Lecture Notes in Networks and Systems

Volume 1326

The series "Lecture Notes in Networks and Systems" publishes the latest developments in Networks and Systems—quickly, informally and with high quality. Original research reported in proceedings and post-proceedings represents the core of LNNS.

Volumes published in LNNS embrace all aspects and subfields of, as well as new challenges in, Networks and Systems.

The series contains proceedings and edited volumes in systems and networks, spanning the areas of Cyber-Physical Systems, Autonomous Systems, Sensor Networks, Control Systems, Energy Systems, Automotive Systems, Biological Systems, Vehicular Networking and Connected Vehicles, Aerospace Systems, Automation, Manufacturing, Smart Grids, Nonlinear Systems, Power Systems, Robotics, Social Systems, Economic Systems and other. Of particular value to both the contributors and the readership are the short publication timeframe and the world-wide distribution and exposure which enable both a wide and rapid dissemination of research output.

The series covers the theory, applications, and perspectives on the state of the art and future developments relevant to systems and networks, decision making, control, complex processes and related areas, as embedded in the fields of interdisciplinary and applied sciences, engineering, computer science, physics, economics, social, and life sciences, as well as the paradigms and methodologies behind them.

Indexed by SCOPUS, EI Compendex, INSPEC, WTI Frankfurt eG, zbMATH, SCImago.

All books published in the series are submitted for consideration in Web of Science.

For proposals from Asia please contact Aninda Bose (aninda.bose@springer.com).

Badal Soni · Poonam Saini · Gyanendra K. Verma · Brij B. Gupta
Editors

Beyond Artificial Intelligence

Select Proceedings of the International Conference, AICTA 2023

Springer

Editors
Badal Soni
Department of Computer Science and Engineering
National Institute of Technology Silchar
Silchar, Assam, India

Poonam Saini
Department of Computer Science and Engineering
Punjab Engineering College
Chandigarh, Punjab, India

Gyanendra K. Verma
Department of Information Technology
National Institute of Technology
Raipur, Chhattisgarh, India

Brij B. Gupta
Department of Computer Science and Information Engineering
Asia University
Taichung, Taiwan

ISSN 2367-3370 ISSN 2367-3389 (electronic)
Lecture Notes in Networks and Systems
ISBN 978-981-96-4169-7 ISBN 978-981-96-4170-3 (eBook)
https://doi.org/10.1007/978-981-96-4170-3

This Springer imprint is published by the registered company Springer Nature Singapore Pte Ltd.
The registered company address is: 152 Beach Road, #21-01/04 Gateway East, Singapore 189721, Singapore

Contents

A Residual Network-Based Approach for Early Diabetic Macular Edema Detection ... 1
Nitigya Sambyal and Sukhpal Singh

Quarter-Car Passive Suspension System Control Using Different Machine Learning Algorithms ... 11
Ashwani Kharola, Syed Farrukh Rasheed, Vishwjeet Choudhary, Arvind Kumar, Tarun Kumar Dhiman, and Abhijat Karn

Industrial Safety System Using IOT ... 21
Shital Sobale, Prathamesh Sawle, Sumeet Sapkal, Satvik Vishnoi, and Saurav Shimare

Design and Implement Security Techniques for Authenticating IoT-Based Smart Homes ... 33
Neha Sharma and Pankaj Dhiman

Analysis of Hybrid Features for Breast Cancer Classification ... 41
Palwinder Kaur and Amandeep Kaur

DCA-IoT: A Distributed Classification Approach for Combating IoT Traffic-Based DDoS Attacks Using Dataset Characterization and Comparison ... 51
Praveen Shukla, C. Rama Krishna, and Nilesh Vishwasrao Patil

Mapping User Perception Towards Ethical Aspects of Artificial Intelligent Systems: A Study of AI Mobile Applications ... 67
Himanshu and Rachita Sambyal

Detection and Mitigation of NDP Attacks for AMI Devices Using Cooja Simulator ... 85
Shelza Dua, Virender Kadyan, Rahul Bhogal, and Mohit Dua

An Efficient CNN Model for COVID-19 Detection Based on X-Ray Images 101
Aditi Kailas, M. Sinthuja, and Neeraj Vaidyamath

Application of Emerging Technology in an Emerging Economy: Case of GovTech Through Blockchain-Based Seeds Traceability in Jharkhand 111
Rameshwar Arora

Recent Patterns of Plant Leaf Disease Characterization with Artificial Intelligence Techniques 123
Apoorva Arora and Vinay Gautam

Mango Maturity Classification Based on LAB Color Features Using Machine Learning 145
Abiban Kumari and Jaswinder Singh

Smart Identification of COVID-19 Patients Among Patients with Other Viral Diseases Based on Analysis of Common Symptoms ... 161
Aditika Tungal, Kuldeep Singh, and Prabhsimran Singh

Affective Driver Emotion Analysis in VANETs Using Heterogeneous Dataset 173
Anwesha Patel, Rishu Chhabra, and C. Rama Krishna

Tomato Leaf Disease Classification Using YOLOv8 189
Ambati Leena Reddy and Surendra Reddy Vinta

Detection of Audio Spoofing Attack Using Integrated Hand-Crafted Features with LSTM 203
Nidhi Chakravarty and Mohit Dua

Skin Cancer Diagnosis and Classification Using VGG 19 with Transfer Learning and Data Augmentation 211
Palak Goyal, Anushka Aggarwal, and Rinkle Rani

Smart Stick Assistant for Visually Impaired People 223
Juiley Sangram Raut, Vaibhav Shamrao Ingale, and Jayashree Ashok Salunkhe

A Review of Deep Learning Techniques for Classification of Alzheimer's Disease Using MRI 235
Nand Kishore and Neelam Goel

A Comparative Study of Recent Biometric CryptoSystems 249
Prabhjot Kaur, Nitin Kumar, and Maheep Singh

Comparative Analysis of Pretrained Model-Based Transfer Learning Techniques for Glaucoma Detection from Fundus Images 263
M. K. Shajila Beegam and Mala Kalra

Crop Type Identification in Indian Agriculture: A Contemporary Review 281
Sanjay Moulik and V. Kethareswaran

Role of Social Media in Disaster Management Using NLP and Deep Learning 293
Pooja Rai, Sudhakar Kumar, Sunil K. Singh, Harkiran Kaur, Simran Jaggi, Varsha Arya, and Brij B. Gupta

FPGA Implementation of AI Algorithms for Image Detection 305
Ishpreet Kaur, Garvit Banga, Vivek Singh, Akshit Muval, and Jyoti Kedia

Yoga Pose Detection Model Using VGG16 329
Hardik Raina and Laxmi Chaudhary

An Augmented Efficient Net Based Deep Learning Model for Diabetic Retinopathy Detection and Classification 341
M. A. Abini and S. Sridevi Sathya Priya

Multi-Model Sarcasm Classification for Online Text: A Comparative Study 357
Umesh Gupta, Manish Agarwal, Pranay Kumar, Ujjwal Yadav, and Aditya Bhardwaj

Moving Object Detection and Segmentation for Various Surveillance Application Using Deep Learning Approaches in Explainable Artificial Intelligence 373
Dipika Gupta, Kashish Bhatia, and Manish Kumar

Design and Exploration of Data-Handling Quantum-Computational Computer Circuits for the Implementation of Ultra-Area and Low-Power Devices 389
Mukesh Patidar, Praveen Bhanodia, Shreyaskumar Patel, Dron Mishra, Rupesh Shukla, Dilip Kumar Sharma, Aaryesh Shukla, and Mayur Tembhurney

Information Retrieval Using Large Language Model for Various Alloys 405
Yash Garg, Akanksha Jaiswal, Daevesh Singh, and Ramkumar Rajendran

Estimating Soil Quality Content of Arid and Semi-Arid Districts of Punjab Using Machine Learning Techniques 415
Mandeep Kaur and Kamal Malik

Implementation of CNN-Based Word Recognition System 425
Bittu Kumar, Gudi Srikanth, Vijay Rao Kumbhare, Amit Kumar Shrivastava, and Ashwini Kumar Varma

Evaluating of Pedestrian Speed Data for Pedestrians Accessibility Within the University Campus Using Data Science 439
Badveeti Adinarayana and A. Krishna Chaitanya

Role of Bio-Inspired Algorithms in Detection of Lung Disease Using Deep Learning 447
Hardeep Saini and Davinder Singh Saini

A Review of Behavioral Biometric Authentication Techniques Using IoT and Deep Learning 459
Narendra Kumar and Shefali Arora

Empowering Smart Healthcare with Low-Power IoT Networks 469
Prabh Deep Singh, Kiran Deep Singh, Shanjal Gupta, and Pardeep Kumar Jindal

Sustainable Farming with Energy-Optimized Internet of Things 477
Kiran Deep Singh, Prabh Deep Singh, Shanjal Gupta, and Pardeep Kumar Jindal

Efficient Resource Allocation and Load Balancing in Cloud Environments with CloudAnalyst 485
Navneet Kumar Rajpoot, Prabh Deep Singh, Bhaskar Pant, and Vikas Tripathi

IoT in Smart Cities: Review of IoT Architecture for Smart Cities, Applications, and Key Challenges 497
Rajwinder Kaur, Jaswinder Singh, Satinder Kaur, and Vishal Malhotra

IoT Based Smart Agriculture for Crop Water and Nutrition Management 511
Talvinder Singh and Shano Solanki

Enhancing Predictive Maintenance for Induction Motors: A 3D Holography-Based Digital Twin Approach with Extended Kalman Filter 531
G. M. Yashaswini, Hari Krishna S. Thota, Anand Kumar Subramaniyan, and Rajkumar Velu

SDN-IOT: DDoS Detection and Mitigation in SDN-Enabled IoT Network 543
Hariom Shukla and Ankit Kumar Jain

Deep Reinforcement Learning for Connected Autonomous Vehicles (CAVs) Task Offloading 555
Aditya Bhardwaj

Activation Functions Impact on Malware Detection: A Multi-layer Perceptron Approach for Android Applications 569
Shruti Brahma, Shweta Sharma, and C. Rama Krishna

ANNDroid: Android Malware Classification and Detection Techniques 583
Pooja, Arvind Mahindru, and Pardeep Kumar Arora

System Identification of Batch Reactor Using Machine Learning Techniques 593
Sivakumar Rajendran, Gowsic Kandaswamy, and Thirunavukkarasu Indiran

Comparative Analysis of Machine Learning Models for Traffic Prediction 601
Tanya Garg, Gurjinder Kaur, and Prashant Singh Rana

A Hybrid Model for Aspect-Based Sentiment Analysis Using Boosting Technique 617
Amit Chauhan, Aman Sharma, and Rajni Mohana

Exploring the Association Between Mental Health and Efficacy Using Student Dataset 625
Rakhi Nagpal, Saravjeet Singh, and Aditi Moudgil

Validating Data Warehouse Requirements Decomposition Metrics Formally by Implementing Zuse's Formal Framework 639
Tanu Singh, Manu Singh, Arvind Panwar, and Urvashi Sugandh

Exploring Genomic Sequence Encoding Methods for 1D Convolutional Neural Networks: A Comparative Study 651
Shikha Sharma, Nidhi Kalra, and Rinkle Rani

Analyzing the Relationship Between Legal Moves in Blind Chess and ELO Ratings Using Speech Recognition and Machine Learning 665
Utkarsh Chauhan, Tushar Singh, Amit Chhabra, Animesh Singh, and Briti Singla

Earthquake Prediction Using Ensemble Regression Models 683
Corinna Mum Norbu, Ramakrishna Challa, and Jagriti Saini

Event-Driven Stream Processing Using Window-Based Query Execution 691
Shruti Arora, Anushka Aggarwal, and Rinkle Rani

YouTube Video Performance: Leveraging Data Analysis and Forecasting with ARIMA Modeling for Optimal Engagement 701
K. S. Hemanth, R. Krishna Murthy, and B. Muralidhara

COVIDWavesSenti: Large Scale Twitter Dataset of Different COVID Waves 713
Jaspal Kaur Saini, Jyoti Maggu, and Pratishtha Verma

Classification of Codling Moth in Apple Orchards with Ensemble Learning ... 727
Kunzang Dolma and Mala Kalra

Effective Advertisement Decision-Making Using Telecom Data Analytics ... 737
Aditya Bhardwaj, Divyansh Palia, Umesh Gupta, and Shreya Chatterjee

Employee Retention and Attrition Prediction Using Machine Learning. Case Study on ITES Company ... 749
Uttam Singh Chaudhary and Darpan Anand

Cold Start Problem in a Recommender System: Application, Evaluation, Performance and Methods ... 763
Rajiv Kumar Nath and Tanvir Ahmad

Malwares for Numerous Platforms: A Deep Learning ... 781
Vinisha Sumra, Naveen Malik, and Sandip Kumar Goyal

Pushing the Boundaries of Mortality Prediction: Advancing High-Risk Sepsis-III Patient Care Through Cutting-Edge Deep Learning Techniques ... 791
Deepak Mahto, Sunil K. Singh, Sudhakar Kumar, Manya, Rashmi, Varsha Arya, Kwok Tai Chui, and Brij B. Gupta

Enhancing Review Authenticity: A Linguistic and Behavioral Approach for Detecting Deceptive Online Reviews Using Machine Learning Models ... 805
Ishtveer Singh Billing, Sudhakar Kumar, Sunil K. Singh, Manraj Singh Gill, Brij B. Gupta, and Varsha Arya

A Residual Network-Based Approach for Early Diabetic Macular Edema Detection

Nitigya Sambyal and Sukhpal Singh

Abstract Diabetic Macular Edema (DME) is a critical complication of diabetes mellitus and is responsible for vision impairment and blindness. Due to an increase in diabetes cases, early and accurate DME detection has become important. This paper presents a deep learning-based solution for DME detection using ResNet50 and ResNet101 architectures. The proposed ResNet50 model achieved an accuracy of 94.56%, sensitivity of 95.98%, specificity of 87.8%, and precision of 97.45%, surpassing existing methods. The ResNet101 model, while slightly lower in accuracy at 90.83%, provides a high sensitivity of 91.83% and precision of 97.45%, with a specificity of 84.38%. This research demonstrates the transformative potential of deep learning in revolutionizing DME diagnosis.

Keywords Macular edema · Deep learning · Fundus images · ResNet50 · ResNet101

1 Introduction

Diabetic Macular Edema (DME) is one of the prevalent complications associated with diabetes mellitus, which affects the blood vessels of the eye, resulting in significant vision impairment and blindness [1]. According to the World Health Organization (WHO), approximately 537 diabetes cases were reported globally in 2021, which are expected to rise to 784 million by 2045 [2]. The increase in diabetes cases has resulted in a growing occurrence of related complications, including DME, specifically among the working-age population.

DME appears as fluid buildup in the macula, a vital region of the retina responsible for precise and detailed vision. This fluid accumulation is caused by the blood-retinal

N. Sambyal · S. Singh (✉)
Department of Computer Science and Engineering, Thapar Institute of Engineering and Technology, Patiala, India
e-mail: ssingh_me21@thapar.edu

N. Sambyal
e-mail: nitigya.sambyal@thapar.edu

B. Soni et al. (eds.), *Beyond Artificial Intelligence*, Lecture Notes in Networks and Systems 1326, https://doi.org/10.1007/978-981-96-4170-3_1

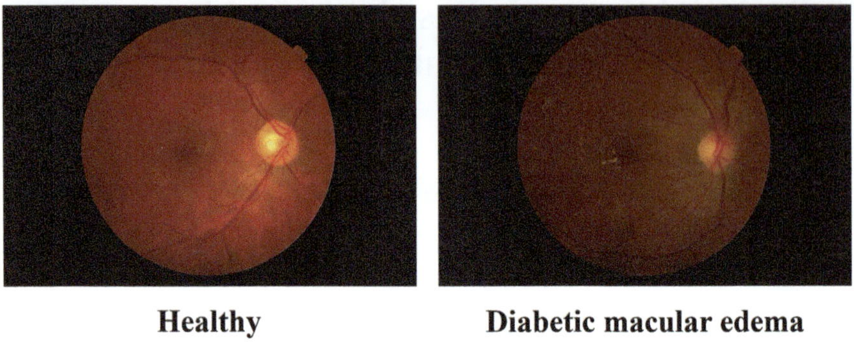

Fig. 1 Healthy and DME affected eye from Messidor dataset

barrier breakdown, primarily driven by chronic hyperglycemia, resulting in fluid and lipid leakage into the retinal tissue [3]. The disruption of the macula hampers the normal function of retinal cells, causing distorted vision and, if untreated, results in severe visual impairment and blindness.

The common symptoms of DME include blindness or blurred vision, color blindness, difficulty in reading or seeing faces, floaters, flashes of light, or dark spots in vision [4]. The causes of macular edema are complex and multifactorial, often stemming from factors such as chronic hyperglycemia, inflammation, and vascular abnormalities [5]. Figure 1 depicts a healthy eye and eye with DME.

DME diagnosis traditionally relies on the expertise of ophthalmologists, who meticulously analyze fundus images. However, this approach has inherent limitations, such as labor-intensive nature, lengthy assessment timelines, and susceptibility to interobserver variability. Due to these challenges, AI based techniques specifically deep learning, has emerged as a promising tool in medical image analysis [6]. The main strength of deep learning lies in its ability to extract valuable insights from complex data swiftly and accurately. In this paper, we propose a deep learning-based solution for DME detection.

Contributions: The main contributions of the research work are as follows:

1. We propose a state-of-the-art deep learning approach for DME detection by utilizing ResNet50 and ResNet101 architectures, which are fine-tuned through transfer learning. This novel methodology demonstrates remarkable accuracy and efficiency in identifying DME cases, potentially revolutionizing the diagnostic process.
2. The proposed ResNet50 model achieves an exceptional accuracy of 94.56%, demonstrating its ability to detect DME accurately.
3. The proposed ResNet-101 model maintains a robust accuracy of 90.83%, sensitivity of 91.83%, precision of 97.45%, and specificity of 84.38%. This showcases the versatility of our approach, catering to different diagnostic requirements.

4. In comparison to existing literature, the proposed ResNet50 model substantially improves accuracy by 3.38%, sensitivity by 15.89%, and precision by 18.25% for DME detection.

Paper Organization: Sect. 2 focuses on existing literature. Section 3 outlines the proposed methodology. Section 4 presents results and discussion along with a comparative analysis of existing methods in the literature. Finally, the conclusion is given in Sect. 5.

2 Literature Review

Over the years, several research efforts have been directed toward the development of automated DME detection methods using deep learning techniques. This section presents an overview of various such research work in the domain of DME detection.

In 2011, Lim et al. [7] proposed an automated method for classifying DME stages using the Early Treatment Diabetic Retinopathy Studies (ETDRS) grading scale. Their approach involved localization of the fovea and marking of macular regions based on the ETDRS scale. The proposed approach achieved 85.2% accuracy on 88 images of the Messidor dataset.

Naguib et al. [8] focused on content-based image retrieval (CBIR) for DME. Their fully automated CBIR system utilized texture discontinuities to represent the lesion content of the retina. The system achieved a precision of 79.2% on the Messidor dataset.

In 2017, Baby and Chandy [9] utilized the dual-tree complex wavelet transform for feature extraction in a content-based retinal image retrieval system. They achieved a precision rate of 78.23% on the Messidor dataset.

Rekhi et al. [10] proposed a method using anisotropic diffusion and adaptive thresholding for segmenting hard exudates from fundus images. Their system employed geometrical, shape, and orientation features for SVM-based classification, achieving an accuracy of 92.13% on DIARETDB1 and 90% on Messidor datasets.

Lim et al. [11] used a modified completed Local Binary Pattern (CLBP) for texture feature extraction in DME images. The researchers utilized a modified Local Binary Pattern (CLBP) technique to extract the local gray level of the image across all RGB channels. The feature vector that was acquired was subsequently fed into a multiclass Support Vector Machine (SVM) for classification, which achieved a sensitivity of 80% and a specificity of 70%.

Li et al. [12] introduced the CANet, a cross-disease attention network that detected diabetic retinopathy and DME. The network utilized disease-specific and disease-dependent attention modules to maximize overall performance. Their method demonstrated promising results on the ISBI 2018 IDRiD challenge and Messidor datasets with 91.2% accuracy and an AUC of 92.4%.

Recent studies, like Lahmar et al. [13], 2023, highlight the importance of computer-aided diagnosis for early diabetic retinopathy (DR) detection. Their evaluation of 28 deep hybrid architectures and seven end-to-end DL architectures favors the hybrid approach, especially with SVM and MobileNet_V2, showcasing superior performance.

Zhang et al. [14] in 2023, introduced OFFNet, a novel feature fusion network for grading DR and DME. Featuring a unique self-attention-based module and cross-attention-based fusion, OFFNet introduced the key-value query concept, which requires only image-level supervision. Experimental results on Messidor and the 2018 IDRiD challenge datasets highlighted OFFNet's superior joint grading accuracy and its capacity for capturing long-distance dependencies.

These studies collectively showcase the evolving landscape of deep learning techniques in DME detection and classification, emphasizing a shift towards more sophisticated and effective methodologies.

3 Methodology

This section elaborates on the methodology adopted for Diabetic Macular Edema (DME) detection using deep learning. Figure 2 illustrates the proposed methodology: the fundus image is initially preprocessed and then given as input to a deep learning model, which performs feature extraction and classification tasks. The results are then analyzed using several performance metrics. The various steps of the methodology are discussed as follows:

3.1 Data Preparation

In this work, we have used the Messidor dataset available at https://www.adcis.net/en/third-party/messidor/. The Messidor dataset contains a collection of fundus images

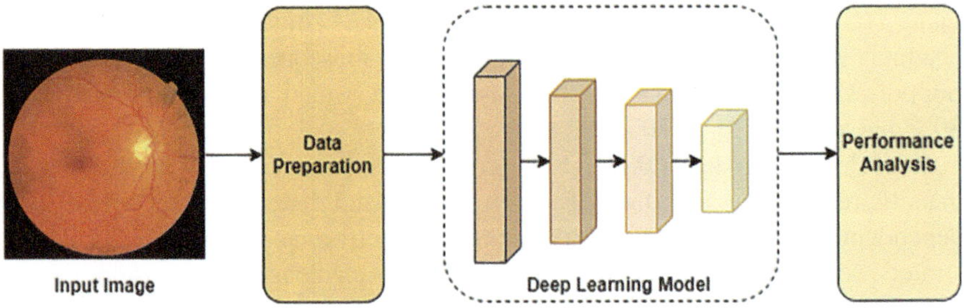

Fig. 2 Proposed methodology for DME detection

Table 1 Messidor dataset details

Class	Images
Healthy	974
Macular	226
Total	1200

consisting of normal cases and instances of DME, the details of which are provided in Table 1.

To ensure compatibility with deep learning models and maintain consistent quality, the images have been adjusted to a resolution of 299 × 299 pixels. This uniform resolution eliminates discrepancies from varying sizes and aspect ratios while retaining crucial details. The images are further augmented using transformations like pixel normalization, horizontal flips, vertical flips, random rotation, zooming up to 1.1x, and lighting variations. The dataset is split into 80:20 ratios for training and validating the model.

3.2 Deep Learning Model

We have employed 50 layered and 101 layered residual networks [15], which have been pre-trained on the ImageNet dataset. These models consist of residual or skip connections, which allow for the training of profound networks by mitigating the vanishing gradient problem. To ensure the optimal performance of DME classification models, we have used novel training concepts like transfer learning [16], cyclic learning [17], and discriminative learning [18]. The proposed fine-tuned models were further trained with cyclic learning techniques using stochastic gradient descent with warm restarts (SGDR) and cosine annealing. Cyclic learning dynamically adjusts learning rates at different intervals using the cosine function, facilitating better convergence and improved generalization on unseen data. Further, we have also employed differential learning by training the model with different learning rates for distinct network blocks. The initial layers, focused on general features, used lower learning rates, while the middle and final layer blocks, responsible for intricate feature extraction, utilized higher learning rates.

3.3 Performance Metrics

In order to evaluate the effectiveness of the proposed DME classification approach, we have evaluated the results on performance metrics like accuracy, sensitivity, specificity, and precision, the formulas of which are given in Eqs. (1, 2, 3, and 4), respectively [19].

$$Accuracy = \frac{(TP + TN)}{(TP + TN + FP + FN)} \quad (1)$$

$$Sensitivity = \frac{TP}{(TP + FN)} \quad (2)$$

$$Specificity = \frac{TN}{(TN + FP)} \quad (3)$$

$$Precision = \frac{TP}{(TP + FP)} \quad (4)$$

True Positives (TP) are the cases where the model correctly identifies unhealthy images as 'Macular Edema'. True Negatives (TN) represent cases where the model accurately identifies healthy images as 'Normal.' False Positives represent the cases where the model incorrectly classifies 'Macular Edema' images as 'Normal'. False Negatives (FN) are the cases where the model erroneously classifies 'Normal' images as 'Macular Edema.'

4 Results and Discussion

This section elaborates on the hyperparameters, and results obtained in this research. The experiments were conducted on a Windows 11 Pro desktop with Intel Core i7-11700 K CPU, Nvidia RTX A6000 GPU, and 64 GB RAM, tailored for demanding deep learning tasks. We have used the fastai library [20] for efficient and reliable experimentation.

4.1 Hyperparameters

The last layers of the ResNet50 model are trained for 10 epochs with a learning rate of, 4×10^{-2} then unfrozen, and the entire model is trained with a differential learning rate of 4×10^{-2} for the initial layers, 3.2×10^{-1} for the middle layers, and 6×10^{-1} for the last layers. The last layers for the proposed ResNet101 model are trained for 10 epochs with a 10^{-3} learning rate. Subsequently, the model is unfrozen, and training is continued with differential learning rates of 10^{-3}, 5.5×10^{-3}, and 10^{-2} for the initial, middle, and last layers, respectively. Table 2 presents the hyperparameter configurations for ResNet50 and ResNet101.

Table 2 Hyperparameters for DME detection models

Parameters	Proposed ResNet50	Proposed ResNet101
Target labels	2	2
Image size	299 × 299	299 × 299
Batch size	16	16
Dropout	0.25	0.30
Initial learning rate	4×10^{-2}	10^{-3}
Epochs with online data augmentation	10	10
After unfreezing the network	50	40
Total epochs	60	50
Learning rate of the initial block	4×10^{-2}	10^{-3}
Learning rate of the middle block	3.2×10^{-1}	5.5×10^{-3}
Learning rate of the last block	6×10^{-1}	10^{-2}

Table 3 Results on the Messidor dataset

Model	Training loss	Validation loss	Accuracy (%)	Sensitivity (%)	Specificity (%)	Precision (%)
ResNet50	0.211042	0.193041	94.56	95.98	87.8	97.45
ResNet101	0.321917	0.317434	90.83	91.83	84.38	97.45

4.2 *Results*

The proposed ResNet50 model achieved 94.56% accuracy, 95.98% sensitivity, 87.8% specificity, and 97.45% precision. On the other hand, ResNet101 achieved 90.83% accuracy, 91.83% sensitivity, 84.38% specificity, and 97.45% precision. The results show that both models effectively distinguish between healthy and DME cases. Table 3 shows the performance metrics values obtained for the proposed model. Also, the confusion matrix of ResNet50 and ResNet101 is shown in Fig. 3. In general, the results emphasize the potential for precise and early DME diagnosis, offering prospects for improved patient outcomes and more efficient healthcare delivery.

4.3 *Comparative Analysis*

The comparison of the proposed DME detection models with the existing literature is shown in Table 4. Our ResNet50 model showed an approximate improvement in accuracy by 3.38%, sensitivity by 15.89%, and precision by 18.25% compared to authors Li et al. [12], Lim et al. [7], and Naguib et al. [8], respectively. Lim et al. [7] obtained high specificity in the existing literature since they used a hybrid approach combining

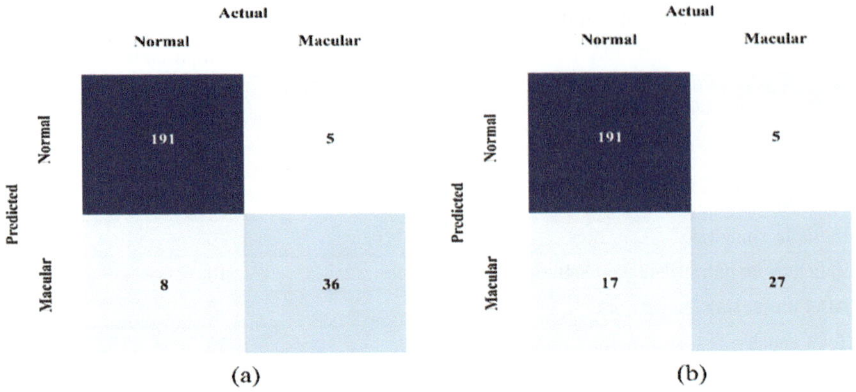

Fig. 3 Confusion matrix for DME classification using deep learning models. **a** ResNet50, **b** ResNet101

Table 4 Performance comparison with existing literature for DME detection

Study	Accuracy	Sensitivity	Specificity	Precision
Lim et al. [7]	85.2%	80.9%	90.2%	–
Naguib et al. [8]	–	–	–	79.2%
Baby and Chandy [9]	–	–	–	78.23%
Rekhi et al. [10]	90%	–	–	–
Lim et al. [11]	–	80%	70%	–
Li et al. [12]	91.2%	–	–	–
Proposed ResNet50	94.56%	95.98%	87.8%	97.45%
Proposed ResNet101	90.83%	91.83%	84.38%	97.45%

localization and classification for DME detection. These substantial results uncover the potential of deep learning in revolutionizing DME diagnosis.

5 Conclusion

In this paper, we focus on deep learning-based automatic detection of diabetic macular edema, a prevalent condition associated with diabetes mellitus that has the potential to cause significant visual impairment. In this study, we have proposed two residual models, ResNet50 and ResNet101, for DME classification. ResNet50 achieved 94.56% accuracy, 95.98% sensitivity, 87.8% specificity, and 97.45% precision. ResNet101 obtained accuracy, sensitivity, specificity, and precision of 90.83%, 91.83%, 84.38%, and 97.45%, respectively. The comparative analysis of the proposed model against existing methods demonstrates substantial improvement in accuracy, sensitivity, and precision. In the future, the focus can be on developing deep learning models for multistage DME classification. This would enable more precise and personalized treatment, potentially improving patient outcomes.

References

1. Diabetic Macular Edema (DME)-Prevent Blindness, https://preventblindness.org/diabetic-macular-edema-dme/. Last accessed 07 Sept 2023
2. Facts & Figures.: World Diabetes Day, https://worlddiabetesday.org/about/facts-figures/. Last accessed 07 Sept 2023
3. Achiron, A., Lagstein, O., Glick, M., Gur, Z., Bartov, E., Burgansky-Eliash, Z.: Quantifying metamorphopsia in patients with diabetic macular oedema and other macular abnormalities. Acta Ophthalmol **93**(8), e649–e653 (2015). https://doi.org/10.1111/AOS.12735
4. Frank, R.N., Glybina, I.: Macular Edema. Encycl Eye 1–12 (2023). https://doi.org/10.1016/B978-0-12-374203-2.00139-1
5. Daruich, A., et al.: Mechanisms of macular edema: beyond the surface. Prog. Retin. Eye Res. **63**, 20–68 (2018). https://doi.org/10.1016/J.PRETEYERES.2017.10.006
6. Muchuchuti, S., Viriri, S.: Retinal disease detection using deep learning techniques: a comprehensive review. J Imaging **9**(4), 84 (2023). https://doi.org/10.3390/JIMAGING9040084
7. Lim, S.T., Zaki, W.M.D.W., Hussain, A., Lim, S.L., Kusalavan, S.: Automatic classification of diabetic macular edema in digital fundus images. In: 2011 IEEE Colloquium on Humanities, Science and Engineering, CHUSER 2011, pp. 265–269 (2011). https://doi.org/10.1109/CHUSER.2011.6163730
8. Naguib, A.M., Ghanem, A.M., Fahmy, A.S.: Content based image retrieval of diabetic macular edema images. In: Proceedings of CBMS 2013-26th IEEE International Symposium on Computer-Based Medical Systems, pp. 560–562 (2013). https://doi.org/10.1109/CBMS.2013.6627877
9. Baby, C.G., Chandy, D.A.: Content-based retinal image retrieval using dual-tree complex wavelet transform. In: International Conference on Signal Processing, Image Processing and Pattern Recognition 2013, ICSIPR 2013, vol. 1, pp. 195–199 (2013). https://doi.org/10.1109/ICSIPR.2013.6497987
10. Rekhi, R.S., Issac, A., Dutta, M.K., Travieso, C.M.: Automated classification of exudates from digital fundus images. In: 2017 International Work Conference on Bio-Inspired Intelligence: Intelligent Systems for Biodiversity Conservation, IWOBI 2017-Proceedings (2017). https://doi.org/10.1109/IWOBI.2017.7985527
11. Lim, S.T., Ahmed, M.K., Lim, S.L.: Automatic classification of diabetic macular edema using a modified completed Local Binary Pattern (CLBP). In: Proceedings of the 2017 IEEE International Conference on Signal and Image Processing Applications, ICSIPA 2017, pp. 6–10 (2017). https://doi.org/10.1109/ICSIPA.2017.8120570
12. Li, X., Hu, X., Yu, L., Zhu, L., Fu, C.W., Heng, P.A.: CANet: cross-disease attention network for joint diabetic retinopathy and diabetic Macular edema grading. IEEE Trans. Med. Imaging **39**(5), 1483–1493 (2020). https://doi.org/10.1109/TMI.2019.2951844
13. Lahmar, C., Idri, A.: Deep hybrid architectures for diabetic retinopathy classification. Comput. Methods Biomech. Biomed. Eng. Imaging Vis. **11**(2), 166–184 (2023). https://doi.org/10.1080/21681163.2022.2060864
14. Zhang, J., Guo, X., Lin, Q., Wang, H., Hu, X., Che, S.: A novel original feature fusion network for joint diabetic retinopathy and diabetic Macular edema grading. Neural Comput. Appl. **35**(9), 6699–6712 (2023). https://doi.org/10.1007/S00521-022-08038-Y/TABLES/8
15. He, K., Zhang, X., Ren, S., Sun, J.: Deep residual learning for image recognition. In: Proceedings of the IEEE Computer Society Conference on Computer Vision and Pattern Recognition, vol. 2016-December, pp. 770–778 (2015). https://doi.org/10.48550/arxiv.1512.03385

16. Day, O., Khoshgoftaar, T.M.: A survey on heterogeneous transfer learning. J. Big Data **4**(1), (2017). https://doi.org/10.1186/S40537-017-0089-0
17. Smith, L.N.: Cyclical learning rates for training neural networks. In: Proceedings-2017 IEEE Winter Conference on Applications of Computer Vision, WACV 2017, pp. 464–472 (2015). https://doi.org/10.1109/WACV.2017.58
18. Zhang, X., et al.: Generative and Discriminative Learning. Encycl. Mach. Learn. 454–455 (2011). https://doi.org/10.1007/978-0-387-30164-8_332
19. Ragan, A.: Taking the Confusion Out of Confusion Matrices. Towards Data Science. https://towardsdatascience.com/taking-the-confusion-out-of-confusion-matrices-c1ce054b3d3e. Last accessed 01 Mar 2023
20. fast.ai-fast.ai—Making Neural Nets Uncool Again, https://www.fast.ai/. Last accessed 03 July 2023

Quarter-Car Passive Suspension System Control Using Different Machine Learning Algorithms

Ashwani Kharola, Syed Farrukh Rasheed, Vishwjeet Choudhary, Arvind Kumar, Tarun Kumar Dhiman, and Abhijat Karn

Abstract This article presents control of quarter-car passive suspension system using an Artificial neural network (ANN) controller. A mathematical model of the proposed system has been developed and simulated in Proportional-integral-derivative (PID) controlled environment. The ANN controller has been designed using the outputs of PID controller and trained using three different training algorithms namely Levenberg–Marquardt (LM), Bayesian-Regularization (BR) and Scaled-Conjugate-Gradient (SCG). The performance of these machine learning algorithms has been analyzed in terms of settling time and maximum percentage overshoot response. The results clearly indicate superior performance of BR-algorithm compared to other proposed algorithms. The settling time taken to stabilize the complete system using BR, LM and SCG algorithm were 4.0 s, 8.0 s and 14.0 s respectively. The maximum percentage overshoot responses obtained for sprung mass displacement using BR, LM and SCG algorithm were 0.7%, 1.4% and 6.7% respectively.

Keywords ANN · PID · Quarter-Car · Passive suspension · Training algorithms

A. Kharola (✉) · S. F. Rasheed · A. Karn
Graphic Era Deemed to Be University, Dehradun, India
e-mail: ashwanidaa@gmail.com

S. F. Rasheed
e-mail: syedfrasheed.me@geu.ac.in

V. Choudhary
Birla Institute of Technology and Science, Pilani, India

A. Kumar
DPS Educational Society, Dehradun, India

T. K. Dhiman
Graphic Era Hill University, Dehradun, India
e-mail: tkdhiman@gehu.ac.in

B. Soni et al. (eds.), *Beyond Artificial Intelligence*, Lecture Notes in Networks and Systems 1326, https://doi.org/10.1007/978-981-96-4170-3_2

1 Introduction

Vehicles are exposed to continuous vibrations due to uneven road surfaces which may damage vehicle parts as well as harm passengers [1]. Thus, advanced suspension systems are necessary for improved driving, vibration reduction and better efficiency. Suspension systems have received a significant amount attention as a part of ongoing effort to improve vehicle ride and comfort [2]. For an instance, Robert et al. [3] analyzed the performances of model-free intelligent fuzzy controller with model-based conventional controller for displacement and acceleration control of quarter-car model. The simulation results confirmed superior performance of model-free fuzzy controller compared to model-based conventional controller. Further, Nguyen [4] proposed a dynamic model having five state variables to mimic the vehicle oscillations based on four excitation criterions. A complicated solution was created using Sliding Mode-Proportional-Integral-Derivative (PID) tuned fuzzy control algorithm. The findings indicate that both acceleration and displacement of car body has been significantly reduced using proposed algorithm. Moreover, Theunissen et al. [5] conducted a review of literature on evolution of suspension systems over the past few decades. The study analyzed various control strategies, most relevant performance indicators and the effect of the road preview time on the performance of suspension systems.

Nagarkar et al. [6] proposed a method for designing a passive suspension system that determines the optimal suspension settings while indicating a feasible performance near an active suspension system. Fuzzy logic control law was used for accomplishing the desired control. The findings indicate that the optimized passive suspension resulted in same suspension travel and velocity as indicated by active suspension system. Moreover, Issa and Samn [7] considered Harris Hawk Optimization (HHO) algorithm to optimize the design of a vehicle passive suspension system. A novel multi-objective function was developed which included different objectives. Tests were conducted on both quarter-car and half-car models. The results indicate superior performance of HHO algorithm compared to several other algorithms. Recently, Tulsian and Dewangan [8] conducted a study to review the use of various types of regenerative suspension systems in generating power and increasing efficiency of daily lives. The study successfully analysed different techniques which could be adopted for converting vibration energy into electrical energy.

This article considered PID and Artificial neural network (ANN) controllers for control of quarter-car passive suspension system model of a vehicle. A mathematical model of the system was developed and simulated in a controlled environment using PID controller. The results of PID controller were used for learning of ANN controller using three different training algorithms. The primary objective of the study is to analyze and compare the performance of different machine learning algorithms namely Levenberg–Marquardt (LM), Bayesian-Regularization (BR) and Scaled Conjugate Gradient (SCG) for supervised control of highly nonlinear quarter-car passive suspension system. The performances of these training algorithms has been analyzed in terms of settling time and maximum overshoot responses. The results clearly indicate superior learning performance of BR-algorithm both in terms settling time and maximum percentage overshoot response. The article has been divided into following sections: Sect. 1 gives a brief introduction and critical review of recent literature related to control of quarter-car passive suspension system. Further, Sect. 2 highlights the mathematical equations and simulink model of the proposed system. The Sect. 3 highlights procedure for designing an ANN controller using results of PID controller. The Sect. 4 demonstrates result, discussions and comparison of different control algorithms. Lastly, Sect. 5 highlights the conclusions drawn from the study and the scope for future work.

2 Mathematical Model of Quarter-Car Passive Suspension System

A Quarter-car model of a vehicle is a two degree of freedom system representing one-fourth of the vehicle mass. The system constitutes a sprung mass ($m_1 = 650$ kg) and an unsprung mass ($m_2 = 60$ kg) attached as shown in Fig. 1. The sprung mass is connected to the unsprung mass through a spring having stiffness ($k_1 = 5750$ N/m) and a damper having damping coefficient ($c_1 = 315$ Ns/m). The stiffness (k_2) and damping coefficient (c_2) of the wheel were 145000 N/m and 1550 Ns/m respectively.

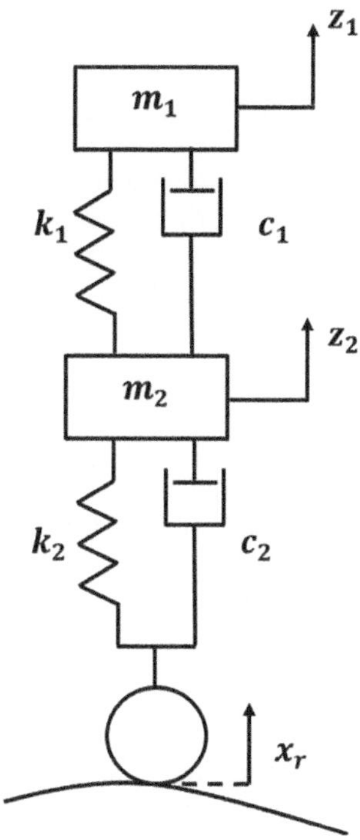

Fig. 1 Model of quarter-car passive suspension system

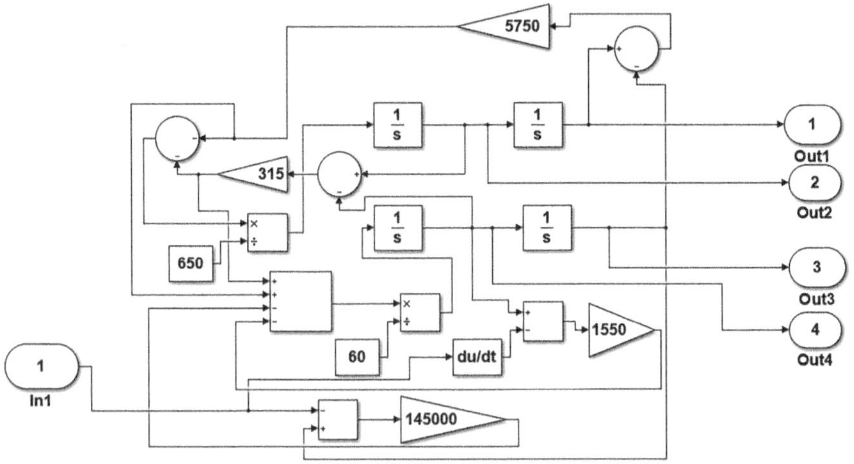

Fig. 2 Simulink model of quarter-car passive suspension system

The expressions obtained for forces acting on the system using Newton's second law are shown with the help of Eq. (1) and Eq. (2) respectively.

$$m_1\ddot{z}_1 = -k_1(z_1 - z_2) - c_1(\dot{z}_1 - \dot{z}_2) \tag{1}$$

$$m_2\ddot{z}_2 = k_1(z_1 - z_2) + c_1(\dot{z}_1 - \dot{z}_2) - k_2(z_2 - x_r) - c_2(\dot{z}_2 - \dot{x}_r) \tag{2}$$

In the above equations, the vertical displacement of the sprung mass, unsprung mass and road surface were represented by z_1, z_2 and x_r respectively. These equations were further used constructing a Simulink model of proposed system as shown in Fig. 2. In the figure, In1 represents the control force, Out1 represents the sprung mass displacement, Out2 represents the sprung mass velocity, Out3 represents the unsprung mass displacement and Out4 represents the unsprung mass velocity respectively.

3 ANN Control of Quarter-Car Model

The passive suspension system of quarter car model was initially simulated using a PID controller and the results were used for learning of ANN controller designed using three different training algorithms namely LM-algorithm, BR-algorithm and SCG-algorithm respectively [9, 10]. Two separate PID controllers were designed for control of the complete system. The PID gains obtained after tuning for both the controllers are shown in Table 1. The studies have revealed that LM-algorithm is the most popular algorithm for various machine learning applications. Therefore, there is an urge to compare the performance of LM-algorithm with some of the similar

Table 1 PID gains obtained for different controllers

PID gains			
Controller	k_p	k_i	k_d
Sprung mass	0.9014	0.3561	0.1893
Unsprung mass	1.3431	0.7183	0.3187

existing algorithms like BR and SCG algorithms particularly for controlling highly nonlinear systems like quarter-car passive suspension system.

The ANN architecture was designed using 20 neurons in the hidden layer and 1 neuron in the output layer as shown in Fig. 3 [11]. The inputs considered for ANN controller were sprung mass displacement, sprung mass velocity, unsprung mass displacement and unsprung mass velocity and the output considered for analysis was control force [12]. A total of 370 data samples were collected and randomly divided into training (258 nos.), validation (56 nos.) and testing samples (56 nos.). The mean square error (*MSE*) and regression (*R*) results obtained after training for different training samples using different algorithms are shown in Table 2. Further, the *R*-plots obtained using different control algorithms are shown in Fig. 4.

It can be observed form the above results that the least *MSE* and best *R*-results were obtained using BR-algorithm followed by LM-algorithm and SCG-algorithm

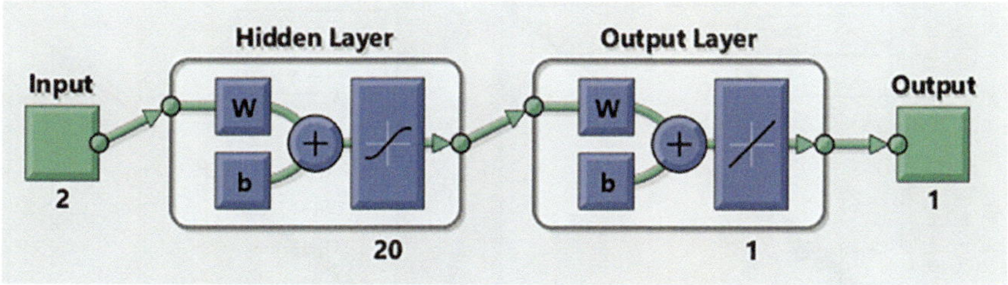

Fig. 3 ANN architecture considered for analysis

Table 2 *MSE* and *R* values obtained for different training algorithms

Sample type			
BR-algorithm	Training	Validation	Testing
MSE	3.90142e-10	0	5.02866e-10
R	9.99999e-1	0	9.99999e-1
LM-algorithm	Training	Validation	Testing
MSE	1.35733e-5	9.44754e-5	2.76503e-5
R	9.99618e-1	9.98321e-1	9.99256e-1
SCG-algorithm	Training	Validation	Testing
MSE	6.08967e-5	4.63330e-4	6.54676e-5
R	9.98418e-1	9.92126e-1	9.98344e-1

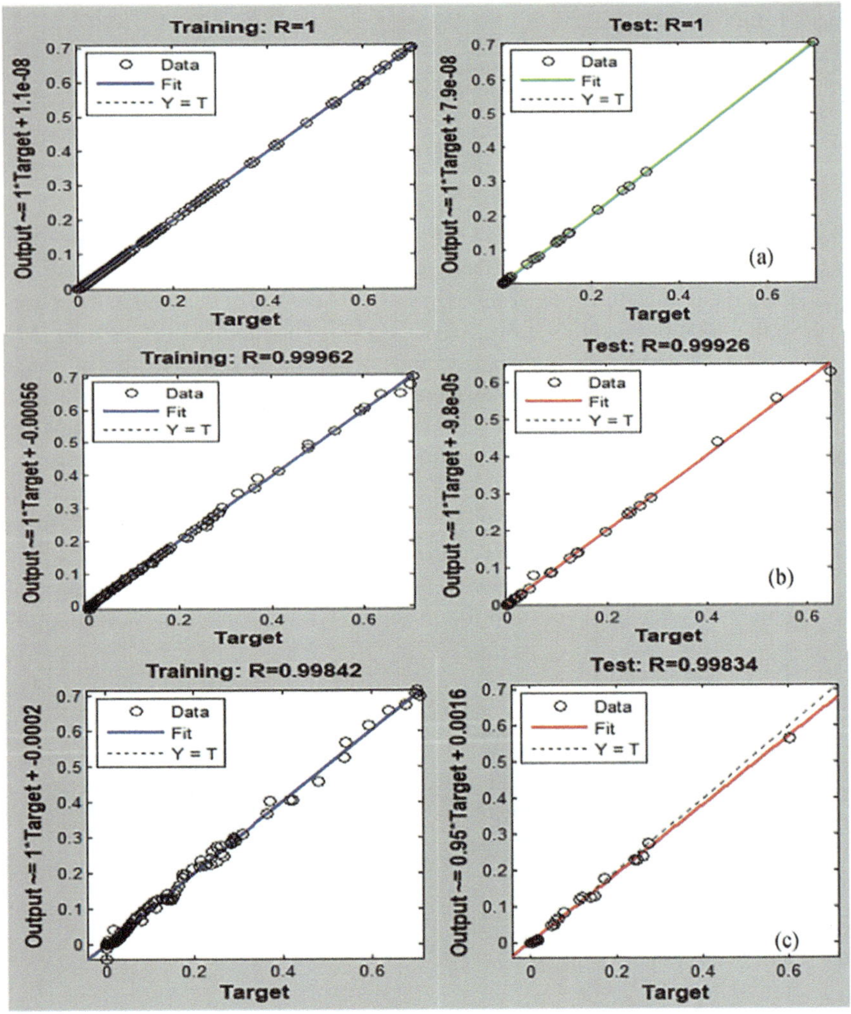

Fig. 4 *R*-plots obtained for training and test samples using **a** BR-algorithm, **b** LM-algorithm, **c** SCG-algorithm

respectively. Further, the error histogram representing histogram of errors along 20 smaller bins using BR, LM and SCG algorithms is shown in Fig. 5. It can be observed from results that using BR-algorithm almost the error samples fall in zero error line within bin value of −0.0048.

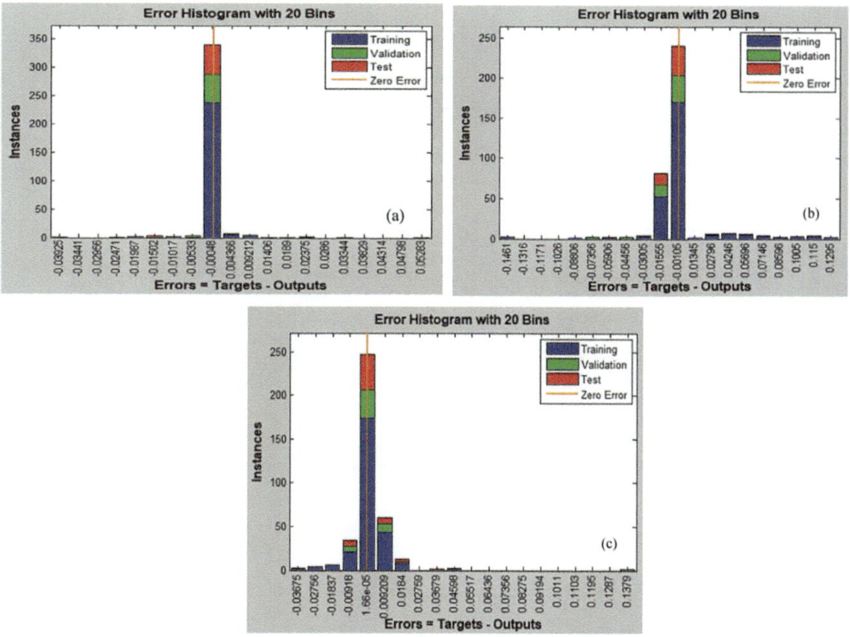

Fig. 5 Error histogram obtained using **a** BR-algorithm, **b** LM-algorithm, **c** SCG-algorithm

4 Results and Discussion

The simulation results obtained for different control attributes i.e., sprung mass displacement, sprung mass velocity, unsprung mass displacement and unsprung mass velocity using different control algorithms are shown with the help of Fig. 6. Further, a comparative analysis of different control algorithms in terms of settling time and maximum percentage overshoot responses are shown with the help of Table 3.

It can be observed from the above results that BR-algorithm provides better responses compared LM and SCG algorithms. The settling time taken for attribute 'sprung mass displacement' using BR, LM and SCG algorithm were 4.0 s, 8.0 s and 14.0 s respectively. Similarly, the settling time taken for attribute 'unsprung mass displacement' using BR, LM and SCG algorithm were 0.5 s, 5.0 s and 8.0 s respectively. Further, the maximum percentage overshoot responses obtained for 'sprung mass displacement' using BR, LM and SCG algorithm were 0.7%, 1.4% and 6.7% respectively. Additionally, the maximum percentage overshoot responses obtained for attribute 'unsprung mass displacement' using BR, LM and SCG algorithm were 0.8%, 1.2% and 4.1% respectively. Lastly, the maximum percentage overshoot responses obtained for attribute 'unsprung mass velocity' using BR, LM and SCG algorithm were 32.5%, 33.0% and 73.0% respectively which were comparatively high compared to overshoot responses obtained for other attributes. The results of this study can be compared with the results of a similar study conducted

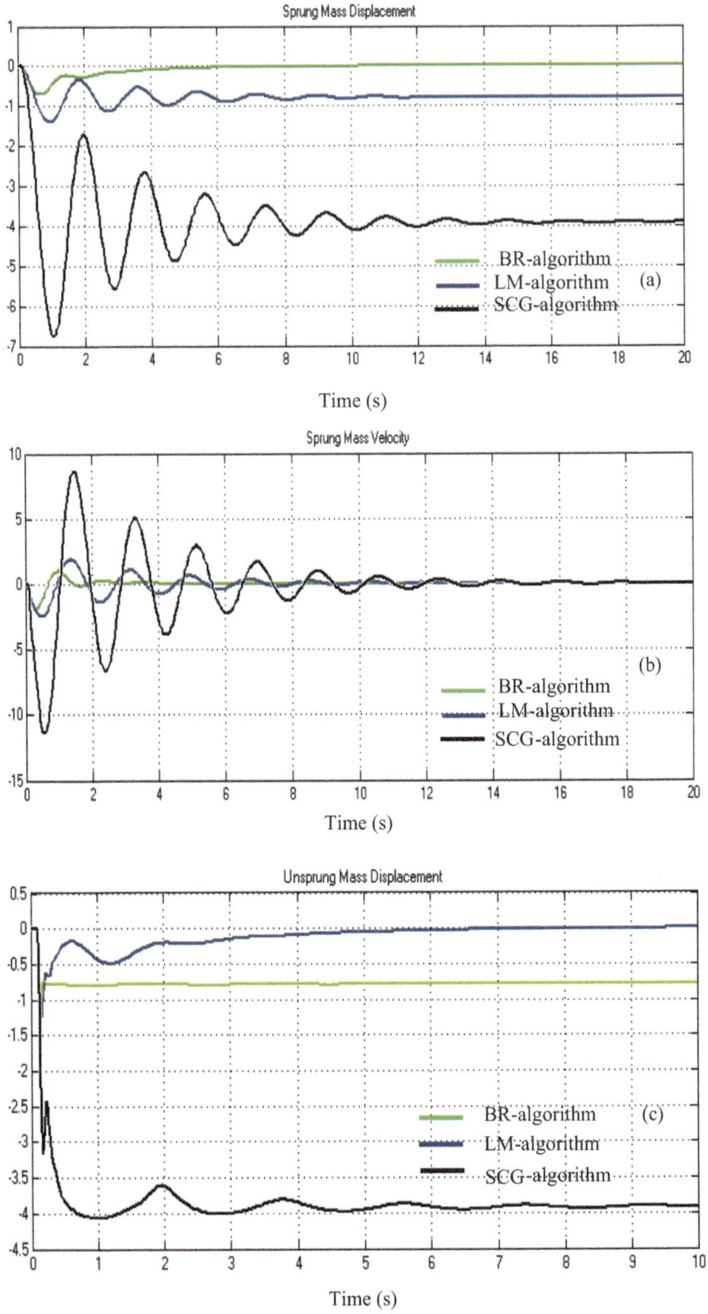

Fig. 6 Simulation responses obtained for: **a** sprung mass displacement, **b** sprung mass velocity, **c** unsprung mass displacement, **d** unsprung mass velocity

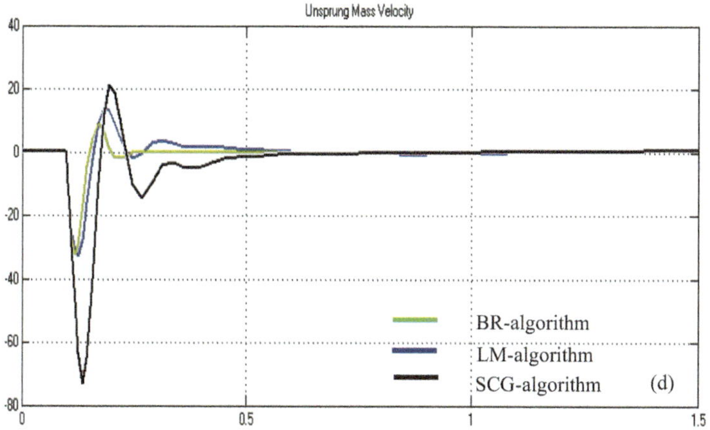

Fig. 6 (continued)

Table 3 Comparison of results for different algorithms

Attribute	Settling time (sec)			Maximum overshoot (%)		
	BR (s)	LM (s)	SCG (s)	BR (%)	LM (%)	SCG (%)
Sprung mass displacement	4.0	8.0	14.0	0.7	1.4	6.7
Sprung mass velocity	3.0	8.0	14.0	2.0	2.5	11.5
Unsprung mass displacement	0.5	5.0	8.0	0.8	1.2	4.1
Unsprung mass velocity	0.3	1.0	2.0	32.5	33.0	73.0

by Ghoniem et al. [13]. In their study the authors considered a semi-active vehicle suspension system controlled using a traditional ANN controller. A comparison of results indicate superior performance of ANN controller designed in this study using BR-algorithm giving a lower settling time response for sprung mass displacement i.e., 4.0 s compared to 5.5 s taken by the traditional passive suspension system controller at different throttle valve openings.

5 Conclusion

The study successfully considers three different machine learning algorithms for control of passive suspension system of quarter-car model. A mathematical model of the system has been initially developed and controlled using a PID controller. The results of PID controller were further used for training of an ANN controller. The results clearly indicate superior performance of BR-algorithm, followed by

LM-algorithm and SCG-algorithm respectively. Additionally, the least *MSE* and *R*-values were obtained using BR-algorithm. Further, the settling time and maximum percentage overshoot responses have been significantly reduced using BR-algorithm. The settling time taken by BR, LM and SCG algorithm to stabilize the complete system were 4.0 s, 8.0 s and 14.0 s respectively. Further, the maximum percentage overshoot responses obtained for sprung mass displacement using BR, LM and SCG algorithm were 0.7%, 1.4% and 6.7% respectively. Similarly, the maximum percentage overshoot responses obtained for unsprung mass displacement using BR, LM and SCG algorithm were 0.8%, 1.2% and 4.1% respectively. Lastly, as an extension for future work efforts can be made to reduce the maximum overshoot percentage responses for attribute unsprung mass velocity by either changing the number of neurons in the hidden layer or by adopting other popular machine learning algorithms like gradient descent, quasi-newton, etc.

References

1. Singh, D.: Passenger body vibration control in active quarter car model using ANFIS based super twisting sliding mode controller. Simul. Model. Pract. Theory **89**, 100–118 (2018)
2. Ahamed, M.M., Natrayan, L.: Structural and dynamic analysis of automotive quarter car model suspension system for different materials. Mater. Today Proc. **62**(Part 4), 2291–2296 (2022)
3. Robert, J.J., Kumar, P.S., Nair, S.T., Moni, Sharne Moni, D.H., Swarneswar, B.: Fuzzy control of active suspension system based on quarter car model. Mater. Today Proc. **66**(Part 3), 902–908 (2022)
4. Nguyen, T.A.: Research on the sliding mode–PID control algorithm tuned by fuzzy method for vehicle active suspension. Forces Mech. **11**, 100206 (2023)
5. Theunissen, J., Tota, A., Gruber, P., Dhaens, M., Sorniotti, A.: Preview-based techniques for vehicle suspension control: a state-of-the-art review. Annu. Rev. Control **51**, 206–235 (2021)
6. Nagarkar, M., Bhalerao, Y., Bhaskar, D., Thakur, A., Hase, V., Zaware, R.: Design of passive suspension system to mimic fuzzy logic control active suspension system. Beni-Suef Univ. J. Basic Appl. Sci. **11**, 109 (2022)
7. Issa, M., Samn, A.: Passive vehicle suspension system optimization using Harris Hawk Optimization algorithm. Math. Comput. Simul. **191**, 328–345 (2022)
8. Tulsian, N., Dewangan, S.: A discussion on energy harvesting through suspension system. Mater. Today Proc. **79**(Part 2), 189–192 (2023)
9. Kansal, V., Jain, U., Pant, B., Kotiyal, A.: Comparative analysis of convolutional neural network in object detection. Lect. Notes Netw. Syst. **520**, 87–95 (2023)
10. Dhiman, G., Kumar, A.V., Nirmalan, R., Sujitha, S., Srihari, K., Yuvaraj, N., Arulprakash, P., Raja, R.A.: Multi-modal active learning with deep reinforcement learning for target feature extraction in multi-media image processing applications. Multimed. Tools Appl. **82**, 5343–5367 (2023)
11. Kharola, A.: Novel adaptive control for avoiding fuzzy rule explosion in nonlinear systems. Int. J. Autom. Control. **17**(4), 377–396 (2023)
12. Kharola, A.: Analyzing dynamics of two-wheel mobile robot for real-time development of Chatbot. Int. J. Appl. Nonlinear Sci. **3**(4), 284–298 (2022)
13. Ghoniem, M., Awad, T., Mokhiamar, O.: Control of a new low-cost semi-active vehicle suspension system using artificial neural networks. Alex. Eng. J. **59**, 4013–4025 (2020)

Industrial Safety System Using IOT

Shital Sobale, Prathamesh Sawle, Sumeet Sapkal, Satvik Vishnoi, and Saurav Shimare

Abstract The purpose of this research article is to offer a prototype for an industrial safety protection system that makes use of Internet of Things (IoT) technologies. Industrial disasters can have devastating results, resulting in loss of life and long-term environmental damage. The lack of adequate protective mechanisms in power plant-based enterprises is a critical issue, and this study presents a remedy. The proposed model incorporates gas sensors, temperature sensors, and flame sensors to continuously monitor the environment, operate exhaust fans, and notify appropriate authorities in the event of a leak. Furthermore, an ESP32-based camera system identifies human activity using machine learning and provides all pertinent data to workers via an LCD panel and buzzer, while higher authorities receive alerts via the cloud. The study demonstrates the usefulness of IoT-based safety prevention systems in minimizing the likelihood and severity of industrial mishaps. The article concludes that the proposed prototype contributes significantly to industrial safety and can be applied to other industries, delivering a safer working environment for all stakeholders.

Keywords Internet of Things · Industry · Safety · MQ2 · Machine learning · Human detection

1 Introduction

The goal of the internet of things is to make life easier and faster by automating boring and repetitive chores that are a part of everyday living. As a result, everything is growing smarter. Because IOT is extremely useful for automating jobs, its benefits may be easily noticed in boasting safety measures in industries.

S. Sobale · P. Sawle (✉) · S. Sapkal · S. Vishnoi · S. Shimare
Department of Multidisciplinary Engineering, Vishwakarma Institute of Technology Pune, Pune, India
e-mail: prathamesh.sawle21@vit.edu

S. Sapkal
e-mail: sumeet.sapkal21@vit.edu

B. Soni et al. (eds.), *Beyond Artificial Intelligence*, Lecture Notes in Networks and Systems 1326, https://doi.org/10.1007/978-981-96-4170-3_3

Many health accidents, such as gas leakage, fuel leakage, and others, can occur in factories, causing catastrophic harm to people and equipment. To avert such tragedies, we can employ a variety of current technologies that detect the possibility of an accident before it occurs. Consumers are looking for more intelligent alternatives to tedious, mechanical, and manual processes due to today's quick pace and severe competition.

As a result of a busy schedule and a lack of time, scheduling gas has become one of the tasks that people choose to forget or postpone. The majority of mishaps in homes and companies are caused by gas leaks and temperature spikes, which result in numerous accidents that can be fatal.

Industrial safety systems based on IOT are designed to monitor and control a wide range of operations in industrial environments, including temperature, humidity, and gas and chemical leaks. The sensors collect data, send it to a central system, and then analyze it for anomalies. When an emergency arises, the system immediately sounds alerts and delivers information to the right people, allowing for quick interventions. The system that we propose incorporates a MQ2 sensor to detect gas leakage, a DS18B20 sensor to measure temperature, and an E265 sensor to detect the presence of flames. It leverages the aforementioned data to notify authorities and management if harmful situations are discovered while performing simple programmable operations such as starting exhaust fans.

2 Literature Review

Author has proposed a IOT industry protection system using NodeMCU controller which is a system to protect industries from losses due to accidents using Internet of Things. The system makes use of temperature and gas sensors to detect fire, gas leakage as well as low lighting to avoid any industrial accidents. This is another paper where the author has suggested a system that offers round-the-clock monitoring, gas leak detection, and database storage for prediction and analysis [1]. In comparison to conventional Gas detector systems, the system is significantly more cost-effective because of the IOT components that are utilized. This paper introduces a microcontroller-based model for the industry's fire safety. It can detect gas leakage or hazardous fire as well as take action to extinguish the fire [2]. Upon detecting suspicious activity, the system disconnects the building's primary power source and shuts the main gas valve. At the same time, it also notifies the manager and the nearby fire station via SMS. This report analyzes and introduces an intelligent fire-fighting system based on the new industry standard, and a smoke-fire detection and alarm device using Nb-IoT technology [3]. It also puts forward solutions to the problem of smart smoke. This study presents an automated human detection system for a complex industrial environment, in which people are monitored/detected from a top view perspective [4]. The system uses IOT technologies like 5G, intelligent sensors and devices to collect data which is then analyzed by using deep learning. Safety++ is an IoT system of inter-connected wearable devices for ensuring safety in energy industries [5]. They have used a user defined approach for designing which takes into consideration the needs of the users to improve product's adaptability. This paper suggests a system design for the augmented data recognising (ADR) method, where each hub contains a computational stage that connects certain mechanical characteristics. The Central BEAGLEBONE controller server has received the considerable data through WiFi, combines it with standard edge readings from various sensor network hubs, and sets the early alerts in the event of a masters' safety violation [6]. WiFi serves as a spine for communication between the BEAGLEBONE and the sensor hub. The system that enables the firm safety association systems, such as observing and announcing using augmented data recognising techniques, is broken down into its hardware and programming components in this study. A survey on the fire safety systems in industries suggests that most of the fire detecting systems detect fire after the accident has taken place [7]. So they have proposed a system for the firecracker industry to find the main cause of the fire and prevent the fire before it gets triggered. There is another paper that proposes an IoT-based fire detection system using wireless sensor networks [8] that can be used to detect and alert about fires. The field of IoT has greatly impacted the industrial sector, leading to the development of various innovative systems for monitoring, controlling, and managing industrial processes. Several research papers have been published on different IoT-based industrial applications such as condition monitoring of equipment, real-time process monitoring and control, energy management, predictive maintenance, and quality control [9]. For instance, Kumar et al. proposed an IoT-based condition monitoring system for industrial equipment, while Kanna et al. presented a real-time [10]. Monitoring and controlling system for industrial processes using IoT. Similarly, Patil and Desai [11] developed an IoT-based monitoring and control system for industrial applications, and Ryu et al. [12]. Proposed a smart manufacturing system based on IoT. Other research papers include IoT-based energy management [13] (Ravikumar) predictive maintenance [14] (Mohapatra) smart factory automation (Pham et al.), wireless sensor networks (Rajalakshmi and Yadav), quality control (Kim et al.), and inventory management (Rajendran and Kumaravel). These studies demonstrate the potential of IoT-based systems in enhancing industrial processes, improving productivity, reducing costs, and increasing efficiency.

3 Methodology

3.1 Purpose and Requirement

To increase workers' and employees' safety and wellbeing at work, an industrial safety detection system is used. It is aimed at enhancing the functionality of industries by allowing them to use cutting-edge technology to monitor and inform workers about various aspects of their work environment.

3.2 Design

This system is concentrated on the design and execution of an industrial safety detection system for workers and employees, with a variety of features to improve worker efficiency, protection, and safety. It contains a variety of sensors. In our system, the MQ2 sensor, which is a gas sensor used to identify potentially dangerous gasses, and the LM35 temperature sensor, which measures industrial temperature, are both utilised. A flame sensor is used to find a fire or flame. The camera module that we utilise as the primary sensor for our system allows us to readily identify people who are present in hazardous areas of industry by detecting them and sending signals in response. In our model, we used an axial fan. We also used an Arduino Uno and an ESP32 cam board. We use the Arduino IDE for the programming portion. Overall, this Industrial safety detection system will help to reduce the risk of accidents in big industries, while also improving productivity of companies and worker, employees' safety.

3.3 Information Model Specification

The system would feature numerous sensors to monitor the environment and the safety of workers, according to the information model standard for such a device. Sensors for temperature, air quality, gas concentration, and flame detection may be included. These sensors would produce information that might be utilized to identify possible risks and give an early warning. The data produced by the sensors would be stored in this system's data storage component. Predictive models could be developed using this data and used for analysis and trend detection (Fig. 1).

3.4 System Architecture

Our proposed industrial safety system consists of various sensors and actuators, which ensure the safety of workers and industry by detecting fire, gas leakage or any kind of human intervention in restricted areas. It consists of a temperature sensor which monitors the temperature of the surroundings and sends an alert if the temperature goes above threshold value. Similarly a gas sensor is also deployed which sends an alert if it senses any harmful gasses in the environment. The flame sensor is installed on the system to detect the presence of flames in case of fire in the industry and alert the user. All of the sensors are connected to the ESP8266 which acts as the microcontroller for our system. The ESP32 camera module is used to detect any human activity in prohibited areas of the industry. The ESP8266 receives the data from the sensors and processes it. Then a message alert is sent to the user on Telegram (Fig. 2).

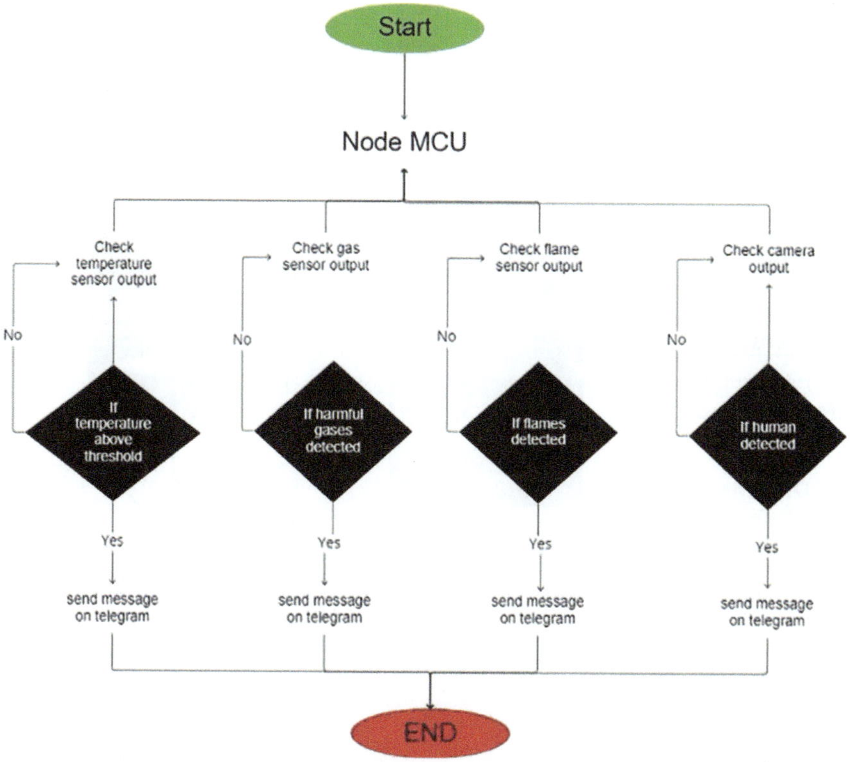

Fig. 1 Process flow

The proposed system provides four main services as:

1. Temperature Monitoring
2. Gas leak detection
3. Fire detection
4. Human detection in prohibited areas.

The Fig. 2 shows service flow diagram for the temperature monitoring. The temperature sensor takes the temperature of the surrounding as an input and checks if the temperature is greater than or less than the threshold value. It gives an output if the temperature is above the threshold value and sends a message to the user on telegram.

The other services also can be represented in a similar way as the Fig. 3.

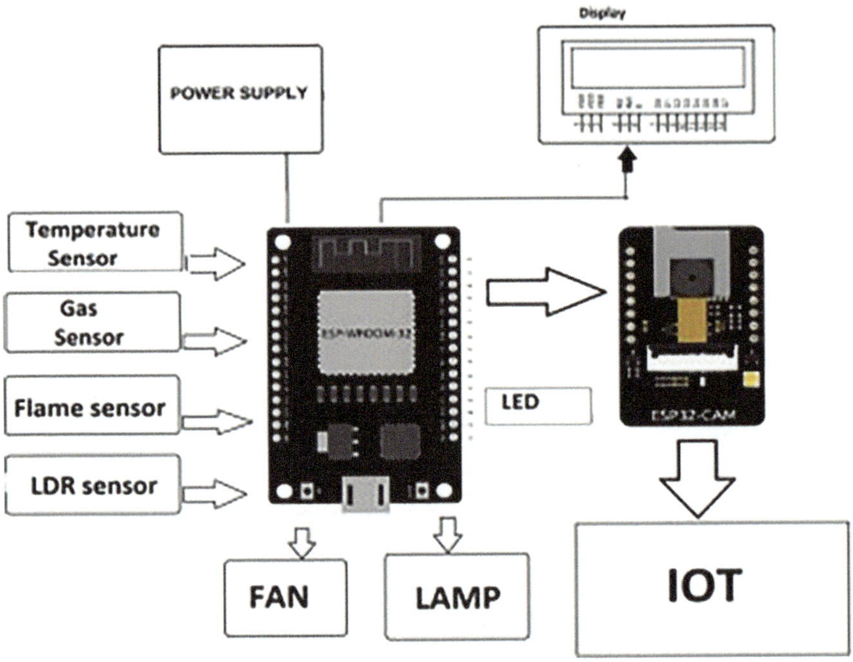

Fig. 2 System architecture

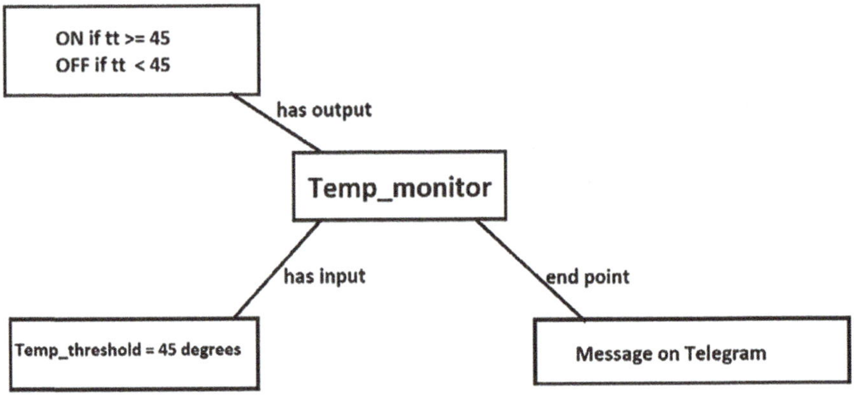

Fig. 3 Service specification

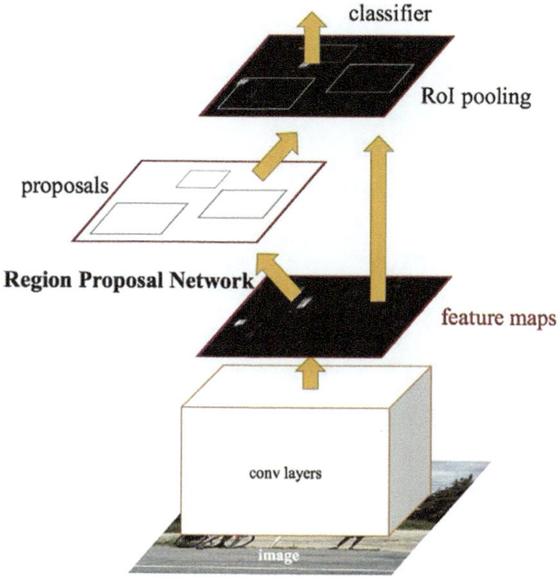

Fig. 4 FRCNN structure

3.5 Human Detection Model

AI is often used with IoT for more effective IoT services. It has also shown great success for various surveillance applications. In our proposed system for detection of humans in prohibited areas of the industry, we are using FRCNN (Faster Region Convolutional Neural Network) to create a neural network. Using that we are detecting the top view along with the top side view. With the help of this we can detect multiple people by considering the head and shoulder data (Fig. 4).

3.6 Description of Sensors

Gas Sensor: A Gas sensor is used to detect Smoke detection in the workplace is done using a gas sensor. An alert is let out when gas is found, telling the worker to leave the area.

Temperature Sensor: A temperature sensor is used to monitor the temperature in the industry. An alert is sent to the worker to let them know what to do if the temperature exceeds a specified threshold.

Buzzer: The worker is informed when an abnormality is found via a buzzer. For instance, the buzzer will sound to inform the worker to take action if smoke or a high temperature is detected.

Fig. 5 Circuit diagram

3.7 Telegram Bot

Telegram bot is a feature of the Telegram messaging app that allows users to automate various tasks and receive information through customized messages. Telegram bots can be created by using the Telegram Bot API. This API provides developers with a set of tools and methods to interact with Telegram servers and perform various tasks, such as sending and receiving messages, managing groups, and monitoring user activity.

4 Results and Discussion

The buzzer is in operation when the flame sensor is active. The fan and led turn on when the temperature sensor is in the on position. The fan and buzzer are active when the gas sensor is in the on position (Figs. 5, 6 and 7).

To increase workers' and employees' safety and wellbeing at work, an industrial safety detection system is used. It is aimed at enhancing the functionality of industries by allowing them to use cutting-edge technology to monitor and inform workers about various aspects of their work environment (Table 1).

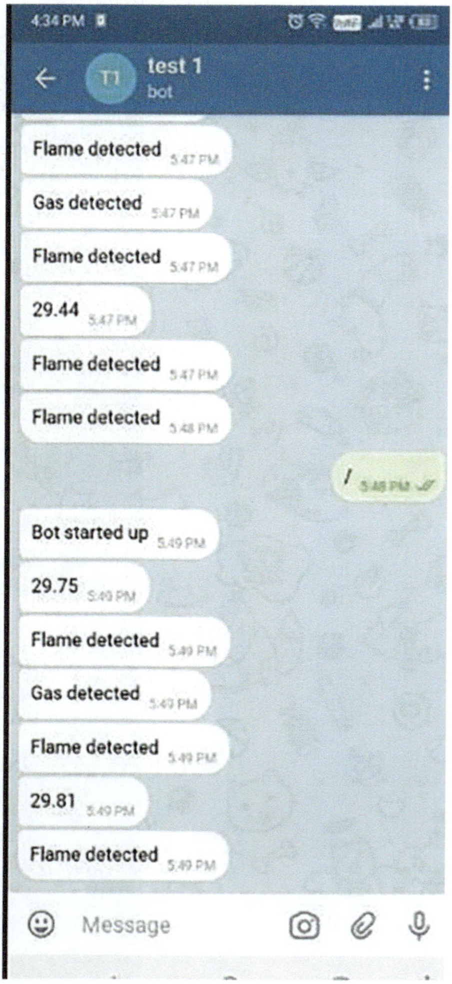

Fig. 6 IOT notification

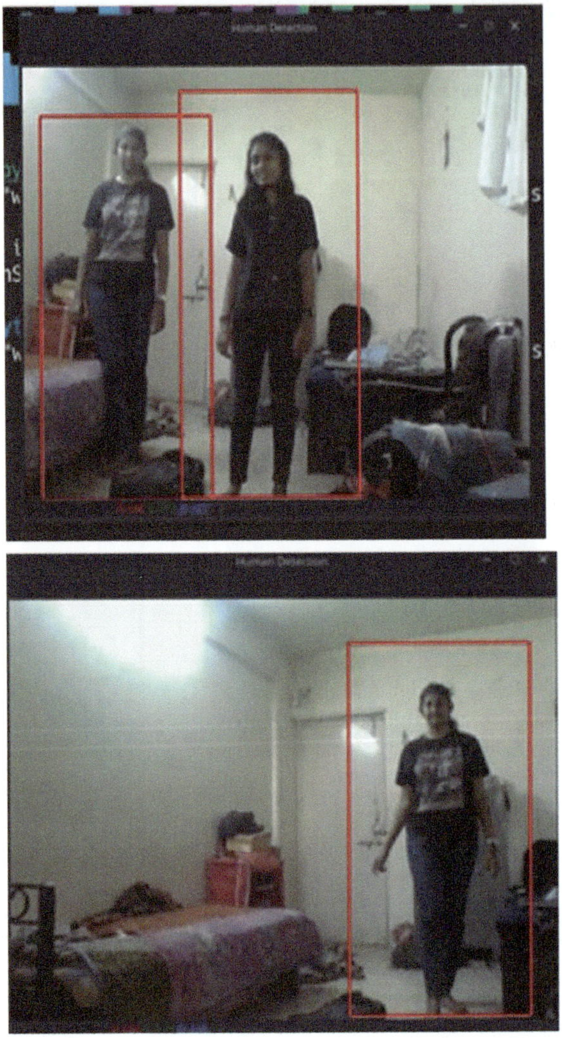

Fig. 7 Human detection model

Table 1 Truth table for input and output

Flame sensor	Gas sensor	Temperature sensor	Fan	Lamp (LED)	Buzzer	Message
0	0	0	0	0	0	0
0	0	1	1	1	0	1
0	1	0	1	0	1	1
0	1	1	1	1	1	1
1	0	0	0	0	1	1
1	0	1	0	1	1	1
1	1	0	0	0	1	1
1	1	1	0	1	1	1

1 = On mode
0 = Off mode

5 Conclusion

In conclusion, the ESP32-based industry protection system with human detection using gas sensor, temperature sensor, and flame sensor is a promising solution for improving industrial safety. Through its ability to detect hazardous conditions, such as gas leaks, abnormal temperature changes, and flames, the system can alert workers to potential dangers in real-time, allowing them to take necessary actions to prevent accidents. Moreover, the human detection feature can significantly reduce the risk of accidents caused by human error.

However, further research is needed to optimize the system's performance and evaluate its effectiveness in real-world industrial settings. Future studies could explore the system's scalability, cost-effectiveness, and integration with other safety technologies to provide a comprehensive solution for industrial safety.

References

1. Wase, V., More, V., Dukre, S., Kawalkar, M., Wankhede, D., Masram, A., Akhtar, F.: IoT Based Industry Protection System for Labor Safe ty
2. Bagwe, C., Ghadi, V., Naik, V., Kunte, N.: IoT Based Gas Leakage Detection System with Database Logging, Prediction and Smart Alerting
3. Saiam, M., Jahan, M.S., Akther, M.S., Alam, M.N.: A Microcontroller Based Fire Protection System for Safety of Industries in Bangladesh
4. Suma, V., Shekar, R.R., Akshay, K.A.: Gas Leakage Detection Based on IoT
5. Joshi, S., Munjal, S., Karanje, U.B.: Gas Leakage Detection and Alert System Using IoT
6. Bernal, G., Colombo, S., Al Ai Baky, M., Casalegno, F.: Safety++, Designing IoT and Wearable Systems for Industrial Safety Through a User Centered Design Approach
7. Palanivelu, R., Srinivasan, P.S.S.: Safety and Security Measurement in Industrial Environment Based on Smart IoT Technology Based Augmented Data Recognizing Scheme
8. Savitha, N., Malathi, S.: A Survey on Fire Safety Measures for Industry Safety Using IoT
9. Sridhar, S., Siva, S.: IoT-Based Fire Detection System Using Wireless Sensor Networks

10. Kumar, A., Kumar, N., Kumar, A.: IoT-Based Condition Monitoring of Industrial Equipment
11. Patil, S.S., Desai, A.S.: IoT-Based Monitoring and Control System for Industrial Applications
12. Ryu, S.-J., Kim, S.-W., Kim, Y.-J.: Smart Manufacturing System Based on IoT
13. Ravikumar, B.R., Savani, V.J., Jivani, B.N.: IoT-Based Energy Management System for Industrial Applications
14. Mohapatra, S.R., Gupta, A.K., Sahoo, S.: IoT-Based Predictive Maintenance for Industrial Equipment

Design and Implement Security Techniques for Authenticating IoT-Based Smart Homes

Neha Sharma and Pankaj Dhiman

Abstract The popularity of IoT-based smart homes is increasing rapidly due to advancements in information and communication technologies (ICT). It is essential to prioritize security measures when using IoT components like sensors, as they can gather sensitive information. A secure, lightweight authentication system can ensure confidentiality and integrity for users and smart devices. The proposed method safeguards the network and provides secure communication. The proposed method was verified informally and formally with the AVISPA tool. It was also compared to other schemes in terms of costs.

Keywords Lightweight authentication protocol · Identity verification and key agreement · Wireless Sensor Networks (WSN) · (GWN) Gateway Node

1 Introduction

The Internet of Things (IoT) is a network of physical sensors, gadgets, and smart devices communicating data through open networks. Smart home systems often use these devices, but their unique characteristics and architectures can pose security risks. It's crucial to prioritize usability while enabling them to make intelligent decisions without human intervention [1, 2]. Smart homes offer cost savings, comfort, and smartphone control while prioritizing safety and environmental protection. Authentication protocols help protect privacy, but data protection concerns persist due to possible breaches [3]. Home networks are vulnerable to security flaws due to the combination of wired and wireless mediums and protocols. Additionally, they

---

N. Sharma (✉) · P. Dhiman
Department of CSE/IT, Jaypee University of Information Technology (JUIT), Waknaghat, Solan 173234, HP, India
e-mail: nehasharma.bu@gmail.com

N. Sharma
Assistant Professor, School of Computing DIT University Dehradun, Dehradun 248009, Uttarakhand, India

B. Soni et al. (eds.), *Beyond Artificial Intelligence*, Lecture Notes in Networks and Systems 1326, https://doi.org/10.1007/978-981-96-4170-3_4

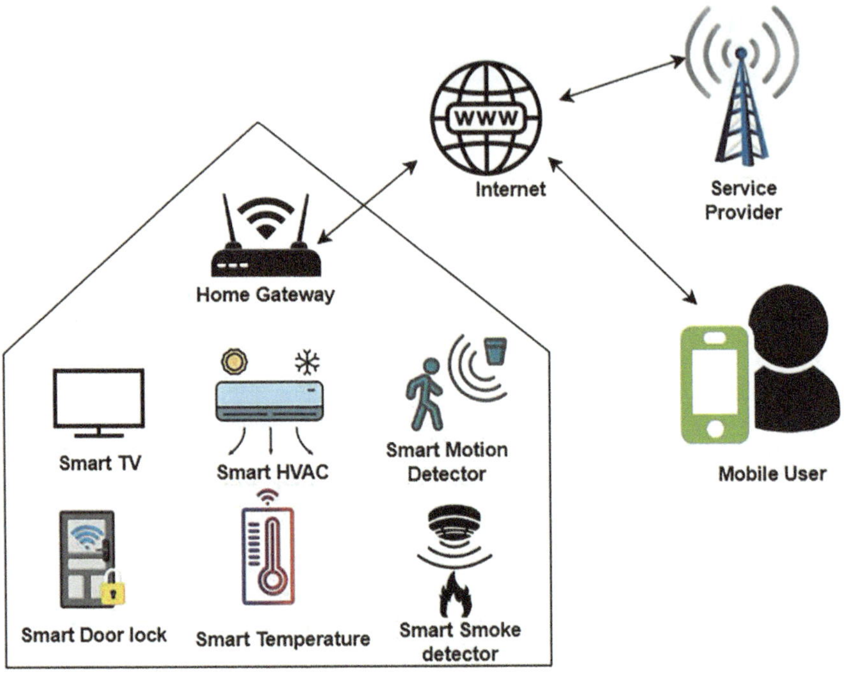

Fig. 1 The system model for a smart home network

struggle to keep up with current network-based cyber threats [4, 5]. We are developing a secure authentication protocol for smart homes with anonymous mutual authentication for safety agreements (Fig. 1).

2 Related Work

To ensure the security of multiple mobile IoT devices on a constantly changing home network, identity verification and risk prevention measures must be implemented [6]. Priv-Home is a system that offers secure data storage, identity verification, access restrictions, and privacy-preserving queries to protect privacy in smart homes [3]. To protect data confidentiality, we use access restrictions and privacy-preserving queries [7]. Our paper proposes a schedule for preventing smart card attacks [8], while the author suggests improving forward security for better communication [9].

3 Attacker Model

In our study, we implemented the Dolev-Yao attack model, which allowed potential attackers to monitor network traffic, create new messages, and manipulate existing ones. Attackers may try to gain unauthorized access to data by exploiting vulnerabilities in smart home devices, whether internal or external to the network [10].

4 Proposed Scheme

Our proposed protocol enables anonymous user authentication in a smart home system consisting of smart device users, IoT device users, and gateways. Figure 2 illustrates the user authentication process and key sharing with GWN.

1. The user's identity is masked as mid_u, which is the same as Id_u, MS, and n_1, also known as sm.
2. The user authenticates by sending ($\mathrm{mid}_u \parallel Ms$) to the IoT device, which then connects to the GWN.
3. The value of the GWN is determined by n_1, that is equivalent to MS Sm, additionally, its value Id_u, that is equivalent to sm, n_1, mid_u.
4. Encrypt n_1 and n_2 for security when connecting with GWN.
5. Users may encrypt data, obtain the session key, and establish n_2 through the GWN's return of n_1.
6. Encrypting n_1 and n_2 ensures message legitimacy and user activity.

Figure 2 depicts the mutually authenticating and confirming each other of an IoT device and a GWN. The following are the steps (Table 1):

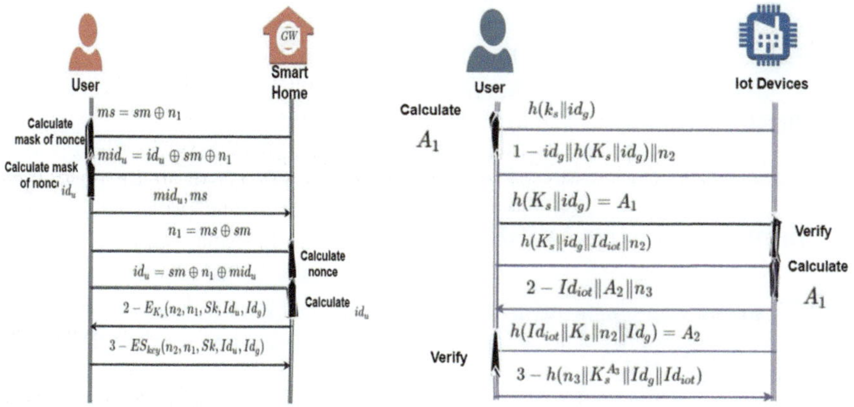

Fig. 2 Phase of user and GWN authentication and GWN and IoT devices Authentication

Table 1 A list of symbols and their descriptions

Notations	Descriptions
Id_u	Identity of the user
Id_{iot}	Device ID for IoT
Id_g	Gateway ID
n_1	User-generated nonce
n_2, n_2'	Nonce produced by home gateway
n_3	IoT device-generated nonce
K_s	Mutually secret key
h(.)	Hash function
S_{Key}	Session Key exchange
mid_u	Mask ID
T	Timestamp
Ms	Mask of nonce
Sm	Secret message
X_{k_s}	Symmetric key encryption between SHS and HG
$\oplus$	xor
$\|$	Concatenation

1. The GWN produces nonces n_2' and sends Id_g towards the IoT device after calculating $A1 = \mathrm{h}(K_s \| Id_g \| n_2' \| Id_g)$.
2. IoT devices validate gateway authenticity by checking A_1. A match with $\mathrm{h}(K_s \| Id_g)$ confirms the request is from the gateway and not a replay attack.
3. The IoT device generates n_3 to ensure the message's freshness and computes $A2 = \mathrm{h}(K_s \| Id_g \| Id_{iot} \| n_2')$ before sending $(\mathrm{Id}_{iot} \| n_3)$ to the GWN.
4. The GWN determines whether $A2 = \mathrm{h}(Id_{iot} \| (K_s \| n_2' \| Id_g)$, in case n_2' has been verified, the IoT device is fresh and authentic.
5. The GWN transmits $A3 = \mathrm{h}(n_3 \| K_s \| Id_g \| Id_{iot})$ to authenticate IoT devices bidirectionally using K_s and n_3, as shown in Fig. 2.

5 Simulation Using AVISPA Tool

AVISPA software automatically analyzes and validates Internet protocols and applications for security [10]. The scheme is validated and secured by the OFMC and CL-AtSe backends. They specify user, gateway, session, environment, and goal roles, as shown in Fig. 3. The scheme is secure against attacks based on OFMC results in

```
%%%%%Role for HG%%%%
Role role_HG (A, HG: agent,
Hash: hash_func,
SK: symmetric key,
SND, RCV: channel (Dy))
Played by HG
Def=
Local State: Nat,
TDi, IDg, Ai, Bi, Alpha: text,
T1, T2, X, Y: text,
V1, V2, CIDi, TK, CI, C2: message,
Hash: hash_func        const
deviceA_hg_Ra, hg_deviceA_Rg,
device_hg_IDi, hg_deviceA_IDg:
protocol_id,
deviceA_hg_T1, hg_deviceA_T2,
sub1, sub2, sub3:protocol_id
Init State: =0
 transition
1. State=0/\RCV (V1', CIDi', C1, T1') =|>
State':=1/\T2':=new ()
/\secret ({Hash(x, y)}, sub2, {A, HG})
/\secret ({MIIDi, IDg}, sub1, {A, HG})
/\Rg':=new ()
/\SK':= Hash (Ra', Rg', MIIDa.IDg.Alpha)
/\xor (Rg', xor (Bi, xor (Hash (Ai, T2'))))
%/\({ Idg, T1, T2} _SK')
/\SND({IDg.T1.T2}_SK,xor(Rg',xor(Bi,xor
(Hash(Ai,T2')))).T2')
/\witness (HG, A, hg_device_T2, T2')
%HG has freshly generated the value of T2 for A
end role
```

```
%%%%%Role for Session%%%%
Role session (A, HG: agent,
Hash: hash_func,
K: symmetric key,
SK: symmetric key)
def=
local AS, AR, HGS, HGR: channel (dy)
Composition
Device A (A, HG, SK, Hash, AS, AR)
/\home gateway (A, HG, SK, Hash, AS, AR)
end role
Role environment ()
def=
Const deviceA, home gateway: agent,
SK: symmetric key,
H: hash_func,
xg,yg,k,alpha,ida,idg,ra,rg,t1,t2:text,deviceA_hg_
Ra,hg_deviceA_Rg,deviceA_hg_MIIDi,
hg_device A_IDg, deviceA_hg_T1,
hg_deviceA_T2, auth1, auth2, auth3: protocol_id
Intruder knowledge = {deviceA, hg, h}
Composition
Session (deviceA, home gateway, h)
/\session (deviceA, i, h)
/\session (home gateway, i, h)
end role goal
secrecy_of sub1,sub2,sub3
authentication on device_hg_ MIIDi
authentication on device_A_IDg
authentication on device_hg_T1
end goal
environment()
```

Fig. 3 Role for user and session and environment

Fig. 4. According to the summary of results under OFMC, the proposed scheme is secure.

6 Performance Evaluation

Our proposed method will be compared to other methods regarding effectiveness, communication, and computational cost.

6.1 Comparison of Communication and Computation Cost

The gateway authenticates with a unique ID before login, as shown in Table 2. Our method uses only 2336 bits and messages for mutual authentication, making

Fig. 4 The AVISPA results utilizing OFMC

```
% OFMC
% Version of 2006/02/13
SUMMARY
SAFE
DETAILS
BOUNDED_NUMBER_OF_SESSIONS
PROTOCOL
/home/span/span/testsuite/results/SHS.if
GOAL
as specified
BACKEND
OFMC
COMMENTS
STATISTICS
parseTime: 0.00s
searchTime: 0.04s
visitedNodes: 12 nodes
depth: 4 plies
```

Table 2 Comparison of communication and computation cost

Scheme	No. of bits	No. of messages	Computation cost	Execution time (milliseconds) (ms)
Fakrron et al. [3]	2510	3	$3T_E/T_D + 2T_{mac} + 3T_{hmac} + T_h$	15.88
Park et al. [8]	4269	2	$4T_E/T_D + 28T_h$	30.12
Santoso et al. [7]	4365	3	$4T_{exp} + 3T_h$	60.52
Wazid et al. [9]	2657	4	$5T_E/T_D + 2T_{fe} + 20T_h$	51.66
Our Scheme (SHS)	2336	2	$2T_E/T_D + 3T_h$	12.6

it easy to compare with other methods. The user sends $(Id_u \oplus sm \oplus n_1)(sm \oplus n_1)$ to the gateway, followed by the gateway sending $n_2, n_1, S_k, Id_u, Id_g K_s$ to the user. The user sends n_2 S_k back to the gateway before it communicates with the outside $(Id_g, h(K_s, Id_g), n_3)$ $u \rightarrow g$:$(Id_u \oplus sm \oplus n_1)(sm \oplus n_1)$,$g \rightarrow U$:$n_2, n_1, S_k, Id_u, Id_g K_s$,$u \rightarrow g$: n_2 S_k,$g \rightarrow o$:$(Id_g, h(K_s, Id_g), n_2$,$g \rightarrow o$:$h(n_3, K_s, Id_g, Id_o)$ require $(128 + 160 + 160 + 160 + 160) = 768$ bits; 256 bits; 256 bits; $(128 + 160 + 160) = 448$ bits. $(128 + 160 + 160) = 448$ bits and 160 bits, respectively.

7 Conclusion

Protecting users' privacy in smart homes is of utmost importance due to the risk of hacking. A secure authentication scheme has been developed to address this issue. It utilizes a specialized protocol for smart homes, ensuring mutual authentication and encryption between the user and the network. This method offers superior security and cost-effectiveness, guarding against threats like essential disclosure and replay attacks.

References

1. Abosata, N.R.A., Kemp, A.H., Razavi, M.: Secure smart-home application based on iotcoap protocol. In: 2019 Sixth International Conference on Internet of Things: Systems, Management and Security (IOTSMS), pp. 13–17. IEEE (2019)
2. Andola, N., Yadav, V.K., Venkatesan, S., Verma, S.: Spychain: a lightweight blockchain for authentication and anonymous authorization in IOD. Wirel. Personal Commun. **119**, 343–362 (2021)
3. Fakroon, M., Alshahrani, M., Gebali, F., Traore, I.: Secure remote anonymous user authentication scheme for smart home environment. Internet Things **9**, 100158 (2020)
4. Gope, P., Hafizul Islam, S.K., Obaidat, M.S., Amin, R, Vijayakumar, P.: Anonymous and expeditious mobile user authentication scheme for glomonet environments. Int. J. Commun. Syst. **31**(2), e3461 (2018)
5. He, J., Xiao, Q., He, P., Pathan, M.S.: An adaptive privacy protection method for smart home environments using supervised learning. Future Int. **9**(1), 7 (2017)
6. Khanpara, P., Lavingia, K., Trivedi, R., Tanwar, S., Verma, A., Sharma, R.: A context-aware internet of things-driven security scheme for smart homes. Secur. Priv. **6**(1), e269 (2023)
7. Santoso, F.K., Vun, N.C.H.: Securing IoT for smart home system. In: 2015 International Symposium on Consumer Electronics (ISCE), pp. 1–2. IEEE (2015)
8. Park, N., Kim, M., Bang, H.-C.: Symmetric key-based authentication and the session key agreement scheme in IoT environment. In: Computer Science and its Applications: Ubiquitous Information Technologies, pp. 379–384. Springer (2015)
9. Wazid, M., Das, A.K., Odelu, V., Kumar, N., Conti, M., Jo, M.: Design of secure user authenticated key management protocol for generic IoT networks. IEEE Int. Things J. **5**(1), 269–282 (2017)
10. von Oheimb, D.: The high-level protocol specification language HLPSL developed in the EU project avispa. In: Proceedings of APPSEM 2005 Workshop, pp. 1–17. APPSEM'05, Tallinn, Estonia (2005)

Analysis of Hybrid Features for Breast Cancer Classification

Palwinder Kaur and Amandeep Kaur

Abstract Early detection of breast cancer plays a vital role in stopping the spread of disease and reducing mortality rates among women. Radiologists face the arduous task of manually interpreting mammograms, which can lead to potential diagnostic errors. Therefore, it is imperative to employ a Computer-Aided Diagnosis (CAD) approach to automatically identify early-stage breast cancer and aid radiologists in their decision-making process. In this paper, we propose a deep learning based automatic breast cancer classification method which utilizes the hybrid features extracted from two different methods Gray-Level Co-occurrence Matrix (GLCM) and Convolutional Neural Networks (CNN). We merge the features before feeding them to the model for classification. Eventually, we evaluate the model performance using the dataset of breast mammography images with masses and compare its performance with state-of-the-art.

Keywords Breast cancer · Mammogram classification · Convolutional Neural Networks · Computer Aided Diagnosis

1 Introduction

Breast cancer is a condition where abnormal cells grow uncontrollably in the breast tissue [1]. These abnormal cells can lead to either benign or malignant cancer. Benign breast conditions are non-cancerous abnormalities that are not considered life-threatening. Malignant breast cancer cells can spread to other parts of the body through the lymphatic system or bloodstream.

Mammograms can identify potential abnormalities in the breast tissue, such as masses, micro-calcifications, or architectural distortions, even before they can be felt during a physical examination. In traditional mammogram classification, radiologists visually inspect the mammography images and assess several characteristics, including shape, size, margins, density, and distribution of abnormalities. They follow

P. Kaur (✉) · A. Kaur
Department of Computer Science and Technology, Central University of Punjab, Bathinda, India
e-mail: palwindergill434@gmail.com

B. Soni et al. (eds.), *Beyond Artificial Intelligence*, Lecture Notes in Networks and Systems 1326, https://doi.org/10.1007/978-981-96-4170-3_5

established guidelines and scoring systems, such as the Breast Imaging Reporting and Data System (BI-RADS) [2], to assign a category or score to the mammogram, ranging from benign to highly suspicious or malignant.

Computerized image classification in the field of healthcare plays a crucial role in various applications, including medical diagnosis, disease detection, and treatment planning. With the progress in technology, computer-aided detection and classification systems have been developed to assist radiologists in the interpretation process. Researchers and medical professionals are continuously working on developing and refining algorithms and techniques to enhance the sensitivity, specificity, and overall performance of mammogram classification systems. These systems employ artificial neural networks, support vector machines, or deep learning models to automatically provide an accurate classification.

The classification of breast cancer mammograms is an active area of research. However, it presents several challenges and limitations that researchers and medical professionals strive to address. Some of the common challenges and limitations include:

- Subjectivity: Mammogram interpretation involves a certain degree of subjectivity, as different radiologists may interpret the same image differently. Variations in expertise, and personal judgement can lead to inconsistencies in classification.
- False positives and false negatives: Mammogram classification can result in false positives, where a benign lesion is incorrectly classified as malignant, leading to unnecessary interventions and anxiety for patients. Conversely, false negatives can occur when a malignant lesion is missed or incorrectly classified as benign, potentially delaying diagnosis and treatment.
- Variability of breast tissue: Breast tissue is complex and exhibits significant variations in density, texture, and appearance. Dense breast tissue can obscure abnormalities and increase the risk of false negatives. Differentiating between benign and malignant lesions in dense breasts can be particularly challenging.
- Overlapping features: Some benign and malignant lesions may exhibit similar radiological features, making it difficult to differentiate between them. For example, benign masses and early-stage tumors can share similar characteristics, leading to potential misclassification.

Addressing these challenges and limitations requires ongoing research, technological advancements, and collaborations between radiologists, researchers, and engineers. By continuously refining classification algorithms, improving imaging techniques, and promoting standardized guidelines, the accuracy and reliability of breast cancer mammogram classification can be enhanced, leading to improved patient outcomes and more effective breast cancer management.

As per The Image Biomarker Satandardization initiative (ISBI) [3] for medical images, we chose to use Gray-Level Co-occurrence Matrix (GLCM) [4] based standard features in addition to Convolutional Neural Networks (CNNs) [5] features in the proposed model. GLCM is a texture analysis technique used to quantify the spatial relationships of pixels in an image. GLCM captures information about the distribution and patterns of gray-level intensities in an image by calculating the frequency of occurrence of different combinations of pixel values or gray-level pairs.

We hybridize the GLCM and CNNs features to fully explore the discriminative features of digital mammograms. To evaluate the model performance, we use an aggregated dataset of three different datasets [6], the Mammographic Image Analysis Society (MIAS), the Digital Database for Screening Mammography (DDSM), and the Inbreast dataset.

The rest of the paper is organized as follows: In Sect. 2, several research papers, their methodologies and results are briefly discussed. Methods and material used for this work are given in Sect. 3. Results are discussed in Sect. 4 and Sect. 5 gives our conclusions.

2 Literature Review

With the advancements in computer science and technology, it has become easy for researchers to develop efficient algorithms and tools to assist healthcare providers. Accurate breast cancer classification helps identify women who may require further diagnostic evaluation or additional imaging studies based on the findings observed on the mammogram. Numerous studies and advancements have been made to develop and improve the performance and applicability of computerized breast cancer classification techniques.

Khaoula et al. [7] used the Entropy thresholding method for pectoral muscle removal and extracted the region of interest (ROI) using the Metaheuristic algorithm Particle Swarm Optimization (PSO). They also extracted Shape and texture features from the abnormalities using the Fourier transform and GLCM respectively. The classification of detected abnormalities is performed using a Support Vector Machine achieving 83.87% accuracy for the Mini-MIAS dataset.

In their study, Kaur et al. [8] employed a combination of Multiclass Support Vector Machine (MSVM) and deep learning techniques for the classification of breast cancer. They trained a CNN using a K-means clustering algorithm and an MSVM classifier. The proposed approach achieved an impressive overall accuracy of 96.9% when applied to the Mini-MIAS dataset.

Hu et al. [9] proposed a method called WMDNet that combines multi-view input and weighted multi-instance learning (MIL) for analyzing mammograms. The multi-view input technique enhanced abnormality detection by leveraging mammograms with different views, capturing more potential features. Weighted MIL was employed to identify the most suspicious lesions, addressing issues related to the small proportion of abnormalities and class imbalance. The WMDNet method achieved 94.7% accuracy using the INbreast dataset and 93.75% accuracy using the MIAS dataset.

Darweesh et al. [10] suggested a hierarchical machine learning (ML) technique for classifying breast cancer using MIAS mammograms. First, they used GLCM for feature extraction and random forest for classification to separate the mammograms into healthy and cancerous mammograms. At the subsequent stage, they used Local Binary Patterns (LBP) for feature extraction and random forest for classification to separate the abnormal mammograms into benign and malignant categories. The suggested method had an overall accuracy of 85%.

Jayesh et al. [11] developed a cancer diagnosis system that combines a CNN and image texture attribute extraction. The system utilized a customized CNN for classification and incorporated texture features from LBP and Gabor filter with dimensionality reduction using Uniform Manifold Approximation and Projection (UMAP). An ensemble algorithm combines the results from each phase to generate the final diagnosis. If any stage indicates malignancy, the overall categorization is considered malignant. The system achieved a testing accuracy of 98% on the MIAS dataset and 98.3% on the DDSM repository.

3 Materials and Methodology

In this research, we are using a master dataset which is an aggregated form of three datasets MIAS, DDSM and Inbreast datasets [6]. The master dataset has a total of 24576 mammograms, of which 13710 are malignant and 10866 are benign.

The proposed model consists of two stages: feature extraction and classification. In the feature extraction stage, two different methods are employed. The first method is GLCM, which allows us to extract standardize texture features from mammograms which are defined in ISBI [3]. The second method is CNN, which extracts more abstract and semantic features from the input images. By combining the features obtained from both methods, we create a comprehensive feature representation for the next stage of the model i.e., classification.

The flow of the proposed work is illustrated in Fig. 1. It consists of multiple steps including data pre-processing, feature extraction, training and testing for classification. First, we resize the input images to a fixed size of 128 $\times$ 128 pixels and then we perform pixel normalization by dividing the pixels by 255. Next, we label the data and split the dataset into 85% for training and 15% for testing. At the feature extraction stage, we extract GLCM and CNN features and concatenate them. To assess the model's performance, we employ stratified 5-fold cross-validation. We take into account the model accuracy and other metrics for further analysis.

In this work, we used the CNN model architecture defined in Kaur and Kaur [12] as a base and modified it to concatenate GLCM features. The architechture of the proposed model is given Table 1. For the efficient classification performance, we used specific parameters and techniques. These included learning rate 0.0001, batch size 64, random state value 42, threshold value greater than 0.5, dropout parameter 0.5, and 150 epochs. To optimize the loss function, we utilized the Adam optimizer. For feature extraction, we used Swish activation function and for classification, we used sigmoid activation function.

To assess and compare the performance of different machine learning models, different performance metrics provide quantitative or qualitative measures. The

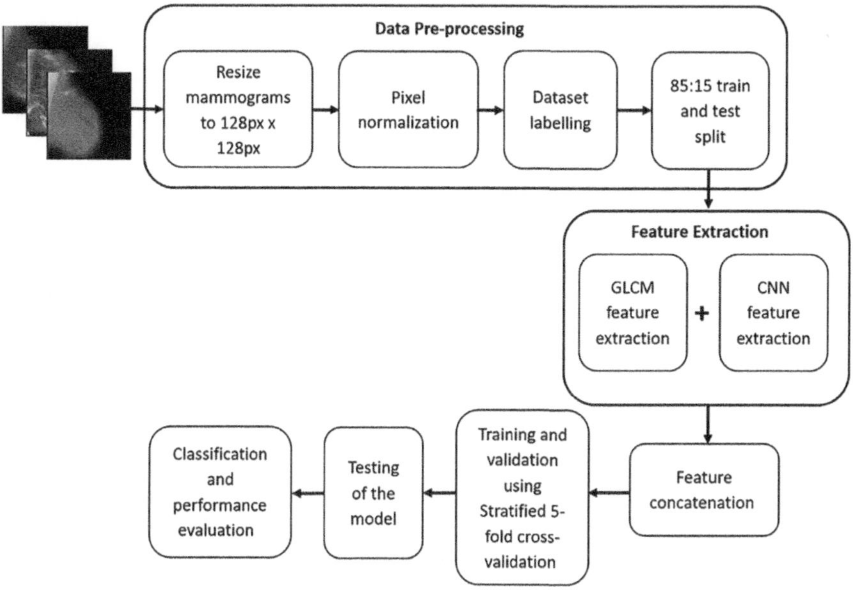

Fig. 1 Workflow of the proposed model

Table 1 Architecture of the proposed model

Layer (Type)	Output shape	Number of parameters
input (InputLayer)	(None, 128, 128, 1)	0
conv2d (Conv2D)	(None, 126, 126, 16)	160
max_pooling2d (MaxPooling2d)	(None, 63, 63, 16)	0
conv2d_1 (Conv2D)	(None, 61, 61, 32)	4640
max_pooling2d_1 (MaxPooling2d)	(None, 30, 30, 32)	0
conv2d_2 (Conv2D)	(None, 28, 28, 64)	18496
max_pooling2d_2 (MaxPooling2d)	(None, 14, 14, 64)	0
dropout (Dropout)	(None, 14, 14, 64)	0
flatten (Flatten)	(None, 12544)	0
input_1 (InputLayer)	(None, 20)	0
concatenate (Concatenate)	(None, 12564)	0
dense (Dense)	(None, 32)	402080
batch_normalization (BatchNormalization)	(None, 32)	128
dropout_1 (Dropout)	(None, 32)	0
dense_1 (Dense)	(None, 1)	33

performance metrices used in the proposed model can be represented mathematically as follows:

$$Accuracy = (TP + TN) \big/ (TP + TN + FP + FN) \tag{1}$$

$$Precision = TP \big/ (TP + FP) \tag{2}$$

$$Sensitivity \big/ Recall = TP \big/ (TP + FN) \tag{3}$$

$$Specificity = TN \big/ (TN + FP) \tag{4}$$

$$F1 - score = 2 * \left((Precision * Recall) \big/ (Precision + Recall)\right) \tag{5}$$

In this context, TP (true positive) signifies the count of subjects accurately predicted as true by the model. TN (true negative) indicates the count of subjects correctly predicted as false. FP (false positive) refers to the count of subjects incorrectly predicted as true. FN (false negative) represents the count of subjects incorrectly predicted as false by the model.

4 Results and Discussion

To assess the model performance, we generate confusion matrices and plot accuracy-loss graphs. Figure 2 presents the confusion matrix and Receiver Operating Characteristics (ROC) for the test phase of the classification experiment. The confusion matrix reveals that there are only a few incorrect predictions that correspond to False Positive and False Negative.

Additionally, Fig. 3 displays the accuracy and loss graphs of the model. These visual representations demonstrate that the proposed model has undergone successful training without being either underfit or overfit.

To evaluate the performance of the proposed model, accuracy, precision, sensitivity, specificity and F1-score are computed for each fold during the training and testing phases. The average outcomes are then considered to evaluate the overall performance of the model. Table 2 presents a comprehensive overview and comparison of the model's performance with state-of-the-art [13].

In the state-of-the-art method, the researchers used a light-weighted CNN model developed from scratch. Using dataset of breast mammography images [6] as a benchmark, we compare state-of-the-art model with the proposed model. Considering the results shown in the Table 2, it comes out that state-of-the-art model performed better than the proposed model. Our research is a potential evidence that CNN alone can automatically learn hierarchical features from images including texture features. By training on large datasets, CNNs can extract high level texture features that are useful

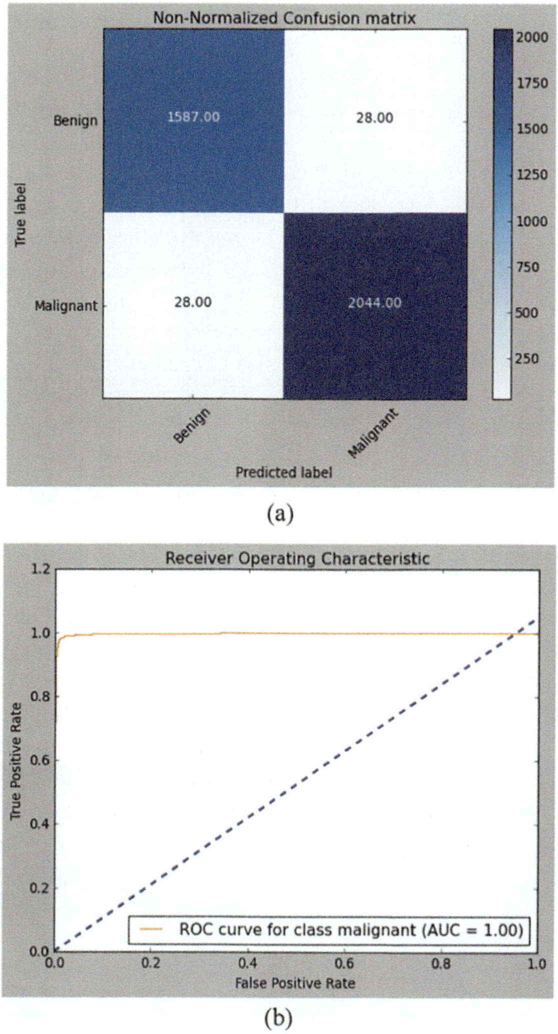

Fig. 2 **a** Confusion matix and **b** ROC of the proposed model

in various applications such as classificaton, segmentation, medical image analysis etc. Furthermore, the combination of CNN and GLCM can turn out useful to evaluate further treatment plans, cancer stage or survival rate.

Fig. 3 **a** Accuracy and **b** Loss graphs of the proposed model

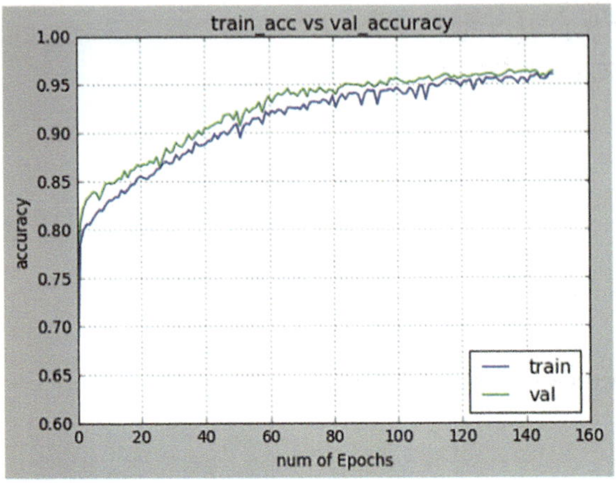

(a)

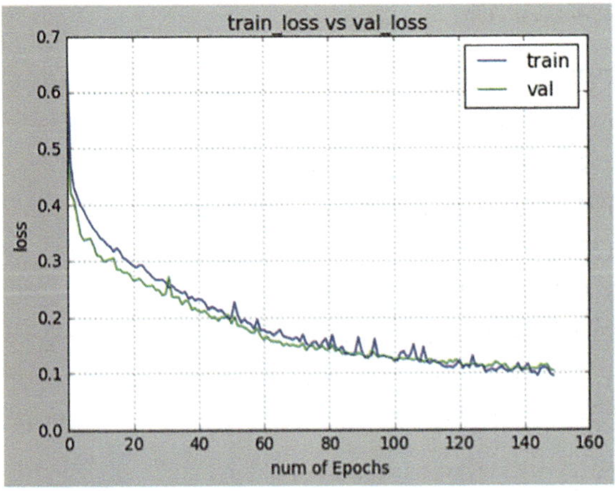

(b)

Table 2 Performance evaluation and comparison of proposed model with state-of-the-art

Model	Process	Precision	F1-score	Sensitivity	Specificity	Accuracy
Kaur et al. [13]	Training	99.15	99.15	99.15	99.02	99.15
	Testing	99.80	99.75	99.70	99.75	99.72
Proposed model	Training	95.73	95.72	95.72	94.93	95.72
	Testing	98.48	98.48	98.64	98.26	98.48

5 Conclusion and Future Scope

The classification of mammograms is essential for determining the appropriate course of action, including further diagnostic tests, biopsies, or treatment plans. Our technique analyzes digital mammograms to aid in breast cancer diagnosis, even without prior knowledge of cancerous lesions. We hybridize GLCM texture features and CNN-extracted features to classify breast mammograms and evaluated the proposed model's performance against state-of-the-art methods using a dataset of mammography images. In this study, we found potential evidence of CNN's capability that CNN alone can extract all the necessary features to differentiate benign and malignant mammograms. In future, the hybridization of GLCM and CNNs features can be evaluated for further treatment plans, cancer stage or survival rate.

References

1. Hassan, N.M., Hamad, S., Mahar, K.: Mammogram breast cancer CAD systems for mass detection and classification: a review. Multimed. Tools Appl. **81**(14), 20043–20075 (2022)
2. D'Orsi, C., Bassett, L., Feig, S.: Breast İmaging Reporting and Data System (BI-RADS). Breast Imaging Atlas, 4th edn. American College of Radiology, Reston (2018)
3. The Image Biomarker Satandardization Initiative, https://ibsi.readthedocs.io/en/latest/index.html. Last accessed 25 Aug 2023
4. Haralick, R.M., Shanmugam, K., Dinstein, I.: Textural features for image classification. IEEE Trans. Syst. Man Cybern. **3**(6), 610–621 (1973)
5. LeCun, Y., Bottou, L., Bengio, Y., Haffner, P.: Gradient-based learning applied to document recognition. Proc. IEEE **86**(11), 2278–2324 (1998)
6. Huang, M.L., Lin, T.Y.: Dataset of breast mammography images with masses. In: Data in Brief, vol. 31, no. 105928 (2020)
7. Soulami, K.B., Saidi, M.N., Tamtaoui, A.: A cad system for the detection of abnormalities in the mammograms using the metaheuristic algorithm particle swarm optimization (pso). In: Advances in Ubiquitous Networking 2: Proceedings of the UNet'16 vol. 2, pp. 505–517. Springer, Singapore (2017)
8. Kaur, P., Singh, G., Kaur, P.: Intellectual detection and validation of automated mammogram breast cancer images by multi-class SVM using deep learning classification. Inform. Med. Unlocked **16**, 100151 (2019)
9. Hu, T., Zhang, L., Xie, L., Yi, Z.: A multi-instance networks with multiple views for classification of mammograms. Neurocomputing **443**, 320–328 (2021)
10. Darweesh, M.S., Adel, M., Anwar, A., Farag, O., Kotb, A., Adel, M., Tawfik, A., Mostafa, H.: Early breast cancer diagnostics based on hierarchical machine learning classification for mammography images. Cogent Eng. **8**(1), 1968324 (2021)
11. Melekoodappattu, J.G., Dhas, A.S., Kandathil, B.K., Adarsh, K.S.: Breast cancer detection in mammogram: combining modified CNN and texture feature based approach. J. Ambient. Intell. Humaniz. Comput. **14**(9), 11397–11406 (2023)
12. Kaur, P., Kaur, A.: Review of progress in diagnostic studies of autism spectrum disorder using neuroimaging. Interdiscip. Sci. Comput. Life Sci. **15**(1), 111–130 (2023)
13. Kaur, P., Kaur, A.: Classification of breast cancer mammographic ımages using a light-weighted Convolutional Neural Network. In: 2023 International Conference on Intelligent and Innovative Technologies in Computing, Electrical and Electronics (IITCEE), pp. 42–46. IEEE (2023)

DCA-IoT: A Distributed Classification Approach for Combating IoT Traffic-Based DDoS Attacks Using Dataset Characterization and Comparison

Praveen Shukla, C. Rama Krishna, and Nilesh Vishwasrao Patil

Abstract The rapid proliferation of Internet of Things (IoT) devices has revolutionized several industries by facilitating automated functionalities. Additionally, it empowers us to experience seamless connectivity and interaction with our surroundings. However, this widespread deployment of less secure IoT devices has also brought multifold security challenges, such as Distributed Denial of Service (DDoS) attacks. This attack may completely crash the victim's online services and systems or make them unavailable to legitimate users. Several detection methods have been proposed in the literature. However, these methods are unable to provide a real-time detection solution for dynamic and high-speed IoT traffic-based attacks. In this paper, we propose a distributed classification approach for detecting DDoS attacks induced by IoT traffic in real-time, named by, DCA-IoT. This study centers around two key aspects: first, identification of the most suitable dataset for implementing the DCA-IoT approach, involving extensive dataset characterization and comparison. Second, development of an efficient DCA-IoT approach using distributed Machine Learning (ML) techniques, empowered by the high-performance Apache Storm stream-processing framework. We deployed the most efficient model on the Apache Storm cluster for analyzing incoming traffic streams quickly. The experimental results reveal that the proposed GBM-based distributed classification model achieves an accuracy of 99.73%.

Keywords IoT devices · Internet of Things · DDoS attacks · Distributed machine learning techniques · Apache Storm stream processing framework · Bot-IoT dataset · HDFS

P. Shukla (✉) · C. R. Krishna
Computer Science and Engineering, NITTTR Chandigarh, Chandigarh, India
e-mail: praveen.cse22@nitttrchd.ac.in

N. V. Patil
Computer Engineering, Government Polytechnic, Ahmednagar, India

B. Soni et al. (eds.), *Beyond Artificial Intelligence*, Lecture Notes in Networks and Systems 1326, https://doi.org/10.1007/978-981-96-4170-3_6

1 Introduction

In this modern world, we find ourselves surrounded by a multitude of IoT applications that are reshaping our daily lives. The impact of IoT technology has been profound, leaving its mark on every industry and society. This technology has increased efficiency and productivity through the power of automation. Numerous factors have fueled the rapid adoption of IoT technology in recent times, with virtually every sector across the globe recognizing and adopting its manifold advantages. Recent statistics show a vivid picture of this trend, projecting a considerable 31 billion connected devices worldwide by 2025 [1]. Furthermore, forecasts indicate that the IoT devices market is poised to surge to a staggering 1.6 trillion US dollars by 2025 [2].

The IoT network refers to a group of thousands of "interrelated or Internet-connected" things/devices. These devices comprise computing machines, mechanical machines, and digital devices, all collaboratively working to sense, collect, process, and analyze data from their surroundings [3], to carry out specific tasks autonomously, without human interference. However, the proliferation of less secure connected devices has exposed many security issues. Consequently, cybercriminals exploit this vulnerability, utilizing the opportunity to compromise additional devices and launch DDoS attacks.

1.1 DDoS Attack

DDoS attacks are one of the most serious cyber threats in the growing landscape of cyber security risks. It is a malicious attempt to disrupt the normal operations of online networks, services, or websites by overwhelming them with a flood of attack traffic. These attacks immediately cause the victim's server, system, or network to crash/slow down [4, 5].

Typically, DDoS attacks are launched using a network of compromised devices, generating an immense volume of fake requests that overwhelm the target system's capacity. Figure 1 depicts the setup for executing large-scale DDoS attacks from a pool of compromised IoT device(s). The primary goal is to render the victim services or network inaccessible to legitimate users, resulting in downtime, financial losses, including the organization's reputation. The increasing number of connected devices is making it difficult to keep track of the growing number of cyber threats. DDoS attacks pose a significant cybersecurity threat and require real-time and effective defense mechanisms to mitigate their impact. In the following section, we presented the primary contributions of the proposed distributed classification technique.

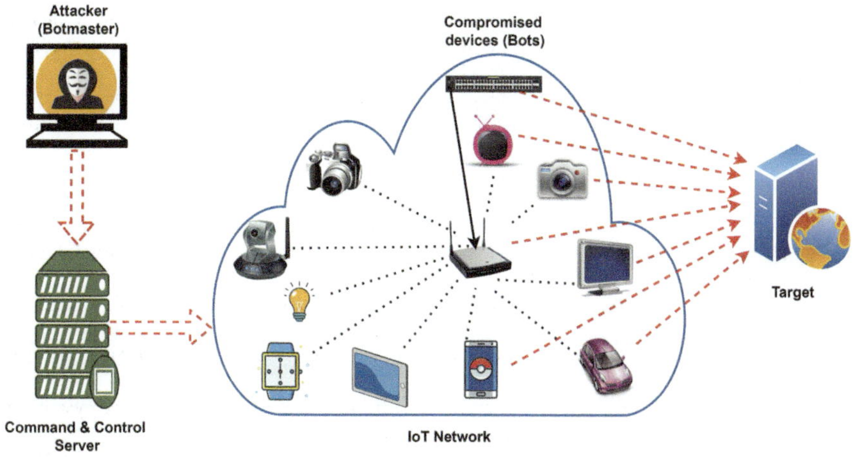

Fig. 1 A standard configuration to execute DDoS attacks by compromised IoT devices [6]

1.2 Contributions

The significant contributions of this paper are presented below:

- A novel distributed classification approach using the Apache Storm framework has been proposed for identifying DDoS attacks in real-time, namely, DCA-IoT.
- Characterize and compare various datasets to identify the most suitable one for implementing the DCA-IoT approach.
- Design and implement a distributed detection mechanism using advanced distributed machine learning algorithms from a highly scalable H2O.ai platform.
- Proposed approach deployed on a Storm cluster within the testing environment, to analyze massive incoming traffic data in real-time and classify it into five distinct classes.
- Proposed DCA-IoT approach is created and validated using the publicly available "Bot-IoT Dataset."
- Proposed DCA-IoT approach significantly strengthens the security of IoT environments by providing a highly scalable, distributed, portable, and real-time defense mechanism.

1.3 Structure of the Paper

The rest of this paper is structured as follows: Sect. 2 offers a concise overview of existing DDoS attack classification methods. In Sect. 3, we present an exhaustive characterization and comparison of various datasets. Section 4 provides a comprehensive explanation of the proposed distributed classification approach designed for

DDoS attack detection. Section 5 delves into the specifics of the practical development environment and the software tools employed to implement the proposed classification technique. Following that, Sect. 6 presents the results and provides an in-depth analysis. Finally, in Sect. 7, we draw conclusions regarding the work presented in this paper including future directions.

2 Related Work

Numerous DDoS attack protection solutions have been proposed in the literature to detect malicious traffic and protect the victim's assets and services. These solutions play a crucial role in reducing the risk of malicious attacks. However, the widespread integration of IoT technology has added complexity and increased the frequency of DDoS attack incidents, making it difficult to defend against them. The existing solutions are not capable to defend the ever-evolving and highly sophisticated DDoS attacks. Therefore, it is essential to develop intelligent DDoS protection methods that can identify and mitigate these threats quickly and effectively. Further research is needed to create more robust and efficient mechanisms for protecting against these types of attacks. In this section, we present a summary of recent approaches to detect DDoS attacks based on IoT traffic.

In the literature, centralized computing architecture-based methods have been proposed in various studies [6–9]. However, the majority of these approaches are incapable of protecting victims during large-scale DDoS attacks. To address this limitation, a few researchers have introduced Distributed Processing Framework (DPF)-based detection methods [10–12]. However, most of them are implemented on the Hadoop framework, which processes data in batches. Further, some researchers have implemented their proposed techniques [10–12] on a Distributed Stream Processing Framework (DSPF), enabling real-time analysis of network traffic data. However, even these solutions have not provided a comprehensive defense against DDoS attacks originating from the IoT environment. In the following, we have listed some limitations of existing approaches:

- Many centralized architecture-based classification methods [6–8] are vulnerable targets, and themselves become victims during large-scale attacks.
- Several AI-based classification techniques [6, 9] have been developed using traditional ML libraries that do not support the scalability feature, leading to scalability issues when deploying these models on DPF/DSPF.
- The majority of the classification approaches are built for batch processing mode. Hence, they could not offer a real-time protection solution that processes data quickly and efficiently.
- Approaches [13, 14] incorporate counter-based techniques for traffic classification but struggle to detect low-rate DDoS attacks effectively.
- Few existing DSPF-based approaches [10–12] detect various DDoS attack types by analyzing a multitude of packets in parallel across a cluster of machines.

However, they fall short of addressing DDoS attacks originating from IoT network traffic.
- Several existing approaches [7, 9] have been designed and evaluated using outdated and limited datasets. Therefore, these approaches are not capable of defending the ever-evolving and highly sophisticated DDoS attacks.

3 Characterization and Comparison of Various Datasets

Artificial intelligence (AI)-based solutions heavily rely on good quality data for solving complex problems. The effectiveness of these solutions is affected by many crucial factors, such as selecting the most relevant features, the nature of data (balance/unbalance), the accuracy of the data, having an adequate number of records, etc. Sometimes, models are designed using asymmetric or inappropriate (non-IoT-specific) datasets. It may be possible for them to perform satisfactorily during training and testing. However, when these models are deployed in a real-time environment, they often struggle to maintain satisfactory performance. In the literature, numerous publicly accessible datasets that can be utilized to create and validate robust DL/ML-based classification models for identifying DDoS attacks. These datasets can be broadly divided into three categories: "synthetic datasets", "benchmark datasets", and "real datasets." Each of these categories serves a distinct purpose in designing and evaluating effective models to counter DDoS attacks.

3.1 Synthetic Datasets

Synthetic datasets are artificially created data that simulate real-world events for training and validating DL/ML models. They are useful when the original data is not available, but they lack the diversity of real situations. However, an important point to note is that the credibility of synthetic datasets cannot always be trusted [15], while they offer a convenient way to test unusual conditions or scenarios, their reliability can be questionable [16]. The examples of synthetic datasets are TON_IoT [17], Edge-IIoTset, and MQTT-IoT-IDS [18].

3.2 Benchmark Datasets

Benchmark datasets are free and standard data for evaluating model performance. They are recommended for researchers and businesses to develop their proposed methods. These datasets play a pivotal role in the creation and validation of suggested methods. One of the most significant advantages associated with benchmark datasets is the ability to directly compare experimental results with those of existing approaches. This comparison facilitates a straightforward evaluation of how well our approach stacks up against existing approaches. The examples of benchmark datasets are NSL KDD, DS2OS [19], IoTID20 [20], IoTDevNet, and IoT Botnet [21].

3.3 Real Datasets

Real datasets are more complex and composed of real-world traffic data. They are collected from actual network traffic and stored in various formats such as PCAP, PCAPNG, TXT, CSV, JSON, etc. They are publicly available and can create accurate models. They are more useful for testing model effectiveness on real data. There are various real datasets are available like FIFA [22], ISP, and N-BaIoT [23].

3.4 A Brief Summary of Published Datasets

In this subsection, we summarize various datasets that researchers recently used to create their proposed DDoS attack detection system.

- **IoTID20 Dataset** [20]: The primary purpose of utilizing the IoTID20 dataset is to create Intrusion Detection Systems (IDS). This dataset simulates contemporary cyber-attacks by connecting two smart devices to a Wi-Fi router. The dataset incorporates 83 distinct features related to network traffic, and it has three more features dedicated to labels: binary, category, and subcategory. The dataset contains instances of five attack categories: Denial of Service (DoS), Mirai, scanning, Man-in-the-Middle (MITM), and legitimate traffic. In total, this dataset comprises a considerable number of 625,783 traces.
- **ToN_IoT Dataset** [17]: ToN_IoT dataset introduces a fresh collection of datasets to assess how well different AI-based cybersecurity applications perform with their reliability. This dataset is referred as labeled ToN IoT because it consists of a diverse range of data originating from IoT and IIoT sensor telemetry, diverse operating systems, and records of IoT network traffic. This dataset incorporates instances related to nine distinct target categories with comprising 46 unique attributes.
- **IoT-23 Dataset** [25]: It is a collection of network traffic data captured from various IoT devices, serving as a testing surface for assessing intrusion detection and security mechanisms. This dataset consists of network traffic data from 23 different IoT devices, ranging from smart thermostats to security cameras. It includes multiple attack scenarios, including DoS, DDoS, reconnaissance attacks, etc. Each device's traffic data is delivered in separate PCAP files, making it simple for the researcher's experimentation. It comprises 21 unique features and 20 target classes. This dataset contributes significantly in providing a framework for designing ML-based intrusion detection mechanisms and assessing their robustness.

- **MQTT-IoT-IDS2020 Dataset** [18]: It is a valuable resource designed for the evaluation and development of the IDS tailored to Message Queuing Telemetry Transport (MQTT)-based IoT attacks. MQTT is a protocol commonly used in IoT communication. This dataset comprises network traffic data captured from simulated MQTT-based IoT devices in a controlled laboratory environment. It contains various real-world attack scenarios like "regular operation", "aggressive scanning", "UDP scanning", "Sparta SSH brute force", and "MQTT brute force", making it a perfect tool for assessing the effectiveness of intrusion detection mechanisms specific to MQTT-based traffic.
- **MedBIoT Dataset** [26]: It has been generated using a medium-sized network testbed with 83 IoT devices, comprising both actual and simulated devices. This dataset is captured from three recognized botnets: "Mirai, BashLite, and Torii". They are deployed for gathering data from various stages of communication with command-and-control (C&C). Security professionals and researchers can leverage the MedBIoT dataset to develop and assess intrusion detection systems specifically tailored for malicious IoT environments. Ultimately, the dataset's utilization contributes significantly to the overall security enhancement of IoT environments.
- **IoTNID Dataset** [27]: This dataset simulates various real-world cyberattacks, comprising diverse network traffic data. It includes 42 PCAPs containing raw network packets captured during different time frames. Within these PCAPs, there are a total number of 825,000 network flows, each trace having seven distinctive attributes. The dataset is structured around five primary categories in addition to eleven subcategories. Among these five main categories, one is dedicated to legitimate transmissions, while the remaining four pertain to various types of cyber-attacks.
- **Bot-IoT Dataset** [24]: It is a significant resource dedicated to the advancement of cyber security research in the field of IoT environments. Specially crafted for discovering malicious activities in IoT environments. It provides a comprehensive collection of network traffic data with benign and synthetic IoT network activity, including various real-world attack scenarios. This dataset offers complete details about captured packets, well-suited labels, and an enormous count of around 72 million records. The dataset's source files are openly accessible in two formats: PCAPs and CSVs, with respective sizes of 69.3 and 16.7 GB. It consists of eleven distinct target categories with 46 attributes.
- **N-BaIoT Dataset** [23]: This dataset incorporates real IoT traffic data originating from nine IoT devices used for commercial purposes. To generate this dataset, two botnets "Mirai" and "Bashlite" were utilized by dataset creators. The malicious traffic in this dataset is broadly classified into two main types and further, subdivided into ten distinct attack categories. Along with these attack categories, there exists one benign class. Notably, this dataset is highly imbalanced in its composition, with a significantly smaller number of benign samples compared to malicious ones.
- **DS2oS Dataset** [19]: This dataset includes packet data collected from the IoT environment. It captures the application layer DDoS attacks. Additionally, it

comprises an IoT middleware component housing data and services for intelligent spaces. The dataset has 347,935 number of network flows. Each containing 13 distinct attributes. These attributes help categorize traces into eight distinctive attack classes: "Normal, Scan, DDoS, Multitious control, Multitious operation, Scan wrong setup, Spying, and Data type probing."

The comprehensive analysis of the various datasets presented here offers valuable insights for cyber security research. We have critically examined their features, traffic types, number of flows, and attack scenarios and presented an in-depth knowledge of each dataset's strengths and limitations. These datasets collectively provide a vital foundation for researchers and practitioners aiming to improve the security of IoT networks. However, the effectiveness of the trained model depends on the quality of the dataset. A good quality dataset should possess certain key attributes. It must be accurately labeled, having a diverse array of contemporary attack traffic scenarios, and its features should contain valuable insights regarding the target variable. It should have a sufficient number of network traces. Regular updates to the dataset further enhance its relevance and reliability. The Bot-IoT dataset has most of these attributes. It has the maximum number of network traces compared to other datasets and also provides contemporary attack traffic with 11 distinct target classes. It is a realistic IoT traffic-specific dataset. Many researchers have already leveraged this dataset to develop and assess their strategies for detecting attacks, as well as to assess the durability of IoT networks against cyber security threats. Therefore, it can be inferred that this dataset is an optimal choice for implementing the proposed distributed classification method.

4 Proposed Research Methodology

This section presents the detailed functioning of the proposed DCA-IoT distributed classification approach for detecting DDoS attacks induced by IoT network traffic. It is a two-stage process: the first stage is building a distributed classification model and the second stage is deploying an efficient distributed classification model. The logical architecture of the proposed DCA-IoT approach is presented in Fig. 2.

4.1 Building a Distributed Classification Model

In this subsection, we build a distributed, scalable, and real-time classification approach using the scalable H2O.ai distributed ML platform. This approach plays a crucial role in advancing the security of IoT networks against cyber threats. The step-by-step process to build a distributed classification model for the DDoS attack is given below. The first stage of the DCA-IoT approach is further divided into several steps:

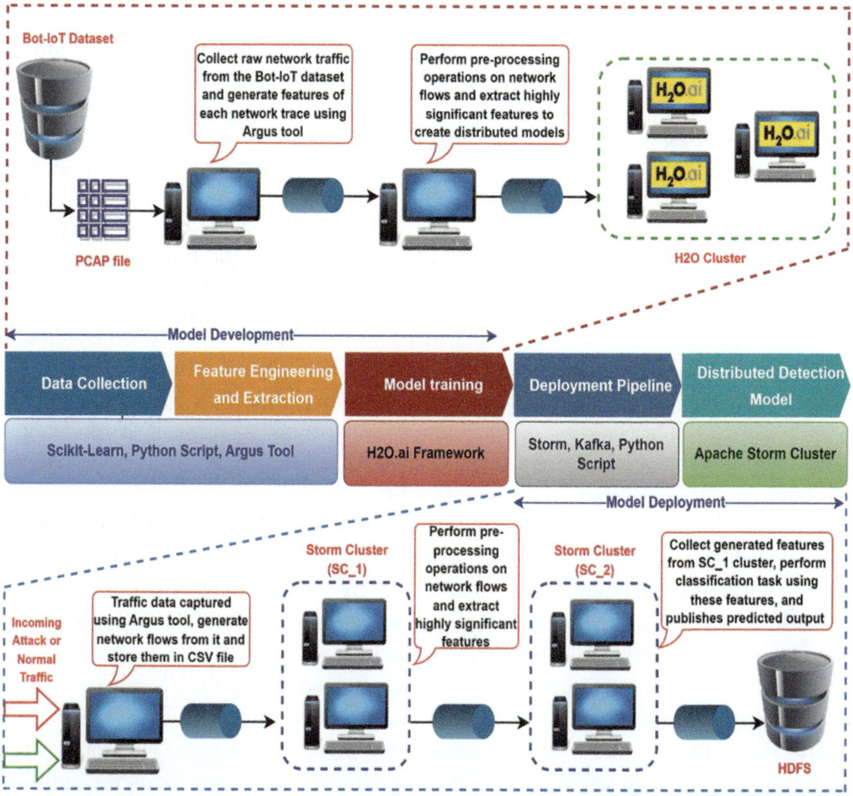

Fig. 2 Logical architecture of the proposed DCA-IoT approach

(i) Collect network flows, (ii) Perform pre-processing tasks, (iii) Extract significant features, and (iv) Build classification models.

Collect Network Flows: In this step, we have gathered raw network traffic in the form of PCAP files with Benign and different attack scenarios: Data Gathering, DDoS, DoS, and Data Theft. After that convert PCAPs into CSVs files using the Argus tool to generate network flows for each network trace. The generated flows comprise 29 features including target labels. For experimentation purposes, we have utilized the .pcap files from the publicly available real-world Bot-IoT dataset [24].

Perform Pre-processing Tasks: This step begins with cleaning the network flows that were generated in the previous step. Initially, we eliminate any inaccurate or incomplete records, such as those having missing or NaN values. There are so many categorical variables in the dataset. Therefore, these attribute values are converted into unique numerical values using label encoding techniques. The next step involves standardizing the data values of attributes. By utilizing the "MinMax method" available from open access and the most prevalent "sklearn library." The objective here is to bring every attribute value to a similar scale (within the range of 0–1). After that, the records undergo normalization, and the network traces are appropriately labeled. Finally, a data balancing procedure is executed because the dataset is highly imbalanced. To make it balanced apply the "Synthetic Minority Over-sampling Technique (SMOTE)", injecting synthetically generated records into the minority class. In our case, the DDoS class contains the highest number of traces. Therefore, the other target classes were up-sampled to 1,926,624 instances. Because an imbalanced dataset detrimentally impacts model accuracy and tends to favor a respective class. The up-sampling process results in a dataset of 9.6 million flows, which will be used for the development of the proposed distributed classification approach.

Extract Significant Features: The selection of features is a pivotal step in the development process of an ML model. It has a significant impact on the model's effectiveness, having a crucial role in its enhanced performance. In addition to trimming dimensions, feature selection serves to lessen the computation expenses related to data mining while avoiding the problem of overfitting. As a result, it contributes to the creation of a cost-effective solution for DDoS attack detection. Firstly, we eliminate all the correlated features, effectively erasing redundancy. Subsequently, we employed the Distributed Random Forest (DRF) technique to estimate the scaled importance of individual features. This dual-step procedure ultimately leads to the identification of the top ten attributes, based on their importance scores.

Build Classification Models: This step involves the creation of multiple classification models for DDoS detection originating from IoT networks. For this, we have employed scalable and distributed ML techniques from the H2O.ai framework, such as "Gradient Boosting Machine (GBM)", "Generalized Linear Model (GLM)", and "Naive Bayes (NB)." These techniques are executed on the Apache Hadoop cluster. Both the H2O.ai framework and the Hadoop cluster offer capabilities for building models in a distributed manner [11].

Before starting model training, it's necessary to split the pre-processed dataset into training and testing samples. We have applied the 80:20 principle to ensure that 80% of data samples are designated for training, while the remaining 20% are allotted for testing. The efficacy of the ML model heavily relies on the nature of the training data. Consequently, a stratified technique is employed to ensure the equitable division of data samples into these two parts.

Subsequently, multiple distributed models are developed using 80% of the data samples on the H2O cluster. These trained models are then evaluated using 20% of the testing data samples, based on manifold performance metrics. The performance evaluation of these distributed models is presented in the Results and Discussion section. The most effective model (in our case: the GBM model) that provides impressive results is stored in persistent storage using the MOJO (Model Object) format. This MOJO model is subsequently implemented in a production setup for validation purposes.

4.2 Deploying an Efficient Distributed Classification Model

The second stage of the proposed DCA-IoT approach is primarily focused on the deployment of an efficient distributed classification model on the SC_2 Storm cluster and classifying ingress network traffic in real-time. In our study, we have identified the GBM-based model as an efficient distributed classification model, as discussed in the earlier section. The step-by-step process of the second stage, describing how incoming network traffic is categorized through the DCA-IoT approach, is given below. To facilitate the task, we have divided the entire second-stage process into two distinct modules: "Packet Decoder" and "Attack Detector." Details of each module are given in the following:

Packet Decoder: This module is implemented on the SC_1 Storm cluster. It centers around capturing incoming network traffic, performing pre-processing tasks on them, and extracting highly relevant features within a 1 ms time frame. The process commences with the Argus tool ingesting raw incoming network traffic and generating network flows from it. These flows are consistently made available on the "ingress_traffic" topic by the Kafka producer agents. This module has various tasks to perform, one of the most significant tasks is the quick extraction of highly relevant features.

Task-1: The primary goal of this task is to capture incoming traffic efficiently. It has been accomplished by configuring the SC_1 cluster to immediately take incoming network flows from the "ingress_traffic" topic, which is continuously populated by producer agents.

Task-2: As a critical element of this module, the SC_1 cluster executes various pre-processing activities on the incoming network flows. These activities include jobs like data cleaning and data standardization, all aimed at preparing the data for subsequent analysis.

Task-3: In this task, ten extremely relevant attributes related to network traffic are extracted from the pre-processed flows quickly. These attributes are then published on the "relevant_flows" topic, which is designed to be consumed by the SC_2 cluster. This task significantly contributes to enhancing the accuracy of predicting malicious traffic.

Attack Detector: This module is implemented within the SC_2 Storm cluster. Its primary objective is to classify incoming network flows and then publish predicted outcomes via Kafka topics, allowing for timely and suitable actions to be performed. Initially, each node of the SC_2 Storm cluster has been deployed with a well-performing, GBM-based distributed classification model. The main purpose of these nodes is to consume highly relevant attributes from the topic named "relevant_flows", published by the SC_1 cluster. Further, these nodes process the ingress network flows in real-time and classify them into five classes based on the ten highly relevant attributes. The subsequent step involves publishing the outcomes produced by the classification model on the "output_data" topic, which initiates the process of taking appropriate security measures. The final step involves merging the classification results with the highly relevant input features and store this record on the HDFS.

This amalgamated record is then stored on the Hadoop Distributed File System (HDFS).

5 Experimental Environment

This section presents comprehensive information about the experimental environment utilized to execute the proposed DCA-IoT approach. We have developed a dedicated testbed, the design of which is illustrated in Fig. 3. In this testbed, we have configured a source network paired with a victim network. The source network serves the dual purpose of generating attack and benign traffic directed towards the victim network. To capture this incoming network traffic, we deployed the Argus tool, which effectively transforms it into network flows.

These network flows are then efficiently managed by two separate Storm clusters, each equipped with two nodes, placed within the victim network. The experiment is supported by a server Ubuntu 18.04.5.LTS, a 64-bit operating system (AMD64) with 64 GB RAM. The processor is an Intel(R)-Xeon(R) Processor E5-2420 v2, operating at a clock speed of 2.20 GHz. The graphics card is llvmpipe (LLVM-6.0 256-bits), while the storage capacity is 984-GB. For creating and evaluating the proposed DCA-IoT approach, the recently published Bot-IoT dataset is utilized. Additionally, a set of open-source software is employed such as Apache Hadoop, HDFS, Apache Storm, Python 3.7.10, and the H2O.ai framework.

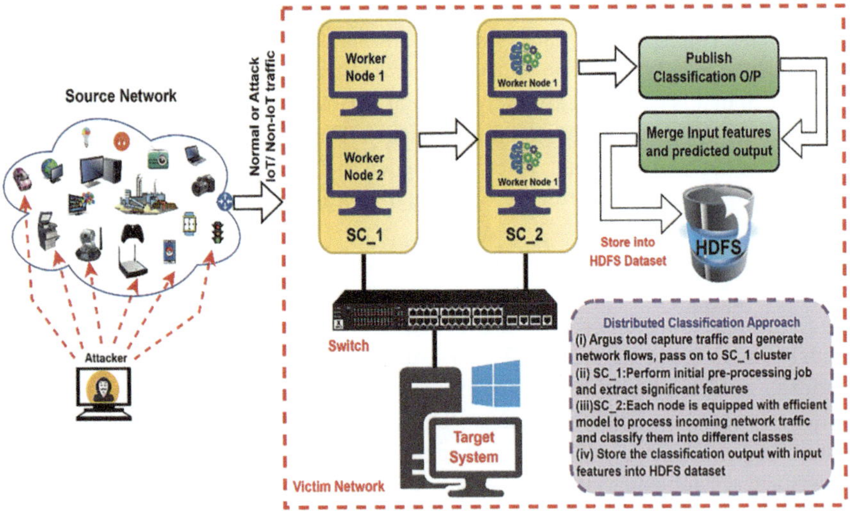

Fig. 3 Experimental setting for the proposed DCA-IoT approach

6 Results and Discussion

In this section, we focus on the evaluation and analysis of the DCA-IoT approach that is presented for detecting DDoS attacks originating from IoT networks. The proposed DCA-IoT distributed classification approach categorizes incoming network traffic into five distinct classes. To illustrate the effectiveness of the model, we measure its classification accuracy along with standard metrics such as precision, recall, and the F1-score. Further, to validate the approach, we utilize the publicly accessible "Bot-IoT" dataset.

This approach involves the development of three distributed ML classification models and the deployment of one highly efficient model into a testbed setting. We employed three distributed models (GBM, GLM, and NB) using distributed ML techniques available in the H2O.ai framework. These models were trained on the ten most relevant features, having been selected based on their scaled importance. Class-wise performance analysis of these three distributed models is shown in Table 1. As part of this analysis, the representation of the confusion matrix enhances the visualization of each model's performance. Within Table 1, the "Accuracy" column displays the classification accuracy of different distributed classification models with the ten most significant features. Notably, the GBM-based model stands out with a significant accuracy of 99.94%, while the GLM and NB, with accuracies of 94.44% and 91.20%, respectively. The GBM-based model takes only 16 s for testing, which is significantly smaller than the testing times of the GLM and NB-based models, which require 31 and 23 s, respectively.

Table 1 Demonstrates class-wise performance analysis of the distributed classification models

Models	Classes	Metrics			
		Precision	Recall	F1-score	Accuracy (%)
GBM	0	01.00	01.00	01.00	99.94
	1	00.99	01.00	01.00	
	2	01.00	01.00	01.00	
	3	00.99	00.99	00.99	
	4	01.00	01.00	01.00	
GLM	0	01.00	01.00	01.00	94.44
	1	00.99	00.99	00.99	
	2	00.92	00.95	00.94	
	3	00.89	00.90	00.89	
	4	00.89	00.84	00.87	
NB	0	00.99	00.98	00.99	91.20
	1	00.99	00.98	00.99	
	2	00.84	00.86	00.84	
	3	00.82	00.83	00.83	
	4	00.85	00.88	00.86	

Notations 0-Benign, 1-DoS, 2-DDoS, 3-Data Gathering, 4-Data Theft

After assessing the performance of the models, it has been observed that the GBM-based distributed model outperforms other models in terms of both effectiveness and speed. As a result, we have implemented the GBM-based distributed model within the Apache Storm cluster SC_2. Furthermore, following the deployment of the GBM-based model, we conducted a trace-driven execution. The proposed approach facilitates the real-time classification of incoming network traffic and achieves better accuracy. However, we can implement a distributed environment using cloud computing with an effective classification approach for real-time detection of DDoS attacks originating from IoT traffic. This approach allows for faster and more efficient detection, particularly in handling modern DDoS attack patterns.

7 Conclusions and Future Directions

In this paper, we proposed a distributed classification approach for real-time detection of DDoS attacks originating from IoT traffic. This study mainly focused on two key aspects: First, we conducted an extensive dataset characterization and comparison to identify the most suitable dataset for implementing the DCA-IoT approach. Second, we developed an efficient DCA-IoT approach using distributed ML techniques, empowered by the high-performance Apache Storm stream processing framework. The entire process is divided into two stages. In the first stage, we collected raw network traffic, pre-processed it, extracted highly relevant features, and then designed detection models using the H2O.ai framework on a Hadoop cluster. The second stage includes deployment of two Strom clusters: SC_1 and SC_2. The SC_1 cluster hosts the "packet decoder" module, responsible for capturing incoming traffic, performing preprocessing tasks, and extracting the ten best features. The "attack detector" module, implemented on the SC_2 cluster, analyzes and classifies incoming traffic into five distinct classes in real-time. Finally, the system saves the ten input features along with classification results into the HDFS. The results show that the proposed DCA-IoT approach gives a better classification accuracy of 99%+ . We aim to improve the proposed approach by using deep learning, and ensemble learning techniques in the future, which will enable us to detect more modern attacks faster and more efficiently.

References

1. Sinha, S.: Number of connected IoT devices 2021. https://iot-analytics.com/number-connected-iot-devices/ (2021)
2. Vailshery, L.S.: Global IoT end-user spending worldwide 2017–2025. https://www.statista.com/statistics/976313/global-iot-market-size/ (2021)
3. Vermesan, O., Friess, P., Guillemin, P., Sundmaeker, H., Eisenhauer, M., Moessner, K., Le Gall, F., Cousin, P.: Internet of Things Strategic Research and İnnovation Agenda, in Internet of Things, pp. 7–151. River Publishers (2022)

4. Patil, N.V., Krishna, C.R., Kumar, K.: Apache hadoop based distributed denial of service detection framework. In: International Conference on Information, Communication and Computing Technology, pp. 25–35 (2019)
5. Patil, N.V., Rama Krishna, C., Kumar, K.: S-DDoS: Apache spark based real-time ddos detection system. J. Intell. & Fuzzy Syst. (Preprint), 1–9 (2020)
6. Shukla, P., Krishna, C.R., Patil, N.V.: Eiot-DDoS: Embedded classification approach for iot traffic-based ddos attacks. Clust. Comput. 1–20 (2023)
7. Kumari, A., Mehta, A.K.: A hybrid intrusion detection system based on decision tree and support vector machine. In: 2020 IEEE 5th International Conference on Computing Communication and Automation (ICCCA), pp. 396–400 (2020)
8. Shurman, M.M., Khrais, R.M., Yateem, A.A.: DoS and DDoS attack detection using deep learning and IDS. Int. Arab J. Inf. Technol. **17**(4A), 655–661 (2020)
9. Shukla, P., Krishna, C.R., Patil, N.V.: A detection approach for IoT traffic-based DDoS attacks. In: International Conference on Artificial Intelligence of Things, pp. 201–214. Springer Nature Switzerland, Cham (2023)
10. Ahmed, A., Hameed, S., Rafi, M., Mirza, Q.K.A.: An intelligent and time-efficient ddos identification framework for real-time enterprise networks: SAD-F: spark based anomaly detection framework. IEEE Access **8**, 219483–219502 (2020)
11. Patil, N.V., Krishna, C.R., Kumar, K.: SS-DDoS: spark-based ddos attacks classification approach. In: Security and Resilience of Cyber Physical Systems, pp. 81–90. Chapman and Hall/CRC (2022)
12. Snehi, M., Bhandari, A.: A novel distributed stack ensembled meta-learning-based optimized classification framework for real-time prolific IoT traffic streams. Arab. J. Sci. Eng. **47**(8), 9907–9930 (2022)
13. Hameed, S., Ali, U.: Efficacy of live ddos detection with hadoop. In: Network Operations and Management Symposium (NOMS), 2016 IEEE/IFIP, pp. 488–494. IEEE (2016)
14. Bhayo, J., Jafaq, R., Ahmed, A., Hameed, S., Shah, S.A.: A time-efficient approach toward DDoS attack detection in IoT network using SDN. IEEE Internet Things J. **9**(5), 3612–3630 (2021)
15. Nonnemaker, J.E.: The Safe Use of Synthetic Data in Classification. ProQuest (2008)
16. Strangmark, G.: Making synthetic data more useful, https://www.rokittastra.com/making-synthetic-data-useful/
17. Booij, T.M., Chiscop, I., Meeuwissen, E., Moustafa, N., Hartog, F.T.: Ton IoT: the role of heterogeneity and the need for standardization of features and attack types in IoT network intrusion data sets. IEEE Internet Things J. **9**(1), 485–496 (2021)
18. Hindy, H., Bayne, E., Bures, M., Atkinson, R., Tachtatzis, C., Bellekens, X.: Machine learning based iot intrusion detection system: an MQTT case study (MQTT-IoT-IDS2020 dataset). In: International Networking Conference, pp. 73–84. Springer (2020)
19. Latif, S., Zou, Z., Idrees, Z., Ahmad, J.: A novel attack detection scheme for the industrial internet of things using a lightweight random neural network. IEEE Access **8**, 89337–89350 (2020)
20. Ullah, I., Mahmoud, Q.H.: A scheme for generating a dataset for anomalous activity detection in IoT networks. In: Canadian Conference on Artificial Intelligence, pp. 508–520. Springer (2020)
21. Islam, N., Farhin, F., Sultana, I., Kaiser, M.S., Rahman, M.S., Mahmud, M., Hosen, A., Cho, G.H.: Towards machine learning based intrusion detection in IoT networks. Comput. Mater. Contin **69**(2), 1801–1821 (2021)
22. FIFA 1998 world cup dataset, https://ita.ee.lbl.gov/html/contrib/WorldCup.html
23. Meidan, Y., Bohadana, M., Mathov, Y., Mirsky, Y., Shabtai, A., Breitenbacher, D., Elovici, Y.: N-baiot: network-based detection of IoT botnet attacks using deep autoencoders. IEEE Pervasive Comput.Comput. **17**(3), 12–22 (2018)
24. Koroniotis, N., Moustafa, N., Sitnikova, E., Turnbull, B.: Towards the development of realistic botnet dataset in the internet of things for network forensic analytics: Bot-IoT dataset. Future Gener. Comput. Syst. **100**, 779–796 (2019)

25. Dutta, V., Choraś, M., Pawlicki, M., Kozik, R.: A deep learning ensemble for network anomaly and cyber-attack detection. Sensors **20**(16), 4583 (2020)
26. Guerra, A., Medina, J., Bahsi, H., Nõmm, S.: Medbiot: generation of an IoT botnet dataset in a medium-sized IoT network. In: ICISSP, pp. 207–218 (2020)
27. Liu, Z., Thapa, N., Shaver, A., Roy, K., Yuan, X., Khorsandroo, S.: Anomaly detection on IoT network intrusion using machine learning. In: 2020 International Conference on Artificial intelligence, Big Data, Computing and Data Communication Systems (icABCD), pp. 1–5. IEEE (2020)

Mapping User Perception Towards Ethical Aspects of Artificial Intelligent Systems: A Study of AI Mobile Applications

Himanshu and Rachita Sambyal

Abstract This study maps user perceptions towards ethical approaches of artificial intelligence (AI) in mobile applications. The study analyzed various AI mobile apps based on different themes of ethics, including privacy, transparency, accountability, safety and security, human control of technology, fairness and nondiscrimination, and promotion of human value. User's perception towards ethical dimensions of apps were analyzed using Binary logistic regression. The finding suggests that understanding of science and technology, gender and understanding of AI play a momentous role in perception towards AI Ethics. The findings also suggests that developers should prioritize improving their policies and practices related AI ethics dimensions.

Keywords Artificial intelligence · Ethics · User perception · AI dimensions · Binary logistic regression

1 Introduction

According to Markets and Markets,[1] global market of artificial intelligence (AI) is projected to reach $1,345.2 billion by 2030 with a CAGR of 36.8% from 2023 to 2030. Businesses around the world are employing AI to improve productivity and efficiencies, reduce time and cost [10]. Due to this AI is becoming top rated resource across the industries. The application of AI span from usage in areas like cyber security, fraud management to customer relationship management like chat bots.

[1] https://www.globenewswire.com/newsrelease/2023/08/17/2727391/0/en/Artificial-Intelligence-Market-Worth-1-345-2-billion-by-2030-Growing-at-a-CAGR-of-36-8-Report-by-MarketsandMarkets.html.

Himanshu · R. Sambyal (✉)
University Institute of Applied Management Sciences, Panjab University, Chandigarh, India
e-mail: rachita@pu.ac.in

Himanshu
e-mail: him3236@gmail.com

B. Soni et al. (eds.), *Beyond Artificial Intelligence*, Lecture Notes in Networks and Systems 1326, https://doi.org/10.1007/978-981-96-4170-3_7

While the widespread usage of AI powered systems have added innovation in the exiting processes, policy makers and researchers have shown grown concern over the ethical challenges attached to it. The idea behind the concern has lead to developing the AI which is more trustworty and human centric. While many researchers have addressed the issues related to AI developmet [6]; its application [11], there is little research investigating public perception on the ethical challenges. The insight from the citizen's perception will help the policy makers and AI developers to design systems in social context [1]. This study investigates users's perception towards ethical principles of AI.

2 Review of Literature

Over the years many researchers have attempted to to provide ethical guidelines for Artificial Intelligent Systems and technologies. This section reviews some of the key literature in the domain of AI and Ethics. Studies defines AI ethics as the principles and guidelines that govern the development and use of artificial intelligence to ensure it is deployed in an ethical and responsible manner [4]. They encompasses values, principles, and techniques to align AI with human values, prevent harm to individuals and society, and ensure that AI systems are transparent, accountable, and reliable. Many advocates, policy makers and researcher working with technologies need to adopt universal framework to ensure ethical use of AI [3]. Eight themes of AI athics namely Privacy, Accountability, Safety and Security, Transparency and Explainability, Fairness and Non-discrimination, Human Control of Technology, Professional Responsibility and Promotion of Human Values were identified. Some studied concluded that [9] three themes of AI design namely user-oriented AI innovation; framing AI as a solution to RI issues; and context-sensitivity are important for building a responsible AI. Researchers have also [5] identified the dimensions of explainable AI from the user's prespective. From the SLR four dimensions of AI emerges i.e., format, completeness, accuracy and currency. Some researhcers attempted to develop and validate an instrument for measuring undergraduate students' attitudes toward the ethics of artificial intelligence (AT-EAI) [8]. Intially they conducted literature review to identify existing instruments for measuring attitudes toward AI ethics, and then use this information to develop their own instrument, the AT-EAI. The study concluded that the AT-EAI is a reliable and valid instrument for measuring undergraduate students' attitudes toward the ethics of AI. Few researchers have also explored the importance of ethical considerations in the development and deployment of AI systems, and discusses the challenges and opportunities that arise in this context. The authors argued that it is crucial to institutionalize ethics in AI development and use, and suggested that this can be achieved through a combination of technical and non-technical measures [2]. Overall, the paper provided valuable insights into the ethical challenges and opportunities of AI, and highlights the importance of addressing these issues in the Some authers [7] discussed user perception towards eight dimensions of AI ethics. They created four

Table 1 Themes of AI ethics

Theme	Definition
Privacy	This includes implementing robust data protection measures, such as encryption and access controls, and ensuring that data is only used for legitimate purposes
Accountability	This includes ensuring that there are clear lines of responsibility and mechanisms for redress in the event that AI systems cause harm
Safety and security	This includes implementing appropriate safeguards and testing procedures to ensure that AI systems are safe and secure
Transparent and explainability	This includes providing clear explanations of the data and algorithms used, as well as the reasoning behind decisions
Fairness and nondiscrimination	AI systems must be developed and used in a way that is fair and does not discriminate against individuals or groups. This includes ensuring that biases are not present in the data and algorithms used, and that decisions are made based on relevant factors rather than irrelevant personal characteristics
Human control of technology	This includes implementing appropriate human oversight and ensuring that AI systems are designed to support rather than replace human decision-making
Professional responsibility	This includes ensuring that AI systems are developed and used in a way that is consistent with these principles and standards, and that appropriate training and education is provided to those working with AI systems
Promotion of human value	This includes ensuring that AI is used to enhance rather than replace human capabilities, and that the benefits of AI are distributed fairly and equitably across society

Source Adapted from the literature review

different scenarios and asked the users about their opinion on the ethical dimensions. The results concluded that age and gender of the user predominantly play an important part during the interplay of ethical dimensions of AI. They also suggested that while making any ethical policy, all dimensions of ethics should be considered.

Based on literature review, this study identified following themes of AI ethics (Table 1).

3 Research Methodology

This section discusses the methodology followed in the study. The data for the study was collected in two phases. In phase I, respondents were delivered lecture on AI and AI ethics. After this, a two week time was given to them to download the selected application programs i.e., Wysa: Anxiety and therapy ChatBot App, Wonder—AI Art Generator, ELSA—Learn English Speaking, FaceApp: Face Editor, Voice Assistant: DataBot AI on their mobile and use it. These mobile apps were selected because they all used AI and ranked highest with the keyword "Top ranked AI apps" in play store (ANNEXTURE II). Efforts were made to ensure apps from different domains namely wellness, word which generates image, english learning platform, photo editing AI, databot used for various purpose were selected to ensure unbiased user review. In phase II, respondents were asked to fill the questionnaire (ANNEXTURE I) containing statements regarding ethical aspects of AI. The questionnaire consisted of a series of multiple-choice questions and Likert scale items, and it was designed to collect information on the participants' personal details, science interests, attitudes toward the ethics of artificial intelligence, and their perceptions and opinions about AI-based application programs. For data analysis, many questions asked were recorded to meet the assumptions of the statistical analysis i.e., binomial logistic regression.

4 Results and Discussions

Table 2 shows the demographic profile of the respondents. Of the sample of 455, 63.7% were males and 36.95% were females. Of the total sample 83.5% had high interest in science and technology (S&T) followed by 8.8% who had medium interest and 7.7% with low interest. 81.3% respondents actively search the internet for the latest information about science and technology. 42.9% respondent found information about S&T easy to understand followed by 53.8% who found it difficult to understand and 3.3% who could not understand it. 28.6% respondents were high users of smartphones followed by 53.8% medium users and 17.6% as low duration users. The results of the Table 2 show that the majority of respondents selected the correct definitions of AI, indicating that they have a good understanding of the technology. To strengthen understanding of AI and its concepts, it is recommended that video lectures and other forms of educational material be made available to the users of AI applications. This could help to clarify any confusion and fill gaps in knowledge about AI. Additionally, ongoing education and awareness campaigns could be developed to keep users up-to-date with the latest advancements and ethical considerations related to AI and their impacts like on ethics, privacy, safety and security etc. Since more emphasis is being laid of development of AI assisted application programs, due attention need to be paid on the AI education, its ethics and its implications.

The result of the Table 3 shows that Voice Assistant: Data AI was top ranked on the bases of Privacy followed by FACEAPP: ELSA, WONDER and WYSA. It has explicitly shown that it does not share data with any third party. The same should be followed by AI powered the mobile application programs. Highest "Accountability" was found to be of Voice Assistant Data BOT AI as it's data is encrypted in transit and data is transferred over a secure connection; developer's need to replicates the accountability policy of it in other application programs.

Highest "Safety and security" was found to be of Wonder AI Generator, as for this you don't have to login and create an id this wonder AI feature of generating an image can be easily used without creating an account that is major thing which respondents are thinking as safety and security for them, developers need to replicates the Safety

Table 2 Demographic profile of the respondents

Category	Options	Responses (%)
Gender	Males	290 (63.7%)
	Females	165 (36.3%)
Science Interests	High	380 (83.5%)
How much are you interested in science and technology?	Medium	40 (8.8%)
	Low	35 (7.7%)
Do you actively search for information about science and technology?	Yes	370 (81.3%)
	No	65 (14.3%)
	Don't know	20 (4.4%)
When you have looked for information about science and technology in the past, have you generally been able to find what you were looking for?	Yes, but it is often difficult to understand	195 (42.9%)
	Yes, but it is often difficult to understand	245 (53.8%)
	No, I often can't find what I am looking for	15 (3.3%)
Approximately how long do you use a computer or smartphone per day?	High	130 (28.6%)
	Medium	245 (53.8%)
	Low	80 (17.6%)
Which of the following options is the most appropriate explanation of AI as of today?	A robot that thinks and acts on its own, without human assistance	115 (25.3%)
	A program that makes decisions based on learning results	290 (63.7%)
	A computer that interacts with people	35 (7.7%)
	A new type of smartphone	15 (3.3%)
Which of the following options is the most appropriate explanation of what AI can do as of today?	It makes moral decisions on its own	95 (20.9%)
	It understands and interprets human languages	275 (60.4%)
	It develops software on its own	55 (12.1%)
	It has free will	30 (6.6%)

(continued)

Table 2 (continued)

Category	Options	Responses (%)
Which of the following options is the most appropriate explanation of AI developers as of today?	The government is developing AI	35 (7.7%)
	Information scientists and researchers are developing AI	315 (69.2%)
	Computer programs are developing AI without human intervention	85 (18.7%)
	Everyone is developing AI using smartphones	20 (4.4%)

Table 3 Response for the eight themes in apps

Themes	WYSA		WONDER		ELSA		FACEAPP		DATABOT	
	Mean	SEM	Mean	SEM	Mean	SEM	Mean	SEM	Mean	SEM
Privacy	2.93	0.12	3.05	0.12	3.05	0.111	3.07	0.125	3.07	0.125
Accountability	2.9	0.113	3.02	0.119	3	0.113	2.88	0.125	2.88	0.125
Safety and security	2.99	0.122	3.26	0.114	3.12	0.108	3.01	0.107	3.01	0.107
Transparency and explainability	3.02	0.109	3.14	0.12	3.19	0.119	3.14	0.119	3.14	0.119
Fairness and non-discrimination	3.3	0.118	3.22	0.116	3.13	0.105	3.08	0.104	3.08	0.104
Human control of technology	3.11	0.114	3.25	0.111	3.2	0.121	3.07	0.108	3.07	0.108
Professional responsibility	3.14	0.119	3.13	0.118	3.13	0.113	3.07	0.107	3.13	0.107
Promotion of human value	3.22	0.11	3.21	0.11	3.26	0.117	3.2	0.115	3.2	0.115

Note SEM = Standard Error of the Mean

and security policy of it in other application programs. Highest "Transparency and Explainability" was found to be of Voice Assistant Data BOT AI, there is transparency and explainability of each policy in detail and it doesn't share the data with third party this is also one of the reason the respondent think it is more transparent and explanibilty is also good.

Developers need to replicates the Transparency and Explainability policy of it in other application programs. Highest ranked app on "Fairness and non discrimination" was Wysa. The reason for the same can be attributed that the app has no discrimination in it and it is easily accessible to all and features of the app for improving your mental health is there you can chat with the AI and it will help you with a fair answer to your problem. The app is easily available to all and no bias is there in the app, developer's needs to replicates the Fairness and non-discrimination policy of it in other application programs. Highest ranked app on "Human Control of Technology" was found to be of Wonder AI Generator, as it gives human the control to give command and generate so it can be the one the reason it is meaner score than others, developers need to replicates the Human Control of Technology policy of it in other application programs. Highest ranked app on "Professional Responsibility" was found to be of Wysa, as it provides a way for you to request that your data be deleted easily. Developers need to replicates the Fairness and non-discrimination policy of it in other application programs. Highest ranked app on "Promotion of Human Value" was found to be of Elsa. The app by giving people a chance to learn English in an interactive way with AI guiding you provides it a human value as you need a guide to learn a language and you are getting it integrated in the app. Developers need to replicates the Promotion of Human Value policy of it in other application programs. Overall, incorporating these recommendations will not only improve the ethics of AI apps but also enhance user trust and satisfaction with the app.

Tables 4, 5 and 6 show the results of logistic regression which was performed to ascertain the effects of gender, duration of Smartphone usage, interest in science and technology and understanding of AI on the likelihood that participants having perception towards themes of AI Ethics.

For each regression, the respective baselines were male, group with high interests in science and technology and not correct understanding of AI was used.

Privacy: The logistic regression model was statistically significant, for privacy with $\chi2(7) = 16.181$, significance value (0.02) which is less than 0.05. The model explained 49.0% (Nagelkerke R2) of the variance in AI privacy and correctly classified 67% of cases. Male participants were 1.408 times more likely to be concerned about privacy as females. Likewise, participants with high interest in science and technology were interested in science were 1.355 times more likely to focus on privacy and participants who used smartphone less were 0.598 time more likely to be concerned about privacy as compared to user who spent more time. Participants who had a good understanding of AI had 1.903time more likely hood of being concerned about privacy.

Accountability: The logistic regression model was statistically significant, for accountability with $\chi2(7) = 22.356$, significance value (0.002) which is less than 0.05. The model explained 66.0% (Nagelkerke R2) of the variance in AI privacy and correctly classified 63.3% of cases. Male participants were 1.408 times more likely to be concerned about accountability as compared to females. Likewise, participants with high interest in science and technology were interested in science were 1.394 times more likely to focus on accountability and Participants who had a good understanding of AI had 1.531 time more likely hood of being concerned about accountability.

Safety and security: The logistic regression model was statistically significant, for accountability with $\chi2(7) = 16.553$, significance value (0.021) which is less than 0.05. The model explained 50.0% (Nagelkerke R2) of the variance in AI privacy and correctly classified 69.2% of cases. Participants who were highly interested in science

Table 4 Results of binomial logistic regression of the responses to the privacy, accountability and safety and security themes of AI ethics

Variable	Privacy					Accountability					Safety and security				
	(95% Confidence interval for B)					(95% Confidence interval for B)					(95% Confidence interval for B)				
	B	Lower	Upper	P	EXP (B)	B	Lower	Upper	P	EXP (B)	B	Lower	Upper	P	EXP (B)
Gender (Men)	0.342	0.912	2.172	0.022*	1.408	0.55	1.128	2.662	0.012*	1.733	0.071	0.604	1.435	0.746	0.931
Duration high	0.238	0.434	1.431	0.434	0.788	0.006	0.545	1.811	0.984	0.994	0.245	0.688	2.37	0.438	1.277
Duration low	0.515	0.351	1.016	0.057*	0.958	0.177	0.492	1.427	0.514	0.838	0.139	0.499	1.518	0.625	0.87
Interest in S&T level high	0.837	0.055	2.279	0.075*	1.355	0.931	0.061	2.529	0.026*	1.394	0.143	0.18	7.401	0.081*	1.153
Interest in S&T level medium	1.903	0.02	1.098	0.062	0.149	1.808	0.023	1.193	0.074	0.164	1.464	0.029	1.871	0.17	0.231
Interest in S&T level low	1.347	0.036	1.851	0.179	0.26	1.457	0.033	1.659	0.146	0.233	0.269	0.107	5.46	0.789	0.764
Understanding of AI (High)	0.352	0.455	1.087	0.013*	1.903	0.633	1.044	1.82	0.004*	1.531	0.374	0.441	1.074	0.010*	1.018

Notes **B** = Unstandardized regression coefficient, **P** = Probability value at a 95% confidence interval, **EXP (B)** = Odds ratio, **Lower and Upper** = Confidence interval boundaries, * Statistically significant at a 95% confidence interval ($p < 0.05$)

Table 5 Results of binomial logistic regression of the responses to the transparency and explainability, fairness and nondiscrimination and human control of technology themes of AI ethics

Variables	Transparency and explainability					Fairness and nondiscrimination					Human control of technology				
	(95% Confidence interval for B)					(95% Confidence interval for B)					(95% Confidence interval for B)				
	B	Lower	Upper	P	EXP (B)	B	Lower	Upper	P	EXP (B)	B	Lower	Upper	P	EXP (B)
Gender (Men)	0.254	0.828	2.009	0.261	1.289	0.157	0.54	1.353	0.503	0.855	0.187	0.781	1.861	0.398	1.206
Duration high	0.311	0.729	2.556	0.331	1.365	0.126	0.128	0.236	0.686	0.252	0.429	2.407	0.005*	1.777	
Duration low	0.061	0.532	1.664	0.834	0.941	0.737	0.273	0.841	0.010*	0.479	0.582	0.325	1.949	0.331	1.555
Interest in S&T level high	0.96	1.031	6.609	0.043*	2.611	0.75	0.833	5.376	0.115	2.116	0.373	0.659	3.199	0.355	1.452
Interest in S&T level medium	0.383	0.445	4.828	0.529	1.466	0.264	0.208	2.836	0.692	0.768	0.639	0.166	1.677	0.279	0.528
Interest in S&T level low	0.385	0.433	1.069	0.094	0.68	0.57	1.346	3.925	0.023*	1.565	0.352	0.452	1.094	0.119	0.703
Understanding of AI (High)	1.876	1.828	2.019	0.001*	1.153	0.985	1.14	1.953	0.018*	1.374	0.7	1.181	2.861	0.039*	1.497

Notes **B** = Unstandardized regression coefficient, **P** = Probability value at a 95% confidence interval, **EXP (B)** = Odds ratio, **Lower and Upper** = Confidence interval boundaries, * Statistically significant at a 95% confidence interval ($p < 0.05$)

Table 6 Results of binomial logistic regression of the responses to the professional responsibility and promotion of human value themes of AI ethics

Variables	Professional responsibility (95% Confidence interval for B)					Promotion of human value (95% Confidence interval for B)				
	B	Lower	Upper	P	EXP (B)	B	Lower	Upper	P	EXP (B)
Gender (Men)	0.036	0.663	1.621	0.005*	1.037	0.022	0.627	1.528	0.925	0.979
Duration high	0.165	0.643	2.163	0.594	1.179	0.357	0.753	2.711	0.024*	1.429
Duration low	0.559	0.325	1.005	0.052	0.572	0.045	0.536	1.706	0.88	0.956
Interest in S&T level high	0.3	0.609	2.993	0.020*	1.35	1.524	1.552	13.582	0.006*	4.592
Interest in S&T level medium	0.242	0.263	2.344	0.664	0.785	0.345	0.145	3.455	0.67	0.708
Interest in S&T Level Low	0.654	0.322	0.839	0.407	0.52	0.549	0.363	0.92	0.021*	0.578
Understanding of AI (High)	0.807	0.663	1.621	0.004*	1.446	2.164	0.627	1.528	0.000*	1.115

Notes **B** = Unstandardized regression coefficient, **P** = Probability value at a 95% confidence interval, **EXP (B)** = Odds ratio, **Lower and Upper** = Confidence interval boundaries, * Statistically significant at a 95% confidence interval ($p < 0.05$)

and technology and had good understanding of AI were more likely to be concerned about safety and security with likelihood ratio of 1.153 and 1.018 respectively.

Transparent and Explainability: The logistic regression model was statistically significant, for Transparent and Explainability with $\chi2(7) = 10.469$, significance value (0.016) which is less than 0.05. The model explained 33.0% (Nagelkerke R2) of the variance in AI privacy and correctly classified 71.6% of cases. Participants who were highly interested in science and technology and had good understanding of AI were more likely to be concerned about safety and security with likelihood ratio 2.611and 1.153 respectively.

Fairness and nondiscrimination: The logistic regression model was statistically significant, for Fairness and nondiscrimination with $\chi2(7) = 19.891$, significance value (0.003) which is less than 0.05. The model explained 64.0% (Nagelkerke R2) of the variance in AI privacy and correctly classified 75.4% of cases. Participants spending less time on smartphone who were highly interested in science and technology and comparatively low interest in S&T and had good understanding of AI were more likely to be concerned about safety & security with likelihood ratio 0.479, 1.565 and 1.374 respectively.

Human control of technology: The logistic regression model was statistically significant, for Fairness and nondiscrimination with $\chi2(7) = 15.116$, significance value (0.019) which is less than 0.05. The model explained 42.3.0% (Nagelkerke

R2) of the variance in AI privacy and correctly classified 69% of cases. Participants spending less time on smart phone who were highly interested in science and technology and had good understanding of AI were more likely to be concerned about Human control of technology with likelihood ratio 1.777, and 1.497 respectively.

Professional responsibility: The logistic regression model was statistically significant, for Fairness and nondiscrimination with $\chi 2(7) = 18.927$, significance value (0.009) which is less than 0.05. The model explained 59.0% (Nagelkerke R2) of the variance in AI privacy and correctly classified 73.7% of cases. Participants who were male, had high interest in S&T and good understanding of AI were more likely to be concerned about Professional responsibility with likelihood ratio 1.037, 1.350 and 1.446 respectively.

Promotion of human value: The logistic regression model was statistically significant, for Fairness and nondiscrimination with $\chi 2(7) = 27.033$, significance value (0.000) which is less than 0.05. The model explained 38.0% (Nagelkerke R2) of the variance in AI privacy and correctly classified 72.1% of cases. Participants who spent more time on smartphones, had high interest in S&T and good understanding of AI were more likely to be concerned about Promotion of human value with likelihood ratio 1.429, 4.592 and 1.115 respectively.

5 Recommendations

Based on data analysis general recommendation for AI developers are provided. Developers should work with policymakers to develop clear and concise AI ethics guidelines that prioritize user privacy, fairness, and accountability. These guidelines should be easily accessible to users and provide transparency on how AI apps operate. Developers should conduct user testing and collect feedback to better understand user needs and preferences. This information can be used to improve the functionality and ethics of AI apps. Various demographic factors like gender, interest in science and technology, understanding of the concepts like ethics, AI should also be considered while designing application programs. Developers should regularly update and maintain AI apps to ensure that they meet current ethical standards and remain relevant to users. Developers, policymakers, and users should work collaboratively to address ethical issues surrounding AI. This collaboration can help ensure that AI is developed and used in a way that benefits society as a whole. Developers should promote education and awareness about AI ethics to users. This can be done through educational campaigns, tutorials, and other forms of outreach to ensure that users understand the ethical implications of AI. By following these recommendations, developers can create AI apps that are not only functional but also ethical and responsible. Overall, this study provides insights into users' understanding of AI technology and highlights the importance of ongoing education and awareness campaigns to promote a better understanding of the technology and its implications. By increasing users' knowledge of AI, developers and policymakers can work together to develop AI solutions that are both functional and ethical, ensuring that AI technology is developed and used in a way that benefits society as a whole.

6 Limitations and Future Scope

Owing to small sample size, this study might not be representative of the AI users. A larger sample size would have provided more accurate and reliable results. Since the study is limited to AI mobile applications only, which may not provide a comprehensive understanding of user perceptions towards AI ethics in all AI applications. The study relied on self-reported data from respondents, which may be subject to social desirability bias or inaccuracies in recall. An expansion of the study could be conducted to explore variances in attitudes towards apps that incorporate artificial intelligence across a multitude of demographic groups, encompassing age brackets, sexes and levels of education. This could help identify any disparities in attitudes and highlight areas where targeted efforts to promote positive attitudes toward AI may be needed. It is plausible for scholars to amass information concerning the usage of mobile applications and juxtapose it with responses gathered through questionnaires so as to ascertain whether favourable dispositions are correlated with heightened utilization rates. Overall, these future scopes could help provide a more comprehensive understanding of public attitudes toward AI apps and identify strategies for promoting positive attitudes and addressing concerns related to the use of AI.

Annexture I

Questionnaire

Artificial Intelligence (AI): Artificial intelligence (AI) is the capacity of a machine to comprehend and pick up knowledge from experiences, making judgements and forecasts based on facts without being explicitly programmed. It can accomplish tasks like recognizing images, processing language, and making decisions because it replicates the cognitive functions of the human brain. It has the power to completely alter how we work, live, and engage with technology.

Personal details.

1. Age (Years): Below 18/19–25/26–35/36–55/56–70/Above 70
2. Gender*: Male/Female/Prefer not to say

Science Interests

3. How much are you interested in science and technology? Please select only one. [Very interested/Quite interested/Neither interested not disinterested/ Not very interested/Not interested at all/Don't know]

4. Do you actively search for information about science and technology? [Yes/No/ Don't know]
5. When you have looked for information about science and technology in the past, have you generally been able to find what you were looking for? Please select the closest answer? [Yes, and it tends to be easy to understand/Yes, but it is often difficult to understand/No, I often can't find what I am looking for/Don't know]
6. Approximately how long do you use a computer or smartphone per day? [Less than 30/30 min to 1/1 h to 2/2 h to 3/3 h to 4/4 h to 6/6 h to 8 h/8 h to 10 h/10 h or more/ I do not use them at all]

***The items 8–10 were randomly presented to the respondents**

(This sentence was not shown to the respondents.)

8. Which of the following options is the most appropriate explanation of AI as of today? [A robot that thinks and acts on its own, without human assistance/ A program that makes decisions based on learning results/ A computer that interacts with people/ A new type of smartphone]
9. Which of the following options is the most appropriate explanation of what AI can do as of today? [It makes moral decisions on its own/it understands and interprets human languages It develops software on its own/It has free will]
10. 10. Which of the following options is the most appropriate explanation of AI developers as of today? [The government is developing AI/Information scientists and researchers are developing AI/Computer programs are developing AI without human intervention/Everyone is developing AI using smartphones]

Attitudes toward the ethics of Artifcial intelligence (AT-EAI).

The purpose of this questionnaire is to investigate post graduate students' attitudes toward AI ethics. Each item is followed by five digits (1, 2, 3, 4, 5), and we would like you to write on the digit that most accurately represents your opinion. 1 means "I strongly disagree with this statement."; 2 means "I disagree with this statement."; 3 means "I neither agree nor disagree with this statement."; 4 means "I agree with this statement."5 means "I strongly agree with this statement."

There are no correct or incorrect responses for these items. The collected data will be kept strictly confidential and will be used solely for research purposes.
Thank you for your participation.

S.NO.	Question	WYSA	Wonder AI	Elsa	FaceApp	Voice assistant
1	I feel comfortable sharing personal information with app					
2	I trust app to keep my personal information secure and confidential					

(continued)

(continued)

S.NO.	Question	WYSA	Wonder AI	Elsa	FaceApp	Voice assistant
3	I am confident in the accuracy and reliability of app responses to my mental health concerns					
4	The App clearly explain how it uses my data and information					
5	I believe that app treats all users fairly and without discrimination					
6	I feel in control of the conversations I have with app					
7	I think that app is a responsible tool for addressing mental health concerns					
8	I believe that app promotes values that are important for mental health and wellbeing					

Annexture-II

Table of eight themes with the applications used

Eight themes	Wysa	Wonder—AI	ELSA	FaceApp	Voice assistant: DataBot AI
Privacy	Data shared	Data shared	No data shared with third parties	Data shared	No data shared with third parties

(continued)

(continued)

Eight themes	Wysa	Wonder—AI	ELSA	FaceApp	Voice assistant: DataBot AI
Data shared and for what purpose	App interactions, Crash logs, Phone number, Health info, Other in-app messages, in-app search history, user-generated content, Device or other IDs	Location, Personalization, User IDs, In-app search history, App info and performance, Crash logs	Location, Personal info, App functionality, Advertising or marketing, Personalization, Account management, User IDs, Account management, Financial info, User payment info, Account management, Crash logs, App functionality, Analytics, Diagnostics, Device or other IDs, App functionality	Location, Financial info, Purchase history, Analytics, App activity, App interactions, Analytics, App info and performance, Diagnostics and other app performance data, Diagnostic Analytics, Purchase history, Analytics, App activity, Crash logs, Analytics, Diagnostics, Analytics, security and compliance, Device or other IDs, Analytics, Advertising or marketing, Fraud prevention, security and compliance	No information
Accountability	Data is encrypted in transit	Data isn't encrypted	Data is encrypted in transit	Data is encrypted in transit	Data is encrypted in transit
Safety and security	– Data is encrypted in transit. – You can request that data be deleted	Data isn't encrypted. Data can't be deleted	Data is encrypted in transit. You can request that data be deleted	Data is encrypted in transit. You can request that data be deleted	Data is encrypted in transit. You can request that data be deleted

(continued)

(continued)

Eight themes	Wysa	Wonder—AI	ELSA	FaceApp	Voice assistant: DataBot AI
Transparent and explainability	The developer has provided this information about how this app collects, shares and handles your data	The developer has not provided this information about how this app collects, shares and handles your data	The developer has provided this information about how this app collects, shares and handles your data	The developer has provided this information about how this app collects, shares and handles your data	The developer has provided this information about how this app collects, shares and handles your data
Fairness and nondiscrimination	The app is easily available to all and no bias is there in the app	The app is easily available to all and no bias is there in the app	The app is easily available to all and no bias is there in the app	The app is easily available to all and no bias is there in the app	The app is easily available to all and no bias is there in the app
Human control of technology	You can request that data be deleted. The developer provides a way for you to request that your data be deleted	The data can't be deleted. The developer doesn't provide a way for you to request that your data be deleted	You can request that data be deleted. The developer provides a way for you to request that your data be deleted	You can request that data be deleted. The developer provides a way for you to request that your data be deleted	You can request that data be deleted. The developer provides a way for you to request that your data be deleted
Professional responsibility	The developer provides a way for you to request that your data be deleted	The developer doesn't provide a way for you to request that your data be deleted	The developer provides a way for you to request that your data be deleted	The developer provides a way for you to request that your data be deleted	The developer provides a way for you to request that your data be deleted

References

1. Berendt, B.: AI for the common good?! Pitfalls, challenges, and ethics pen-testing. Paladyn J. Behav. Robot. **10**(1), 44–65 (2019). https://doi.org/10.1515/pjbr-2019-0004
2. Dignum, V.: Ethics in artificial intelligence: introduction to the special issue. Ethics Inf. Technol. **20**(1), 1–3 (2018). https://doi.org/10.1007/s10676-018-9450-z

3. Fjeld, J., Achten, N., Hilligoss, H., Nagy, A., Srikumar, M.: Principled Artificial Intelligence: Mapping Consensus in Ethical and Rights-Based Approaches to Principles for AI (2020)
4. Floridi, L., Cowls, J., Beltrametti, M., Chatila, R., Chazerand, P., Dignum, V., Luetge, C., Madelin, R., Pagallo, U., Rossi, F., Schafer, B.: AI4People—an ethical framework for a good AI society: opportunities, risks, principles, and recommendations. Mind. Mach. **28**(4), 689–707 (2018). https://doi.org/10.1007/s11023-018-9482-5
5. Haque, A.B., Islam, A.K.M.N., Mikalef, P.: Explainable Artificial Intelligence (XAI) from a user perspective: a synthesis of prior literature and problematizing avenues for future research. Technol. Forecast. Soc. Chang. **186**(PA), 122120 (2023). https://doi.org/10.1016/j.techfore.2022.122120
6. Hasan, M.M., Islam, M.U., Sadeq, M.J., Fung, W.K., Uddin, J.: Review on the evaluation and development of artificial intelligence for COVID-19 containment. Sensors **23**(1), 1–35 (2023). https://doi.org/10.3390/s23010527
7. Ikkatai, Y., Hartwig, T., Takanashi, N., Yokoyama, H.M.: Octagon measurement: public attitudes toward AI ethics. Int. J. Hum.-Comput. Interact. **38**(17), 1589–1606 (2022)
8. Jang, Y., Choi, S., Kim, H.: Development and validation of an instrument to measure undergraduate students' attitudes toward the ethics of artificial intelligence (AT-EAI) and analysis of its difference by gender and experience of AI education. In: Education and Information Technologies, vol. 27. Springer, US (2022). https://doi.org/10.1007/s10639-022-11086-5
9. Lukkien, D.R., Nap, H.H., Buimer, H.P., Peine, A., Boon, W.P., Ket, J.C., Minkman, M.M., Moors, E.H.M.: Toward responsible artificial intelligence in long-term care: a scoping review on practical approaches. Gerontologist **63**(1), 155–168 (2023). https://doi.org/10.1093/geront/gnab180
10. Perifanis, N.A., Kitsios, F., Haque, A.B., Islam, A.K.M.N., Mikalef, P., Lukkien, D.R.M., Uddin, J.: Artificial intelligence ethics by design. Evaluating public perception on the importance of ethical design principles of artificial intelligence. Gerontologist **186**(1), 60 (2023). https://doi.org/10.1177/20539517221092956
11. Yüksel, N., Börklü, H.R., Sezer, H.K., Canyurt, O.E.: Review of artificial intelligence applications in engineering design perspective. Eng. Appl. Artif. Intell. **118**(November 2022), 105697 (2023). https://doi.org/10.1016/j.engappai.2022.105697

Detection and Mitigation of NDP Attacks for AMI Devices Using Cooja Simulator

Shelza Dua, Virender Kadyan, Rahul Bhogal, and Mohit Dua

Abstract For the smart grid and smart metering systems to monitor and effectively manage the amount of electricity consumed by consumers and industrial areas, an AMI (Advanced Metering Infrastructure) is essential. Cyberattacks on the smart meter made it vulnerable to the electricity meter and consumer privacy after we switched to smart metering. Smart metering services can be disrupted by cyberattacks, and this attack primarily targets AMI devices that are used in smart meters. Cyberattacks have resulted in inaccurate readings, incorrect billing, etc. In this paper, we seek to identify and counter (Neighbor Discovery Protocol) NDP attacks on the AMI devices of the smart meter system using the Routing Protocol for Low-Power and Lossy Networks (RPL) in the Cooja simulator. The detection of cyberattacks from PCAP file output into the Cooja simulator's implementation of the RPL protocol. This includes putting RPL protocol countermeasures like route diversion, packet filtering, and rate limiting into practice. The effectiveness of these mitigation strategies is assessed based on their capacity to preserve the reliability and availability of AMI data as well as their effect on the performance of the entire network. The proposed approach is validated through extensive simulations in the Cooja simulator that mimic real-world AMI scenarios. The simulations take into account various attack scenarios and assess how well the detection and mitigation mechanisms perform in terms of detection precision and the capacity to lessen the effects of an NDP attack.

S. Dua
Department of Electronics and Communication Engineering, Punjab Engineering College, Chandigarh, India

V. Kadyan
School of Computer Science, University of Petroleum and Energy Studies, Dehradun, India
e-mail: vkadyan@ddn.upes.ac.in

R. Bhogal · M. Dua (✉)
Department of Computer Engineering, National Institute of Technology, Kurukshetra, India
e-mail: er.mohitdua@nitkkr.ac.in

R. Bhogal
e-mail: 32213205@nitkkr.ac.in

B. Soni et al. (eds.), *Beyond Artificial Intelligence*, Lecture Notes in Networks and Systems 1326, https://doi.org/10.1007/978-981-96-4170-3_8

Keywords NDP (Neighbour Discovery Protocol) · AMI (Advanced Metering Infrastructure) · RPL (Routing Protocol for Low-Power and Lossy Networks) · Cooja simulator

1 Introduction

Modern electricity distribution networks now include advanced metering infrastructure (AMI) as a necessary component due to the widespread adoption of smart grid technologies. Utility providers can collect real-time data on energy consumption, improve grid management, and provide consumers with cutting-edge services thanks to AMI devices like smart meters. As shown in Fig. 1. However, the addition of these components to the electrical grid poses new cybersecurity difficulties [1], making them vulnerable to a range of malicious actions, such as Network Denial of Service NDP attacks.

NDP attacks pose a serious risk to the consistent performance of AMI devices. By flooding the network with a lot of malicious traffic, these attacks try to stop it from operating normally. Attackers can clog up the network, deplete its resources, and disrupt services by sending an excessive number of packets, which will result in inaccurate measurements of the amount of energy used and compromised data integrity. These attacks can have serious consequences, including the possibility of financial losses for both utility companies and customers as well as a reduction in the overall dependability of the power grid. This paper concentrates on developing and implementing a detection and mitigation strategy using the Routing Protocol for Low-Power and Lossy Networks (RPL) in the Cooja simulator to address the growing threat of NDP attacks in AMI systems. Since AMI devices frequently operate

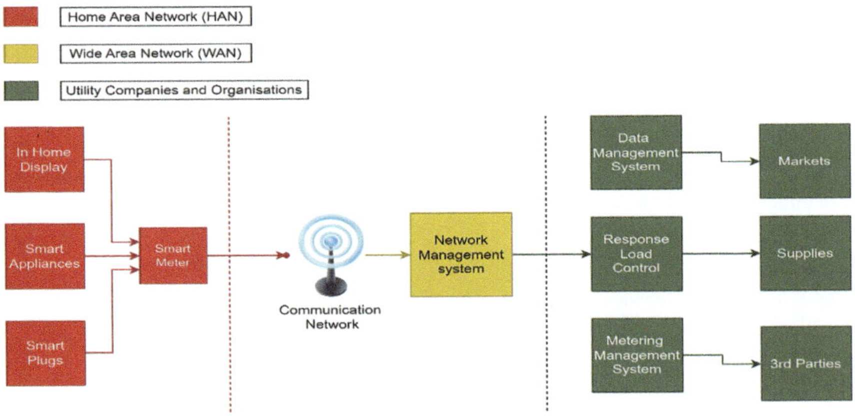

Fig. 1 Advanced metering infrastructure block diagram

with constrained power and processing resources, RPL is well-suited for resource-constrained networks. By effectively detecting and mitigating NDP attacks, we hope to improve the security of AMI systems by utilizing the features of RPL.

A suitable setting for testing and assessing the suggested strategy is the Cooja simulator. It provides a realistic picture of the AMI system by allowing us to simulate various attack scenarios and network conditions. By running simulations, we can evaluate the performance metrics, determine how well the detection and mitigation technique's function, and learn how the system responds to various attack intensities and network topologies. The main goal of this paper is to create a thorough framework for detection and mitigation that can recognize NDP attacks with accuracy and act quickly to lessen their effects. The suggested method entails including anomaly detection methods in the implementation of the RPL protocol. These algorithms continuously track patterns of network traffic, examining elements including packet rates, message integrity, and routing behaviors to spot outliers that might be signs of an NDP attack. The RPL protocol's mitigation measures take effect as soon as an NDP attack is discovered. These tactics are designed to stop and stop the attack, protecting the availability and integrity of AMI data. Route diversion, packet filtering, rate restriction, and traffic prioritization are a few examples of mitigation strategies. To reduce their negative effects on operational efficiency, these strategies are chosen and tuned taking into account the resource limitations of AMI devices.

We will evaluate the effectiveness and efficiency of the suggested technique using exact simulations and performance measurements. Accuracy of detection, false positive rates, the efficiency of neutralizing NDP attacks, and the effect on overall network performance are key performance indicators. The outcomes of these assessments will give important information about the scalability, resilience, and viability of the suggested strategy in actual AMI deployments. The findings of this investigation have applications for utility companies and system administrators in charge of ensuring the security of AMI devices. In order to better defend smart meter systems against the changing threat landscape of NDP attacks, the proposed detection and mitigation approach can help. AMI device cybersecurity is a growing topic, and the knowledge gathered from this research can contribute significantly to that sector's knowledge, enabling the creation of smart grid infrastructures that are more reliable and resilient. The methodology, system design, implementation specifics, and outcomes of the simulations and evaluations will all be covered in detail in the next sections of this report, which will also include a summary of the suggested strategy. We will also talk about the paper's limits and offer suggestions for future improvements and research paths in an effort to support ongoing efforts to protect AMI devices and guarantee the trustworthy and secure operation of smart meter systems.

2 Motivation

Modern utility systems now have a completely new way to monitor and regulate electricity use thanks to the implementation of Advanced Metering Infrastructure (AMI). A key part of AMI, smart meter devices give real-time data on energy consumption, allowing utilities to optimize distribution, increase billing accuracy, and improve operational efficiency overall. These smart meter devices, however, have changed from possible targets for cyberattacks to potential targets because of their increased connection and dependency on communication networks.

The NDP assault, also known as a network denial of service attack, is one of the biggest risks to AMI devices. By flooding AMI devices with a large amount of malicious traffic or by taking advantage of flaws in the communication protocols, an NDP assault seeks to prevent AMI devices from operating normally. Inaccurate meter readings, incorrect billing, and interruptions to consumer services are just a few of the serious effects of such attacks. The development of efficient mitigation and detection methods that are especially suited to NDP assaults is essential to ensuring the dependable and secure operation of AMI devices. Due to its suitability for resource-constrained devices and support for effective communication within the network, the implementation of the Routing Protocol for Low-Power and Lossy Networks (RPL) in AMI systems offers a potential strategy.

The goal of this paper is to address the critical need for effective NDP attack detection and mitigation measures that target AMI devices in a smart meter system. We are able to analyse and assess the efficacy of different detection and mitigation measures in a controlled setting by utilizing the capabilities of RPL and simulating the system using the Cooja simulator. The effective development of such methods may result in a number of important advantages. The integrity of energy consumption data is protected, and accurate invoicing for customers is ensured. Initially, it aids utility companies in defending their infrastructure against potential cyber threats. Second, it improves smart meter systems' overall availability and reliability, avoiding service interruptions and upholding client satisfaction. Last but not least, it makes a contribution to the larger field of cybersecurity for AMI devices by offering the industry useful insights and best practices to embrace and enhance the security of their systems.

Through the completion of this study, we hope to expand our understanding of AMI device cybersecurity and offer workable solutions to reduce the dangers brought on by NDP attacks. The results of this paper will help utility companies, system administrators, and researchers put effective security measures into place, leading to a more secure and robust AMI architecture.

3 Literature Survey

The use of the smart grid system allows us to monitor, analyze, and control the mechanism of two-way communication with the supply chain to improve efficiency and reduce the consumption and cost of electricity [2]. The smart grid uses information technology to provide energy to the end-users through two-way communications, which is changing the infrastructure for transmitting energy in terms of efficiency, scalability, and interoperability [3]. The smart meter is a consumer product that is outfitted with electronic hardware and software for connectivity and real-time measurement [4]. At the user end, smart meters are installed and use the wireless area network to connect with the host. Due to the widespread adoption of smart meters, the supplier must maintain cyber security in order to improve the infrastructure and add cyberattack defenses.

Smart metering communication network, the communication network is in a hierarchical structure, where layers can be categorized into three classical networks based on the range [5]. The use of information technology by the smart grid to supply energy effectively while incorporating green technologies is the main topic of this article. However, the system is now vulnerable to a number of security risks due to communication technology flaws. Current survey articles emphasize safeguards based on availability, confidentiality, and integrity but frequently ignore responsibility. In order to identify and neutralize these dangers, the paper evaluates security standards, discusses serious cyber-attacks, and suggests a thorough cyber-security plan. Future research directions are also recommended [6]. As the below show the communication network topologies are:

- Home Area Network (HAN)
- Neighborhood Area Network (NAN)
- Wide Area Network (WAN).

The most crucial element of a smart grid is advanced metering infrastructure (AMI), which connects all other elements to boost efficiency [4]. Smart meters, a data management system, automatic meter readings, transmission networks, and other components make up AMI, which enables two-way communication between customers and service providers. The groupings that can be used to categorize attacks and other parties using smart metering data are listed below [7]:

- Utility companies: To produce invoices and track electricity usage.
- Law enforcement: To find harmful or unlawful behavior on the part of the consumer.
- Data management companies and marketers: to gather personal data for marketing.
- Curious eavesdroppers: to snoop on their neighbors' goings-on.
- Passive attackers: To showcase their hacking capabilities and penetration tactics.
- Consumers: Stealing power to avoid paying as much on bills.
- Unintentional listeners: to determine which rich families to target and to determine when a break-in is most likely to occur.

- Aggressive assailants: To carry out massive terrorist assaults on the utility or the power grid.

Information security has become more important in all spheres of life as a result of our growing reliance on computer networks and the internet. Institutions must, however, strengthen their security infrastructure due to dangers and cyber-attacks. This study suggests creating a simulated intelligent instrument to solve the problem of unreliable replies in conventional questionnaire research. In contrast to theoretical surveys, the instrument will leverage gamification in experimental research to assess users' behaviors toward cybersecurity rules [8]. The article talks about how challenging it is to defend against Distributed Denial-of-Service (DDoS) attacks, which continue to be a severe threat to network security. The proposed system uses a hybrid approach that combines feature selection strategies with machine learning classifiers in order to rapidly and effectively identify DDoS attacks on IoT devices. The study employs feature selection methods such as chi-square, Extra Tree, and ANOVA on classifiers including Random Forest, Decision Tree, k-nearest Neighbors, and XGBoost. Experimental results on the CICDoS2019 dataset demonstrate the superiority of the proposed method, which aids in the early detection of DDoS attacks on IoT devices with 98.34% accuracy using ANOVA for XGBoost and a superb 82.5% feature reduction ratio [9, 10].

The Service Unavailability Cyber Impact Factor (SUCIF), which rates the severity of breaches on a scale from 0 to 1, is the first index. Systems with little knowledge of cyber-physical infrastructure can nonetheless use it. Time-based cyber impact factor (TBCIF), the second index, employs graph-based bad node identification and takes intrusion discovery time into account when calculating impact. TBCIF is appropriate for smart grid networks and well-known test systems. These indices are used in the research to compare the effects of hacking on the Saudi Aramco Oil and Gas Company and the Ukrainian Power Grid, as well as on several IEEE test systems to validate TBCIF [11]. The Neighbour Discovery Protocol (NDP) Security mechanism, a critical component of IPv6, provides address resolution, router location, and duplicate address detection in a smart metre network [12]. A process in NDPs will presume the dependability of every node linked to the network. This leaves the network open to assault, and there are many different kinds of attacks, including Denial of Service on Duplicate Address Detection (DAD) attacks, address resolution-based attacks, router advertising (RA) attacks, and redirect attacks.

The concept of Low Power and Lossy Networks (LLNs) is presented, along with the need for trustworthy and resource-conserving routing protocols for Point-to-Point (P2P) communication in various LLNs applications. The difficulties of energy utilization and reliability are recommended to be resolved by the special region-based routing protocol known as ER-RPL. Since it only employs a small subset of nodes for route discovery, ER-RPL uses substantially less energy than traditional protocols. Numerous simulations demonstrate that the benchmark protocols ER-RPL outperform RPL and P2P-RPL [13]. The goal of the Internet of Things is to link every physical object to the Internet. A distinct Internet Protocol address should be displayed for each device. The terms IPv4 and IPv6 refer to the two IP versions. The assumption is that IPv4 will cover all network interfaces. Since the emergence of IoT, the number of linked devices has drastically expanded, and IPv4 cannot support such massive numbers [14]. Compared to IPv4, IPv6 improve security in comparison to IPv4 and also as well as the quality of services, but IPv6 continues to face various type of attacks likewise Denial of services and Man-in-the-Middle attacks.

In many IoT-based networks, RPL is a low-power, energy-intensive communication method. The term "Internet of Things" refers to a network that links many heterogeneous things. IoT devices include smart appliances, sensors, and actuators. Control messages are sent between nodes in RPL acceptance and transmission packets. In a graph known as a (Destination-Oriented Directed Acyclic Graph) DODAG [15], these messages will be sent and received. It explores the two case studies TinyRPL and ContikiRPL and presents early simulation results using the COOJA tool. The conversation then shifts to the use of RPL in precision agriculture to fulfil application-specific requirements. The paper concludes by summarizing recent findings and offering prospective solutions for technology issues in the future [16]. The security of Wireless Sensor Networks (WSNs) is the main topic of this study. In a 20-node WSN with 10% hostile nodes, it uses RPL for IDS. Cooja with InstantContiki2.7 and NS2 with the AODV protocol are used in simulations to monitor node mobility, identify network intrusions, and find lost packets during IDS in WSN [17].

In the above literature survey explain the different strategies which are work on the different data sets and techniques that also uses machine learning or deep learning. But now this paper presents the security mechanism for AMI devices (RPL) with the Contiki-os-based Cooja simulator.

4 Proposed Approach

A systematic and thorough technique using the Routing Protocol for Low-Power and Lossy Networks (RPL) in the Cooja simulator is the recommended strategy for the identification and mitigation of Network attacks on NDP assaults against AMI devices in a smart meter system.

4.1 Cooja Simulation Setup

The Cooja simulation setup process is detailed has been described in Kurniawan [18]. Now lets we discuss more about the Cooja simulation in the following section.

Here we use the Linux—os (ubuntu) to perform the Cooja simulation setup, to start the Cooja simulation software, first open the terminal window and enter the following commands:

- cd contiki/tools/cooja
- ant run.

After that, we click on the file and create a new simulation. The settings on this need not change, in deployment of the Unit Disk Graph Model (UDGM). These are added by clicking on motes, adding motes, creating a new mote type, and then sky mote.

After adding the motes, we observed the different functionalities:

- Addresses: IP or Rime
- Log output: printf()'s
- LED's
- Radio Traffic
- Position
- Mote type
- Mote attributes
- Radio environment (UDGM).

More detailed Cooja simulation setup and implementation part has been discussed in Sect. 5.

4.2 Analysis of Vulnerabilities

The analysis of the AMI devices' particular vulnerabilities in an RPL-based network is the first step in the investigation. AMI devices, such as smart meters, frequently have restricted processing power and memory due to resource constraints. Because of this, they are vulnerable to many types of assaults, including NDP attacks that try to stop these devices' communication and operation. Studying potential attack vectors such as packet flooding, resource fatigue, compromised nodes, and protocol vulnerabilities is part of the analysis. The unique characteristics of AMI systems, such as the network's hierarchical structure, reliance on wireless connection, and the significance of the data being sent, must be taken into account throughout the analysis phase.

4.3 Detection System Development

The creation of a reliable detection system that can spot irregularities suggestive of NDP attacks is the next step in the suggested methodology. The RPL protocol is implemented in the Cooja simulator with the inclusion of anomaly detection methods. As part of its ongoing monitoring of network traffic patterns, the detection system examines variables including packet rates, message integrity, and routing behavior to spot outliers in behavior that might be signs of an NDP attack. On the basis of thresholds and statistical models, statistical analysis can find anomalies, whereas machine learning algorithms can discover anomalous patterns by learning from previous data. To find specific attack signatures, rule-based systems can use predetermined rules.

Accurate and timely detection must be achieved by the careful selection and adjustment of suitable anomaly detection techniques. To determine which algorithm or method is best for the AMI system, many can be compared and assessed in terms of detection accuracy, false positive rates, and processing demands.

4.4 Mitigation Strategy Implementation

When an NDP attack is discovered, the paper's main focus is on putting good mitigation measures in place to lessen the effects on the smart meter system. Assuring the integrity and availability of AMI data, mitigation mechanisms are implemented inside the RPL protocol to thwart and neutralize the assault.

Some common mitigation techniques that can be implemented within the RPL protocol include:

- Route Diversion: When an NDP attack is detected, alternative paths can be established to bypass the affected nodes. This helps in maintaining communication and ensuring the delivery of critical AMI data.
- Packet Filtering: The detection system can implement packet filtering mechanisms to identify and discard malicious traffic. Filtering can be based on specific packet attributes, such as source IP address, payload content, or abnormal packet rates.
- Rate Limiting: Controlling the flow of packets is important to prevent resource exhaustion and ensure the stability of the network. Rate-limiting mechanisms can be implemented to restrict the rate at which packets are processed and transmitted.
- Traffic Prioritization: In the presence of an NDP attack, it is crucial to prioritize critical AMI data to ensure its timely delivery. Traffic prioritization mechanisms can be used to give priority to important packets, such as energy consumption data while limiting the impact of the attack on non-critical traffic.

To guarantee that they may be applied effectively without having a substantial negative impact on the operating performance of AMI devices, these mitigation approaches should take into account the resource limitations of such devices.

4.5 Simulation and Performance Evaluation

The Cooja simulator is used to run numerous simulations that are used to thoroughly test. With several attack scenarios and network variables taken into account, simulations mimic real-world AMI scenarios. On the basis of detection precision, false positive rates, the efficiency with which NDP attack impacts are mitigated, and the influence on overall network performance, the effectiveness of the detection and mitigation techniques is evaluated.

The behavior of the suggested technique under various assault intensities, network topologies, and communication patterns can be better understood through simulations. The simulations shed light on the suggested approach's scalability and resilience in large-scale AMI installations. The percentage of attacks that are effectively mitigated, the rate of attack detection, the time it takes to detect an attack, and the impact on network performance, including latency and throughput, are some examples of performance evaluation metrics. The effectiveness and efficiency of the suggested approach can be evaluated quantitatively using these indicators.

4.6 *Optimization and Refinement*

The research focuses on optimizing and improving the detection and mitigation mechanisms based on the outcomes of the simulation and the performance evaluation. This iterative procedure seeks to raise the proposed strategy's general efficacy and efficiency.

The algorithms for anomaly detection may need to be adjusted as part of optimization in order to increase precision and decrease false positives. Incorporating domain knowledge with the detection algorithms and examining the NDP attack features unique to the AMI system can accomplish this. By taking into account the trade-off between security and resource use, the mitigation measures can also be improved. In order to minimize resource exhaustion and preserve effective network communication, for instance, rate-limiting techniques might be changed to achieve the ideal balance. In order to make sure that the suggested approach is workable and doable in actual deployments, it is also crucial to optimize the resource usage of AMI devices, such as memory, computing power, and energy consumption. It is possible to investigate methods to reduce resource overhead while preserving security, including data compression, lightweight cryptographic algorithms, and energy-efficient routing protocols. The paper's goal is to increase the security of smart meter systems and guarantee the consistent performance of AMI devices by continually improving and refining the suggested strategy.

5 Implementation/Observations

Simulation tool: Cooja Network simulator, the **Trial Network**: Trial Network with rpl-border-router mote and udp-sender mote to understand the simulator.

We used /contiki/examples/ipv6/rpl-border-router/border-router.c as a Contiki firmware for compiling to create a mote "mote 1". We again used/home/user/contiki/examples/ipv6/rpl-collect/udp-sender.c for compiling to create other 24 corresponding motes in the network panel. "Serial Socket (SERVER)" of Sky 1 will create a serial port on the Border Router which is accessible via UDP port number

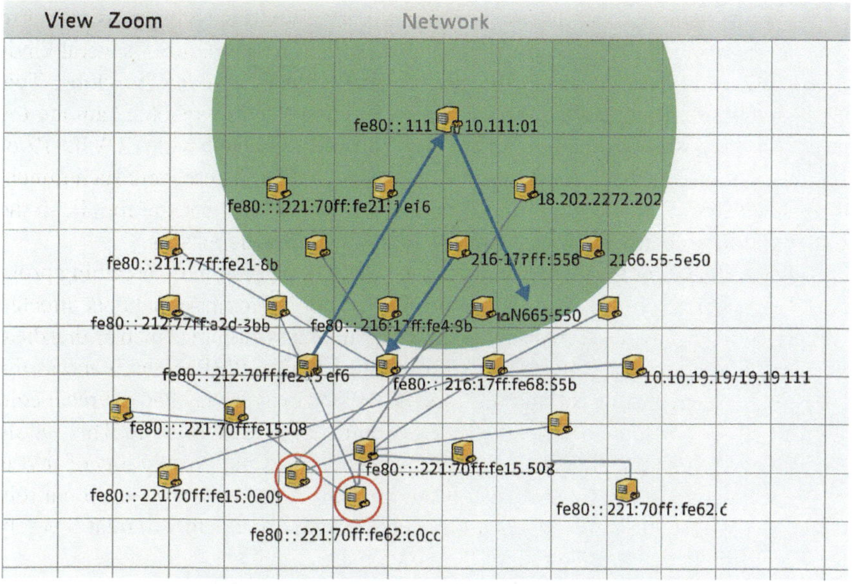

Fig. 2 Network creation between border-router and udp-sender motes

60001 on the local machine [18, 19]. Figure 2 shows the network which is create with the help of rpl-border and udp-server.

A Terminal window is showing the following commands, in Fig. 3, which will establish the bridge to the Border Router. This will return a bridge that has been established and IPv6 addresses assigned with the prefix aaaa::/64. The Border Router's IPv6 address of aaaa::212:7401:1:101 can be seen. While the simulation is running, the bridge can be used to Ping motes in the network as follows, although not all are successful due to the volatility of the network.

Figure 3 Terminal window output of neighbours motes.

Finally, it is also possible to display the Border Router's routing table by opening up a web page and entering the IPv6 address of the Border Router as in the following figure. This shows routes established. The approach of simulation:

Step 1: Create an NDP network between the udp-server mote and udp-client mote and implement the proposed attack scenario.

Step 2: Create an NDP network between the udp-server mote and udp-client mote and implement the proposed mitigation algorithm.

Step 3: Comparative study between Step 1 and Step 2.

Now we want to see the output of each particular mote, so use the Wireshark to see the output, then we want to save the output in the form of a PCAP file for future requirements. Figure 4 shows the output of the above network PCAP file.

As a result, the suggested method offers a methodical framework for identifying and thwarting NDP attacks in AMI devices utilizing RPL in the Cooja simulator. By employing this strategy, the paper seeks to improve the security of smart meter

(a)

(b)

Fig. 3 **a** Ping to the neighbours motes, **b** Ping to the neighbours motes

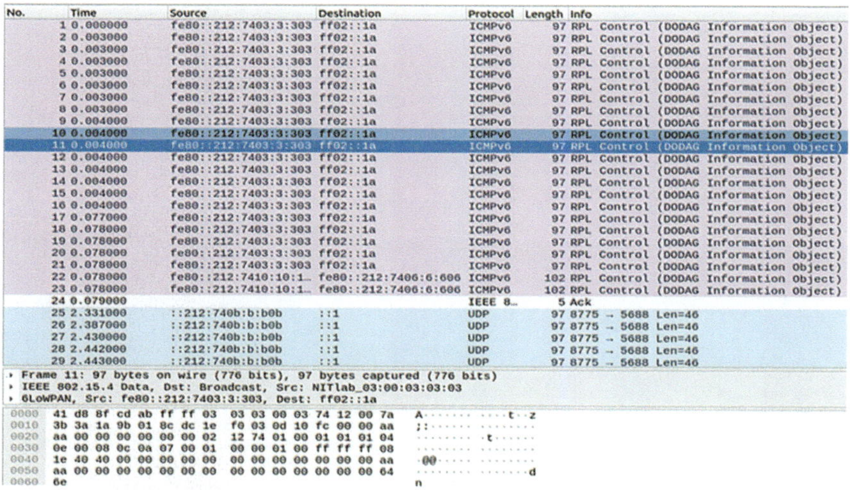

Fig. 4 Pcap file output of neighbor's motes

systems, guarantee the dependable operation of AMI devices, and safeguard the integrity of data on energy use. The findings from this study make a contribution to the field of cybersecurity for AMI devices by giving utility companies and system administrators useful tips and tricks for quickly identifying and thwarting NDP assaults on their smart meter systems. In this experiment we create a network of 1 mote is rpl-border-router and 24 mote is rpl-collect/udp-sender.c and create a (Destination-Oriented Directed Acyclic Graph) DODAG network, messages were transmitted and received in a network, following Fig. 5 shows the output of the transmitted and received message in the network.

6 Conclusion and Future Plan

The conclusion is, that we using Cooja simulator to identify the attacker mote in the network by seeing the PCAP file and the Cooja simulator mote output window shows us the output of each particular mote. This paper mainly focuses on AMI devices to protect from malicious mote who try to disrupt the services of the server to clients. We all know the smart meter system comes under the AMI devices so using Cooja simulator by rpl-border-router to server and rpl-client to client mote who send the data to the server.

By using this technique, we try to detect and mitigate the NDP attack on the AMI devices which is used in smart meter also. In this paper, we are trying to explain the method and idea of how to protect our smart meter from the malicious mote in the network. In the smart meter system, we generally use home appliances, and after calculating all the appliances our home panel where all data is calculated which is

Time	Mote	Message
00:00.618	ID:4	Rime started with address 0.18.116.4.0.4.4.4
00:00.627	ID:4	MAC 00:12:74:04:00:04:04:04 Contiki 3.0 started. Node id is set to 4.
00:00.635	ID:4	nullsec CSMA ContikiMAC, channel check rate 8 Hz, radio channel 26, CCA threshold -45
00:00.646	ID:4	Tentative link-local IPv6 address fe80:0000:0000:0000:0212:7404:0004:0404
00:00.648	ID:4	Starting 'UDP client process'
00:00.651	ID:4	UDP client process started
00:00.652	ID:1	Rime started with address 0.18.116.1.0.1.1.1
00:00.655	ID:4	Client IPv6 addresses: aaaa::212:7404:4:404
00:00.658	ID:4	fe80::212:7404:4:404
00:00.660	ID:1	MAC 00:12:74:01:00:01:01:01 Contiki 3.0 started. Node id is set to 1.
00:00.663	ID:4	Created a connection with the server :: local/remote port 8765/5678
00:00.669	ID:1	nullsec CSMA ContikiMAC, channel check rate 8 Hz, radio channel 26, CCA threshold -45
00:00.680	ID:1	Tentative link-local IPv6 address fe80:0000:0000:0000:0212:7401:0001:0101
00:00.682	ID:1	Starting 'UDP server process'
00:00.683	ID:1	UDP server started
00:00.687	ID:1	created a new RPL dag
00:00.692	ID:1	Server IPv6 addresses: aaaa::212:7401:1:101
00:00.694	ID:1	aaaa::ff:fe00:1
00:00.697	ID:1	fe80::212:7401:1:101
00:00.703	ID:1	Created a server connection with remote address :: local/remote port 5678/8765
00:00.985	ID:5	Rime started with address 0.18.116.5.0.5.5.5
00:00.994	ID:5	MAC 00:12:74:05:00:05:05:05 Contiki 3.0 started. Node id is set to 5.
00:01.002	ID:5	nullsec CSMA ContikiMAC, channel check rate 8 Hz, radio channel 26, CCA threshold -45
00:01.013	ID:5	Tentative link-local IPv6 address fe80:0000:0000:0000:0212:7405:0005:0505
00:01.015	ID:5	Starting 'UDP client process'
00:01.017	ID:5	UDP client process started
00:01.022	ID:5	Client IPv6 addresses: aaaa::212:7405:5:505
00:01.025	ID:5	fe80::212:7405:5:505
00:01.030	ID:5	Created a connection with the server :: local/remote port 8765/5678
00:01.169	ID:3	Rime started with address 0.18.116.3.0.3.3.3
00:01.178	ID:3	MAC 00:12:74:03:00:03:03:03 Contiki 3.0 started. Node id is set to 3.
00:01.186	ID:3	nullsec CSMA ContikiMAC, channel check rate 8 Hz, radio channel 26, CCA threshold -45
00:01.197	ID:3	Tentative link-local IPv6 address fe80:0000:0000:0000:0212:7403:0003:0303
00:01.199	ID:3	Starting 'UDP client process'
00:01.202	ID:3	UDP client process started
00:01.206	ID:3	Client IPv6 addresses: aaaa::212:7403:3:303
00:01.209	ID:3	fe80::212:7403:3:303
00:01.214	ID:3	Created a connection with the server :: local/remote port 8765/5678
01:09.941	ID:5	DATA send to 1 'Hello 1'
01:09.962	ID:2	DATA recv 'Hello 1 from the client' from 5
01:17.977	ID:3	DATA send to 1 'Hello 1'
01:17.998	ID:2	DATA recv 'Hello 1 from the client' from 3
01:37.388	ID:4	DATA send to 1 'Hello 1'
01:37.410	ID:2	DATA recv 'Hello 1 from the client' from 4
02:22.894	ID:4	DATA send to 1 'Hello 2'
02:22.916	ID:2	DATA recv 'Hello 2 from the client' from 4
02:50.214	ID:5	DATA send to 1 'Hello 2'
02:50.236	ID:2	DATA recv 'Hello 2 from the client' from 5
02:51.533	ID:3	DATA send to 1 'Hello 2'
02:51.554	ID:2	DATA recv 'Hello 2 from the client' from 3
03:04.254	ID:4	DATA send to 1 'Hello 3'
03:04.277	ID:2	DATA recv 'Hello 3 from the client' from 4
03:20.664	ID:3	DATA send to 1 'Hello 3'
03:20.686	ID:2	DATA recv 'Hello 3 from the client' from 3
03:49.558	ID:5	DATA send to 1 'Hello 3'

Fig. 5 Mote output of neighbor's mote

calculated our electricity consumption, and transfer the data to the server located at the provider's end, and then the provider generates the bill. In this process, if a malicious mote is trying to manipulate and send false data to the provider and malicious trying to flood the server of the provider by sending the false data so many times, this thing disrupts services and this problem we solve by using Cooja simulator. Route diversion, packet filtering, and rate restriction are a few of the mitigating techniques used in the RPL protocol that have been successful in reducing the negative effects of NDP attacks on the smart meter system.

References

1. Goel, L., Chawla, H., Dua, M., Dua, S., Dhingra, D.: Novel hybrid chaotic map-based secure data transmission between smart meter and HAN devices. Smart Sci. **12**(1), 27–42 (2024)
2. Tak, L., Yadav, A.K., Singh, N.K., Mahajan, V.: (2020) Reliability assessment of smart grid with renewable energy sources, storage devices, and cyber intrusion. In: International Conference on Smart Technologies for Energy, Environment, and Sustainable Development, pp. 525–542. Springer Nature Singapore, Singapore
3. Khoei, T.T., Slimane, H.O., Kaabouch, N.: A Comprehensive Survey on the Cyber-Security of Smart Grids: Cyber-Attacks, Detection, Countermeasure Techniques, and Future Directions. arXiv:2207.07738 (2022)
4. Prabhakar, P., Arora, S., Khosla, A., Beniwal, R.K., Arthur, M.N., Arias-Gonzáles, J.L., Areche, F.O.: Cyber Security of Smart Metering Infrastructure Using Median Absolute Deviation Methodology. Security and Communication Networks (2022)
5. Win, L.L., Tonyalı, S.: Security and privacy challenges, solutions, and open issues in smart metering: a review. In: 2021 6th International Conference on Computer Science and Engineering (UBMK), pp. 800–805. IEEE (2021)
6. El Mrabet, Z., Kaabouch, N., El Ghazi, H., El Ghazi, H.: Cyber-security in smart grid: survey and challenges. Comput. Electr. Eng. **67**, 469–482 (2018)
7. Bačnar, D., Leytner, L., Prenc, R., Jardas, V., Lerga, J.: On security and privacy in smart metering systems. In: 2022 7th International Conference on Smart and Sustainable Technologies (SpliTech), pp. 1–6. IEEE (2022)
8. Alissa, K.A., AlDeeb, B.A., Alshehri, H.A., Dahdouh, S.A., Alsubaie, B.M., Alghamdi, A.M., Alsmadi, M.K.: Developing a simulated intelligent instrument to measure user behavior toward cybersecurity policies. Int. J. Commun. Netw. Inf. Secur. **13**(1), 82–91 (2021)
9. Gaur, V., Kumar, R.: Analysis of machine learning classifiers for early detection of DDoS attacks on IoT devices. Arab. J. Sci. Eng. **47**(2), 1353–1374 (2022)
10. Osanaiye, O.A., Alfa, A.S., Hancke, G.P.: Denial of service defence for resource availability in wireless sensor networks. IEEE Access **6**, 6975–7004 (2018)
11. Singh, N.K., Mahajan, V.: Analysis and evaluation of cyber-attack impact on critical power system infrastructure. Smart Sci. **9**(1), 1–13 (2021)
12. Al-Ani, A., Al-Ani, A.K., Laghari, S.A., Manickam, S., Lai, K.W., Hasikin, K.: NDPsec: neighbor discovery protocol security mechanism. IEEE Access **10**, 83650–83663 (2022)
13. Zhao, M., Ali, G. M. N., Lu, R., Chong, P.H.J.: An energy-efficient and self-regioning based RPL for low-power and Lossy networks. In: 2016 IEEE 84th Vehicular Technology Conference (VTC-Fall), pp. 1–6. IEEE (2016)
14. Abusafat, F., Pereira, T., Santos, H.: Roadmap of security threats between IPv4/IPv6. In: 2021 IEEE International IOT, Electronics and Mechatronics Conference (IEMTRONICS), pp. 1–6. IEEE (2021)
15. Behal, A., Sandhu, J.K., Gupta, G.: Using the Cooja Simulator, Analysing The Routing Protocol (RPL) For Low Power And Lossy Networks In IoT. In 2023 IEEE International Students' Conference on Electrical, Electronics and Computer Science (SCEECS), pp. 1–4. IEEE (2023)
16. Chen, Y., Chanet, J.P., Hou, K.M.: RPL Routing Protocol a case study: precision agriculture. In: First China-France Workshop on Future Computing Technology (CF-WoFUCT 2012), pp. 6-p (2012)
17. Kipongo, J., Swart, T.G., Esenogho, E.: Design and implementation of intrusion detection systems using RPL and AOVD protocols-based wireless sensor networks. Int. J. Electron. Telecommun. 309–318 (2023)
18. Thomson, C., Romdhani, I., Al-Dubai, A., Qasem, M., Ghaleb, B., Wadhaj, I.: Cooja Simulator Manual (2016)
19. Kurniawan, A.: Practical Contiki-NG (2018)

An Efficient CNN Model for COVID-19 Detection Based on X-Ray Images

Aditi Kailas, M. Sinthuja, and Neeraj Vaidyamath

Abstract The first COVID-19 instances were reported in December 2019 by Wuhan City, Hubei Province, China (Xu et al.). Since the COVID-19 worldwide pandemic affects health care and lifestyles all over the world, early diagnosis is crucial to stopping the spread of cases and lowering mortality. Backward transcription The Polymerase Chain Reaction (RT PCR) is the gold standard in diagnostic testing; yet, due to its exorbitant costs and lengthy turnaround times, alternate quick and readily accessible diagnostic approaches are necessary. Current deep learning models (CNN) are used in this paper's method to analyze the images and classify them as positive or negative for COVID-19. This method was motivated by recent research that connects the presence of COVID-19 to findings in Chest X-ray images. The suggested system includes a preprocessing stage that includes lung segmentation, eliminating the surrounds that do not provide pertinent information for the job and may create biased findings; Performance evaluation of the classification model using a confusion matrix and performance measures comes after this initial phase in which it was trained using a transfer learning technique. The most precise models had a 97% accuracy in detecting COVID-19. The improvement from the other models is then demonstrated by comparing our model with the current model for COVID-19 detection.

Keywords CNN (Convolutional Neural Networks) · COVID-19

A. Kailas · N. Vaidyamath
Department of Information Science and Engineering, M.S. Ramaiah Institute of Technology, Bengaluru, India
e-mail: kaadi22@gmail.com

N. Vaidyamath
e-mail: neerajvaidyamath1@gmail.com

M. Sinthuja (✉)
School of Computer Science and Engineering (SCOPE), Vellore Institute of Technology, Vellore, India
e-mail: sinthuja.m@vit.ac.in

B. Soni et al. (eds.), *Beyond Artificial Intelligence*, Lecture Notes in Networks and Systems 1326, https://doi.org/10.1007/978-981-96-4170-3_9

1 Introduction

The severe acute respiratory syndrome coronavirus, or SARS-CoV-2, was first discovered in late 2019. Currently, a virus that originated in China is to blame for. The World Health Organization (WHO) confirmed the COVID-19 illness, which is brought on by the corona virus, to exist. A pandemic of sickness will start in March 2020. Global healthcare authorities' most recent research indicates. The pandemic had an effect on millions of individuals, including federal globally. Symptoms of COVID-19's most serious illness, pneumonia, are lung- related. Involves dyspnea, a fever, a runny nose, and a cough, however the illness's symptoms might vary. These circumstances occur most frequently. Examined by looking at the anomalies on chest X-ray pictures. The motivation for this project came from the experience of the COVID-19 pandemic. The RT-PCR kits were an accurate method of detecting COVID-19, but they were expensive and often unavailable in remote places. The project explores the application of deep learning to detect COVID-19 via X-rays. The goal is to be better prepared for the future by providing an alternative method of detection that is more accessible and affordable. The diagnosis of COVID-19 relied mainly on molecular diagnostic tests such as the reverse transcription-polymerase chain reaction (RT-PCR) test. However, the RT-PCR test has 1 some limitations, including long turnaround times, supply shortages, and the need for specialized laboratory equipment and trained personnel, making it challenging to scale up testing to meet the demand.

Additionally, the RT-PCR test has been reported to have a high rate of false negative results, which can lead to delayed diagnosis and increased transmission of the virus. The development of an accurate and reliable model that can identify between individuals with COVID-19 and those who do not in their chest radiographs (X-rays). This problem is essential in the ongoing fight against the COVID-19 pandemic because quick treatment and disease containment can be improved by early illness detection and diagnosis. Where Paul et al. [1] AI is a vast field with several sub-fields that may be used to address difficult issues in our daily lives.

Learning, planning, representing information, and seeking are some of these sub areas. The objective is to develop a deep learning model that can accurately classify X-rays as positive or negative for COVID-19 in order to assist medical professionals in making rational judgements. To identify unique COVID-19 infection characteristics in the massive dataset of X-ray images, a deep learning system is built using convolutional neural networks (CNNs) or other deep learning architectures.

2 Literature Review

Provide an overall summary of the potential usage of machine learning and deep learning techniques to tackle COVID-19, with a focus on how these tools might be used to address additional issues like societal and economic repercussions. Fatima et al. [2] has conducted a survey-based study with the goal of assessing the efficiency of deep learning methods for COVID-19 detection. However, it does not include further features of sickness [3]. Compared different X-ray images using CNN, Transfer learning and Deep learning techniques but failed to determine the generalizability of the models on different datasets or populations [4]. Used transfer learning to fine-tune the pre-trained VGG-16, ResNet-50, and InceptionV3 models.

Accuracy, precision, recall, and F1-score were among the performance indicators used to assess the models, although the size of the data set was small and there were no external validations [5]. Convolutional neural network (CNN) provides the foundation for a deep learning model. He did not carry out a clinical study of the suggested model's performance in actual clinical settings, but instead trained and tested it on the dataset and measured its performance using metrics like accuracy, precision, recall, and F1-score. Sethi et al. [6] compared various X-ray images using CNN, Transfer learning, and Deep learning approaches and issued a proposal for Deep Learning based Diagnosis of COVID-19 utilizing Chest X-rays.

Images; however, his advice only included two (CNNs) and a classifier. The classifier predicts the diagnosis while the CNNs are utilized to extract characteristics [7]. Deep learning framework was used to classify anomalous traces in the lung's CT scan and detect lung abnormality, however it was vulnerable to visual bias, which increased unreliability. Convolutional neural network (CNN) provides the foundation for a deep learning model. The dataset was used to train and test the model, and measures like recall, precision, accuracy and F1-score were used to assess the model's performance but just like [5] the study did not conduct a clinical evaluation of the proposed model's effectiveness in real-world medical settings.

3 Problem Statement

The development of an accurate and reliable model that can identify between individuals with COVID-19 and those who do not in their chest radiographs (X-rays). This problem is essential in the ongoing fight against the COVID-19 pandemic because quick treatment and disease containment can be improved by early illness detection and diagnosis. The objective is to develop a deep learning model that can accurately classify X-rays as positive or negative for COVID-19 in order to assist medical professionals in making rational judgements. To identify unique COVID-19 infection characteristics in the massive dataset of X-ray images, a deep learning system is built using convolutional neural networks (CNNs) or other deep learning architectures.

4 Proposed Model

4.1 Data Preprocessing

Input images undergo normalization to ensure consistent scale and brightness, enhancing model performance and convergence during training. Resizing images to a standard size, such as 224 × 224 pixels, Panwar et al. [8] promotes computational efficiency and facilitates model compatibility. Which is the usual size used for model. A training set, a validation set, and a testing set are created from the dataset. The training set is utilized for model parameter optimization, the validation set aids in hyperparameter tuning and model selection, while the testing set assesses final model performance on unseen data.

4.2 Data Augmentation

Data augmentation is an essential phase in the preprocessing chain that is used to enhance the training dataset by applying various picture transformations. This method improves the model's ability to make generalizations to previously unseen data and reduces the possibility of overfitting. The model uses random zooming, rotation, and flipping to simulate varying perspectives and improve robustness. Zooming aids in the recognition of characteristics at various scales, rotation introduces variety in object orientation, and flipping mirrors images, allowing learning regardless of orientation, particularly in non-classic scenarios.

4.3 Convolutional Layers

Convolutional layers in CNNs use learnable filters to capture spatial hierarchies and patterns in images. These filters distinguish between edges, textures, and shapes. The activation function Rectified Linear Unit (ReLU) introduces non linearity, boosting sparsity and efficient learning. When convolutional layers are combined with ReLU activation, the model may learn hierarchical representations, making it useful for image-related tasks. This architecture allows for accurate predictions and meaningful visual information extraction.

Hence, we use four convolutional layers with a ReLU activation function are part of the model. A max pooling layer is added after each convolutional layer to down sample the feature maps.

4.4 Fully Connected Layers

Three completely connected layers receive flattened output from the final pooling layer. The final output layer includes a sigmoid activation function to output a probability of whether the input image contains COVID-19 or not. The first two fully connected layers feature a ReLU 2 activation function.

4.5 Training

The model is trained using the binary cross-entropy loss function and the Adam optimizer, a loss function designed specifically for issues with binary classification. The goal is to decrease the difference between anticipated and actual binary labels, directing the model to produce predictions that correspond to true class labels. The Adam optimizer provides adaptive learning rates and dynamic parameter modifications, enabling the model to explore complex high dimensional domains effectively. Exposure to a wide range of examples within the training set improves the model's capacity to make correct predictions on new data, hence improving its performance on real-world binary classification challenges.

4.6 Evaluation

The model's performance is evaluated on the testing set using accuracy, precision, recall, and F1-score as these are critical evaluation measures, particularly in COVID-19 detection. These metrics provide a complete picture of the model's performance by taking into consideration of the confusion matrix. A confusion matrix is included to provide a more nuanced insight of the model's strengths and limitations, displaying specific errors such as false positives and false negatives. This complete examination is required for accurate COVID-19 diagnosis and the reduction of false positives and false negatives. The confusion matrix correlates with interpretability, resulting in increased trust in real-world healthcare circumstances.

5 Implementation

The Initial step is to gather a dataset of X-ray images that are labeled as COVID-19 positive or negative. There are multiple publicly available datasets that can be used, such as the COVID-19 Radiography Database or the COVIDx dataset. Once the dataset is available, the images must be preprocessed by being resized to a standard size, having their pixel values normalized, and being divided into training, validation, and testing sets. The next step is to construct a CNN model using any deep learning library such as TensorFlow or PyTorch to build the CNN model. Convolutional layers, pooling layers, and fully connected layers make up a standard CNN architecture. In this model, we begin with a straightforward architecture that includes a few convolutional layers, pooling layers, and entirely interconnected layers. To determine which architecture works best for the dataset, various ones are tested.

The prepared dataset is subsequently used to train the CNN model. The algorithm gains the ability to spot patterns in the X-ray images that identify COVID-19 positive from negative images throughout training. To reduce the loss function and update the model parameters, you can use several optimization techniques like stochastic gradient descent or Adam optimizer. The performance of the CNN model is assessed on the testing set after training. Measures of the model's performance, including as accuracy, precision, recall, and F1-score. The confusion matrix is also visualized in the project to show the quantity of accurate positive, accurate negative, accurate positive, and accurate inaccurate forecasts. Once the model's performance has been validated, it is used to categories fresh X-ray images. Then, a web application is developed that accepts input from X-ray Images.

5.1 Framework and System Design

The model used is Sequential method is a way to build a model in Keras. It allows you to build a model layer by layer. Internal Layer: sequential layer, convolutional 2D layer, max pooling layer, flatten layer, drop out, dense layer. Output Layer: It is a fully connected layer, where the input from the other layers is flattened and sent so as the transform the output into the number of classes as desired by the network [9]. Figure 1 depicts the layered structure.

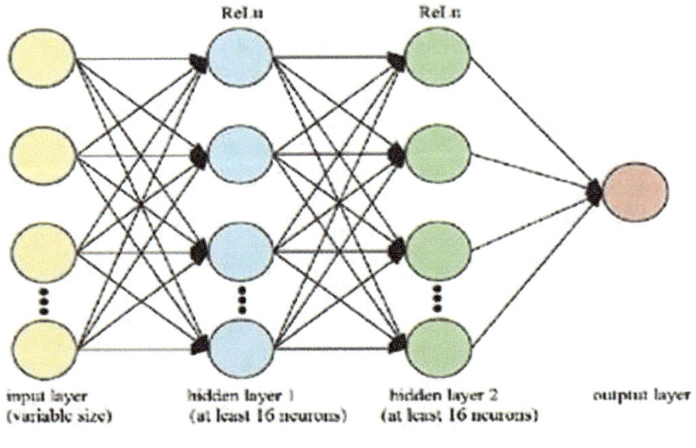

Fig. 1 Layered structure [9]

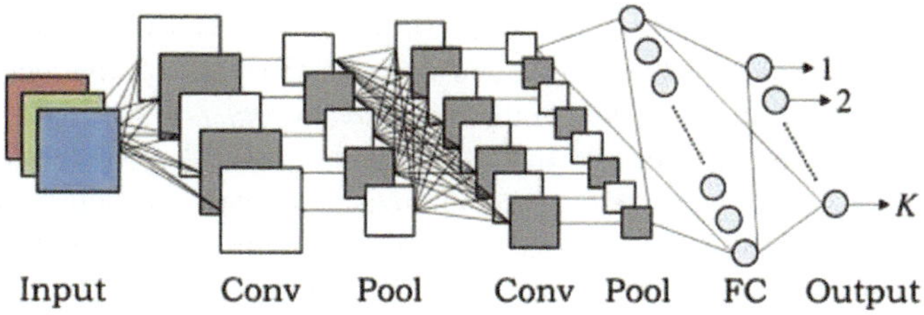

Fig. 2 CNN flowchart [10]

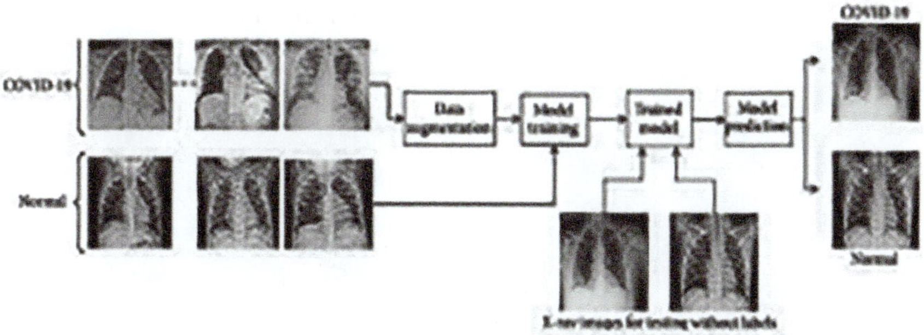

Fig. 3 System architecture

In order to extract features and produce predictions, the proposed CNN model feeds an input image via a number of convolutional layers, pooling layers, and fully connected layers. A task-specific performance optimization strategy can be applied to the CNN model's specific architecture, which includes the number of filters, filter sizes, pooling regions, and neurons in the fully connected layers. Model of CNN is shown in Palanisamy et al. [10] Fig. 2. Architecture of the model is illustrated in Fig. 3.

5.2 Experimentation and Result

The proposed model achieved an accuracy of 0.97 precision of 1 and recall of 0.944 in detecting COVID-19 from X-rays. The model accurately classified 279 images out of 288 images as COVID-19 or Normal images. The models high ac curacy and sensitivity indicate its potential as a reliable tool for detecting COVID-19 from lungs X-ray images. However further testing and validation on larger datasets are required to confirm the model's performance and reliability.

Figure 4 shows the performance evaluation and Fig. 5 shows the confusion matrix. The y-axis represents the True classification and the X-axis Predicted classification where 0 indicates COVID-19 Positive and 1 indicates COVID-19 Negative. As we can see the model predicted all the true COVID-19 negatives samples correctly, were as coming to COVID-19 positives samples it predicted 37 cases correctly but gave 2 false predictions saying that COVID-19 cases as negatives.

Graphical analysis is displayed in Fig. 6. It is a graphical representation of the accuracy and loss curves for training and test datasets, as well as their interrelationship. We can see how, as the loss decreases, the system's accuracy increases with each epoch, and how, as the epoch increases, the accuracy increases as well. After a certain epoch, the accuracy tends to become saturated (Table 1).

The model in this project is compared with various other existing models implied to test the test COVID-19 using X-rays.

	precision	recall	f1-score	support
Covid	0.97	1.00	0.99	35
Normal	1.00	0.97	0.99	37
accuracy			0.99	72
macro avg	0.99	0.99	0.99	72
weighted avg	0.99	0.99	0.99	72

Precision score: 0.9730
Recall score: 1.0000
F1 Score: 0.9863

Fig. 4 Performance evaluation

Fig. 5 Confusion matrix

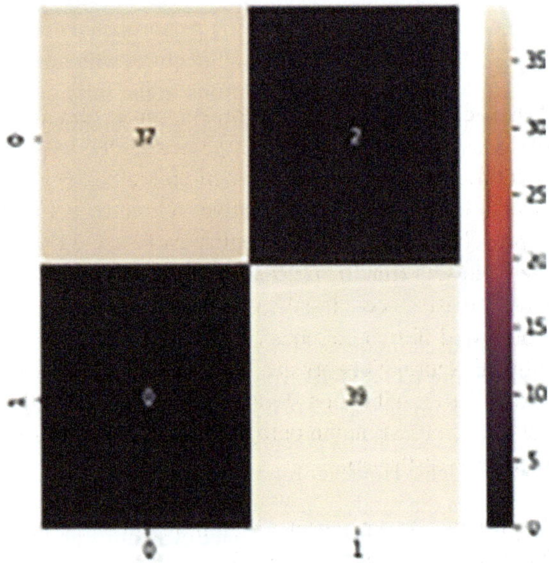

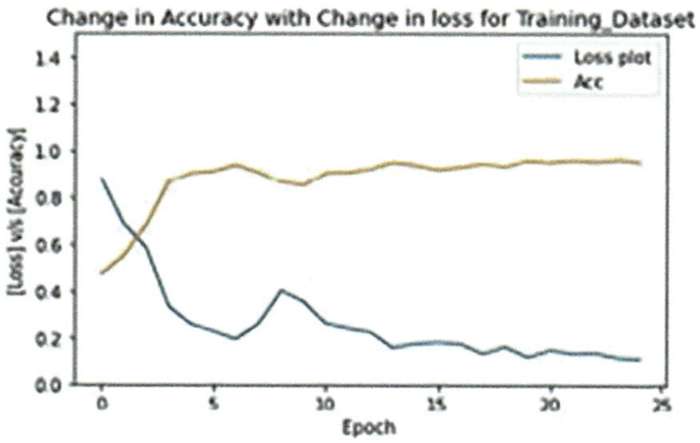

Fig. 6 Graphical analysis

Table 1 Model comparison

Model	Accuracy	Precision
RF	0.95	0.96
GBM	0.91	0.92
SVC	0.93	0.95
LR	0.92	0.93
KNN	0.93	0.99
CNN	0.97	0.92

6 Conclusion and Future Work

In conclusion, the implementation provides encouraging results for the COVID-19 detection model for detecting COVID-19 in chest X-ray pictures. With four convolutional layers, two pooling layers, and three fully connected layers, the proposed model makes use of a CNN architecture. A dataset from two independent sources is used to train the model, which has an accuracy rate of about 93%. It should be mentioned, nonetheless, that the effectiveness of the model significantly depends on the caliber of the input data. Chest X-ray pictures can occasionally be generic and overlap with other respiratory disorders, therefore utilizing them as a sole diagnostic tool for COVID-19 has other drawbacks.

The proposed model should not be utilized in place of a professional medical examination; rather, it should be used as an additional tool to enhance clinical decision-making. Overall, this model represents a promising advancement in COVID-19 identification and diagnosis, and it has the potential to be further enhanced through more thorough testing and validation on a variety of datasets.

There are several future scopes for the COVID-19 detection model proposed. Some of them are: • Expansion of the dataset: The model's accuracy and robustness could be further proved by expanding the dataset to include more diverse X-ray images from different sources. • Improving the accuracy: Changing the architecture of the current model to make it more efficient and explore more diverse ways to come across errors and enhance the model for various other diseases • Detection of other respiratory diseases: The proposed model could be adapted to detect other respiratory diseases besides COVID-19, such as pneumonia, tuberculosis, and lung cancer. • Integration with electronic health records (EHR): Integration of the model with electronic health records (EHR) could help in automating the detection and diagnosis process and making it more efficient. • Deployment on cloud platforms: The model can be deployed on cloud platforms like AWS, Azure or Google CloudPlatform, making it more accessible to users worldwide. • Integration with mobile applications: Integration of the model with mobile applications could allow for easier access and usage, particularly in resource-constrained settings. Overall, the COVID-19 detection model proposed in the code has the ability to be further developed and improved to assist medical practitioners in diagnosing COVID-19 and other respiratory diseases more accurately and efficiently.

References

1. Paul, S.G., Saha, A., Biswas, A.A., Zulfiker, M.S., Arefin, M.S., Rahman, M.M., Reza, A.W.: Combating COVID-19 using machine learning and deep learning: applications, challenges, and future perspectives. Array **17**, 100271 (2023). https://www.sciencedirect.com/science/article/pii/S2590005622001047
2. Fatima, I., Saeed, S., Rehman, A., Khan, M.A., Hussain, M., Khan, S.U., Iqbal, S.: Automatic detection of COVID-19 using deep learning techniques: a review. Front. Artif. Intell. **5**, 912022 (2022)
3. Tomar, M., Arora, T., Sabitha, A.S., Hasteer, N.: Comparative study of deep learning models for COVID-19 diagnosis. In: 2021 2nd International Conference on Secure Cyber Computing and Communications (ICSCCC), pp. 507–512 (2021)
4. Saeed, A.Y.A., Ba Alawi, A.E.: COVID-19 diagnosis model using deep learning with focal loss technique. J. Healthc. Eng. **2021**, 1–9 (2021)
5. Hişam, D., Hişam, E.: Deep learning models for classifying cancer and COVID-19 lung diseases. J. Ambient. Intell. Hum. Comput. 1–9 (2021)
6. Sethi, R., Mehrotra, M., Seth, D.: Deep learning based diagnosis recommendation for COVID-19 using chest X-rays images. Int. J. Inf. Technol. & Decis. Mak. **20**(6), 1727–1746 (2021)
7. Bhandary, A., Prabhu, G.A., Rajnikanth, V.: Deep learning framework to detect lung abnormality. Int. J. Mach. Learn. Cybern. **15**, 1–14 (2021)
8. Panwar, H., Gupta, P., Siddiqui, M.K.: Deep learning and grad-cam based color visualization approach for fast detection of COVID-19 cases using chest X-ray and CT-scan images. IEEE Sens. J. **21**(22), 24248
9. Alexandru, T.G., Pupăză, C.: Machine learning generalization of lumped parameter models for the optimal cooling of embedded systems. Stud. Inform. Control **29**, 169–177 (2020). https://doi.org/10.24846/v29i2y202003
10. Palanisamy, P., Urooj, S., Arunachalam, R., Lay-Ekuakille, A.: A novel prognostic model using chaotic CNN with hybridized spoofing for enhancing diagnostic accuracy in epileptic seizure prediction. Diagnostics **13**(21), 3382 (2023). https://doi.org/10.3390/diagnostics13213382

Application of Emerging Technology in an Emerging Economy: Case of GovTech Through Blockchain-Based Seeds Traceability in Jharkhand

Rameshwar Arora

Abstract A strong agriculture sector plays an important role in ensuring the food security of a country. In addition to the ecological and social importance of seeds, the seeds' quality holds the utmost priority in the agricultural supply chain, thus governments are employing tactics focused on seed quality, and traceability. Governments have been deploying emerging technological tools to monitor the various phases of the agricultural supply chain. This paper analyzes the role of GovTech in agricultural seed-traceability. I examine the same by understanding the deployment of blockchain-based techniques by the government of the Indian state of Jharkhand. Thus, this paper demonstrates the responsible action undertaken by the government to ensure transparency, efficiency, and immutability in the agricultural supply chain data which leads to effective seed traceability and eventually quality. The paper presents a potential guide for policymakers and government institutions in reconfiguring the agricultural sector.

Keywords GovTech · Agriculture · Traceability · Blockchain · India

1 Introduction

With an 18.8% share in Gross Value Added (GVA), the agriculture sector is the largest employment generator in India [1]. The traceability of seeds in the agricultural supply chain contributes to the mission of food safety and quality concerns [2]. This is of prime concern to agriculture departments of governments marred by issues such as trust and ease of communication of quality information to the farmers [2]. The issue becomes more complicated owing to the prevalence of informal seed markets and fragmented farmer population with almost all efforts proven inefficient for seed traceability, especially in emerging economies.

R. Arora (✉)
Jindal Global Business School, O.P. Jindal Global University, Sonipat, India
e-mail: f14rameshwara@iimidr.ac.in

B. Soni et al. (eds.), *Beyond Artificial Intelligence*, Lecture Notes in Networks and Systems 1326, https://doi.org/10.1007/978-981-96-4170-3_10

In this research study, I strive to answer the following research question: "How does emerging economies government employ emerging technologies to ensure seed traceability?". The authors join prior work [3] in highlighting the issues faced in terms of the quality of seeds which is marred due to the traceability issues. Hence, in this paper, I employ the case of the implementation of blockchain, an emerging technology, in a bid to alleviate the issues faced due to seed traceability concerns. I do so by using the frame of reference of the most important actor, the government and thus highlight the phenomena of GovTech. For this paper, I adopt the definition proposed by Bharosa [4], wherein, GovTech is defined as "socio-technical solutions—that are developed and operated by private organizations—intertwined with public sector components for facilitating processes in the public sector" (p. 3).

I investigate the phenomena of seed traceability governed by the government of one of the Indian states, Jharkhand by employing blockchain technology. I apply the GovTech framework proposed by Bharosa [4] and find that multiple actors originating from private and public spaces come together to accomplish "Citizen-centered public service delivery" (p. 3) to facilitate quality seed distribution and traceability. In this paper, I focus on the GovTech for public service consumption, and public service delivery using data-driven policymaking. This paper contributes to highlighting that GovTech is an effective policy tool for intelligence, monitoring, and supervision of public service delivery and execution.

The rest of the paper is divided as follows, Sect. 2 discusses the literature on the seeds distribution system, emerging issues and challenges in the distribution of seeds, blockchain and agriculture, and finally the GovTech Model. Furthermore, Sect. 3 deals with the research methodology, and Sect. 4 discusses the findings from the data sources. Finally, the conclusion of the study is produced in Sect. 5.

2 Related Literature

This section illustrates the extant literature on various key concepts explored and investigated in this paper.

2.1 Blockchain and Agriculture

Blockchain provides the features of decentralization, immutability, and cryptographic links. Multiple blockchain applications have been proposed. Blockchain applicability in agriculture has been explored in academic literature for quite some time [5–7]. For instance, blockchain has been applied to disburse subsidies [8], credit schemes for farmers [9], food supply chains, agricultural insurance [10], smart farming, and transactions of agricultural products [7].

The studies, by and large, propose one or the other framework or method to adopt blockchain in one or the other processes of agriculture (for instance the reviews in Bharosa [4], Chen et al. [11] and Katsikouli et al. [5]). Further, the authors have tested the propositions of their respective studies by employing the same over a smaller facility to showcase experimental validity and encourage potential widespread scaled-up adoption. The existing studies seldom discuss a case that has independently employed blockchain.

2.2 GovTech Model

The fast-mover advantage of private players in the technology space has emerged as a clear partner for the governments. Governments across the world have been struggling to cope with the technological advancements and at the same time aspire to harness the potential of digital technology to provide public service more efficiently than can be achieved on their own. The GovTech model (Fig. 1) aims at easing technology adoption by public services [4]. This also facilitates cutting down the red tape and reduces the turnaround time.

GovTech aims to achieve two types of goals, viz., citizen-centered public service delivery and data-driven government. While some services try to accomplish both goals, not all achieve more than one simultaneously.

The plethora of existing scholarly literature demonstrates how governments have used or envisioned employing technology such as artificial intelligence (AI) technologies to highlight the importance of emerging technologies [12, 13]. For instance,

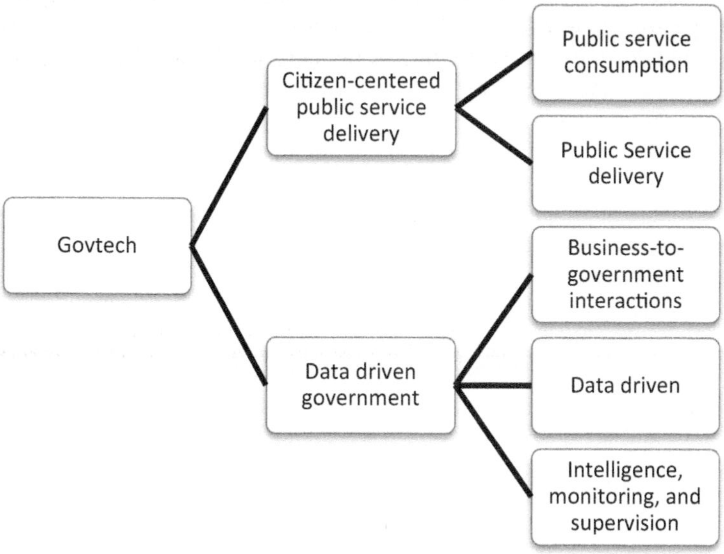

Fig. 1 Flow chart illustrating the concept of "Govtech". *Source* Bharosa [4]

emerging technologies such as AI can improve the efficiency, quality, and accessibility of public services [14] such as AI in public transport [15], agriculture using IoT [16], AI in healthcare, education, social security, etc. [9, 17, 18].

Some of the examples of GovTech could be artificial intelligence and machine learning (AI/ML) based chatbots [11], automated business registration and application processes [18], etc.

This section highlights the vast scholarly literature on governments considering the importance of different technologies (such as AI/ML, IoT, etc.) to deliver services or achieve data-driven governance. This poses a fertile ground to study the impact of other emerging technologies such as blockchain, which I pursue as a case study in this research study. I use the GovTech model proposed by Bharosa [4] as it encompasses a marriage of "technology, social science, public values, and private innovation" (pg. 4) which serves as a great perspective lens for our research question.

2.3 Seed Distribution System

Typically, there are two types of seed distribution systems, viz., informal and formal seed systems. The formal system is the one, wherein the value chain (series of activities leading to the availability of a valuable product to the end user) includes government-led seed and agricultural research institutes, complemented with state-run and private-sector distribution agencies. The formal system is characterized by genetically pure, standardized, and uniform seeds. On the other hand, the informal system, as the name suggests, involves the farmers and local networks to disseminate the locally produced seeds, thus escaping the stringent seed quality assurance practices laid down by the policymakers and enforced through various authorities. This also faces higher traceability issues as compared to the formal system.

While the informal system is marred with no or low assurance of proper breeding of seeds in terms of their genetic purity, their penetration and reach among small farmers is way more than seeds distributed in formal systems [19, 20]. The lesser checks, leading to sub-optimal seed quality affect the yields. This has a definite impact on the economic condition of the farmers and impacts their financial planning both in the short and long term.

The informal system has demonstrable demerits over the formal system [20–22], however, despite the shortcomings, the informal system cannot be completely phased out, especially, since they are deep-rooted and understand their importance from resource-poor farmers [23]. It is paramount for policymakers to strengthen the informal system and integrate the informal system into the mainstream formal system [24]. For a country such as India with a highly fragmented farm system in terms of many small farmers and a lack of awareness of governmental policies amongst the farmers, informal sectors play a crucial role and take precedence in seed supply chain management. The steps undertaken in cropping system in agriculture and potential blockchain implementation in one of the step, viz., sourcing of seed is demonstrated in Fig. 2.

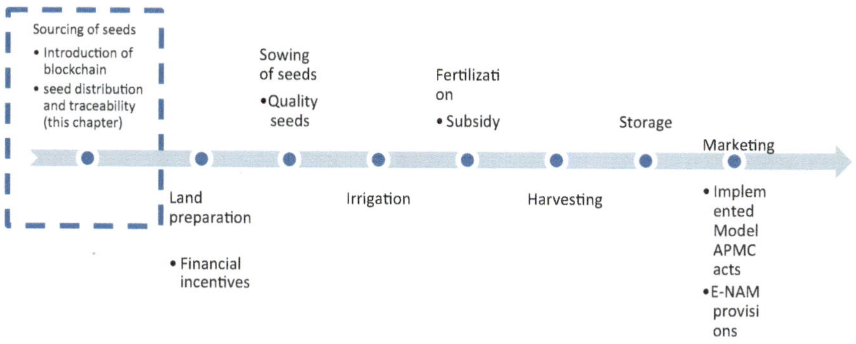

Fig. 2 Cropping system and potential Blockchain implementation

3 Research Methodology

The concepts of GovTech and Blockchain are novel and contemporary. The concept of Blockchain is an emerging technology that has been gaining tremendous practitioner and research interest in different sectors. The concept of GovTech is novel to the extent that the number of research and practitioners' articles on the topic is further limited. The novelty of the concepts under investigation warrants an exploratory study. The case-based methodology is the best-suited research methodology for exploratory studies [25]. The case study methodology is appropriate for deep diving into the phenomena of the government's adoption of blockchain technology in the seed traceability aspect of the agriculture supply chain. Thus, the collection of data is from different secondary sources catering to the various components of the research objective. The data source comprises LexisNexis, and Google News for news articles. Furthermore, the databases such as Google Scholar, and Scopus.com to scope out research literature on the topic of interest.

Owing to the qualitative nature of the data obtained in the form of texts, pictures, and videos, to tease out themes from the unstructured data, I employed the widely used thematic analysis. The government websites and the public tender information from third-party websites, along with the information from the third-party vendor that facilitated the process were also employed for triangulation. The case study-based approach provides rich insights into the "how" aspect of phenomena, which bodes well with the research question of this study. The overall research methodology adopted in this chapter is summarised in Table 1.

Table 1 Research methodology

Research type	Exploratory
Research method	Single case
Keywords	Blockchain, Government, Agriculture, Public sector
Data source	LexisNexis, Google News, YouTube, Press reports, Government websites, Google Scholar, Scopus.com
Analysis	Thematic analysis

4 Result Analysis

Jharkhand is an eastern state in the Republic of India. It is a predominantly agrarian state with around 80% population of the state directly or indirectly dependent on agriculture. Since Jharkhand's formation by splitting away from Bihar state, its economy has grown tremendously with the agriculture sector given major impetus by the state government owing to its important contribution to the state's economy [26]. Despite the majority of small land holdings by farmers, the state has registered exceptional growth in state-wise yield of food grains year-on-year starting from 2004 [27, 28].

Jharkhand government has played a phenomenal enabling role in strengthening all the aspects of the agriculture sector within the state [29]. This can be substantiated by their adoption of several reforms in the field of agricultural marketing, such as the establishment of private market yards, the establishment of direct purchase of agricultural produce from agriculturists, to promotion and permit e-trading, establishment of farmers/consumers market managed by a person other than a market committee, registration of contract farming sponsor, single point levy of market fee, adoption of e-NAM, an e-market platform which provides ease to farmers to participate in National Agricultural market thus providing better opportunities to farmers beyond the State's markets [29]. Similarly, the state government announced the provision of subsidies for seeds and fertilizers [30].

4.1 GovTech Model in Jharkhand

In 2022, Jharkhand became the first state of India to implement blockchain in seed distribution (Fig. 3). The government-issued tender outsourced the implementation to an international firm, SettleMint, a blockchain platform-as-a-service company. SettleMint developed web and app-based interfaces for farmers and the end to ensure the timely availability of seeds to distributors, retailers and, the farmers.

The entities involved in distribution, micro-financing, and retail activities such as collectives such as Large Area Multipurpose Societies (LAMPS), Primary Agricultural Credit Societies, Farmer Producer Organizations along with the distributors and retailers are riddled with middlemen who restrain the trickling down of profits.

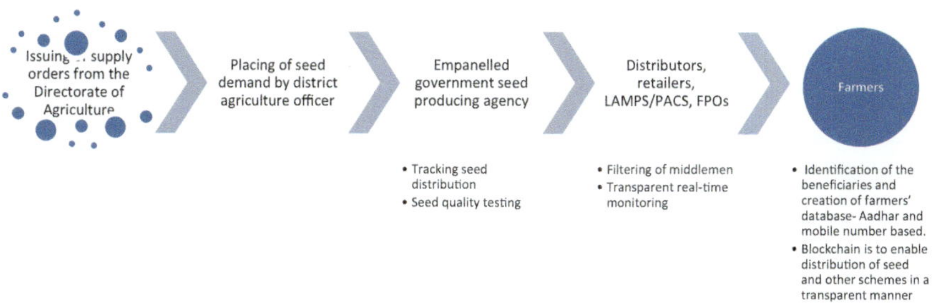

Fig. 3 Seed distribution system aided by blockchain technology in Indian state of Jharkhand. *Source* Authors compilation from various data sources

Furthermore, the system doesn't have provisions in place for real-time monitoring of activities such as quantity/variety/frequency of seed purchase. Blockchain-enabled technology developed by the external vendor, SettleMint, envisages filtering out middlemen by registering the transactions and ease of connection of the buyers and the seller's body and can provide real-time monitoring of activities transparently.

The end beneficiaries of the quality seeds, that is, the farmers, need to be validated and identified using the Indian Government's 12-digit individual identification number known as Aadhaar. This also led to the creation of a beneficiary database which is purported to disseminate information on agri-related social welfare schemes organized by the Jharkhand Government. The Aadhar number is linked to a mobile number and OTP-based verification which supports the user authentication module and is supported by MySQL® data analysis. Any occurrence of transaction for any Aadhar number is fed into the blockchain-based ledger system for transaction traceability and transparency.

I use the GovTech design and governance framework [4] and interpret our findings as a part of the four elements of the framework (Table 2), viz., governance, human-centered design, institutional, and technological design. Below I explain the four components and supplant the case facts accordingly.

The first element of the GovTech framework is governance which entails the decision-making structures including but not limited to bodies and mandates focused on achieving GovTech goals that participate in the change process. In the context of this case, the Department of Agriculture under the aegis of the Ministry of Agriculture of the state of Jharkhand assumes this role. This began by issuing a tender inviting participation from external and private sector bodies to usher in the changes as envisaged by the policy-making bodies [31]. The tender was ultimately awarded to Settlemint India, a three-year-old Indian subsidiary of Settlemint Belgium [32].

The next element is institutional design, which ensures the existence of a regulatory framework for the smooth conduct of the change process envisaged as a part of the GovTech plan. In terms of the adoption of this technology by the Jharkhand State Agriculture Department, includes the development, implementation, and maintenance of a seed traceability platform using Blockchain Technology. This also ensures smooth fungible, and acceptable technology across all the plans related to farmer

Table 2 Effective Seed Traceability Measures Implementation in Agriculture Jharkhand based on author's data using GovTech design and governance framework of Bharosa [4]

Elements of GovTech framework	Description	Implementation	Issues
Governance	Decision-making structures	Tender floated in April 2021 to get a private partner on board	Experience with domestic/international suppliers
Institutional	Existence of regulatory frameworks	An agreement signed; Aadhaar-based data protection policies	Data Privacy/ confidentiality
Human-centered design	End-user centric designs	Settle Mint app-Jharkhand Seed Website & App	Adoption by consumers;
Technological design	Interoperable digital designs	Aadhar Based Identification, Database management system, Blockchain based transactions	Ease of adoption; Willingness for end-users;

welfare. This includes the electronic know your customer (e-KYC), an Aadhar-linked mobile number to generate one time password (OTP) used for verification of farmers. This also helps in building a robust farmers' database.

Thirdly, the human-centered design incorporates the need to design the product as per the requirements of the end-users, that is, the farmers of the Jharkhand state. This includes the development of a website and mobile [33, 34] and making the farmers accustomed to the usage and features of the digital channels. Finally, the technological design, which includes the designing, development, and connection of interoperable digital technologies which are based on the other three components. The database's creation ensures the efficient delivery of agri-related schemes to the farmers. The backbone is blockchain-based which allows for replication of transactions distributed across the intended participants. The Hyperledger BESU private blockchain has demonstrable performance [35] and the same has been deployed by Settlemint for the Jharkhand seed-traceability project.

The framework also advocates for managing the interdependencies of the four design areas, lying under the purview of the governance mechanisms. Thus, the responsibility lies with the governing body (Department of Agriculture, Jharkhand, and the Ministry of Agriculture) to ensure the smooth operations of the project and to pave the way to ensure the smooth conduct of the tasks. This also includes onboarding of all stakeholders such as suppliers, distributors, famers and training them to adopt the technology by focusing on its benefits.

5 Contribution

Through this study, I intend to contribute to the understanding of GovTech, agriculture, and seed traceability. The contributions are based on the analysis of existing literature and data collected and triangulated through several sources.

In this paper, I analyzed the adoption of blockchain technology in seed distribution traceability by the state government of Jharkhand, India. This study advances the understanding of emerging technologies in the agriculture sector, by bringing to the fore a data source for researchers to situate their theoretical and hypothetical research on traceability [3, 36]. This study contributes to the growing literature of data-driven policymaking in the public sector [4]. Furthermore, this study allows us to deep dive into the concept of GovTech [4] and understand the practical implication of data-driven governance and public service delivery through the adoption of technology. I build on the extant literature of GovTech and unpack its various aspects, shedding light on the transformative potential of socio-technical solutions.

This study demonstrates a case of willpower of the government to successfully make use of emerging technology such as Blockchain to reduce the role of the middleman and enable managing the seed distribution process thus enabling effective traceability. The intention of the government to ensure the continued dissemination of high-quality seeds for the greater good is reflected in their endeavor to adopt emerging digital technologies. Thus, this study contributes to the importance of policy-level changes toward the successful transformation of a sector. Lastly, the study contributes a potential solution to the problem posed by informal seed distribution [17] in the form of emerging technology-based traceability.

References

1. Key Highlights of the Economic Survey 2021–22 (n.d.). https://pib.gov.in/PressReleasePage.aspx?PRID=1793829. Last accessed 08 Aug 2023
2. Feng, H., et al.: Applying blockchain technology to improve agri-food traceability: a review of development methods, benefits and challenges. J. Clean. Prod. **260** (2020). https://doi.org/10.1016/j.jclepro.2020.121031
3. Sandeep Chaitanya, N., et al.: A modern approach to seed quality check and it's traceability in agriculture BT. In: Reddy, A.B., et al. (eds.) Proceedings of Second International Conference on Advances in Computer Engineering and Communication Systems, pp. 117–129. Springer Nature Singapore, Singapore (2022)
4. Bharosa, N.: The rise of GovTech: trojan horse or blessing in disguise? A research agenda. Gov. Inf. Q. **39**(3), 101692 (2022). https://doi.org/10.1016/j.giq.2022.101692
5. Katsikouli, P., et al.: On the benefits and challenges of blockchains for managing food supply chains. J. Sci. Food Agric. **101**(6), 2175–2181 (2021)
6. Krithika, L.B.: Survey on the applications of blockchain in agriculture. Agriculture **12**(9), (2022). https://doi.org/10.3390/agriculture12091333
7. Xiong, H., et al.: Blockchain technology for agriculture: applications and rationale. Front. Blockchain. **3**(February), 1–7 (2020). https://doi.org/10.3389/fbloc.2020.00007

8. Bakare, S., et al.: A blockchain-based framework for Agriculture subsidy disbursement. IOP Conf. Ser. Mater. Sci. Eng. **1110**(1), 012008 (2021). https://doi.org/10.1088/1757-899x/1110/1/012008
9. Patel, N., et al.: KRanTi: blockchain-based farmer's credit scheme for agriculture-food supply chain. Trans. Emerg. Telecommun. Technol. **December 2020**, 1–16 (2021). https://doi.org/10.1002/ett.4286
10. Manoj, T., et al.: A trusted IoT data sharing and secure oracle based access for agricultural production risk management. Comput. Electron. Agric. **204**, 107544 (2023). https://doi.org/10.1016/j.compag.2022.107544
11. Chen, T., et al.: The adoption and implementation of artificial intelligence chatbots in public organizations: evidence from US state governments. Am. Rev. Public Adm. 02750740231200522 (2023)
12. Misuraca, G., et al.: The use of AI in public services: rResults from a preliminary mapping across the EU. ACM Int. Conf. Proc. Ser. 90–99 (2020). https://doi.org/10.1145/3428502.3428513
13. de Sousa, W.G., et al.: How and where is artificial intelligence in the public sector going? A literature review and research agenda. Gov. Inf. Q. **36**(4), 101392 (2019)
14. Reis, J., et al.: Artificial intelligence in service delivery systems: a systematic literature review. Trends Innov. Inf. Syst. Technol. **1**(8), 222–233 (2020)
15. Abduljabbar, R., et al.: Applications of artificial intelligence in transport: an overview. Sustainability **11**, 1 (2019). https://doi.org/10.3390/su11010189
16. Malhotra, C., Anand, R.: Accelerating public service delivery in India: application of internet of things and artificial intelligence in agriculture. In: Proceedings of the 13th International Conference on Theory and Practice of Electronic Governance, pp. 62–69 (2020)
17. Matuschke, I., Qaim, M.: Seed market privatisation and farmers' access to crop technologies: the case of hybrid pearl millet adoption in India. J. Agric. Econ. **59**(3), 498–515 (2008). https://doi.org/10.1111/j.1477-9552.2008.00159.x
18. Newman, J., et al.: Digital technologies, artificial intelligence, and bureaucratic transformation. Futures **136**, 102886 (2022)
19. Akpo, E., et al.: Analyzing pathways of nurturing informal seed production into formal private ventures for sustainable seed delivery and crop productivity: experiences from Ethiopia. Sustainability **12**(17), 6828 (2020)
20. Biemond, P.C., et al.: Does the informal seed system threaten cowpea seed health? Crop Prot. **43**, 166–174 (2013)
21. McGuire, S., Sperling, L.: Seed systems smallholder farmers use. Food Secur. **8**(1), 179–195 (2016)
22. Wattnem, T.: Seed laws, certification and standardization: outlawing informal seed systems in the Global South. J. Peasant Stud. **43**(4), 850–867 (2016)
23. Tripp, R.: Seed regulatory frameworks and resource-poor farmers: a literature review. Netw. Pap. Adm. (Res. Ext.) Netw. (U. K.) (1995)
24. Sperling, L., McGuire, S.: Understanding and strengthening informal seed markets. Exp. Agric. **46**(2), 119–136 (2010)
25. Eisenhardt, K.M., Graebner, M.E.: Theory building from cases: opportunities and challenges. Acad. Manag. J. **50**(1), 25–32 (2007)
26. Pandey, G., Kumari, S.: Dynamics of agricultural growth and diversification in Eastern India. J. Agribus. Dev. Emerg. Econ. **11**(2), 105–120 (2021). https://doi.org/10.1108/JADEE-06-2020-0131
27. Deshpande, T.: State of agriculture in India. PRS Legis. Res. **53**(8), 6–7 (2017)
28. Singh, R.P., Prakash, J.: Transformation of agriculture in Jharkhand. Agro-Economist **4**(1), 45–54 (2017)
29. Reforms in agricultural marketing (n.d.). https://pib.gov.in/newsite/printrelease.aspx?relid=137359
30. Barve, D.: Jharkhand Govt to offer 90% subsidy on seeds and fertilizer for Kharif crops, https://agrospectrumindia.com/2020/04/25/jharkhand-govt-to-offer-90-subsidy-on-seeds-and-fertilizer-for-kharif-crops.html

31. Request for Proposal (RfP) for Design, Development Implementation and Maintenance of Seed Traceability Platform Using Blockchain Technology, Ranchi, Jharkhand Tender (n.d.). https://www.tendersontime.com/india/details/request-proposalrfp-design-development-implementation-and-maintenance-seed-traceability-platform-32a0c85/, last accessed 2023/08/08.
32. Jharkhand Introduces Blockchain Technology in Seed Distribution to Farmers (n.d.). https://agriculturepost.com/farm-inputs/seeds/jharkhand-introduces-blockchain-technology-in-seed-distribution-to-farmers/. Last accessed 09 Sept 2023
33. Jharkhand Seed-Google App (n.d.)
34. Jharkhand Web App (n.d.). https://jharkhandseed.com/. Last accessed 07 Aug 2023
35. Fan, C., et al.: Performance analysis of hyperledger besu in private blockchain. In: Proceedings of the 4th 2022 IEEE International Conference on Decentralized Applications and Infrastructures (DAPPS) September 2022, pp. 64–73 (2022). https://doi.org/10.1109/DAPPS55202.2022.00016
36. Kamble, S.S., et al.: Modeling the blockchain enabled traceability in agriculture supply chain. Int. J. Inf. Manage. **52**(April), 1–16 (2020). https://doi.org/10.1016/j.ijinfomgt.2019.05.023

Recent Patterns of Plant Leaf Disease Characterization with Artificial Intelligence Techniques

Apoorva Arora and Vinay Gautam

Abstract Agriculture is the base for the growth of the economy of a country. Maintaining the health of crops is a significant part of it. Early identification of plant leaf diseases can help in dealing with and eradicating the various challenges. Using deep learning in detecting leaf diseases is a crucial and essential research area in the agriculture field. Such kind of research focuses on seventeen percent of the total gross domestic product (GDP) is contributed by agriculture in India. Enhancing the growth of crops can lead to growth in farmer's profit, which further boosts the economy of the country. This paper presents a comprehensive review of various research works being carried out in the field of plant disease detection using deep-learning-based techniques. Using handcrafted feature-based approaches can be challenging. Models based on deep learning give an edge over such handcrafted feature-based techniques. The research shows that various deep learning approaches show significant accuracy on particular data sets, however, the accuracy declines when tested on field image conditions or on a separate data set. While reviewing it was found that models like GoogleNet and InceptionV3 have shown better results. Using transformer-based techniques could be a potential approach to detect plant leaf diseases.

Keywords Plant disease · Deep learning · Transfer learning · Image segmentation · Feature extraction

1 Introduction

Agriculture is the major contributor when the economy is taken into consideration in India. Just like humans, plants also undergo various health issues, affecting the healthy growth of the plant [1]. The diseases can affect any part of the plant like flowers, leaf root, and fruit. As the scale of agriculture is humongous in the country, the number of diseases is also numerous [2]. A timely diagnosis helps save plant life from loss in qualitative and quantitative forms [3–5]. Most of the farmers do not have

A. Arora · V. Gautam (✉)
Chitkara University Institute of Engineering and Technology, Chitkara University, Rajpura, India
e-mail: vinaykkr@gmail.com

B. Soni et al. (eds.), *Beyond Artificial Intelligence*, Lecture Notes in Networks and Systems 1326, https://doi.org/10.1007/978-981-96-4170-3_11

the precise knowledge that helps in detecting plant diseases [6]. Manual detection of diseases can have various shortcomings and limited accuracy. On the other hand, automated identification of diseases minimizes human effort and enhances accuracy [7].

Today techniques based on machine learning and deep learning are used effectively to monitor health and diseases in plants. Maintaining a standard and quality in real-time applications is the major outcome of this kind of system. Minimized time and cost, along with precise results are an addition.

The agriculture field provides ample opportunities for researchers to use machine learning techniques as an application to multiple aspects like plant identification, disease detection, etc. Here, we are studying the diseases that occur in the leaves of plants, which can be detected based on shape, texture, or color using various deep-learning models.

In recent times, automatic detection of plants has been a frequent topic of research. To measure plant diseases quantitatively, precise and accurate identification is required [8]. Authors around the world have surveyed diseases related to various plants like potatoes, tomatoes, and even rice [9–11]. The careful analysis involved, using a classifier, a dataset, the classification of diseases into classes, and pre-processing techniques followed by segmentation.

A survey was conducted by Prajapati et al. [12] using machine learning techniques regarding the classification of a cotton plant disease. The classification of citrus plants was surveyed by Iqbal et al. [13]. Similarly, a survey was also conducted by Kaur et al. [14] which helped in identifying and classifying plant leaf diseases through images.

The above survey and studies have been focused on handcrafted features, but to do so, there are certain steps to be taken first. Preprocessing, along with segmentation, and extraction of features from the images are a few steps. This way is more time-consuming, and a lot of hard work is required. With technology advancing every day, artificial intelligence using machine learning has come up with various new techniques and models using computer vision [15]. Deep learning has been expanding its arms in various fields like voice recognition [16, 17], and image recognition [18]. Deep learning has made an impact in developing automated cars, machine interpretation, and many other complex applications. Using deep learning to detect plant leaf diseases is still a new technology with a lot of scope [19, 20].

Recently researchers have been paying attention to models and data sources, along with various preprocessing techniques, to detect plant diseases using deep learning [21, 22]. The above papers have mainly focused on discussing the various deep learning techniques, majorly CNN models used to detect leaf diseases.

Here in this research paper, the focus is on several methods to detect leaf diseases in plants using deep learning feature-based identification. Here we try and bring forth various pros and cons of identification techniques. We also address various shortcomings of deep learning approaches discussed and various challenging issues faced during the identification process.

The description of complete paper has been laid down in various sections. Section 2 covers the recent advancement in plant disease classification and detection.

The result and analysis have been presented in Sect. 3 and last Sect. 4 is conclusive section.

2 State-of-the-Art

2.1 Plant Leaf Disease Identification

To effectively identify plant diseases several steps are required. Majorly these are data collection and pre-processing followed by extraction of features. At the end, these features are classified using various classifiers.

Data Collection

The availability of online data sets has made it easy to form a database of images to detect leaf diseases in plants. Some of them are the PlantVillage dataset [23], the Cassava dataset [24, 25], the Hops dataset [26], and the rice disease data set [27, 28]. Now let's talk about these data sets in detail.

Plant Village data set comprises 14 different species of plants categorized into 38 different classes. The species include cherry, orange, peach, pepper, strawberry, tomato, etc. The diseases identified in these plants are cherry healthy, peach bacterial spot, pepper bacterial spot, and so on. Every image in the data set was carefully captured carefully in laboratory conditions.

The Cassava disease dataset comprises images classified into 5 major classes. Another important factor about the data is that the images are captured in the field in real-time.

Diseases included are cassava mosaic disease, cassava healthy, and so on. Hops data set has a unique property. It provides a non-uniform background condition to the five classes it uses to classify the data. Another data set is the Rice data set, which consists of 4 different classes, again captured in the field. Some of the diseases include brown spot and tungro.

Preprocessing

The next important step towards identifying plant diseases is pre-processing. From resizing an image to noise removal and color transformation to morphological operations, the list of pre-processing techniques is endless.

Further, several filtering techniques are used like Wiener Filter, Gaussian filter [29], and median filter. It helps make images of disease-affected plants noise-free. To enhance the process different color spaces like RGB, HSV, $CIEL * a * b$, and many more are used. And then to narrow down the region of interest, several segmentation techniques are used like color threshold [30, 31], Otsu segmentation [32, 33], and the Sobel Edge detector [34].

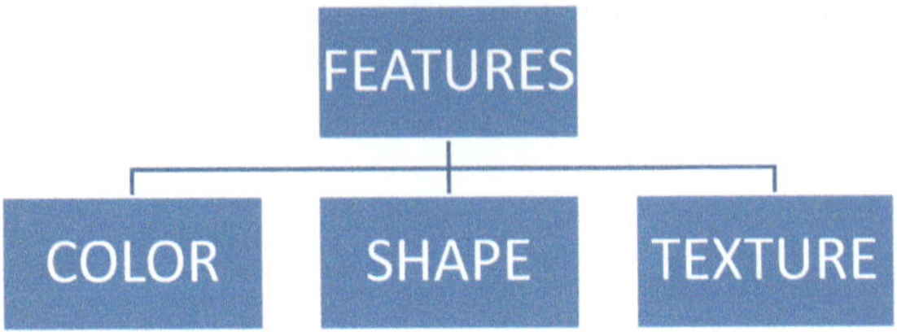

Fig. 1 Features used in detecting plant disease

Feature Extraction

This is a significant step in the image detection process using deep learning techniques. Providing a mathematical form of the disease information makes classification an easier process. Classification can be done using various types of features like color, shape [29, 35], texture [34–36], and deep learning-based features. Using the color feature helps in detecting the disease region. Similarly, shape features like major and minor axis length, and parameter help in pinpointing the disease area. Identifying the leaves with the disease can be also texture-based. For instance, Local Binary Pattern (LBP) [37], Gray Level Co-occurrence Matrix (GLCM) [33], and Gabor texture features [38] are commonly used for identification purposes (Fig. 1).

Classification

The next step is classifying the images into various categories based on the identification techniques used before. It helps categorize and organize data of leaf images by numerically analyzing the various image features. K-nearest neighbor (KNN) [38], Logistic Regression (LR), decision tree (DT) [34], Artificial Neural Networks (ANN) [37], and Support Vector Machine (SVM) [32, 33, 39] are some of the commonly used techniques for classification.

2.2 *Literature Studied and the Present State of Art Regarding Identified Problem*

Researchers have been using various methods to identify plant diseases using machine learning techniques like color, texture, and shape as we discussed earlier. These steps being hand-crafted take a lot of energy and time. The results are delayed and at times not completely accurate. On the other hand, Deep learning techniques have gained a lot of success in this field (Fig. 2).

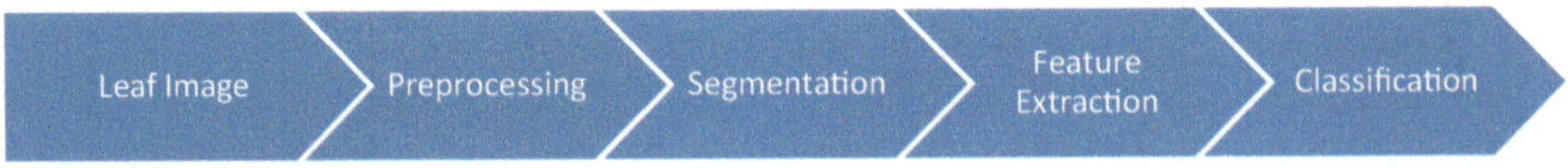

Fig. 2 Steps for identification of diseases in leaves

2.3 Deep-Learning-Based Identification of Diseases

Due to its vast scope and outstanding performance in image recognition, deep learning has grown exponentially as far as computer vision is concerned. Implementing pattern recognition has now become much simpler and more accurate e.g., DL Models performing the ImageNet Challenge. Extending image recognition to identify leaf diseases in plants [2, 40–42], weed detection [43], and fruit identification [44, 45] has been made possible through deep learning.

Deep learning is an automated version of machine learning, all segmentation and feature extraction is done without any human intervention.

Kawasaki et al. [46] used CNN to identify two separate types of diseases found in cucumber, namely melon yellow spot virus (MYSV) and zucchini yellow mosaic virus (ZYMV). In their research, a large number of images were used in rotation to enhance the data size and further conclude that a larger database is effective in increasing performance. The DL model acquired an accuracy of 94.9%.

Further in Fujita et al. [47] authors took the previous work to the next level and found seven different types of diseases in cucumber plants. This was accomplished with the help of an enormous database and two CNN architectures. The various considerations while using the data set were various image aspects like angle, background, distance, and uneven lighting conditions. To enhance the size of the data set, techniques like rotation and shifting were used for augmentation. Image mirroring was also used to enhance image quality.

VGG Net [48] and Caffe Framework [49] were based on the grounds of CNN. CNN-1 showed diminished performance in comparison to CNN-2 in the case of a poor image condition. The reason is, that images in good as well as bad condition were used to train CNN-2. The accuracy rate achieved was 82.3%.

Sladojevic et al. [50] identified about 13 separate diseases found in plants using a fine-tuned and pre-trained CaffeNet Model. To train the model and evaluate its performance, an enormous number of about 4483 internet-downloaded images was used. Along with that, a technique for data augmentation was used to multiply the size of the data. Evaluating the model using a 10-fold cross-validation technique made it reach an accuracy level of 96.3%.

Rangarajan et al. [51] made use of VGG16 Net to identify six types of diseases in the leaves of tomato plants. The accuracy attained in classifying the types was about 97.29% for VGG16 Net, whereas the AlexNet architecture reached an accuracy level of 97.49%.

Using two profound architectures of CNN, namely AlexNet and GoogleNet, Mohanty et al. [40] classifies 26 distinguished diseases in 14 species of crops. The task was accomplished using 54,306 images. The data set was divided into multiple training sets. However, 3 different datasets were used, and RGB images could achieve the maximum training accuracy of 99.35% using GoogLeNet.

The above process had some limitations as well. For instance, the complete detection process was carried out in a laboratory setup. Which led to a major decrease in performance when the model was tested on real-time images. Another constraint was that the classification for models was done on front-facing single leaves on similar background images.

Identification of apple leaf diseases of six different types was achieved by Nachitigall et al. [52] using several CNN architectures. AlexNet model [53] resulted in an enhanced accuracy of 97.3%. Using a CNN model gave good results too. So to compare the results an MLP—Multi-Layer Perceptron was used, which further helped in achieving higher levels of recognition accuracy in classifying images [54]. The CNN architecture was designed using tools like the Caffe [49] and the DIGITS tools.

Amara et al. [55] used a deep learning CNN architecture, LeNet [56] to find leaf diseases in banana plants. The promising thing about this approach was that it was found to be effective under challenging situations. For instance, a complex background, illumination, resolutions, orientations, sizes, and poses of real-life images. To check the performance levels both greyscale and RGB images were used. They recorded a high accuracy rate of 98.61% for color images and 94.44% for greyscale images.

Another research to identify olive leaf diseases was done by our next set of authors [57]. Here a combination of texture and shape-based features, like Hu's moments [58] and Zernike's moments [59] was used along with CNN. After undergoing the process for a long duration the model achieved an accuracy of 98.61%. Atole et al. [60] used a finely tuned AlexNet architecture to look for diseases in rice plants and achieved an accuracy level of 91.23%.

Another author Ferentinos et al. [2] made use of 5 separate CNN architectures (AlexNet, AlexNetOWTBn, GoogLeNet, VGG, and Overfeat) to find plant diseases using images of leaves. The data set was huge, containing 58 different image classes, which helped achieve top-notch results with an accuracy level of 99.53% with the help of VGA [48] architecture. The reason behind such sky-reaching results was the use of the world's largest dataset available on plant disease. It included images of leaves with partial shading as well as multiple objects like fingers, hands, shoes, and so on. The training of these models was time-consuming. However, the major issue was expanding the current database, to enhance the plant species variety, to detect more diseases. The testing dataset belonged to the same database that became the base of the training data set, which became one of the main problems.

The research performed by Rangaranjan et al. [61] was on eggplant diseases of five different forms. A trained VGG16 network was used for this purpose. To test their results, they used YCbCr, HSV, and grayscale color spaces. The accuracy results achieved are 99.4% the important thing to consider here is that the images used were taken both in laboratory and real-life conditions. The research in Rangarajan Aravind and Raja [62] used six separate pre-trained deep learning models to find 10 categories of diseases found in various crops—beans, lime, eggplant, etc. Although many models were used, VGG16 has the best results, giving an accuracy of 90%. Before moving to further research let's look at what we have studied so far.

- To minimize or completely remove the pre-processing steps involved in feature extraction from an image for identifying a disease, a pre-trained model is necessary.
- A pre-trained model works better if a large data set is used.
- Approaches like fine-tuning and transfer learning elevate the accuracy of results.
- When using coloured images, RGB gives best results than any other form of image format.
- Using architectures like LeNet, AlexNet, VGG, and GoogLeNet, demand longer computation duration because of the number of parameters used.
- The training time required by these models is much higher, which also requires high-power GPUs for model training.

Moving further, author Lee et al. [63] wrote a paper to compare the performances of a few models namely, VGG16, InceptionV3, and GoogLeNetBN. The purpose of using these models was to identify the diseases in plants using these models but also to find the best model using the transfer learning approach. The result of this diagnosis was that the overfitting impact could be minimized by pre-training of models.

The work done by authors in Johannes et al. [64] was extended by Picon et al. [65] to find three different types of diseases found in wheat crops. For this purpose, a modified ResNet50 architecture was used. Initially, a 7×7 convolution layer of ResNet50 was taken place by two 3×3 consecutive convolutional layers. After that, the sigmoid function was introduced in place of the softmax activation function. This helped in multiple disease detection on the same leaf. The accuracy achieved was about 87%, which was an improved version achieved using the superpixel segmentation approach. This approach used artificial background while training the model.

Another research for the detection of diseases in tomato plants was done by Fuentes et al. [41]. The approach was practical and could be used to detect diseases in tomato plants as well as recognize pests in real-time. The system used deep learning methods for robust functioning and accurate results. Moreover, the model was able to take care of complex conditions like illumination, background change, and also changes in object dimensions. The architecture used were Faster R-CNN, SSD, and R-FCN models. These are meta-architectures of CNN used for extracting features. In addition to that VGG16 [48], ResNet-50 [66], and ResNet-101 models were also used. The accuracy obtained by the model was 83%. The model included the current stage of the disease a plant is suffering from along with the position of its occurrence.

Finding the diseases in cassava plants was done by Ramcharan et al. [24]. An image-based deep-learning technique was used. The task was to identify three types of diseases and two pest damages. For this transfer, a learning approach was used on the InceptionV3 model. Here two data sets were used, the original cassava dataset which included several leaves on one image, whereas the other was, leaflet cassava dataset, which included singular leaf images. The leaflet database showed some improvement in the results. The evaluation of results was done using softmax layer SVM and KNN classifiers. The accuracy obtained was 98%, with the help of an SVM classifier.

Ahmed et al. [67] found separate types of tomato diseases. The method used pre-trained deep learning models like VGG16, and VGG19. Other models used were ResNet, and InceptionV3. The network was fine-tuned, to achieve the best results. Two data sets were used, one in which all the laboratory-generated images were used and another was a collection of real-time field images. The results showed that the laboratory images gave promising results compared to the other dataset. As far as the DL models are concerned InceptionV3 results were outstanding. The accuracy results achieved for laboratory and field images were 99.60% and 93.70% respectively. The results were achieved after fine-tuning the parameters.

Five types of cassava plant diseases were detected by Oyewola et al. [25]. The task was achieved using a plain convolution neural network (PCNN). In addition to that a deep residual network (DRNN) was used. The results showed how the results of DRNN overshadowed the results of PCNN by a margin of 9.25%.

Elhassouny et al. [68] used a MobileNet CNN-based model, to identify ten commonly found tomato diseases, the result of the method was 90.3% accurate.

Another author named Li et al. [69] identified diseases in ginkgo leaf diseases. The models used were VGG16 and InceptionV3. The images used were taken both in laboratory conditions and in real-time field conditions. The results were separate for both models. VGG16 worked better on laboratory images, while InceptionV3 gave higher accuracy on field images.

Yong et al. [70] used InceptionResNetV2 architecture to find diseases in various crops along with insect pets found in these crops. The purpose was to set up platforms in remote areas like mountains to inspect diseases. To make that possible Internet of Things (IoT) platforms were built. The accuracy rate achieved by this model was 86.1%.

Another author DeChant et al. [71] used a pipeline CNN model to detect diseases in maize plants. The training of CNNs included 3 layers. The first layer contained training for the classification of small sections of images containing lesions. The next layer was a combination of all the predictions from the first layer bundled up together. The last and final layer helps feed the images into the CNN layer resulting in the classification of disease-affected plants from healthy ones.

Ozuguven et al. [72] used a model named Faster R-CNN to detect sugar beet diseases. The method could achieve an accuracy rate of 95.48%.

Using deep CNN Oppenheim et al. [73] found a way to identify four types of diseases in potato crops. The database used potato images with varied shapes and sizes, with expert labeling of images. The overfitting problem was solved with the help of multiple dropout layers. An accuracy of 96% was achieved.

A CNN network used by Geetharamani et al. [74] had nine layers. To identify various plant diseases various techniques like data augmentation along with a large dataset can help enhance accuracy. The techniques used in this paper are noise rejection, scaling, gamma correction, and many more. The result was as anticipated and the accuracy levels increased from 91.43 to 96.46%. Certain comparisons were made which showed that deep learning approaches gave better results than traditional approaches.

Wang et al. [75] identify leaf diseases in apple crops with the help of fine-tuned CNN network. The author compared two different approaches in this paper. One architecture is built into a shallow network from scratch. It also contains a convolution layer, two completely connected layers, and also a softmax layer, which helps anticipate the output. Another approach is the architecture built using transfer learning. This approach is very useful in case of limited data. It works by fine-tuning the pre-trained network parameters of a large data set. ImageNet [76] can be used as an example in this case.

Our next author [77] identified various types of leaf diseases found in maize plants. The architectures used were GoogLeNet and Cifar10. The data size was enhanced with the help of specific augmentation techniques. Further pooling combinations and rectified linear unit functions were used to improve hyperparameters. The accuracy was 98.9% in GoogLe Netmodel and 98.8% in Cifer10 [78].

Another author Liu et al. [79] found four types of apple leaf diseases using images of leaves. The models used were AlexNet and GoogLeNet. The major point in this model is that a convolution layer along with an inception layer was replaced by a fully connected layer of AlexNet. This resulted in lowering the model parameter by a large margin and enhanced the accuracy level to 97.62%. the model also used an accelerator *named* Nesterov's accelerated gradient (NAG).

Ramcharan et al. [80] who built a mobile-based cassava disease detection system. The model used was a single-shot multi-box (SSD) model. It further included a MobileNet detector and a pre-trained classifier on a COCO dataset [81]. The evaluation included images and video files as well of the disease leaves. The accuracy obtained was 80.6% for images and 70.4% for video files.

Architecture built by Toda et al. [82] was a deep CNN with InceptionV3. The approach used an attention map and altered various layers as per their contribution to the identification process. The alteration process proves to be a success as it did not affect the accuracy but rather reduces the no, of parameters to about 75%. The accuracy achieved was 97.1%.

Another model proposed by Lu et al. [83] was to find 10 types of diseases in rice crops. For best results, various pooling operations and convolution filter sizes were used. This led to a high accuracy rate of 95.48% as a result of stochastic pooling and with a 16×16 convolutional filter, the accuracy was found to be 93.29%. The main advantage of this model is, with the rise of no. of layers the time of computation decreases.

Another architecture was used by Durmus et al. [84] namely SqueezeNet [85]. The model was built to find diseases in the leaves of tomato crops. A robot was used to detect diseases in plant leaves in the field as well as in the greenhouse. The accuracy level reached through this autonomous detection was 94.3%.

Gensheng et al. [86] used a modified version of the Cifer10 quick CNN model. The architecture was used to identify various tea leaf diseases. Here a depth-wise separable convolution was used in place of standard convolution. As a result, the number of parameters was diminished. The results were compared with the other models like LeNet-5 [87], AlexNet, and VGG16 [75]. The acquired accuracy was 92.5%.

A mobile deployable model was prepared with minimum cost by Bi et al. [88], to find various types of diseases in apple leaves. The models used were MobileNet, ResNet152, and InceptionV3. The results of these were compared to the data set collected by agricultural aces. The accuracy achieved was 73.50%, 77.65%, and 75.59% for the above models respectively. The results showed that using MobileNet the resultant handling time was significantly lower than InceptionV3.

Chen et al. [42] identified rice and maize leaf diseases with the help of the INC-VGGN approach. A few alterations were made in the existing models, like replacing the last convolution layer with a couple of inception layers and adding a global average pooling layer. The inception layer helped in extracting the high-dimensional features. The accuracy level achieved was 92% for rice and 80.38% for maize.

The architecture used by Atila et al. [89] was to detect various diseases in plants. The performance of EfficientNet architecture was held in comparison with a few different CNN models like AlexNet, ResNet50, VGG16, and InceptionV3. The results were completely bent toward EfficientNet architecture. The accuracy achieved was 99.91%. Another advantage was that the number of parameters required to train was less and hence the time required was less too.

A hybrid CNN approach was followed by Tuncer [90] to detect plant leaf diseases. The author tries to reduce the number of parameters and the cost of computation with the help of an inception network along with a separate convolution network. The model shows an accuracy level of 99.27% by using a k-fold cross-validation technique.

Our next set of authors [91] used deep learning techniques to detect plant leaf diseases. The images used were gathered from different datasets, 15% from the PlantVillage dataset, and other images captured in real-time. The models used were CNN, VGG, and Inception architectures. Soyabean leaf diseases were detected using pre-trained AlexNet and GoogleNet [92] by authors in Jadhav et al. [57]. The method was to use healthy leaf images with a few modified hyperparameters like max epoch and bias learning rate. The accuracy rate is 95%.

Classification of diseases in maize leaves was done by authors in Arora et al. [93]. The method was similar to the above because they too used healthy images with modified hyperparameters related to the forest like the number of trees and grains and so on. The accuracy achieved was 96%.

Sibiya et al. [94] built a fully connected CNN model to identify maize leaf diseases. The accuracy level reached was 92.85% in finding 3 types of maize leaf diseases. Singh [95] used a multiple-layered CNN architecture to detect leaf diseases in mango crops and achieved an accuracy of 96%.

Authors in Türkoğlu and Hanbay [96] used various CNN architectures like AlexNet, VGG16, VGG19, and ResNet, to detect leaf diseases in plants. The model was trained using images of eight different diseases, which too captured through the camera. Various classifiers were used as well, like KNN, SVM, and ELM. The accuracy rate was 97.86% in the case of ResNet architecture. Also, 93.82% accuracy was achieved by a NASNet-based CNN architecture [97] to detect leaf diseases in general.

Yang Li et al. [98] used an SCNN (shallow CNN) for the detection of plant leaf diseases in apples, maize, and grape. The method used CNN with SVM and RF classifiers. As per them, the combination of machine learning techniques and shallow CNN made the results better and could overcome overfitting.

Enkvetchakul and Surinta [99] came up with a CNN model using a transfer learning approach for a couple of plant diseases. The models used were pre-trained NASMobileNet and MobileNetV2. NASMobileNet provided the most accurate results. Data augmentation techniques used in this model helped reduce the over-fitting technique. The data sets used were iCassava 2019 and other leaf disease data sets. The accuracy achieved was 84.51%.

Singh et al. [100] used CNN techniques to detect a wide range of diseases in cotton plants. The illness ranges from bacterial to nutritional deficiency. The data set used has 22 leaf disease types. Further, data augmentation has been used to boost performance levels of the models. The accuracy achieved was 99.39%. The proposed approach is efficient and can be implemented in real-time leaf disease detection in cotton plants.

Peanut being an essential food and oil seed crop, it's imperative to deal with low yield and destruction of this crop due to leaf diseases. Xu et al. [101] proposed a model which combines enhanced X-ception and a couple of attention augmented branches. The accuracy obtained is 99.69%. The model gave excellent results on other crops like apple, rice corn, cucumber, and wheat, for leaf disease detection. The accuracy achieved was 99.61%.

Binnar et al. [102] uses CNN models, AlexNet, MobileNet, simple sequential model, and Inception V3 to detect plant leaf diseases. The data set used was the New plant disease data set, with 38 distinct classes including healthy as well as diseased plant leaf images. The model acts as a prediction tool for plant lead disease detection and can be very well extended as real-world integrated plant leaf detection system. The best model with a 99.07% training accuracy and 97.52% validation accuracy was MobileNet (Tables 1, 2, 3 and 4).

3 Result and Analysis

The following table shows the various limitation that was found while surveying the work done by various authors. It includes details like the number of images included in data sets, and consideration of complex backgrounds while collecting images. The table also includes which authors could detect multiple images on the same image along with the accuracy of testing on-field images.

Table 1 Summary of state of the art study

Name of the author	Year of research	Crop name	Architecture used	Accuracy (%)
Binnar et al. [102]	2023	All plant leaves	AlexNet, MobileNet, simple sequential model, and Inception V3	0.9752
Liu et al. [103]	2023	Tomato	NanoSegmenter Model based on Tranformer structure	98
Thai et al. [104]	2023	Cassava plant leaves	LeIAP and SPMM alogorithms	0.9752
Singh et al. [100]	2023	Cotton plant	CNN, Data Augmetation	0.9939
Xu et al. [101]	2023	Peanut	X-ception and a couple of attention augmented branches	0.9969
Li et al. [105]	2022	Sugarcane	Shuffle Convolution LVit	98.87 & 87.64
Enkvetchakul and Surinta [99]	2022	2 Plants	NASMobileNet and MobileNetV2	84.51
Atila et al. [89]	2021	Several	EfficientNet	99.91
Tuncer et al. [90]	2021	Several	Hybrid CNN	99
Oyewola et al. [25]	2021	Cassava	DRNN	96.75
Shirivastava et al. [106]	2021	Rice plant	Pre-trained CNN with SVM classifier	0.9137
Sethy et al. [28]	2020	Rice	Different deep learning model with SVM	98.38 (F1-score)
Arora et al. [93]	2020	Maize	Deep forest	96.25
Bi et al. [88]	2020	Apple	MobileNet	73.5
Turkoglu et al. [96]	2019	Eight types of plant diseases	Modified deep learning model with SVM, ELM, KNN	95.5 (ALexNet), 95 (VGG16)
Singh et al. [95]	2019	Mango	Multilayer CNN	97.13
Ozguven et al. [72]	2019	Beet	Faster R-CNN	95.48
Adedoja et al. [97]	2019	Multiple	NASNet	93.8
Gensheng et al. [86]	2019	Tea	Modified Cifar10	92.5
Elhassouny et al. [68]	2019	Tomato	MobileNet	90.3

(continued)

Table 1 (continued)

Name of the author	Year of research	Crop name	Architecture used	Accuracy (%)
Ramcharan et al. [80]	2019	Cassava	MobileNet	80.6
Liu et al. [107]	2018	Apple	AlexNet	97.62
Atole et al. [60]	2018	Rice	AlexNet	91.23
Mohanty et al. [40]	2016	Several	AlexNet, GoogLeNet	99.35
Nachtigall et al. [52]	2016	Apple	AlexNet	97.3
Sladojevic et al. [50]	2016	Grape and apple	CNN architecture	96.3
Ramcharan et al. [24]	2017	Cassava	Inception V3, GoogLeNet	98
DeChant et al. [71]	2017	Maize	Multi-layered CNN architecture	96.7
Lu et al. [83]	2017	Rice	CNN with multiple stages	95.48
Wang et al. [75]	2017	Apple	VGG 16	90.4
Fuentes et al. [41]	2017	Tomato	Faster R-CNN (ResNet)	83

4 Conclusion

The study presents a nutshell the importance of using deep learning for the detection of plant diseases with the help of leaf images. Just like humans, plants undergo various health issues, and timely detection of leaf diseases can help in better growth. Ignoring early signs of any disease in the case of plants, can be disastrous, leading to a diminished world economy as a result of losses in food crops [110]. Moreover, detecting these diseases manually can be time-consuming and misleading as well [111]. Here, in this report, we discussed various deep-learning techniques for the identification of these diseases. After the survey and comparison of the performance of various techniques, it was observed that the ResNet50, InceptionV3, and DenseNet201 architectures are best suited for plant leaf disease detection. We used various data sets in our survey as well, and it can also be concluded that as the disease changes in the plant body, it becomes difficult to detect the disease using identification techniques.

Table 2 Summary of literature survey

Name of the authors	Durmus et al. [84]		Cruz et al. [108]		Amara et al. [55]		Ferentinos et al. [2]	
Year of research	2017		2017	2017	2017	2017	2018	2018
Crop name	Tomato		Olive		Banana		Several	
Architecture used	Alexnet, SqueezeNet		Lenet hybridized with shape, edge, Hu's moments	Zernike moments	LeNet		AlexNetOWTBn, VGG	
Accuracy (%)	95.65 (AlexNet)	94.3 (SqueezeNet)	98.6		98.61 (RGB image)	94.44 (grayscale image)	99.49 (AlexNet)	99.53 (VGG)
Number of images in the data set	Tomato images	Plant village data set	299		3700		87848	

Table 3 Detailed related work

Name of the authors	Rangarajan et al. [51]		Lee et al. [63]		Zeng et al. [109]			
Year of research	2018	2018	2020	2020	2020	2020	2020	2020
Crop name	Tomato		Several		Rice			Cucumber
Architecture used	AlexNet, VGG16		VGG16	InceptionV3	Transfer learning and	Training from scratch	SACNN	
Accuracy (%)	97.29 (ALexNet)	97.49 (VGG16)	99.09 (GoogLeNetBN)	99.00 (VGG16)	99.35 (GoogLeNet)		95.33 (AES-CD9214)	98.00 (MK-D2)
Number of images in the data set	13262		54306		AES-CD9214, MK-D2			

Table 4 Limitations in the literature surveyed

Limitations in work of different authors						
Name of author	Large data set used	Vast variations in species used	Testing field images—accuracy	Same image with multiple diseases	Complex background consideration	Different data sets are used for testing and training
Mohanty et al. [40]	✓	✓	Less	×	×	×
Ferentinos et al. [2]	✓	✓	Less	×	×	×
Geetharamani et al. [74]	✓	✓	×	×	✓	×
Barbedo et al. [110]	✓	✓	Low	✓	×	×
Cruz et al. [108]	×	×	×	×	×	×
Sladojevic et al. [50]	×	✓	×	×	✓	×
Brahimi et al. [111]	✓	✓	×	×	✓	×
Ozguven et al. [72]	×	×	×	×	×	×
Wang et al. [75]	×	×	×	×	×	×
Oyewola et al. [25]	×	✓	×	×	✓	×
Ramcharan et al. [24]	×	✓	×	×	✓	×
Liu et al. [107]	×	×	×	×	×	×
Amara et al. [55]	×	×	×	×	×	×
Fuentes et al. [41]	×	×	✓	✓	✓	×
DeChant et al. [71]	×	×	×	×	×	×
Ramcharan et al. [80]	×	✓	✓	×	✓	✓

References

1. Chouhan, S.S., Kaul, A., Singh, U.P., Jain, S.: Bacterial foraging optimization based Radial Basis Function Neural Network (BRBFNN) for identification and classification of plant leaf diseases: an automatic approach towards plant pathology. IEEE Access **6**, 8852–8863 (2018)

2. Ferentinos, K.P.: Deep learning models for plant disease detection and diagnosis. Comput. Electron. Agric. **145**, 311–318 (2018)
3. Bharate, A.A., Shirdhonkar, M.: A review on plant disease detection using image processing. In: Proceedings of the 2017 International Conference on Intelligent Sustainable Systems (ICISS), pp. 103–109. Palladam, India (2017)
4. Das, R., Pooja, V., Kanchana, V.: Detection of diseases on visible part of plant—a review. In: Proceedings of the 2017 IEEE Technological Innovations in ICT for Agriculture and Rural Development (TIAR), pp. 42–45. Chennai, India (2017)
5. Bock, C., Poole, G., Parker, P., Gottwald, T.: Plant disease severity estimated visually, by digital photography and image analysis, and by hyperspectral imaging. Crit. Rev. Plant Sci. **29**, 59–107 (2010)
6. Tlhobogang, B., Wannous, M.: Design of plant disease detection system: a transfer learning approach work in progress. In: Proceedings of the 2018 IEEE International Conference on Applied System Invention (ICASI), pp. 158–161. Chiba, Japan (2018)
7. Mutka, A.M., Bart, R.S.: Image-based phenotyping of plant disease symptoms. Front. Plant Sci. **5**, 734 (2015)
8. Nutter, F.W., Esker, P.D., Netto, R.A.C.: Disease assessment concepts and the advancements made in improving the accuracy and precision of plant disease data. Eur. J. Plant Pathol. **115**, 95–103 (2006)
9. Munyaneza, J.E., Crosslin, J.M., Buchman, J.L., Sengoda, V.G.: Susceptibility of different potato plant growth stages to purple top disease. Am. J. Potato Res. **87**, 60–66 (2010)
10. Díaz-Pendón, J.A., Cañizares, M.C., Moriones, E., Bejarano, E.R., Czosnek, H., Navas-Castillo, J.: Tomato yellow leaf curl viruses: Ménage à trois between the virus complex, the plant and the whitefly vector. Mol. Plant Pathol. **11**, 441–450 (2010)
11. Shah, J.P., Prajapati, H.B., Dabhi, V.K.: A survey on detection and classification of rice plant diseases. In: Proceedings of the 2016 IEEE International Conference on Current Trends in Advanced Computing (ICCTAC), pp. 1–8. Bangalore, India (2016)
12. Prajapati, B.S., Dabhi, V.K., Prajapati, H.B.: A survey on detection and classification of cotton leaf diseases. In: Proceedings of the 2016 International Conference on Electrical, Electronics, and Optimization Techniques (ICEEOT), pp. 2499–2506. Chennai, India (2016)
13. Iqbal, Z., Khan, M.A., Sharif, M., Shah, J.H., ur Rehman, M.H., Javed, K.: An automated detection and classification of citrus plant diseases using image processing techniques: a review. Comput. Electron. Agric., **153**, 12–32 (2018)
14. Kaur, S., Pandey, S., Goel, S.: Plants disease identification and classification through leaf images: a survey. Arch. Comput. Methods Eng. **26**, 507–530 (2019)
15. Lee, S.H., Chan, C.S., Wilkin, P., Remagnino, P.: Deep-plant: plant identification with convolutional neural networks. In: Proceedings of the 2015 IEEE International Conference on Image Processing (ICIP), pp. 452–456. Quebec City, QC, Canada (2015)
16. Hinton, G., Deng, L., Yu, D., Dahl, G.E., Mohamed, A., Jaitly, N., Senior, A., Vanhoucke, V., Nguyen, P., Sainath, T.N., et al.: Deep neural networks for acoustic modeling in speech recognition: the shared views of four research groups. IEEE Signal Process. Mag. **29**, 82–97 (2012)
17. Wen, T., Zhang, Z.: Deep convolution neural network and autoencoders-based unsupervised feature learning of EEG signals. IEEE Access **6**, 25399–25410 (2018)
18. Ciresan, D.C., Meier, U., Masci, J., Gambardella, L.M., Schmidhuber, J.: Flexible, high performance convolutional neural networks for image classification. In: Proceedings of the Twenty-Second International Joint Conference on Artificial Intelligence, pp. 1237–1242. Barcelona, Spain (2011)
19. Carranza-Rojas, J., Goeau, H., Bonnet, P., Mata-Montero, E., Joly, A.: Going deeper in the automated identification of Herbarium specimens. BMC Evol. Biol. **17**, 181 (2017)
20. Yang, X., Guo, T.: Machine learning in plant disease research. Eur. J. BioMed. Res. **3**, 6–9 (2017)
21. Nagaraju, M., Chawla, P.: Systematic review of deep learning techniques in plant disease detection. Int. J. Syst. Assur. Eng. Manag. **11**, 547–560 (2020)

22. Li, L., Zhang, S., Wang, B.: Plant disease detection and classification by deep learning—a review. IEEE Access **9**, 56683–56698 (2021)
23. Hughes, D., Salathé, M.: An open access repository of images on plant health to enable the development of mobile disease diagnostics. arXiv:1511.08060 (2015)
24. Ramcharan, A., Baranowski, K., McCloskey, P., Ahmed, B., Legg, J., Hughes, D.P.: Deep learning for image-based cassava disease detection. Front. Plant Sci. **8**, 1852 (2017)
25. Oyewola, D.O., Dada, E.G., Misra, S., Damaševicˇius, R.: Detecting cassava mosaic disease using a deep residual convolutional neural network with distinct block processing. PeerJ Comput. Sci. **7**, e352 (2021)
26. Hops Disease Dataset, https://www.kaggle.com/scruggzilla/hops-classification
27. Rice Disease Dataset, https://www.kaggle.com/shayanriyaz/riceleafs
28. Sethy, P.K., Barpanda, N.K., Rath, A.K., Behera, S.K.: Deep feature based rice leaf disease identification using support vector machine. Comput. Electron. Agric. **175**, 105527 (2020)
29. Camargo, A., Smith, J.: Image pattern classification for the identification of disease causing agents in plants. Comput. Electron. Agric. **66**, 121–125 (2009)
30. Islam, M., Dinh, A., Wahid, K., Bhowmik, P.: Detection of potato diseases using image segmentation and multiclass support vector machine. In: Proceedings of the 2017 IEEE 30th Canadian Conference on Electrical and Computer Engineering (CCECE), pp. 1–4. Windsor, ON, Canada (2017)
31. Chuanlei, Z., Shanwen, Z., Jucheng, Y., Yancui, S., Jia, C.: Apple leaf disease identification using genetic algorithm and correlation based feature selection method. Int. J. Agric. Biol. Eng. **10**, 74–83 (2017)
32. Anthonys, G., Wickramarachchi, N.: An image recognition system for crop disease identification of paddy fields in Sri Lanka. In: Proceedings of the 2009 International Conference on Industrial and Information Systems (ICIIS), pp. 403–407. Peradeniya, Sri Lanka (2009)
33. Yao, Q., Guan, Z., Zhou, Y., Tang, J., Hu, Y., Yang, B.: Application of support vector machine for detecting rice diseases using shape and color texture features. In: Proceedings of the International Conference on Engineering Computation 29–31, pp. 79–83. Vancouver, BC, Canada (2009)
34. Sabrol, H., Satish, K.: Tomato plant disease classification in digital images using classification tree. In: Proceedings of the 2016 International Conference on Communication and Signal Processing (ICCSP), pp. 1242–1246. Melmaruvathur, India (2016)
35. Padol, P.B., Yadav, A.A.: SVM classifier based grape leaf disease detection. In: Proceedings of the 2016 Conference on Advances in Signal Processing (CASP), pp. 175–179. Pune, India (2016)
36. Padol, P.B., Sawant, S.: Fusion classification technique used to detect downy and Powdery Mildew grape leaf diseases. In: Proceedings of the 2016 International Conference on Global Trends in Signal Processing, Information Computing and Communication (ICGTSPICC), pp. 298–301. Jalgaon, India (2016)
37. Pujari, J.D., Yakkundimath, R., Byadgi, A.S.: Image processing based detection of fungal diseases in plants. Procedia Comput. Sci. **46**, 1802–1808 (2015)
38. Prasad, S., Peddoju, S.K., Ghosh, D.: Multi-resolution mobile vision system for plant leaf disease diagnosis. Signal Image Video Process **10**, 379–388 (2016)
39. Kaur, S., Pandey, S., Goel, S.: Semi-automatic leaf disease detection and classification system for soybean culture. IET Image Process **12**, 1038–1048 (2018)
40. Mohanty, S.P., Hughes, D.P., Salathé, M.: Using deep learning for image-based plant disease detection. Front. Plant Sci. **7**, 1419 (2016)
41. Fuentes, A., Yoon, S., Kim, S., Park, D.: A robust deep-learning-based detector for real-time tomato plant diseases and pests recognition. Sensors **17**, 2022 (2017)
42. Chen, J., Chen, J., Zhang, D., Sun, Y., Nanehkaran, Y.A.: Using deep transfer learning for image-based plant disease identification. Comput. Electron. Agric. **173**, 105393 (2020)
43. Farjon, G., Krikeb, O., Hillel, A.B., Alchanatis, V.: Detection and counting of flowers on apple trees for better chemical thinning decisions. Precis. Agric. **21**, 503–521 (2019)

44. Bresilla, K., Perulli, G.D., Boini, A., Morandi, B., Corelli Grappadelli, L., Manfrini, L.: Single-shot convolution neural networks for real-time fruit detection within the tree. Front. Plant Sci. **10**, 611 (2019)
45. Trong, V.H., Gwang-hyun, Y., Vu, D.T., Jin-young, K.: Late fusion of multimodal deep neural networks for weeds classification. Comput. Electron. Agric. **175**, 105506 (2020)
46. Kawasaki, Y., Uga, H., Kagiwada, S., Iyatomi, H.: Basic study of automated diagnosis of viral plant diseases using convolutional neural networks. In International Symposium on Visual Computing, pp. 638–645. Springer, Berlin (2015)
47. Fujita, E., Kawasaki, Y., Uga, H., Kagiwada, S., Iyatomi, H.: Basic investigation on a robust and practical plant diagnostic system. In: Proceedings of the 2016 15th IEEE International Conference on Machine Learning and Applications (ICMLA), pp. 989–992. Anaheim, CA, USA (2016)
48. Simonyan, K., Zisserman, A.: Very deep convolutional networks for large-scale image recognition. arXiv:1409.1556 (2014)
49. Jia, Y., Shelhamer, E., Donahue, J., Karayev, S., Long, J., Girshick, R., Guadarrama, S., Darrell, T.: Caffe: convolutional architecture for fast feature embedding. In: Proceedings of the 22nd ACM International Conference on Multimedia, pp. 675–678. Orlando, FL, USA (2014)
50. Sladojevic, S., Arsenovic, M., Anderla, A., Culibrk, D., Stefanovic, D.: Deep neural networks based recognition of plant diseases by leaf image classification. Comput. Intell. Neurosci. **2016**, 3289801 (2016)
51. Rangarajan, A.K., Purushothaman, R., Ramesh, A.: Tomato crop disease classification using pre-trained deep learning algorithm. Procedia Comput. Sci. **133**, 1040–1047 (2018)
52. Nachtigall, L.G., Araujo, R.M., Nachtigall, G.R.: Classification of apple tree disorders using convolutional neural networks. In: Proceedings of the 2016 IEEE 28th International Conference on Tools with Artificial Intelligence (ICTAI), pp. 472–476. San Jose, CA, USA (2016)
53. Krizhevsky, A., Sutskever, I., Hinton, G.E.: Imagenet classification with deep convolutional neural networks. Adv. Neural Inf. Process. Syst. **25**, 1097–1105 (2012)
54. Hara, Y., Atkins, R.G., Yueh, S.H., Shin, R.T., Kong, J.A.: Application of neural networks to radar image classification. IEEE Trans. Geosci. Remote Sens. **32**, 100–109 (1994)
55. Amara, J., Bouaziz, B., Algergawy, A.: A deep learning-based approach for banana leaf diseases classification. In: Proceedings of the Datenbanksysteme für Business, Technologie und Web (BTW 2017)-Workshopband. Stuttgart, Germany (2017)
56. LeCun, Y., Boser, B., Denker, J.S., Henderson, D., Howard, R.E., Hubbard, W., Jackel, L.D.: Backpropagation applied to handwritten zip code recognition. Neural Comput. **1**, 541–551 (1989)
57. Jadhav, S.B., Udupi, V.R., Patil, S.B.: Identification of plant diseases using convolutional neural networks. Int. J. Inf. Technol. **13**, 2461–2470 (2020)
58. Hu, M.K.: Visual pattern recognition by moment invariants. IRE Trans. Inf. Theory **8**, 179–187 (1962)
59. Zhenjiang, M.: Zernike moment-based image shape analysis and its application. Pattern Recognit. Lett. **21**, 169–177 (2000)
60. Atole, R.R., Park, D.: A multiclass deep convolutional neural network classifier for detection of common rice plant anomalies. Int. J. Adv. Comput. Sci. Appl. **9**, 67–70 (2018)
61. Rangarajan, A.K., Purushothaman, R.: Disease classification in eggplant using pre-trained VGG16 and MSVM. Sci. Rep. **10**, 1–11 (2020)
62. Rangarajan Aravind, K., Raja, P.: Automated disease classification in (Selected) agricultural crops using transfer learning. Automatika Casopis za Automatiku Mjerenje Elektroniku Racˇunarstvo i Komunikacijeˇ **61**, 260–272 (2020)
63. Lee, S.H., Goëau, H., Bonnet, P., Joly, A.: New perspectives on plant disease characterization based on deep learning. Comput. Electron. Agric. **170**, 105220 (2020)
64. Johannes, A., Picon, A., Alvarez-Gila, A., Echazarra, J., Rodriguez-Vaamonde, S., Navajas, A.D., Ortiz-Barredo, A.: Automatic plant disease diagnosis using mobile capture devices, applied on a wheat use case. Comput. Electron. Agric. **2017**(138), 200–209 (2017)

65. Picon, A., Alvarez-Gila, A., Seitz, M., Ortiz-Barredo, A., Echazarra, J., Johannes, A.: Deep convolutional neural networks for mobile capture device-based crop disease classification in the wild. Comput. Electron. Agric. **161**, 280–290 (2019)
66. He, K., Zhang, X., Ren, S., Sun, J.: Deep residual learning for image recognition. In: Proceedings of the IEEE Conference on Computer Vision and Pattern Recognition, pp. 770–778. Las Vegas, NV, USA (2016)
67. Ahmad, I., Hamid, M., Yousaf, S., Shah, S.T., Ahmad, M.O.: Optimizing pretrained convolutional neural networks for tomato leaf disease detection. Complexity **2020**, 8812019 (2020)
68. Elhassouny, A., Smarandache, F.: Smart mobile application to recognize tomato leaf diseases using convolutional neural networks. In: Proceedings of the 2019 International Conference of Computer Science and Renewable Energies (ICCSRE), pp. 1–4. Agadir, Morocco (2019)
69. Li, K., Lin, J., Liu, J., Zhao, Y.: Using deep learning for image-based different degrees of ginkgo leaf disease classification. Information **11**, 95 (2020)
70. Ai, Y., Sun, C., Tie, J., Cai, X.: Research on recognition model of crop diseases and insect pests based on deep learning in harsh environments. IEEE Access **8**, 171686–171693 (2020)
71. DeChant, C., Wiesner-Hanks, T., Chen, S., Stewart, E.L., Yosinski, J., Gore, M.A., Nelson, R.J., Lipson, H.: Automated identification of northern leaf blight-infected maize plants from field imagery using deep learning. Phytopathology **107**, 1426–1432 (2017)
72. Ozguven, M.M., Adem, K.: Automatic detection and classification of leaf spot disease in sugar beet using deep learning algorithms. Phys. A Stat. Mech. Its Appl. **535**, 122537 (2019)
73. Oppenheim, D., Shani, G.: Potato disease classification using convolution neural networks. Adv. Anim. Biosci. **8**, 244 (2017)
74. Geetharamani, G., Pandian, A.: Identification of plant leaf diseases using a nine-layer deep convolutional neural network. Comput. Electr. Eng. **76**, 323–338 (2019)
75. Wang, G., Sun, Y., Wang, J.: Automatic image-based plant disease severity estimation using deep learning. Comput. Intell. Neurosci. **2017**, 2917536 (2017)
76. Deng, J., Dong, W., Socher, R., Li, L.J., Li, K., Fei-Fei, L.: Imagenet: a large-scale hierarchical image database. In: Proceedings of the 2009 IEEE Conference on Computer Vision and Pattern Recognition, pp. 248–255. Miami, FL, USA (2009)
77. Zhang, X., Qiao, Y., Meng, F., Fan, C., Zhang, M.: Identification of maize leaf diseases using improved deep convolutional neural networks. IEEE Access **6**, 30370–30377 (2018)
78. Salimans, T., Goodfellow, I., Zaremba, W., Cheung, V., Radford, A., Chen, X.: Improved techniques for training GANs. arXiv:1606.03498 (2016)
79. Liu, B., Zhang, Y., He, D., Li, Y.: Identification of apple leaf diseases based on deep convolutional neural networks. Symmetry **10**, 11 (2018)
80. Ramcharan, A., McCloskey, P., Baranowski, K., Mbilinyi, N., Mrisho, L., Ndalahwa, M., Legg, J., Hughes, D.P.: A mobile-based deep learning model for cassava disease diagnosis. Front. Plant Sci. **10**, 272 (2019)
81. Lin, T.Y., Maire, M., Belongie, S., Hays, J., Perona, P., Ramanan, D., Dollár, P., Zitnick, C.L.: Microsoft coco: common objects in context. In: European Conference on Computer Vision, pp. 740–755. Springer, Berlin (2014)
82. Toda, Y., Okura, F.: How convolutional neural networks diagnose plant disease. Plant Phenomics **2019**, 9237136 (2019)
83. Lu, Y., Yi, S., Zeng, N., Liu, Y., Zhang, Y.: Identification of rice diseases using deep convolutional neural networks. Neurocomputing **267**, 378–384 (2017)
84. Durmuş, H., Güneş, E.O., Kırcı, M.: Disease detection on the leaves of the tomato plants by using deep learning. In: Proceedings of the 2017 6th International Conference on Agro-Geoinformatics, pp. 1–5. Fairfax VA, USA (2017)
85. Iandola, F.N., Han, S., Moskewicz, M.W., Ashraf, K., Dally, W.J., Keutzer, K.: SqueezeNet: AlexNet-level accuracy with 50x fewer parameters and <0.5 MB model size. arXiv:1602.07360 (2016)
86. Hu, G., Yang, X., Zhang, Y., Wan, M.: Identification of tea leaf diseases by using an improved deep convolutional neural network. Sustain. Comput. Inform. Syst. **24**, 100353 (2019)

87. LeCun, Y., Bottou, L., Bengio, Y., Haffner, P.: Gradient-based learning applied to document recognition. Proc. IEEE **86**, 2278–2324 (1998)
88. Bi, C., Wang, J., Duan, Y., Fu, B., Kang, J.R., Shi, Y.: Mobilenet based apple leaf diseases identification. Mob. Netw. Appl. **27**, 172–180 (2020)
89. Atila, Ü., Uçar, M., Akyol, K., Uçar, E.: Plant leaf disease classification using EfficientNet deep learning model. Ecol. Inform **61**, 101182 (2021)
90. Tuncer, A.: Cost-optimized hybrid convolutional neural networks for detection of plant leaf diseases. J. Ambient. Intell. Humaniz. Comput. **12**, 8625–8636 (2021)
91. Chohan, M., Khan, A., Chohan, R., Katpar, S.H., Mahar, M.S.: Plant disease detection using deep learning. Int. J. Recent Technol. Eng. **9**, 909–914 (2020)
92. Szegedy, C., Liu, W., Jia, Y., Sermanet, P., Reed, S., Anguelov, D., Erhan, D., Vanhoucke, V., Rabinovich, A.: Going deeper with convolutions. In: Proceedings of the IEEE Conference on Computer Vision and Pattern Recognition, pp. 1–9. Boston, MA, USA (2015)
93. Arora, J., Agrawal, U., Sharma, P.: Classification of maize leaf diseases from healthy leaves using deep forest. J. Artif. Intell. Syst. **2**, 14–26 (2020)
94. Sibiya, M., Sumbwanyambe, M.: A computational procedure for the recognition and classification of maize leaf diseases out of healthy leaves using convolutional neural networks. AgriEngineering **1**, 119–131 (2019)
95. Singh, U.P., Chouhan, S.S., Jain, S., Jain, S.: Multilayer convolution neural network for the classification of mango leaves infected by anthracnose disease. IEEE Access **7**, 43721–43729 (2019)
96. Türkoglu, M., Hanbay, D.: Plant disease and pest detection using deep learning-based features. Turk. J. Electr. Eng. Comput. Sci. **27**, 1636–1651 (2019)
97. Adedoja, A., Owolawi, P.A., Mapayi, T.: Deep learning based on nasnet for plant disease recognition using leave images. In: Proceedings of the 2019 International Conference on Advances in Big Data, Computing and Data Communication Systems (icABCD), pp. 1–5. Winterton, South Africa (2019)
98. Li, Y., Nie, J., Chao, X.: Do we really need deep CNN for plant diseases identification? Comput. Electron. Agric. **178**, 105803 (2020)
99. Enkvetchakul, P., Surinta, O.: Effective data augmentation and training techniques for improving deep learning in plant leaf disease recognition. Appl. Sci. Eng. Prog **15**, 3810 (2022)
100. Singh, P., Singh, P., Farooq, U., et al.: CottonLeafNet: cotton plant leaf disease detection using deep neural networks. Multimed. Tools Appl. (2023)
101. Xu, L., Cao, B., Ning, S., et al.: Peanut leaf disease identification with deep learning algorithms. Mol. Breed. **43**, 25 (2023)
102. Binnar, V., Sharma, S.: Plant leaf diseases detection using deep learning algorithms. In: Doriya, R., Soni, B., Shukla, A., Gao, XZ. (eds.) Machine Learning, Image Processing, Network Security and Data Sciences. Lecture Notes in Electrical Engineering, vol. 946. Springer, Singapore (2023)
103. Liu, Y., Song, Y., Ye, R., Zhu, S., Huang, Y., Chen, T., Zhou, J., Li, J., Li, M., Lv, C.: High-precision tomato disease detection using nanosegmenter based on transformer and lightweighting. Plants **12**, 2559 (2023)
104. Thai, H.-T., Le, K.-H., Nguyen, N.L.-T.: FormerLeaf: an efficient vision transformer for cassava leaf disease detection. Comput. Electron. Agric. **204**, 107518 (2023). ISSN 0168-1699
105. Li, X., Li, X., Zhang, S., Zhang, G., Zhang, M., Shang, H.: SLViT: shuffle-convolution-based lightweight Vision transformer for effective diagnosis of sugarcane leaf diseases. J. King Saud Univ. Comput. Inf. Sci. **35**(6), 101401 (2023). ISSN 1319-1578
106. Shrivastava, V.K., Pradhan, M.K.: Rice plant disease classification using color features: a machine learning paradigm. J. Plant Pathol. **103**, 17–26 (2021)
107. Liu, T., Chen, W., Wu, W., Sun, C., Guo, W., Zhu, X.: Detection of aphids in wheat fields using a computer vision technique. Biosyst. Eng. **141**, 82–93 (2016)
108. Cruz, A.C., Luvisi, A., De Bellis, L., Ampatzidis, Y.: X-FIDO: an effective application for detecting olive quick decline syndrome with deep learning and data fusion. Front. Plant Sci. **8**, 1741 (2017)

109. Zeng, W., Li, M.: Crop leaf disease recognition based on self-attention convolutional neural network. Comput. Electron. Agric. **172**, 105341 (2020)
110. Upadhyay, S.K., Kumar, A.: A novel approach for rice plant diseases classification with deep convolutional neural network. Int. J. Inf. Technol. **14**, 185–199 (2022)
111. Dawod, R.G., Dobre, C.: Upper and lower leaf side detection with machine learning methods. Sensors **22**, 2696 (2022)

Mango Maturity Classification Based on LAB Color Features Using Machine Learning

Abiban Kumari and Jaswinder Singh

Abstract Mango is an important horticulture crop with high economic value. The manual classification and postharvest handling cause high postharvest losses. The classification of mango using emerging machine learning techniques offers an advanced solution to reduce postharvest losses. The fruits were classified into three different categories; raw, intermediate, and ripe, using L*, a*, and b* (LAB) color characteristics. The non-invasive classification was performed using Support Vector Machine (SVM), Decision Tree (DT), Random Forest (RF), and K-Nearest Neighbour (K-NN). The K-fold cross-validation (CV) was used to assess the model's performance. Further, the precision, recall, Root Mean Square Error (RMSE), F1 measure, and confusion matrices were used to assess model performance. The SVM resampling using 10-fold cross-validation and the tuning parameter "sigma = 1.577 and C = 1.0" showed the highest accuracy of 98.67% among all four algorithms used across the cross-validation set. Thus, suggesting its efficient potential in classifying mangoes with varying maturity levels.

Keywords Artificial intelligence · Machine learning · Fruit classification · Mango · LAB color

1 Introduction

Mango (*Mangifera indica* L.) is a commercially important and widely consumed fruit globally. It is considered a king of fruits, grown in tropical and sub-tropical climates. India is the leading producer of mango with 10.99 million tonnes of annual production in a 1.23-million-hectare cultivation area, with >55% of the total global mango production [1]. As climacteric in nature, the rapid ripening of the fruit continues even after harvesting. During the ripening process, mangoes experience multiple physiochemical changes which make fruit highly susceptible to perishability, and

A. Kumari (✉) · J. Singh
Department of Computer Science and Engineering, Guru Jambheshwar University of Science and Technology, Hisar, India
e-mail: abibanlangaya@gmail.com

B. Soni et al. (eds.), *Beyond Artificial Intelligence*, Lecture Notes in Networks and Systems 1326, https://doi.org/10.1007/978-981-96-4170-3_12

reduce postharvest shelf life, thus ultimately affecting the fruit storage, transportation, and export. Thus there is huge losses take place after harvesting in the supply chain [2–5]. This is attributed mainly due to the poor grading, sorting, and shelf-life prediction after harvesting. The major challenges which affect mango trades are the high perishability, short postharvest shelf-life [5], high susceptibility to chilling injury, high post-harvest losses and consumer demand for high-quality fruit [3].

The efficient prediction of a postharvest shelf of the fruit will ultimately help to reduce postharvest losses and effective management during the supply chain. The determination of the fruit ripening stage is very important, as the ripening stages of the fruit are directly related to metabolic activity. The color of the fruit is the prime quality attribute which can be used for the classification of climacteric fruit. Because, during ripening the degradation of chlorophyll takes place and synthesis of the anthocyanin and carotenoids takes place which is responsible for yellow [3] and red color [6]. Thus, the color of the fruits helps to determine the storage time before consumption. Therefore, the application of non-destructive classification of the fruit based on their color using emerging techniques helps in the reduction of postharvest losses [4, 7–9].

Color is considered an important quality attribute and can be correlated with the fruit quality attributes. The color attributes to predict fruit quality were explored very limited, thus opening the opportunity among researchers. In industries, an online imaging system is used to monitor and control fruit quality [10]. However, only limited studies are available on color-based transformation and modelling kinetics. To improve the process, engineers required quantitative models. Therefore, in the present investigation color attributes were used as a rapid and nondestructive substitute for precise quality index prediction. Previous studies showed the color based maturity classification in different fruits using Artificial Neural Networks (ANN); Fuji apple maturity classification using SVM and Linear Discriminant Analysis (LDA); classification of fruit CNN-RNN-Long-Short Term Memory (LSTM), mango plantation mapping using SVM and Random Forest (RF), counting and classification of pomegranate using SVM and K-Nearest Neighbour (KNN) and blueberry fruit maturity classification using KNN [11–17]. Further, different types of algorithms have been trained for the classification and shelf life prediction of fruit such as SVM, DT, KNN, and Artificial Neural Networks (ANN) etc. [18–20]. In most of the previous studies, the images of the fruits have been used for the classification of fruits. The developed model showed a reduced classification accuracy in the range of 73–94% [21–23].

It is generally accepted that color characteristics are first-order spatial statistics that are dependent on the values of individual pixels [24]. Earlier studies have reported the use of high-quality images in order to achieve a more accurate classification of fruit. Furthermore, the major drawback of earlier studies is that the processing of high-resolution images, such as those produced by hyperspectral and multispectral imaging systems, requires a significantly high amount of computational time and results in a significantly high cost of investment. In addition to this, the Red, Green and Blue (RGB) color space is device dependent color space which means the RGB values depend upon the instrument. The different instrument produces different RGB values, and thus does not represent the color perceived by human eyes [24, 25].

The main novelty in the present study is the utilization of the color spaces values (L*, a* and b*) for fruit quality prediction. Because, the LAB color parameters are less affected by the sensor variation in comparison to the RGB. Thus, represents the actual color as perceived by the human eyes. To the best of our knowledge, the classification of mango was not performed in previous studies using the LAB color spaces. Therefore, in the present study LAB color data was directly used for fruit classification. The study focused on the classification of mango fruits into three different maturity classes based on their color characteristics using machine learning.

2 Material and Methods

2.1 Material

The mango (Chausa) was used in the present study for quality classification. The freshly harvested mango was obtained from the local market, Hisar, Haryana. The fruits were washed and used for non-destructive analysis.

2.2 Data Acquisition

The color is the first sign of the maturity level of the fruits (Fig. 1). Color features of the mango fruits such as L* (lightness), a* (red-green) and b* (yellow-blue) were extracted by colorimeter (Hunter Lab colorimeter, Color Flex EZ, USA) [26, 27]. The measurement was taken from the different sides of the mango by rotating the fruit to acquire uniform color measurements.

Fig. 1 Images of the mango at different stages (raw, intermediate and ripe mango)

Table 1 Color characteristics of mango

Color characteristics	Minimum value	Maximum value
L*	38.5	71.4
a*	−9.3	12.6
b*	28.5	57.1

L*: Black to white (0 to 100)
a*: red to green (+120 to −120)
b*: yellow to blue (+120 to −120)

A total of 1500 instances (for L*, a*, and b*) of the color spaces dataset were obtained from mangoes. An intelligent model was built for the classification of the maturity level of mango based on the color feature values [28]. The maximum and minimum values of the mangoes color attributes (LAB) are presented in Table 1.

3 Machine Learning Procedures

The selection of a classifier is a critical step because the same set of features can give diverse results. In the present study supervised learning was used to predict fruit grade. The four different classifiers (SVM, KNN, RF, and DT) were used, and their performance was compared.

Table 2 shows the generalized process for ML algorithms used in this study. The mango dataset was split into 0.70 and 0.30 for both the training phase and testing phase, respectively. The k-times cross-validation was used for the model verification. The training data was divided into K partitions and the model was verified K times. Specifically, a 10-fold cross-validation was conducted, and the average performance was computed. Subsequently, the model's performance was assessed using the testing dataset. These evaluation and validation processes aid in identifying potential overfitting of the model. There are two evaluation metrics were employed: classification accuracy and confusion matrix metrics including precision, recall, and F-score.

The programming languages used in the study were Python, IPython, Anaconda environments, and jupyter notebook editor terminal. Further, the Python modules were used for the deep network training process.

3.1 Support Vector Machine (SVM)

SVMs are based on the principle that two classes can be linearly separated. The hyperplane categorizes all data points into two groups. In addition to being useful for classifying easily separable data, SVM classifiers can also be applied to the data which

Table 2 The generalized process for machine learning algorithms

 # Initialization
Import Libraries' (Pandas, Numpy, Sklearn, etc.)
Load and split the data features and predictor which is (x, y)
Perform Data Preprocessing (ex. Labelencoder)
 # Repeat training and testing of the select MLA's
for an algorithm to MLA's
Shuffle splitting of x and y datasets into train and test datasets
 # 10-times training Cross-Validation (CV) using KFold
for 1 to 10-times
train the models using MLA's
Evaluation of the trained model's performance
training and CV score of MLA's
 # Testing the model
Test the models using MLA's
Evaluation of the tested model's performance
 # Result
Classification Report, Accuracy, Precision, recall, F1 score
Confusion metric, RMSE, R^2

is not easily separable. The researchers showed the application of SVM for transforming the data space into a higher dimensional space, where classes can be separated linearly [29]. With kernel functions, the data space can be altered in an implicit manner. Following the determination stage, SVMs significantly rely on the training data samples provided to assess the necessary parameters (support vectors). The radial basis functions (RBF), polynomial, and sigmoid have been reported as kernel functions in earlier studies. The reliability of the regression model was examined by comparing many kernel functions [30].

During the SVC training process, the input kernel parameters (Sigma) and C (cost) were adjusted to accomplish the intended trade-off between training and generalization performance. The kernel parameters sigma and cost were varied within their respective ranges to determine which set of parameters generated the most accurate SVM classifier.

3.2 Random Forest (RF)

Tree-based classifiers are the basis of RF, an ensemble classification method that combines the outputs of several nodes. The algorithm outperforms a single classifier while also being more robust against bagging/boosting-based noise. In the random forest (RF) classifier, initially, a training dataset was generated by independent sampling from the original training data and thus increasing classification trees diversity [19]. Further, each pixel was assigned to a class based on the highest votes class from all the tree predictors in the forest. The missing data from the bootstrap sample was used to assess the significance of the variables and the degree of misclassification. The most common implementation of RF builds a decision tree based on the Gini contamination measure, which decreases when the split selection is performed [16].

3.3 Decision Tree (DT)

It is a nonparametric statistical approach used to construct a decision tree that specifies the dependent variable as a function of independent variables [31]. The tree was built by repeatedly slicing subsets into descendant subsets during the construction process. The single independent variable was used to define every split. To construct a tree model, the root node was assigned to all objects and optimum features were selected. The training data was divided into sub-nodes for better classification under assigned conditions.

The tree was constructed when all the sub-nodes belonged to leaf nodes, otherwise, the split process continued until all sub-nodes belonged to leaf nodes [32]. The node impurity was measured using three methods that were gini index, information ratio, and information gain. The Gini impurity for a node in a decision tree is calculated as follows:

$$Gini(node) = 1 - \sum_{i=1}^{c} (\mathrm{P(i|node)})2$$

where P(i|node) represents the probability of class i at the current node.

The success of DT is proportional to how well it is built on the training set.

3.4 K Nearest Neighbour (KNN)

KNN is also a non-parametric machine learning method used for classification and regression, where K defines the number of neighbour. The training samples were used as input and output was the categorized instances number. Generally, KNN classifies the samples using the distance between the samples and the training set. Distances between the test sample and the training data were analyzed as part of the process of developing KNN decision criteria [33].

To differentiate the test sample, the Euclidean distance was computed to determine the distances between both training and testing data. The distance metric (Euclidean distance) was used to find neighbours.

3.5 Model Performance Evaluation

In order to determine the model performance, four parameters were employed which include; precision, recall, accuracy, and F1 score. Each of these values ranges between 0 and 1. If the value of the proposed model was nearest to 1 then the model was declared as successful and robust. Further, the model's performance was measured by using the Eqs. (1–4).

$$Accuracy = \frac{TP + TN}{TP + TN + FP + FN} \tag{1}$$

Precision refers to the exactness of the classifiers, lies in [0, 1] and is computed as:

$$Precision(P) = \frac{TP}{TP + FP} \tag{2}$$

Recall indicates the completeness of a classifier, it lies in [0, 1] and is calculated as

$$Recall(R) = \frac{TP}{TP + FN} \tag{3}$$

The F1 score is the harmonic mean of precision and recall, it lies in [0, 1] and is computed as:

$$F1\ score = \frac{2.P * R}{P + R} \tag{4}$$

The four main notations were explained as; (A) The numbers of positive class predictions (True Positives—TP); (B) The numbers of negative class predictions (True Negatives—TN): (C) The numbers of false positive (False Positives—FP) and (D) the numbers of false negatives (False Negative—FN) class predictions.

The multiple parameters essential for each model were fine-tuned using tenfold cross-validation (CV) on the training set. This optimization process focused on enhancing precision, recall, and accuracy. The accuracy, recall, precision, and F1 score as well as the differences between models for fruit class classification is presented in Table 3. The coefficient of determination (R2) and root mean squared error (RMSE) were also used to determine the variances between the true and predicted values. The parameters were determined as given below:

Table 3 Overall performance of mango maturity classification

MLA's	Parameters	Classes	Precision	Recall	FI-score	Accuracy (%)
SVM	Cost = 1.0	Intermediate	0.98	0.98	0.98	98.67
	Sigma = 1.577	Raw	0.98	0.98	0.98	
		Ripe	1	1	1	
Decision tree	Depth = 4	Intermediate	0.96	0.84	0.90	92
		Raw	0.85	1	0.92	
		Ripe	0.97	0.95	0.96	
Random forest	Number of trees = 40	Intermediate	0.98	0.97	0.98	98
	Depth = 8	Raw	0.96	0.98	0.97	
		Ripe	1	1	1	
K-nearest neighbours	K(neighbours) = 2	Intermediate	0.95	0.98	0.97	97.33
		Raw	0.98	0.96	0.97	
		Ripe	1	0.97	0.99	

$$RMSE = \sqrt{\frac{1}{n}\sum_{i=1}^{n}(P_i - A_i)^2}$$

$$R^2 = 1 - \left(\frac{\sum_{i=1}^{n}(P_i - A_i)^2}{\sum_{i=1}^{n}(A_i)^2}\right)$$

where n is the total number of test data, P_i concerns the predicted values and A_i the actual values.

4 Results

The color is the first sign of the maturity level of the fruits. The result showed variation in the LAB color spaces of the mangoes. The L* values (light to dark) of the mango varied from 38.5 to 71.4. The a* values (green to red) varied from −9.3 to 12.6. The value in the negative showed the greenness in the mangoes. The b* values varied from 28.5 to 57.1. The increase in the b* value was due to the yellowness of the fruits which revealed the ripeness of the mango as shown in Fig. 1. In climacteric fruits, chlorophyll degradation and synthesis of carotenoids and anthocyanin are responsible for yellow and red color [34]. The box plot of color values (L*, a*, and b*) of the mangoes are shown in Fig. 2a, b, and c respectively.

The parameters tuning was performed for training and testing accuracy in different model (Fig. 3).

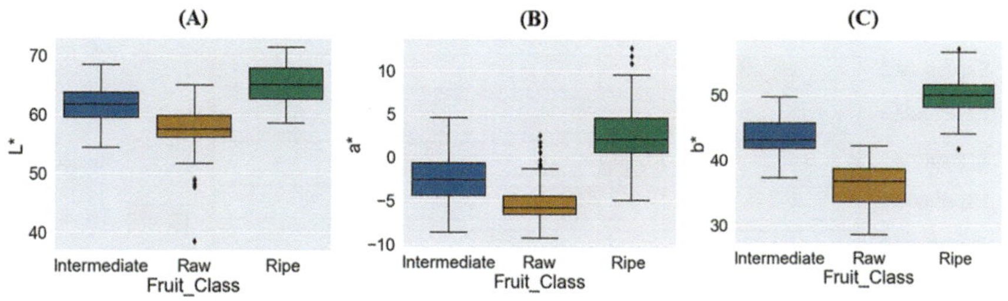

Fig. 2 Box plot of **a** L* (light to dark), **b** a* (green to red) and **c** b* (blue to yellow) value of mango

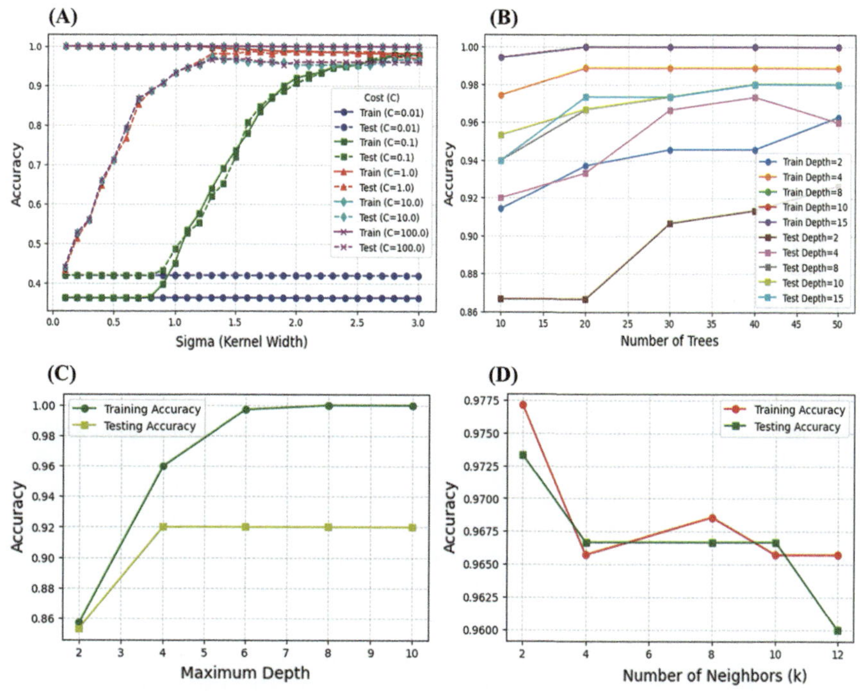

Fig. 3 **a** Parameter tuning (sigma and cost) for SVC, **b** Parameter tuning (number of trees and depth) for RF, **c** Parameter tuning (number of depth) for DT, and **d** Parameter tuning (numbers of neighbours) for KNN classifiers

The different models with multiple parameters (Cost, sigma, depth, number of trees and number of neighbours) were used to determine the training and testing accuracy for better classifier as shown in Fig. 3. Further, in random forest, the optimal model was obtained by using 40 trees and 8 depths with an accuracy of 98%. Furthermore, in modelling with DT, the optimal model is obtained by using depth = 4 with an accuracy of 92%. Lastly, in modeling with KNN, the optimal model is obtained by k = 2 with an accuracy of 97.33%

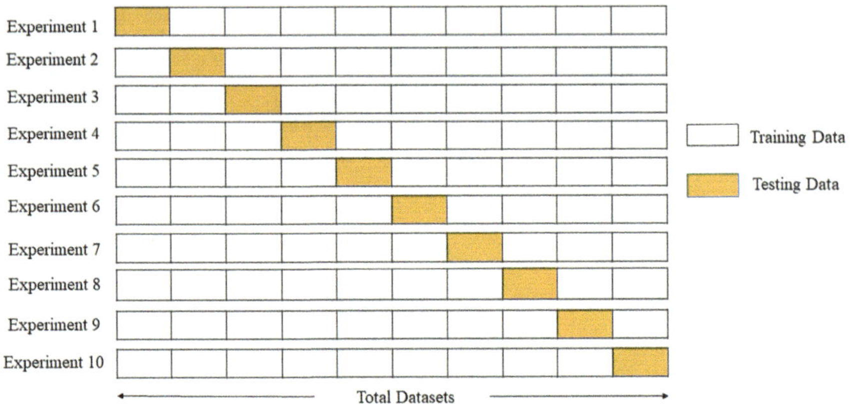

Fig. 4 Presentation of 10-fold cross validation

A supervised ML model was used to predict the maturity grade of mangoes. A dataset derived from mango was used to determine the relationship among the various variables and the model was fit to achieve the most efficient performance. To determine the most efficient model for mango classification, algorithms such as SVM, KNN, RF, and DT models were employed and compared. The classification was performed using 10-fold cross-validation Fig. 4 and average accuracy was determined.

The performance of the model was assessed using the cross-validation technique. The method selects one observed case out of all cases for data validation and uses other data to obtain the predictive capabilities. The procedure was carried out for each and every observation. Therefore, all samples were used for data validation. To calibrate and validate the model, the maximum amount of available datasets were utilized. The results obtained from different classifiers were calculated and compared with their performance. Further, the classifier training and validation scores are shown in Fig. 5.

Furthermore, the performance of the projected model was assessed by employing a range of color spaces and classification approaches. This was done to showcase the most resilient, appropriate, and effective model for the dataset. To evaluate the performance of the proposed models, the confusion matrix was employed as a foundational tool. These matrices offer insights into the predicted classes, taking into account the fruit's maturity level. The level of maturity was classified into three categories which include: Raw, Intermediate, and Ripe. The confusion matrices for the different classes of mango are presented in Fig. 6.

The highest accuracy score (>0.9) was obtained in the SVC while the DT showed the lowest accuracy score among all the algorithms used in the study.

The study showed an accuracy in the range of 92–98.67%. Among all the tested models, the SVM resampling using 10-fold cross-validation and the tuning parameter "sigma = 1.577 and C = 1.0" showed the highest accuracy of 98.67% with high average values of recall (0.98), F1 measure (0.98) and precision (0.98). On

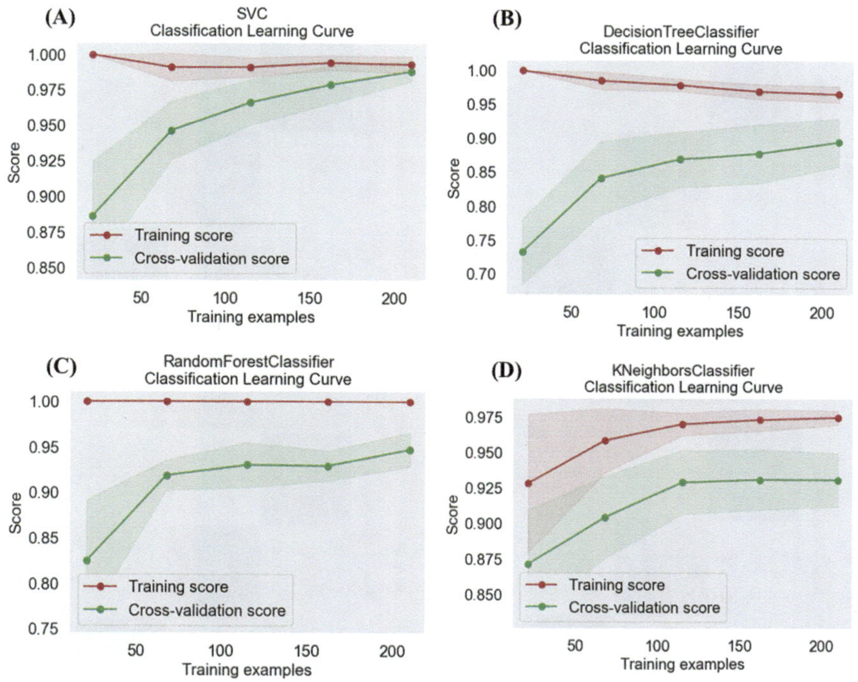

Fig. 5 The learning curve of different models (**a**-Support vector classifier, **b**-Decision Tree, **c**-Random Forest, **d**-K nearest neighbour)

the contrary, the lowest performance was obtained in the DT model with the values of recall, precision, F1 measure, and accuracy of 0.9276, 0.9227, 0.9234, and 92% respectively. Overall, the results (Table 3) suggested that the SVM model outclassed the RF, DT, and KNN models predicting the different mango classes. It was observed that the SVM classifier outperforms all the other models at an accuracy of 98.67%. Results were classified based on the three different hyperplanes such as hp1, hp2, and hp3. The kernel function was the key factor of SVM because the mangoes were categorized into different classes using hyperplanes generated from different kernels. In this study, a kernel function is a type of mathematical tool that is frequently employed in SVM, one type of ML technique performed. To make data simpler to categorize, it is used to turn it into a higher-dimensional space. The performance was evaluated using a confusion matrix that provides information on the predicted ripeness level by comparing it with the actual level. The maturity of mango was categorized into three grades: raw mango, ripe mango and medium mango. The confusion matrix showed the performance statistics of the SVM model using precision, F1 score criteria, recall and accuracy. The classification results reached an accuracy of 0.98 and a recall of 0.98 in the test data set. Thus, the developed model is declare successful and reliable if the value is nearest to 1.

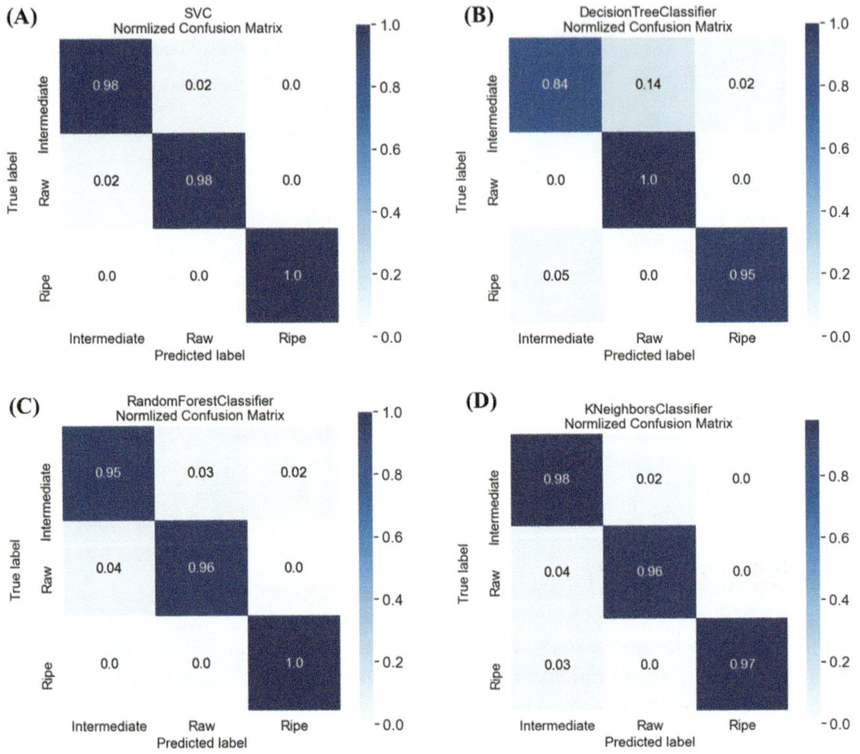

Fig. 6 Confusion matrices score of different models (**a**-Support vector classifier, **b**-Decision Tree, **c**-Random Forest, **d**-K nearest neighbour)

4.1 Discussion

The mango is an important horticulture crop and the determination of the ripeness level in fruit has a significant impact on the reduction of postharvest losses in the mango. The classification of fruit will help in the extended storage and transportation. In the present study, classification models were developed to distinguish the different levels of maturity (raw, intermediate and ripe). According to the aim of the study, the model represents the favorable performance with high accuracy (98.67%). The mode showed high accuracy as compared to the accuracy (94.4%) obtained using the random forests classification technique for the quality grading of mango. Similarly, in another study, 93.33% accuracy was obtained for mango grading using deep learning [9, 19]. The high-performance accuracy in the present study showed that the projected model can be used as an effective and efficient model as compared to the previously developed models. The R^2 value of the different models (SVC, DT, RF and KNN) was in the range of 0.748–0.979. The SVC classifier showed the highest R^2 value (0.979). The RMSE value for the SVC classifier was 0.115.

In a previous study, the neural network and computer vision-based system was used for mango maturity classification using color-based features. The developed model was effective in the classification of mango classes (ripe and unripe) with reduced (94%) accuracy [23]. Further, vision-based system was also proposed for the grading of mangoes based on the maturity level using SVM. The study showed that developed model was effective for mangoes grading with nearly 90% of performance accuracy and 96% maturity prediction accuracy [22]. The mangoes were classified according to their geographic area cultivation practice for their mineral contents. Data obtained from the atomic absorption spectrometry (AAS) was used for discriminating purposes. Nearly, 73% of classification performance accuracy was obtained using Linear discriminant analysis (LDA) and 93% of prediction ability was obtained using support vector machines (SVM) [21]. Therefore, the present study showed that the SVC classifier showed a high accuracy as compared to the previous studies.

5 Conclusion

The maturity-based classification of mangoes is very important to reduce postharvest losses during supply chain and storage. Based on the results obtained from different models, the study concluded that all the machine learning algorithms used in the study were performed satisfactorily with the given L*, a*, and b* values. The classification accuracy percentage in the different models varied in the range of 9–98.67%. Among all the ML algorithms used in the study, the SVC classifier model was found more effective in the maturity classification of mango with an accuracy of 98.67%. Since the maturity classification of mango is important to determine the storage time, the algorithm based on the color could be employed at the commercial level for the detection of maturity.

Acknowledgements This present work did not receive any specific grant from funding agencies in public, commercial and not for profit sector.

Declarations The authors declare that they have no conflict of interest in relation to this research work.

Abbreviations

ANN	Artificial Neural Network
AAS	Atomic Absorption Spectrometry
CV	Cross-Validation
DT	Decision Tree
FP	False Positives
FN	False Negative
K-NN	K-Nearest Neighbour

LAB	L a*b*
LDA	Linear Discriminant Analysis
ML	Machine Learning
MLA's	Machine Learning Algorithm's
R^2	Coefficient of determination
RBF	Radial Basis Functions
RF	Random Forest
RMSE	Root Mean Square Error
SVM	Support Vector Machine
SVC	Support Vector Classifier
TP	True Positives
TN	True Negative

References

1. Taria, S., Alam, B., Rane, J., Kumar, M., Babar, R., Singh, N.P.: Deciphering endurance capacity of mango tree (*Mangifera indica* L.) to desiccation stress using modern physiological tools. Sci. Hortic. (Amst.) **303**, 111247 (2022)
2. Le, T.D., Viet Nguyen, T., Muoi, N. Van, Toan, H.T., Lan, N.M., Pham, T.N.: Supply chain management of mango (Mangifera indica L.) fruit: a review with a focus on product quality during postharvest. Front. Sustain. Food Syst. **5**, 799431 (2022)
3. Ali, U., Kanwar, S., Yadav, K., Basu, S., Mazumder, K.: Effect of arabinoxylan and β-glucan stearic acid ester coatings on post-harvest quality of apple (Royal Delicious). Carbohydr. Polym. **209**, 338–349 (2019)
4. Ali, U., Basu, S., Mazumder, K.: Improved postharvest quality of apple (Rich Red) by composite coating based on arabinoxylan and β-glucan stearic acid ester. Int. J. Biol. Macromol. **151**, 618–627 (2020)
5. Ali, U., Maurya, R., Basu, S., Mazumder, K.: Indian apple quality characteristics as affected by controlled atmosphere storage. J. Appl. Hortic. **24**, 269–275 (2023)
6. Usman, A., Kumar, P.: Effect of soxhlet and ultrasound assisted extraction on antioxidant activity of pomegranate peel extract. Int. J. Food Nutr. Sci. **3**, 265–270 (2015)
7. Mahanti, N.K., Pandiselvam, R., Kothakota, A., Ishwarya S., P., Chakraborty, S.K., Kumar, M., Cozzolino, D.: Emerging non-destructive imaging techniques for fruit damage detection: image processing and analysis. Trends Food Sci. Technol. **120**, 418–438 (2022)
8. Ismail, N., Malik, O.A.: Real-time visual inspection system for grading fruits using computer vision and deep learning techniques. Inf. Process. Agric. **9**, 1–14 (2021)
9. Bhole, V., Kumar, A.: A transfer learning-based approach to predict the shelf life of fruit A transfer learning-based approach to predict the shelf life of fruit. Intel. Artif. **24**, 102–120 (2021)
10. Sanaeifar, A., Bakhshipour, A., De La Guardia, M.: Prediction of banana quality indices from color features using support vector regression. Talanta **148**, 54–61 (2015)
11. Piedad, E., Larada, J.I., Pojas, G.J., Ferrer, L.V.V.: Postharvest classification of banana (Musa acuminata) using tier-based machine learning. Postharvest Biol. Technol. **145**, 93–100 (2018)
12. Septiarini, A., Sunyoto, A., Hamdani, H., Kasim, A.A., Utaminingrum, F., Hatta, H.R.: Machine vision for the maturity classification of oil palm fresh fruit bunches based on color and texture features. Sci. Hortic. (Amst.) **286**, 110245 (2021)
13. Zhang, Y., Wang, S., Ji, G., Phillips, P.: Fruit classification using computer vision and feedforward neural network. J. Food Eng. **143**, 167–177 (2014)

14. Zhang, M., Zhang, B., Li, H., Shen, M., Tian, S., Zhang, H., Ren, X., Xing, L., Zhao, J.: Determination of bagged 'Fuji' apple maturity by visible and near-infrared spectroscopy combined with a machine learning algorithm. Infrared Phys. Technol. **111**, 103529 (2020)
15. Gill, H.S., Khehra, B.S.: An integrated approach using CNN-RNN-LSTM for classification of fruit images. Mater. Today Proc. **51**, 591–595 (2021)
16. Luo, H.X., Dai, S.P., Li, M.F., Liu, E.P., Zheng, Q., Hu, Y.Y., Yi, X.P.: Comparison of machine learning algorithms for mapping mango plantations based on Gaofen-1 imagery. J. Integr. Agric. **19**, 2815–2828 (2020)
17. Tan, K., Lee, W.S., Gan, H., Wang, S.: Recognising blueberry fruit of different maturity using histogram oriented gradients and colour features in outdoor scenes. Biosyst. Eng. **176**, 59–72 (2018)
18. Gururaj, N., Vinod, V.: Predictive model for optimum fruit maturity grading. Int. J. Innov. Technol. Explor. Eng. **9**, 3567–3571 (2019)
19. Sun, J., Li, S., Yao, X.: A novel method for multi-feature grading of mango using machine vision. J. Comput. **31**, 65–77 (2020)
20. Mon, T.O., ZarAung, N.: Vision based volume estimation method for automatic mango grading system. Biosyst. Eng. **198**, 338–349 (2020)
21. Hernández-Sánchez, C., Luis, G., Moreno, I., Cameán, A., González, A.G., González-Weller, D., Castilla, A., Gutiérrez, A., Rubio, C., Hardisson, A.: Differentiation of mangoes (*Magnifera indica* L.) conventional and organically cultivated according to their mineral content by using support vector machines. Talanta **97**, 325–330 (2012)
22. Nandi, C.S., Tudu, B., Koley, C.: Computer vision based mango fruit grading system. Int. Conf. Innov. Eng. Technol. 1–5 (2014)
23. Yossy, E.H., Pranata, J., Wijaya, T., Hermawan, H., Budiharto, W.: Mango fruit sortation system using neural network and computer vision. In: Procedia Computer Science, pp. 596–603. Elsevier B.V. (2017)
24. Ireri, D., Belal, E., Okinda, C., Makange, N., Ji, C.: A computer vision system for defect discrimination and grading in tomatoes using machine learning and image processing. Artif. Intell. Agric. **2**, 28–37 (2019)
25. Quevedo, R.A., Aguilera, J.M., Pedreschi, F.: Color of salmon fillets by computer vision and sensory panel. Food Bioprocess Technol. **3**, 637–643 (2010)
26. Ali, U., Kaur, P., Kanwar, S., Kumar, V., Maurya, R., Bishnoi, M., Basu, S., Mazumder, K.: Application and toxicity studies of arabinoxylan and β-d-glucan stearic acid ester composite coatings in extending postharvest storage of peach. Sci. Rep. **11**, 1–11 (2021)
27. Ali, U., Bijalwan, V., Mazumder, K., Basu, S., Kesarwani, A.K.: Effect of β-glucan-fatty acid esters on microstructure and physical properties of wheat straw arabinoxylan films. Carbohydr. Polym. (2016)
28. Hitesh Singh, W.: Experimental RGB and CIE lab colour space analysis and comparison for fruits and vegetables. JETIR **4**, 1–4 (2017)
29. Dhakshina Kumar, S., Esakkirajan, S., Bama, S., Keerthiveena, B.: A microcontroller based machine vision approach for tomato grading and sorting using SVM classifier. Microprocess. Microsyst. **76**, 103090 (2020)
30. Ghazal, S., Qureshi, W.S., Khan, U.S., Iqbal, J., Rashid, N., Tiwana, M.I.: Analysis of visual features and classifiers for Fruit classification problem. Comput. Electron. Agric. **187**, 106267 (2021)
31. Goel, N., Sehgal, P.: Fuzzy classification of pre-harvest tomatoes for ripeness estimation-an approach based on automatic rule learning using decision tree. Appl. Soft Comput. **36**, 45–56 (2015)
32. Yang, X., Zhang, R., Zhai, Z., Pang, Y., Jin, Z.: Machine learning for cultivar classification of apricots (Prunus armeniaca L.) based on shape features. Sci. Hortic. (Amst.) **256**, 108524 (2019)

33. Syazwani, R.W.N., Asraf, H.M., Amin, M.M.S., Dalila, K.N.: Automated image identification, detection and fruit counting of top-view pineapple crown using machine learning. Alex. Eng. J. **61**, 1265–1276 (2022)
34. Kapoor, L., Simkin, A.J., Doss, C.G.P., Siva, R.: Fruit ripening: dynamics and integrated analysis of carotenoids and anthocyanins. BMC Plant Biol. 1–22 (2022)

Smart Identification of COVID-19 Patients Among Patients with Other Viral Diseases Based on Analysis of Common Symptoms

Aditika Tungal, Kuldeep Singh, and Prabhsimran Singh

Abstract A dramatic increase in the number of instances of any disease invariably creates concern among common people. COVID-19 is a recent outbreak that has troubled the world. Due to its high fatality and dissemination rates, people were terrified of getting infected with it. This covid phobia has prompted individuals to misdiagnose themselves due to symptom similarities with other popular diseases viz., common cold, flu, and allergy. Thus, this study introduces a machine learning-driven approach to discern whether an individual presenting as a suspect is genuinely infected with COVID-19 or possibly affected by other common diseases. On evaluating multiple machine learning models, the Naive Bayes (NB) model exhibited superior performance with an accuracy of 95.41% for precise identification. As a result, this research can be used to differentiate between COVID-19 and other common diseases that have similar symptoms. Future studies can widen the scope to include more diseases having COVID-19-like symptoms.

Keywords COVID-19 · Machine learning · Pandemic · Smart Healthcare · Symptoms

A. Tungal · P. Singh (✉)
Department of Computer Engineering and Technology, Guru Nanak Dev University, Amritsar, India
e-mail: prabhsimran.dcet@gndu.ac.in; prabh_singh32@yahoo.com

A. Tungal
e-mail: aditikacet.rsh@gndu.ac.in

K. Singh
Department of Electronics Technology, Guru Nanak Dev University, Amritsar, India
e-mail: kuldeep.ece@gndu.ac.in

B. Soni et al. (eds.), *Beyond Artificial Intelligence*, Lecture Notes in Networks and Systems 1326, https://doi.org/10.1007/978-981-96-4170-3_13

1 Introduction

COVID-19 is the recent global public health emergency to emerge and trap the 7.8 billion people worldwide. It first appeared towards the end of December 2019 in the Chinese city of Wuhan and quickly decimated the entire world [1]. This latest COVID-19 outbreak generated by the SARS-CoV-2 virus has frenzied the sphere into disarray due to its frighteningly high transmission rate [2, 3]. In response to the prevailing circumstances, the World Health Organization (WHO) was forced to designate it as a Global Pandemic on the 12th of March, 2020 [4]. As the number of COVID-19 instances skyrocketed, governments began to manage the infection by imposing lockdowns and emergencies to minimize COVID-19 transmission by authorizing social distancing measures [5].

During the initial phase of this disease, everyone was unaware of the impending catastrophic outbreak that would kill and shatter people's physical, emotional, and mental health. Countries all over the world were forced to suffer financial losses, which led to a severe global recession [6]. The brutal reality and worst impacts of COVID-19 have so far been seen by the world in three waves, each of which saw a different variation in the disease's symptoms [7]. While everyone was trying to confront the pandemic and adjust to the "new normal pandemic life," several kinds of COVID-19 variants emerged that affected individuals all over the globe. This whole scenario created a fear of terror and a feeling of "coronaphobia" among the masses as they were not sure whether they had been infected with COVID-19 or some variant of it or had a common cold or flu-like disease. There were such minute differences among all these and mostly the symptoms were the same for all, so the fear of getting infected has raised the stress levels among the masses.

COVID-19 symptoms include a high fever, coughing, sleepiness, breathing problems, loss of smell or taste, and phlegm [8, 9]. In addition to these, patients infected with COVID-19 may develop a variety of other clinical symptoms (such as diarrhea, pink eyes/conjunctivitis, muscle discomfort, and so on) [10]. People all across the world were so terrified of contracting COVID-19 that even if they had a typical cold and cough, they assumed that COVID-19 might be the reason for it. This whole scenario created panic among the citizens across nations and created lots of myths and misconceptions among the general public. This horrible disease even entirely disrupted the current public healthcare systems, prompting doctors to treat severely or critically ill patients at the risk of their own lives.

Although RT PCR (aka Real-Time Reverse Transcription Polymerase Chain Reaction) or Rapid Antigen tests can be performed to confirm COVID-19 instances, [11]. The population of the twenty-first century has not recently experienced such a fatal disease, so those who were suspected of having it barricaded themselves in their homes out of fear to avoid being sent to government isolation wards, etc. Many patients' medical conditions worsened as a result of this; some of them recovered, while others were unable to return home. This was the worst aspect of this "life-taking" pandemic. Therefore, it is essential to establish a system so that individuals can determine whether they have a typical cold, some flu, a form of allergy, or are COVID-19 positive so that they can be treated accordingly without wastage of time.

Nevertheless, the recent pandemic-prone disease called COVID-19 has utterly traumatized the entire world, and even if someone had a regular cough or fever, they feared they were gripped with COVID-19. So, in order to correctly comprehend the condition based on the symptoms of the patients, it is critical to differentiate the diseases so that patients can receive appropriate treatment. In regard to this, this study aimed to establish a machine learning-based automated system that can determine and readily identify whether a person is infected with COVID-19, a cold, flu, or any type of allergy based on their symptoms. Among different machine learning (ML) models, the Naive Bayes classification model performed the best, with an overall accuracy of 95.41%, followed by Random Forest (95.32%), Stochastic Gradient Descent (93.63%), K-nearest neighbor (89.64%), and Linear Discriminant Analysis (87.26%). As a result, the proposed method can be utilized to readily identify the patient's disease among the cold, COVID-19, flu, or allergy symptoms.

The following sections comprise the current work: Sect. 1 describes an overview of the current situation with the novel pandemic-prone disease and how it has caused widespread anguish. In addition, the existing work done in the field of COVID-19 identification using symptom analysis has been illustrated in Sect. 2. Similarly, Sect. 3 addresses the suggested machine learning-based approach for accurately identifying 'viral' diseases among allergies, cold, covid, and flu and explains the step-by-step framework of the proposed methodology. Furthermore, Sect. 4 summarizes and examines the acquired results, while Sect. 5 incorporates the concluding remarks.

2 Related Work

With the passage of time, there has been significant progress in the medical field. High-tech gadgets and tools have fundamentally revolutionized the treatment process, making it more advanced and less time-consuming. Many researchers have worked tirelessly to provide methods for eradicating the COVID-19 pandemic using cutting-edge ML approaches [12]. In the same regard, Wubineh et al. [13] developed an expert system for diagnosing COVID-19 individuals by assessing their symptoms in order to detect the disease early. The designed system attained an accuracy of 92.47% when cross-validation techniques were used to identify patients based on their symptoms. Furthermore, Khanday et al. [14] used feature engineering measures on ensemble ML-based classifiers for the detection of clinically based COVID-19 data. Several machine learning algorithms produced encouraging results for accurately identifying the disease. Similarly, Otoom et al. [15] created a monitoring mechanism to detect COVID-19 in a prompt manner. Symptom-based data was collected from patients and various ML techniques were used. In diagnosing SARS-Cov-2 cases, this study achieved an accuracy of more than 90%.

In the same regard, Solayman et al. [16] employed various ML methods on the Israeli dataset, finding that the Hybrid Convolution Neural Network- Long Short-Term Memory (CNN-LSTM) yielded the highest accuracy at 96.34%. Correspondingly, Malik et al. [17], utilized the same dataset, determining COVID-19 patients with Naïve Bayes and Decision Tree models achieving an accuracy of 93.70%. Arowolo et al. [18] investigated the San Francisco COVID-19 dataset, employing an ML-based IoT framework, with ABC-Q-SVM techniques recording the highest accuracy of 97% across COVID-19, viral, and non-viral categories. Additionally, Podder et al. [19] utilized multiple classification algorithms, revealing XGBoost's accuracy of 92.67% in COVID-19 patient detection. Likewise, Chadaga et al. [20] collected data from two Manipal hospitals, employing preprocessing techniques followed by a multi-level stacked model that showcased a remarkable accuracy of 96%. Table 1 outlines cutting-edge research endeavors employing diverse machine learning and deep learning techniques in identifying COVID-19 patients.

3 Proposed Methodology

In the current scenario, computational technologies and high-tech advanced solutions have been widely used to tackle the various challenges posed by the 'viral' diseases in today's world. With the development of AI and IoT-based methodologies for implementing Industry 4.0 in the healthcare domain, the landscape of healthcare systems has undergone a complete revolution [21, 22]. There has been a drastic change in the current public healthcare systems because of these cutting-edge advanced solutions. Figure 1 depicts the framework of the suggested strategy, in which data is initially acquired from an open public source and a data preparation technique in the form of label encoding is implemented. Following that, specific machine learning approaches were utilized to assess the accuracy of the built model for reliably recognizing 'popular' diseases such as allergy, cold, covid, and flu.

3.1 Data Collection

The open-source dataset used in this study was obtained from the **Covid, Flu, Cold Symptoms Dataset** [23], which is a symptom-based dataset with four output classes: Allergy, Cold, Covid, and Flu. It is an extensive data set with approximately 44,453 entries identifying various symptoms as 0 or 1, with 0 representing the absence of those symptoms and 1 signifying the presence of those symptoms. Cough, muscular pain, dizziness, sore throat, a runny nose, stuffy nose, a high fever, vomiting, nausea, diarrhea, shortness of breath, difficulty in breathing, loss of appetite, loss of sensation of smell, itchy nose, itchy eyes, itchy mouth, itchy inner ear, sneezing, and redness in the eyes are common symptoms in the studied dataset [24].

Table 1 State-of-the-Art literature review on COVID-19 detection using machine learning and deep learning techniques

Author, year	No. of input features and output classes	Methodology	Classifier	Accuracy (%)
Wubineh et al. [13], (2023)	Input: 8 features Output class: COVID-19 positive/ negative	Use of acquired symptom data for developing the rule-based system	Part-time rule-based algorithm	92.47
Khanday et al. [14], (2023)	Input: 24 features Output class: COVID, ARDS, SARS, Both (COVID, ARDS)	Term frequency/inverse document frequency and bag of words-based features	Logistic regression and Naive Bayes	96.2
Otoom et al. [15], (2020)	Input: 20 features Output class: COVID-19 positive/ negative	Selection of appropriate features using Spectral score, Information score, Pearson correlation, Intra-class distance, Interquartile range, and Variance feature weighting techniques	Support vector machine and decision table	92.95
Solayman et al. [16], (2023)	Input: 8 features Output class: COVID-19 positive/ negative	Data processing by dropping null values, feature engineering, and synthetic oversampling (SMOTE) for data balancing	Hybrid CNN-LSTM	96.34
Malik et al. [17], (2022)	Input: 10 features Output class: COVID-19 positive/ negative	Handling of missing values and Correlation-based feature selection	Naive Bayes and decision tree	93.70
Arowolo et al. [18], (2023)	Input: Multiple features Output class: COVID-19, viral, non-viral acute respiratory diseases	Artificial Bee Colony (ABC) technique-based feature selection	ABC-Q-SVM	97

(continued)

Table 1 (continued)

Author, year	No. of input features and output classes	Methodology	Classifier	Accuracy (%)
Podder et al. [19], (2021)	Input: 25 features Output class: COVID-19 positive/ negative	Data Balancing and feature selection using the univariate method	XGBoost	92.67
Chadaga et al. [20], (2023)	Input: 22 features Output class: COVID-19 positive/ negative	Handling of missing values using mean, median, and mode; data balancing; data scaling using Standardization and normalization; and feature selection using Grey Wolf optimizer	Multi-level stacked model	96

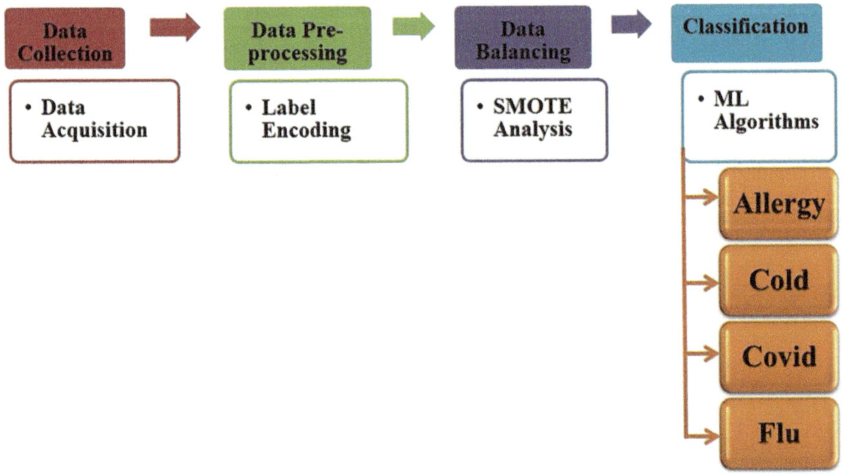

Fig. 1 Step-by-step process of the proposed methodology

3.2 Data Pre-processing

Data pre-processing is regarded as the most important stage in machine learning and data mining [25]. For this study, the four output classes were converted from categorical to integer values using the Label encoding technique, resulting in 0 for Allergy, 1 for Cold, 2 for Covid, and 3 for Flu. Label encoding makes the processing of data much easier for further processing into machine learning models [26].

3.3 Data Balancing

Furthermore, dealing with unbalanced data is a fundamental difficulty in the processing of machine learning models that must be handled to obtain accurate results. SMOTE (which stands for the Synthetic Minority Oversampling Technique) is an effective method for addressing the issue of inequalities between classes [27, 28]. It works by creating synthetic samples to improve the representation of the minority class in a dataset [29]. The original dataset had the values as Allergy: 16381, Cold: 1024, Covid: 2048, Flu: 25000, thus SMOTE was used and all columns were balanced up to 25000. This is done to improve the efficiency and effectiveness of the machine learning classifiers.

3.4 Classification

After balancing all of the data, the next critical activity is to train the machine learning models. For this study, around eighty percent of the data was used to train the model, while the other 20% was used to test the model. Various machine learning models, including the Naive Bayes algorithm (NB), the random forest model (RF), SGD (Stochastic Gradient Descent), the KNN model (K-nearest Neighbors), and LDA in particular (Linear Discriminant Analysis), were trained in the current study to accurately classify whether the infected individual has flu, allergy, cold, or covid.

4 Results and Discussions

This section provides and examines the observations from the current study, which were based on multiple ML models such as Naive Bayes (also known as NB), the random forest (RF), SGD (also referred to as Stochastic Gradient Descent), KNN (k-nearest-neighbors), and LDA (also known as Linear Discriminant Analysis). Table 2 displays the evaluation results of the various ML classifiers employed in the present study using the metrics of sensitivity, specificity, accuracy, F1 scores, and FDR.

According to the current study, the NB model beats other classification models with a maximum accuracy of 95.41%, followed by RF with an accuracy of 95.32%, and SGD models with a total accuracy of 93.63%. Following that, the KNN model has an accuracy of 89.64%, accompanied by the LDA model, which has an accuracy of 87.26%. Figure 2 graphically depicts the statistical data from Table 2.

Table 2 also shows the overall sensitivity, specificity, F1-score, and FDR values in percentage obtained by the above-mentioned classifiers, where NB shows the best results with the highest values as 95.38% sensitivity, 98.47% specificity, F1-score value as 95.43%, and FDR values of only 4.27%. Similarly, the performance metrics for other classifiers are listed in Table 2. Furthermore, the class-wise performance of

Table 2 Performance assessment of the given ML classifiers for the accurate identification of the 'viral' disease

Classifiers	Accuracy (%)	Sensitivity (%)	Specificity (%)	F1-score (%)	FDR (%)
NB	95.41	95.38	98.47	95.43	4.27
RF	95.32	95.30	98.44	95.34	4.39
SGD	93.63	93.61	97.88	93.65	5.97
KNN	89.64	89.59	96.54	89.56	8.74
LDA	87.26	87.21	95.74	87.34	9.85

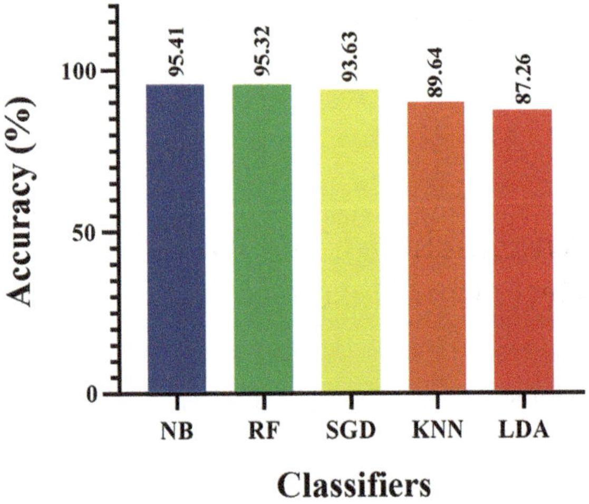

Fig. 2 Accuracy plot of various classifiers

the aforementioned classifiers is effectively displayed in Figs. 3, 4, 5, and 6, where the blue bar represents the Allergy class, the green bar represents the Cold class, the yellow bar represents the Covid class, and the red bar represents the Flu class. The results of the current investigation show that the NB model outperforms the other classifiers in terms of class-wise performance, with sensitivity values of 95.93, 100, 95.03, and 90.58%, specificity values of 100, 96.42, 97.45 and 100%, F1 score values of 97.92, 94.98, 93.75 and 95.06%, and FDR values of 0, 9.56%, 7.5% and 0.

Overall, the proposed system is capable of correctly identifying patients based on their symptoms. These types of systems can be adjusted with upcoming or future diseases based on their symptoms and can be used in the medical field by healthcare experts while diagnosing patients with the disease. Following that, future research will concentrate on improving the model's performance by using various feature selection tactics, as well as updating the model and symptoms to reflect the future trend and nature of the disease.

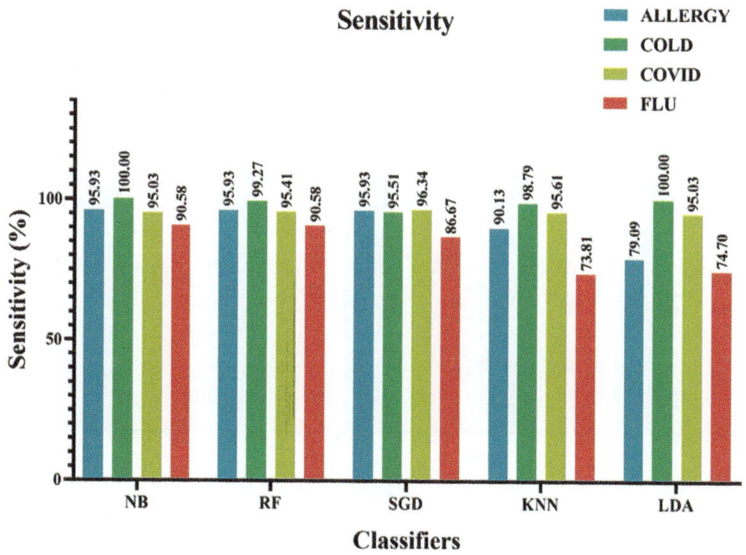

Fig. 3 Class-wise performance in terms of sensitivity of various ML models

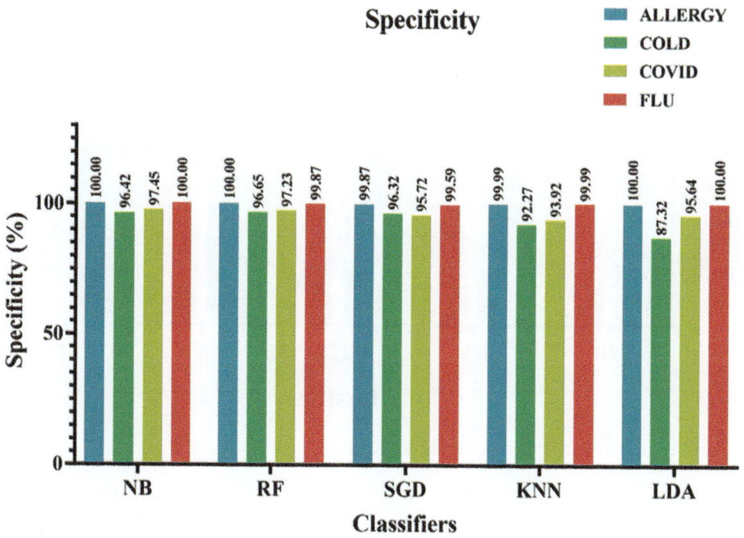

Fig. 4 Class-wise performance in terms of specificity of various ML models

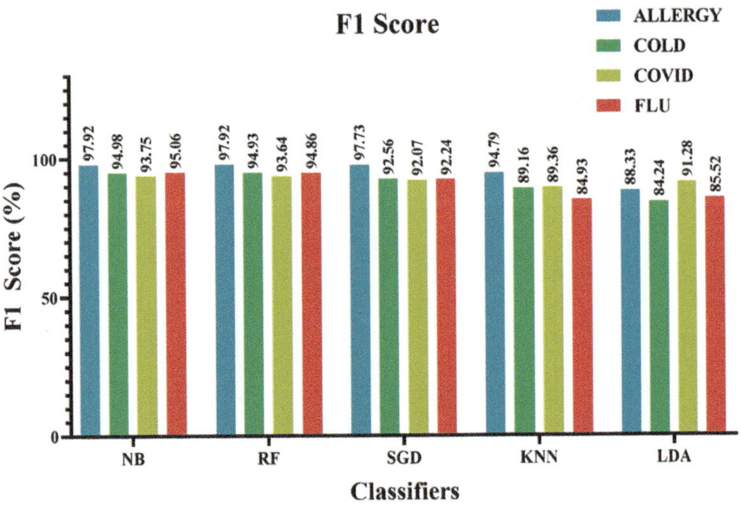

Fig. 5 Class-wise performance in terms of F1-score of various ML models

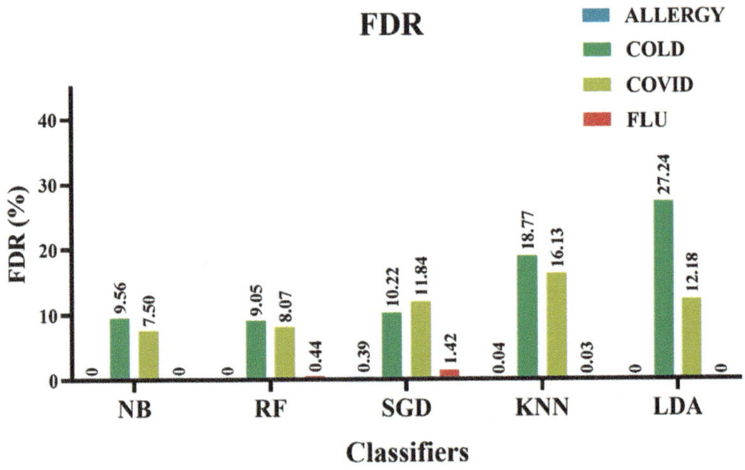

Fig. 6 Class-wise performance in terms of FDR of various ML models

5 Conclusions

The viral propagation of any disease has had an enormous effect on an individual's psychological well-being, and in today's era, a COVID-19-like outbreak has completely ruined their overall physical and mental health. The resemblance between COVID-19 symptoms and those of flu, cold, or allergies caused widespread concern, leading to a pervasive 'corona phobia' among the public. Therefore, it is critical to clear the delusions from people's minds so that those who do have

COVID-19 and require urgent care can be treated on time and those with other diseases should be diagnosed accordingly. As a result, the current study focused on developing machine learning-based models that can effectively determine whether a person is infected with a cold, flu, allergies, or covid. Multiple machine learning models were constructed, encompassing NB, RF, SGD, KNN, and LDA, among which NB exhibited superior performance, achieving an overall accuracy rate of 95.41%. The proposed model is validated by medical experts for real-time usage in reliably identifying diseases based on their symptoms. Overall, technological breakthroughs and modernization can help to improve patient diagnosis and treatment. Subsequently, upcoming research endeavors will prioritize augmenting the model's performance through the utilization of diverse feature selection methodologies. Additionally, there will be a focus on refining the model and symptoms to align with the anticipated evolutionary trends and nature of the disease in the future.

Acknowledgements Aditika Tungal is grateful to the Department of Science and Technology (DST), for providing assistance, in the form of INSPIRE Fellowship (IF210299).

References

1. Kumar, A., Singh, R., Kaur, J., Pandey, S., Sharma, V., Thakur, L., Sati, S., Mani, S., Asthana, S., Sharma, T.K., Chaudhuri, S.: Wuhan to world: the COVID-19 pandemic. Front. Cell. Infect. Microbiol. **11**, 596201 (2021)
2. Kaye, A.D., Okeagu, C.N., Pham, A.D., Silva, R.A., Hurley, J.J., Arron, B.L., Sarfraz, N., Lee, H.N., Ghali, G.E., Gamble, J.W., Liu, H.: Economic impact of COVID-19 pandemic on healthcare facilities and systems: International perspectives. Best Pract. Res. Clin. Anaesthesiol. **35**(3), 293–306 (2021)
3. Baig, A.M.: Neurological manifestations in COVID-19 caused by SARS-CoV-2. CNS Neurosci. Ther. **26**(5), 499 (2020)
4. Mahase, E.: Covid-19: WHO declares pandemic because of "alarming levels" of spread, severity, and inaction. BMJ **368**(8), 1036 (2020)
5. Panchal, U., Salazar de Pablo, G., Franco, M., Moreno, C., Parellada, M., Arango, C., Fusar-Poli, P.: The impact of COVID-19 lockdown on child and adolescent mental health: systematic review. Eur. Child & Adolesc. Psychiatry **32**(7), 1151–1177 (2023)
6. Yeyati, E.L., Filippini, F.: Social and Economic Impact of COVID-19. Brookings Institution (2021)
7. Poland, G.A., Issa, M., Sundsted, K.: Year three of COVID-19: harsh truths, brutal realities, and glimmers of hope. In: Mayo Clinic Proceedings. Elsevier (2022)
8. Klein, H., Asseo, K., Karni, N., Benjamini, Y., Nir-Paz, R., Muszkat, M., Israel, S., Niv, M.Y.: Onset, duration and unresolved symptoms, including smell and taste changes, in mild COVID-19 infection: a cohort study in Israeli patients. Clin. Microbiol. Infect. **27**(5), 769–774 (2021)
9. Alimohamadi, Y., Sepandi, M., Taghdir, M., Hosamirudsari, H.: Determine the most common clinical symptoms in COVID-19 patients: a systematic review and meta-analysis. J. Prev. Med. Hyg. **61**(3), E304 (2020)
10. Glover, H.A., Davis, A.B.: A case of multisystem inflammatory syndrome in children following SARS-CoV-2 infection in a rural emergency department. Adv. Emerg. Nurs. J. **43**(2), 114 (2021)

11. Sinan, H., Uzunoğlu, E., Uğur, M., Avci, E., Akdemir, C., Direkel, Ş: Rapid antigen tests for COVID-19: are their specificity, sensivity and accuracy sufficient? Middle Black Sea J. Health Sci. **9**(3), 387–393 (2023)
12. Kwekha-Rashid, A.S., Abduljabbar, H.N., Alhayani, B.: Coronavirus disease (COVID-19) cases analysis using machine-learning applications. Appl. Nanosci. **13**(3), 2013–2025 (2023)
13. Wubineh, B.Z., Salau, A.O., Braide, S.L.: Knowledge based expert system for diagnosis of COVID-19. J. Pharm. Negat. Results 1242–1249 (2023)
14. Khanday, A.M.U.D., Rabani, S.T., Khan, Q.R., Rouf, N., Mohi Ud Din, M.: Machine learning based approaches for detecting COVID-19 using clinical text data. Int. J. Inf. Technol. **12**, 731–739 (2020)
15. Otoom, M., Otoum, N., Alzubaidi, M.A., Etoom, Y., Banihani, R.: An IoT-based framework for early identification and monitoring of COVID-19 cases. Biomed. Signal Process. Control **62**, 102149 (2020)
16. Solayman, S., Aumi, S.A., Mery, C.S., Mubassir, M., Khan, R.: Automatic COVID-19 prediction using explainable machine learning techniques. Int. J. Cogn. Comput. Eng. **4**, 36–46 (2023)
17. Malik, M., Waseem Iqbal, M., Shahzad, S.K., Mushtaq, M.T., Naqvi, M.R., Kamran, M., Khan, B.A., Tahir, M.U.: Determination of COVID-19 patients using machine learning algorithms. Intell. Autom. & Soft Comput. **31**(1) (2022)
18. Arowolo, M.O., Ogundokun, R.O., Misra, S., Agboola, B.D., Gupta, B.: Machine learning-based IoT system for COVID-19 epidemics. Computing **105**(4), 831–847 (2023)
19. Podder, P., Bharati, S., Mondal, M.R.H., Kose, U.: Application of machine learning for the diagnosis of COVID-19. In: Data Science for COVID-19, pp. 175–194. Academic Press (2021)
20. Chadaga, K., Prabhu, S., Bhat, V., Sampathila, N., Umakanth, S., Chadaga, R.: A decision support system for diagnosis of COVID-19 from non-COVID-19 influenza-like illness using explainable artificial intelligence. Bioengineering **10**(4), 439 (2023)
21. Chataut, R., Phoummalayvane, A., Akl, R.: Unleashing the power of IoT: a comprehensive review of IoT applications and future prospects in healthcare, agriculture, smart homes, smart cities, and Industry 4.0. Sensors **23**(16), 7194 (2023)
22. Rani, S., Kataria, A., Kumar, S., Tiwari, P.: Federated learning for secure IoMT-applications in smart healthcare systems: a comprehensive review. Knowl.-Based Syst. 110658 (2023)
23. Covid, Flu, Cold Symptoms Dataset, https://www.kaggle.com/datasets/walterconway/covid-flu-coldsymptoms?resource=download
24. Mayo Clinic Site, https://www.mayoclinic.org/diseases-conditions/coronavirus/in-depth/covid-19-cold-flu-and-allergies-differences/art-20503981
25. Obaid, H.S., Dheyab, S.A., Sabry, S.S.: The impact of data pre-processing techniques and dimensionality reduction on the accuracy of machine learning. In: 2019 9th Annual Information Technology, Electromechanical Engineering and Microelectronics Conference (IEMECON), pp. 279–283. IEEE (2019)
26. Sharma, B., Sharma, L., Lal, C., Roy, S.: Anomaly based network intrusion detection for IoT attacks using deep learning technique. Comput. Electr. Eng. **107**, 108626 (2023)
27. Xu, S., Li, Z., Yuan, B., Yang, G., Wang, X., Li, N.: A no parameter synthetic minority oversampling technique based on finch for imbalanced data. In: International Conference on Intelligent Computing, pp. 367–378. Springer Nature Singapore, Singapore (2023)
28. Chawla, N.V., Bowyer, K.W., Hall, L.O., Kegelmeyer, W.P.: SMOTE: synthetic minority over-sampling technique. J. Artif. Intell. Res. **16**, 321–357 (2002)
29. Alex, S.A., Nayahi, J.J.V.: Classification of imbalanced data using SMOTE and AutoEncoder based deep convolutional neural network. Int. J. Uncertain. Fuzziness Knowl.-Based Syst. **31**(03), 437–469 (2023)

Affective Driver Emotion Analysis in VANETs Using Heterogeneous Dataset

Anwesha Patel, Rishu Chhabra, and C. Rama Krishna

Abstract Driving is referred to as a cognitively challenging task that involves key-decision process. However, this decision process can be impeded due to the dynamic changes in the emotions of the driver on the drive, which not only impact their own driving but can also pose a threat to the living elements surrounding the driver which are all a part of a Vehicle Ad-Hoc Network (VANET). Therefore, integration of the affective automotive driver emotion detection devices with the VANETs can improve the dynamic driving experience of all the drivers in the VANETs. This paradigm of driving introduced can help the drivers in the connected networks of the VANETs to not only be vigilant of the physical on-road driving conditions around them, but also helps in the discerning the driving patterns of the co-drivers through communications in the VANETs. Such a system of vigilant driving through emotion detection can result in a cooperative and adaptive driving environment which can benefit the drivers by increasing their safety, reducing the driving-associated risk, and improvising the driver's driving experience. Thereby enhancing both the driver's safety and experience on the road. This paper implements the various state-of-art deep learning models in detecting and interpreting human emotions for affective automotive emotion detection systems in VANETs and it has been observed that the SqueezeNet model exhibits the highest accuracy of 99.91%.

Keywords Deep learning · Machine learning · VANET · Valence

A. Patel (✉) · C. R. Krishna
Department of Computer Science and Engineering, National Institute of Technical Teachers Training and Research (NITTTR), Chandigarh, India
e-mail: anwesha.patel@gmail.com

R. Chhabra
Chitkara University Institute of Engineering and Technology, Chitkara University, Punjab, India

B. Soni et al. (eds.), *Beyond Artificial Intelligence*, Lecture Notes in Networks and Systems 1326, https://doi.org/10.1007/978-981-96-4170-3_14

1 Introduction

A lot of research and innovation is happening nowadays in the field of autonomous vehicles. There is continual research and innovation to improve the safety of the user driver and to enhance the driving experience of the driver. There is a continual effort to avoid the risk and accidents of the driver on the road using autonomous vehicles. Even after some casualties, the popularity and the demand for autonomous vehicles have increased and will be exponentially increasing in the upcoming years which is depicted in Fig. 1 [1].

Driving is a continuous, decision-making process which involves continual inputs involving high risk. Driver emotion can directly affect their actions. According to Yerkes-Dodson Law, the medium level of arousal is the safest area of driving experience while a high level of arousal either in positive or negative valence is dangerous [2]. Positive emotional valence implies a relatively safer driving experience while a negative emotional valence implies a higher accident risk. Emotion inference enables in-vehicle applications to focus on driver safety and improve well-being and comfort. It is often found that the negative experience is often associated with traffic behaviour while the positive experience is associated with enjoying their independence and a mode of rebellion for teenagers. As a consequence of negative experiences on the road, participants wished for a better information management and a higher degree of automation while drivers with positive experience with automotive devices wished for a back to basic mode in cars which can give them a thrill of driving experience done by themselves [3]. So in particular understanding the driver's emotion through their body state and activities becomes essential and relevant for the upcoming age of highly automated driving particularly while switching in between the various levels of automation.

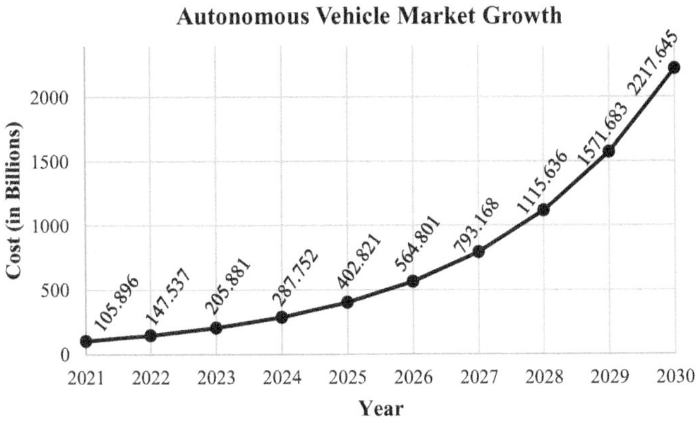

Fig. 1 Size of the global autonomous vehicle market in 2021 and 2022, with a forecast through 2030

Hence, to have this handover or switching between automated, manual, or semi-automated driving systems, the Advanced Driver Assistance System (ADAS) and the Driver Monitoring System (DMS) should have an appropriate and proper in-depth understanding of their driver to support this switch. Since emotional states are likely to be expressed differently depending on the difference in the cultural state-of-affairs/backdrop/frame-of-reference depending on Power Distance, Individualism, and Uncertainty Avoidance. According to Hofstede [4], people in the Asian and African subcontinent have higher uncertainty avoidance, lower Individualism, and lower or medium uncertainty avoidance would rather withhold their emotion and not emote properly. However, in American, European, and Australian subcontinents, people are more willing to emote their emotions without holding back. In the major half of the developing subcontinents like Asia and Africa, solutions are needed to be developed that are more integrable and assimilable without disrupting the current vehicular systems. For this smartphone-based solutions are more feasible as they are less expensive alternatives [5, 6].

One of the major research i.e., going on in the automatic state and emotion recognition is affective computing whose main goal is to sense, interpret, adapt and potentially respond to human emotions. Researchers in this automotive field have pinned their focus on finding effective systems to increase the traffic safety [2]. One of the recommended systems is to prevent negative emotional state, which are statistically correlated with unsafe driving behavior [7]. This paper also proved that a higher arousal of the emotion related to happiness is another source of accidents.

Ensuring the safety of the driver in the VANETs through an affective automotive emotion detection system poses several challenges for the researchers [8, 9]. In this paper, we have presented various state-of-the-art driver emotion detection techniques through deep learning models with the key performance parameters like accuracy, precision, recall, F1-score, loss, and area under the curve. The artificial intelligence models have been trained over a custom heterogeneous dataset addressing ethnicity, pose, illumination, partial occlusion, age factors, and environment-related issues.

Section 2 presents a literature review of the works done in the detection of positive and negative emotion detection techniques. A detailed discussion of the datasets used and the work done is presented in Sect. 3. Our work has been concluded in Sect. 4 with suggestions for future research.

2 Literature Review

In today's landscape along with fully and semi-automated vehicles, manual vehicles are also in operation. Detecting the driver's emotional valence is necessary since it not only affects the driver but also impacts other drivers on drive in VANETs. In Eyben et al. [10], the authors have discussed the various underlying factors behind the emotions influencing the driving of the drivers on the road wherein aggressiveness, sadness, and fatigue are reported to have caused maximum causalities. The authors have not only discussed the detection techniques but have also presented a broad perspective on the various mitigation strategies. In Huo et al. [11], the authors have mapped the user experience according to Kansei engineering that allowed to accurately describe each of the user's emotionality based on their own individualism which can help to improvise the user's experience overall. Table 1 summarizes the various negative and positive emotion detection techniques.

It is evident from the literature review that there are various methodologies using which the emotions are detected for the driver. The most prominent are the deep learning techniques which detects the emotion from the face in an unobtrusive manner. We have therefore used facial emotion for detecting the emotions of the driver. The following section presents information about the heterogeneous datasets used and the results derived using various deep-learning techniques from the dataset.

3 Methodology Implemented

In order to develop tools that can assist the driver in identifying emotions and assist in the self-attuning to drive properly, the basic step is to recognize their emotional valence. According to available literature [19–21], valence can be identified into two major categories as negative and positive valence. In order to carry out the proposed work, we have curated our own dataset by appending crucial information classes from two different standard datasets. As each of the considered datasets have their own features, therefore fusion of the datasets generated a diverse dataset that comprises various ethnicities and a wide assortment of age group from 18–50 (captured under both controlled laboratory and real-time on-road in-vehicle conditions). We have trained the curated dataset over eight state-of-art deep learning models.

3.1 Datasets

By fusing two standard datasets namely, extended Cohn-Kanade database (CK+) and Keimyung University Facial Expression of Drivers (KMU-FED) a custom dataset has been compiled for this work. The class distribution is balanced in our dataset. The dataset created considers three main emotional classes namely, anger, sad, and happy, classified under two valence categories namely, negative and positive. Table 2 shows the configuration of the datasets used.

3.2 Proposed Framework

The experiments in this study were performed over the Kaggle kernels using the Keras and Tensorflow libraries. This study encompasses eight state-of-art CNN models namely Inception V1, Xception, MobileNet, DenseNet, SqueezeNet, NasNetLarge,

Table 1 Negative and Positive Emotion Detection of drivers on the road

Paper references and year	Methodologies used	Situations influencing driver emotions	Valence		Inferences	Disadvantages
			Negative	Positive		
Braun et al. [3], (2018)	Questionnaire analysis	Accident, Driving patterns of drivers in proximity, Personal driving expertise	✓	✓	While in-car interfaces are designed, driver emotion should be considered Design recommendations like warning, positive reinforcement, and others are assessed for their feasibility of being incorporated	A demographic considered during the survey is from North America and Europe and no respondents from a developing subcontinent were considered where implementing these changes is an actual challenge
Braun et al. [12], (2018)	Emotion induction using music and autobiographic recollection	Vehicle UI interactions	✓	✓	Autobiographic recollection performs best in initial emotion inducement Music playback during the driving task proved to prolong the induced emotional state	The emotion elicitation techniques should be more diverse and inclusive of other techniques
Izquierdo-Reyes et al. [13], (2018)	Frontal facial expression, intrusive physiological signals	Accident and near-accident scenarios	✓	✓	ADAS was integrated with emotion detection and an ADMAS model was introduced to study driver emotion mainly through facial analysis done using the Gaussian Mixture Model	Used complicated systems and costly hardware
Braun et al. [14], (2019)	Speech, questionnaire analysis	In-car voice assistant		✓	Personalization has a positive effect on trust and likability if the voice assistant character matches the user's personality	The study participants belonged to BMW Group with a minuscule representation of women respondents

(continued)

Table 1 (continued)

Paper references and year	Methodologies used	Situations influencing driver emotions	Valence		Inferences	Disadvantages
			Negative	Positive		
Kowalczuk et al. [15], (2019)	Frontal facial expression	Face captures of the driver on the road	✓	✓	Facial recognition and speed of change in emotion using Action Units and Kalman Filter is performed	The actual measured value of emotion may not reflect the actual emotional state of the driver
Braun et al. [2], (2020)	Questionnaire, facial emotion, and camera to track the road situation	Vehicle UI interactions, Road conditions	✓	✓	The feedback from respondents reported a high demand for affective automotive UIs having emotional awareness, data privacy, and a careful consideration of sociocultural aspect is in demand	A fixed demographic dividend is considered, Germans and Chinese, and only a premium car was considered for an experiment
Du et al. [16], (2020)	Facial expression and Heart rate analysis	Emotions were simulated in lab controlled environment	✓	✓	Due to the introduction of heart rate analysis along with facial expression, the accuracy increased by 2–8% for various emotions	The authors have used Kinect V2 which is a costly device and can pose difficulty for mobility. The negative emotion of fear was detected with 59.09% accuracy

(continued)

Table 1 (continued)

Paper references and year	Methodologies used	Situations influencing driver emotions	Valence		Inferences	Disadvantages
			Negative	Positive		
Sukhavasi et al. [17], (2022)	Frontal facial expression	Emotions using the heterogeneous standard datasets namely CK + , FER 2013, KDEF, and KMU-FED datasets	✓	✓	A combination of Viola-Jones and Kanade Lucas Thomasi is robust for illumination and occlusion issues while Multi-Task cascaded Convolutional Neural Network is robust for pose issue	The negative emotions like anger, and afraid were less accurate in the detection strategies implemented by the authors
Alsaid et al. [18], (2023)	Facial expressions	Aggressive, moderate, and conservative automation styles, intersections, traffic flow	✓	✓	The driver valence and arousal were affected at the intersections	NADS MiniSim was used for this experiment which is a costly simulator device

Table 2 Details of our custom dataset compiled

Dataset	Respondents	Ethnicity	Changes	Image capture environment	Images size	Negative valence	Positive valence
CK+	210	American African Europian	Pose	Controlled laboratory environment	48 × 48 (PNG)	219	207
KMU-FED	12	Asian	Illumination Partial occlusions	In-vehicle, on-road real-time environment	1600 × 1200 (JPG)	376	210
						595	417
					Total:	1012	

NasNetMobile, and ConvNeXtXLarge. For all the CNN models, the knowledge stored in the weights of a pre-trained model from the extensive ImageNet dataset is leveraged, and thereby, the weights are initialized from the ImageNet. This weight initialization benefits the models by hierarchical and abstract representations learnt during the pre-training phase. Subsequently, the entire architecture is fine-tuned over a carefully curated dataset tailored to our specific emotional valence recognition task for 150-epoch run. This fine-tuning process helps the model to adapt the learnt features to the intricacies present in our target dataset, enabling it to capture emotional expressions more effectively. The last layer of the network is reshaped to produce output relevant to the emotional valence. This hybrid approach of combining the strength of pre-trained weights with dataset-specific fine-tuning and customization, has proven instrumental in achieving a model that inherits prior valuable information and also excels in discerning emotional nuances within our target domain. The following Fig. 2 represents the overall implementation framework for all the CNN Models to be studied and the Table 3 represents the implemented hyper parameters.

4 Result and Discussion

The proposed work compares the performance for the driver's facial emotion expression using eight different approaches of Convolution Neural Networks (CNN). Table 4 presents the comparative analysis of CNNs used for the facial emotion detection. The different techniques have been evaluated using accuracy, precision, recall, F1-score, loss, and Area Under Curve (AUC) as performance metrics. Figures 3, 4, 5, 6, 7 and 8 shows the results for all the performance metrics, namely, accuracy, precision, recall, F1-score, loss, and AUC for the validation dataset. It is evident from the results that the highest accuracy, precision, and recall is achieved by the CNN model SqueezeNet, followed by Xception and DenseNet model, and for F1-score SqueezeNet performed the best followed by Inception V1 and Xception. For Loss,

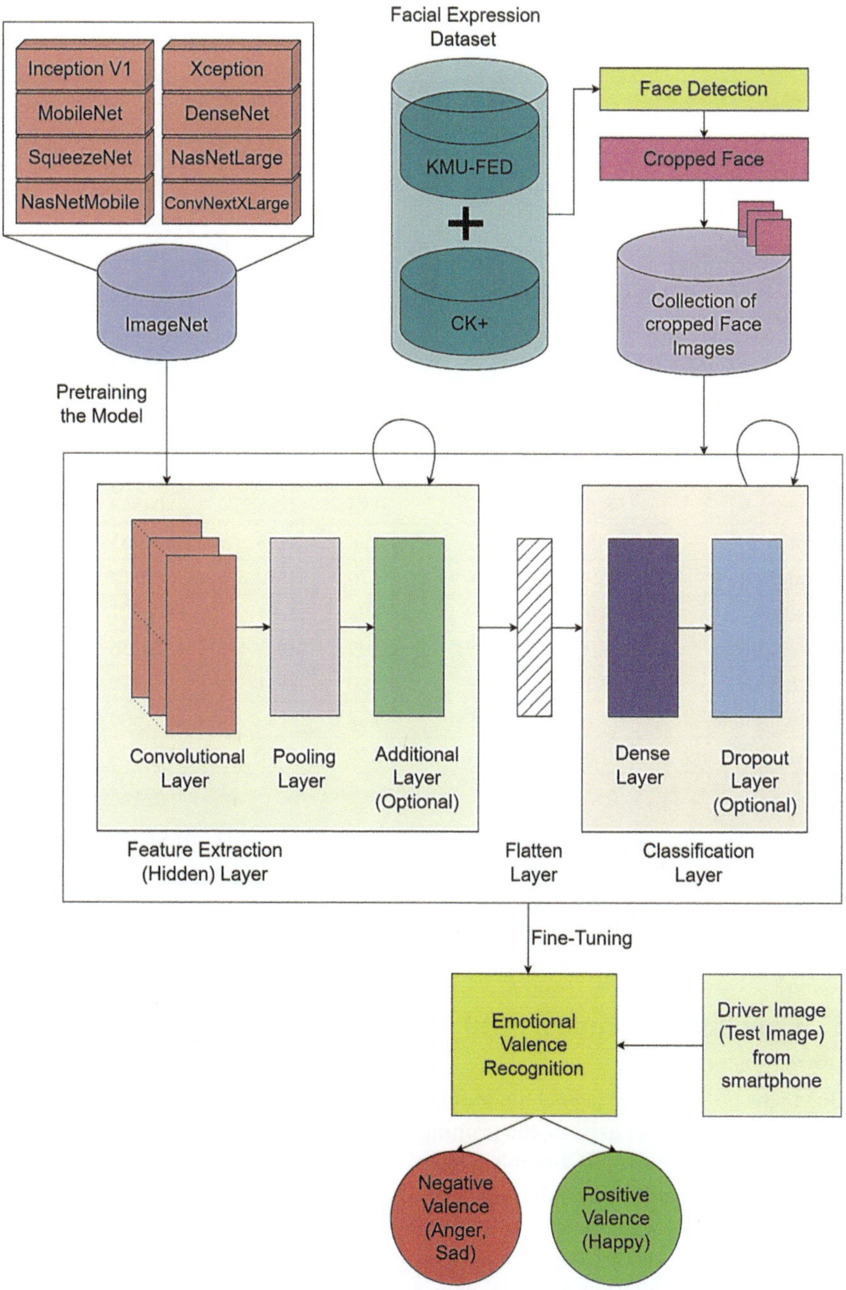

Fig. 2 Proposed framework for the overall implementation of all the CNN models

Table 3 Implemented hyper parameters for the models

Implementation parameters	Implementation value
Weights	Initialized to ImageNet
Batch size	128
Learning rate	0.0001
Optimiser	RMSprop
Loss function	Binary cross entropy
Activation function	Softmax
Regularization	Batch normalization

Table 4 The various CNN models used along with the performance metrics

Algorithm/ performance metrics	Accuracy	Precision	Recall	F1-score	Loss	AUC
Inception V1	98.52	0.98	0.98	0.9855	0.04563	0.9998
Xception	99.507	0.99	0.98	0.98	0.04951	0.9900
MobileNet	72.906	0.74	0.70	0.70	0.50712	0.8354
DenseNet	99.01	0.99	0.98	0.91	0.05321	0.9999
SqueezeNet	99.907	0.99	0.99	0.99	0.29156	0.9799
NasNetLarge	83.251	0.88	0.80	0.81	0.36770	0.9316
NasNetMobile	89.163	0.90	0.88	0.89	0.27600	0.9636
ConvNeXtXLarge	90.640	0.91	0.90	0.90	0.28878	0.9611

Inception V1 performed the best followed by Xception and DenseNet and for AUC, DenseNet performed the best followed by Inception V1 and Xception.

Figures 3–8 Comparative Analysis of CNN Models.

5 Concluding Remarks and Future Directions

Due to the complex driving behavior of the drivers and the unacquainted environment that can pose over the drivers, the training and validation of the driver's facial image can be a challenge. In this study, we have compared eight state-of-art CNN models for driver emotion recognition using heterogeneous dataset based on six performance parameters. To elevate the model's training and learning, we have simultaneously curated a heterogeneous dataset addressing issues of ethnicity, pose, illumination, partial occlusion, age factors, and environment in which the shot was captured in both negative and positive valence. However, there are several research challenges including the privacy of the driver which can be addressed by using depth-based camera. Also, as future research work, the proposed work can be expanded for accommodating compound emotion expressions. Further research can be carried out

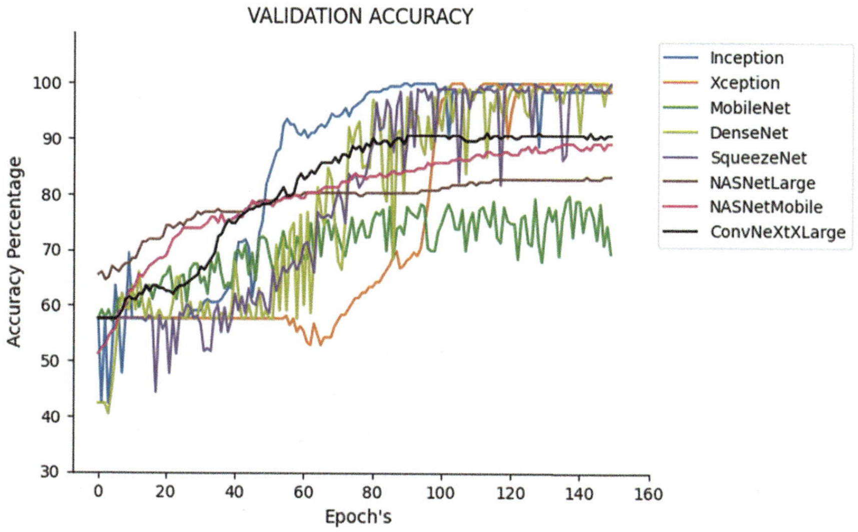

Fig. 3 Validation accuracy for the implemented models

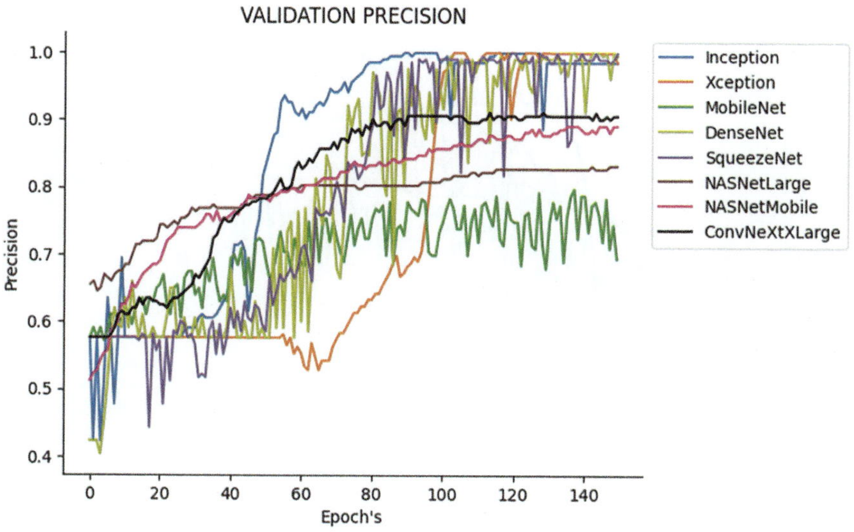

Fig. 4 Validation precision for the implemented models

to devise next mitigation strategy post emotion detection. It generates warning for the drivers by developing haptic processing based on various triggers which can be of spatial, warning, communication, information, or interaction type to ensure a two-layer safety for the drivers on the road and to overall enhance the driver's driving experience.

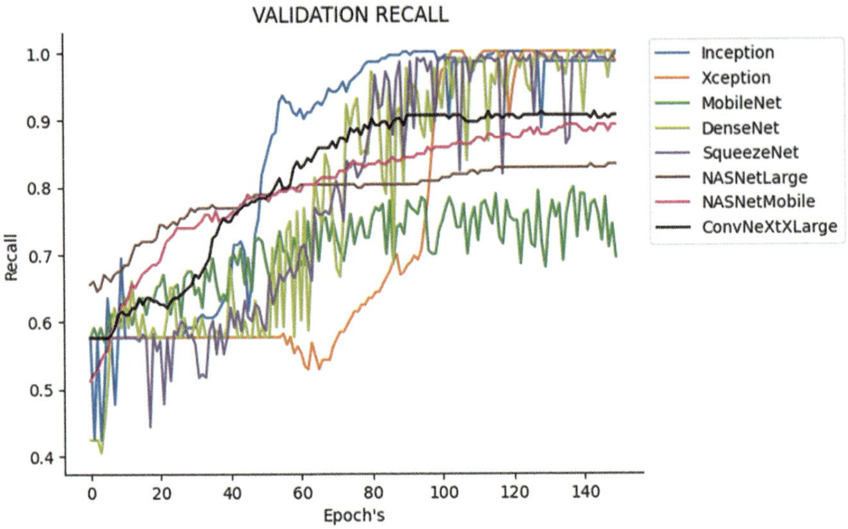

Fig. 5 Validation recall for the implemented models

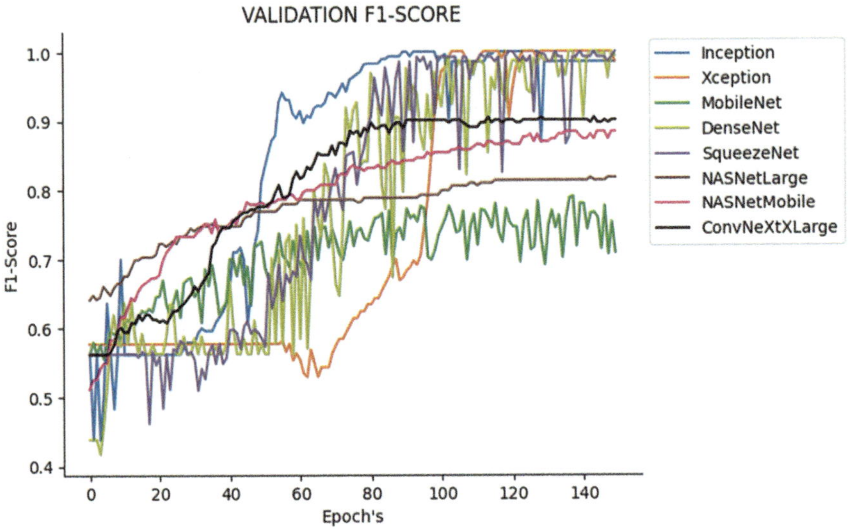

Fig. 6 Validation F1-score for the implemented models

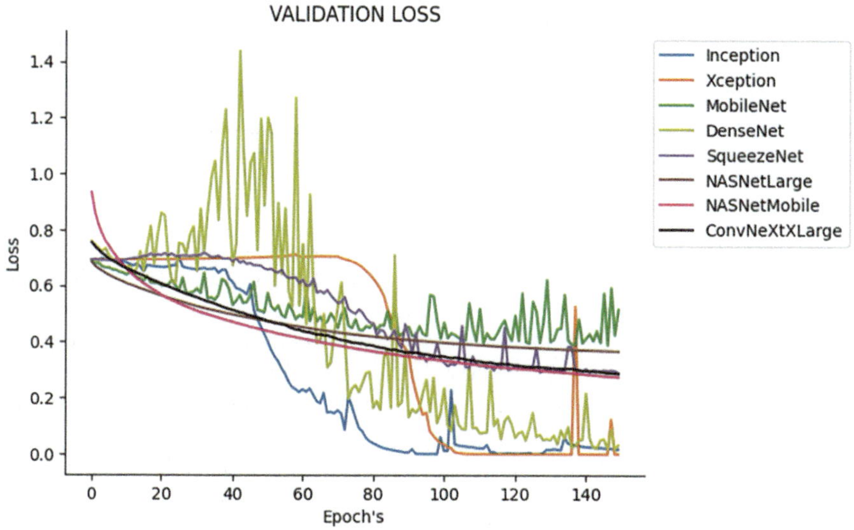

Fig. 7 Validation loss for the implemented models

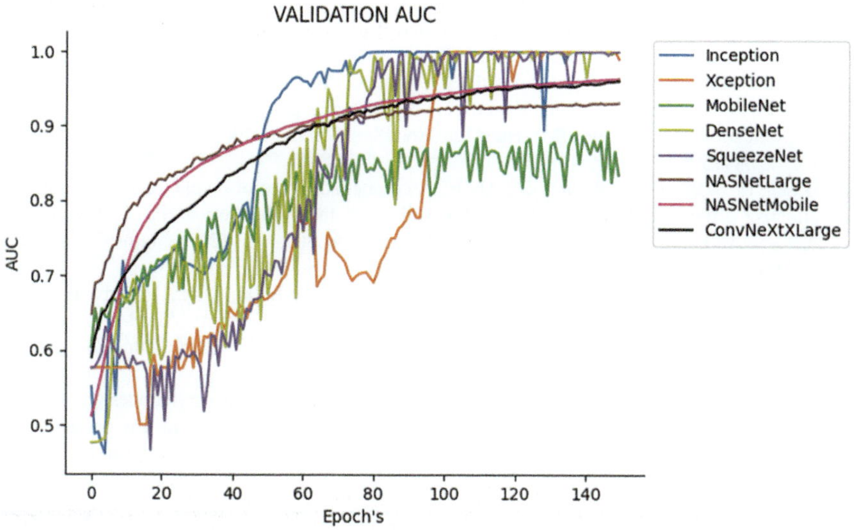

Fig. 8 Validation AUC for the implemented models

References

1. T. & L. Vehicles & Road Traffic.: Size of the global autonomous vehicle market in 2021 and 2022, with a forecast through 2030. Statista (2023). https://www.statista.com/statistics/1224515/av-market-size-worldwide-forecast/. Last accessed 19 Sept. 2023)

2. Braun, M., Li, J., Weber, F., Pfleging, B., Butz, A., Alt, F.: What if your car would care? Exploring use cases for affective automotive user interfaces. In: 22nd International Conference on Human-Computer Interaction with Mobile Devices and Services (2020). https://doi.org/10.1145/3379503.3403530
3. Braun, M., Pfleging, B., Alt, F.: A survey to understand emotional situations on the road and what they mean for affective automotive UIs. Multimodal Technol. Interact. **2**(4) (2018). https://doi.org/10.3390/mti2040075
4. Hofstede, G., Hofstede, G.J., Minkov, M.: Cultures and Organizations: Software of the Mind, 3rd edn. (2010)
5. Chhabra, R., Krishna, C.R., Verma, S.: Smartphone based context-Aware driver behavior classification using dynamic Bayesian network. J. Intell. & Fuzzy Syst. **36**(5) (2019). https://doi.org/10.3233/JIFS-169995
6. Chhabra, R., Krishna, C.R., Verma, S.: Augmenting driver's situational awareness using smartphones in VANETs. Arab. J. Sci. Eng. **47**(2), 2271–2288 (2022). https://doi.org/10.1007/s13369-021-06159-5
7. Jeon, M., Walker, B.N., Bin Yim, J.: Effects of specific emotions on subjective judgment, driving performance, and perceived workload. Transp. Res. Part F: Traffic Psychol. Behav. **24** (2014). https://doi.org/10.1016/j.trf.2014.04.003
8. Rani, S., Koundal, D., Kavita, F., Ijaz, M.F., Elhoseny, M., Alghamdi, M.I.: An optimized framework for WSN routing in the context of Industry 4.0. Sens. (Basel, Switz.) **21**(19) (2021). https://doi.org/10.3390/s21196474
9. Verma, K., Bhardwaj, S., Arya, R., Ul Islam, M.S., Bhushan, M., Kumar, A., Samant, P.: Latest tools for data mining and machine learning. Int. J. Innov. Technol. Explor. Eng. **8** (2019). https://doi.org/10.35940/ijitee.I1003.0789S19
10. Eyben, F., Wöllmer, M., Poitschke, T., Schuller, B., Blaschke, C., Färber, B., Nguyen-Thien, N.: Emotion on the road-Necessity, acceptance, and feasibility of affective computing in the car. Adv. Hum.-Comput. Interact. **2010** (2010). https://doi.org/10.1155/2010/263593
11. Huo, F., Zhao, Y., Chai, C., Fang, F.: A user experience map design method based on emotional quantification of in-vehicle HMI. Hum.Ities Soc. Sci. Commun. **10**(1) (2023). https://doi.org/10.1057/s41599-023-01761-4
12. Braun, M., Weiser, S., Pfleging, B., Alt, F.: A comparison of emotion elicitation methods for affective driving studies. In: Adjunct Proceedings of the 10th International Conference on Automotive User Interfaces and Interactive Vehicular Applications (2018). https://doi.org/10.1145/3239092.3265945
13. Izquierdo-Reyes, J., Ramirez-Mendoza, R.A., Bustamante-Bello, M.R., Navarro-Tuch, S., Avila-Vazquez, R.: Advanced driver monitoring for assistance system (ADMAS): based on emotions. Int. J. Interact. Des. Manuf. (IJIDeM) **12**(1), 187–197 (2018). https://doi.org/10.1007/s12008-016-0349-9
14. Braun, M., Mainz, A., Chadowitz, R., Pfleging, B., Alt, F.: At your service: designing voice assistant personalities to improve automotive user interfaces a real world driving study. In: Proceedings of the 2019 CHI Conference on Human Factors in Computing Systems (2019). https://doi.org/10.1145/3290605.3300270
15. Kowalczuk, Z., Czubenko, M., Merta, T.: Emotion monitoring system for drivers. IFAC-PapersOnLine **52**(8) (2019). https://doi.org/10.1016/j.ifacol.2019.08.071
16. Du, G., Wang, Z., Gao, B., Mumtaz, S., Abualnaja, K.M., Du, C.: A convolution bidirectional long short-term memory neural network for driver emotion recognition. IEEE Trans. Intell. Transp. Syst. **22**(7) 2020. https://doi.org/10.1109/TITS.2020.3007357
17. Sukhavasi, S.B., Sukhavasi, S.B., Elleithy, K., El-Sayed, A., Elleithy, A.: Deep neural network approach for pose, illumination, and occlusion invariant driver emotion detection. Int. J. Environ. Res. Public Health **19**(4) (2022). https://doi.org/10.3390/ijerph19042352
18. Alsaid, A., Lee, J.D., Noejovich, S.I., Chehade, A.: The effect of vehicle automation styles on drivers' emotional state. IEEE Trans. Intell. Transp. Syst. **24**(4), 3963–3973 (2023). https://doi.org/10.1109/TITS.2023.3239880

19. Russell, J.A.: A circumplex model of affect. J. Pers. Soc. Psychol. **39**(6) (1980). https://doi.org/10.1037/h0077714
20. Ekman, P.: Are there basic emotions? Psychol. Rev. **99**(3), 550–553 (1992). https://doi.org/10.1037/0033-295X.99.3.550
21. Plutchik, R.: The nature of emotions: human emotions have deep evolutionary roots, a fact that may explain their complexity and provide tools for clinical practice. Am. Sci. **89**(4) (2001)

Tomato Leaf Disease Classification Using YOLOv8

Ambati Leena Reddy and Surendra Reddy Vinta

Abstract Tomato is one of the finest vegetable which is famous all around the world most of the farmers cultivate the potatoes all around the world and export them as goods. Tomato have a great finite taste which can be a best part of side dish to food item. The Tomatoes are roots with lot if minerals hidden in them but all this nutrients can be lost with the attack of pests which will spoil the crop. The leaf are the main part of the plant for the growth without proper leaves the plant cannot be grown for the further nutrients. Tomatoes are a cornerstone of agricultural production, with farmers cultivating and exporting them as key commodities due to their rich flavor and versatile culinary applications. Despite their nutritional value, tomatoes are vulnerable to pests, particularly leaf diseases, which jeopardize crop health and yield. Early disease detection is critical for mitigating losses and ensuring sustainable agricultural practices. To address the imperative of timely disease detection, this study employs the cutting- edge YOLOv8 (You Only Look Once) algorithm, a Deep Learning model built on basis of (CNNs). The most known alternaria is a leaf disease in Tomato called as early blight and the late blight disease, leaf mold, sepotria leaf spot, bacterial spot. Detection of diseases in early stage of the crop field with this project can decrease the loss and early detection of the leaf affected and try to change and adapt to chemicals used by farmers. The YOLOv8, an improved version of the YOLOv5, stands out for its superior performance, particularly in the segmentation of classified datasets for object detection tasks. This study asserts that YOLOv8 outperforms existing methodologies in disease detection, demonstrating increased accuracy. By detecting diseased tomato leaves at their prior stage, this algorithm will farmers to effectively adapt and implement targeted chemical interventions, reducing disease's impact on crop quality as well as yield. Therefore, we use a Deep Learning algorithm YOU ONLY LOOK ONCE is used for Tomato Leaves Plant Disease for Identification Software using Convolutional Network. The convolutional neural network are used for mainly image classification for predicting the best accuracy

A. L. Reddy · S. R. Vinta (✉)
School of Computer Science and Engineering (SCOPE), VIT-AP University, Amaravati, India
e-mail: vsurendra.cse@gmail.com

A. L. Reddy
e-mail: leena.21BCE7018@vitapstudent.ac.in

B. Soni et al. (eds.), *Beyond Artificial Intelligence*, Lecture Notes in Networks and Systems 1326, https://doi.org/10.1007/978-981-96-4170-3_15

Of 98%. The YOLOv8 is new version of YOLOv5, the YOLOv8 mostly used in segmentation of the classified dataset for the object detection required tasks. So, YOLOv8 performs better than the existing methods as its accuracy is high.

Keywords Tomato leaf disease · YOLOv8 (You Only Look Once) · Late blight · Early blight · Loss function · Classification · Segmentation

1 Introduction

Tomatoes are sensitive vegetable and it is one of the popular finest alrounder one extensively farmed crop, a significant the beginning of nutrients for humans. Tomato plants, on the other hand, are subject to a broad range of illnesses, which can drastically impair output and quality. Tomato leaf disease is one of a list of the most widespread illnesses that harm tomato plants. Tomato leaf are more important to get a best vegetable but affecting of the leaves can also affect the crop and there are many problem which are affecting the crop yield and reduce the quality. There are many of diseases which are affecting the crop that are the above mentioned list of the most widespread illnesses that can harm tomato plants. Tomato leaf infections are a typical issue that many gardeners and farmers experience. These diseases, which can harm the parts of plant and fruit of tomato plants, that are due to a range of micro organisms, diseases, consist fungus, tiny bacteria's, and viruses.

Tomato leaf disease is a catch-all name for a variety of fungal, bacterial, and viral diseases that harm tomato leaves. These diseases can cause yellowing, withering, spotting, and curling of the leaves, among other symptoms. In severe situations, the leaves may fall off, reduction of the plant's capacity to capture rays from sun and produce fruit. Environmental circumstances, plant genetics, and cultural practices are all variables that involve in generating to the improvement in tomato leaf disease. High humidity, moderate temperatures, and damp leaves, for example, can provide excellent circumstances for the establishment and spread of fungal and bacterial infections. Similarly, certain tomato types may be more vulnerable to disease than others, and cultural techniques like as irrigation and fertilization can also influence disease susceptibility. Crop rotation, trimming, and cleanliness are examples of cultural management strategies that can assist prevent disease transmission. Fungicides and bactericides are used in chemical control procedures to kill or inhibit disease development. Biological control approaches include the employment of helpful microorganisms such as fungus and bacteria to inhibit disease development.

Despite the availability of these management measures, tomato leaf disease continues to be a major issue for tomato farmers worldwide. As a result, additional study is required to better our understanding of the condition and create more effective control techniques. We will evaluate the present level of knowledge on tomato leaf disease and propose new research avenues for the future in this study.

Tomato leaf diseases can cause discoloration, spotting, withering, and deformation of the leaves, resulting in a reduction in plant growth and output. To prevent tomato leaf diseases, effective cultural beliefs such as crop rotation, cleanliness, correct watering must be combined with the use of chemical and biological controls. It is critical for survival to prevent and manage tomato leaf diseases. Implement efficient supervision strategies, and limit crop losses. Tomato leaf diseases are frequently prevented using a combination of cultural techniques such as adequate plant spacing, watering, and fertilization, as well as the use of immune to disease tomato peoples as well as pesticides or bactericides. The use of computerized vision and algorithmic learning tricks could be applied to automate the characterization and discovery of tomato leaf inflammation, increasing the effectiveness and long-term viability of tomato production.

2 Literature Survey

Mehdhar et al. [1] utilizing transfer learning and feature concatenation, tomato leaf disease categorization a common ailment that can damage tomato plants' leaves, stems, and fruit are tomato leaf diseases. Transfer learning may be used to study how various diseases are the reason to influence tomato plants.

Sabrol and Satish [2] severity of tomato leaf diseases were examined in this study across conventional, organic, and integrated pest management (IPM) agricultural techniques in 2016.

Basavaiah and Arlene Anthony [3] in his article published in 2020, Jagadeesh Basavaiah stated that there are several approaches. The primary goal of the suggested strategy is to provide a system for identifying tomato plant leaf disease by increasing classification accuracy and decreasing computing time.

Attallah [4] the genomic regions associated to disease resistance that the researchers uncovered might be utilized to create disease- resistant tomato varieties.

Singh et al. [5] in this study, genetic markers were used to pinpoint the genes that confer resistance to bacterial spot, early blight, and late blight, among other tomato leaf diseases.

Trivedi et al. [6] in this study, the effectiveness of several fungicides and bio-control agents for treating tomato leaf diseases in the field was examined.

Shoaib et al. [7] according to Researcher, The use of this project was to identify quantitative trait loci associated with tomato plant resistance to bacterial wilt.

Basavaiah [8] the Methods used by the researchers demonstrated that certain fungicides and biocontrol agents are useful in reducing disease severity and improving tomato yield.

Ramalakshmi [9] it was shown that when compared to conventional production methods, IPM and organic production methods had lower incidence and severity of tomato leaf diseases.

Nawaz et al. [10] a strong deep neural network method for identifying and systematization tomato plant leaf diseases. For the development of tomato cultivars that are resistant, study is essential.

Sharma et al. [11] author in 2021 used molecular markers to locate the QTLs and found several genomic regions associated with disease resistance.

Muhammad et al. [12] this study investigated the efficacy of machine learning methods for recognition and organization tomato leaf illness using image processing techniques. The investigators found that a combination of Convolutional Neural Networks models were able to accurately recognize illnesses in foliage from tomatoes.

Wu et al. [13] this review article discusses the prevalent ailments of tomato leaves, their symptoms, and treatment options, which include the use of indigenous practices, chemical surveillance, and biological inhibition.

Vinay [14] in order to visualize future diseases for categorization purposes, the first symptoms have been removed.

Brahimini and Boukhalfa [15] the paper from the year 2021 describes this technique. Each of the aforementioned illnesses provided a wealth of information into the molecular mechanisms behind disease resistance as well as the origin, symptoms, and treatment approaches.

Saranya et al. [16] they discovered that early issue mitigation requires precise and quick diagnosis of plant diseases.

Zaki et al. [17] the analysis and comparison of Fuzzy SVM, CNN, and R-CNN classification techniques using R-CNN aims to identify the best precise algorithm for plant disease forecasting.

Nagamani and Sarojadevi [18, 19] to minimize future losses, early illness identification is required, according to Sarojadevi H's article from the year 2022.

Aishwarya and Pushpa [20] with 2-Max Assembling covers and a completely related layer, a 2-Dimensional neural network convolution modelling technique Compared to SVM, VGG16, Inception V3, and the Mobile Net CNN model, among other classification models

Liu and Wang [21] in Researcher study from two years ago, the YOLOv3 model was combined with mobilenet for maximum accuracy.

Chauhan and Mohana [22] the model by enhancing the accuracy and F1 score of the existing model by fine-tuning the technique.

By studying all the references Tomato is one of the finest vegetable with many diseases which require a large dataset of image capturing to train and validate the model. So, we used the most recent article that is YOLOv8 which is the new version of the YOLOv5 with more acquired features and showing the highest accuracy while classifying the data.

Series No.	References	Model used	Accuracy (%)
1	Al-Gaashani Mehdhar	SVM, RF, Mulitinomial regression	97
2	H. Sabrol and K. Satish	Supervised-learning technique	97.3
3	Jagadeesh Basavaiah	Random forest	90
4	Omneya Atta	KNN and SVM	99.2

(continued)

(continued)

Series No.	References	Model used	Accuracy (%)
5	Suneet Gupta	Convolution-neural network	91.2
6	Hani Moaiteq Aljahdali	CNN	98.49
7	Ihsan Ullah	U-Net and Modified U-Net, Inception	99
8	Jagadeesh Basavaiah	Decision tree classifier	90
9	K. Ramalakshmi	RNN-CNN	99.64
10	Marriam Nawaz	Convolutional-block attention module	99.97
11	Rakesh Sharma	Machine learning, CNN, Res Net 50	88.44
12	Nabil Ibtehaz	DenseNet201	97.99
14	Vinay and Pranay	Neural network	97
15	Kamel Boukhalfa	Convolutional neural network	99.8
16	V. Saranya	ECNN	90
17	Zulkifley	MobileNetv2	90
18	Nagamani HS	Deep learning techniques	96
19	Sarojadevi H's	(Fuzzy-SVM), CNN, (R-CNN)	96.735
20	Pushpa BR, Aiswarya VV	CNN	96
21	Xuewei's	MobileNetv2-YOLOv3	93.24
22	Amit Chauhan and Rajni Mohana	Bert model	85.02

2.1 Data Set

In our model data has been loaded from robo flow site. The dataset is nothing but the collection of group of related information which are particularly kept apart. In our dataset the image type of dataset which has been recruited and this dataset is in three different deviations that are train, Val, test in this way the entire dataset is splited and in each and every part the different type of images has been placed.

In our dataset we have collected images for each of this diseases that are 2127 pictures of the target location as well as a bacterial spot 952 images depict the final blight, leafy mold, and the Lycopersicon esculentum related yellowy leaf curl virus. Early blight are having 1000 Spider mites and their visuals Two, noticed a spider mite this both are mixed with 1676 as images, Tomato healthy are 1591 images, Septoria leaf spot are 1771 images (Figs. 1 and 2).

```
Number of images in 'Tomato_Spider_mites_Two_spotted_spider_mite' are 1676.
Successfully processed 'Tomato_Spider_mites_Two_spotted_spider_mite' class having a total of 1676 images.
Number of images in 'Tomato_Leaf_Mold' are 952.
Successfully processed 'Tomato_Leaf_Mold' class having a total of 952 images.
Number of images in 'Tomato_Bacterial_spot' are 2127.
Successfully processed 'Tomato_Bacterial_spot' class having a total of 2127 images.
Number of images in 'Tomato_Septoria_leaf_spot' are 1771.
Successfully processed 'Tomato_Septoria_leaf_spot' class having a total of 1771 images.
Number of images in 'Tomato_Late_blight' are 1908.
Successfully processed 'Tomato_Late_blight' class having a total of 1908 images.
Number of images in 'Tomato_healthy' are 1591.
Successfully processed 'Tomato_healthy' class having a total of 1591 images.
Number of images in 'Tomato_Early_blight' are 1000.
Successfully processed 'Tomato_Early_blight' class having a total of 1000 images.
```

Fig. 1 Total number of images and classes

Fig. 2 Image dataset of diseases

3 Methodology

Our model is YOLOv8 which is new version of YOLOv5 which is increased in its performance and in predicting the accuracy. It is used to predict the condition of the leaf whether it is healthy or it is affected by any diseases (Fig. 3).

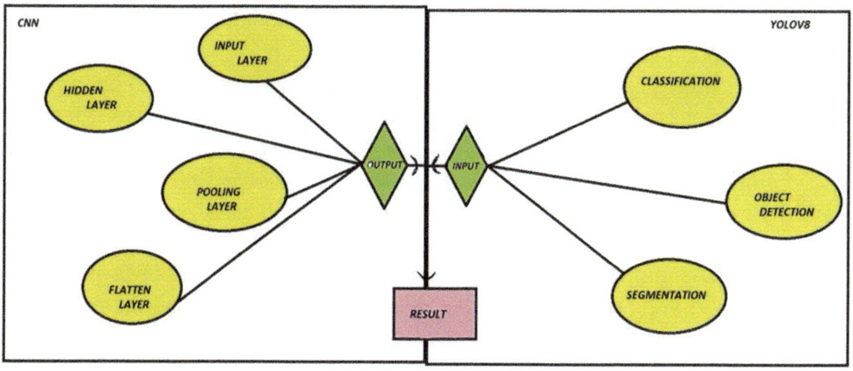

Fig. 3 Architecture of the CNN and YOLOv8

3.1 Flow Diagram

3.2 Methods

The neural networks based on con a type of artificially generated neural network, are used to categorize photographs. This is a demonstration of a deep learning algorithm. When comparison is made within the related neural networks the CNN is the most higher secured with data that contain images, sounds, or audio information. They have three separate types of layers: convolution layer, dipping level, and fully linked layer. The convolutional level is the initial level of a Semantic Network. The fully-linked tier is the final level, albeit one that may be preceded by additional convolutional levels or levels that pool information. As the layer's complexity rises as a result of additional images being detected, the CNN will diverge even more. Early layers emphasize basic components like hues and edges. The CNN layers will enhance the data depicts as they navigate the pixel data and seek to identify the item.

The layer of convolution, composed of the data being input, a filter, plus a characteristic map, is the basic element of a CNN. A 3D color imagine with dimensions of height, width, and depth is used for the input. To determine if a feature is present, the feature locater, which is additionally referred to as a kernel or filtering system, wipes over the image's receptive areas. After applying the filter onto the picture, the resultant dots is computed for the data provided and boundaries. Feature maps, activation maps, or convolved features are what is produced that take their the greatest form.

Rectified Linear Unit (ReLU) or other activation functions are used by a CNN after each convolution operation to adjust the feature map for data nonlinearity. The first convolution layer can be followed by another convolution layer to produce a hierarchical structure. By applying an aggregation function to the data in the field of reception, the layers of pooling, which is frequently referred to as down sampling, reduce heterogeneity. The pooling comprises two submodules: the ultimate pooling and average pooling.

The fully linked layer, which works categorization based on traits, connects each node in the last layer directly to a node in the layer that came before it retrieved from earlier layers and filters. While FC layers often employ an activation function with softmax activation for organizing data, convolutional as well as pooling layers typically utilize the ReLU algorithm functions which will classify the inputs that are approximate.

The YOLOv8 models have code that is comparable to the YOLOv5 models, but they have an entirely novel design that makes it feasible to utilize the exact same code routines for labeling, segmentation of instances, and recognition of objects task types. The data source format and configuration of models retain to use the application YAML format.

4 Results

After evaluating the result of several Deep neural network models to recognize disorders of leaves of tomatoes based on various features, our analysis indicates that the CNN model with a pertained architecture YOLOV8 gives a 98% accuracy which is easy to run the epochs with less time requiration (Figs. 4, 5 and 6).

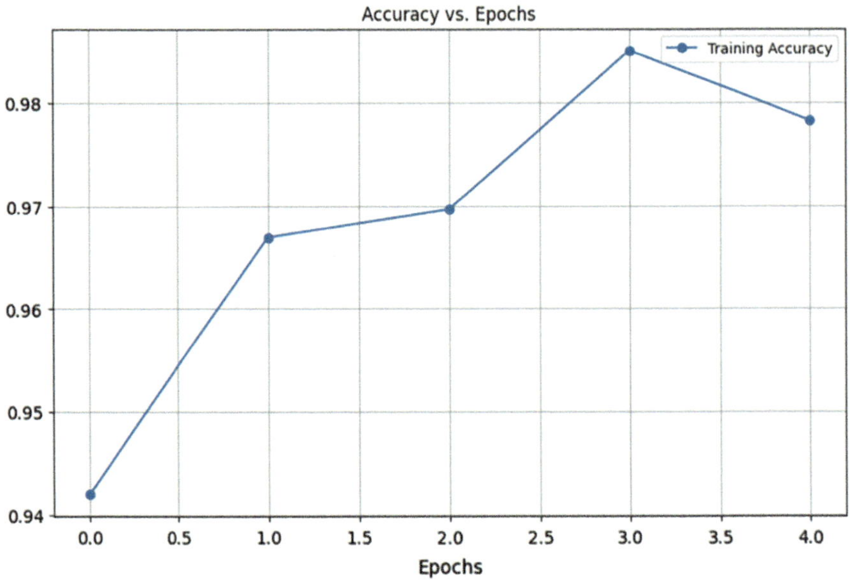

Fig. 4 Accuracy versus graph

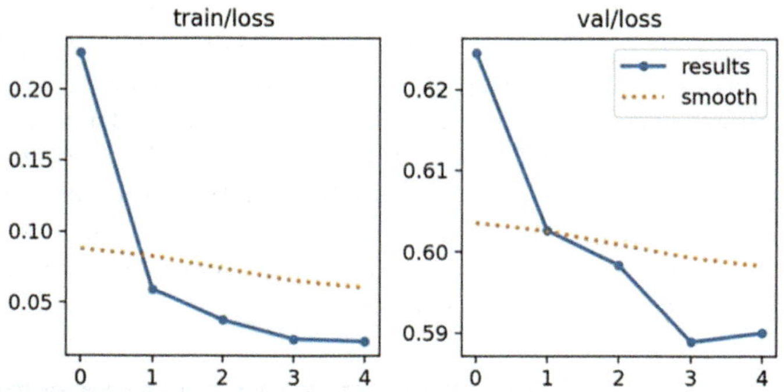

Fig. 5 Train versus loss and Val versus Loss graphs

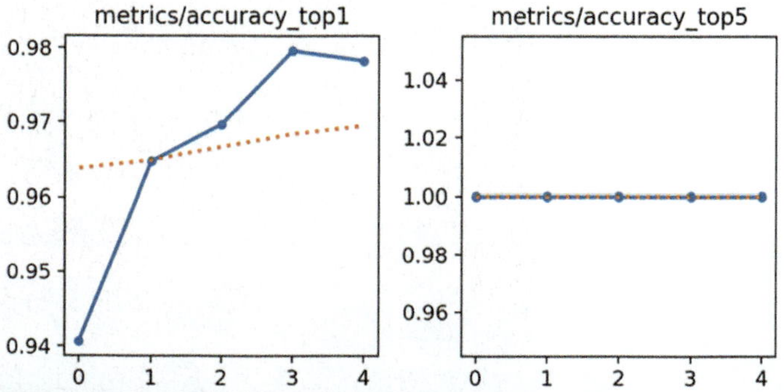

Fig. 6 Metrics versus Accuracy_top1 and Accuracy_top5 graphs

The above figure presents the Metrics and accuracy are shown in the graph that follows along with instruction and verification losses. For the YOLOv8 Classification model on Tomato leaf Disease, trained for 10 epochs (Fig. 7).

The pictures shows the prediction on some of the leaves according to the training done by YOLOv8 classification model.

Fig. 7 Classified Images from dataset

4.1 Score

```
validation_metrics = model.val(data = os.path.join(os.getcwd(), 'tomato_leaf_disease'),
                               model = os.path.join(os.getcwd(), 'runs', 'classify', 'train', 'weights', 'best.pt'))
print(f"Accuracy of the model is {validation_metrics.results_dict['metrics/accuracy_top1'] * 100:.2f}% on the validation dataset.")

Ultralytics YOLOv8.0.140 🚀 Python-3.10.6 torch-2.0.1+cu118 CUDA:0 (Tesla T4, 15102MiB)
YOLOv8n-cls summary (fused): 73 layers, 1443847 parameters, 0 gradients
               classes   top1_acc   top5_acc: 100%|██████████| 139/139 [00:03<00:00, 35.05it/s]
                   all      0.991          1
Speed: 0.1ms preprocess, 0.7ms inference, 0.0ms loss, 0.0ms postprocess per image
Results saved to runs/classify/val2
Accuracy of the model is 99.09% on the validation dataset.
```

Comparison table for used model and our model.

Series No.	Models	Published	Accuracy (%)
1	YOLOv8	–	98

(continued)

(continued)

Series No.	Models	Published	Accuracy (%)
2	CNN	2017	71.41
3	KNN & DT & SVM	2020	89.27
4	CNN & YOLOV5	2022	95.90
5	VGG16 & VGG19	2016	96

5 Limitations

Limited dataset: The consummation of model YOLOv8 the quantity and the caliber of the training information affect the efficiency of this project

Classes and images are limited: We concentrated on identifying types of tomato leaves, many additional diseases can harm tomato plants, and the model's effectiveness against these illnesses is uncertain, images variability found in real-world conditions. Computational requirements: The YOLOv8 model is a computationally expensive approach that requires a strong GPU.

6 Conclusions

Finally, utilizing an enormous collection of leaf photos, the YOLOv8 object identification approach finds and categorizes tomato leaf illnesses. The results demonstrate YOLOv8's excellent success rate in recognizing and classifying various diseases, proving the power of deep learning approaches for accurate and speedy diagnosis.

Deep learning, as shown by YOLOv8, has the potential to transform tomato leaf disease discovery and grouping. Producers and researchers gain greatly from the accuracy and quickness with which this technology diagnoses diseases. This research could aid in the formation of successful pest control measures, which could lead to increased tomato crop profits and quality.

These results also imply that expanding the amount and variety of the information set could increase the accuracy and adaptability capabilities of the YOLOv8 model. As we focus on advanced-stage illness detection, we realize the difficulty of early-stage disease detection when signs are modest. Additional research may investigate the difficulties of recognizing disorders in their infancy.

In conclusion, the YOLOv8 object identification method proves to be a useful tool for accurate recognition and diagnosis of Lycopersicon esculentum leaf diseases. Further research into the applicability of deep learning-based models in addressing inflammations in other plant species and crops is warranted, paving the way for innovative advances in agricultural disease management. Based on this our findings indicate that YOLOv8 can be a highly successful technique for detecting and diagnosing tomato leaf diseases. More study is required to examine the possibilities of deep learning-based technical models for other plant inflammations and crops.

7 Future Scope

The future scope of this research paper is upcoming future generations will high advanced technology will improve the power of detection in more difficult circumstances. Furthermore, researching pathogen genetic diversity and variability, as well as host- pathogen interactions, might give insights into disease management techniques. Holistic treatment approaches that integrate several therapies and account for ecological dynamics can be examined. Finally, using omics techniques to study the molecular basis of disease resistance in tomatoes has the potential to benefit marker-assisted breeding and genetic engineering initiatives.

The future direction of tomato leaf illness classification research holds huge potential for advancement, particularly with the anticipated evolution of advanced technologies. Integrating cutting-edge technologies like advanced machine learning algorithms and sophisticated imaging systems could significantly improve the detection capabilities of models like YOLOv8 in difficult conditions. This could entail investigating novel sensor technologies, such as hyperspectral imaging, to capture subtle spectral variations associated with early-stage disease symptoms, allowing for more robust and early detection.

The ongoing development of high-tech tools such as advanced machine learning algorithms and imaging systems is expected to significantly improve the detection capabilities of models such as YOLOv8. Continued research into these technologies may lead to more efficient and accurate disease detection, even in difficult environmental conditions.

Pathogen genetic diversity research and understanding host-pathogen interactions remains a promising avenue. Omics techniques such as genomics, transcriptomics, proteomics, and metabolomics are at the forefront of future research. These avenues have the potential to improve our ability to detect, understand, and manage tomato leaf diseases, which will benefit agricultural practices and crop yields in the long run.

Tomato leaf disease classification research will expand into advanced technology, genetic exploration, integrated disease management, and omics-based molecular studies in the future. These paths not only promise to improve disease detection accuracy, but also hold the key to developing sustainable and resilient agricultural practices in the face of changing challenges.

References

1. Al-Gaashani Mehdhar, S., Shang, F., Muthanna, M.S., Khayyat, M.: Tomato Leaf Disease Classification by Exploiting Transfer Learning and Feature Concatenation (2022). https://ietresearch.onlinelibrary.wiley.com/doi/full/https://doi.org/10.1049/ipr2.12397
2. Sabrol, H., Satish, K.: Tomato Plant Disease Classification in Digital Images using Classification Tree (2016). https://www.ius.edu.ba/sites/default/files/users/files/user765/07754351.pdf
3. Basavaiah, J., Arlene Anthony, A.: Tomato Leaf Disease Classification using Multiple Feature Extraction Techniques (2020). https://www.researchgate.net/profile/Jagadeesh-Basavaiah/publication/342248958_Tomato_Leaf_Disease_Classification_using_Multiple_Feature_Extraction_Techniques/links/5f46911592851c6cfddd4d88/Tomato-Leaf-Disease-Classification-using-Multiple-Feature-Extraction-Techniques.pdf
4. Attallah, O.: Tomato Leaf Disease Classification via Compact Convolutional Neural Networks with Transfer Learning and Feature Selection (2023). https://www.mdpi.com/2311-7524/9/2/149
5. Agarwal, M., Singh, A., Arjaria, S., Sinha, A., Gupta, S.: Tomato Leaf Disease Detection using Convolution Neural Network (2020). https://www.sciencedirect.com/science/article/pii/S1877050920306906
6. Trivedi, N.K., Gautam, V., Anand, A., Aljahdali, H.M.: Early Detection and Classification of Tomato Leaf Disease Using High-Performance Deep Neural Network (2021). https://www.ncbi.nlm.nih.gov/pmc/articles/PMC8659659/
7. Shoaib, M., Hussain, T., Shah, B., Ullah, I.: Deep Learning-Based Segmentation and Classification of Leaf Images for Detection of Tomato Plant Disease (2022). https://www.frontiersin.org/articles/https://doi.org/10.3389/fpls.2022.1031748/full
8. Basavaiah, J.: Tomato Leaf Disease Classification Using Multiple Feature Extraction Techniques (2020). https://www.researchgate.net/publication/342248958_Tomato_Leaf_Disease_Classification_using_Multiple_Feature_Extraction_Techniques
9. Ramalakshmi, K.: Literature Review of Disease Detection in Tomato Leaf Using Deep Learning Techniques (2021). https://www.researchgate.net/publication/352123881_Literature_Review_of_Disease_Detection_in_Tomato_Leaf_using_Deep_Learning_Techniques
10. Nawaz, M., Nazir, T., Javed, A., Masood, M.: A Robust Deep Learning Approach for Tomato Plant Leaf Disease Localization and Classification (2022). https://www.nature.com/articles/s41598-022-21498-5
11. Sharma, R., Panigrahi, A., Garanayak, M.: Tomato Leaf Disease Detection Using Machine Learning (2022). https://ceur-ws.org/Vol-3283/Paper115.pdf
12. Chowdhury, M.E., Rahman, T., Khandakar, A., Ibtehaz, N.: Tomato Leaf Diseases (2021). https://www.intechopen.com/chapters/76494
13. Wu, Y., Xu, L., Goodman, E.D.: Tomato Leaf Disease Identification and Detection Based on Deep Convolutional Neural Network (2021). https://www.techscience.com/iasc/v28n2/42061/html
14. Vinay P.K.: Classification of Tomato Plant Leaf Disease Using Neural Network (2021). https://papers.ssrn.com/sol3/papers.cfm?abstract_id=3747953
15. Brahimini, M., Boukhalfa, K.: Deep Learning for Tomato Diseases: Classification and Symptoms Visualization (2017). https://www.tandfonline.com/doi/full/https://doi.org/10.1080/08839514.2017.1315516
16. Saranya, C.P., Palanivel Rajan, D., Mythili, M., Pushpalatha, K.., Saranya, V.: Enhanced Convolution Neural Network for Tomato Leaf Disease Classification (2021). https://www.ijrte.org/wp-content/uploads/papers/v8i6/F8970038620.pdf
17. Zaki, S.Z.M., Zulkifley, M.A.: Classification of Tomato Leaf Diseases Using MobileNetv2 (2021). https://www.semanticscholar.org/paper/Classification-of-tomato-leaf-diseases-using-v2-Zaki-Zulkifley/ed082a98b01694d917ac68542541e08a5c916ae5
18. Nagamani, H.S.: Tomato Leaf Disease Detection using Deep Learning Techniques (2022). https://thesai.org/Downloads/Volume13No1/Paper_38-Tomato_Leaf_Disease_Detection.pdf

19. Nagamani, H.S., Sarojadevi, H.: Tomato Leaf Disease Detection using Deep Learning Techniques (2022). https://thesai.org/Publications/ViewPaper?Volume=13&Issue=1&Code=IJACSA&SerialNo=38
20. Aishwarya, V.V., Pushpa, B.R.: Tomato Leaf Disease Detection and Classification Using CNN (2022). https://www.philstat.org/index.php/MSEA/article/download/853/486/1385
21. Liu, J., Wang, X.: Early Recognition of Tomato Gray Leaf Spot Disease Based on MobileNetv2-YOLOv3 Model (2022). https://plantmethods.biomedcentral.com/articles/https://doi.org/10.1186/s13007-020-00624-2
22. Chauhan, A., Mohana, R.: Improving Bert Model Accuracy for Uni-Modal Aspect-Based Sentiment Analysis Task (2023). https://www.scpe.org/index.php/scpe/article/view/2444/806, https://doi.org/10.12694/scpe.v24i3.2444

Detection of Audio Spoofing Attack Using Integrated Hand-Crafted Features with LSTM

Nidhi Chakravarty and Mohit Dua

Abstract Spoofing attacks, advanced text-to-speech (TTS) mechanisms, and voice conversion (VC) technologies pose significant security risks to Automatic Speaker Verification (ASV) systems by generating realistic-sounding synthetic speech and spoofed voices. To protect ASV systems from unauthorized access, it is essential to detect such synthetic speech and spoofed voices effectively. An Audio Spoof Detection (ASD) system typically comprises two main components: front-end feature extraction and back-end classification. In this paper, we introduce a novel approach to address this challenge by combining hand-crafted features with graph-based features during the front-end feature extraction stage. Specifically, we utilize Acoustic Ternary Pattern (ATP), Mel Frequency Cepstral Coefficient (MFCC), Gammatone Cepstral Coefficient (GTCC), and a new graph-based feature called Graph Frequency Cepstral Coefficient (GFCC). For the back-end classification, we employ Long Short-Term Memory (LSTM) networks. To evaluate the effectiveness of our proposed system, we develop six distinct configurations. The first three systems use ATP, MFCC, and GTCC features individually, while the remaining three systems combine GFCC with ATP, MFCC, and GTCC features, respectively. We conduct our experiments using the ASVspoof 2019 Physical Access (PA) evaluation datasets.

Keywords ASV · Replay attack · ATP · MFCC · GTCC · LSTM

1 Introduction

In recent years, the widespread use of speech-based authentication technologies has heralded a new era of ease and security. These systems provide a seamless method of user authentication by analyzing the unique qualities of a person's speech.

N. Chakravarty (✉) · M. Dua
Department of Computer Engineering, National Institute of Technology, Kurukshetra, India
e-mail: nidhichakravarty03@gmail.com

M. Dua
e-mail: er.mohitdua@nitkkr.ac.in

B. Soni et al. (eds.), *Beyond Artificial Intelligence*, Lecture Notes in Networks and Systems 1326, https://doi.org/10.1007/978-981-96-4170-3_16

However, as the use of such systems has grown, a critical weakness has emerged: the vulnerability to many attacks such as Impersonation attack [1] and Replay attacks [2].

ASD systems have two important components; Frontend and Backend. At frontend feature extraction has been done. Front-end feature extraction techniques used in audio spoofing are crucial for distinguishing genuine audio from fabricated audio. These approaches include extracting discriminative characteristics from audio signals. Several front-end approaches have been utilized for feature extraction in Audio Spoof Detection systems (ASD), including Mel-Frequency Cepstral Coefficients (MFCCs) [3], Linear Frequency Cepstrum Coefficient (LFCC) [4] and Perceptual Linear Prediction (PLP) [5]. Gammatone Frequency Cepstral Coefficients (GTCCs) [6] and Constant Q Cepstral Coefficients (CQCC) [7] are the two other feature extraction techniques that have gained popularity in recent years. However, like MFCC, CQCC does not perform well in noisy environments [8]. On the other hand, GTCC [9] performs well in noisy conditions [10]. Also, researchers try to improve these existing methods to improve their performance against different types of attacks. For instance, Authors in [11] have introduced modified CQCC and LFCC features by adding mean and variance to the log power spectrum [10]. In this work we have introduced hybrid features by combining graph-based features with cepstral features.

Similarly, back-end classification has seen substantial progress in recent years. Various machine learning methods are now employed for this purpose. For instance, Malik et al. [12] employed K-Nearest Neighbours (KNN), Naïve Bayes (NB), and Support Vector Machines (SVM) to train a binary classifier. Additionally, deep learning techniques such as Convolutional Neural Networks (CNN), Long Short-Term Memory (LSTM), and their upgraded versions play a key role in constructing effective classification models for ASV systems [13]. This work's contributions include the following:

1. This study develops six distinct ASD models to address the problem. Initially, three models are constructed using the audio features MFCC, GTCC, and ATP individually. Subsequently, an additional set of three models is created by combining the novel Graph Frequency Cepstral Coefficient (GFCC), a feature based on Graph Signal Processing (GSP), with each of the existing features. This approach allows for a comprehensive analysis of the effectiveness of both individual audio features and their combinations.
2. The study tests these models using the PA datasets from ASVspoof2019 and evaluates performance using metrics such as Precision, F1-score, Accuracy, Equal Error Rate (EER), and Recall.

The rest of the paper is organized as follows: Subsection 1.1 reviews related work, Sect. 2 thoroughly explains the proposed system architecture, Sect. 3 presents the experiments and findings, and Sect. 4 concludes the paper.

1.1 Related Work

This section provides the overview of existing countermeasures developed by authors in recent years. Authors [14] have developed two low-level characteristics from the audio's high-frequency spectrum: CQCC and our suggested High-Frequency Cepstral Coefficients (HFCC) and with an EER of 11.5%, which is 53% lower than the baseline GMM applied to CQCC, combining both features, HFCC + CQCC—and the Deep Neural Network (DNN)-Support Vector Machine (SVM) classifier. Chen et al. [15] have presented a solution for replay attack detection. The proposed model has implemented the frontend part using CQCC and MFCC features, and for the backend, DNN, Gaussian Mixture Models (GMM), and residual network have been used. Fusion of MFCC-ResNet, CQCC-GMM, and CQCC ResNet has produced a minimum tdcf of 13.30 and an EER of 2.58%.

The CQCC [16], MFCC and Mel Filter bank Slope (MFS) characteristics are used to construct three separate GMM-based replay detection systems. With the fusion of MFCC + MFS, EER was 16.3%, accuracy was 69.6%, and for non-voiced and full utterance, EER was 19.16%, and accuracy was 63.20%. The suggested method outperforms the best baseline system, which has a 37% relative improvement in Equal Error Rate (EER).

Chakravarty et al., [17] used data augmentation techniques, hybrid MFCC-GTCC features [18], and two deep learning models and two machine learning models, Long Short-Term Memory (LSTM), Gated Recurrent Unit (GRU), Random Forest (RF) and Support Vector Machine (SVM).

In Ref. [19] authors have proposed a new approach using a segment-based hybrid Feature Extraction method MFCC, CQCC and their hybrid features. The hybrid features were trained using various deep learning networks, including DenseNet, Long Short-Term Memory (LSTM). Joshi et al. [20] have proposed an ASV system that uses GTCC for feature extraction and LSTM for classification. The model is trained and evaluated using the VSDC dataset. The proposed system has achieved an EER of 2.38% and 2.72% in 1PR and 2PR attacks, respectively, and an accuracy of 98% and 97.26% in the same attacks [21, 22].

From literature, it can be concluded that handcrafted features like ATP, MFCC, and GTCC at the front end, and LSTM models at the back end, outperform traditional feature extraction and classification methods for speech verification tasks. Motivated by these works, we have employed ATP, GTCC, and MFCC features at the front end and LSTM at the back end.

2 Proposed Methodology

The proposed ASD system has been implemented by developing two models, as illustrated in Fig. 1. In both models, the first phase involves feature extraction to obtain information from the audio, while the second phase uses acoustic models to classify the audio. Figure 1 shows the proposed model, which processes speech signals from

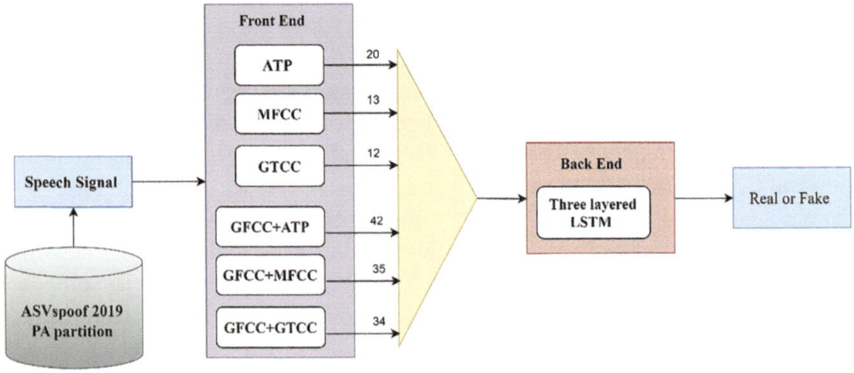

Fig. 1 Proposed architecture

the PA partition of the ASVspoof 2019 dataset. At the front end, three single features and three dual features are created using ATP, MFCC, and GTCC. Additionally, a new feature, GFCC, inspired by Graph Signal Processing (GSP), is created and combined with the existing features. At the back end, a deep learning-based LSTM acoustic model is employed to classify the audio.

2.1 Feature Extraction

This section provides details on the feature extraction step, which has been carried out using ATP, MFCC, and GTCC feature extraction techniques. Function 1 describes the algorithm for producing the ATP feature matrix. It takes an audio file in '.flac' format as input.

Function 1: *generate_ATP ()*

Input: audio file in .flac format

```
1.  generate_ATP (audio_file)
2.  {
3.     neighbour = x(i + (r − (P/2)));
4.     ltp_lower(1,k) = 1 ∗ lower_frame(1,k) + 2 ∗ lower_frame(2,k) + 4 ∗
       lower_frame(3,k) + 8 ∗ lower_frame(4,k) + 16 ∗ lower_frame(5,k) + 32 ∗
       lower_frame(6,k) + 64 ∗ lower_frame(7,k) + 128 ∗ lower_frame(8,k);     //
       converts the binary histogram to a decimal histogram for the lower ATP binary
5.     ltp_uper(1,k) = 1 ∗ uper_frame(1,k) + 2 ∗ uper_frame(2,k) + 4 ∗
       uper_frame(3,k) + 8 ∗ uper_frame(4,k) + 16 ∗ uper_frame(5,k) + 32 ∗
       uper_frame(6,k) + 64 ∗ uper_frame(7,k) + 128 ∗ uper_frame(8,k)//
       upper ATP binary histogram
6.     k = k + 1;
7.     ltp_hist_uper = hist(ltp_uper, 10);
8.     ltp_hist_lower = hist(ltp_lower, 10);
9.     ltp_full = [ltp_hist_uper ltp_hist_lower];
10.    return ltp_full
11. }
```

Output: *ATP feature matrix for each audio*

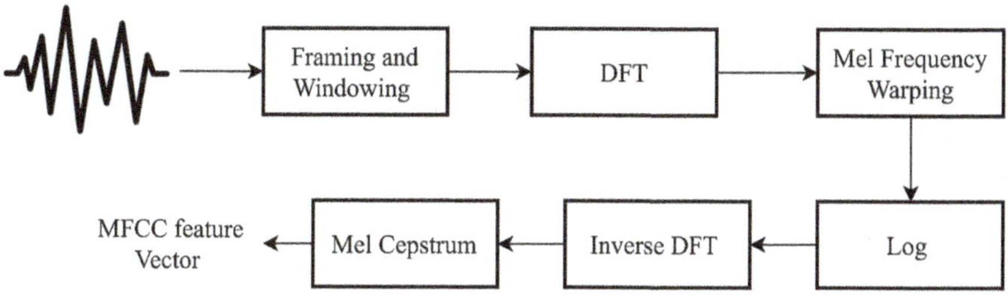

Fig. 2 Steps for extracting MFCC features

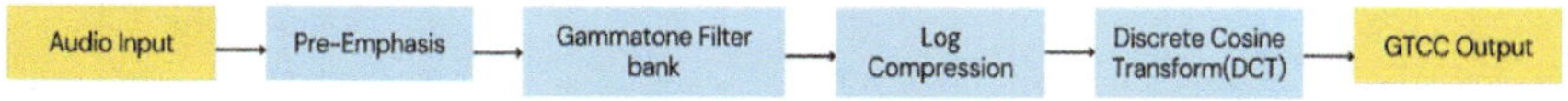

Fig. 3 Steps to generate GTCC features

Mel-Frequency Cepstral Coefficients (MFCC) are a set of coefficients used to represent the short-term power spectrum of an audio signal. They are derived from either the linear cosine transforms or the log spectrum. The steps to construct MFCC are given in Fig. 2.

Gammatone Cepstral Coefficients (GTCC) [23] processing is similar to MFCC. Figure 2 shows the steps involved in extracting GTCC features. GTCC utilizes the properties of the Gammatone Filter, which mimics the human auditory response to various frequencies. Figure 3 shows the steps to get GTCC features.

Graph signal processing (GSP) is a field focused on the analysis, processing, and manipulation of signals defined on graphs. In GSP, signals are represented as functions on graph nodes, capturing phenomena such as social interactions, traffic flows, or brain activity. Figure 4 illustrates the process of creating GFCC features. Compared to the fast Fourier transform (FFT), the GFT has been found to more accurately represent the structural relationships of speech samples.

2.2 *Classification*

The suggested method uses an LSTM-based acoustic model [24] to categorize sounds as real or fake, which serves as the mode's backend. Figure 5 depicts the proposed LSTM network, which is composed of 50 LSTM layers, each with its unique ReLU activation function. In this configuration, a feature matrix is sent into the LSTM as input. The model has 50 units in each of the three levels of the three-layered LSTM, followed by a dense output layer. To prevent overfitting, a 20% dropout layer is utilized, which disables some neurons at random. The output from the dropout layer is then sent to a dense layer with sigmoid activation.

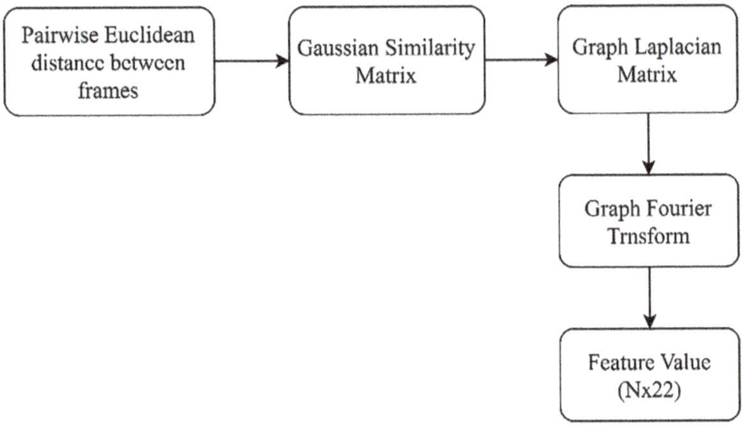

Fig. 4 Steps involved in Graph Signal Processing

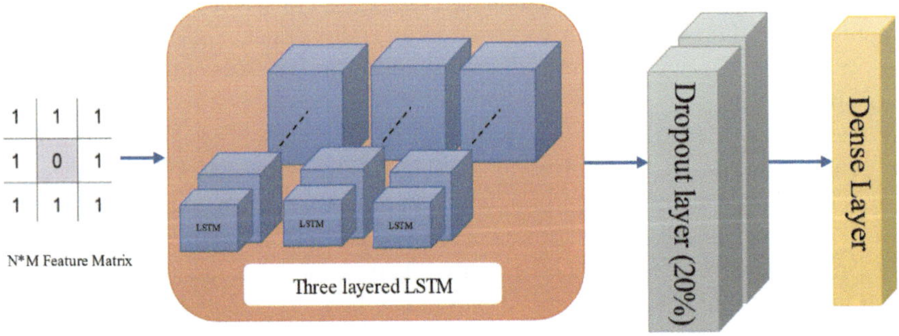

Fig. 5 Architecture of proposed LSTM model

3 Experimental Setup and Results

- **Dataset**: The ASVspoof 2019 PA dataset has been used to train and evaluate the proposed system. This portion of the dataset consists of recorded instances of replay attacks in English, with audio samples ranging from 2 to 8 s in length. Table 1 provides a detailed breakdown of the training, validation, and evaluation audio samples.

Table 1 ASVspoof2019 PA Partition

Samples	Physical Access
Training	38,880
Validation	9720
Evaluation	116,640

Table 2 Results of proposed models over PA dataset

Feature set	Accuracy (%)	Recall (%)	F1-score (%)	Precision (%)	EER (%)
ATP	83	84	85	81	7.53
MFCC	91	91	91	91.3	8.76
GTCC	98	99	98	98	2.82
GFCC + ATP	92	92	92	92.2	5.3
GFCC + MFCC	94	94.5	94	94	6.4
GFCC + GTCC	99	99	99	99	1.85

- **Performance Analysis of Model:** Table 2 presents the outcome for the proposed ASD system, which employs ATP, MFCC, and GTCC audio features, as well as their combination with GFCC, at the front end, and LSTM at the backend. Additionally, Table 2 shows the results for the suggested ASD system that uses six sets of audio features and LSTM at the backend for the PA dataset. The results indicate that the combination of GFCC and GTCC features achieves the highest EER of 1.85%. From result it can be concluded that the GFCC and GTCC feature combination outperforms the other five proposed feature sets.

4 Conclusion

The proposed work implements ASV models, with the first three systems utilizing ATP, MFCC, and GTCC features individually. The remaining three systems combine GFCC with ATP, MFCC, and GTCC features, respectively. The best results were achieved with the combination of GFCC and GTCC features along with the LSTM acoustic model for the ASVspoof 2019 PA dataset, obtaining an EER of 1.85%. This result indicates that combining handcrafted features with graph-based features can enhance speaker verification systems across various evaluation metrics. The proposed approach can be further expanded by incorporating more refined spectrogram-based features and advanced acoustic modeling techniques at the backend to handle real-time noisy scenarios.

References

1. Tan, C.B., et al.: A survey on presentation attack detection for automatic speaker verification systems: State-of-the-art, taxonomy, issues and future direction. Multimed. Tools Appl. **80**(21–23), 32725–32762 (2021)
2. Chakravarty, N., Dua, M.: A lightweight feature extraction technique for deepfake audio detection. Multimed. Tools Appl. pp 1–25 (2024)
3. Todisco, M., Delgado, H., Evans, N.W.D.: A new feature for automatic speaker verification anti-spoofing: constant Q cepstral coefficients. Odyssey **2016**, 283–290 (2016)

4. Fan, X., Hansen, J. H. L.: Speaker identification with whispered speech based on modified LFCC parameters and feature mapping. In: 2009 IEEE international conference on acoustics, speech and signal processing, pp. 4553–4556 (2009). https://doi.org/10.1109/ICASSP.2009.4960643
5. Hermansky, H.: Perceptual linear predictive (PLP) analysis of speech. J. Acoust. Soc. Am. **87**(4), 1738–1752 (1990)
6. Valero, X., Alías, F.: Gammatone cepstral coefficients: biologically inspired features for non-speech audio classification. Multimedia IEEE Trans. **14**, 1684–1689 (2012). https://doi.org/10.1109/TMM.2012.2199972
7. Todisco, M., Delgado, H., Evans, N.: Constant Q cepstral coefficients: a spoofing countermeasure for automatic speaker verification. Comput. Speech Lang. 45 (2017). https://doi.org/10.1016/j.csl.2017.01.001
8. Qian, Y., Chen, N., Dinkel, H., Wu, Z.: Deep feature engineering for noise robust spoofing detection. In: IEEE/ACM Trans. Audio Speech Lang. Process. 25(10), 1942–1955 (2017)
9. Fedila, M., Bengherabi, M., Amrouche, A.: Gammatone filterbank and symbiotic combination of amplitude and phase-based spectra for robust speaker verification under noisy conditions and compression artifacts. Multimed. Tools Appl. **77**, 16721–16739 (2018)
10. Malik, K., Javed, A., Malik, H., Irtaza, A.: A light-weight replay detection framework for voice controlled IoT devices. IEEE J. Sel. Top. Signal Process. PP (2020). https://doi.org/10.1109/JSTSP.2020.2999828
11. Mingrui, H. E., Ali, Z. S. F., Mianxin, T., Zhiyong, S., Zhengru, J., Longting, X. U.: Modified cepstral feature for speech anti-spoofing. J. Donghua Univ. 40(2) 2023
12. Mittal, A., Dua, M.: Static–dynamic features and hybrid deep learning models based spoof detection system for ASV. Complex Intell. Syst. 1–14 (2021)
13. Aravind, P. R., Nechiyil, U., Paramparambath, N.: Audio spoofing verification using deep convolutional neural networks by transfer learning. arXiv Prepr. arXiv2008.03464 (2020)
14. Wickramasinghe, B., Irtza, S., Ambikairajah, E., Epps, J.: Frequency domain linear prediction features for replay spoofing attack detection. In: Interspeech 661–665 (2018)
15. Chen, Z., Xie, Z., Zhang, W., Xu, X.: ResNet and model fusion for automatic spoofing detection (2017)
16. Cooper, D.: Speech detection using gammatone features and one-class support vector machine (2013)
17. Chakravarty Mohit, N. D.: Data augmentation and hybrid feature amalgamation to detect audio deep fake attacks. Phys. Scr. (2023) [Online]. Available: http://iopscience.iop.org/article/https://doi.org/10.1088/1402-4896/acea05
18. Chakravarty, N., Dua, M.: Noise robust ASV spoof detection using integrated features and time delay neural network. SN Comput. Sci. **4**(2), 127 (2022)
19. Huang, L., Gan, Y., Ye, H.: Audio-replay attacks spoofing detection for automatic speaker verification system. IEEE International Conference on Artificial Intelligence and Computer Applications (ICAICA) **2019**, 392–396 (2019)
20. Joshi, S., Dua, M.: LSTM-GTCC based approach for audio spoof detection. In: 2022 international conference on machine learning, big data, cloud and parallel computing (COM-IT-CON), vol. 1, pp. 656–661 (2022)
21. Dua, M., Sadhu, A., Jindal, A., Mehta, R.: A hybrid noise robust model for multireplay attack detection in Automatic speaker verification systems. Biomed. Signal Process. Control **74**, 103517 (2022). https://doi.org/10.1016/j.bspc.2022.103517
22. Chakravarty, N., Dua, M.: An improved feature extraction for Hindi language audio impersonation attack detection. Multimed. Tools Appl. 1–26 (2024)
23. Chakravarty, N., Dua, M.: Feature extraction using GTCC spectrogram and ResNet50 based classification for audio spoof detection. Int. J. Speech Technol. 1–13 (2024)
24. Chakravarty, N., Dua, M.: Spoof detection using sequentially integrated image and audio features. Int. J. Comput. Digital Syst. **13**(1), 1–1 (2023)

Skin Cancer Diagnosis and Classification Using VGG 19 with Transfer Learning and Data Augmentation

Palak Goyal, Anushka Aggarwal, and Rinkle Rani

Abstract Skin cancer, one of the most prevalent types of cancer in the world, has gradually on the rise for the past several decades. The VGG-19 deep convolutional neural network is used in this study to introduce a reliable method for the diagnosis and classification of skin cancer that is enhanced by transfer learning and data augmentation approaches. Further, we fine-tune the network to improve its diagnostic abilities using a pre-trained VGG-19 model on a substantial skin cancer image dataset. In addition, data augmentation technique is used to improve model generalisation and address the overfitting issue at the same time. Our approach exhibits astounding precision in the identification and classification of different forms of skin cancer, offering a viable tool for early diagnosis and accurate classification. This study makes a substantial contribution to the discipline of dermatology and offers an important resource for improving skin cancer diagnosis and patient care.

Keywords Skin cancer · VGG-19 · Transfer learning · Data augmentation

1 Introduction

A multitude of factors, such as lifestyle decisions like smoking, drinking, and eating as well as environmental changes like sun exposure and other kinds of radiation as well as viruses, have lately contributed to the recent rise in the number of cancer diagnoses [1]. Out of all, Skin cancer is among the most typical cancers. Unusual cell swellings might appear as a symptom of skin cancer. According to World Health

P. Goyal (✉) · A. Aggarwal · R. Rani
Department of Computer Science and Engineering, Thapar Institute of Engineering and Technology, Patiala, Punjab 147001, India
e-mail: pgoyal60_phd18@thapar.edu

A. Aggarwal
e-mail: aaggarwal_be22@thapar.edu

R. Rani
e-mail: raggarwal@thapar.edu

B. Soni et al. (eds.), *Beyond Artificial Intelligence*, Lecture Notes in Networks and Systems 1326, https://doi.org/10.1007/978-981-96-4170-3_17

Organisation studies, skin cancer affects one in three persons worldwide. One in five Americans, according to the Skin Cancer Foundation's projections [2–4], may get skin cancer at some point in their lives.

Skin cancer is a common and possibly fatal condition that continues to be a worldwide health challenge. Each year, many new cases are recorded each year, and the incidence of skin cancer has been rising over time [5]. Effective therapy and better patient outcomes depend on early and precise diagnosis [6]. Deep learning and artificial intelligence (AI) advancements recently have showed considerable promise for improving the diagnostic capacities of healthcare systems. This study employs the VGG-19 deep CNN architecture along with transfer learning and data augmentation approaches to build and evaluate a novel strategy for the diagnosis and classification of skin cancer.

The three most typical skin cancer subtypes are basal cell carcinoma, squamous cell carcinoma, and melanoma [7]. It is essential to quickly and accurately distinguish between these subtypes because their prognoses and treatment options differ greatly. Dermatologists typically use dermatoscopy, a non-invasive imaging technique, in conjunction with a visual examination to identify skin lesions. These techniques, however, may be subjective and sensitive to interobserver variability [8]. These conventional methods may be supplemented by the application of AI and deep learning models to dermatology, which would result in more accurate and effective diagnostic instruments.

Deep learning models, particularly CNNs, have proven to be exceptionally adept at tasks requiring image processing, such as the interpretation of medical images. Skin cancer diagnosis is a good fit for the VGG-19 architecture, which is renowned for its depth and efficacy in feature extraction [9]. Additionally, transfer learning enables us to use models that have already been pre-trained on sizable datasets, making them suitable for tasks requiring medical image classification with scant labelled data [10]. Additionally, data augmentation is essential for improving the generalization and resilience of deep learning models [11, 12]. The algorithm may learn to recognize skin lesions under various settings by artificially expanding the dataset with variants of existing images, thereby increasing its diagnostic accuracy.

In order to diagnose and categorize skin cancer, this research article will examine the performance of the VGG-19 architecture using transfer learning and data augmentation. We predict that this strategy will greatly increase the reliability and precision of skin cancer diagnosis, enabling early detection and effective treatment.

The rest of this essay is structured as follows: A summary of related research in the field of skin cancer diagnosis is given in Sect. 2, the technique and dataset used in our study are described in Sect. 3, the experimental findings and analysis are presented in Sect. 4, and the consequences and future directions of this research are covered in Sect. 5.

2 Related Work

On the basis of a brief assessment, some of the many studies conducted on a variety of issues and challenges connected to the detection, categorization, and diagnosis of skin cancer are reviewed.

On the basis of deep learning techniques, various research for skin cancer have emerged in recent years. A deep neural network for skin cancer classification was first introduced by [13], performing at a dermatologist-level. A 72.1% accuracy rate was attained by their model. Their work was limited, nevertheless, by the requirement for a sizable quantity of labelled data in order to successfully train deep learning models. Additionally, a convolution neural network (CNN)-based system for melanoma identification was proposed in the work of [14], reporting an accuracy of 76.7%. Large datasets are important, but as their analysis showed, there is a risk that dataset biases would impair model performance.

Additionally, [15] highlighted the most recent developments in the field while discussing the ISBI challenge on skin lesion analysis. The dependence on publicly accessible datasets, which might not accurately reflect real-world clinical circumstances, was a constraint. Innovation was sparked by the problem, yet the models still need improvement. Deep neural networks were used in a study by [16] to detect melanoma and show the promise of AI in identifying skin cancer with sensitivity and specificity of 88.0% and 76.1%, respectively. Their study had the drawback that the quality and diversity of the dataset can have an impact on the model's performance.

Another work by [17] investigated the use of ensemble deep learning methods for melanoma detection in dermoscopy images. This study demonstrated the value of combining numerous models and the improvement in classification performance that comes from ensembling as opposed to using a single model. The computational complexity of ensemble methods was a drawback. Furthermore, [18] reported an accuracy of 86.4% in a study that suggested a deep learning architecture for classification. The necessity for interpretability in medical settings was a drawback of their strategy, which included data augmentation.

Regardless of the studies mentioned above that classified skin cancer using deep learning algorithms, [19] unveiled a smartphone application for melanoma detection that makes use of deep learning. The application aimed to increase early melanoma detection, yet particular accuracy figures were not provided. Their work demonstrates the usefulness of AI in dermatology by bringing skin cancer detection to mobile platforms. However, a constraint was the requirement to resolve privacy issues and guarantee correct performance on different smartphone cameras.

Another investigation by [20] looked into the application of AI melanoma detection algorithms in clinical settings. The classification of skin lesions using a group of CNNs was the focus of their study. According to the study, accuracy was 86.6%. The requirement for thorough clinical validation and integration into healthcare workflows was one possible barrier. They examined the use of an ensemble of CNNs for skin lesion categorization in their previous study [21], attaining an accuracy of

80.3%. The requirement for substantial generalization across various skin types and lesion variations was a restriction.

In the last, [22] carried out a prospective investigation of a convolutional neural network for skin cancer diagnosis and showed how the incorporation of deep learning models into clinical practice could improve diagnostic accuracy. They could not disclose exact accuracy numbers, but their findings imply that AI can improve diagnostic precision. The necessity for additional research in actual clinical settings was a restriction.

Collectively, these findings highlight the promise of deep learning approaches in skin cancer classification problem. They noted several constraints, including the necessity for a variety of and representative datasets, interpretability in clinical practice, thorough clinical validation, and integration into healthcare workflows, even though they reported accuracy rates that were encouraging. By addressing some of these issues and investigating the use of VGG-19 with transfer learning and data augmentation for better skin cancer diagnosis, this research study makes a contribution to this developing field.

3 Proposed Approach

This study provides a thorough method that makes use of deep learning network for the diagnosis and categorization of skin cancer. To improvise the precision and robustness of the model, the study specifically uses the VGG-19 architecture along with transfer learning and data augmentation. Skin cancer treatment success depends on a prompt and accurate diagnosis. This condition is still a major global health concern. The suggested strategy is to enhance the potential for early cancer detection and the diagnostic capabilities of automated systems, making them an important tool in dermatology. Figure 1 presents the workflow for the proposed model and the details of the methodology is provided below.

3.1 Data Collection

The dataset used in this study was obtained from ISIC Repository (https://www.isic-archive.com/#!/topWithHeader/wideContentTop/main). It contains data on 3297 people, that are separated into two groups of patients: Benign and Malignant moles. Additionally, the Benign class consists of 1800 patients, while the Malignant class consists of 1497 patients. The data used in this investigation are summarized in Table 1, along with details like the number of patients in each class, which was further divided into a train and test folder.

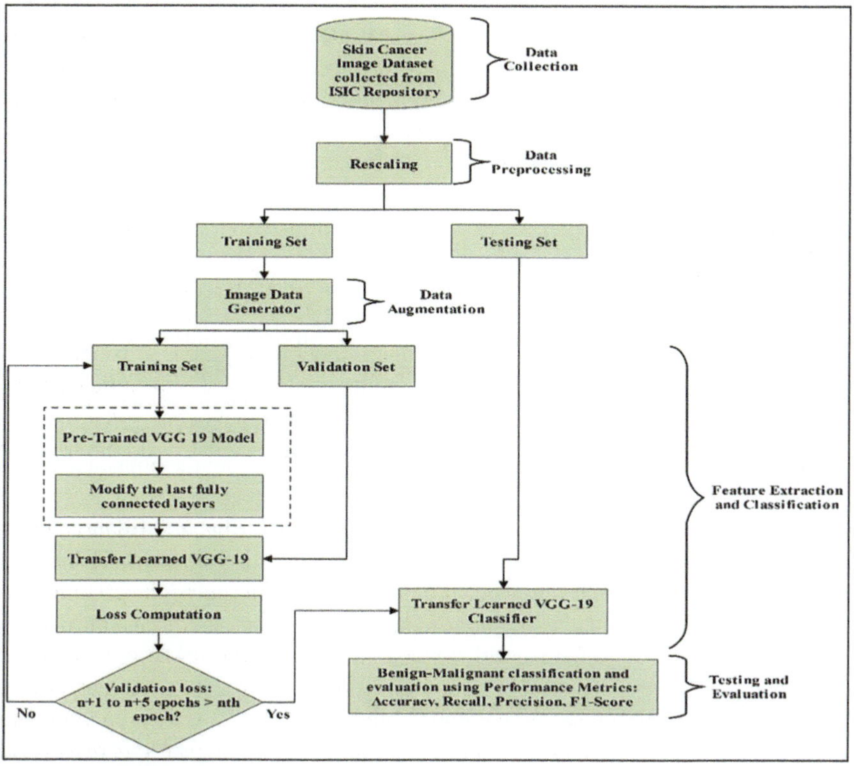

Fig. 1 Proposed approach

Table 1 Dataset description

Criteria	Benign patients	Malignant patients
Count of patients	1800	1497
Train folder	1440	1197
Test folder	360	300

3.2 Data Preprocessing

The range of pixel values or voxel intensity levels is changed in the following phase as a preprocessing approach to reduce noise in the data. In order to prevent the value of one or more pixels from becoming overpowering, it reduces or eliminates important data variations. Various normalization techniques exist, including the zero mean and unit standard deviation method, [0,1] rescaling, [−1,1] rescaling, etc. In this paper, the unit standard deviation and zero mean techniques are used. A final normalized image has pixel values with a standard deviation of one and an average of zero, on which further steps are applied.

Table 2 Augmented dataset details

Criteria	Benign patients	Malignant patients
Count of patients	2360	2300
Train folder	2000	2000
Test folder	360	300

3.3 Data Augmentation

Through the wide representation of skin moles, the skin cancer identification system must be strengthened to withstand various scenarios. More data must be gleaned from the training of the dataset in order to decrease the likelihood of overfitting. It can be accomplished utilizing a number of image alteration techniques, i.e., Data Augmentation techniques. In this study, Image Data Generator method has been applied for data augmentation in which the images are altered without changing the boundary boxes using the methods: Padding, flipping (Vertical and Horizontal), Rotating, Rescaling and Zooming. The rotated and flipped images along with balanced dataset have enhanced the model's detection performance and durability. In addition, Table 2 describes the statistics of the augmented dataset used in the study.

4 Feature Extraction and Classification

This step entails the extraction of features and classification on the original as well as augmented dataset for the classification of skin cancer. In this study, model was build using the VGG-19 deep convolutional neural network architecture using pre-trained weights from an image dataset (like ImageNet) to initialize the network. Then, the model is fine-tuned using appropriate loss functions and optimization parameters and transfer learned by retraining the fully connected layers of VGG-19 model so that it can work according to the current problem, i.e., to classify the skin cancer and at the same time, to prevent overfitting. Further, Fig. 2 presents the fine-tuned and transfer learned deep convolutional network model used in the study.

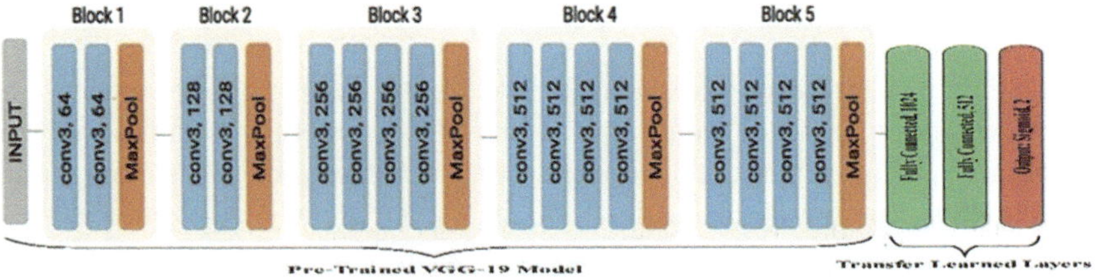

Fig. 2 Transfer learned VGG-19 model

5 Testing and Performance Evaluation

In this step, the trained models are compared to test data in order to evaluate their performance. Performance evaluation measures like recall, accuracy, precision, and F1-Score are computed during this step. Further, the results of the aforementioned models is compiled and analyzed using confusion matrices and the aforementioned performance evaluation measures.

6 Results and Discussions

The Anaconda platform is used for this study's experimentation in Jupyter Notebook. Additionally, the following standards were used for performance evaluation:

Accuracy: It is demonstrated by taking fur values: true positive (t_p), true negative (t_n), false positive (f_p) and false negative (f_n) as:

$$Acc = \frac{t_p + t_n}{t_p + t_n + f_n + f_p} \tag{1}$$

Precision: It's the portion of the outcomes that apply to the search. This is how it can be characterized:

$$Precision = \frac{t_p}{t_p + f_p} \tag{2}$$

Recall: The part of the data that is pertinent to the query is the data that is successfully obtained; calculated as:

$$Recall = \frac{t_p}{t_p + f_n} \tag{3}$$

F1-Score: The following definition identifies it as a measure that combines recall and precision:

$$F1 - Score = \frac{2 * Precision * Recall}{Precision + Recall} \tag{4}$$

To measure the performance of models using the aforementioned metrics, data must be processed effectively to remove noise and enhance classifier performance. To decide which preprocessing techniques should be applied to eliminate noise from the data, data analysis is necessary first. For this, the images are analyzed and rescaling was done to reduce the noise from the dataset. Then, after the preprocessing step, augmentation was done on the dataset using Image Data Generator to enhance the dataset size so as to increase the generalization performance and prevent overfitting

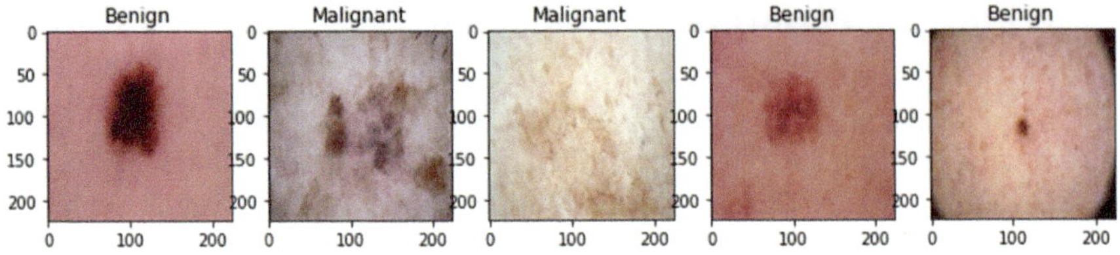

Fig. 3 Samples of the benign and malignant class

simultaneously. Further, the samples of the final dataset after preprocessing and augmentation is shown in Fig. 3.

After the aforementioned steps, final augmented dataset as well as original dataset is used for evaluating and classifying the skin cancer patients; and results are compared for both the datasets. Firstly, to compare the training set performance, four parameters named Training Accuracy, Validation Accuracy, Training Loss and Validation Loss are calculated for both the datasets and Fig. 4 presents the learning curves for the datasets.

After the anaysis of curves, the most optimal iteration is recorded and the model corresponding to that parameters is further used for testing and the values of four parameters for the aforementioned iteration is depicted in Table 3.

In the next step, the testing data was pre-owned to evaluate the fine-tuned and transfer learned VGG19 model for the optimal iteration weights. The performance metrics values for the datsets is shown in Table 4. The corresponding confusion matrices for both the datasets on the testing set is shown in Fig. 5 respectively.

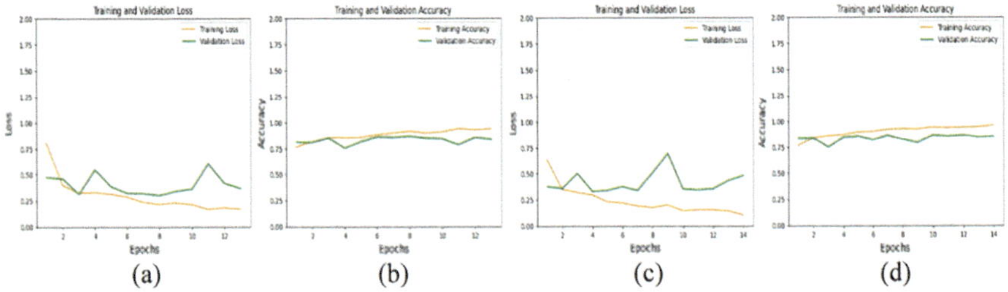

Fig. 4 Training and validation curves: **a** Loss curve without augmentation **b** Accuracy Curve without augmentation **c** Loss curve with augmentation **d** Accuracy curve with augmentation

Table 3 Training and validation loss and accuracy values for optimal iteration

Dataset	Training loss	Training accuracy	Validation loss	Validation accuracy
Without augmentation	0.232347	0.896159	0.318534	0.856061
With augmentation	0.293242	0.868125	0.330437	0.84625

Table 4 Proposed Model Results

Dataset	Testing accuracy	Precision	Recall	F1-Score
Without augmentation	0.8227	0.8343	0.8314	0.8328
With augmentation	0.8636	0.8622	0.8636	0.8629

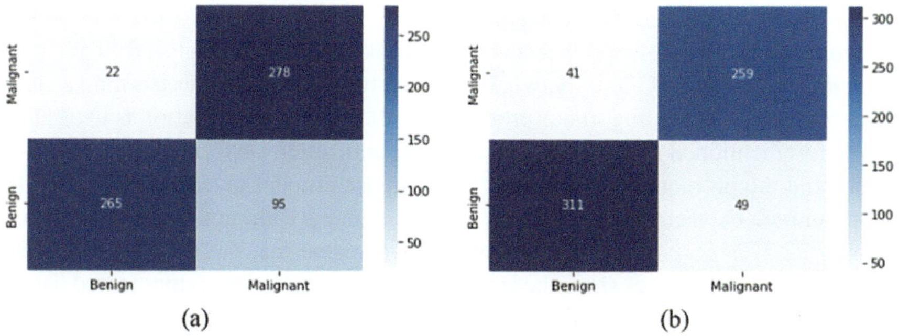

Fig. 5 Confusion matrices: **a** dataset without augmentation **b** dataset with augmentation

As seen in Tables 3 and 4, the difference between the training, validation and testing accuracy is very large for the original dataset without augmentation as compared to dataset with augmentation; which proves that original dataset gets overfitted for the deep learning model. Further, the testing accuracy is large for the augmented dataset as compared to the original dataset as seen in Table 4 and Fig. 5 and the comparison for them is shown in Fig. 6 using bar graphs. Finally, it can be concluded that the suggested deep learning model, along with transfer learning and data augmentation, improves generalisation performance for skin cancer diagnosis without overfitting the outcomes.

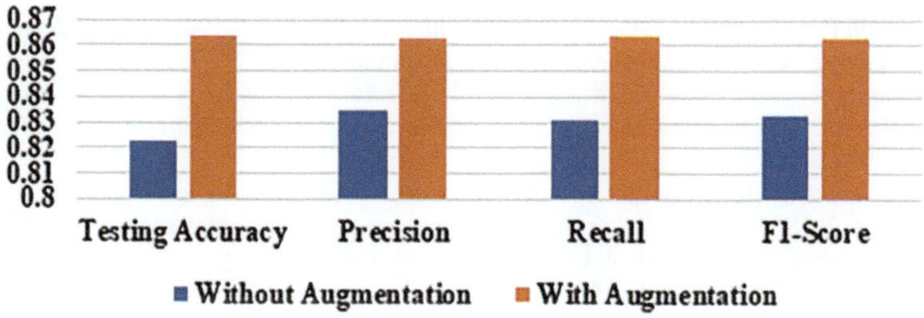

Fig. 6 Performance comparison of datasets: without augmentation and with augmentation

7 Conclusion

Skin cancer is among the most prevalent and dangerous cancers. Dermatologists primarily use ocular examinations to diagnose this illness. However, this is a difficult task because of the little variation in the appearance of mole in different classes. On the contrary, in current research, deep neural networks have performed better than dermatologists at skin cancer categorization. In this study, the effectiveness of deep convolution networks to predict and classify skin cancer is analyzed. This study uses the pre-trained VGG-19 network with fine tuning and transfer learning so as to make it capable according to our problem. Further, data augmentation is used along with aforementioned model to enhance the generalization performance as well as to overcome the problem of overfitting. The proposed model given 86.36% accuracy on augmnted dataset which is higher when compared with non-augmented dataset having 82.27% accuracy. Further, the results revealed that the training, validation and testing performances show the large difference in case of non-augmented dataset due to overfitting as compared to augmented dataset. In the end, we can say that the proposed model can be beneficial in generalization for skin cancer detection and classification. As can be observed, this suggested approach is only tested against a dataset related to skin cancer, but it may eventually be broadly applicable to other clinical research disciplines. Also, other deep learning methods such as MobileNet, ResNet etc. can be used to improve the classification results.

References

1. Dorj, U.O., Lee, K.K., Choi, J.Y., Lee, M.: The skin cancer classification using deep convolutional neural network. Multimedia Tools Appl. **77**, 9909–9924 (2018)
2. World Health Organization.: Radiation: ultraviolet (UV) radiation and skin cancer—how common is skin cancer. https://www.who.int/news-room/q-a-detail/radiation-ultraviolet-(uv)-radiation-and-skin-cancer (2017), Accessed 12th Jun 2021
3. Piccialli, F., Di Somma, V., Giampaolo, F., Cuomo, S., Fortino, G.: A survey on deep learning in medicine: why, how and when? Informat. Fusion **66**, 111–137 (2021)
4. Razmjooy, N., Ashourian, M., Karimifard, M., Estrela, V.V., Loschi, H.J., Do Nascimento, D., Vishnevski, M.: Computer-aided diagnosis of skin cancer: a review. Curr. Med. Imaging **16**(7), 781–793 (2020)
5. Bray, F., Ferlay, J., Soerjomataram, I., Siegel, R. L., Torre, L. A., Jemal, A.: Global cancer statistics 2018: GLOBOCAN estimates of incidence and mortality worldwide for 36 cancers in 185 countries. CA: Cancer J. Clinic. 68(6), 394–424 (2018)
6. Balch, C.M., Gershenwald, J.E., Soong, S.J., Thompson, J.F., Atkins, M.B., Byrd, D.R., Scolyer, R.A.: Final version of 2009 AJCC melanoma staging and classification. J. Clin. Oncol. **27**(36), 6199–6206 (2009)
7. Guy, G.P., Machlin, S.R., Ekwueme, D.U., Yabroff, K.R.: Prevalence and costs of skin cancer treatment in the US, 2002–2006 and 2007–2011. Am. J. Prev. Med. **48**(2), 183–187 (2015)
8. Argenziano, G., Soyer, H.P., De Giorgi, V., et al.: Dermoscopy of pigmented skin lesions: results of a consensus meeting via the Internet. J. Am. Acad. Dermatol. **48**(5), 679–693 (2018)
9. Simonyan, K., Zisserman, A.: Very deep convolutional networks for large-scale image recognition. arXiv preprint arXiv:1409.1556 (2015)

10. Litjens, G., Kooi, T., Bejnordi, B.E., Setio, A.A.A., Ciompi, F., Ghafoorian, M., Sánchez, C.I.: A survey on deep learning in medical image analysis. Med. Image Anal. **42**, 60–88 (2017)
11. Shorten, C., Khoshgoftaar, T.M.: A survey on image data augmentation for deep learning. J. Big Data **6**(1), 60 (2019)
12. Arora, S., Rani, R.: Real-time classification of cardiac events in arrhythmia disease. In 2022 IEEE international conference on current development in engineering and technology (CCET), 1–5 (December, 2022)
13. Esteva, A., Kuprel, B., Novoa, R.A., et al.: Dermatologist-level classification of skin cancer with deep neural networks. Nature **542**(7639), 115–118 (2017)
14. Codella, N., Cai, J., Abedini, M., Garnavi, R., Halpern, A., Smith, J.R.: Deep learning, sparse coding and SVM for melanoma recognition in dermoscopy images. In: International workshop on machine learning in medical imaging, 118–126 (2017)
15. Menegola, A., Tavares, J.M.R.S., Fornaciali, M., et al.: Skin lesion analysis toward melanoma detection: a challenge at the 2017 International Symposium on Biomedical Imaging (ISBI), hosted by the International Skin Imaging Collaboration (ISIC). In: International symposium on biomedical imaging (ISBI), 2017 IEEE 14th International Symposium, 168–172 (2017)
16. Tschand, P., Rinner, C., Apalla, Z., et al.: Human-computer collaboration for skin cancer recognition. Nat. Med. **24**(9), 1–5 (2018)
17. Sun, L., Liang, Y., Zhang, X., et al.: Ensemble deep learning for melanoma recognition in dermoscopy images. Comput. Biol. Med. **101**, 1–9 (2018)
18. Yu, L., Chen, H., Dou, Q., et al.: Multi-class skin cancer classification during deep learning features. Comput. Biol. Med. **99**, 81–88 (2018)
19. Haenssle, H.A., Fink, C., Schneiderbauer, R., et al.: Melanoma detection using deep learning: a smartphone application. Dermatol. Pract. Conceptual **8**(1), 1–5 (2018)
20. Brinker, T.J., Hekler, A., Enk, A.H., et al.: Deep learning outperformed 136 of 157 dermatologists in a head-to-head dermoscopic melanoma image classification task. Eur. J. Cancer **113**, 47–54 (2019)
21. Brinker, T.J., Hekler, A., Utikal, J.S., et al.: Deep ensemble of convolutional neural networks for skin lesion classification. Eur. J. Cancer **119**, 24–31 (2019)
22. Gessert, N., Parisi, F., Giorgi, V.D., et al.: A prospective study of a convolutional neural network for skin cancer diagnosis: The New Era of Total Body Dermatoscopy? J. Eur. Acad. Dermatol. Venereol. **35**(6), 1315–1322 (2021)

Smart Stick Assistant for Visually Impaired People

Juiley Sangram Raut, Vaibhav Shamrao Ingale, and Jayashree Ashok Salunkhe

Abstract The existing solutions for the visually impaired (VI) include usage of guide dogs, walking sticks and devices for obstacle detection. The devices detect the obstacle and inform the user about the location of the obstacle via audio or vibration. The walking sticks and devices using hats, belts, etc. are the extra items that the VI Person need to carry while going out. Also, these aids do not detect the object type i.e. if it is any animal, human or any object and if object then what object it is. The problems such as having to carry extra items for aid, no object identification, etc. Should be eliminated and the VI should be able to walk around like any other normal person. The proposed system provides smart assistant to the visually impaired people through stick. The main objectives of the proposed systems are: 1. To design and develop a "Smart Stick for VI people". 2. To help visually impaired people in navigation. 3. To design a system which will recognize the obstacles in the route of VI person The system identified the obstacles/objects which comes in front of the VI people. Also identify object types as well as number of objects. Authors used object detection algorithm.

Keywords Object detection · Smart assistant · Visually impaired · Smart stick

J. S. Raut (✉)
Veermata Jijabai Technological Institute, Mumbai, India
e-mail: juiley.raut@gmail.com

V. S. Ingale
Punjab Engineering College, Chandigarh, India
e-mail: vaibhavshamraoingale.phd21cs@pec.edu.in

J. A. Salunkhe
Governmant Polytechnic, Thane, India

B. Soni et al. (eds.), *Beyond Artificial Intelligence*, Lecture Notes in Networks and Systems 1326, https://doi.org/10.1007/978-981-96-4170-3_18

1 Introduction

Robots that can complete intended task in unstructured conditions without ongoing human supervision are called autonomous robots. Many different types of robots have some level of autonomy. Robots can indeed be autonomous in a variety of ways. A high level of autonomy is especially desired in industries like space exploration and deep-sea research where communication hiccups or delays are inevitable. People had dreams of exploring the deep waters before the twentieth century. They were exposed to risks from the ocean since they were unaware that there was crude oil under the sea. They had the technological means to look into shipwrecks. Nowadays, due to technological advancements and the creation of autonomous underwater vehicles, biological resources in rivers may be exploited for object detection in surveillance and reconnaissance, critical area safety, pipeline welding, mining, fishing, archaeology, and environmental pollution. The creation of a low power, compact gadget capable of taking pictures at extremely low light levels was made possible by the ease with which still or video images can now be captured in the age of digital electronics. The study of the environment of aquatic creatures benefits greatly from collections of underwater pictures and movies. A managed autonomous underwater vehicle is one that carries out underwater tasks. AUVs are often compact and straightforward gadgets that are solely used for monitoring, to take pictures and do certain measurements via underwater cameras [1].

Deep learning models differs in term of the structural and functional characteristics of biological brains (especially human brains), which renders each other inconsistent to neuroscience proof. Deep learning models are only loosely inspired by information communication and processing patterns in biological nervous systems.

Deep learning is a part of machine leaning which learns representation of data instead of some task-based algorithms. It is also called as deep structured learning or hierarchical learning. Unsupervised, semi-supervised, and supervised learning are all possible [2].

While older machine learning models cease advancing after reaching a saturation threshold, deep learning models often function well with the volume of data.

2 Literature Review

A system with five ultrasonic sensors that constantly trigger sequential manner and are positioned as left, right, and front was introduced by Laubhan et al. in [8]. The impediments are found by the ultrasonic sensors, which then compare them to the threshold. The white walking cane that Mahmud, et al. [9] exhibited system white has ultrasonic sensors and vibrators attached to it. When an impediment is present and within 70 cm of the user, the system's built-in speech and vibration feedback informs them. The Aljahdali et al. [10] approach is mostly used for elderly adults over the age of 65. Hardware and software are used by the system. The system is an Internet of Things-based system that connects to the server online immediately. According to Aymaz et al. [11], the system consists of a headset with two membranes: the left membrane detects impediments on the left, while the right membrane detects obstacles on the right. Two ultrasonic sensors that can detect obstructions to the right and left are placed on each membrane. The ultrasonic distance sensors are 60° apart. Patil et al. [12] describe a wearable, portable, and user-friendly technology that people may utilize in their daily lives. For simulating the authentic sensation produced by the classic white cane, this system comprises of the belt and a modernized cane. The belt is equipped with a Raspberry Pi 3, Lithium battery, PIR, Ultrasonic, and PIR sensors. The basic goal, according to Muthu Senthil, et al. [13], is to transform a regular text with various typefaces to an audio output. The suggested method employs a camera module to capture a picture of the text and an ultrasonic sensor to find the correct distance to the item. The system's ultrasonic sensor determines the object's distance first, and then a camera module, which takes a picture of the text inside, completes the process. Redmon et al. [14] offer an object identification model from which the pixels of the image to the corresponding coordinates of boxes and confidence score in a single regression scenario. You can tell what things are available and where they have been placed by only glancing at a picture once (YOLO). A neural network guesses several dimensions of frames and a class probability for each box simultaneously. Sharma et. al. [15], offer a system that employs ultrasonic sensors to perform obstacle avoidance. Sensor automatically transmits eight pulses at 40 kHz and checks to see if a pulse signal has returned. The time from sending ultrasonic to returning is measured in terms of high output I/O duration if the signal is returned. The user will simply reach its target using the Mohanty, et al. [16] technique described in the study. To carry out their further implementation with the aid of an Android application, the system suggests using shoes. An input must be sent from source to destination by the user. The system presented by Rastogi, et al. [17] uses 2 components. The shoes serve as the system's principal module, while the cane serves as its secondary module. The main module has two IR proximity sensors that can identify an object's front and left side. The IR proximity sensors function similarly to an ultrasonic sensor to identify impediments. According to Murthy et al. [18], the system includes several hardware elements, including eyeglasses, a waistbelt, and shoes. Nine ultrasonic sensors are used altogether, three for each item. All of the sensors on each item operate in three separate directions in order to locate the impediments. To ensure that there are no obstacles or objects remaining to recognize, the three devices are put on the user's three separate body parts. The technology was introduced by Mustapha, et al. [19] and fundamentally operates wirelessly and securely. The suggested system detects the item using infrared and ultrasonic sensors. With the aid of a force sensing resistor, both sensors are turned on. The other sensors are prevented from being constantly triggered by the use of Force Sensor Resistors. Khursade et al. [20] found IOT system. The system involves sensors and vibrators to detect impediments in the immediate surroundings and to provide the blind person with input on their location. The entire mechanism was fitted inside the shoes. In order to help visually impaired persons , avoid impediments and improve their ability to walk about, Khampachua et al. [21] devised a system that provides a wrist-mounted wearable gadget. The accelerometer in the smartphone and the ultrasonic sensor on the watchstrap are the sources of the data used in the detection method.

3 Proposed System

The proposed system (Fig. 1) of the Smart Stick for Visually Impaired with Object Detection and Classification represents a cutting-edge assistive technology designed to enhance the mobility and safety of individuals with visual impairments. This innovative device employs a combination of advanced sensors, a camera module, and machine learning algorithms to detect and identify objects in the user's vicinity. Here's a detailed breakdown of its key components and functionalities (Table 1).

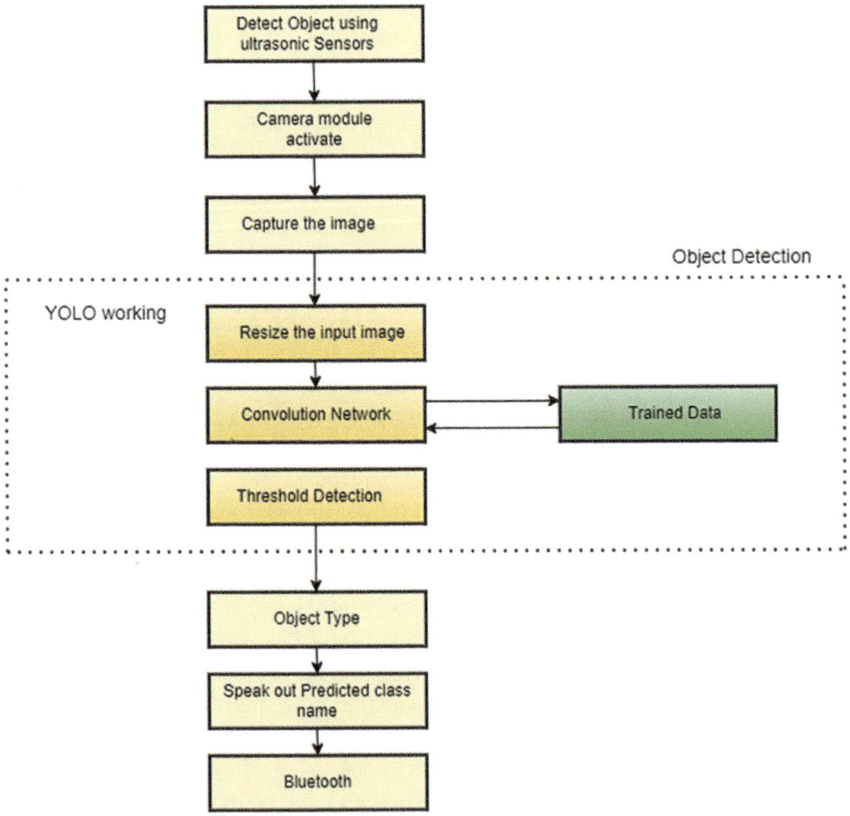

Fig. 1 Proposed system

Table 1 Comparison of different systems for VI

S. No	Title	Advantage	Limitation
1	"A Wearable Portable Electronic Travel Aid for Blind" [3]	Does not use walking stick, uses wireless Bluetooth for informing user about the obstacle via audio	It does not classify the object ahead i.e. either living or non-living
2	"Vibration and Voice Operated Navigation System for Visually Impaired Person" [4]	Tells the direction of the object whether left, right etc. via audio and vibrations	The system uses a white cane and it does not detect the object type
3	"IoT Based Assistive Walker Device for Frail & Visually Impaired People" [5]	It navigates the user towards the particular destination according to maps	It requires Internet. Does not detect the type of obstacle and distance of that obstacle
4	"Ultrasonic Assistive Headset for Visually Impaired People " [6]	Ultrasonic sensors detect the objects and inform the user via headset. Uses Solar panels and hence is battery less	The placement of ultrasonic sensors on the headphones are not able to detect the obstacles on the ground level
5	"Environment Sniffing Smart Portable Assistive Device for Visually Impaired Individuals" [7]	Detects the signs, human using PIR and obstacles such as stairs	Does not differentiate between the human and other animals. Requires the use of Walking stick
6	"You Only Look Once: Unified, Real-Time Object Detection" [8]	We can directly use full images to train YOLO. Fast YOLO is one of the fastest in object detectors	YOLO shows poor performance while detecting small objects like flock of birds
7	"Smart Assistance for Blind People using Raspberry Pi "[9]	Implements OCR Recognizes the Text using image processing and converts it to audio output	Only detects the text but has no use in navigation of the blind person
8	"Designof Microcontroller Based Virtual Eye for The Blind "[10]	Detects the obstacle and its distance using the ultrasonic sensors	Uses Hat, Shoes as well as cane to guide through obstacles. Does not detect the object type
9	"Smart Shoe for Visually Impaired" [11]	Provides Navigation Assistancewhile travelling using android application	RequiresInternet. Environment changes dynamically hence the predefined path cannot always be safest route
10	"Smart Assistive Shoes and Cane: Sole mates For the Blind People" [12]	It doesn't need any external help/assistant to walk on streets	It uses walking cane for navigation aid. Does not detect the object type

Ultrasonic Sensor: The system begins with an Ultrasonic Sensor as its primary input. This sensor emits high-frequency sound waves, which bounce off objects in the user's path and return as echoes. By measuring the time it takes for the echoes to return, the system can calculate the distance between the user and the objects, providing essential information about obstacles in the surrounding environment.

Camera Module: Upon detecting an object within a certain range through the ultrasonic sensor, the Camera Module is activated. This module consists of a high-resolution camera that captures real-time images or video feed of the detected object or objects.

YOLO (You Only Look Once) Module: The captured images or video feed are processed through the YOLO (You Only Look Once) module, which is a state-of-the-art object detection algorithm. YOLO is capable of identifying and locating multiple objects within an image or video frame simultaneously, making it an ideal choice for real-time object recognition.

Object Type and Counting: The YOLO module identifies the type of objects present in the user's environment. It recognizes common objects such as pedestrians, vehicles, obstacles, and more. Additionally, it counts the number of each type of object present in the field of view.

Audio Feedback: Once the object types and counts are determined, the Smart Stick provides audio feedback to the visually impaired user. It communicates the results through text-to-speech technology, describing the detected objects and their quantities in a clear and concise manner. For example, it may say, "Two pedestrians and one vehicle detected in front."

4 Implementation

This consists of information on the working of the system, system flow and step wise procedure of working of the implemented system.

4.1 Working

For the purpose of assisting blind and VI individuals with their navigation, the system creates a model for object detection and identification. The system that has been created comprises of two modules in total: obstacle detection and object detection. Three HC-SR04 ultrasonic sensors are used to identify obstacles.

By detecting the amount, the time that has passed between sending and receiving an ultrasonic pulse, an ultrasonic sensor may determine the distance from a target. The distance will be compared to the predetermined threshold after receiving the ultrasonic pulse. The Raspberry Pi will trigger the Camera Module Rev. 1.3 to activate if the threshold value is crossed. The YOLO (You Only Look Once) algorithm will be used by the camera module to detect objects in real-time. The YOLO approach first resizes the image before running a convolution network and predicting the object's class. The system then converts the predicted object class from text-to-speech in order to tell the user of the discovered object. The Bluetooth Headset allows the user to hear the audio output. The raspberry pi 4 b+ handles all of the object detection processing.

4.2 *Hardware*

The Hardware components used to implement the implemented system are as follows:

4.2.1 Raspberry 4 Model B+

The Raspberry Pi 4 Model B+ can handle PoE thanks to an additional PoE HAT, which has a 64-bit four core CPU running at 1.4 GHz, dual-band 2.4 GHz and 5 GHz wireless LAN, Bluetooth 4.2/BLE, faster Ethernet, and PoE capability. Because of the modular compliance certification for the dual-band wireless LAN, the board may be used in completed products with substantially less wireless LAN compliance testing, lowering both cost and time to market.

4.2.2 Camera Module Rev 1.3

This Pi Camera Module was created specifically as an accessory for the Raspberry Pi. It connects to the Raspberry Pi via one of the two tiny connections on the top of the board. This interface makes use of the specific CSI interface that was created just for connecting to cameras. The CSI bus just transports pixel data and is capable of carrying exceptionally high data speeds.

4.2.3 Ultrasonic Sensors HC-SR04

Ultrasonic sensors use ultrasonic waves to measure distance. An ultrasonic wave is sent by the sensor head, which then picks up the wave that the target reflects back to it. The time elapsed between the emission and reception is measured by ultrasonic sensors to determine the target's distance.

The formula below can be used to get the distance.

The formula to calculate distance:

$$\text{Distance L} = 1/2 \times \text{T} \times \text{C}$$

where L represents the distance, T represents the time passed between emission and reception, and C represents the speed (The value is divided by partial since T is the time required time to return the distance.)

4.3 Circuit Diagram

A circuit diagram (Fig. 2) is a graphical depiction of a wiring schematic. A graphical circuit diagram uses basic representations of elements, whereas a circuit diagram shows the circuit's elements and interconnections using defined symbolic representations. Figure 2 depicts the circuit for the implemented system. The ultrasonic sensors are wired to the raspberry pi. All ultrasonic sensors' Vcc is combined and linked to Raspberry Pi pin number 2. All ultrasonic sensors' Trig pins are shot together and linked to Raspberry Pi pin number 12. In addition, the Gnd pins of all ultrasonic sensors are pooled and linked to Raspberry Pi pin number 6. Each ultrasonic sensor's Echo is linked to Raspberry Pi pins 16, 18, and 22 independently. The CSI Camera Connector is attached to the camera module. Pin 17 is linked to the vibration motor.

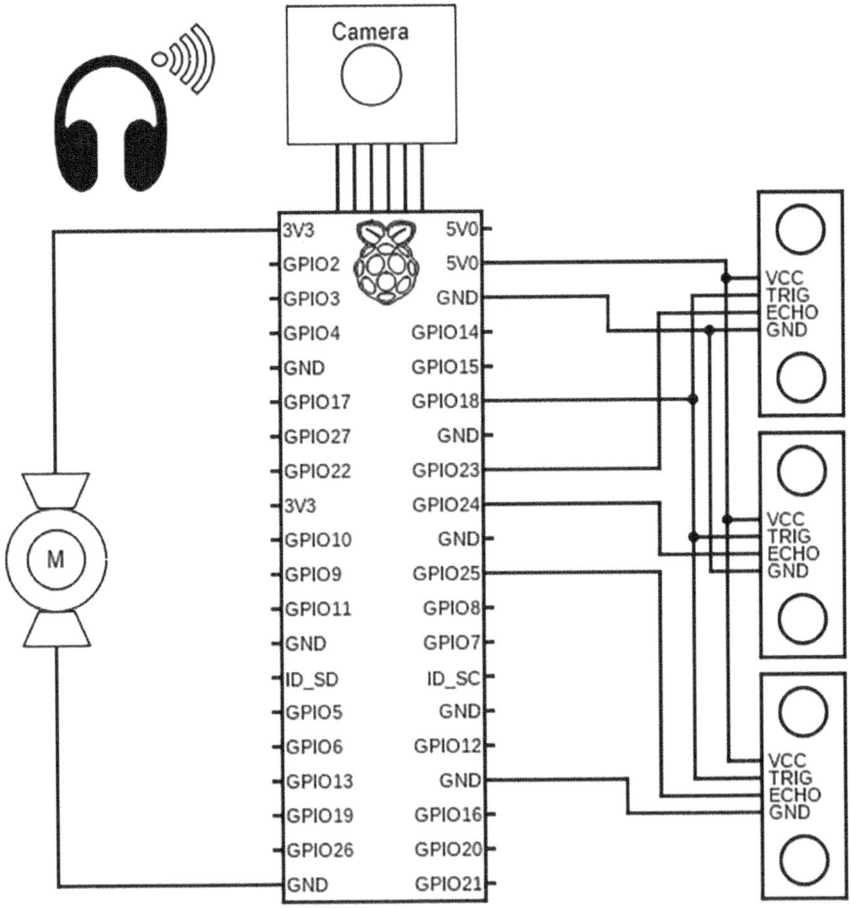

Fig. 2 Circuit diagram

Fig. 3 Implemented smart stick

5 Results and Conclusion

Implemented Fig. 3 shows the actual hardware implementation of the proposed system. It involves of ultrasonic sensors, raspberry pi, camera module, headphones as well as jumper pins for connections. The Entire system is mounted on a stick. The audio output can be heard through any Bluetooth device. In this case, Bluetooth headphones (Figs. 4 and 5).

The implemented system will detect the object and uses audio to tentatively alert the user. The real-time object detection is done. Machine learning has grown in popularity in recent years has witnessed a growth in object detection. The technique of locating occurrences of actual things in pictures or movies is known as object detection. The application of object detection to assist the blind and VI is seen to be advantageous. The technologies that are accessible for their assistance use extra objects to carry such walking canes, caps, etc. These technologies simply identify the existence of an item at a precise distance and place. However, they do not inform the user if an item, a person, an animal, etc., is present.

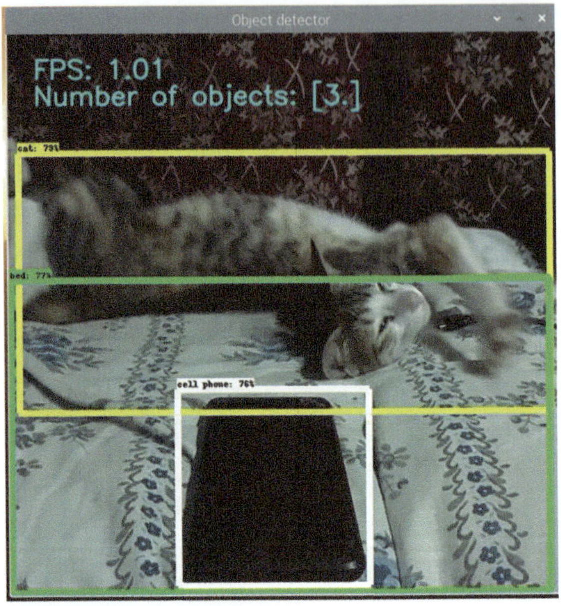

Fig. 4 Output of the smart stick (Detected three objects)

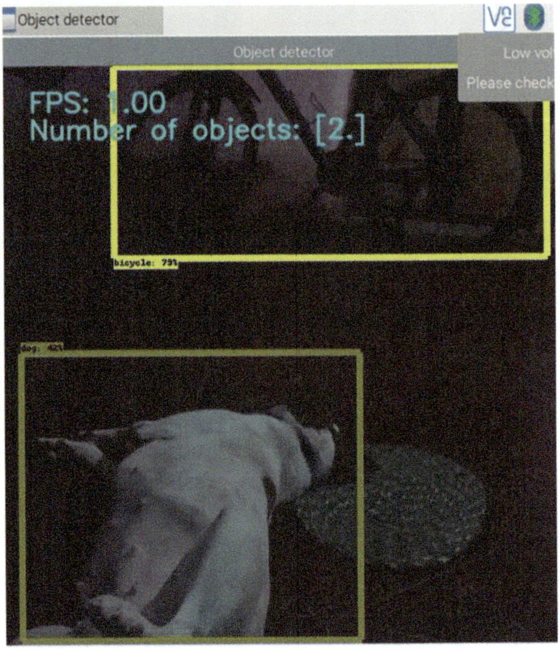

Fig. 5 Output of the smart stick (Detected two objects)

References

1. https://www.forbes.com/sites/bernardmarr/2018/10/01/what-is-deep-learning-ai-a-simple-guidewith-8-practical-exampes/#69f549588d4b
2. https://machinelearningmastery.com/what-is-deep-learning/
3. Laubhan, K., Trent, M., Root, B., Abdelgawad, A., Yelamarthi, K.: A wearable portable electronic travel aid for blind. In: International conference on electrical, electronics, and optimization techniques (ICEEOT), pp. 1999–2003 (2016)
4. Mahmud, N., Saha, R. K., Zafar, Bhuian, M. B. H., Sarwar, S. S.: Vibration and voice operated navigation system for visually impaired person. In: 3rd international conference on informatics electronics & vision (2014)
5. Aljahdali, M., Abokhamees, R., Bensenouci, A., Brahimi, T., Bensenouci, M. A.: IoT based assistive walker device for frail &visually impaired people. IEEE 171–177 (2018)
6. Aymaz, S., Çavdar, T.: Ultrasonic assistive headset for visually impaired people. IEEE 388–391 (2016)
7. Patil, P., Sonawane, A.: Environment sniffing smart portable assistive device for visually impaired individuals. In: International Conference on Trends in Electronics and Informatics ICEI, pp. 317–321 (2017)
8. Redmon, J., Divvala, S., Girshick, R., Farhadi, A.: You only look once: unified, real- time object detection. arXiv, pp. 1–10 (2015)
9. Muthusenthil, B., Joshuva, J., Kishore, S., Narendiran, K.: Smart assistance for blind people using raspberry Pi. In: International Journal of Advance Research, Ideas and Innovations in Technology, (Volume 4, Issue 2), pp. 884–891 (2018)
10. Sharma, P., Shimi, S. L., Chatterji, S.: Design of microcontroller based virtual eye for the blind. In: International Journal of Scientific Research Engineering & Technology (IJSRET), Vol. 3, Issue 8, pp. 1137–1142 (2014)
11. Mohanty, S., Karunan, M., Sayyad, I., Khursade, S., Gite, B. B.: Smart shoe for visually impaired. Int. J. Adv. Res. Comput. Commun. Eng. 6(11), 244–246 (2017)
12. Rastogi, S., Sharma, P., Dhall, P., Agarwal, R., Thakur, S.: Smart Assistive Shoes and Cane: Solemates for the Blind People. Int. J. Adv. Res. Electron. Commun. Eng. (IJARECE) **6**(4), 334–345 (2017)
13. Murthy, R. N., Sudha, P. N.: Smart navigation system for visually challenged people. Int. J. Ind. Electron. Electr. Eng. pp. 45–48 (2016), ISSN: 2347–6982
14. Mustapha, B., Zayegh, A., Begg, R. K.: Reliable wireless obstacle detection system for elderly and visually impaired people with multiple alarm units. In: IEEE 2014 international conference on computer, communication, and control technology, pp. 271–276 (2014)
15. Khursade, S., Karunan, M., Sayyad, I., Mohanty, S., Gite, B.B.: Smart shoes: a safe future for the blind. Int. J. Innov. Res. Comput. Commun. Eng. 6(5), 5808–5814 (2018)
16. Khampachua, C., Wongrajit, C., Waranusast, R., Pattanathaburt, P.: Wrist-mounted smartphone-based navigation device for visually impaired people using ultrasonic sensing. In: 2016 Fifth ICT International Student Project Conference (ICT-ISPC), pp. 93 (2016)

A Review of Deep Learning Techniques for Classification of Alzheimer's Disease Using MRI

Nand Kishore and Neelam Goel

Abstract The Alzheimer's disease is a progressive neurological disorder that gradually impairs cognitive abilities and behavior, affecting daily functioning. Current treatments for this medical condition are insufficient in addressing its unpredictability. Due to the intricate structure of the brain, it is extremely challenging to identify Alzheimer's disease beyond its early stages. Deep learning-based computer-aided diagnosis (CAD) methods utilizing MRI imaging data have been proposed as accurate tools for identifying Alzheimer's disease during its initial phases. In this study assessed the effectiveness of deep learning techniques on magnetic resonance imaging (MRI) scans in diagnosing Alzheimer's disease. Furthermore, this paper offers an extensive overview of related studies comparing Alzheimer's disease (AD) patients versus cognitively normal (CN) individuals. Although deep learning displays promise in diagnosing AD, several concerns persist. Deep learning models hold great significance for managing Alzheimer's disease. Developing a thorough diagnostic framework encompassing all forms of dementia—not only Alzheimer's disease—is essential to advance clinical practices and improve patient outcomes in the long run.

Keywords Alzheimer's disease · Deep learning · Classification · MRI · Convolutional neural networks

1 Introduction

Alzheimer's disease (AD) is a neurodegenerative disorder characterized by the accumulation of amyloid and tau proteins in the brain, leading to cognitive and functional decline, as well as behavioral abnormalities [1]. AD can cause memory, thinking, and movement difficulties. People with AD may struggle with learning, planning, attention, spatial awareness, and daily activities. Alzheimer's disease has no known cure, but symptoms can be managed with certain medications and therapies.

N. Kishore · N. Goel (✉)
Department of Information Technology, University Institute of Engineering and Technology, Panjab University, Chandigarh, India
e-mail: erneelam@pu.ac.in

B. Soni et al. (eds.), *Beyond Artificial Intelligence*, Lecture Notes in Networks and Systems 1326, https://doi.org/10.1007/978-981-96-4170-3_19

Alzheimer's disease (AD) is spreading at an alarming rate as the world's population ages [2]. There are an estimated 6.2 million individuals with Alzheimer's disease dementia in the United States, and 50 million worldwide [3]. It is estimated that 60% of individuals who have the condition are currently residing in countries with lower and middle-income economies. These countries often have limited resources and access to treatment for the affected individuals. By 2050, global population is projected to reach 150 million, with 12.7 million in the US. It is crucial to discover new medications that can prevent, delay the progression of, or improve the symptoms of Alzheimer's disease, especially given these epidemiologic predictions.

There are three stages of AD are early or mild, middle or moderate, and late or severe [4]. Early stage of AD can be challenging to retain recent information or acquire new knowledge when one is struggling with this condition, as it can impact one's cognitive abilities and memory function. Middle stage, as individuals age, they may experience a decline in their capabilities, making it challenging to perform daily activities without assistance. Struggling to remember specific words may result in using the wrong words frequently, which can lead to difficulties with speech. During the advanced stages of the condition, caregivers have the responsibility of providing all necessary support for those who are affected. Alzheimer's disease can cause a decline in muscle mass and mobility, leading to a loss of independence in daily tasks and eventually being bedridden and unable to feed oneself.

Early detection of Alzheimer's disease (AD) can be aided by biomarkers, which have both diagnostic and prognostic value [5]. Research is currently ongoing on a wide range of biomarkers that are associated with Alzheimer's disease. Neuroimaging has the potential to predict the transition from MCI to AD. Various brain-imaging techniques are commonly used to study neuropathological processes, and morphological and functional changes in AD. Neuroimaging techniques can help diagnose Alzheimer's disease and distinguish it from other neurodegenerative diseases. Magnetic resonance imaging (MRI) is a non-invasive medical imaging techniques that produce a detailed images of almost every internal structure in the human body, including organs, bones, muscles, and blood vessels. MRI scanners use a powerful magnet and radio waves to create detailed images of the inside of the body. MRI exams do not produce ionizing radiation, unlike X-rays. These images provide essential information for your physician to diagnose your medical condition accurately and plan an effective course of treatment.

Deep learning is the superior machine learning method for identifying complex patterns in high-dimensional data [6]. The exceptional performance of this technology in computer vision is proven by its innovative approach. These methods have greatly improved machine learning algorithms, enabling them to perform complex tasks with high accuracy. This has made deep learning a crucial area of research for academia and industry. The techniques used for diagnosing Alzheimer's disease have gained significant attention in research since 2014, the quantity of published papers in this field is growing quickly. Recent studies indicate that deep learning techniques exhibit superior performance in accurately diagnosing Alzheimer's disease when compared to traditional machine learning models.

The following are the key contributions:

- An analysis of deep learning methods for detecting Alzheimer's disease.
- The objective is to research the effectiveness of MRI technology in identifying Alzheimer's disease.
- The performance of existing methods in distinguishing Alzheimer's disease (AD) from cognitively normal (CN) stages.

In Sect. 2, the primary goal of the research was to differentiate between cognitively normal (CN) and Alzheimer's disease (AD) classifications. Section 3, Results and discussion. Section 4, conclusion and future scope.

2 MRI Based Deep Learning Methods for AD Diagnosis

Numerous advanced deep learning techniques have been developed to significantly improve the accuracy of computer-aided diagnoses for Alzheimer's disease. The framework for Landmark-based Deep Multi-Instance Learning (LDMIL) used to accurately diagnose brain disorders is discussed in [7]. LDMIL method makes it easy to develop an MR image classifier that captures both local and global structural information. It works by using landmark patches to identify local features, and all landmarks were used to provide a complete understanding of the image. Deep learning followed by an activity monitoring framework that used body-worn inertial sensors to monitor the subjects' daily activities is discussed in [8]. Activity monitoring can be highly advantageous to patients. It not only offers a well-organized approach to help them carry out their daily activities, but also evaluates their vulnerability based on their level of physical activity. The diagnosis for Alzheimer's disease achived an 82% accuracy compared to traditional methods.

3D-Convolutional Neural Networks and trained 3D-autoencoder technology to detect Alzheimer's disease at its earliest stages is described in [9]. The accuracy of AD classification was improved to 98.54% through the use of ensemble techniques. AD versus CN is classified by unsupervised learning utilizing convolution autoencoders, while supervised transfer learning was used for pMCI versus sMCI classification [10]. The models have proven to be superior in AD and pMCI classification, achieving remarkable accuracies of 86.60% and 73.95% respectively. The results showed that the temporal and parietal lobes play a crucial role in brain classification.

Three 3D convolution networks process the gray matter (GM), white matter (WM), and cerebrospinal fluid (CSF) tissues individually, which were later combined into a single representation as discussed in [11]. In the first scenario, numerous tissue scales are fused together; in the second, multiple tissue areas are fused. XGBoost to prevent overfitting by reducing feature dimensions through boosting before classification. The research yielded a conclusive finding that implementing single-sample methods leads to the most accurate predictions [12]. While the suggested approaches with three orthogonal planes of data (top) produce the most accurate overall AD versus MCI predictions of 97.1% while MCI versus CN was 92.6%. Advanced research using deep learning and brain MRI images has led to significant progress in accurately diagnosing Alzheimer's disease [13]. Training deep learning architectures such as CNNs can be challenging due to the significant amount of data required to ensure proper model training. The VGG-16 architecture is an effective feature extractor in transfer learning for solving classification problems. The approach utilized for validating the set of multiple classes achieved a score of 95.73%.

Alzheimer's disease can be diagnosed with confidence by utilizing a combination of time-series analysis of structural magnetic resonance imaging and convolutional and recurrent neural networks. Convolutional Neural Networks, or CNNs, expertly classify MR images by analyzing their spatial features [14]. The outputs from CNN were processed with Recurrent Neural Networks at different time intervals. To identify disease patterns over a period, three layers of bidirectional gated recurrent units were combined to extract relevant features for disease classification. Spatial and longitudinal data were successfully acquired without the need for feature extraction during classifier training. The classifier for detecting diseases demonstrated exceptional efficacy and remarkable outcomes in examining longitudinal MR images. It attained an accuracy rate of 91.33% in distinguishing AD from CN and 71.71% in discriminating pMCI from sMCI, which is quite promising. Hierarchical fully convolutional networks were utilized in whole-brain sMRI to detect discriminative local patches and regions. These were then combined to form hierarchical classification models for diagnosing AD [15].

The classification of Alzheimer's disease can be made by using sMRI to assess the size and shape of the hippocampus as discussed in [16]. It is worth noting that these traits might not fully capture the complexity of the data and could potentially have segmentation issues. In addition, these characteristics are obtained without relying on the classification model, which may result in subpar performance. A deep learning multi-model approach for Alzheimer's disease classification using MRI scans. An automated hippocampus partitioning and AD classification method based on multi-model deep learning and convolution neural networks (CNN) may be used with structural MRI. Features of the CNN and Dense Net models that have been learned work together to diagnose AD. The distinction between AD and CN patients had a promising accuracy rate of 88.9% and an equally impressive AUC rate of 92.5%. Conversely, the accuracy rate for MCI patients versus CN patients was 76.2% with an AUC rate of 77.5%.

Integrated deep feature extraction and identification of Alzheimer's disease cortical areas with iterative sparse deep learning (ISDL) is a deep learning approach as discussed in [17]. 62 cortical areas were analysed using deep feature extraction. The sparse regression module found important brain circuits in the DFE module, excluding unnecessary prefrontal cortex and improving accuracy.

To achieve accurate AD classification from brain MR images, the recommended approach is to implement a fully connected convolution neural network that incorporates correlation attention mechanisms. The advanced technique will facilitate the acquisition of multi-level features, leading to highly effective results as discussed in [18]. By using dense neural networks, the extracted features from pre-processed images at various scales. As a result, it was able to transform MR images into hierarchical structures that effectively utilized related features from various levels through connection-wise attention mechanisms. The approach has an accuracy of 97.35% in differentiating AD versus CN, 87.82% in distinguishing MCIc versus CN, and 78.79% in distinguishing MCIc versus MCInc.

It is important to identify and develop prognostic biomarker signatures to address challenges in AD drug research [19]. To detect Alzheimer's disease, deep learning technology was used to analyze 2D anatomical slices obtained through magnetic resonance imaging [20]. The application of AI algorithms has permitted more accurate diagnosis of Alzheimer's disease [21]. Research into early prediction is still needed so that treatment can start with either medication or a change in lifestyle. The study developed a method for diagnosing AD using 3D MRI scans. This method involved both a fuzzy hyperplane-based least square twin support vector machine and a deep learning model to extract features at all levels from the MRI slices.

It is possible to effectively train models using the publicly available ADNI dataset. The construction of the hyperplane for classification was accomplished through the use of the triangle fuzzy function. The approach provided the most precise categorizations among existing systems. A convolution neural network was used in conjunction with T1w-MRI to segment Alzheimer's disease into three categories [22]. A systematic examination of three computational neuroimaging methodologies is presented in this study (3D patch-based, 3D subjects' level, 3D slice-based). Texture, edge, appearance, and strength feature extraction were combined with clinical data as discussed in [23]. The rotation forest was used to generate the subset of training features utilized by the Deep Enhanced Stacked Autoencoder classifier. The CAD system has shown better performance than the baseline system, with improved accuracy, sensitivity, and specificity for multi-class problems. The accuracy, sensitivity, and specificity rates were 98.68%, 97.72%, and 97.19%, respectively.

A network called Multi-Relation Reasoning Network (MRN) was developed to effectively learn diverse brain areas multi-relation knowledgeable representations using sMRI data for diagnosing Alzheimer's disease is presented in [24]. It took into account into account spatial correlations and topological information. A graph was generated for each scan, highlighting brain regions that have experienced abnormal changes through comparison between groups. The nodes in the graph represent these regions, while the edges between them show their semantic or spatial relation. The integration of deep features extracted from brain MR images and transformations is achieved through two fusion methods [1]. The decision-level and feature-level fusion techniques, are highly effective in integrating data from multiple sources. They are designed to ensure accurate and comprehensive analysis of the data. It is evident from the findings that the most accurate approach can be achieved through the amalgamation of EfficientNet-B7 and an Artificial Neural Network (ANN). Table 1 presented a comprehensive summary of cutting-edge deep learning strategies that employ MRI scans to classify AD.

Table 1 MRI based Deep learning Methods for Alzheimer's disease diagnosis

References	Modality	Methods	Description/Pros	Limitation	Classification	Acc (%)
[7] 2018	T1-Weighted MRI	Multi-instance landmark based learning	• Landmark based deep multi-instance learning • Anatomical landmarks • Capturing local and global information	• Patches of MR images challenges problem • Voxel level representations more time consuming	CN versus AD	91.09
[8] 2019	MRI	CNN	• Deep feature extraction • Deep learning achieves-good performance in Medical-imaging	• CNN lack of orientation, size, position etc • High cost • Complex classification task • Require a lot of training data • Weight initialization is difficult	CN versus AD	98.74
[9] 2019	MRI	3D Caps Nets and Convolution Neural Networks	• Caps-Nets is efficient spatial relationship features of images changes in shape, size, position, orientation, luminance • Cover more of the image feature and regions than CNN • High level feature extraction of image • Low computational time	• Limited amount of labeled data and class imbalance	CN versus AD	92.98
[10] 2019	MRI	Convolutional Autoencoder	• Dimensionality reduction • Discriminative representation in a lower dimensional space can improve the performance of classification • Reduce dependency on fully connected layer	• Unsupervised learning • Random initialization of trainable weights difficult task • Small sample size • Voxel wise saliency score compute	CN versus AD	86.60

(continued)

Table 1 (continued)

References	Modality	Methods	Description/Pros	Limitation	Classification	Acc (%)
[11] 2019	MRI	3D Multiscale Deep Convolutional Network	• Multi-level feature fusion • Extracts image level and semantic level features • Reduce the dimension of features • Only features with high importance are chosen for the final classification	• Overfitting problem • High dimension number of features vector • Large memory requirement • More training time	CN versus AD	99.67
[12] 2019	MRI	Convolution Neural Networks	• 2D CNNs use 2D convolutional Kernels to predict the segmentation map for one slice • Segmentation maps can be more accurate when voxel information from neighboring slices is used	• Limited sample size • Segmentation maps are generated By predicting one slice at time for the entire volume	CN versus AD	89.15
[13] 2019	T1-Weighted MRI	2D-Convolution Neural Networks	• CNN model is an incredibly robust tool for extracting features • In the medical field, CNNs have achieved remarkable results for computer vision tasks in the past couple of year	• To achieve optimal training results, a substantial amount of data is essential • A lot of computational resources	CN versus AD	99.14
[14] 2019	T1-Weighted MRI	Recurrent Neural Networks (RNN)	• Model a collection of records (i.e., time collection) so that each pattern can be assumed to be dependent on previous ones • It gives an effective pixel neighborhood with the help of convolutional layers	• Complex training • Difficulty to process very long process • Gradient vanishing and exploding	CN versus AD	91.33

(continued)

Table 1 (continued)

References	Modality	Methods	Description/Pros	Limitation	Classification	Acc (%)
[15] 2020	MRI	Hierarchical fully convolutional network (HFCN)	• Automatically identify all discriminative locations • Reduce the computational complexity	• A lot of training data is needed • High cost • Complex classification task • Size of input image patches was fixed	CN versus AD	90
[16] 2020	T1-Weighted MRI	3D DenseNet	• Reduce the number of parameters • Reduce the feature dimension and classify disease status • Alleviate the vanishing-gradient problem	• The feature maps of each layer are spliced with the previous layer • Data is replicated multiple times	CN versus AD	88.9
[17] 2021	T1-Weighted MRI	Iterative sparse and Deep learning	• Deep feature extraction • Efficient handling of data • Better classification performance	• More computational resource is permitted • Complex network	CN versus AD	95.32
[18] 2021	T1-Weighted MRI	Densely connected neural network	• Reduce the number of parameters • Reduce the feature dimension and classify disease status • Alleviate the vanishing-gradient problem	• The feature maps of each layer are spliced with the previous layer • Data is replicated multiple times	CN versus AD	97.35
[19] 2021	T1-Weighted MRI	VGG-16	• Reduce computational complexity • Memory requirements • Overfitting	• Complex architecture • More training time	CN versus AD	82.5
[20] 2021	MRI	2D-CNN	• Prevent overfitting • Feature extraction • 3D-patch and sliced based approach	• Single image produces the great number of patches • It suffers from overfitting • Information loss	CN versus AD	99.30

(continued)

Table 1 (continued)

References	Modality	Methods	Description/Pros	Limitation	Classification	Acc (%)
[21] 2022	sMRI	FLS-TWSVM	• Generalization performance • Less computational cost • Optimization • More stable for re-sampling	• Not suitable for large data sets • Classes are overlapping	CN versus AD	97.15
[22] 2022	T1-Weighted MRI	Volumetric Convolution Neural Networks	• Reduce the parameter size • Volume metric data	• Computation overhead • Large memory requirement	CN versus AD	97.39
[23] 2022	MRI	Hybrid rotation forest deep neural network	• Hybrid feature extraction • High order statistical texture features • Additional output layer Softmax regression	• Computational burden • Overfitting	CN versus AD	98.68
[24] 2023	MRI	Graph convolutional Neural network	• Multi-Relation Reasoning Network for brain regions • Developing unique node representations is crucial for building brain graphs	• Computation overhead • Large memory requirement	CN versus AD	92.57
[1] 2023	MRI	Deep convolutional networks	• Extract features from each matrix and combine them into one feature vector using the same CNNs • Identify key factors for classifying dementia severity	• Complex architecture • More training time	CN versus AD	90.3

AD: Alzheimer's disease; CN: cognitively normal; 3D CNN: 3 dimensional convolutional neural networks; CNN: convolutional neural networks; 3D DenseNet: 3D Densely Connected Convolutional Networks; VGG-16: Visual Geometry Group; DCNN: deep convolution neural network; FLS-TWSVM: fuzzy hyperplane based least square twin support vector machine; NIFTI: Neuroimaging Informatics Technology Initiative; JPEG: joint photographic experts groups

3 Results and Discussion

The present study entails an analysis of 19 papers that are concerned with the classification of Alzheimer's disease using MRI. Among these, 17 papers employed deep learning techniques, whereas one paper of transfer learning and another explored machine learning. Figure 1 shows that using deep learning-based MRI can help diagnose Alzheimer's disease (AD) is more accurately and efficiently. It is a promising approach. Figure 2 demonstrated that Deep Learning is a recent and emerging field of study and research. The Fig. 2 suggests that more research is needed to Transfer learning-based MRI for diagnosing and treating AD.

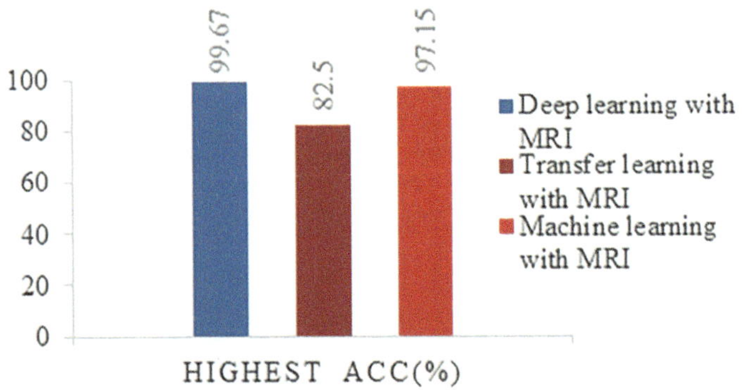

Fig. 1 MRI based classifications of Alzheimer's disease

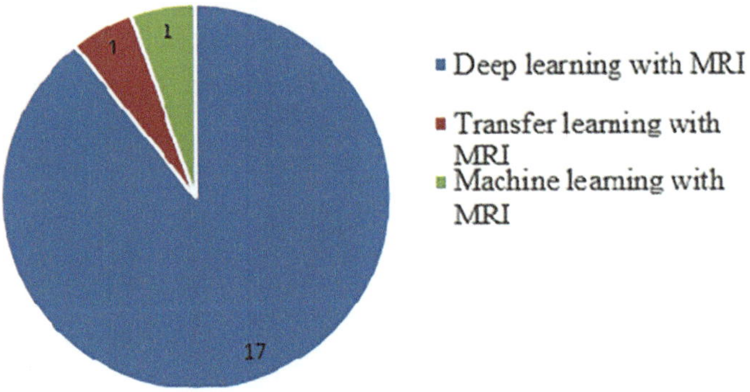

Fig. 2 No of studies analysis for detection of AD

3.1 Discussion

In recent years, various Deep learning methods have been proposed to accurately diagnose patients with Alzheimer's disease classification through MRI scans. Directly comparing these methods can be challenging as they use diverse datasets and pre-processing techniques. Table 1 presented an effective summary of the classification between CN and AD.

Although CNNs have achieved promising outcomes in medical computer vision tasks in recent times, training these massive architectures from scratch has its limitations. Training deep neural networks properly requires a significant amount of annotated data. This can be particularly challenging in the medical imaging field, where obtaining sufficient data can be expensive and difficult. The training process for these types of structures demands a significant amount of computational resources. Some of these studies have an accuracy rate of 99% [11, 13, 20] while a few others have an accuracy rate above 80% [19].

Achieving optimal performance in deep learning requires meticulous and thorough hyper-parameter tuning. Neglecting this crucial step can lead to detrimental consequences such as overfitting or underfitting, which in turn can result in unsatisfactory overall performance. Researchers have developed transfer learning as a successful alternative approach to solve these issues. Transfer learning is a powerful deep learning technique that uses an existing model to develop a new one for a specific task. The use of deep learning shows great potential in detecting Alzheimer's disease. The study suggests that the use of deep learning and MRI methods can reliably detect early signs of Alzheimer's disease.

4 Conclusion

The impact of Alzheimer's disease (AD) on the elderly demo graphic is a significant and far-reaching issue that affects populations worldwide. Computer-based methods are utilized to improve the diagnosis and treatment planning of AD, as detecting it in its early stages can be challenging. In recent years, deep learning has received significant attention for its ability to detect Alzheimer's disease. It has also provided valuable insights into various pre-processing methods and deep learning architectures, such as CNN and RNN, which are used to classify AD. In classification algorithms, imbalanced data can lead to unreliable predictions and overfitting. In future studies, it is imperative for researchers to place their focus on the early detection and diagnosis of AD for individuals who are at a high risk of developing the condition.

References

1. Asgharzadeh-Bonab, A., Kalbkhani, H., Azarfardian, S.: An Alzheimer's disease classification method using fusion of features from brain Magnetic Resonance Image transforms and deep convolutional networks. Healthc Anal. **4**, 100223 (2023). https://doi.org/10.1016/j.health.2023.100223
2. Butynski, T. M.: Report at a glance. A rapid biol assess lokutu. Democr. Repub. Congo 1–5 (2022). https://doi.org/10.1896/978-1-934151-04-4.7
3. WHO GUIDELINES: Risk reduction of cognitive decline and dementia: WHO guidelines (2019)
4. Sheng, J., Shao, M., Zhang, Q., et al.: Alzheimer's disease, mild cognitive impairment, and normal aging distinguished by multi-modal parcellation and machine learning. Sci. Rep. **10**, 1–10 (2020). https://doi.org/10.1038/s41598-020-62378-0
5. International AD, University M: World Alzheimer Report 2021. Alzheimer's Dis. Int. 2–314 (2021)
6. Zhao, X., Ang, C.K.E., Acharya, U.R., Cheong, K.H.: Application of Artificial Intelligence techniques for the detection of Alzheimer's disease using structural MRI images. Biocybern. Biomed. Eng. **41**, 456–473 (2021). https://doi.org/10.1016/j.bbe.2021.02.006
7. Liu, M., Zhang, J., Adeli, E., Shen, D.: Landmark-based deep multi-instance learning for brain disease diagnosis. Med. Image Anal. **43**, 157–168 (2018). https://doi.org/10.1016/j.media.2017.10.005
8. Raza, M., Awais, M., Ellahi, W., et al.: Diagnosis and monitoring of Alzheimer's patients using classical and deep learning techniques. Expert Syst. Appl. **136**, 353–364 (2019). https://doi.org/10.1016/j.eswa.2019.06.038
9. Kruthika, K.R., Rajeswari, M.H.D.: Erratum: CBIR system using Capsule Networks and 3D CNN for Alzheimer's disease diagnosis (Informatics in Medicine Unlocked (2019) 14 (59–68), (S235291481830176X), (10.1016/j.imu.2018.12.001)). Informatics Med Unlocked **16**, 59–68 (2019). https://doi.org/10.1016/j.imu.2019.100227
10. Oh, K., Chung, Y.C., Kim, K.W., et al.: Classification and visualization of Alzheimer's disease using volumetric convolutional neural network and transfer learning. Sci. Rep. **9**, 1–16 (2019). https://doi.org/10.1038/s41598-019-54548-6
11. Ge, C., Qu, Q., Gu, I.Y.H., Jakola, A.S.: Multi-stream multi-scale deep convolutional networks for Alzheimer's disease detection using MR images. Neurocomputing **350**, 60–69 (2019). https://doi.org/10.1016/j.neucom.2019.04.023
12. Bi, X., Li, S., Xiao, B., et al.: Computer aided Alzheimer's disease diagnosis by an unsupervised deep learning technology. Neurocomputing **392**, 296–304 (2020). https://doi.org/10.1016/j.neucom.2018.11.111
13. Jain, R., Jain, N., Aggarwal, A., Hemanth, D.J.: Convolutional neural network based Alzheimer's disease classification from magnetic resonance brain images. Cogn. Syst. Res. **57**, 147–159 (2019). https://doi.org/10.1016/j.cogsys.2018.12.015
14. Todd, L.: Accepted manuscript Development Accepted manuscript. 1–33 (2015)
15. Lian, C., Liu, M., Zhang, J., Shen, D.: Hierarchical fully convolutional network for joint atrophy localization and Alzheimer's disease diagnosis using structural MRI. IEEE Trans. Pattern Anal. Mach. Intell. **42**, 880–893 (2020). https://doi.org/10.1109/TPAMI.2018.2889096
16. Liu, M., Li, F., Yan, H., et al.: A multi-model deep convolutional neural network for automatic hippocampus segmentation and classification in Alzheimer's disease. Neuroimage 208 (2020). https://doi.org/10.1016/j.neuroimage.2019.116459
17. Chen, Y., Xia, Y.: Iterative sparse and deep learning for accurate diagnosis of Alzheimer's disease. Pattern Recognit. **116**, 107944 (2021). https://doi.org/10.1016/j.patcog.2021.107944
18. Zhang, J., Zheng, B., Gao, A., et al.: A 3D densely connected convolution neural network with connection-wise attention mechanism for Alzheimer's disease classification. Magn. Reson. Imaging **78**, 119–126 (2021). https://doi.org/10.1016/j.mri.2021.02.001

19. Yang, L., Wang, X., Guo, Q., et al.: Deep learning based multimodal progression modeling for Alzheimer's disease. Stat Biopharm Res **13**, 337–343 (2021). https://doi.org/10.1080/19466315.2021.1884129
20. Al-Khuzaie, F. E. K., Bayat, O., Duru, A. D.: Diagnosis of Alzheimer disease using 2D MRI slices by convolutional neural network. Appl. Bionics. Biomech. 2021 (2021). https://doi.org/10.1155/2021/6690539
21. Sharma, R., Goel, T., Tanveer, M., Murugan, R.: FDN-ADNet: Fuzzy LS-TWSVM based deep learning network for prognosis of the Alzheimer's disease using the sagittal plane of MRI scans. Appl. Soft Comput. **115**, 108099 (2022). https://doi.org/10.1016/j.asoc.2021.108099
22. Goenka, N., Tiwari, S.: AlzVNet: A volumetric convolutional neural network for multiclass classification of Alzheimer's disease through multiple neuroimaging computational approaches. Biomed. Signal Process. Control **74**, 103500 (2022). https://doi.org/10.1016/j.bspc.2022.103500
23. Raghavaiah, P., Varadarajan, S.: A CAD system design for Alzheimer's disease diagnosis using temporally consistent clustering and hybrid deep learning models. Biomed. Signal Process. Control **75**, 103571 (2022). https://doi.org/10.1016/j.bspc.2022.103571
24. Zhang, J., He, X., Qing, L., et al.: Multi-relation graph convolutional network for Alzheimer's disease diagnosis using structural MRI. Knowledge-Based Syst. **270**, 110546 (2023). https://doi.org/10.1016/j.knosys.2023.110546

A Comparative Study of Recent Biometric CryptoSystems

Prabhjot Kaur, Nitin Kumar, and Maheep Singh

Abstract Feature detection and description are fundamental tasks in computer vision and image processing. In this paper, we have performed experiments using six existing Feature detection and description techniques *namely,* Scale-Invariant Feature Transform (SIFT), Binary Robust Independent Elementary Features (BRIEF), Speeded-Up Robust Features (SURF) and its variant for Biometric CryptoSystems *namely,* BRIEF based fuzzy vault scheme in Biometric CryptoSystem (BRIEFBCS), SURF based fuzzy vault scheme in Biometric CryptoSystem (SURFBCS) and SIFT based fuzzy vault scheme in Biometric CryptoSystem (SIFTBCS) on MMU and YALE datasets. We present a comprehensive comparative analysis of these widely used methods. These methods are evaluated on the basis of various parameters such as: feature extraction time, feature description time, overall execution time, security component, rotation invariance, scale invariance, illumination invariance, descriptor type, key construction time, robustness, etc. Through a systematic evaluation, we aim to provide insights into the strengths and weaknesses of each method, aiding practitioners in choosing the most suitable approach for their specific applications.

Keywords SIFTBCS · SURFBCS · BRIEFBCS · Biometric CryptoSystem

P. Kaur (✉) · M. Singh
Department of CSE, National Institute of Technology Uttarakhand, Srinagar, India
e-mail: prabhjot.kaur@dituniversity.edu.in

M. Singh
e-mail: maheepsingh@nituk.ac.in

P. Kaur
School of Computing, DIT University, Makka Wala, India

N. Kumar
Department of CSE, Punjab Engineering College (Deemed to be University), Chandigarh, India
e-mail: nitik@pec.edu.in

B. Soni et al. (eds.), *Beyond Artificial Intelligence*, Lecture Notes in Networks and Systems 1326, https://doi.org/10.1007/978-981-96-4170-3_20

1 Introduction

In the realm of computer vision and image processing, the ability to extract robust and distinctive features from images forms the bedrock of various applications, such as object recognition, image registration, scene understanding, and augmented reality. These applications heavily rely on the efficiency and accuracy of feature extraction techniques. During feature extraction process, feature detection and description are essential steps, allowing for the identification and matching of distinctive points or regions in images. As a result, researchers and practitioners have developed a plethora of algorithms to address this fundamental challenge. This work provides a comparative analysis of six such algorithms. In this paper, we explore six prominent feature detection and description algorithms: SIFT [7], SURF [1], BRIEF [2], SIFTBCS [4], SURFBCS [6], and BRIEFBCS [3]. We provide an overview of each algorithm's principles, followed by a comprehensive comparative analysis based on various criteria, including computational efficiency, robustness, and accuracy. The SIFT, SURF and BRIEF are the primary components for feature extraction and description in SIFTBCS, SURFBCS and BRIEFBCS respectively. Figure 1 shows the schematic diagram that shows the use of each technique at different levels in a system. There are two coloured components in the diagram *viz.* pink box and green box. The pink box showcases the use of SIFT, SURF and BRIEF techniques for feature extraction in the biometric images. The green box showcases the usage for SIFTBCS, SURFBCS and BRIEFBCS techniques which are considered as the part of umbrella term Biometric CryptoSystem (BCS) [5]. Lets understand the working as shown in Fig. 1 by considering SIFTBCS technique.

The working of SIFTBCS technique first involves the use of Scale Invariant Feature Transform method to extract the keypoints, followed by their description. Then, these features are encoded using Reed-Solomon encoding scheme to generate a reference template (R) and stored in the database and a biometric key (κ) is also generated using these feature descriptors. This key is stored as one component in fuzzy vault (v). The other component of fuzzy vault are the set of feature points amalgamated with the chaff points. The chaff points are the dummy points that added to the gen-

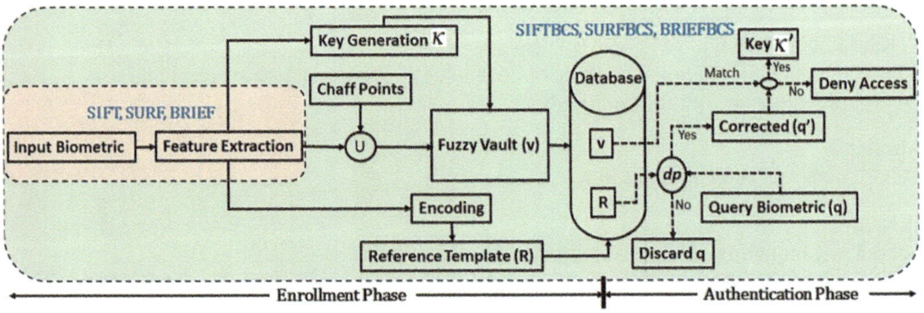

Fig. 1 Components of SIFT, SURF, and BRIEF in BCS: SIFTBCS, SURFBCS, and BRIEFBCS

uine points to make it indistinguishable to identify for the intruders. The v is stored in the database along with R. This ends with the first phase *i.e.* enrollment phase. During the authentication phase, the query biometric is presented to the system and its errors are corrected using Reed-Solomon decoding algorithm and the value of decision parameter (dp) is set. Based on dp, a decision is made whether the number of to construct corrected query biometric (q') or to discard q. In case of, burst error correction is done for q to construct q', then its matched with the vault points (v). In case of successful match, key (κ') is derived using Lagrange's Interpolation, an in case of non-match, access is denied. The Fig. 1 clearly shows that SIFT, SURF and BRIEF are used only upto feature extraction and description. These three form the subset of SIFTBCS, SURFBCS and BRIEFBCS techniques respectively. Further subsections explain the fundamental concepts of the aforementioned six techniques.

The rest of this paper will encompass the following aspects: Sect. 2 will elucidate the contributions, Sect. 3 will delve into state-of-the-art feature extraction techniques, Sect. 4 will encapsulate the methodology, Sect. 5 will explore experimental findings and discussions, and Sect. 6 will furnish the concluding insights.

2 Contribution

Our primary contribution centres around conducting an extensive comparative analysis of the six techniques, namely SIFT, SURF, BRIEF, SIFTBCS, BRIEFBCS, and SURFBCS. This analysis goes beyond merely listing their features and capabilities; it involves a rigorous examination of their performance metrics, computational efficiency, robustness in various scenarios, and potential trade-off. By presenting a comprehensive overview of these techniques, we aim to equip researchers and practitioners with valuable guidance for making informed decisions when selecting the most appropriate method for the specific image processing and computer vision tasks.

3 State-of-the-Art Feature Extraction Techniques

This section includes state-of-the-art feature extraction techniques that widely embraced within the field of BCS. These state-of-the-art methodologies represent the latest advancements in the extraction of discriminative patterns and meaningful information from data, playing a pivotal role in enhancing the accuracy and efficiency of BCS systems.

3.1 Scale Invariant Feature Transform (SIFT)

SIFT, introduced by David G. Lowe in 2004 , marked a significant milestone in computer vision by providing a framework for the detection and description of distinctive features that are invariant to scale, rotation, and illumination changes. Its robustness and distinctiveness made it a de facto choice for various computer vision applications, yet it comes with computational costs that have spurred the development of more computationally efficient alternatives.

3.2 Speeded Up Robust Features (SURF)

Herbert Bay et al., introduced the concept of Speeded Up Robust Features in 2006. SURF brings the features of robustness and efficiency by utilizing integral images and approximations in its calculations. This combination allows it to maintain strong performance even under challenging conditions while remaining computationally efficient, positioning it as a versatile choice for various computer vision tasks.

3.3 Binary Robust Independent Elementary Features (BRIEF)

BRIEF, in contrast, embodies a the property of computational efficiency. Introduced by Michael Calonder et al. in 2010. This technique offers a different approach to feature description by representing feature descriptors as binary strings. This binary representation drastically reduces the computational complexity and memory requirements while maintaining a fair level of robustness. BRIEF's emphasis on speed and simplicity has made it an attractive option for real-time applications and resource-constrained environments.

Adding a captivating layer of complexity to our exploration is the emergence of SIFTBCS, which stands for "Scale Invariant Features Transform based Fuzzy Vault Scheme in Biometric CryptoSystem". This innovative application of SIFT represents a convergence of feature extraction and cryptographic principles in the domain of biometric security. SIFTBCS introduces enhancements to the core SIFT algorithm, tailoring it for the specific demands of Biometric CryptoSystems, particularly in the context of secure biometric key generation and verification.

3.4 SIFTBCS (Scale-Invariant Feature Transform Based Fuzzy Vault Scheme in Biometric CryptoSystem)

SIFTBCS extends the remarkable scale-invariant feature extraction capabilities of SIFT into the domain of biometric security. In essence, it leverages the distinctive SIFT features to create a secure and robust Biometric CryptoSystem, ensuring the confidentiality and integrity of biometric templates. The heart of the SIFTBCS scheme lies in the creation of a "fuzzy vault" that protects the biometric template using SIFT-based cryptographic keys. SIFT features, known for their invariance to scale and rotation, provide a stable foundation for representing biometric characteristics, making them an ideal choice for secure template storage. Within the fuzzy vault, the biometric data is securely encrypted and bound to the SIFT-derived cryptographic keys. This ensures that only individuals possessing the correct biometric data, and thereby the corresponding SIFT keys, can access and unlock the vault. SIFTBCS thus achieves a high level of security while preserving the uniqueness and distinctiveness of the biometric template.

3.5 SURFBCS (Speeded-Up Robust Features Based Fuzzy Vault Scheme in Biometric CryptoSystem)

SURFBCS represents a pioneering fusion of robust feature extraction with cryptographic principles, tailored explicitly for the realm of biometric security. This innovative approach aims to secure sensitive biometric data while simultaneously enabling secure authentication and key generation in applications like fingerprint recognition, iris scanning, and facial verification. In the SURFBCS scheme, the robust feature extraction capabilities of the original SURF algorithm serve as the foundation for the secure biometric vault. These features, which exhibit resilience to variations in scale, rotation, and illumination, are uniquely suited to represent biometric characteristics. SURFBCS enhances the SURF framework with cryptographic techniques to create a secure environment, commonly referred to as a "fuzzy vault," where biometric data is stored and protected. The core idea behind the fuzzy vault concept is to create a secure vault that can only be unlocked or accessed using the correct biometric data. In the context of SURFBCS, this involves leveraging the distinctive SURF features as cryptographic keys to safeguard the biometric template. The biometric data is combined with additional security elements, creating a cryptographic binding that ensures the integrity and confidentiality of the stored biometric information. SURFBCS offers an unprecedented level of security, making it suitable for applications that demand robust protection of biometric data. By integrating the strengths of SURF with cryptographic techniques, it addresses the critical challenges of biometric security, including template protection and secure authentication.

3.6 BRIEFBCS (Binary Robust Independent Elementary Features Biometric CryptoSystem)

BRIEFBCS represents a convergence of computational efficiency and security within the realm of biometric cryptosystems. Building upon the efficiency of BRIEF in feature extraction, BRIEFBCS introduces cryptographic techniques to safeguard biometric data and enable secure biometric authentication and key generation. In BRIEFBCS, the binary feature representation of BRIEF serves as the basis for secure template storage. BRIEF features, which are highly compact and efficient, are well-suited for encoding biometric characteristics. These features are then used as cryptographic keys to create a secure "fuzzy vault" where the biometric template is stored. The fuzzy vault concept ensures that only individuals with the correct biometric data and corresponding BRIEF keys can access and unlock the secure vault. BRIEFBCS thus combines the computational efficiency of BRIEF with the security required for biometric applications, making it an attractive choice for real-time and resource-constrained environments. Applications of BRIEFBCS extend to various biometric recognition systems, including voice authentication, palmprint recognition, and behavioural biometrics, where the need for both efficiency and security is paramount.

In the subsequent sections of this comparative analysis, we will delve deeper into the inner workings and performance characteristics of SIFT, SURF, BRIEF, SURFBCS, SIFTBCS, and BRIEFBCS, exploring their strengths, limitations, and real-world applications in the domain of biometric security.

4 Methodology

The first three core feature extraction and description techniques *namely,* SIFT, SURF, BRIEF are explained in this section followed by the extended techniques SIFTBCS, SURFBCS and BRIEFBCS. The SIFT, SURF and BRIEF differ mathematically in the way of feature extraction and description. With the amalgamation of Biometric CryptoSystem into SIFT, SURF and BRIEF, a new category of BCS systems *namely,* SIFTBCS, SURFBCS and BRIEFBCS are constructed. Thus, within the SIFT method, to acquire the feature vector of the enrolled individual, the most discriminating keypoints need to be detected. The keypoints are detected and described using SIFT method. The basic idea here is to use normalized Laplacian of Gaussian (L_η) to obtained the image sub-scales, which is computed from the equations below. Here, η_σ denotes the Gaussian filter with initial scale. The first order derivative (∇) w.r.t. x and y is computed. The Laplacian of Gaussian (L) is computed by applying Gaussian filter to the second order derivative of Laplacian (∇^2). This is further normalized by multiplying it with σ^2 to obtain a normalized Laplacian of Gaussian (L_η).

$$\eta_\sigma = \frac{1}{2\pi\sigma^2} e^{\frac{-(x^2+y^2)}{2\sigma^2}} \tag{1}$$

$$\nabla = \frac{\partial}{\partial x} + \frac{\partial}{\partial y} \tag{2}$$

$$\nabla^2 = \frac{\partial^2}{\partial x^2} + \frac{\partial^2}{\partial y^2} \tag{3}$$

$$L = \nabla^2 \eta_\sigma \tag{4}$$

$$L_\eta = \sigma^2 \nabla^2 \eta_\sigma \tag{5}$$

Lets understand this concept using biometric image $\mathbf{B}_i$ of ith person. First of all, the biometric image is smoothed to remove noise by applying Gaussian filter (n_σ) with varied scales, where σ denotes the scale of filter. The scale of the Gaussian filter can be changed as given in equation below, where σ_0 is the initial scale and s is the constant multiplier.

$$\sigma_d = \sigma_0 s^d, d = 0, 1, 2, 3.... \tag{6}$$

Once, the Gaussian filter is applied to the image at varied scales, a stack of smoothed images known as scale space (Γ_{B_i}) is obtained, whereby image blurring increase with higher value of σ. Next, a location $\mathbf{l} = (a, b)$ is selected in all the images to check for extremum by computing its normalized second order derivative of Gaussian w.r.t. to the location. If there exists an extremum corresponding to the selected location in any scaled image, then that point is a potential candidate for keypoint and corresponding scale (σ) is its characteristic scale.

$$\Gamma_{B_i} = L_\eta * \mathbf{B}_i \tag{7}$$

The normalized Laplacian of Gaussian can be approximated by another technique known as difference of Gaussian (**D**) as proposed by Lowe [7]. The difference of Gaussian is computed by taking the difference of any two consecutive Gaussian. The **D** produces the results that approximate to $L_\eta \cdot \beta$, where β is a varying scale factor as given below.

$$\mathbf{D} \approx L_\eta \cdot (\beta - 1) \tag{8}$$

$$\mathbf{D}_{B_i} = \Gamma_u - \Gamma_v \tag{9}$$

Here, u, v represents consecutive scales. Thus, instead of convolving image points with L_η, the **D** is computed to determine potential candidates for keypoints. The construction of keypoint descriptor is obtained by diving the 16×16 cluster area around each keypoint into 4×4 groups. The gradient orientation of each 4×4 group is computed and a histogram is obtained for each group. The histogram of these four groups is concatenated to construct a 64− dimensional feature descriptor for respective keypoint. Similarly, feature descriptors are created for all the selected keypoints. The collection of feature descriptors (**f**) around selected keypoints make feature vector ($\mathbf{F}_i$) of ith person. Once, the descriptor is created, then these are secured using fuzzy vault scheme as given in Algorithm 1. These descriptors are also exploited to generate the biometric key. The key and descriptor are formed as components of the fuzzy vault and subsequently stored in the database. The security component is achieved using SIFTBCS technique. Similarly, in SURFBCS, the keypoints are obtained by exploiting the concept of image integral. The integral ($\mathbf{I}_{int}$) of biometric image (**I**), having size $r \times c$ of an image is computed as given below:

$$\mathbf{I}_{int} = \sum_{j=0}^{j \leq r} \sum_{k=0}^{k \leq c} \mathbf{I}(j, k) \tag{10}$$

In case of BRIEFBCS, the keypoints are detected using Features from Accelerated Segment Test (FAST) [8]. Once, the images are convolved and a scale space is obtained, then keypoints are extracted using FAST method. In this method, any point is selected in the smoothed image as a potential candidate for keypoint and we name it as candidate point for better understanding, i.e., a point which is not yet selected as a keypoint. A Bresanham's circle (**cr**) of 16 points is created around each candidate point and the pixel intensities of all these 16 pixels is compared one-by-one with the candidate point. If the results of consecutive 12 pixel intensities is either greater than or lower than the intensity value of candidate point, then the candidate point is selected as a keypoint. To alleviate the problem of multiple interest points in adjacent locations, Non-maximum suppression algorithm is applied. Considering an example, if a location l has multiple interest points. Say **h** is the set of w overlapping interests points. Then, for each point, compute the score (ω) by computing the sum of difference between each interest point and the corresponding circle having 16 pixels. Consider the following equation for computing sum score (ω_j) for jth overlapping interest point having intensity as $In(\mathbf{h}_j)$. Here, $In(\mathbf{cr}_k)$ is the intensity value of kth circle pixel. The score (ω_j) is computed for each overlapping interest point.

$$\omega_j = \sum_{k=1}^{16} \left(|In(\mathbf{h}_j) - In(\mathbf{cr}_k)| \right) \tag{11}$$

Once, the score (ω_j) is computed for each overlapping interest point then a comparison is made to select the maximum scoring interest point one as the keypoint. Now, construct a patch (ρ) around each keypoint. Then, a test (τ) is defined on ρ as given in equation below, where $\rho(x)$ denotes the pixel intensity used for comparison. Then, a distribution is formed from the spread of $0.04 * \omega^2$ around each keypoint to select random pair values for both x and y pixels. A feature descriptor (**f**) is constructed out the results of τ.

$$\tau(\rho; x, y) = 1 \; if x < y, 0 \; otherwise \tag{12}$$

The steps for enrolling the person and authenticating the enrolled user in the system are given in Algorithm 1 and 2 respectively.

5 Experimental Results and Discussion

The experiments using six techniques are performed on two publicly available datasets: MMU [9] and YALE [10]. MMU dataset consists of 450 iris images while YALE dataset includes 165 face images of different subjects. The experiments are performed on both these datasets using SIFT, SIFTBCS, SURF, SURFBCS, BRIEF and BRIEFBCS. Table 3 shows the results obtained using these methods on various parameters such as keypoint extraction time, number of extracted keypoints, overall execution time and the accuracy.The time parameters have the metric unit of millisecond (ms) and the unit for accuracy is percentage (%). Similarly, Table 4 showcases various properties that these six methods exhibit. Considering the scale and rotation invariance property, except BRIEF and BRIEFBCS, all the methods possess it. Illumination is handled only in SIFTBCS and SIFT methods. With the advent of IoT, the devices need to be connected to the network, thus security becomes utmost concern. Thus, out of all six methods, SIFTBCS, SURFBCS and BRIEFBCS methods offer the security features. The portable devices have low processing power, thus making BRIEFBCS and BRIEF the most suitable among all. Similarly, real time applications

Table 1 Algorithm for enrollment phase

Input: Biometric images from face/iris dataset to be used for training

Output: Extracted features, biometric key κ, reference template R, fuzzy vault υ

The following steps are repeated for all the biometric images used for training the model. For simplicity, the following steps are defined for a single image:

1. The keypoints are detected from the biometric image using all the six methods.

1.1 In case of SIFT and SIFTBCS, the difference of Gaussian is computed from the scale space of images created using image convolution.

1.2 In case of SURF and SURFBCS, the determinant of Hessian matrix is computed on scale space of images created by applying box filters.

1.3 In case of BRIEF and BRIEFBCS, the Features from Accelerated Segment Test (FAST) are used to compute interest points, which are further used to compute descriptors.

2. A biometric key (κ) is generated from these points using SIFTBCS, SURFBCS and BRIEFBCS methods.

3. These features are also encoded using error correcting codes to construct a reference template (R), which is stored in the database.

4. Chaff points are generated which are later amalgamated with the extracted features to construct the fuzzy vault (υ).

5.The κ is also formed a part of fuzzy vault which is extracted later during successful authentication.

6. The υ and R are stored in the database.

Table 2 Algorithm for authentication phase

Input: Query biometric image q, υ, R
Output: Biometric key κ'
The following steps are performed if a user wants to access the system
1. The error correction code is applied to the q using R, if upto j error(s) can be corrected, then only q is considered for further processing, otherwise q is discarded
2. The keypoints are detected using SIFT, SURF, BRIEF, SIFTBCS, SURFBCS and BRIEFBCS in the similar manner as performed during enrollment phase
3. The extracted features are matched with the stored points in υ
4. If there is a match then the κ' is also extracted from the υ using Lagrange's Interpolation. The authentication is denied in cases of a non-match

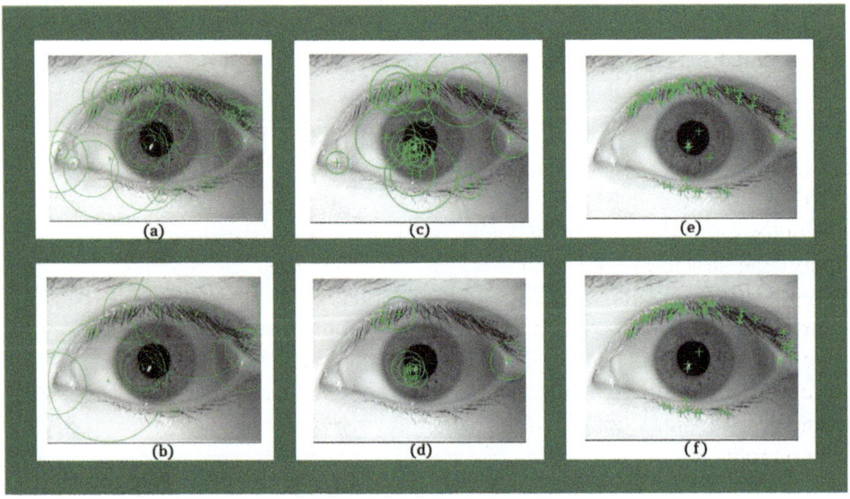

Fig. 2 Features extracted using **a** SIFT **b** SIFTBCS **c** SURF **d** SURFBCS **e** BRIEF **f** BRIEFBCS on MMU dataset

focus on speed with accuracy and BRIEFBCS and BRIEF offer speed component by moderating the accuracy.

All these six methods compute descriptor around each keypoint. The BRIEFBCS and BRIEF computes binary type descriptor, where as SURF and SIFT based techniques computes integral and gradient based descriptors. Gradient based descriptors are more computationally intensive and require more memory, but are more robust to noise and blur. As, descriptors are computed around each keypoint, where keypoints are detected from extrema, Hessian matrix and corners (using FAST) in SIFTBCS, SURFBCS and BRIEFBCS methods. Figures 2 and 3 depict the qualitative analysis of feature descriptor components on MMU and YALE datasets respectively using the aforementioned six techniques.

Table 3 Quantitative analysis using several performance parameters on MMU (iris) and YALE (face) datasets on all six methods *Note* Time is in ms and Accuracy is in percentage

Database	Performance parameter	SIFTBCS [4]	SURFBCS [6]	BRIEFBCS [3]	SIFT [7]	SURF [1]	BRIEF [2]
MMU	Keypoint extraction time	0.22	0.18	0.9	0.24	0.20	0.11
	Number of keypoints	50	68	55	78	110	70
	Overall execution time	0.38	0.32	0.27	0.31	0.29	0.18
	Accuracy	89.4	93.3	92.4	85.2	90.6	91.4
YALE	Keypoint extraction time	0.14	0.07	0.3	0.18	0.20	0.04
	Number of keypoints	30	62	18	52	109	25
	Overall execution time	0.26	0.31	0.09	0.29	0.46	0.11
	Accuracy	82.7	91.5	78.3	81.3	90.1	79.6

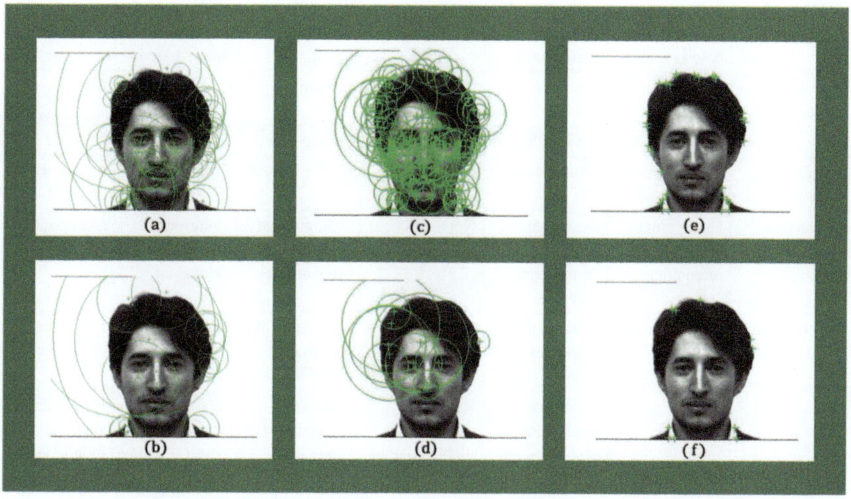

Fig. 3 Features extracted using **a** SIFT **b** SIFTBCS **c** SURF **d** SURFBCS **e** BRIEF **f** BRIEFBCS on YALE dataset

Table 4 Comparative analysis on all six methods based on various properties

Property	SIFTBCS	SURFBCS	BRIEFBCS	SIFT	SURF	BRIEF
Scale invariance	✓	✓	×	✓	✓	×
Rotation invariance	✓	✓	×	✓	✓	×
Illumination invariance	✓	×	×	✓	×	×
Security component	Present	Present	Present	Absent	Absent	Absent
Suitability for portable devices	Low	Medium	High	Low	Medium	High
Suitability for real-time applications	Low	Medium	High	Low	Medium	High
Memory requirement	High	Medium	Low	High	Medium	Low
Robustness to blur	Medium	Medium	Low	Medium	Medium	Low
Robustness to noise	High	Medium	Low to medium	High	Medium	Low to medium
Computational speed	Low to medium	Medium to high	High	Low to medium	Medium to high	High
Descriptor type	Gradient-based	Integral Image-based	Binary	Gradient-based	Integral image-based	Binary
Descriptor nature	Continuous	Continuous	Binary	Continuous	Continuous	Binary
Scale space	Octave-based	Pyramid-based	No scale space	Octave-based	Pyramid-based	No scale space
Keypoint detection	From extrema	From Hessian	From corners	From extrema	From Hessian	From corners
Orientation assignment	Histogram of gradients and orientations	Dominant orientation from Haar wavelets	Not applicable	Histogram of gradients and orientations	Dominant orientation from Haar wavelets	Not applicable
Keypoint localization	Difference of Gaussian pyramid	Hessian matrix responses	FAST corner detection	Difference of Gaussian pyramid	Hessian matrix responses	FAST corner detection
Use cases	Object recognition, image stitching	Real-time applications, object tracking	Real-time applications, mobile vision	Object recognition, image stitching	Real-time applications, object tracking	Real-time applications, mobile vision

6 Conclusion

In this paper, we conducted a thorough comparative analysis of six prominent feature detection, description and security algorithms: SIFT, BRIEF, SURF, SIFTBCS, SURFBCS and BRIEFBCS. Our evaluation encompassed various aspects, including computational efficiency, accuracy, memory requirement, real-time applicability etc. The choice of algorithm depends on the specific requirements of the application, with SIFT excelling in robustness, BRIEF in computational efficiency, and SURF offering a balanced approach and additional security component is provided by Biometric CryptoSystem leading to SIFTBCS, SURFBCS and BRIEFBCS. The results presented in this paper provide valuable insights for researchers and practitioners in the field of computer vision, aiding in the selection of the most suitable algorithm for particular application. Consequently, conducting a comparative analysis of these algorithms becomes imperative for researchers and engineers to make informed decisions when choosing an appropriate method for their tasks. As technology and computing power continue to advance, new feature detection and description algorithms are likely to emerge. Future research could focus on developing new approaches or enhancing the existing algorithms to further improve their performance in various applications.

References

1. Bay, H., Tuytelaars, T., Gool, L.V.: Surf: speeded up robust features. In: European Conference on Computer Vision, pp. 404–417. Springer (2006)
2. Calonder, M., Lepetit, V., Ozuysal, M., Trzcinski, T., Strecha, C., Fua, P.: Brief: computing a local binary descriptor very fast. IEEE Trans. Pattern Anal. Mach. Intell. **34**(7), 1281–1298 (2011)
3. Kaur, P., Kumar, N.: Briefbcs: binary robust independent elementary features based fuzzy vault scheme in biometric cryptosystem. Cluster Computing, pp. 1–14 (2023)
4. Kaur, P., Kumar, N.: Siftbcs: scale invariant feature transform based fuzzy vault scheme in biometric cryptosystem. Multimedia Tools and Applications, pp. 1–22 (2023)
5. Kaur, P., Kumar, N., Singh, M.: Biometric cryptosystems: a comprehensive survey. Multimedia Tools Appl. **82**(11), 16635–16690 (2023)
6. Kaur, P., Kumar, N., Singh, M.: Surfbcs: speeded up robust features based fuzzy vault scheme in biometric cryptosystem. J. Supercomput. **79**, 12292–12316 (2023)
7. Lowe, D.G.: Distinctive image features from scale-invariant keypoints. Int. J Comput. Vis. **60**(2), 91–110 (2004)
8. Rosten, E., Porter, R., Drummond, T.: Faster and better: a machine learning approach to corner detection. IEEE Trans. Pattern Anal. Mach. Intell. **32**(1), 105–119 (2008)
9. University, M.: Mmu2 iris dataset. https://mmuexpert.mmu.edu.my/ccteo
10. University, Y.: Yale face dataset. http://cvc.cs.yale.edu/cvc/projects/yalefaces/yalefaces.html

Comparative Analysis of Pretrained Model-Based Transfer Learning Techniques for Glaucoma Detection from Fundus Images

M. K. Shajila Beegam and Mala Kalra

Abstract Glaucoma is a disorder of the eyes for which there is no recognized treatment and which, if unchecked, can result in irreversible visual loss. The early and accurate diagnosis of glaucoma, a primary contributor to irreversible blindness, is of paramount importance in ophthalmology. Fundus images have become a valuable resource for aiding in the diagnosis of this condition. In recent years, transfer learning, a powerful approach in the field of deep learning, has gained significant attention for its potential to enhance glaucoma detection accuracy. Transfer learning involves passing the pre-trained model's weights to the new task, where it can be adjusted or used as a feature extractor. This survey paper provides a comprehensive examination of transfer learning approaches employed in glaucoma detection from fundus images. This study compares pre-trained models like VGG16, ResNet, GoogLeNet, MobileNet, DenseNet, and Inception providing insights into their performance, strengths and limitations in glaucoma detection using various datasets and methodologies.

Keywords Glaucoma detection · Vision loss · Fundus image · Pre-trained model · Cup-to-Disc ratio · Transfer learning

1 Introduction

Lack of qualified professionals and modern imaging technology hinders accurate disease screening, especially in developing countries. Automatic detection and diagnosis systems, including medical image processing, could greatly improve routine clinical checkups and therapy planning for various ailments. Without treatment, glaucoma, an irreversible and progressive optic neuropathy, results in visual loss [1].

M. K. Shajila Beegam (✉) · M. Kalra
National Institute of Technical Teachers Training & Research, Chandigarh, India
e-mail: shajila.cse21@nitttrchd.ac.in

M. Kalra
e-mail: malakalra@nitttrchd.ac.in

B. Soni et al. (eds.), *Beyond Artificial Intelligence*, Lecture Notes in Networks and Systems 1326, https://doi.org/10.1007/978-981-96-4170-3_21

Glaucoma is a condition characterized by the degeneration of retinal ganglion cells, often linked to elevated intraocular pressure due to the presence of aqueous humor [2]. Glaucoma has two basic categories, open-angle glaucoma [3] and angle-closure glaucoma [4]. As per the World Health Organization's assessment, glaucoma holds the second position as a main contributor to blindness worldwide. Around 60 million people worldwide suffer from glaucomatous optic neuropathy, with 8.4 million blind, and it is predicted to afflict 111.8 million by 2040 [5]. Retinal Ganglion Cell(RGC) deterioration is a significant contributor to irreversible blindness in individuals over 60 years and this degradation results in two significant problems: (a) the Optic Nerve Head (ONH) and the Retinal Nerve Fibre Layer (RNFL) structural abnormalities and (b) problems with Visual Field (VF) functions. This vision loss starts as barely noticeable blind spots towards the borders of the visual field and proceeds to tunnel vision and eventually blindness [6].

Glaucoma is described as the silent assailant of vision as the majority of glaucomatous patients do not experience any early signs or pain [7]. So, an early diagnosis and detection method is urgently needed to effectively follow up the disease to stop its progression [8].

Color fundus imaging, which highlights distinct retinal regions by varying the intensity of the R, G, and B bands, can be used to identify glaucoma [9]. Figure 1, depicts the fundus of a healthy eye [10]. It clearly shows the Optic Nerve Head. The optic nerve's inner ring is the Optic Cup, the outer ring is the Optic Disc, and the area between the inner and outer ring is the neuroretinal rim. Both the disc and cup serve as the cornerstone for segmenting the ONH and detecting glaucoma. Manual estimation of ONH anatomy, such as rim area, disc diameter, inferior, superior, nasal, and temporal (ISNT) rule, and the Cup-to-Disc Ratio (CDR), are the possible diagnostic markers for glaucoma monitoring [11].

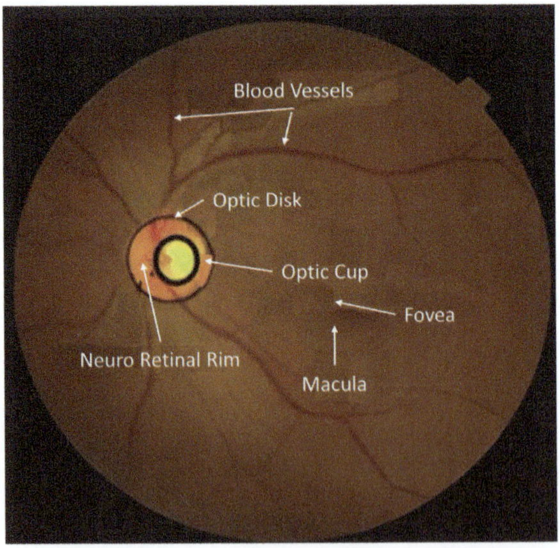

Fig. 1 Fundus Image denoting the healthy eye

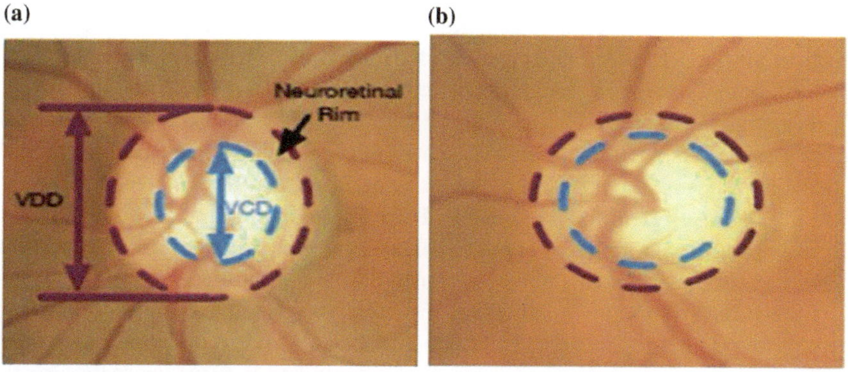

Fig. 2 **a** Normal eye, **b** Glaucoma eye

Figure 2a and b show zoom-in regions of normal disc and glaucoma disc respectively [12]. Finding the ratio between the cup diameter and the disc diameter is the first step in calculating the cup-to-disc ratio. One indicator that is most frequently used in the clinical evaluation of glaucoma is the CDR [13]. In clinical practice, a CDR of 0.65 or above is regarded as glaucoma-indicative [14].

Technological breakthroughs have brought more attention to computer-assisted diagnostics for early disease detection [15]. In this situation, retinal fundus imaging has been used to identify glaucoma utilising deep learning (DL) [16]. Convolutional Neural Network (CNN) models that have already been trained are used in these methods to classify data. The Residual Network (ResNet), Visual Geometry Group Network (VGGNet), and GoogLeNet are some of the models [15].

The study is structured as follows. Initially, an introduction to glaucoma, the main types, the use of fundus imaging technology, and the need for deep learning with transfer learning (TL) is pointed out in Sect. 1. An overview of the pre-trained model and transfer learning is discussed in Sect. 2. The architectures of pre-trained models are described in Section three. Section 4 delves into the performance metrics utilized in various research studies. A comparative analysis of pre-trained models is outlined in Sect. 5. Section 6 concludes the study.

2 Pretrained Models and Transfer Learning

Pretrained models and transfer learning are closely linked concepts in machine learning and deep learning domains. Pretrained models often serve as a foundational component of transfer learning. A description of both is given below.

Pretrained Models: Pretrained models are neural networks that have undergone training using extensive and varied data to excel at a designated task. Through their initial training, these models acquire feature representations, parameters, and knowledge from the source task. Typical instances of pre-trained models encompass architectural designs such as VGG, ResNet, BERT, GPT-3, and numerous others, each pre-trained for specific domains like image classification, natural language understanding, or speech recognition.

Transfer Learning: TL is a machine learning (ML) method where knowledge from a source task is transferred to a related target task to enhance performance. It entails repurposing the pre-trained model by modifying it for the intended task.

Leveraging Pretrained Models for Transfer Learning: In TL, trained models act as the knowledge source. For the target task, these models start with the features, representations, and parameters they learned during their initial training.

Fine-tuning: This process involves modifying a pre-trained model to fit a new task. Using data unique to the target job entails altering and updating some or all of the model's parameters. The model can be fine-tuned to focus on addressing the objective problem while still maintaining the skills and traits it learned during its first training.

Benefits of Transfer Learning: The utilization of pre-trained models in TL has the advantage that they offer a solid framework and a set of high-level features that can capture general patterns and data from the source domain. When there is little data available that is specific to the target task, this knowledge might help enhance the model's performance and generalization on the task.

3 Architecture of Pre-Trained Models

The integration of pre-trained models in artificial intelligence and natural language processing has revolutionized various applications like text generation, picture analysis, and problem-solving. Some of the common models are explained here.

3.1 VGG-16 Architecture

VGG16 is a VGG family model developed by the University of Oxford, renowned for its efficacy and ease of use in image categorization tasks. The following layers make up VGG 16 model [17] as shown in Fig. 3.

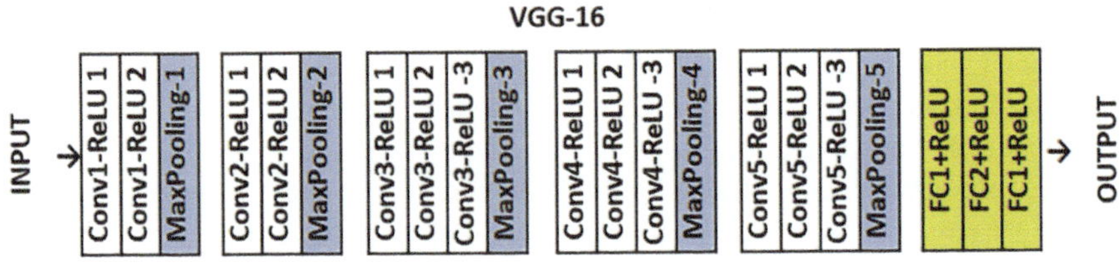

Fig. 3 VGG-16 architecture [17]

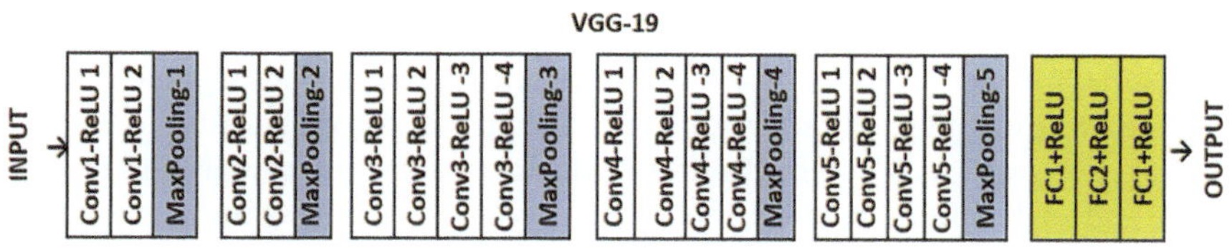

Fig. 4 VGG-19 architecture [17]

Input: It accepts an input image of size 224 × 224 pixels. There are 13 **Convolutional Layers** each with a small 3 × 3 kernel and a Rectified Linear Unit (ReLU) activation function. The convolutional layers gradually increase the number of filters and learn low-level and mid-level features from the input image. VGG16 includes a **Max-Pooling Layer** with a 2 × 2 window and a stride of 2 which helps reduce the spatial dimensions of the feature maps and retains the most important information. Following the convolutional layers and max-pooling layers, there are three **Fully Connected (FC) Layers** or dense layers, which ultimately produce the classification output. The **Softmax Layer** is the final layer that creates the probability distribution over various classes for the input image.

3.2 Architecture of VGG-19

The VGG19 architecture [17] as shown in Fig. 4, is similar to the VGG-16 model, except that it has a total of nineteen layers, which comprise sixteen convolutional layers, and three fully connected layers. It also has five max-pooling layers and one softmax layer for classification.

3.3 The ResNet50 Architecture

Residual Network-50 (ResNet50) [18] as shown in Fig. 5, consists of 50 layers that include both convolutional and residual layers. The architecture is explained below.

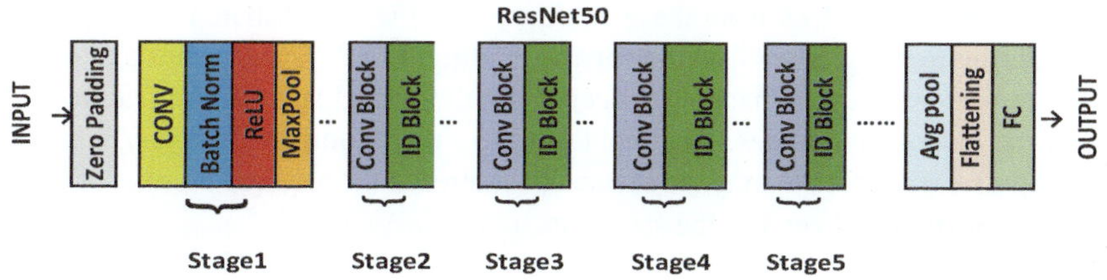

Fig. 5 ResNet50 architecture [18]

The **Input Layer** accepts an RGB image with a 224 × 224 pixels resolution. ResNet50's initial layers include **Convolutional Layers** and **max-pooling layers** for feature extraction. These layers can gather basic knowledge from the source image. ResNet50 uses several **Residual Blocks** to address the vanishing gradient issue and enable the model to be much deeper without sacrificing training accuracy. There are skip connections, in a residual block that omit one or more convolutional layers. The gradient can move directly across the network thanks to these skip links without vanishing. Typically, each residual block has several convolutional layers, batch normalization, and a ReLU activation function. After the convolutional layers and residual blocks, ResNet50 uses **Global Average Pooling (GAP)** to shrink the spatial dimensions of the feature maps to a predetermined size. To build a single vector of feature values for each channel, this technique entails calculating the average of the feature values inside each feature map. A **Fully Connected layer and the Softmax Layer** perform the same function as in VGG 16.

3.4 Architecture of Inception V3

Inception V3 [19], also known as GoogLeNetV3, is a complex convolutional neural network design created especially for applications involving object recognition and picture categorization. Inception V3 as shown in Fig. 6, features the following layers: **The Input Layer** accepts an RGB image with 299 × 299 pixels resolution. To accomplish initial feature extraction and dimension reduction, the network begins with a **Stem Layer** made up of a few **convolutional** and **max-pooling** layers. **Inception modules:** These modules that make up the backbone of Inception v3 give the network the ability to gather features of varied complexity and scales. This module creates a range of features by merging feature maps from different convolutional filters and kernel sizes. Each Inception module typically consists of max-pooling, 1 × 1, 3 × 3, and 5 × 5 convolutions, all performed in parallel. From minute details to larger patterns, these processes capture a variety of aspects. Batch normalization and ReLU activation features are also included here.

At the intermediate layers, Inception v3 has two **Auxiliary Classifiers** to aid gradient flow during training and contribute to regularization. It consists of average pooling, a completely connected layer, and a softmax activation function, collectively responsible for predicting class labels. To help the network develop more accurate representations, the loss from these auxiliary classifiers contributed to the primary loss during training. The **Global Average Pooling (GAP)** layer creates a fixed-size feature vector for each channel by decreasing the spatial sizes of the feature maps after the Inception modules. The final layer is a **Fully Connected Layer**, with 1000 neurons, which serves to map the extracted features to the required number of output classes. **Softmax Layer** has the same function as in VGG 16.

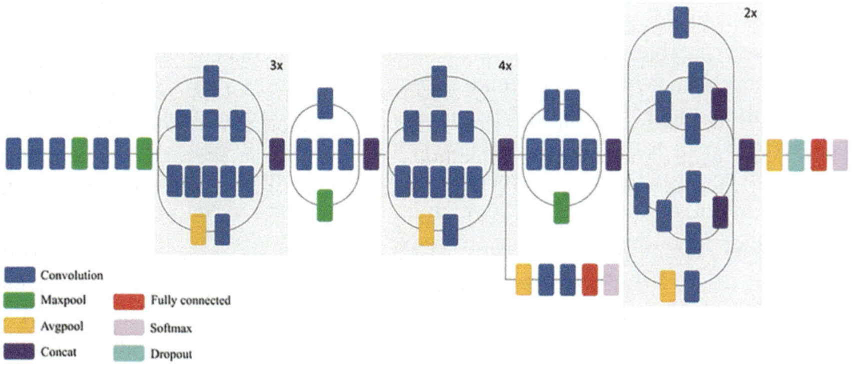

Fig. 6 Inception V3 architecture [19]

3.5 Architecture of Xception (Extreme Inception)

Xception, [20] which stands for Extreme Inception, is a deep convolutional neural network model made for picture categorization and object detection applications. The various layers as shown in Fig. 7, are explained here.

Input Layer. This layer accepts an RGB image with a 224 × 224 pixels resolution. **Entry Flow.** Xception's first stage uses **depthwise separable convolutional layers**, combining depthwise convolution and 1 × 1 **pointwise convolution** for efficient spatial information capture across channels. This combination decreases the parameter count and complexity of computation while retaining efficient feature extraction. Skip connections, which are comparable to residual networks, are also included in

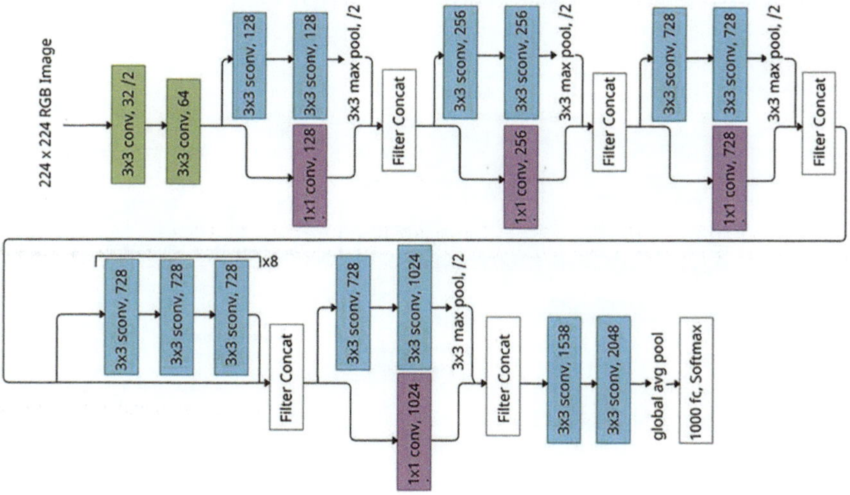

Fig. 7 Xception architecture [20]

the entry flow to aid with gradient flow and feature learning. **Middle Flow.** Repeated depth-wise separable convolution blocks make up the middle section of Xception. The network's capacity to recognize hierarchical features steadily improves as a result of these blocks' additional feature extraction and refinement from the input data. **Exit Flow.** The exit flow is identical to the entry flow, except it has more convolutional layers. It is intended to record high-level patterns and features. It also has a **Global average pooling (GAP)** which functions as in ResNet50 and a **Softmax Layer** which functions similarly as in ResNet50.

3.6 MobileNet Architecture

MobileNet is a family of convolutional neural network models optimized for efficient, lightweight deep learning on mobile and embedded devices, achieving.

reasonable performance in computer vision tasks. Here is an overview of the MobileNet architecture, accompanied by a diagram depicted in Fig. 8.

Depthwise Separable Convolutions: Here this layer's function is the same as in Xception architecture. **Width Multiplier and Resolution:** The width multiplier parameter, designated as alpha is frequently included in MobileNet models. This parameter controls how many channels are allowed in each network layer, making it simple to trade off model size and performance. Also, to balance accuracy and computational cost, MobileNet models can be adapted to various input resolutions. Lower resolutions lead to the development of smaller models that require fewer computational resources. MobileNet commonly uses **Global Average Pooling (GAP)** after the convolutional layers. This results in a fixed-size feature vector for each channel and minimizes the feature maps' spatial dimensions. **Fully Connected Layer and Softmax** layers perform the same function as in VGG16.

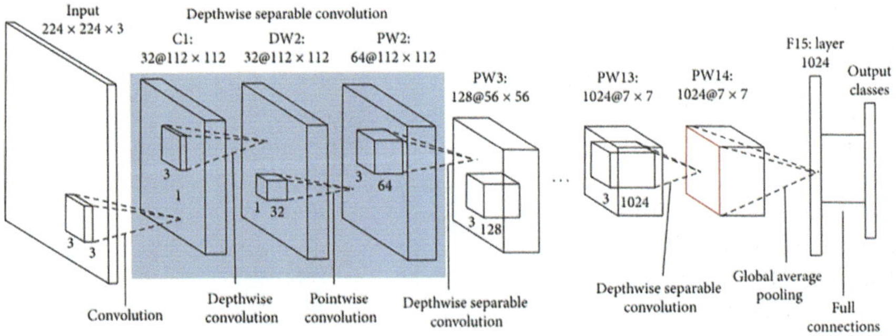

Fig. 8 Architecture of MobileNet [21]

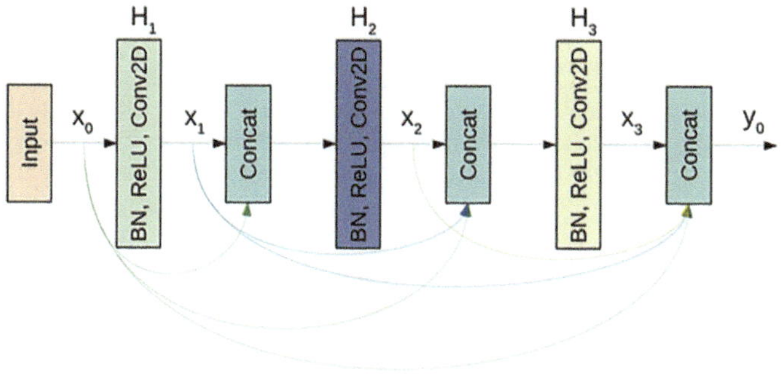

Fig. 9 DenseNet architecture [22]

3.7 DenseNet Architecture

Researchers Huang, Liu, and van der Maaten developed the DCNN architecture known as DenseNet in 2017 as shown in Fig. 9, to enhance data exchange and feature reuse, which in turn enhanced training and classification accuracy by utilizing dense connectivity between layers.

The main elements of the DenseNet architecture are explained below.

Input Layer. This layer accepts an RGB image with a 224 × 224 pixels resolution. The input image is passed through an initial **Convolution Layer** to extract initial features. **Dense Blocks**, the core of DenseNet, consist of convolutional layers that execute operations like 3 × 3 convolution, batch normalization, and ReLU activation, enhancing discriminative learning through feedforward connectivity.

Transition layers between dense blocks control network growth and parameter reduction using batch normalization, convolutional layers, and average pooling, while a **GAP** layer reduces spatial magnitudes and feature vectors.

Fully Connected Layer and Softmax layers have the same function as in previous models.

Table 1 displays a comparison of different pre-trained models, providing descriptions of each, along with their respective strengths and weaknesses. Table 2 depicts the various performance measures utilized in the studies.

4 Performance Metrics

The following performance measures as in Table 2 are utilised in the studies.

Table 1 Comparison of pretrained models

Model	Description	Pros	Cons
VGG16/VGG19 [24–26]/[24, 26]	Simple and deep architectures with 16 or 19 weight layers	Performance and understandability are good and are widely used for image classification	Relatively large model size, slower inference on resource-constrained devices
ResNet (ResNet50) [23–28]	Deep architecture with residual blocks to address the vanishing gradient problem	Excellent performance, relatively easier to train deep networks, good for image classification and transfer learning	Larger model size compared to some newer architectures
Inception (InceptionV3) [23, 25, 26, 28, 29]	Utilizes inception modules for feature extraction and efficient utilization of computation	High accuracy, efficient architecture, suitable for image classification and object detection	Slightly more complex architecture
Xception [24, 27–29]	Based on depth-wise separable convolutions	High accuracy, efficient design, suitable for various computer vision tasks	May require more computational resources
MobileNet (MobileNetV2) [28, 29]	Designed for mobile and embedded devices with depth-wise separable convolutions	Extremely efficient and lightweight, ideal for mobile applications and real-time tasks	May have lower accuracy compared to larger models
DenseNet (DenseNet121) [25, 26, 28, 29]	Features densely connected layers to encourage feature reuse	Strong performance, efficient use of parameters, and well-suited for image classification	Slightly larger model size compared to MobileNet

5 Comparative Analysis of Pretrained Models

This study compared pre-trained models through a literature review of relevant papers, analyzing studies by various researchers, as presented in Tables 3 and 4.

M. Christopher et al. [23] tested deep learning methods to recognize glaucomatous optic neuropathy (GON) from 14,822 fundus photos. ResNet50 with TL was the outstanding model, considering the physical traits of the optic disc. Here, transfer learning improved performance and reduced training time. Serener et al. [27] used GoogLeNet and ResNet50 models for automatic glaucoma diagnosis from fundus images, with GoogLeNet outperforming ResNet-50 for early-stage and progressive glaucoma differentiation.

Table 2 Performance metrics

Metric	Equation	Explanation
Sensitivity (True Positive Rate) (TPR)	TP/(TP + FN)	Sensitivity computes the ratio of actual positive cases that are precisely identified as such. In the context of glaucoma detection, the test can correctly identify those with the disease [23, 24, 27, 29–31]
Specificity (True Negative Rate) (TNR)	TN/(TN + FP)	Specificity computes the ratio of actual negative instances that are precisely identified as such. In the context of glaucoma detection, the test can correctly identify those without the disease [27, 30, 31]
Precision (Positive Predictive Value)	TP/(TP + FP)	It is the proportion of correct identifications. In the context of glaucoma detection, it is how many of the detected positive cases have the disease [24, 29]
Accuracy	(TP + TN)/(TP + TN + FP + FN)	It is the ratio of total correct predictions. It measures how well a test correctly identifies or excludes a condition [24–31]
F1 Score	2 × (Precision × Sensitivity)/(Precision + Sensitivity)	The F1 score, measuring test accuracy, reflects both precision and sensitivity. It is a harmonic mean value derived from precision and sensitivity [29, 31]
AUC	TPR against FPR	AUC (the abbreviation for Area Under Receiver Operating Characteristic Curve) is a plot of True Positive Rate (TPR) against False Positive Rate (FPR) at various classification levels [23, 27–31]

Xu et al. [30] propose a Transfer Induced Attention Network (TIA-Net) for automated glaucoma detection under bounded supervision, integrating maximum mean discrepancy and channel-wise attention. To determine the optimal number of layers for training, they conducted experiments using two clinical datasets and performed TL emphasizing the key features of the fundus image.

Utilizing pre-trained CNN models like VGG11, VGG16, VGG19, AlexNet, ResNET-18, ResNet50, ResNet-101, ResNet-152 and GoogleNet, Sallam et al. [24] introduce an automated strategy using the Large-scale Attention-based Glaucoma (LAG) dataset. Here the ResNet-152 model excelled with commendable accuracy, recall, and precision, each measuring at 86.9% in early detection of glaucoma.

Latif et al. [25] use transfer learning on DCNNs to improve glaucoma diagnosis, with inception V3, ResNet152 and EfficientNetB7 achieving accuracies 96.47%, 99.65% and 100%. Chayan et al. [26] compared their transfer learning model to ResNet50, VGG-16, DenseNet121, InceptionV3, and VGG-19, all optimized with Adam optimizer. ResNet50 achieved the highest validation accuracy at 94.7% after 50 epochs. They also incorporated explainability using Local Interpretable Model-Agnostic Explanations (LIME).

To address the lack of annotated data and the diversity and asymptomaticity of glaucoma, Yi Li et al. [31] introduce GMNNnet, a multimodal neural network for

Table 3 Performance of various models in general prediction

Paper	Dataset	Performance measures (%)							
		Models	AUC	Accuracy	Sensitivity	Specificity	F1 Score	Loss	Precision
Sreng et al. [28]	REFUGE	GoogLeNet	88.69	92.50	–	–	–	–	–
		Inception V3	87.26	90.25	–	–	–	–	–
		Xception	85.29	89.00	–	–	–	–	–
		ResNet50	92.97	93.50	–	–	–	–	–
		MobileNet	93.04	94.50	–	–	–	–	–
		DenseNet	94.64	93.00	–	–	–	–	–
	RIMONE	GoogLeNet	91.03	94.74	–	–	–	–	–
		Inception V3	76.92	71.05	–	–	–	–	–
		Xception	88.14	81.58	–	–	–	–	–
		ResNet50	98.08	92.11	–	–	–	–	–
		MobileNet	99.36	92.11	–	–	–	–	–
		DenseNet	99.04	94.74	–	–	–	–	–
	ACRIMA	GoogLeNet	96.49	91.51	–	–	–	–	–
		Inception V3	99.36	93.87	–	–	–	–	–
		Xception	99.80	98.11	–	–	–	–	–
		ResNet50	99.56	95.75	–	–	–	–	–
		MobileNet	99.96	98.58	–	–	–	–	–
		DenseNet	99.98	99.53	–	–	–	–	–
	ORIGA	GoogLeNet	77.03	71.79	–	–	–	–	–
		Inception V3	76.86	75.38	–	–	–	–	–
		Xception	81.68	77.44	–	–	–	–	–
		ResNet50	80.19	75.90	–	–	–	–	–

(continued)

Table 3 (continued)

Paper	Dataset	Performance measures (%)							
		Models	AUC	Accuracy	Sensitivity	Specificity	F1 Score	Loss	Precision
		MobileNet	81.54	80.51	–	–	–	–	–
		DenseNet	73.92	77.44	–	–	–	–	–
	Drishti-GS1	GoogLeNet	79.89	70.00	–	–	–	–	–
		Inception V3	65.08	70.00	–	–	–	–	–
		Xception	66.67	66.67	–	–	–	–	–
		ResNet50	78.31	73.33	–	–	–	–	–
		MobileNet	73.02	76.67	–	–	–	–	–
		DenseNet	78.31	73.33	–	–	–	–	–
X. Xu et al. [30]	Pvt	TIA-Net	92.9	85.7	84.9	86.9	–	–	–
	ORIGA	TIA-Net	83.50	76.6	75.3	77.2	–	–	–
J. Latif et al. [25]	Pvt	EfficientNetB7	–	100	–	–	–	69.3	–
		DenseNet201	–	0.12	–	–	–	69.3	–
		Inception V3	–	96.47	–	–	–	69.3	–
		VGG16	–	26.32	–	–	–	69.3	–
		NASNetMobile	–	26.56	–	–	–	69.3	–
		ResNet50	–	56.64	–	–	–	69.3	–
		ResNet101V2	–	7.87	–	–	–	69.3	–
		ResNet152	–	99.65	–	–	–	69.3	–
Yi Li et al. [31]	Pvt	GMNNnet	93.90	95.1	93.2	88.6	94.3	–	–
M.T Islam et al. [29]	BEH and ACRIMA	DenseNet	84.17	86.96	73.81	–	80.52	–	88.57
		GoogLeNet	95.24	96.52	90.48	–	95.00	–	100
		MobileNetV2	90.98	93.04	83.33	–	89.74	–	97.22

(continued)

Table 3 (continued)

Paper	Dataset	Performance measures (%)							
		Models	AUC	Accuracy	Sensitivity	Specificity	F1 Score	Loss	Precision
		EfficientNet b3	95.74	96.52	92.86	–	95.12	–	97.50
	BEH and ACRIMA (vessel segmented)	MobileNetV2	70.38	76.52	47.62	–	59.70	–	80.00
		MobileNetV3	84.46	83.48	88.10	–	79.57	–	72.55
	HRF	**MobileNetV2/V3**	83.33	83.33	100	–	–	–	75.00
A. Sallam et al. [24]	LAG	VGG-11	–	80	80	–	–	0.4	80
		VGG-16	–	82.2	82	–	–	0.3	82
		VGG-19	–	80.9	80.9	–	–	0.4	80.9
		Alex Net	–	81.4	81.5	–	–	0.4	81.8
		ResNet-18	–	86.7	86.7	–	–	0.3	86.7
		ResNet-50	–	85.6	85.7	–	–	0.3	85.6
		ResNet-101	–	86.2	86.2	–	–	0.3	86.2
		ResNet-152	–	86.9	86.9	–	–	0.2	86.9
		GoogleNet-V1	–	82.9	83	–	–	0.3	82.9
T. I. Chayan et al. [26]	Pvt	ResNet50	–	94.71	–	–	–	12.22	–
		InceptionV3	–	86.42	–	–	–	35.79	–
		DenseNet121	–	86.81	–	–	–	24.10	–
		VGG-16	–	88.63	–	–	–	27.92	–
		VGG-19	–	93.31	–	–	–	14.94	–

Table 4 Performance of various models in severity prediction

Paper	Dataset		Architecture	Performance measures			
				AUC	Accuracy	Sensitivity	Specificity
Christopher et al. [23]	Pvt	Mild	ResNet 50 with TL	0.89	–	–	–
			Inception with TL	–	–	0.82	0.82
		Any Glaucoma	VGG-16	0.83	–	–	–
			VGG-16 with TL	0.89	–	–	–
			Inception	0.83	–	–	–
			Inception with TL	0.9	–	0.84	0.83
			ResNet 50	0.89	–	–	–
			ResNet 50 with TL	0.91	–	0.84	
		Moderate to severe	ResNet 50 with TL	0.97	–	–	–
			Inception with TL	–	–	0.92	0.93
A Serener et al. [27]	Pvt	Mild	ResNet 50	0.84	0.9	0.42	0.94
			GoogLeNet	0.91	0.91	0.17	0.98
		Severe	ResNet 50	0.74	0.86	0.21	0.93
			GoogLeNet	0.75	0.85	0.29	0.91
		Average	ResNet 50	0.79	–	–	–
			GoogLeNet	0.83	–	–	–

glaucoma prognosis and classification, which uses a multimodal dataset of high-resolution medical photographs, electronic medical records, and five significant glaucoma label types. GMNNnet architecture, featuring Catboost classifier, Grad-CAM strategy, and transfer learning, effectively analyzes patient metadata, segments glaucoma, and processes images, excelling in multimodal data.

Islam et al. [29] applied multiple DL models (EfficientNet, MobileNet, DenseNet, and GoogLeNet) to identify glaucoma from fundus images. The EfficientNet-b3 model outperformed on a private dataset, while the MobileNetV2/V3 models demonstrated satisfactory test accuracy and F1-score on the HRF dataset.

Sreng et al. [28] present a dual-step automatic glaucoma screening system using the DeepLabv3 + architecture and multiple deep convolutional neural networks across five datasets. The DenseNet secured the best results with accuracy of 0.9953 and an AUC of 0.9998 on the ACRIMA dataset.

6 Conclusion

In this work, we compared 9 studies to assess the effectiveness of the aforementioned seven pre-trained learning models in diverse data processing scenarios. In the first study [23], Resnet50 with transfer learning achieved an AUC of 0.91 for general glaucoma prediction and 0.97 for moderate to severe prediction. In the second study [27] the model GoogLeNet secured an AUC of 0.91 for mild glaucoma prediction. Using two datasets, the Transfer Induced Attention Network (TIA-Net) obtained AUCs of 0.929 and 0.835 respectively in the next study [30]. In the fourth study [24] which used the LAG dataset, ResNet50 achieved an accuracy of 85.6%. Study 5 [25] proved inception V3, ResNet152 and EfficientNetB7 as the best models with accuracies 96.47%, 99.65% and 100%. ResNet50 achieved 94.71% accuracy in the next work [26]. The seventh work [31] gives an accuracy of 95.1% with a novel multimodal neural network for glaucoma prognosis and classification, GMNNnet. The eighth study [29] projects EfficientNet b3 and MobileNetV2/V3 as the best models. The final study [28] proved DenseNet as the best model with the ACRIMA dataset.

References

1. Davis, B.M., Crawley, L., Pahlitzsch, M., Javaid, F., Cordeiro, M.F.: Glaucoma: the retina and beyond. Acta Neuropathol. **132**(6), 807–826 (2016). https://doi.org/10.1007/s00401-016-1609-2
2. Goel, M., Picciani, R.G., Lee, R.K., Bhattacharya, S.K.: Aqueous humor dynamics: a review. Open Ophthalmol. J. **4**(1), 52–59 (2010). https://doi.org/10.2174/1874364101004010052
3. Dey, A., Bandyopadhyay, S.: Automated glaucoma detection using support vector machine classification method. Br. J. Med. Med. Res. **11**(12), 1–12 (2016). https://doi.org/10.9734/bjmmr/2016/19617
4. Wright, C., Tawfik, M.A., Waisbourd, M., Katz, L.J.: Primary angle-closure glaucoma: an update. Acta Ophthalmol. **94**(3), 217–225 (2016). https://doi.org/10.1111/aos.12784
5. Tham, Y.C., Li, X., Wong, T.Y., Quigley, H.A., Aung, T., Cheng, C.Y.: Global prevalence of glaucoma and projections of glaucoma burden through 2040: a systematic review and meta-analysis. Ophthalmology **121**(11), 2081–2090 (2014). https://doi.org/10.1016/j.ophtha.2014.05.013
6. de Carvalho Junior, A. S. V., Carvalho, E. D., de Carvalho Filho, A. O., de Sousa, A. D., Corrêa Silva, A., Gattass, M.: Automatic methods for diagnosis of glaucoma using texture descriptors based on phylogenetic diversity. Comput. Electr. Eng. 71(June), 102–114 (2018). https://doi.org/10.1016/j.compeleceng.2018.07.028
7. Abdull, M.M., Chandler, C., Gilbert, C.: Glaucoma, 'the silent thief of sight': patients' perspectives and health seeking behaviour in Bauchi, northern Nigeria. BMC Ophthalmol. **16**(1), 1–9 (2016). https://doi.org/10.1186/s12886-016-0220-6
8. Costagliola, C., Dell'Omo, R., Romano, M.R., Rinaldi, M., Zeppa, L., Parmeggiani, F.: Pharmacotherapy of intraocular pressure part II. Carbonic anhydrase inhibitors, prostaglandin analogues and prostamides. Expert Opin. Pharmacother. **10**(17), 2859–2870 (2009). https://doi.org/10.1517/14656560903300129
9. Gour, N., Khanna, P.: Automated glaucoma detection using GIST and pyramid histogram of oriented gradients (PHOG) descriptors. Pattern Recognit. Lett. **137**, 3–11 (2020). https://doi.org/10.1016/j.patrec.2019.04.004

10. Ramani, R.G., Shanthamalar, J.J.: Improved image processing techniques for optic disc segmentation in retinal fundus images. Biomed. Signal Process. Control **58**, 101832 (2020). https://doi.org/10.1016/j.bspc.2019.101832
11. Guo, J., Azzopardi, G., Shi, C., Jansonius, N.M., Petkov, N.: Automatic determination of vertical cup-to-disc ratio in retinal fundus images for glaucoma screening. IEEE Access **7**, 8527–8541 (2019). https://doi.org/10.1109/ACCESS.2018.2890544
12. Fu, H., Cheng, J., Xu, Y., Liu, J.: Glaucoma detection based on deep learning network in fundus image. Adv. Comput. Vis. Pattern Recognit. (September), 119–137 (2019). https://doi.org/10.1007/978-3-030-13969-8_6
13. Jiang, Y., Duan, L., Cheng, J., Gu, Z., Xia, H., Fu, H.: JointRCNN: a region-based convolutional neural network for optic disc and cup segmentation. IEEE Trans. Biomed. Eng. **67**(2), 335–343 (2020). https://doi.org/10.1109/TBME.2019.2913211
14. Akram, M.U., Tariq, A., Khalid, S., Javed, M.Y., Abbas, S., Yasin, U.U.: Glaucoma detection using novel optic disc localization, hybrid feature set and classification techniques. Australas. Phys. Eng. Sci. Med. **38**(4), 643–655 (2015). https://doi.org/10.1007/s13246-015-0377-y
15. Joshi, S., Partibane, B., Hatamleh, W. A., Tarazi, H., Yadav, C. S., Krah, D.: Glaucoma detection using image processing and supervised learning for classification. J. Healthc. Eng. 2022, (2022). https://doi.org/10.1155/2022/2988262
16. Guergueb, T., Akhloufi, M.A.: A review of deep learning techniques for glaucoma detection. SN Comput. Sci. **4**(3), 274 (2023). https://doi.org/10.1007/s42979-023-01734-z
17. Deng, M., Goy, A., Li, S., Arthur, K., Barbastathis, G.: Probing shallower: perceptual loss trained Phase Extraction Neural Network (PLT-PhENN) for artifact-free reconstruction at low photon budget. Opt. Express **28**(2), 2511 (2020). https://doi.org/10.1364/oe.381301
18. Choi, H.S., Kim, J.S., Whangbo, T.K., Kim, K.H.: Transfer learning for effective urolithiasis detection. Int. Neurourol. J. **27**(Suppl 1), S21–S26 (2023). https://doi.org/10.5213/inj.2346110.055
19. Mahdianpari, M., Salehi, B., Rezaee, M., Mohammadimanesh, F., Zhang, Y.: Very deep convolutional neural networks for complex land cover mapping using multispectral remote sensing imagery. Remote Sens. 10(7), (2018). https://doi.org/10.3390/rs10071119
20. Srinivasan, K., Garg, L., Datta, D., Alaboudi, A.A., Jhanjhi, N.Z., Agarwal, R.: Performance comparison of deep cnn models for detecting driver's distraction. Comput. Mater. Contin. **68**(3), 4109–4124 (2021). https://doi.org/10.32604/cmc.2021.016736
21. Wang, W., Li, Y., Zou, T., Wang, X., You, J., Luo, Y.: A novel image classification approach via dense-mobilenet models. Mob. Inf. Syst. 2020, (2020). https://doi.org/10.1155/2020/7602384
22. Alshazly, H., Linse, C., Abdalla, M., Barth, E., Martinetz, T.: COVID-Nets: deep CNN architectures for detecting COVID-19 using chest CT scans. Peer J. Comput. Sci. **7**(July), 1–40 (2021). https://doi.org/10.7717/peerj-cs.655
23. Christopher, M., Belghith, A., Bowd, C., Proudfoot, J.A., Goldbaum, M.H., Weinreb, R.N.: Performance of deep learning architectures and transfer learning for detecting glaucomatous optic neuropathy in fundus photographs. Sci. Rep. **8**(1), 1–13 (2018). https://doi.org/10.1038/s41598-018-35044-9
24. Sallam, A., Gaid, A. S. A., Saif, W. Q. A., Kaid, H. A. S., Abdulkareem, R. A., Ahmed, K. J. A.: Early detection of glaucoma using transfer learning from pre-trained CNN models. In: 2021 Int. Conf. Technol. Sci. Adm. ICTSA 2021 (2021). https://doi.org/10.1109/ICTSA52017.2021.9406522
25. Latif, J., Tu, S., Xiao, C., Rehman, S.U., Sadiq, M., Farhan, M.: Digital forensics use case for glaucoma detection using transfer learning based on deep convolutional neural networks. Secur. Commun. Netw. **2021**, 4494447 (2021). https://doi.org/10.1155/2021/4494447
26. Chayan, T. I., Islam, A., Rahman, E., Reza, M. T., Apon, T. S., Alam, M. G. R.: Explainable AI based glaucoma detection using transfer learning and LIME, no. iii, 2022 [Online]. Available: http://arxiv.org/abs/2210.03332
27. Serener, A., Serte, S.: Transfer learning for early and advanced glaucoma detection with convolutional neural networks. TIPTEKNO 2019—Tip Teknol. Kongresi, pp. 1–4, (2019). https://doi.org/10.1109/TIPTEKNO.2019.8894965

28. S. Sreng, N. Maneerat, K. Hamamoto, S., Win, K. Y.: Deep learning for optic disc segmentation and glaucoma diagnosis on retinal images. Appl. Sci. 10(14), (2020). https://doi.org/10.3390/app10144916
29. Islam, M.T., Mashfu, S.T., Faisal, A., Siam, S.C., Naheen, I.T., Khan, R.: Deep learning-based glaucoma detection with cropped optic cup and disc and blood vessel segmentation. IEEE Access **10**, 2828–2841 (2022). https://doi.org/10.1109/ACCESS.2021.3139160
30. Xu, X., Guan, Y., Li, J., Ma, Z., Zhang, L., Li, L.: Automatic glaucoma detection based on transfer induced attention network. Biomed. Eng. Online **20**(1), 39 (2021). https://doi.org/10.1186/s12938-021-00877-5
31. Li, Y., Han, Y., Li, Z., Guo, Y. Z. Z.: A transfer learning-based multimodal neural network combining metadata and multiple medical images for glaucoma type diagnosis. www.nature.com/scientificreports/, 13(12076), 1–13 (2023). https://doi.org/10.1038/s41598-022-27045-62

Crop Type Identification in Indian Agriculture: A Contemporary Review

Sanjay Moulik and V. Kethareswaran

Abstract Accurate crop identification is crucial for India's diverse agricultural landscape, impacting resource allocation, decision-making, and sustainable development. This paper reviews crop identification techniques, encompassing traditional and modern methods and highlighting their strengths and weaknesses. It emphasizes the significance of integrating remote sensing data and crop modelling for yield estimation. Successful literature studies from India illustrate yield prediction's potential benefits in supply chain management, market predictions, and policy formulation. Challenges persist, particularly in regions like North-East India, characterized by complex terrain and limited infrastructure. Addressing these challenges and fostering research in crop identification is crucial for harnessing their agricultural potential.

Keywords Crop-identification · Remote sensing · Machine learning · North-East · Tea plantation

1 Introduction

Spanning across 28 states and 8 union territories, India stands as a vast nation characterized by a rich tapestry of demographics. Boasting a population exceeding 1.3 billion individuals, it ranks as the world's most populous country. Within its demographic expanse, India houses an intricate mosaic of humanity, encompassing upwards of 2,000 distinct ethnic groups and a linguistic spectrum comprising over 1,600 languages. The geographical fabric of India is equally multifaceted, featuring the towering expanse of the Himalayan mountain range to the north, the expansive Thar Desert gracing the west, and lush tropical forests adorning its southern reaches.

S. Moulik (✉) · V. Kethareswaran
Indian Institute of Information Technology Guwahati, Guwahati, India
e-mail: sanjay@iiitg.ac.in

V. Kethareswaran
e-mail: v.kethareswaran@iiitg.ac.in

B. Soni et al. (eds.), *Beyond Artificial Intelligence*, Lecture Notes in Networks and Systems 1326, https://doi.org/10.1007/978-981-96-4170-3_22

The nation witnesses an array of climatic variations, spanning from the humid subtropical conditions of the north to the tropical environs of the south. This intricate interplay of diverse terrains and climates has indelibly shaped the distinctive cultural heritage and customs that flourish across India's disparate regions.

Crop identification refers to the process of discerning and categorizing different types of crops cultivated in a given area. This is typically accomplished through the analysis of various data sources such as satellite imagery, remote sensing, and sensor data, often employing machine learning techniques. The goal is to accurately classify and map the specific crops present, aiding in agricultural management, yield prediction, and land-use planning. Crop identification holds immense significance for India due to its diverse agricultural landscape and the critical role of agriculture in the country's economy and livelihoods. With over 60% of its population directly or indirectly dependent on agriculture, accurate crop identification is fundamental for efficient resource allocation, informed decision-making, and sustainable development.

India's vast and varied geographical features, ranging from the Himalayan foothills to coastal plains, contribute to a wide array of climatic conditions that support the cultivation of an extensive range of crops (Fig. 1). From grains like rice and wheat to cash crops like cotton and sugarcane, each region specializes in specific crops suited to its ecological conditions. Accurate identification of these crops assists in tailoring agricultural practices to the local environment, optimizing water usage, fertilization, and pest control. This not only enhances productivity but also minimizes resource wastage and environmental impacts. Moreover, precise crop identification is pivotal in ensuring food security by aiding in estimating potential yields, facilitating distribution, and managing supply chains effectively. It also enables policymakers to formulate targeted strategies for agricultural development, trade, and investment, fostering growth and resilience in the sector.

In a country dominated by smallholder farmers, crop identification becomes a tool for financial inclusion. Accurate data substantiates farmers' activities, enhancing their access to credit, insurance, and government support programs. This empowers farmers to make informed choices, adopt modern techniques, and mitigate risks effectively. Furthermore, given India's vulnerability to climate change and natural disasters, timely and accurate identification of crop failures due to extreme weather events is crucial for disaster response and recovery efforts.

Contribution: Crop identification is fundamental to agriculture and resource management in India, especially in a context as diverse as North-East India. The significance of accurate crop identification spans from informing precision agriculture to ensuring food security and disaster management. In this paper, we discuss the diverse methodologies employed for crop identification, ranging from traditional methods to modern techniques like machine learning, deep learning, and computer vision. We emphasize the importance of integrating remote sensing data and crop modeling for yield estimation, with literature studies showcasing their relevance in real-world applications.

Our contribution lies in highlighting the challenges faced in crop identification in regions like North-East India, characterized by complex terrain, limited infrastructure, and linguistic diversity. This region has often been neglected in research due to data constraints and unique challenges. We emphasize the need for increased research focus in such areas to harness their agricultural potential fully. The summary of our contribution is given below:

- Review of diverse crop identification techniques, emphasizing their strengths and weaknesses.
- Highlighting the significance of addressing challenges in under-represented regions like North-East India.
- Emphasizing the need for increased research in crop identification for informed agricultural practices.

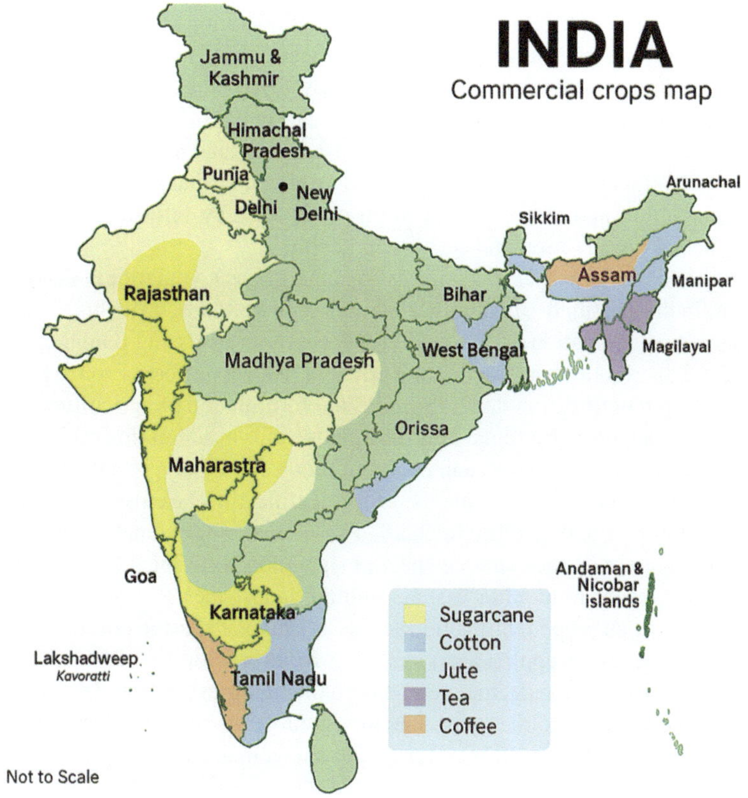

Fig. 1 A commercial crops map [10]

2 Applications Based on Crop Classification

Crop identification plays a pivotal role across various applications within the realms of agriculture, remote sensing, and environmental management. Its significance lies in its ability to provide fundamental insights into the composition of cultivated crops within a given geographical area. This foundational information serves as a crucial step, yielding a multitude of practical applications that span diverse sectors.

At the forefront of these applications is precision agriculture, where accurate identification of crops forms the bedrock of tailored farming practices. Armed with knowledge about the specific crops present, farmers can fine-tune their strategies for irrigation, fertilization, and pest control, optimizing resource utilization while minimizing wastage. Furthermore, crop identification serves as the linchpin for predicting yields. Yield estimation models, integrating historical data with crop identification insights, empower stakeholders to anticipate potential harvests. Such forecasts are invaluable for supply chain management, market predictions, and formulation of pricing strategies, contributing to efficient market operations.

Beyond the agricultural realm, crop identification finds application in land use planning and natural resource management. By discerning which areas are under cultivation, governments and urban planners can make informed decisions about urban expansion, conservation efforts, and sustainable land management. Additionally, crop identification aids in monitoring environmental changes like deforestation, land degradation, and shifts in vegetation cover, thereby informing strategies for environmental preservation and ecosystem health.

During times of crisis, crop identification emerges as a vital tool in disaster management. In the aftermath of natural calamities such as floods, droughts, or wildfires, the accurate identification of affected crops assists in assessing damage, planning targeted aid distribution, and facilitating efficient recovery efforts. From a policy perspective, crop identification data influences the formulation of agricultural policies, subsidies, and support programs. This information aids in addressing regional disparities, promoting specific crops, and ensuring food security at a broader level. Incorporating technological advancements, crop identification heavily relies on remote sensing techniques and satellite imagery. These technologies enable the classification of crops based on their unique spectral signatures, offering a cost-effective and non-intrusive means to monitor vast agricultural landscapes.

We may say that crop identification serves as a pivotal starting point that cascades into a diverse array of applications spanning from precision agriculture and yield prediction to disaster management and environmental preservation. Its foundational insights empower stakeholders across sectors, facilitating informed decision-making, sustainable practices, and effective resource management.

3 Techniques Used for Crop Identification

This section delves into the diverse methodologies employed to accurately identify crops, encompassing both conventional and contemporary approaches.

Traditional Methods: Visual interpretation involves human experts analyzing satellite or aerial imagery to identify crops based on visual characteristics such as color, texture, and shape. This method relies on the expertise of individuals who are trained to recognize distinct features associated with different crops. While accessible and intuitive, visual interpretation can be subjective, prone to errors, and limited in its ability to differentiate subtle variations between crops. Additionally, it might not be suitable for large-scale or real-time applications.

Field surveys involve on-site visits by experts who physically inspect crops to determine their types. This hands-on approach provides ground truth data for validating remote sensing-based identification methods. However, field surveys can be time-consuming, resource-intensive, and logistically challenging, especially in large and diverse agricultural landscapes.

Modern Techniques: Machine learning techniques involve training algorithms on large datasets containing spectral and spatial information extracted from satellite imagery. These algorithms learn patterns and relationships that differentiate various crops, enabling automated classification [17]. Machine learning models can handle a wide range of data features and are capable of scaling to process vast amounts of data efficiently.

Deep learning, a subset of machine learning, utilizes neural networks to automatically extract intricate patterns from data. It excels at capturing complex relationships in imagery, making it suitable for distinguishing crops with nuanced differences [3]. Deep learning models require substantial computational resources for training but can achieve high accuracy levels.

Computer vision techniques focus on image processing and feature extraction. By identifying specific visual attributes in images, such as leaf shape or arrangement, these methods contribute to crop classification [19]. Computer vision is particularly valuable when working with images captured at different resolutions or in challenging lighting conditions.

Comparison of Strengths and Weaknesses: Traditional methods are straightforward but can be time-consuming, costly, and prone to subjectivity. In contrast, modern techniques offer efficiency, scalability, and reduced human bias. Machine learning and deep learning excel in handling complex data relationships, while computer vision is adept at feature extraction. However, modern techniques demand substantial training data, technical expertise, and can be sensitive to variations in lighting, weather, and data quality.

We may summarize that traditional methods provide familiarity but suffer from limitations in efficiency and scalability. Modern techniques leverage the power of data-driven algorithms, offering automation and scalability, albeit with prerequisites of data volume and technical proficiency. The choice between these methods depends on the project's goals, available resources, and the desired level of accuracy and scalability.

4 Literature Survey

Researchers have been working on the problem of crop identification for a very long time. However, we are focusing on the works that have been carried out in the last couple of decades since these works lose their significance with time.

Various techniques have been used to address the crop identification problem. Among them, remote sensing based approaches have been the primary ones. The main contribution of [16] lies in investigating the effectiveness of phenology combined with an object-based image analysis approach for accurate rice mapping. The research underscores the potential of integrating crop phenology and multispectral data for precise rice classification on a larger scale. The authors of [4] address challenges in using satellite remote sensing technology for horticultural crops in India, given their small field sizes and short duration. Focusing on major fruit crops (Banana, Mango, and Citrus) across four states, the study develops a research methodology for accurate area assessment. Pixel-based hybrid classification and object-based classification methods were used with high-resolution remote sensing data. The study's accuracy and methodology were endorsed by State Horticulture Departments and archived in ISRO's geoportals. In [8], the authors have used remote sensing and GIS techniques for accurate paddy field determination in Uttar Pradesh's Haidergarh Canal Command Area. The study addresses the need for swift and economical methods for generating minimum support prices, ensuring benefits for farmers, crop insurance, and efficient storage. By leveraging Sentinel-1 C-band microwave satellite data, the research overcomes cloud cover limitations and estimates a paddy cropped area of 66,287.25 hectares for the year 2018. In [12], Murthy et al. proposed an alternative basis for assessing jute crop yield loss using the 'Crop Health Factor' (CHF), derived from satellite-derived indices. The study effectively correlates CHF with yield, enabling accurate assessment of crop loss and insurance payouts. This approach harnesses remote sensing data and has implications for innovative and parametric crop insurance solutions, especially in the context of jute cultivation in West Bengal.

Machine learning-based approaches have also been used to address this problem. The authors demonstrate the necessity of multi-temporal data analysis for wheat crop delineation in [13]. It highlights the limitations of Maximum Likelihood Classification (MLC) with composite multi-temporal data due to a large null set. To address this, two approaches were evaluated: i. MLC variations and ii. Artificial Neural Networks (ANN) with back-propagation. These methods were applied to Indian Remote Sensing satellite (IRS)-1B images for wheat classification in two different Indian regions. Mathur et al. [11] introduced an efficient approach to supervised classification using Support Vector Machines (SVM). This approach was demonstrated in agricultural classification in Punjab, India, where a small, intelligently selected training dataset produced comparable accuracy (over 91%) to a larger, conventionally acquired one. The intelligently defined set contained a higher proportion of support vectors (0.70) compared to the conventional approach (0.41). Musande et al. [14] investigated the use of temporal remote sensing data to discriminate cotton crops. It explores the impact of various spectral indices in identifying cotton crops from multi-spectral images. The study employs soft computing techniques to address mixed-pixel problems and uses a possibilistic fuzzy classification approach.

There have also been many works that have incorporated different techniques together to address crop identification problems. Bhosle et al. [3] explored the use of deep learning Convolutional Neural Networks (CNN) for both Land Use Land Cover (LULC) classification and crop identification. The Indian Pines dataset from the AVIRIS sensor was employed for LULC classification, while a dataset from the EO-1 Hyperion sensor in Phulambri, Aurangabad, MH, India, was used for crop classification. The accuracy of the CNN model was assessed by optimizing various parameters such as the optimizer, activation function, filter size, learning rate, and batch size. The authors in [1] explored hyperspectral imaging for crop mapping in agriculture, emphasizing band selection considering both reflectance properties and biological conditions. Bands are categorized based on wavelength boundaries and then selected using metrics like entropy, NDVI, and MNDWI. A CNN is designed to classify crops, tested on Indian Pines and Salinas datasets, achieving high overall accuracies. Lee et al. [9] addressed the challenge of accurately mapping sugarcane cultivation in India's Bhima Basin, a critical task for targeted agricultural policies. To achieve this, the researchers employed crowdsourced smartphone data as an alternative ground truth, applying both supervised machine learning and unsupervised classification methods. Key findings include the effectiveness of machine learning with high-resolution satellite imagery for sugarcane mapping and a notable discrepancy between the estimated sugarcane area in the Bhima Basin and government statistics, indicating potential for improving mapping approaches in other regions.

Many researchers have extended the technique of crop identification to provide an estimate of the crop yield. In [5], the authors aimed to predict sugarcane yield at the district level in India using a remote sensing-based approach that leverages the Vegetation Condition Index (VCI). Historical data spanning 13 years (2003–2015) of the Normalized Difference Vegetation Index (NDVI) from the MODIS instrument on the Terra satellite was used to calculate the VCI during the peak growing period. Empirical models were developed through stepwise regression between VCI and historical sugarcane yield data for 52 major sugarcane-growing districts in five Indian states. These models were found to be statistically significant for all districts. Tripathy et al. [18] tried to enhance wheat yield forecasting in Punjab by incorporating remotely sensed data into the World Food Studies crop simulation model. Spatial weather data were generated through interpolation, while sowing dates were estimated using normalized difference vegetation index data from the SPOT satellite. Remotely sensed leaf area index (LAI) data were utilized in the WOFOST model to predict spatial wheat yields. The study in [2] focused on estimating rice yields in India with high spatial resolution using a machine learning technique called Gradient Boosted

Regression (GBR). The researchers used MODIS LAI data and district-level yield observations for calibration. The GBR models were then applied to downscale district yields to a finer spatial resolution of 500 meters. The results showed good agreement with reported district-level observations, with improved performance when separate models were used for different rice cropping densities.

We can observe from the above discussion that crop identification is an important and complex problem, and researchers have tried to propose solutions using various approaches. However, it is a continuous process since most of the solutions are time-sensitive and lose their relevance with time.

5 North-East India and Its Challenges for Crop Identification

North-East India, comprising Arunachal Pradesh, Assam, Manipur, Meghalaya, Mizoram, Nagaland, and Tripura, is characterized by its diverse climate and varied landforms Fig. 2. From the rugged terrain of Arunachal Pradesh to the fertile plains of Assam, each state boasts a unique climate and topography. This region's agricultural practices are mainly based on rice cultivation and tea plantation. The North-Eastern States account for 53% of the total tea production of India. However, only a handful of works have focused research on it. In [6], the authors examine the influence of climatic variables on tea yield in north-east India. Utilizing a garden-level dataset and statistical models, it identifies the negative impact of warming temperatures exceeding 26.6 °C and intense precipitation variability on tea production. The work in [15] employs LISS-III and Cartosat-1 images to monitor tea replantation phases in North East India's Dooars region. Wavelet techniques, including Haar, Daubechies, and Symlet, identify patterns at varying scales. Results highlight the efficacy of levels 3 and 4 in wavelet decomposition for information extraction. Spatial correlations aid in selecting informative patterns, indicating wavelets' utility in revealing spatial variations during replantation stages. The authors in [7] utilize Sentinel-2 data and a random forest algorithm to map aboveground biomass (AGB) in Assam's Barak Valley tea agroforestry systems. It reveals 4.78% (330.69 km^2) of the region under tea cultivation, with AGB predictions ranging from 30 to 124 Mg ha^{-1}. Other works related to the tea plantation in this region have focused on the effect of climate, deforestation, etc., on the tea plantation. However, the region needs more focused research to increase productivity in the region.

The topological diversity presents a series of challenges for the crop identification process. The complex topography creates difficulties in acquiring consistent remote sensing data across the region, affecting the accuracy of identification models. Additionally, the remote and often inaccessible nature of certain areas hampers data collection efforts, leading to limited ground truth data for model validation. The North-East's linguistic and ethnic diversity poses challenges in conducting surveys

Fig. 2 The map of North-East India having 8 states

and collecting accurate data, potentially impacting the development of reliable identification algorithms. Furthermore, challenges like limited infrastructure, variable weather patterns, and the predominance of smallholder farming further compound the task of crop identification. Heavy rainfall and cloudy conditions can affect the quality of satellite imagery, while the lack of awareness and technical expertise among farmers and communities may hinder the adoption of modern identification methods. Addressing these challenges is crucial to effectively harness the agricultural potential of North-East India while ensuring sustainable and informed farming practices.

6 Conclusion

Accurate crop identification is vital for India's agricultural sector, enabling informed decisions and efficient resource allocation. This paper highlights the importance of precision in crop identification, discussing both traditional and modern methods and their relevance. Integrating remote sensing data and crop modeling for yield estimation, as showcased in Indian literature studies, offers significant advantages for supply chain management, market predictions, and policy formulation. Challenges, especially in regions like North-East India, demand attention due to complex geography and infrastructure limitations. Overcoming these challenges and promoting research in crop identification in such areas is essential to maximize their agricultural potential. Continuous research and a focus on underrepresented regions are essential to meet the evolving needs of Indian agriculture.

References

1. Agilandeeswari, L., Prabukumar, M., Radhesyam, V., Phaneendra, K., Farhan, A.: Crop classification for agricultural applications in hyperspectral remote sensing images. Appl. Sci. **12**(3), 1670 (2022)
2. Arumugam, P., Chemura, A., Schauberger, B., Gornott, C.: Remote sensing based yield estimation of rice (Oryza Sativa l.) using gradient boosted regression in India. Remote Sens. **13**(12), 2379 (2021)
3. Bhosle, K., Musande, V.: Evaluation of deep learning CNN model for land use land cover classification and crop identification using hyperspectral remote sensing images. J. Indian Soc. Remote. Sens. **47**(11), 1949–1958 (2019)
4. Chaudhari, K., Nishant, N., Upadhyay, G., More, R., Singh, N., Vyas, S.P., Bhattacharya, B.K.: Crop inventory of orchard crops in India using remotely sensed data. Int. Arch. Photogramm. Remote. Sens. Spat. Inf. Sci. **42**, 269–275 (2019)
5. Dubey, S.K., Gavli, A., Yadav, S., Sehgal, S., Ray, S.S.: Remote sensing-based yield forecasting for sugarcane (saccharum officinarum l.) Crop in India. J. Indian Soc. Remote. Sens. **46**, 1823–1833 (2018)
6. Duncan, J.M., Saikia, S., Gupta, N., Biggs, E.: Observing climate impacts on tea yield in Assam, India. Appl. Geograp. **77**, 64–71 (2016)
7. Kalita, R.M., Nandy, S., Srinet, R., Nath, A.J., Das, A.K.: Mapping the spatial distribution of aboveground biomass of tea agroforestry systems using random forest algorithm in barak valley, northeast india. Agrofor. Syst. **96**(8), 1175–1188 (2022)
8. Krishna, A., Singh, A., Singh, P., Upadhyay, R.K.: Paddy crop classification using microwave satellite data in 23 down haidergarh canal command system, Uttar pradesh, India. Int. Res. J. Eng. Technol. (IRJET) **7**, 3432–3436 (2020)
9. Lee, J.Y., Wang, S., Figueroa, A.J., Strey, R., Lobell, D.B., Naylor, R.L., Gorelick, S.M.: Mapping sugarcane in central India with smartphone crowdsourcing. Remote Sens. **14**(3), 703 (2022)
10. Ltd., M.D.P.: Commercial crops map (2022). https://www.mapsofindia.com/indiaagriculture/commercialcrops.htm
11. Mathur, A., Foody, G.M.: Crop classification by support vector machine with intelligently selected training data for an operational application. Int. J. Remote. Sens. **29**(8), 2227–2240 (2008)
12. Murthy, C., Poddar, M., Choudhary, K.K., Srikanth, P., Pandey, V., Ramasubramanian, S., Kumar, G.S.: Remote sensing based crop insurance for jute (corchorus olitorius) crop in India. Remote Sens. Appl.: Soc. Environ. **26**, 100717 (2022)
13. Murthy, C., Raju, P., Badrinath, K.: Classification of wheat crop with multi-temporal images: performance of maximum likelihood and artificial neural networks. Int. J. Remote Sens. **24**(23), 4871–4890 (2003)
14. Musande, V., Kumar, A., Kale, K.: Cotton crop discrimination using fuzzy classification approach. J. Indian Soc. Remote Sens. **40**, 589–597 (2012)
15. Singh, A., Dutta, R., Stein, A., Bhagat, R.M.: A wavelet-based approach for monitoring plantation crops (tea: Camellia Sinensis) in North East India. Int. J. Remote Sens. **33**(16), 4982–5008 (2012)

16. Singha, M., Wu, B., Zhang, M.: An object-based paddy rice classification using multi-spectral data and crop phenology in Assam, North East India. Remote Sens. **8**(6), 479 (2016)
17. Stournaras, S., Loukatos, D., Arvanitis, K.G., Kalatzis, N.: Crop identification by machine learning algorithm and sentinel-2 data. Chem. Proc. **10**(1), 20 (2022)
18. Tripathy, R., Chaudhari, K.N., Mukherjee, J., Ray, S.S., Patel, N., Panigrahy, S., Parihar, J.S.: Forecasting wheat yield in Punjab state of India by combining crop simulation model Wofost and remotely sensed inputs. Remote Sens. Lett. **4**(1), 19–28 (2013)
19. Wu, Z., Chen, Y., Zhao, B., Kang, X., Ding, Y.: Review of weed detection methods based on computer vision. Sensors **21**(11), 3647 (2021)

Role of Social Media in Disaster Management Using NLP and Deep Learning

Pooja Rai, Sudhakar Kumar, Sunil K. Singh, Harkiran Kaur, Simran Jaggi, Varsha Arya, and Brij B. Gupta

Abstract In the era of technological advancement, social media has emerged as a vital player in disaster management. It facilitates the dis-semination of information by both responders and victims, offering a direct channel for communication with authorities—a feature notably absent in traditional media. The diversity of languages, including regional ones, introduces complexity in comprehending and extracting valuable insights from text-based social media data. Recent progress in Natural Language Processing (NLP), coupled with Machine Learning (ML) and Deep Learning (DL), has been applied to address the challenges presented by disasters on social media. This paper conducts a thorough review of NLP-based research studies focused on social media data related to natural disaster management, particularly those conducted since 2016. The paper delves into the findings and concludes with reflections and suggestions for future research in this field.

Keywords Natural language processing · Disaster management · Social media texts · Machine learning · Deep learning

---

P. Rai
Department of Computer Science, New Alipore College, Kolkata, India

S. Kumar (✉) · S. K. Singh · H. Kaur · S. Jaggi
Chandigarh College of Engineering and Technology, Chandigarh, India
e-mail: sudhakar@ccet.ac.in

V. Arya
Department of Business Administration, Asia University, Taichung, Taiwan

B. B. Gupta
Department of Computer Science and Information Engineering, Asia University, Taichung 413, Taiwan

V. Arya
Hong Kong Metropolitan University, Hong Kong, Hong Kong

B. B. Gupta
University of Economics and Human Science, Warsaw, Poland

B. Soni et al. (eds.), *Beyond Artificial Intelligence*, Lecture Notes in Networks and Systems 1326, https://doi.org/10.1007/978-981-96-4170-3_23

1 Introduction

Social network analysis (SNA) [1, 2] can have a significant impact on social media in the context of disaster management. SNA [3, 4] is the process of examining the relationships and interactions among individuals or entities within a network, and it can provide valuable insights. Indeed, SNA can be enhanced through various technologies and methods like deep active learning [5] and secure blockchain [6] to address the challenges of fake news detection, compromised social network accounts in the context of disaster management.

Any hazardous incident leading to threat and damage to susceptible communities is known as a disaster [7]. Disasters can be caused by natural or man-made that threaten the vulnerability of a community. Effective disaster management calls for intelligent systems to carry out several activities to cope with the disaster [8]. With the emergence of social networks, people can easily provide essential information that can be useful for enduring disasters. Online social platforms can play a vital role in providing situational awareness during and after natural disasters [9]. The massive data with wide variance from social platforms led to the possibility of extracting information for disaster management, at different levels.

Handling complex and high-volume social media data urges automated tools and systems for extracting meaningful information that caters to the needs of disaster management. With the advancement of technology, there have been several research works on the disaster-related social media data and Phengsuwan et al. [9] have presented a taxonomy of the data found in social platforms, as shown in Fig. 1. Figure 1 shows the taxonomy presented in the survey made in [9]. The elements of the taxonomy are as follows:

- Data source: the sources of data in social platforms.
- Information dimension: aspects of information in social platform.
- Methodology: the algorithms used in Data Management and Data Analysis.
- Application: using data on social networks for the management of data
- Language: the language used for writing posts on social platforms.

Nazer et al. [10] have pointed out that social media data reflects the following stages of disaster management: Warning, Impact, Response, Relief as shown in Fig. 2. Figure 3 shows the languages used in social platform posts. Various regional languages can also be seen in such posts. With such language varieties, it becomes challenging to understand text-based information produced by social media. Systematic extraction of vital information and semantics out of the social media text using artificial intelligence aids in disaster management. Hence, NLP with Machine Learning [11, 12] plays a major role in comprehending and unfolding insights from text available on social media, as highlighted in Ref. [13].In this survey, the application of NLP, and ML/DL [14–16] models in the context of social media and disaster management, are reviewed.

Fig. 1 Taxonomy of Disaster management related social media data [9]

Fig. 2 Social media affected disaster management stages

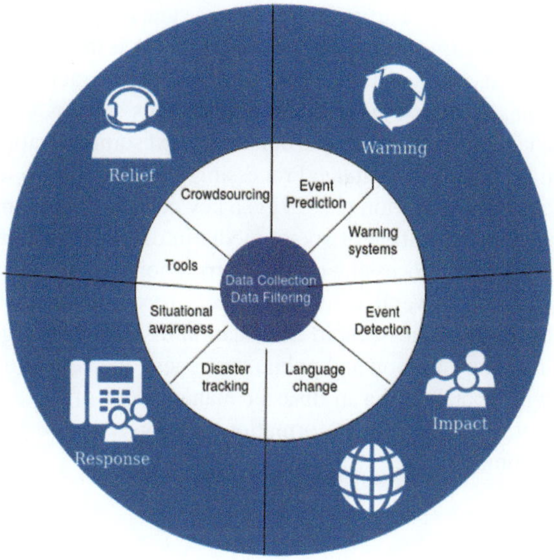

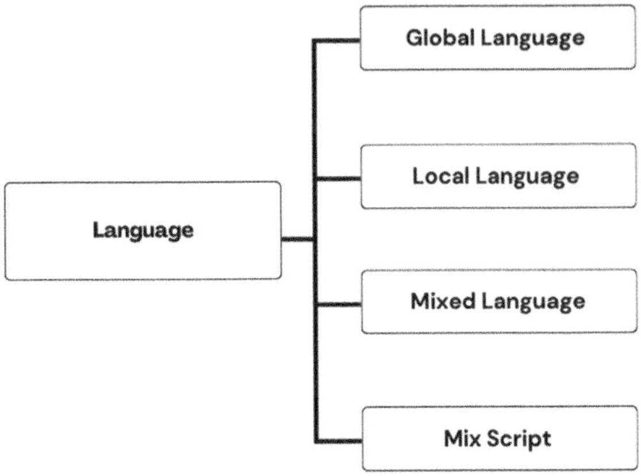

Fig. 3 Language types in social media text

2 NLP and ML for Disaster Management in the Recent Literature

By combining artificial intelligence with technologies like natural language processing it is possible to discover meaningful information from the massive amount of unstructured data from social media. Natural language processing (NLP) belongs to the family of artificial intelligence that aids computers in comprehending as well as manipulating human language. This work is the analysis of the application of existing NLP techniques to social media data for extracting information useful for disaster management. This includes applications such as geo-location detection, sentiment analysis, entity and topic detection, text summarization, and word vector representation. Natural Language Processing (NLP) and deep learning often go hand in hand, and parallelization [17–19] is a key concept in making these technologies efficient.

The utilization of social media in disaster management, combined with Natural Language Processing (NLP), carries noteworthy implications for various industries. Here are several essential aspects to contemplate like early warning and situational awareness, rapid information dissemination, geotagged data analysis, monitoring, evaluation and public awareness and education. To sum up, integrating NLP with social media in disaster management offers a robust set of tools for prompt warnings, effective information sharing, resource allocation, and engagement with communities.

2.1 NLP-Based Disaster Monitoring

A model for disaster monitoring was developed by Gopal et al. [20] by using e-news related to various hazard emergencies. ML-based techniques have been used to separate important information (Text classification) from irrelevant data. Domala et al. [21] have reported work on combining NLP and ML models to improve crisis management using news data. The spider-scraper tool was used for scraping online news. This research [22] focuses on creating a social media analytics system that employs NLP, ML, and Deep learning techniques to analyze social media data. The system's primary goal is to detect and monitor emergency events, including disasters, using data from Twitter. It provides a framework for real-time information extraction and emergency response in crisis management, demonstrating its effectiveness in disaster monitoring and early response. A discussion in Ref. [23] emphasizes how automating the monitoring of tweets can assist in identifying warning signs of impending disasters. Many disaster relief organizations and news agencies hence have been aiming to develop technologies and programs related to tweet analysis. In Ref. [24], the aim is to create a disaster detection system that is automated and uses sentiment analysis and location intelligence powered by AI to gather important data from social media posts. The system employs deep learning and support vector machines monitoring disasters, including heatwaves, cyclones, earthquakes, floods, and bushfires. It highlights the value of comprehending location-specific sentiments towards crisis scenarios for political leaders and decision-makers, making it a valuable tool for early disaster monitoring and response. A hybrid ML model [25] has been used for identifying the aftereffects of disasters from Twitter posts. A graph-based clustering was used for identifying the credible information, and a transformer (with BERT) was used for classification. The locations specified in the posts have been validated by the named entity recognition (NER) model. Furthermore, as demonstrated in [26], the application is useful for post-disaster response and disaster reduction.

2.2 NLP-Based Damage Assessment

Sit et al. [37] used machine learning, deep learning, and natural language processing to analyse Hurricane Irma data to determine the post-disaster effect. Long Short-Term Memory (LSTM) based classification showed the best results.

Stowe et al. [27] formulated an annotation schema of data on twitter related to Hurricane Sandy. The meaningful tweets were categorized into seven categories namely, Reporting, Sentiment, Information, Action, Preparation, Movement. The aim of this work was to ameliorate risk communication as well as precautionary measures amid disasters.

The semantic feature of the textual part of the social media posts has been utilized in [28] for analyzing the information dimension (spatial or temporal) using Latent Dirichlet Allocation (unsupervised topic modeling approach).The model has been successful in accurate predictions of earthquakes as well as other nature-related disasters.

2.3 NLP-Based Post-Disaster Response

Social media is often found to be noisy and unstructured. So, this problem has been addressed by O'Neal et al. [38]. They took data related to the 2017 Hurricane Harvey disaster and used Google Vision API for attribute detection. The concept of domain adaptation has been applied by Li et al. [29] for analyzing tweets (data from CrisisLexT6)for post-disaster response. The model is the ensemble of Naive Bayes (NB) and (NB-ST). The performance of NB-ST (with 86.91% accuracy)was found to be the best compared to other supervised ML classifiers such as SVM, LR, Naive Bayes, and Random Forest in recognizing disaster-relevant tweets. A combination of textual and visual data from social data has been utilized in [34], to design deep learning models for categorizing relevant posts for disasters. These data are again combined with geotagging of posts and several models were designed using Machine learning algorithms namely Linear Regression, Random Forest, Linear SVM, SVM with RBF, and, Sigmoid SVM. Linear Regression outperformed all other models with an accu-racy of 95.2%. The tweet posts related to the earthquake in Nepal and some regions of the Indian subcontinent (2015) as well as Italy (2016) have been uti-lized by Basu et al. [31] for identifying resource needs and availability. They have proposed the model with an unsupervised neural network [39, 40], leverag-ing the strength of embedding, word-level as well as character-level. However, the proposed model performed with 0.57% accuracy and 0.191 F1 score for the Nepal earthquake tweets. Identification of informative data for critical situations in disasters has been reported in [30]. They have used Twitter data from the CrisisLexT26 dataset. They proposed models with Naive Bayes (NB) classifier as well as deep learning models (ANN and CNN), by taking features like polarity, details of the user, and BOW features, independently as well as taking these features as a set. However, the deep learning models outperformed the model with the Naive Bayes (NB) classifier. Filtering of relevant data was done by SVM and LR and applied transfer learning model (BERT) for detecting the outage types. The challenges of scheduling effective rescue during disasters have been addressed in [32]. A model consisting of an attention LSTM (bi-LSTM) and CNN, feature engineering, was proposed to classify Twitter data from Hurricanes Harvey and Irma, CrisisNLP, and CrisisLex. Targeting the rescue operations, they have proposed a hybrid scheduling model too.

3 Discussion

A summarization of the technical analysis of Disaster Response/Damage assessment using NLP and ML/DL on social media data has been illustrated in Table 1. The analysis consists of the NLP and ML/DL techniques used, the disaster type, and performance metrics for evaluating the model.

Models have been designed with NLP-based techniques, including Text classification (40%), Topic modeling (20%), Bag of Words (BoW) based techniques (4%),

Table 1 Analysis of disaster response/damage assessment using NL

Article	Disaster type	NLP technique	ML/DL technique	Performance
[27], 2016	Hurricane	Text classification	SVM	Acc. = 85.6%
[28], 2018	Flood	Topic modeling	LDA	Acc. = 86.45%
[29], 2018	General	Topic modeling	NB-ST	Acc. = 86.91%
[30], 2018	General	BoW based text classification	CNN, RNN, NB	F1 score = 89.5%
[31], 2019	Earthquake	Information retrieval	ANN	Acc. = 57%
[32], 2019	Hurricane	Text classification	Bi-LSTM and CNN	Acc. = 93.7%
[33], 2020	Hurricane	Information retrieval	KNN, LR, SVM, RF, CNN	for BERT:
			LSTM, GRU, AttenLSTM, BERT	Acc. = 88%
[34], 2020	Flood	Topic modeling	CNN	Acc. = 95%
[20], 2020	Landslide, flood, rain	Text classification	SVM, K-NN, Na ıveBaye, MPL, LR	Acc. = 68%
[25], 2020	Hurricane	Named entity recognition	BERT transformer	Acc. = 95.55%
[21], 2020	General	Text classification	Linear classifier, logistic Reg., ANN	Acc. = 89%
[22], 2021	General	Topic modeling, Info. retrieval	Clustering classification	Acc. = 91.78%
[24], 2022	General	Sentiment analysis	SVM	Acc. = 97%
[35], 2023	Hurricane, earthquake,	Information retrieval	CNN with Transfer Learning	Acc. = 98.14%
	Flood, wildfire	Text & image classification		
[36], 2023	General	Text classification	Bi-LSTM, GCN	Acc. = 93%

Information retrieval (20%), Sentiment analysis (8%), Named Entity Recognition (4%), and Multi-label classification (4%). These models are primarily focused on disaster monitoring, damage assessment, and post-disaster response. They have been applied to various disasters, with the following distribution: Hurricanes (32.14%), general disasters (21.42%), floods (21.42%), earthquakes (10.71%), tornadoes (3.57%), rain-related disasters (3.57%), landslides (3.57%), and wildfires (3.57%).

It is seen from Fig. 4 that research has been carried out using NLP and ML/DL, targeting the management of disasters of various types. It illustrates the distribution of research works based on different NLP techniques used for managing the disaster, in terms of percentage. Text classification has been mostly used (40%), followed by Topic modelling (20%). Other NLP techniques namely, Information retrieval, Sentiment analysis, and Named Entity Recognition have also been used by various research works.

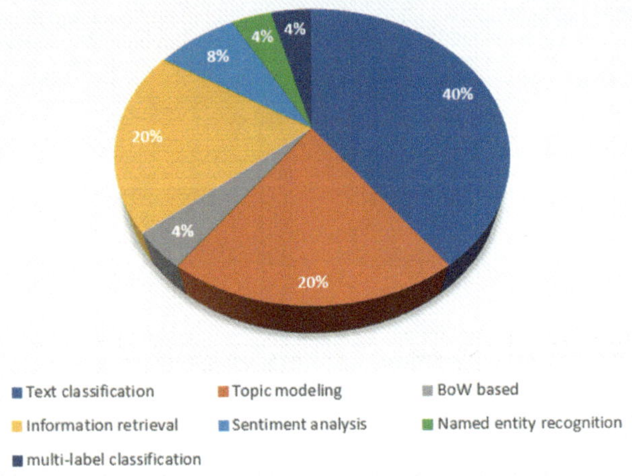

Fig. 4 NLP-based research works by type of NLP given in terms of percentage

Studies based on the type of disasters have also been shown in Fig. 5 in terms of percentage. The highest number of studies have been found for Hurricanes (32.14%), earthquakes (10.71%) being the next. Studies have been there for any disaster type (general) (21.42%), rain (3.57%), flood (21.42%), and landslides (3.57%). However, inclusion and discussion of all related research works may not have been included in this review work due to the keyword-based search-ing of the literature.

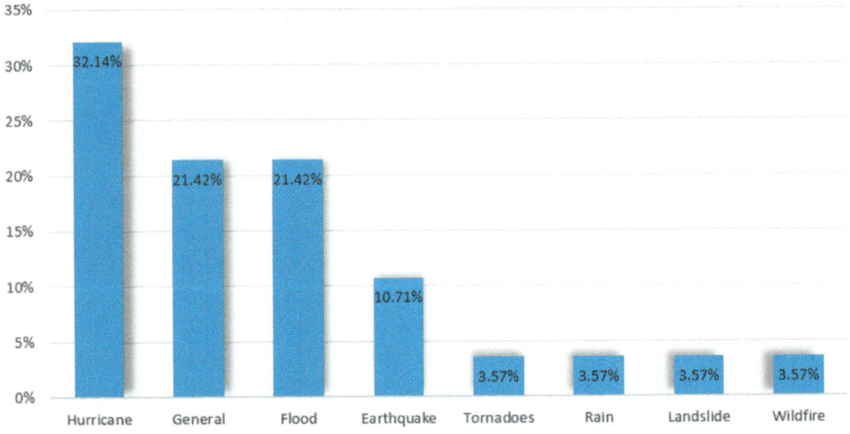

Fig. 5 NLP-based research works by type of disaster given in terms of percentage

4 Conclusion

Disaster management involves the strategic organization of resources to mitigate or abate the harm caused by disasters. With the advancement in technology, social media is one of the most convenient and useful communications in emergencies during disasters. Responders as well as victims can broadcast information on social media and unlike traditional media, feedback can be provided by the people can also to officials or concerned authorities. However, handling such massive and complex social media data for extracting valuable insights in times of disaster management is quite challenging and demands automated techniques for doing so. The complexity of catastrophes has been handled more and more by developments in NLP, ML, and DL. We have analyzed how NLP with ML and DL techniques aid in disaster management. To this end, recent research works (since 2016) for disaster management (mostly Twitter posts), have been used. In terms of the performance of the models, CNN with transfer learning has outperformed all, achieving an accuracy of 98.14%. Furthermore, NLP with ML and DL may be leveraged for other sub-stages of disaster management and improving the current systems by fetching more data for the model.

References

1. Zhang, Z., Gupta, B. B.: Social media security and trustworthiness: overview and new direction. Future Gener. Comput. Syst. 86, 914–925 (2018)
2. Sahoo, S. R., Gupta, B. B.: Real-time detection of fake account in twitter using machine-learning approach. In: Advances in computational intelligence and communication technology: proceedings of CICT 2019, pp. 149–159. Springer (2021)
3. Sahoo, S. R., Gupta, B. B.: Classification of various attacks and their defence mechanism in online social networks: a survey. Enterprise Inform. Syst. 13(6), 832–864 (2019)
4. Sahoo, S. R., Gupta, B. B.: Popularity-based detection of malicious content in facebook using machine learning approach. In: First international conference on sustainable technologies for computational intelligence: proceedings of ICTSCI 2019, pp. 163–176. Springer (2020)
5. Ren, P., Xiao, Y., Chang, X., Huang, P. Y., Li, Z., Gupta, B. B., Chen, X., Wang, X.: A survey of deep active learning. ACM Comput. Survey (CSUR) 54(9), 1–40 (2021)
6. Nguyen, G. N., Le Viet, N. H., Elhoseny, M., Shankar, K., Gupta, B. B., Abd El-Latif, A. A.: Secure blockchain enabled cyber–physical systems in healthcare using deep belief network with resnet model. J. Parallel Distribut. Comput. 153, 150–160 (2021)
7. Wiegmann, M., Kersten, J., Senaratne, H., Potthast, M., Klan, F., Stein, B.: Opportunities and risks of disaster data from social media: a systematic review of incident information. Nat. Hazard. **21**(5), 1431–1444 (2021)
8. Adeel, A., Gogate, M., Farooq, S., Ieracitano, C., Dashtipour, K., Larijani, H., Hussain, A.: A survey on the role of wireless sensor networks and IoT in disaster management, pp. 57–66. Springer Singapore, Singapore (2019)
9. Phengsuwan, J., Shah, T., Thekkummal, N. B., Wen, Z., Sun, R., Pullarkatt, D., Thirugnanam, H., Ramesh, M. V., Morgan, G., James, P., Ranjan, R.: Use of social media data in disaster management: a survey. Future Internet 13(2) (2021)
10. Nazer, T. H., Xue, G., Ji, Y., Liu, H.: Intelligent disaster response via social media analysis a survey. SIGKDD Explor. Newsl. 19(1), 46–59 (2017)

11. Singh, I., Singh, S. K., Singh, R., Kumar, S.: Efficient loop unrolling factor prediction algorithm using machine learning models. In: 2022 3rd international conference for emerging technology (INCET), pp. 1–8. IEEE (2022)
12. Khade, G., Kumar, S., Bhattacharya, S.: Classification of web pages on attractiveness: a supervised learning approach. In: 2012 4th international conference on intelligent human computer interaction (IHCI), pp. 1–5. IEEE (2012)
13. Aggarwal, K., Singh, S. K., Chopra, M., Kumar, S.: Role of social media in the covid-19 pandemic: A literature review. Data mining approaches for big data and sentiment analysis in social media, pp. 91–115 (2022)
14. Mengi, G., Singh, S. K., Kumar, S. K., Mahto, D., Sharma, A.: Automated machine learning (automl): The future of computational intelligence. In: International conference on cyber security, privacy and networking, pp. 309–317. Springer (2021)
15. Pen˜alvo, F. J. G., Maan, T., Singh, S. K., Kumar, S., Arya, V., Chui, K. T., Singh, G. P.: Sustainable stock market prediction framework using machine learning models. Int. J. Software Sci. Comput. Intell. (IJSSCI) 14(1), 1–15 (2022)
16. Aggarwal, K., Singh, S. K., Chopra, M., Sudhakar Kumar, and Colace, F.: Deep learning in robotics for strengthening industry 4.0.: opportunities, challenges and future directions. Robotics and AI for Cybersecurity and Critical Infrastructure in Smart Cities, pp. 1–19 (2022)
17. Sudhakar Kumar, S. K., Singh, S. K., Aggarwal, N., Aggarwal, K.: Efficient speculative parallelization architecture for overcoming speculation overheads. In: International conference on smart systems and advanced computing (Syscom-2021), vol. 3080, pp. 132–138 (2021)
18. Kumar, S., Singh, S. K., Aggarwal, N., Gupta, B. J., Alhalabi, W., Band, S. B.: An efficient hardware supported and parallelization architecture for intelligent systems to overcome speculative overheads. Int. J. Intell. Syst. 37(12), 11764–11790 (2022)
19. Kumar, S., Kr Singh, S., Aggarwal, N.: Speculative parallelism on multicore chip architecture strengthen green computing concept: a survey. In: Advanced computer science applications, pp. 3–16. Apple Academic Press
20. Gopal, L. S., Prabha, R., Pullarkatt, D., Ramesh, M. V.: Machine learning based classification of online news data for disaster management. In: 2020 IEEE global humanitarian technology conference (GHTC), pp. 1–8 (2020)
21. Jayashree Domala, Manmohan Dogra, Vinit Masrani, Dwayne Fernandes, Kevin D'souza, Delicia Fernandes, and Tejal Carvalho. Automated identification of disaster news for crisis management using machine learning and natural language processing. In: 2020 international conference on electronics and sustainable communication systems (ICESC), pp. 503–508 (2020)
22. Asif, A., Maruf Hasan, M., Abdou, S., Elsayed, K. M., Rashwan, M., Khatoon, S., Alshamari, M. A.: Development of social media analytics system for emergency event detection and crisis management. Comput., Mater. Continua 68(3), 3079–3100 (2021)
23. Chopra, M., Singh, S. K., Aggarwal, K., Gupta, A.: Predicting catastrophic events using machine learning models for natural language processing. In: Data mining approaches for big data and sentiment analysis in social media, pp. 223–243. IGI Global (2022)
24. Linardos, V., Drakaki, M., Tzionas, P., Karnavas, Y.L.: Machine learning in disaster management: recent developments in methods and applications. Machine Learn. Knowledge Extract. **4**(2), 446–473 (2022)
25. Fan, C., Wu, F., Mostafavi, A.: A hybrid machine learning pipeline for automated mapping of events and locations from social media in disasters. IEEE Access 8, 10478–10490 (2020)
26. Sharma, A., Singh, S. K., Badwal, E., Kumar, S., Gupta, B. B., Arya, V., Chui, K. T., Santaniello, D.: Fuzzy based clustering of consumers' big data in industrial applications. In: 2023 IEEE international conference on consumer electronics (ICCE), pp. 01–03 (2023)
27. Stowe, K., Paul, M. J., Palmer, M., Palen, L., Anderson, K.: Identifying and categorizing disaster-related tweets. In: Proceedings of the fourth international workshop on natural language processing for social media, pp. 1–6, Austin, TX, USA. Association for Computational Linguistics (2016)

28. Resch, B., Usl¨ander, F., Havas, C.: Combining machine-learning topic models and spatiotemporal analysis of social media data for disaster footprint and damage assessment. Cartogr. Geogr. Inform. Sci. 45(4), 362–376 (2018)
29. Li, H., Caragea, D., Caragea, C., Herndon, N.: Disaster response aided by tweet classification with a domain adaptation approach. J. Contingen. Crisis Manage. **26**(1), 16–27 (2018)
30. Neppalli, V. K., Caragea, C., Caragea, D.: Deep neural networks versus naive bayes classifiers for identifying informative tweets during disasters. In: International conference on information systems for crisis response and management (2018)
31. Basu, M., Shandilya, A., Khosla, P., Ghosh, K., Ghosh, S.: Extracting resource needs and availabilities from microblogs for aiding post-disaster relief operations. IEEE Trans. Comput. Social Syst. 6(3), 604–618 (2019)
32. Kabir, M. Y., Madria, S.: A deep learning approach for tweet classification and rescue scheduling for effective disaster management. In: Proceedings of the 27th ACM SIGSPATIAL international conference on advances in geographic information systems, SIGSPATIAL '19, pp. 269–278, New York, NY, USA. Association for Computing Machinery (2019)
33. Paul, U., Ermakov, A., Nekrasov, M., Adarsh, V., Belding, E.: Outage: detecting power and communication outages from social networks. In: Proceedings of the web conference 2020, WWW '20, pp. 1819–1829, New York, NY, USA. Association for Computing Machinery (2020)
34. Huang, X., Li, Z., Wang, C., Ning, H.: Identifying disaster related social media for rapid response: a visual-textual fused cnn architecture. Int. J. Digital Earth **13**(9), 1017–1039 (2020)
35. Vishwanath, T., Deepak Shirwaikar, R., Jaiswal, W. M., Yashaswini, M.: Social media data extraction for disaster management aid using deep learning techniques. Remote Sensing Appl. Soc. Environ. 30, 100961 (2023)
36. Li, S., Wang, Y., Huang, H., Zhou, Y.: Study on the rumor detection of social media in disaster based on multi-feature fusion method (2023)
37. Sit, M. A., Koylu, C., Demir, I.: Identifying disaster-related tweets and their semantic, spatial and temporal context using deep learning, natural language processing and spatial analysis: a case study of hurricane IRMA. Int. J. Digital Earth 12(11), 1205–1229 (2019)
38. O'Neal, A., Rodgers, B., Segler, J., Murthy, D., Lakuduva, N., Johnson, M., Stephens, K.: Training an emergency-response image classifier on signal data. In: 2018 17th IEEE international conference on machine learning and applications (ICMLA), pp. 751–756 (2018)
39. Inderpreet Singh, Sunil Kr Singh, Sudhakar Kumar, Aggarwal, K.: Dropoutvgg based convolutional neural network for traffic sign categorization. In: Congress on intelligent systems: proceedings of CIS 2021, Vol. 1, pp. 247–261. Springer (2022)
40. Gupta, S., Agrawal, S., Singh, S. K., Kumar, S.: A novel transfer learning-based model for ultrasound breast cancer image classification. In: Computational vision and bio-inspired computing: proceedings of ICCVBIC 2022, pp. 511–523. Springer (2023)

FPGA Implementation of AI Algorithms for Image Detection

Ishpreet Kaur, Garvit Banga, Vivek Singh, Akshit Muval, and Jyoti Kedia

Abstract Nowadays, Deep neural networks (DNNs) including QNN (Quantised Neural Network) and BNN (Binary Neural Network) are widely used in wide range of artificial intelligence (AI) applications including image processing, object detection and robotics. However, implementing these deep neural networks on embedded devices is a major bottleneck due to the massive requirement of computation and storage. Therefore, use of techniques which can carry out efficient processing of DNNs in AI systems and improve inference time and execution time keeping into consideration the critical performance parameters such as accuracy or hardware cost is very important. In this paper, an efficient implementation and analysis of QNN- and BNN-based pattern recognition techniques have been done. The FPGA PYNQ-Z2 has been used for QNN and BNN algorithms and has shown improved performance in terms of inference time, efficiency and hardware. The comparative analysis of implementation of BNN and QNN has been carried out on the basis of accuracy, weight bit error, RoC curve, and execution speed. The proposed Image detection algorithms has produced significant results on identifying images with a percentage decrease in inference time of about 12×, 7× increase in the accuracy, 95% decrease in execution time in comparison to the implementation on software for both the above stated parameters. The plotted Roc Curve indicated good performance levels.

I. Kaur · G. Banga · V. Singh · A. Muval · J. Kedia (✉)
Electronics And Communication Department (B.Tech), Punjab Engineering College, Chandigarh, India
e-mail: jyotikedia@pec.edu.in

I. Kaur
e-mail: ishpreetkaur.bt19ece@pec.edu.in

G. Banga
e-mail: garvitbanaga.bt19ece@pec.edu.in

V. Singh
e-mail: viveksingh.bt19ece@pec.edu.in

A. Muval
e-mail: akshitmuval.bt19ece@pec.edu.in

B. Soni et al. (eds.), *Beyond Artificial Intelligence*, Lecture Notes in Networks and Systems 1326, https://doi.org/10.1007/978-981-96-4170-3_24

Keywords Deep learning neural network (DNN) · Field programmable gate array (FPGA) · Inference time · Receiver operating characteristics (Roc)

1 Introduction

Field programmable gate arrays have emerged as a powerful hardware platform for the implementation of artificial intelligence algorithms for carrying out tasks like image detection and its processing and object detection etc. Image detection [1, 2] which is a part of computer vision which identifies objects, patterns and features within images. This hardware implemented algorithms can be used for variety of applications such as autonomous vehicles surveillance systems and medical imaging etc. Implementation of these algorithms on FPGA offers numerous advantages such as low latency, high throughput and hardware efficiency that makes them suitable for real time image detection. In addition to the advantages offered by FPGA, it also requires the hardware description languages for coding. Further FPGAs may have limited resources in terms of logic gates or memory for implementing complex AI models. However, the use of PYNQ-Z2 boards allow the users to implement the complex algorithms using Python language.

Image detection is a fundamental task in computer vision, in which aim is to recognize and locate objects or patterns of interest within an image. The algorithms are designed to identify patterns and regularities in data, which can then be used to make predictions or classifications. The Convolutional Neural Networks (CNNs) are the algorithms specifically designed for recognition and pattern classification of images. These are highly appreciated and most commonly used for such tasks [3–6].

Although the convolutional neural networks are effective for wide range of image and object detection but they do suffer from certain disadvantages. In case of hardware implementation particularly for limited resources devices the convolutional neural networks are less efficient. Further they demand high computational capacity of the hardware along with significant memory requirement and energy consumption which leads to the increase in complexity and latency of the implemented system.

However, there are few neural network architectures such as Binary Neural Networks and quantized neural networks which have emerged as an alternative to conventional convolution neuron networks these architectures Require less precision of weights and activation functions which addresses to the disadvantages offered by CNN.

Therefore, this paper presents the hardware implementation of Image detection algorithm i.e. QNN and BNN. The hardware used in the work is Xillinx PYNQ-Z2 which is a low power architecture and supports Python. This device can also be attached to the cloud, that increases its compatibility with devices such as mobile etc.

The major contributions of this work are briefly described below:

- Design of a hardware algorithm for Image detection using AI and Ml algorithms in wide spectrum imaging.
- Verification and performance analysis of the implemented system based on parameters like accuracy, weight bit error rate, RoC curve, and execution time on the PYNQ Z2 FPGA board.

This paper is organised as follows: Section A and B of introduction explains the fundamentals of Convolutional Neural Network, Binary Neural Network and Quantized Neural Network. The detailed literature review is in Section II followed by the methodology of proposed system in Section III. The results and analysis of the designed system are presented in Section IV followed by Conclusion and References.

1.1 Convolutional Neural Network (CNN)

CNNs are highly effective deep learning model designed for image and video processing. They work in a similar way as a human brain processes the visual signals. The different layers or the steps involved in CNNs are as shown in Fig. 1 [8]:

- **Convolutional Layers:** These layers apply convolution operations to input images, extracting features like edges, textures, and patterns.
- **Pooling Layers:** Pooling layers downsample feature maps, reducing computational complexity but at the same time preserves important information.
- **Fully Connected Layers:** These layers perform classification or regression tasks based on extracted features.

- **Activation Functions:** Non-linear functions like ReLU (Rectified Linear Unit) introduce non-linearity to the model

The cleaning of CNN is carried out on large data sets with labelled images to make the model learn about the required features and objects these trained CNN algorithms can be fine-tuned for specific projects as they are commonly used in image detection.

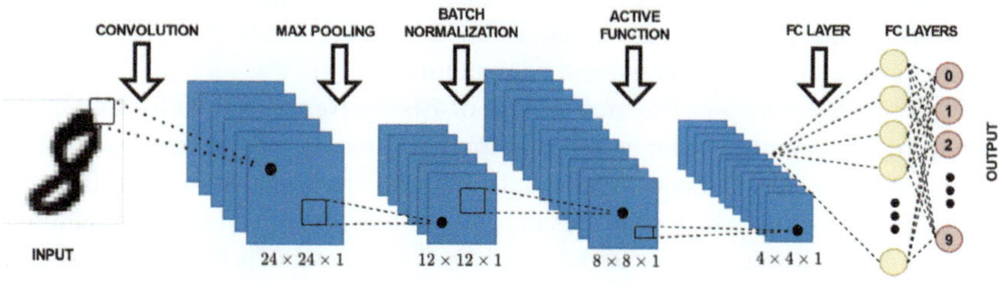

Fig. 1 Architecture of CNN

While Convolutional Neural Networks (CNNs) are highly popular for image detection tasks, however, CNNs require intensive computations which makes it difficult to be implemented on portable mobile interface device. The performance of the image detection along with less complexity is very difficult to maintain and will offer significant challenges on the mobile platforms for the given algorithm. Therefore, there is a need to implement the efficient image detection algorithm on hardware offering less complexity without sacrificing on accuracy. [7].

Recently the Deep Neural Networks (DNN) which include Quantised Neural Network (QNN) And Binary Neural Network (BNN) have been used in place of conventional CNN, due to their advantages of less memory usage and computational resources.

1.2 Binary Neural Network (BNN)

Binary neural networks are a class of artificial neural networks which uses the binary values that is once and 0 S for network activation functions and weights of the neural network which is in contrast to the conventional neural networks which use the real valued weights and activations. BNN offer highly efficient hardware implementation because of the less memory and computational requirement [8, 9]. Because of this advantage of BNN, these network architectures have gained attention in last decade.

Binary Activation Functions:

BNNs use binary activation functions that produce binary outputs. The most common binary activation function is the Sign function:

$$\mathrm{sign(x)} = \{\ 1 \text{ if } \mathrm{x}\ >=\ 0;\ 0 \text{ if } \mathrm{x}\ <\ 0\}$$

Unlike the conventional neural networks which carry out operations such as multiplication and addition using floating point numbers, the BNNs are computationally efficient, as they eliminate costly operations.

In BNNs, the weights of the neural network are also binary. Rather than using continuous values for weights, BNNs represent them as binary values, typically -1 and 1. This binary weight representation further reduces memory requirements and allows for more efficient storage and computation.

Figure 2 shows the general architecture of Binary Neural Network. As it can be seen from the diagram, the input image is applied to binary convolution layer. At this step, the features of the images are scaled down and are passed to the next layer that is activation function which classifies the pattern in the image. The output pattern or object is detected when it is passed through the last layer that is batch normalisation layer.

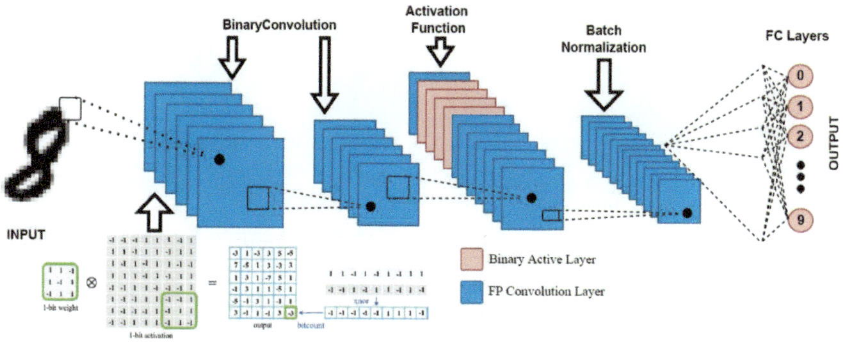

Fig. 2 Architecture of BNN

1.3 Quantized Neural Network (QNN)

Quantized neural network uses the principle of quantization to model the parameters such as weight and activation functions. Instead of using integers or floating-point numbers. QNNs integrate quantum gates with the architecture of neural networks. They consist of quantum layers and classical layers. Quantum layers process data using qubits and quantum gates.

Quantum computing leads to exponentially fast operations. This characteristic of quantum neural networks can provide a significant enhancement in speed of machine learning algorithms which require complex computations and high memory requirements. However, quantization of the parameters also give rise to reduced accuracy therefore development of QNNs is to create a balance between the accuracy of the model and its computational efficiency these characteristics of QNN make them fit for the hardware implementation with limited resources devices such as mobile, IOT devices etc.

Figure 3 shows a QNN architecture. The architecture of a Quantum Neural Network (QNN) combines elements of classical neural networks with quantum computing components. Similar to classical neural networks, a QNN starts with an input layer where data is fed into the network where data is converted to quantum states. The core of the QNN consists of one or more quantum layers, where quantum operations are applied to the data and data is processed using Quantum interference and entanglement effects. After the quantum processing, the quantum states are measured, and the measurement outcomes are mapped to classical values. These layers include linear transformations, activation functions [10–12], and other classical operations etc. the design of QNN architectures [13, 14] offers challenges due to the need to manage quantum resources efficiently and reduce the errors associated with quantum operations.

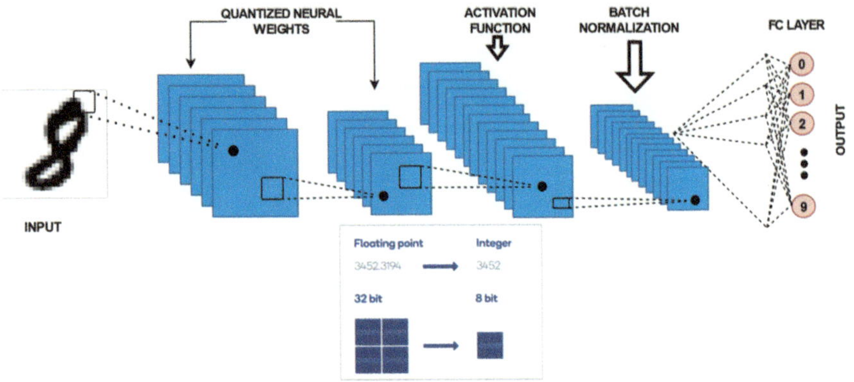

Fig. 3 Architecture of QNN

Advantages of BNN and QNN Over CNN:

1. **Hardware Efficiency:** Well-suited for hardware implementations like Field-Programmable Gate Arrays (FPGAs because the binary or quantized operations can be implemented efficiently because of the integer values leading to reduced power consumption and lower latency.
2. **Memory Efficiency:** The binary and quantized weights significantly reduce the memory requirement of the neural network, making these algorithms suitable for hardware devices.
3. **Faster inference and Energy Efficient:** Binary or quantized activations and weights need significantly less computations during both training and inference, which can result in faster execution times and hence they are energy efficient for limited battery-operated devices.

In spite of the significant advantages of BNN and CNN algorithms as mentioned above but the reduced accuracy and loss of information always remain a challenge. In the presence of these challenges the proposed algorithms are important for applications where Energy efficiency is paramount with limited resources.

2 Related Work

Many researchers have carried out the work related to hardware implementation of various neural networks such as convolutional neural network quantized neural network and binary neural network. The researchers have showed that the implementation of BNN on a hardware platform can significantly increase the speed even by a factor of 200 [5] which has improved further with the use of advanced modern computing platforms.

Recently, a researcher proposed a novel binary neural network with fixed sign to be used on the hardware. The full precision results with better accuracy and faster computation were presented with the proposed algorithm for standard datasets such as CIFAR-10 and an imageNet.

Another realization of binary weighted convolutional neural networks on an FPGA was proposed with low energy resources [15]. With the aim to reduce bandwidth requirement of earth observation images, a hardware implementation of BNN and QNN was proposed by Reiter et al. [16].

A new method called CLIP-Q was proposed [6] to address to the problems such as memory complexity and computations involved of full precision CNN. The developed system thereafter was validated on the CIFAR-10 and CIFAR-100 datasets.

Few latest research works carried out for the design and implementation of QNN algorithms has been reviewed extensively [9, 3, 17–20]. A novel method in which weights of neural networks are quantized with more computational efficiency, and the reduced quantization error, that were used for multilayer perceptron (MLP) and CNN [17]. Likewise, a mixed-precision QNN having [3] progressively decreasing bitwidth has been proposed which was used on commonly used CNNs.

Another experimental work conducted on an Xilinx ZCU102 platform by using QNN for network compression [20].

In the presence of all the recently developed classification and detection algorithms [20, 8], the implementation of a simple and efficient neural network model with the low power and higher computing efficiency is still very challenging.

3 Proposed Methodology

The proposed Image Detection algorithm contains four steps i.e. Binary Neural Weights or Quantized Neural Weights, Activation Function, Batch Normalisation and evaluation of various performance measures.

- In the Neural Weights layer of the algorithm, it transforms input data within the network's hidden layers. This layer identifies and fixes the initial values for all the parameters of the neural network before the model is trained on a data set.
- In the second step the output of the neural network is determined using the activation layer.
- Finally batch normalisation is used to adjust and scale the inputs so that they have standard deviation of one and mean value of zero. It is a crucial component in modern neural networks, contributing to faster training, improved convergence, and the ability to effectively train deep and complex models.

The implementation of Binary Neural Networks and quantized neural networks Xilinx, PYNQ-Z2 boards which has parametrizable Intellectual Property Cores and are easier to use. It uses Python libraries for efficient implementation to make use of the programmable logic and building microprocessor to create the efficient electronic systems.

In the analysis phase of the algorithm various performance measures were analyzed for the DNN classifiers. The weights of a neural network in general are initialized with very small values.

3.1 Neural Weights Layer

Neural networks weight layer is the first step of the model which alters the input data within the hidden layers of the network. The product of the weights of the network layer and the input image is then transferred to next layer in the neural network. For Binary Neural Networks these values are binarized with the help of an activation function following are the two functions used to binarize the weights and the activation:

3.1.1 Sign Function

The sign function is a deterministic function that requires random bit generation during quantization hence sign function is preferred mostly with the exception of the case of certain activation in some experiments.

$$x^b = Sign(x) = \begin{cases} +1 & if\ x > 0 \\ -1 & otherwise \end{cases} \tag{1}$$

3.1.2 Stochastic binarization

The stochastic binarization is non-deterministic function which uses the hard sigmoid.

$$x^b = Sign(x) = \begin{cases} +1 & with\ probability\ p = \sigma(x) \\ -1 & with\ probability\ 1 - p \end{cases} \tag{2}$$

x^b in Eqs. (1) and (2) represents the binarized value of the real valued variable x. In the proposed model the sign function was applied in case of Binary Neural Network as seen in Fig. 4.

As seen in Fig. 4, after passing the image through the Binary Neural Layer the algorithm identifies the Road Sign in the input image and locates it by demarcating the area with a rectangular boundary. This adds weights to the focused area for the successive layers.

Unlike the Binary Neural Networks layer in BNN, in QNN models, the quantization layer saves the desired information in continuous variables and is very important

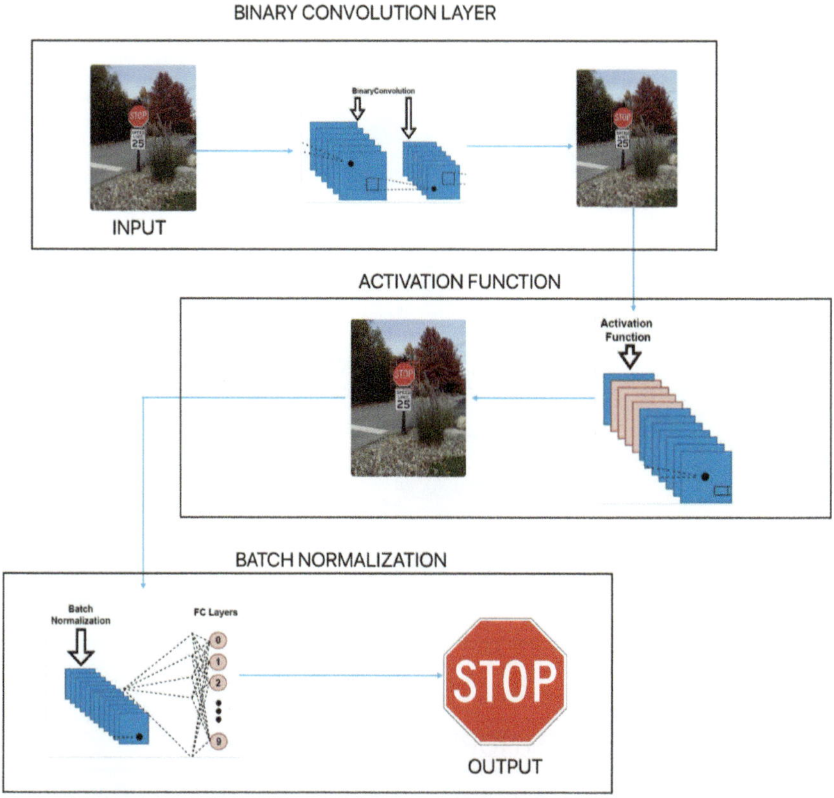

Fig. 4 BNN model

for network performances. The quantization function used in the stated algorithm is described by Eq. (3).

$$m = 1 + \left[log_2(max|x_i|)\right] \; with \; 1 < i < N \tag{3}$$

This converts from a 32-bit floating point to a fixed 8 bit point, required for quantization and m represents the required number of bits for the unsigned integer part.

The Quantized neural layer aids with the identification of features in image and add weights to the pixels wherein the further layers work so as to classify the image on the applied weights. As shown in Fig. 5, the features of the input image were extracted and focused upon after passing it though the first layer. With the adjustment of the weights one can employ more feature extraction which not only leads to better prediction but more concrete and peculiar features to be extracted of the given input dataset.

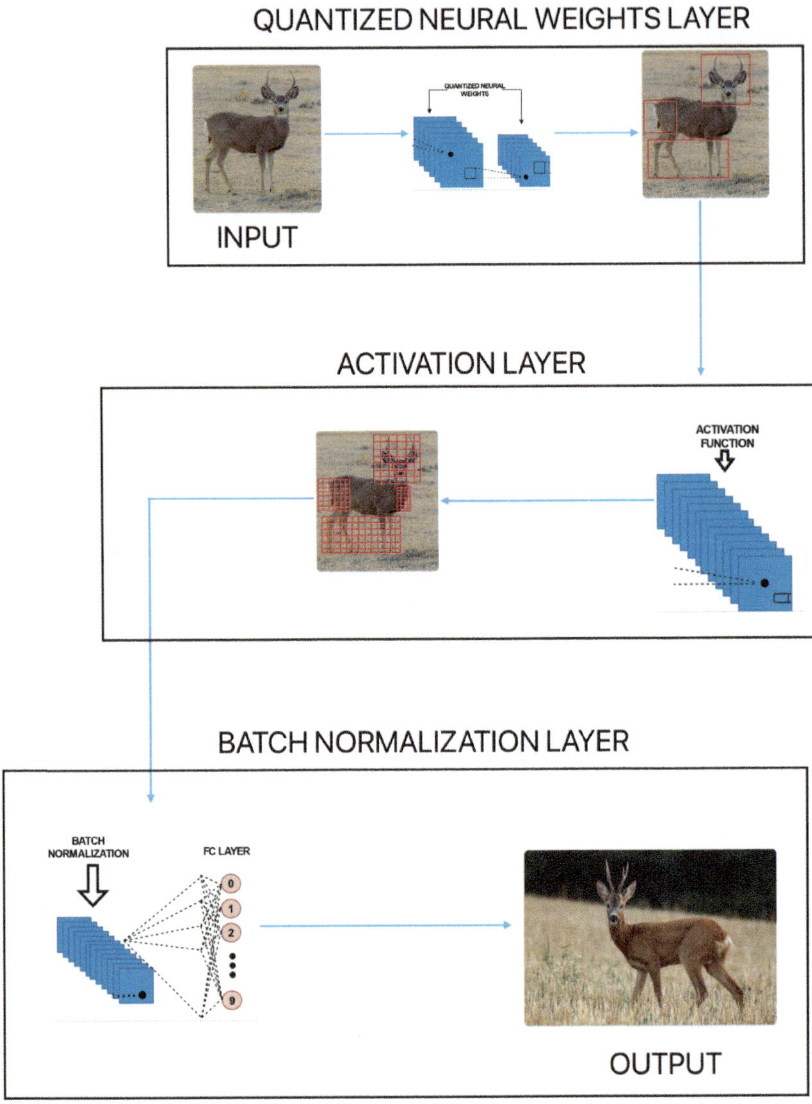

Fig. 5 QNN model

3.2 Activation Layer

Activation layer helps the neural network to learn complex patterns in the dataset. It will restrict the value or amount to a certain limit. The most important feature of the activation layer in the proposed algorithm is of adding non linearity to the model. In the proposed model, the Activation layer possesses the following qualities helping to reduce the inference time by a large amount.

3.2.1 No Vanishing Gradient Problem

By the use of effective datasets and proper number of epochs, the vanishing gradient problem was significantly reduced which led to the gradient not being shifted to zero.

3.2.2 Zero Centered

The selected activation function is symmetrical at zero therefore the gradients did not shift to any particular direction.

3.2.3 Computational Expense

After each layer, the activation function was applied which reduced the computational time of the layer tremendously by the used of PYNQ-Z2 board.

For the BNN layer, a linear activation function was used which is proportional to the input image (as shown in Eq. 4). It simply added up the weight total of the inputs and returned the result.

$$f(x) = x \tag{4}$$

In case of BNN the input image after passing through the activation layer was more rigorously evaluated (as shown in Fig. 4) for feature extraction. The rectangular area marked given as the inout to the layer was focused upon and thoroughly analysed to learn complex patterns. It closely looks into each feature and adds non linearity towards the training of the model.

The QNN Activation Layer can have various values and hence proves to be more resourceful at the cost of computational expense. The activation used for the stated model was a sigmoid function (shown in Eq. (5)) because it has a smooth gradient and has a better classification than most of the activation function available.

By varying the activations (as shown in Fig. 5) the rankings of the input image appear to be closer to the actual class of the image.

$$f(x) = \frac{1}{1 + e^{-x}} \tag{5}$$

The activation algorithm takes a lot of execution time in case of software due to the slow processing RAM and the multiple chain rule for feature extraction. However, with the aid of a hardware device the activation process becomes fast and efficient due to the use of digital gates which aid the process of feature extraction in the model.

3.3 Batch Normalisation

Batch Normalisation is the final layer in case of any neural network and supervises the learning method for normalizing the interlayer outputs. This layer significantly stabilizes the training and advances the generalization ability of the model. The features and other dimensionalities that have been extracted in the previous layers were vigorously studied and analysed by applying mean and standard deviation to the datasets so that batches could be made more uniformly leading to a better training and testing of the model.

In BNN and QNN the output of this layer would be the correct prediction of the input image with high accuracy rates. This layer effectively normalises all the ensuring that no activation value is too high or too low.

The overall process (shown in Fig. 6) for the proposed algorithm can be summed up with the first step being importing the datasets, then setting their parameters required for model training, loading the classifier, performing model training on both software and hardware, finally analysing the model on various parameters.

3.4 Performance Measures

The following performance parameters were analyzed for both BNN and QNN using efficient libraries. All the performances measures give a broad overview of how the algorithm caters to need of having efficient training and test of the models and subsequently analysing the results.

3.4.1 Prediction Accuracy

Accuracy is one of the most important performance parameters for assessing the various classification models. A**ccuracy** can be defined as the fraction of predictions a designed model finds correctly. Formally, the accuracy is given by:

$$Accuracy = \frac{number\ of\ correct\ predictions}{total\ number\ of\ predictions}$$

While training and testing the model and having the input image classified the main aim was not to compromise with accuracy because of the use of hardware devices. Various learning rates were implied and tested to find the prediction accuracy.

Fig. 6 Overall process for the model

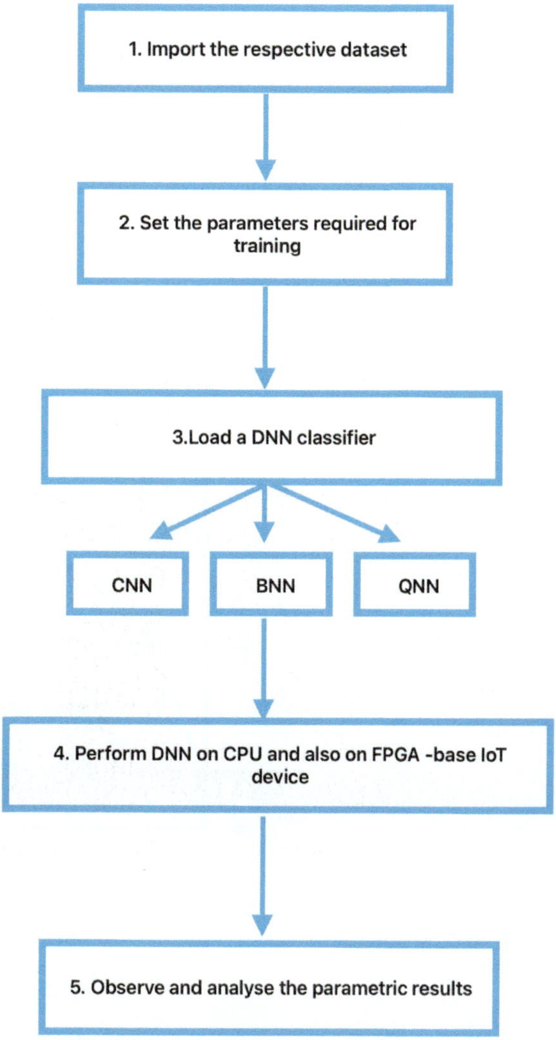

3.4.2 Weight Bit error rate

The error basically signifies how well the network is performing on a certain (training/testing/validation) set. The error occurs in the Weight Neural layer and affects the validation accuracy of the model,

3.4.3 RoC Curve

An ROC curve (receiver operating characteristic curve) demonstrates the performance of a neural network model at all classification thresholds. As shown in Fig. 7, this curve plots True Positive Rate (TPR) and False Positive Rate (FPR).

An ROC curve plots TPR vs. FPR at different thresholds of classification. With the lower values of this threshold, more items are classified as positive that leads to increase in both False Positives and True Positives.

The ROC curve of any sample random classifier with the random performance level (as shown in Fig. 8) always shows a straight line. This ROC curve acts as a tool to assess the performance of a classifier. The plot demarcated by RoC curve gives an estimation about the good or bad performance of thew classifier.

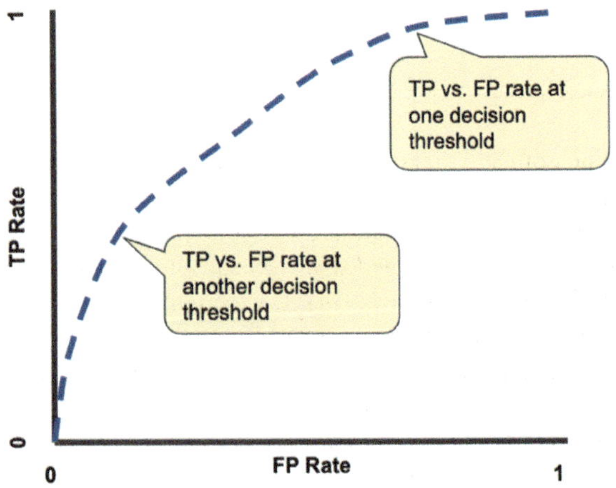

Fig. 7 RoC curve

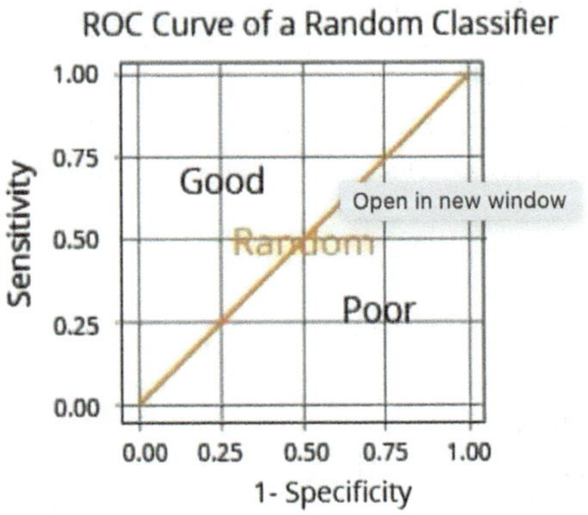

Fig. 8 RoC curve of random classifier

3.4.4 Execution Time

The Execution time overall represents the total time taken by the system for carrying out the model computations, including data splitting, data pre-processing, and model evaluation.

3.4.5 Inference Time

The Inference time is the time taken by a given model to predict a new data. As compared to this the training time is the time taken by the system to train the model on a dataset.

4 Results and Analysis

The BNN and QNN models have been analysed and validated on an FPGA board by Xilinx PYNQ-Z2. The board is compatible with Python and can be connected through the cloud. The board offers low power for its operation [8].

The experimentation was performed on a CPU and then implemented on the FPGA hardware. The comparative analysis of the model's performances in both the cases was caried out on the basis of parameters like accuracy, execution time, weight bit error rate, and RoC curve etc.

The performance measure for BNN and QNN were analysed using the stated algorithm both on software and hardware leading to satisfactory results.

4.1 Performance Measures for BNN

4.1.1 Prediction Accuracy

Figure 9 shows the accuracy of the Road sign Dataset on the hardware with the varying learning rates. As can be seen from figure that the learning rate with 0.01 gives a low prediction rate of about 48% whereas 0.5 learning rate had a high prediction rate of 80% for BNN. The following graph depicts that even with the apt learning rates, the prediction accuracy decreases with increasing noisy labels. With a low learning rate of 0.01 the prediction accuracy can drop upto 10% and in case of learning rate of 0.5 we will have at least 65% of prediction accuracy.

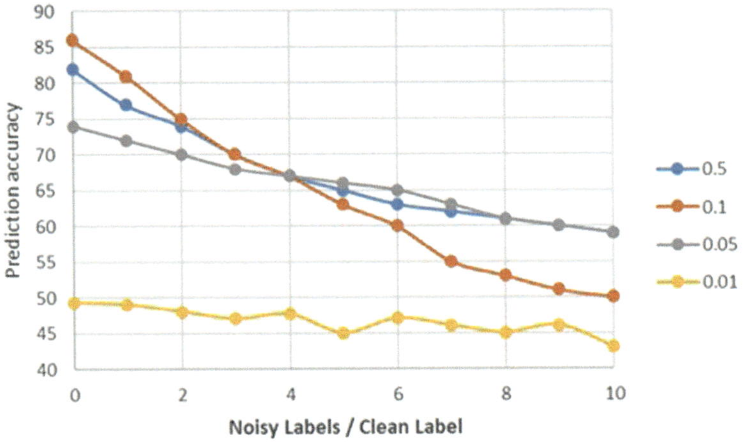

Fig. 9 Prediction accuracy for BNN

4.1.2 Weight Bit Error Rate

The weight bit error rate curve is shown in Fig. 10. As it can be inferred from the graph the weight bit error rate (which signifies how well the network is performing on a certain (training/testing/validation) set) for the model did not have a value exceeding 0.08 and even in the case of such an error the accuracy did not fall below 80%. This makes and efficient model having both low bit error rate and high validation accuracy.

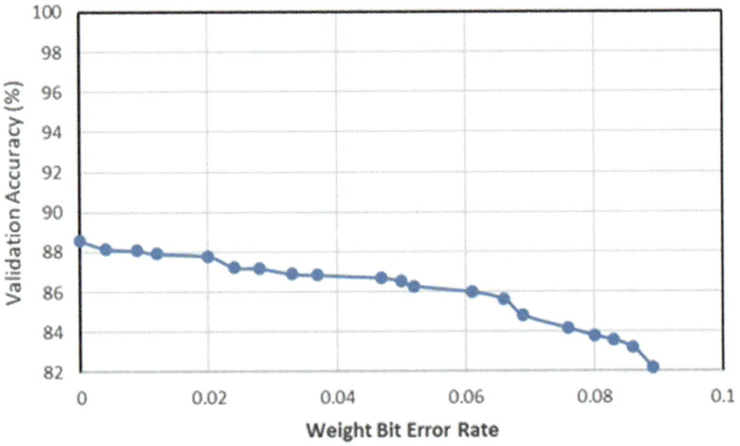

Fig. 10 Weight bit error rate versus validation accuracy for BNN

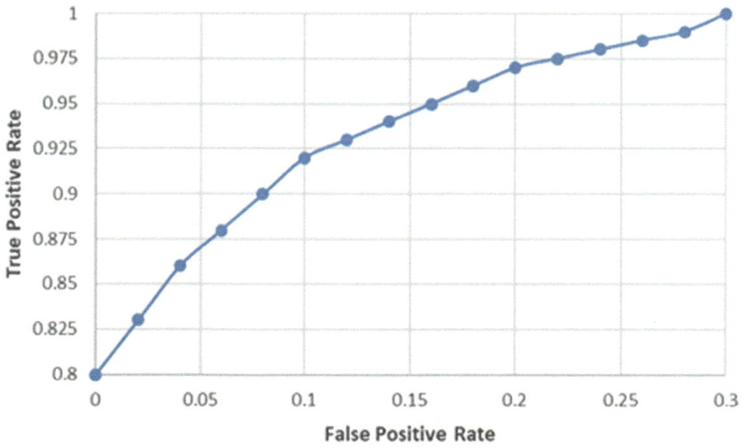

Fig. 11 RoC curve for BNN

Table 1 Accuracy and execution time for software and hardware BNN

Parameters	PYNQ-Z2	Software
Accuracy	97.71%	90.8%
Execution Time	0.2 ms	4.1 ms

4.1.3 Roc Curve

The RoC curves for BNN is shown in Fig. 11. As it is evident from the graph that the Roc Curve falls under the area at the top-left corner thus indicating good performance levels.

4.1.4 Execution Time and Accuracy

As shown in the Table 1 the use of PYNQ-Z2 board led to a 7× increase in the accuracy of object detection along with 95% reduction in execution time. This makes the use of hardware devices apt as it did not lead to a decrease in accuracy and significantly reduced the execution time.

4.1.5 Inference Time

As shown in Table 2, the total inference time along with the inference time calculated in micro second per image is about 12× less for Hardware than software i.e. the PYNQ Z2 board takes little time to process new data and make predictions accurately.

Figure 12 is an indicative of how the use of efficient hardware reduces the inference time by a large amount.

Table 2 Inference time for hardware and software BNN

Inference time	PYNQ-Z2	Software
Total (msec)	2743.00	4,789,107.00
Per image (micro second)	914.33	1,596,369.00

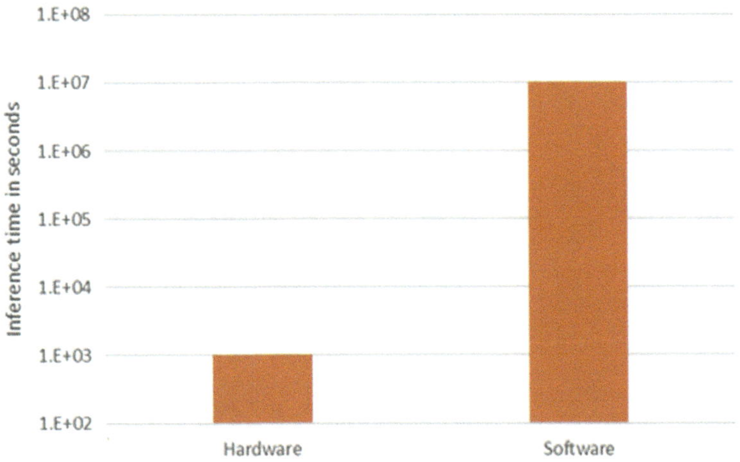

Fig. 12 Inference time graph for BNN

4.2 *Performance Measures for QNN*

4.2.1 Prediction Accuracy

Figure 13 shows the prediction accuracy of the CIFAR-10 Dataset with the varying learning rates. The learning rate with 0.01 showed a low prediction rate of about 48%, and 0.5 had a high prediction rate of 80% for BNN. The following graph depicts that even with the apt learning rates the prediction accuracy decreases with increasing noisy labels. With a low learning rate of 0.01 the prediction accuracy can drop to 10% and in case of learning rate of 0.5 we can have a 65% of prediction accuracy.

4.3 *Weight Bit Error Rate*

The weight bit error rate is shown in Fig. 14. As can be inferred from the graph, the weight bit error rate for the model did not have a value exceeding 0.08 and even in the case of such an error the accuracy did not fall below 94%. This makes and efficient model having both low bit error rate whose maximum value is unto 0.08 and high validation accuracy with a maximum value of 98%.

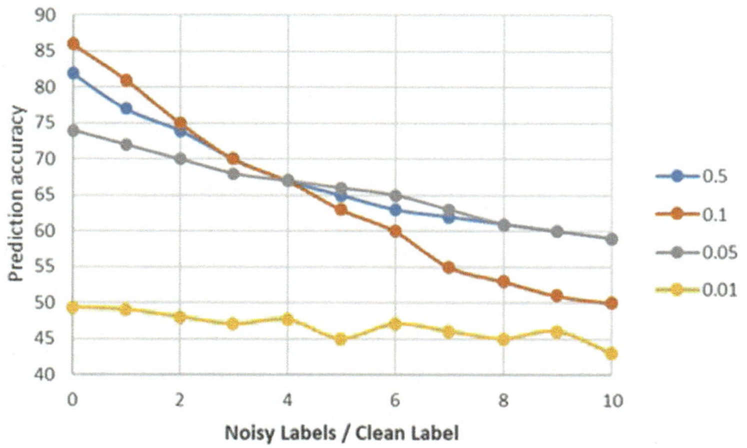

Fig. 13 Prediction accuracy for QNN

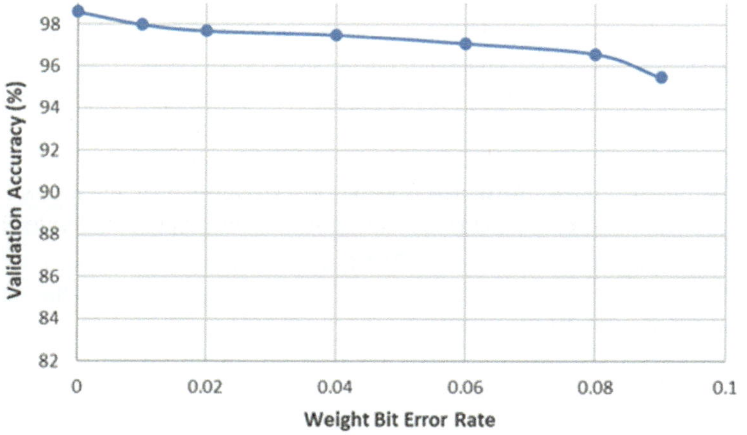

Fig. 14 Weight bit error for QNN

4.4 RoC Curve

The RoC curve for QNN is shown in Fig. 15.

As it is evident from the graph the Roc Curve falls under the area at the top-left corner thus indicate good performance levels.

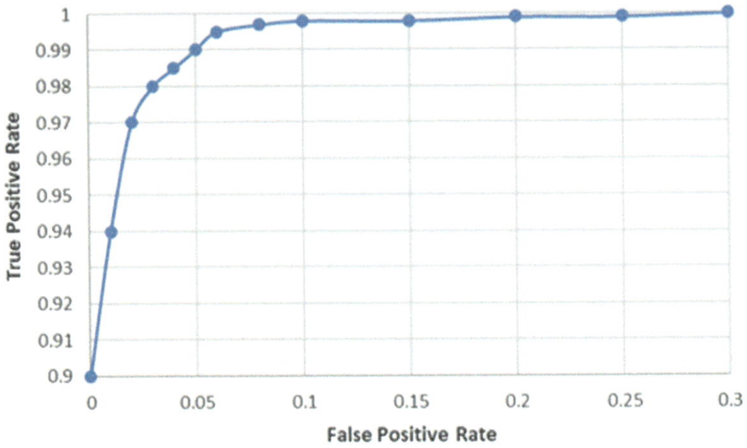

Fig. 15 ROC curve for QNN

Table 3 Accuracy and execution time for hardware and software QNN

Parameters	PYNQ-Z2	Software
Accuracy	97.51%	98.6%
Execution time	0.2 ms	2.7 ms

4.5 *Execution Time and Accuracy*

As shown in the Table 3, the use of PYNQ-Z2 board increase the accuracy 11× times along with the 92% reduction in execution time in comparison to the software implementation of QNN.

4.6 *Rankings*

The different rankings for the recognition of a particular image are plotted (Fig. 16) that gives an idea about the ranking or the probability with which the algorithm predicts the class of the image. (The image to be classified was of a deer).

- Case 1: W1A1 = Weights 1 Activation 1

  In this case of QNN the model trains with 1 Weights layer and 1 Activation Layer. The rankings which are predicted for different classes are shown in Table 4, in which the proposed algorithm predicts the probability of the image to be deer as the highest but the rankings for the other classes are still comparable.
- Case 2: W1A2 = Weights 1 Activation 2

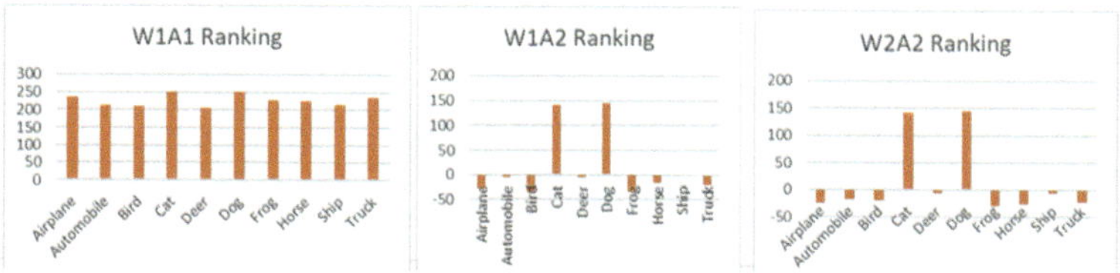

W1A1	
Class	**Ranking**
Airplane	234
Automobile	211
Bird	209
Cat	248
Deer	203
Dog	249
Frog	226
Horse	224
Ship	213
Truck	233

W1A2	
Class	**Ranking**
Airplane	-30
Automobile	-6
Bird	-36
Cat	140
Deer	-6
Dog	144
Frog	-35
Horse	-16
Ship	-3
Truck	-20

W2A2	
Class	**Ranking**
Airplane	-24
Automobile	-18
Bird	-20
Cat	141
Deer	-6
Dog	144
Frog	-29
Horse	-27
Ship	-6
Truck	-23

Fig. 16 Rankings for different cases of QNN model

Table 4 Inference time for hardware and software QNN

Inference time (Total sec)	PYNQ-Z2	Software
W1A1	1583	1,594,724
W1A2	1628	5,678,736
W2A2	4866	9,588,836

In this case the model undergoes training once for the weight layer and twice for the activation layer. The probability of the image to be that of deer increases along with a decrease in the rankings of other classes.

- Case 3: W2A2 = Weights 2 Activation 2

  This the case when the QNN model passes 2 weights and 2 activation layer in each epoch. The ranking in this case of the deer class is the highest which makes this model of the algorithm the best choice.

  As it is inferred from the above graphs (Fig. 16) with the increase in the weights and activation layer the prediction probability of the image to be identified increases from 0.82 (Case-1) to 0.976 (Case-2) and the rankings also are the highest in the third case i.e. Weights 2 Activation 2 (W2A2). This clearly depicts that the weights and activation function adds parameters, biases and functioning to the respective layers which increase the features extraction capability of the model leading to better predictions and more accurate detection of the images to be classified.

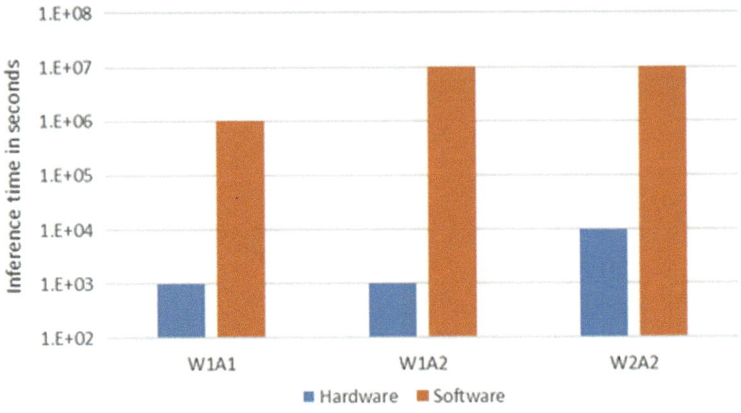

Fig. 17 Hardware versus software inference time QNN

4.7 *Inference Time*

As clearly indicates in Fig. 17 and Table 4, the total inference time calculated in micro second image is 99.9% less for Hardware than software i.e. the PYNQ Z2 board in all the three cases takes little time to process new data and make prediction.

5 Conclusion

The hardware implementation of Binary Neural Network and Quantized Neural Network for image detection has been carried out on low power FPGA hardware i.e. PYNQ-Z2 from Xilinx. A detailed comparative analysis of the various performance parameters such as prediction accuracy, weight bit error rate, ROC curve, and execution time has been carried out for both the neural networks for their implementation on software as well as hardware.

The proposed Image detection algorithms has produced notable results on identifying images when implemented on hardware with a percentage decrease in inference time of about 12×, 7× increase in the accuracy, 95% decrease in execution time in comparison to the implementation on software for both the above stated parameters. The plotted Roc Curve indicated good performance levels.

Funding and/or Conflicts of Interests/competing Interests The authors declare that they have no conflict of interest. The datasets used for the current study are CIFAR-10 and CIFAR-100 datasets (toronto.edu).

References

1. Kousik, N., Natarajan, Y., Raja, R.A., Kallam, S., Patan, R., Gandomi, A.H.: Improved salient object detection using hybrid convolution recurrent neural network. Expert Syst. Appl. **166**, 114064 (2021)
2. Pérez-Hernández, F., Tabik, S., Lamas, A., Olmos, R., Fujita, H., Herrera, F.: Object detection binary classifiers methodology based on deep learning to identify small objects handled similarly: Application in video surveillance. Knowl.-Based Syst. 2020, 194 (105590)
3. Reiter, P., Karagiannakis, P., Ireland, M., Greenland, S., Crockett, L.: FPGA acceleration of a quantized neural network for remote-sensed cloud detection. In: Proceedings of the 7th international workshop on on-board payload data compression, virtual, 21–23 September (2020)
4. Cheng, W., Lin, C., Shih, Y.Y.: An efficient implementation of convolutional neural network with CLIP-Q quantization on FPGA. IEEE Trans. Circuits Syst. I Regul. Pap. 69, 4093–4102 (2022)
5. Lybrand, E., Saab, R.: A greedy algorithm for quantizing neural networks. J. Mach. Learn. Res. **22**, 156–161 (2021)
6. Chen, J., Liu, L., Liu, Y., Zeng, X.: A learning framework for n-bit quantized neural networks toward FPGAs. IEEE Trans. Neural Netw. Learn. Syst. **32**, 1067–1081 (2020)
7. Doon, R., Rawat, T.K., Gautam, S.: Cifar-10 classification using deep convolutional neural network. In: Proceedings of the 2018 IEEE Punecon, Pune, India, 30 November–2 December 2018, pp. 1–5 (2018)
8. Deng, L.: The mnist database of handwritten digit images for machine learning research [best of the web]. IEEE Signal Process. Mag. **29**, 141–142 (2012)
9. Belabed, T., Quenon, A., Da Silva, V.R.G., Sakuyama, C.A.V., Souani, C.: Full python interface control: auto generation and adaptation of deep neural networks for edge computing and hardware applications FPGA-based acceleration. In: Proceedings of the 2021 international conference on innovations in intelligent systems and applications (INISTA), Kocaeli, Turkey, pp 1–6 (2021)
10. Novac, P.E., Hacene, G.B., Pegatoquet, A., Miramond, B., Gripon, V.: Quantization and deployment of deep neural networks on microcontrollers. Sensors **21**, 2984 (2021)
11. Chu, T., Luo, Q., Yang, J., Huang, X.: Mixed-precision quantized neural networks with progressively decreasing bitwidth. Pattern Recognit. 111, 107647 (2021)
12. Qin, H., Gong, R., Liu, X., Bai, X., Song, J., Sebe, N.: Binary neural networks: a survey. Pattern Recognit. **105**, 107281 (2020)
13. Austin, J., Kennedy, J.: Presence, a hardware implementation of binary neural networks. In Proceedings of the International Conference on Artificial Neural Networks, Skövde, Sweden, 2–4 September 1998; Springer: London, UK, pp 469–474 (1998)
14. Kumar, V., Baskar, D., Arunsi, M., Jenova, R., Majid, M.: VLSI design and implementation of High-performance Binary-weighted convolutional artificial neural networks for embedded vision based Internet of Things (Hardware). Procedia Comput. Sci. **163**, 639–647 (2019)
15. Yu, H., Han, Q., Li, J., Shi, J., Cheng, G., Fan, B.: Search what you want: barrier panelty nas for mixed precision quantization. In: Proceedings of the European conference on computer vision, glasgow, UK, 23–28 August 2020; Springer: Cham, Switzerland, pp 1–16 (2020)
16. Bartan, B., Pilanci, M.: Training quantized neural networks to global optimality via semidefinite programming. In: Proceedings of the international conference on machine learning, virtual, 18–24 July, pp 694–704 (2021)
17. Wang, Z., Lu, J., Tao, C., Zhou, J.; Tian, Q.: Learning channel-wise interactions for binary convolutional neural networks. In: Proceedings of the IEEE/CVF conference on computer vision and pattern recognition, Long Beach, CA, USA, 15–20 June 2019; pp 568–577 (2019)
18. Jain, D.K.: An evaluation of deep learning based object detection strategies for threat object detection in baggage security imagery. Pattern Recognit. Lett. **120**, 112–119 (2019)

19. Ariji, Y., Yanashita, Y., Kutsuna, S., Muramatsu, C., Fukuda, M., Kise, Y., Nozawa, M., Kuwada, C., Fujita, H., Katsumata, A., et al.: Automatic detection and classification of radiolucent lesions in the mandible on panoramic radiographs using a deep learning object detection technique. Oral Surg. Oral Med. Oral Pathol. Oral Radiol., 128, 424–430 (2019)
20. Novac, P.E., Hacene, G.B., Pegatoquet, A., Miramond, B., Gripon, V.: Quantization and deployment of deep neural networks on microcontrollers. Sensors 21, 2984 (2021).

Yoga Pose Detection Model Using VGG16

Hardik Raina and Laxmi Chaudhary

Abstract Yoga has surged in popularity as a comprehensive method for enhancing physical, mental, and spiritual well-being. With this growing interest, there is an increasing demand for effective techniques to supervise and monitor personal yoga sessions. In this study, we introduce a Yoga Pose Detection (YPD) method that harnesses VGG16, a Convolutional Neural Network (CNN) model, to accurately recognize yoga poses. Prevailing methods for yoga pose recognition often face challenges in complex real-world contexts due to their reliance on manually engineered features derived from controlled environments. The proposed YPD model effectively overcomes this challenge by fine-tuning the pre-trained VGG16 model using a new dataset specially curated for yoga poses. The efficacy of the proposed approach is validated on the publicly available Yoga-82 dataset, attaining an impressive accuracy of 97.83%. Additionally, the proposed model surpasses numerous state-of-the-art image classification models, including ResNet 50, Inception V3, Inception ResNet-V2, Xception, Classic CNN, and Transfer Learning & Stochastic Gradient Descent (SGD) in terms of accuracy, precision, recall, and F1 score.

Keywords Yoga pose recognition · Pose estimation · Deep learning · Transfer learning · And neural networks

1 Introduction

Yoga, an ancient practice with its roots tracing back millennia, continues to exhibit its efficacy in enhancing comprehensive well-being. Yoga has a good effect on a variety of aspects of physical and mental health, according to research [1, 2]. Regular yoga practise has been linked to lower stress levels, improved mood, increased muscle

H. Raina · L. Chaudhary (✉)
Department of Computer Science & Engineering and Information Technology, Jaypee Institute of Information Technology, Noida, Uttar Pradesh, India
e-mail: laxmichaudari.iet@gmail.com

H. Raina
e-mail: hardikraina079@gmail.com

B. Soni et al. (eds.), *Beyond Artificial Intelligence*, Lecture Notes in Networks and Systems 1326, https://doi.org/10.1007/978-981-96-4170-3_25

strength, enhanced flexibility, and an overall higher quality of life [3]. Notably, yoga has emerged as a promising therapeutic exercise for individuals grappling with psychiatric conditions like depression and anxiety. According to studies, adding yoga to treatment plans significantly reduces symptoms and improves general mental health. In addition, yoga has gained recognition as a successful adjunct treatment for persistent lower back pain, providing pain relief and improving functional mobility [4].

While the benefits of yoga for health are well known, it is important to stress the importance of performing yoga poses correctly and precisely. During yoga practice, incorrect posture or alignment can result in physical harm or strain. This emphasizes the importance of adequate guidance and supervision during yoga sessions [5]. However, in light of contemporary lifestyles and restricted access to yoga instructors, there's an escalating need for alternative solutions that can provide real-time feedback and guidance to practitioners.

To meet these requisites, the field is witnessing a surge in the development of Artificial Intelligence (AI) systems capable of recognizing yoga poses and offering personalized feedback and recommendations. These AI-powered technologies can precisely detect particular yoga postures by analyzing body alignment and movement. This technology has the potential to greatly increase access to yoga instruction by allowing people to practise yoga at their convenience in a safe and efficient manner [6, 7].

To improve user engagement and experience, posture recognition is crucial in the field of home-human interfaces. With a growing elderly population and a shortage of healthcare resources, there is an increased need for remote independence-supporting gadgets for vulnerable people. A healthy lifestyle depends on maintaining good posture, which affects how people hold their bodies and arrange their limbs. However, sedentary lifestyles and reduced physical activity have led to muscle strength decline. The negative effects of this sedentary behaviour on the human body include neck, back, and shoulder pain.

Therefore, in this research work, we provide a YPD model to tackle the aforementioned challenges of classifying and identifying yoga poses. Through the utilization of transfer learning with the VGG16 model, the proposed model overcomes these obstacles while acknowledging the importance of monitoring and preserving proper posture for the well-being and safety of individuals.

The paper is organized as follows: Section 2 presents an overview of the related work, highlighting previous research and its contributions. Section 3 outlines data pre-processing steps and introduces the architecture of the YPD model. Section 4 presents experimental results, encompassing evaluation metrics and a performance analysis of the proposed model. Ultimately, Section 5 concludes the paper by deliberating potential future research directions and unexplored domains in the realm of yoga pose detection.

2 Related Work

Yoga posture classification is a new area of study in deep learning, and researchers have devoted a lot of effort and time to resolve the challenges involved in correctly classifying yoga poses. Chen et al. [7] introduced a self-training paradigm for yoga practice identification based on distinguishing characteristics. In this method, body maps are created by extracting user body outlines using a Kinect device. Star skeletons, a quick skeletonization method, was employed to quickly delineate human positioning and serve as a descriptor. Furthermore, Agrawal et al. [8] used a variety of Machine Learning (ML) techniques to recognise yoga positions. In order to extract real-time skeleton images, they built a dataset with 5,500 photographs of ten different yoga positions and used the tf-pose estimation technique. Particularly, random forest classifiers performed better on this dataset.

A hybrid strategy for posture recognition was developed by Liaqat et al. [6] by combining deep neural networks and conventional machine learning methods. The most accurate class prediction was achieved by combining weights acquired from the deep learning model with predictions from the traditional model. Kumar et al. [9] employed OpenPose for keypoint extraction and adopted a hybrid CNN-LSTM architecture for classification. Utilizing 88 videos capturing six yoga poses, this model underwent training. However, a limitation surfaced in the method's focus on specific datasets, suggesting potential efficacy variations when applied to diverse or expansive datasets.

Byeon et al. [10] investigated the possibility of ensemble deep models in various home settings, particularly in the context of older adults. In this approach, the results shed light on the efficacy of home exercise systems and matched the popularity of relaxing home workouts, especially in the post-COVID era. A neural network model designed for accurate classification of five different yoga positions was introduced in the work of Ashraf et al. [11]. To improve categorization performance, various architectural components and training methods were combined. Furthermore, only a limited number of complex real-world scenarios were used to evaluate the models' performance.

Upadhyay and Ananthakrishnan [12] demonstrated a deep learning method for yoga position recognition. This model efficiently classified yoga poses by utilizing deep neural networks and multi-scale spatial and depth parameters. A framework was developed by Chasmai et al. [13] that can categorize yoga poses regardless of the viewpoint or perspective used to capture the pose. This framework cleverly combines machine learning and image processing approaches, producing reliable and precise categorization results. Lastly, Anand et al.'s [5] research focused on developing an AI-driven system that rates yoga poses and provides constructive criticism. This method, which relied on deep neural networks, accurately measured yoga postures and gave users pointers for improving form.

3 Proposed Model

In this paper, we have proposed YPD, a new method for classifying and detecting yoga poses. We merged the VGG16 model [15] with Transfer Learning [14]. We have improved accuracy and dependability by applying transfer learning and making use of the learnt characteristics from the pre-trained model. We have modified and added layers to the VGG16 model to better suit the objective of yoga pose recognition. The Yoga82 dataset [16], which has been enhanced to improve balance and robustness, is used to train the proposed model. It uses the five most prominent classes from the dataset. In addition to the VGG16 model, the proposed model also contains two extra layers called Flatten and Dense. The parameters used to train the proposed YPD model are listed in Table 1, and Fig. 1 depicts the proposed YPD model's architecture.

1. **VGG16 model:** The basis of the proposed YPD architecture rests upon the VGG16 model. It accepts images with dimensions of (224, 224, 3) and acts as a pre-trained CNN, producing an output tensor with the dimensions (None,

Table 1 Trainable parameters of the proposed YPD model

Layers of the proposed YPD	Trainable parameters
VGG16	21,138,757
Flatten	0
LeakyRelu	0
Dense	125,445

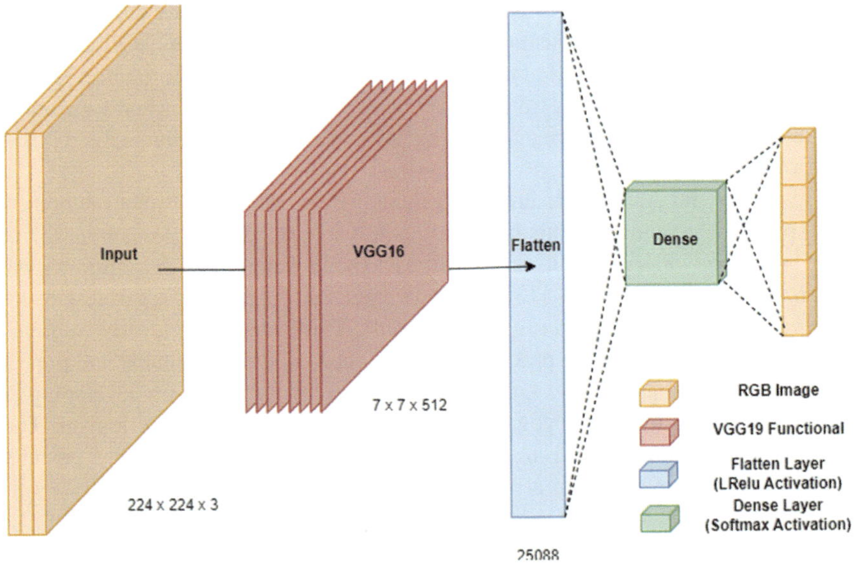

Fig. 1 Proposed YPD model Architecture

7, 7, 512). VGG16 has acquired valuable intricate features from a wide variety of images through its training on the ImageNet dataset. By harnessing these acquired features, the proposed model excels at classifying image data within this framework when applied to novel datasets.
2. **Flatten Layer:** The Flatten layer is added after the VGG16 segment to transform the output into a one-dimensional tensor with dimensions of (None, 25,088). Within the CNN, this transition plays a crucial role in enabling complex computations on input data with significant dimensions, especially images. The Flatten layer is crucial in helping the model grasp complex patterns and relationships as it creates links with completely integrated layers. It effectively creates the output from the previous layers for subsequent analytical processes.
3. **Activation function:** The Leaky Rectified Linear Unit (LeakyReLU) [17], a commonly employed activation function in neural networks, shares a relationship with the Rectified Linear Unit (ReLU) while displaying distinct characteristics. Unlike ReLU, LeakyReLU introduces a minor gradient (typically set at 0.1 in our context) for negative input values, as opposed to outputting zero. This slight gradient guarantees that even when a neuron isn't actively firing, it still contributes to the gradient during backpropagation. This ongoing contribution enables the neuron to continue learning and adjusting its weights based on available information. Furthermore, LeakyReLU's effectiveness in enhancing neural network performance, particularly in image classification tasks, has been noted. By incorporating a slight gradient for negative inputs, LeakyReLU aids in counteracting overfitting by introducing slight noise to the output. This infusion of noise enhances the network's overall resilience, empowering it to effectively manage variations and generalize more effectively to unfamiliar data. The LeakyReLU is given as follows:

$$Leaky\mathrm{Re}LU(x) = \max(ax, x) \tag{1}$$

where x is the input value and a is the leakage coefficient, a positive constant typically set to a small value like 0.01.

4. **Dense Layer:** The main objective of a dense layer is to recognize complex patterns and connections within the data. Each neuron within this layer receives input from all neurons in the preceding layer and produces a singular output value. This output subsequently becomes the input for the following layer. The neurons within the dense layer possess the capability to learn and recognize diverse combinations of features and patterns present in the data, equipping them to make predictions or classifications. This adaptability empowers the dense layer to extract more advanced representations from the input data.

The softmax function is utilized on the output values of the neurons within the dense layer, converting them into probabilities corresponding to each possible class label. These probabilities signify the chances of the input belonging to each class, facilitating the model's predictions based on the highest probability. The softmax function is particularly advantageous in situations involving multi-class classification. It holds a pivotal role during training by motivating the model to produce more accurate predictions. This is accomplished through the implementation of a loss function, such as categorical cross-entropy, which penalizes inaccurate predictions more rigorously.

Training Strategy for the Proposed YPD Model

The training strategy of the proposed YPD architecture are as follows:

A pre-trained VGG16 network used in the YPIM architecture is intended to extract features from the input data. Subsequently, a Flatten layer is employed to transform these features into a one-dimensional tensor. A LeakyReLU activation function is used to introduce nonlinearity and handle vanishing gradients. Ultimately, a Dense layer equipped with a softmax activation function is employed to produce the class probabilities, serving as the final output.

Adam optimization [18] is a widely utilized algorithm known for its efficient optimization of neural network parameters. By merging the strengths of AdaGrad and RMSprop, it adjusts the learning rate of each parameter based on both the gradient's first and second moments. This adaptability empowers Adam to effectively manage various data scenarios, even those involving sparse or noisy gradients. As a result, it ensures quicker convergence and heightened optimization performance. In our implementation, we have employed the Adam optimizer with a smaller learning rate of 0.0001, as opposed to the conventional 0.001. This modification has demonstrated its advantages by improving generalization, reducing the risk of overshooting optimal parameter values, and facilitating a smoother learning curve.

LeakyReLU is used to handle any unwanted noise or unimportant data that may be present in the input. This noise may be caused by a variety of factors, including inaccurate measurements, inconsistent data gathering, and inherent changes in the phenomenon being modeled in reality. Noise can interfere with learning by introducing misleading patterns and distorting relationships between variables when it enters the system. As a result, the model may exaggerate insignificant features or make erroneous generalizations. To address this, we carefully selected an appropriate "alpha" value in LeakyReLU to filter out excessive noise that could lead to model misinterpretation while minimizing the loss of potentially useful information.

Effective model training requires finding a balance between tolerating noise and avoiding overfitting. For this reason, we chose a greater value of 0.1 to allow some noise into our suggested model. With this modification, a more significant leakage has been included, enabling a higher tolerance for negative input values. This adaptation aids in maintaining an optimal stability by preventing excessive information suppression. The proposed model is able to effectively handle input fluctuations due to the slightly increased "alpha" value, ensuring that important information is retained without unnecessary haste.

During the training process, the model is trained in batches of 16 samples to work within hardware constraints. Stochastic Gradient Descent (SGD) [19] is used as the optimization technique, and it involves updating the network weights after each batch to account for the gradient of the loss function for that batch.

4 Experiments and Results

The experiments are conducted on a system with an Intel(R) Xeon(R) CPU @ 2.20GHz, 12 GB of RAM, and 2.2 GHz clock speed, using Google Colab as the development environment. The proposed model's architecture includes a pretrained VGG16 network for feature extraction. Then, a one-dimensional tensor is created from these extracted features. The LeakyReLU activation function has been used to overcome the difficulties of vanishing gradients and introduce non-linearity. Through the integration of a softmax activation function in a dense layer, the final class probabilities are produced. The Adam optimizer, a stochastic optimization technique that decouples weight decay from gradient updates, facilitates optimization throughout the training phase. The categorical cross-entropy is used as a loss function for multi-class classification. Notably, model training is carried out in batches of 16 to accommodate hardware limitations. Additionally, the SGD optimization approach is used, which updates the network weights after each batch while taking the gradient of the loss function for the most recent batch into account.

4.1 Data Description

We employed the Yoga-82 dataset [16] as the basis for implementing the proposed YPD model. This dataset was curated by Verma et al. [16] by manually selecting images from the internet using the Bing search engine. The dataset originally included a wide range of yoga poses that were later grouped into super classes. For our proposed model, we have classified these images into five distinct super classes representing different yoga poses. These super classes consist of Adho Mukha Svanasana, Dev Mudra, Phalakasana, Virabhadrasana, and Vrikshasana. The goal of our experiment is to successfully train the proposed model to recognize the super class that corresponds to each yoga pose. Additionally, each image underwent the necessary preprocessing steps to create a standardized dimension with RGB channels in order to guarantee uniformity. Subsequently, random images have been divided into training and testing sets, ensuring representation from all five classes in both sets.

The proposed YPD approach is illustrated in Fig. 2, along with the data pre-processing steps applied to the Yoga-82 dataset. The data pre-processing steps utilized for the dataset in the proposed YPD model are as follows:

Class Selection: During the initial phase of data preprocessing, we have chosen the most uniform categories from the Yoga 82 Dataset, namely Adho Mukha Svanasana, Dev Mudra, Phalakasana, Virabhadrasana, and Vrikshasana. These classes are chosen to increase the training dataset's diversity and encourage balance.

Image Resizing: The subsequent step entails resizing the image. All the images in the dataset are scaled to a standardized format of (255, 255, 3), which ensures consistency and simplifies model usage. In addition to maintaining uniformity in

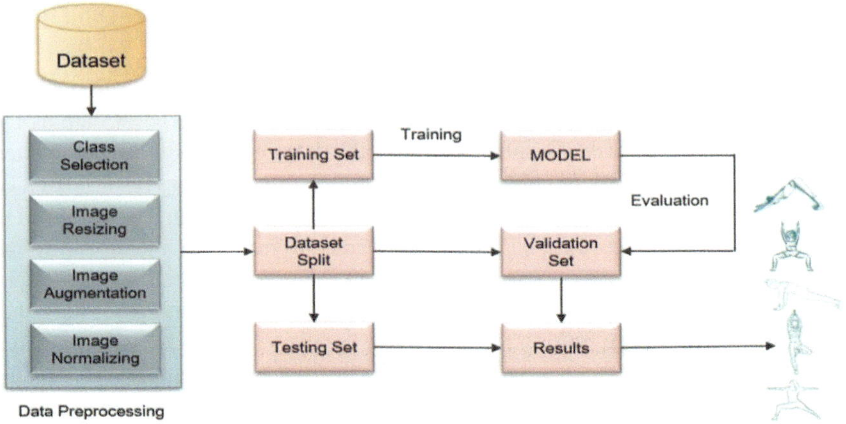

Fig. 2 The Proposed YPD model preprocessing

image size, this scaling process also reduces computational load during both the training and testing phases.

Image Augmentation: Image augmentation is a technique utilized to broaden the training dataset by creating modified versions of the original images. This strategy is employed to address slight disparities in class distribution within the dataset. The images have been enriched through methods such as rotation, flipping, and zooming, generating diverse iterations of the initial dataset. This improves the balance of the dataset and strengthens its general robustness.

Image Normalization: Image normalization reduces the impact of variations in brightness, contrast, and color distribution between various images by converting the image's pixel values into a standardized interval (for the purposes of the suggested model, often ranging from 0 to 1). This process helps to reduce the impact of input variations and allows the network to focus on recognizing important characteristics rather than irrelevant image statistics. Image normalization further speeds up the network's convergence and eliminates the dominance of high-intensity features, hence reducing computational complexity and memory requirements during training.

4.2 Performance Metrics and Existing Methods for Comparison

In Table 2, the performance metrics of different contemporary architectures, including the proposed YPD model, in addressing the classification task using the Yoga-82 dataset are displayed. To evaluate the effectiveness of the proposed YPD approach, we have employed metrics such as accuracy, precision, recall, and F1 score. In the context of classification, a True Positive (TP) signifies the model's accurate prediction of the positive class. Conversely, a True Negative (TN) signifies the model's correct prediction of the negative class. Conversely, an incorrect positive class prediction by the model is categorized as a False Positive (FP), and an erroneous negative class prediction is labeled as a False Negative (FN).

The mathematical formulations for these metrics are provided below.

$$Accuracy = \frac{TP + TN}{TP + TN + FP + FN}$$

$$\Pr ecision = \frac{TP}{TP + FP}$$

$$\text{Recall} = \frac{TP}{TP + FN}$$

$$F1\,Score = \frac{TP + TN}{TP + TN + FP + FN}$$

Table 2 provides details of the hyperparameters used in the proposed YPD model. The hyperparameters are elucidated as follows.

Input image shape: It defines the size of the input image for the proposed model. In this instance, the image dimensions are 224 pixels in width, 224 pixels in height, and consist of 3 channels representing RGB color information.

Channels: It indicates the count of color channels within the input image. A value of 3 denotes that the proposed model is handling images in the RGB format.

Batch size: It signifies the quantity of training samples processed in each iteration of the training process, involving both the forward and backward passes. With a batch size of 16, the model weights are updated collectively using 16 images.

Epoch: An epoch is the complete path through the training data set. The count of epochs specifies the frequency at which the model is trained on the complete dataset. In this case, the proposed model runs 25 training iterations on the dataset.

Optimizer: During training, the optimizer adjusts the model's weights to minimize the loss function. In the proposed model, the Adam optimizer is employed to adapt the learning rate of each weight based on previous gradients.

Table 2 Comparison of the proposed YPD model

Architecture	Depth	Number of parameters	Accuracy	Precision	F1 score
Resnet	50	5,751,813	91.52	91.82	91.48
Inception V3	159	23,966,885	86.44	90.05	87.14
Xception	126	23,025,581	89.83	90.21	89.79
Inception Resnet-V2	572	55,976,549	81.35	81.64	81.29
Classic CNN	8	22,245,957	81.06	81.72	81.01
Transfer Learning & SGD	20	22,227,493	90.63	91.12	90.64
Proposed	**19**	**21,138,757**	**93.19**	**93.4**	**93.08**

Loss function: This is the objective function that measures the difference between the proposed model's predicted result and the actual target during training. Categorical cross-entropy is selected to handle the multiclass classification.

Learning rate: It manages weight adjustment pace in training. The proposed model uses a low 0.0001 rate for precise, gradual weight updates, ensuring accurate convergence and generalization.

Through multiple experiments conducted on the proposed YPD, it has been consistently observed that the proposed model outperforms existing models such as ResNet50, Inception V3, Xception, Inception-ResNet V2, classic CNN, and Transfer learning & SGD. The outcomes demonstrate our proposed model's better accuracy in comparison to several well-known existing architectures for image classification, which vary in depth and parameters. The proposed YPD architecture, with 21 million parameters, outperforms Classic CNN and Transfer Learning & SGD with more parameters, in terms of accuracy, precision, recall, and F1 score, consistently achieving 93.19%, 93.40%, 93.19%, 93.08% respectively. Table 3 indicates accuracy improvement with increased architecture depth but at the expense of computational demands. To balance performance and resources, we developed the YPD model. The learning curve of the proposed model is illustrated in Fig. 3.

Table 3 Hyperparameters of the proposed YPD model

Hyperparameters	Values
Input image shape	224 × 224 × 3 (width × height × channels)
Channel	3
Batch size	16
Epochs	25
Optimizer	Adam
Loss function	Categorical Cross entropy
Learning rate	0.0001

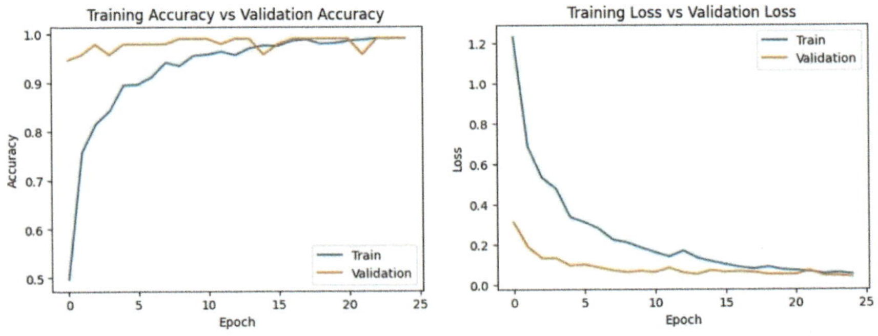

Fig. 3 Learning curve of the YPD model

5 Conclusion

To conclude, this study introduces the YPD approach, which utilizes Transfer Learning and the VGG16 CNN model to achieve precise yoga pose recognition. Unlike existing methods dependent on manually crafted features and controlled conditions, our YPD method fine-tunes the VGG16 model on a novel dataset, ensuring robust performance in complex real-world datasets. The proposed model is evaluated on the Yoga-82 dataset and showcases an impressive accuracy of 93.19%, surpassing several state-of-the-art image classification models including ResNet50, Inception V3, Xception, Inception-ResNet V2, classic CNN, and Transfer Learning & SGD. The proposed architecture effectively extracts spatial and depth features from input images, seamlessly integrating them into the classification process to enhance yoga pose estimation accuracy.

For future work, we propose investigating the following directions for future research. First, optimizing hyperparameters might improve YPD's performance. Additionally, adding more complex and diverse datasets to the model may improve its ability for generalization. In conclusion, the YPD shows promising results in accurately recognizing yoga poses, paving the way for subsequent exploration and advancements in this field.

References

1. Basavaraddi, I. V.: Yoga: its origin, history and development. Government of India, Ministry of External Affairs, 23 ((2015)
2. Balasubramaniam, M., Telles, S., Doraiswamy, P.M.: Yoga on our minds: a systematic review of yoga for neuropsychiatric disorders. Front. Psych. **3**, 117 (2013)
3. Ross, A., Thomas, S.: The health benefits of yoga and exercise: a review of comparison studies. J Alternat Complemen. Med. **16**(1), 3–12 (2010)
4. Sherman, K.J., Cherkin, D.C., Erro, J., Miglioretti, D.L., Deyo, R.A.: Comparing yoga, exercise, and a self-care book for chronic low back pain: a randomized, controlled trial. Ann. Intern. Med. **143**(12), 849–856 (2005)
5. Anand Thoutam, V., Srivastava, A., Badal, T., Kumar Mishra, V., Sinha, G. R., Sakalle, A., Bhardwaj, H. & Raj, M.: Yoga pose estimation and feedback generation using deep learning. Comput. Intell. Neurosci. (2022)
6. Liaqat, S., Dashtipour, K., Arshad, K., Assaleh, K., Ramzan, N.: A hybrid posture detection framework: Integrating machine learning and deep neural networks. IEEE Sens. J. **21**(7), 9515–9522 (2021)
7. Chen, H. T., He, Y. Z., Hsu, C. C., Chou, C. L., Lee, S. Y., & Lin, B. S. P.: Yoga posture recognition for self-training. In: MultiMedia Modeling: 20th anniversary international conference, MMM 2014, Dublin, Ireland, January 6–10, 2014, Proceedings, Part I 20 (pp. 496–505). Springer International Publishing (2014)
8. Agrawal, Y., Shah, Y., & Sharma, A.: Implementation of machine learning technique for identification of yoga poses. In: 2020 IEEE 9th international conference on communication systems and network technologies (CSNT), pp. 40–43. IEEE (2020)
9. Kumar, D., Sinha, A.: Yoga pose detection and classification using deep learning. LAP LAMBERT Academic Publishing (2020)

10. Byeon, Y.H., Lee, J.Y., Kim, D.H., Kwak, K.C.: Posture recognition using ensemble deep models under various home environments. Appl. Sci. **10**(4), 1287 (2020)
11. Ashraf, F.B., Islam, M.U., Kabir, M.R., Uddin, J.: YoNet: a neural network for yoga pose classification. SN Comput. Sci. **4**(2), 198 (2023)
12. Upadhyay, A., Basha, N. K., & Ananthakrishnan, B.: Deep learning-based yoga posture recognition using the Y_PNMSSD model for yoga practitioners. In: Healthcare, Vol. 11, No. 4, p 609. MDPI (2023)
13. Chasmai, M., Das, N., Bhardwaj, A., Garg, R.: A view independent classification framework for yoga postures. SN Comput. Sci. **3**(6), 476 (2022)
14. Bansal, M., Kumar, M., Sachdeva, M., Mittal, A.: Transfer learning for image classification using VGG19: Caltech-101 image data set. J. Ambient Intell. Human. Comput. 1–12 (2021)
15. Simonyan, K., Zisserman, A.: Very deep convolutional networks for large-scale image recognition (2014). arXiv preprint arXiv:1409.1556
16. Verma, M., Kumawat, S., Nakashima, Y., Raman, S.: Yoga82: a new dataset for fine-grained classification of human poses. In: Proceedings of the IEEE/CVF conference on computer vision and pattern recognition workshops, pp. 1038–1039, (2020)
17. Xu, J., Li, Z., Du, B., Zhang, M., Liu, J.: Reluplex made more practical: Leaky ReLU. In 2020 IEEE symposium on computers and communications (ISCC), pp. 1–7. IEEE (2020)
18. Kingma, D. P., Ba, J.: Adam: A method for stochastic optimization (2014). arXiv preprint arXiv:1412.6980
19. Bottou, L.: Stochastic gradient descent tricks. In: Neural networks: tricks of the trade: Second Edition, pp. 421–436. Berlin, Heidelberg: Springer Berlin Heidelberg (2012)

An Augmented Efficient Net Based Deep Learning Model for Diabetic Retinopathy Detection and Classification

M. A. Abini and S. Sridevi Sathya Priya

Abstract Over the last few years, there has been a rise in the prevalence of Diabetic Retinopathy (DR), one of the most severe consequences of diabetes that can cause lifelong blindness. Analysis of diabetic retinopathy (DR) using color fundus images requires experienced physicians to determine the presence of DR. In order to detect issues in DR and its numerous stages utilizing retinal scans, many different artificial intelligence and deep learning techniques have been developed. The present study aims to develop a robust system for the automated identification and categorization of diabetic retinopathy (DR). A Convolutional Neural Network (CNN) model called Augmented Efficient Net is used to detect diabetic retinopathy from fundus images in this work. The classification of diabetic retinopathy was performed using four different Efficient Net models, namely Efficient Net B0, Efficient Net B1, EfficientNetB3 and EfficientNet-B4 and tested on publicly available fundus image datasets APTOS 2019. Each model performance was assessed, and the best model was found among the four models. We use these models to solve the difficulties we have in transfer learning along with preprocessing. Data augmentation and hyper parameter optimization to make our models stronger and more accurate. Experiments showed that on the APTOS 2019 datasets, The training accuracy for EfficientNet-B4 was 99.03%, while the validation and testing accuracy were 87% respectively.

Keywords Efficient Net · Deep learning · Convolutional Neural Networks · Diabetic Retinopathy

M. A. Abini (✉) · S. S. S. Priya
Karunya Institute of Technology and Sciences, Coimbatore, India
e-mail: abinima87@gmail.com

S. S. S. Priya
e-mail: sridevi@karunya.edu

B. Soni et al. (eds.), *Beyond Artificial Intelligence*, Lecture Notes in Networks and Systems 1326, https://doi.org/10.1007/978-981-96-4170-3_26

1 Introduction

Diabetic retinopathy (DR), which damages the retina and makes it difficult to visualize, is often caused by diabetes mellitus [1]. If untreated, hyperglycemia can weaken and destroy the blood vessels in the retina, causing progressive vision loss and ultimate blindness. Although diabetic retinopathy does not cause many symptoms in its early stages, damage to the nerve cells in the retina and changes to the blood vessels cannot be seen by the naked eye. Therefore, people with diabetes should undergo regular eye examinations because it is important to identify problems early and treat them immediately. A rapid, non-invasive imaging technique that is extensively used is fundus photography [2]. Ophthalmologists use fundus images to visualize retinal lesions at a high resolution so that they can diagnose diabetic retinopathy and determine how bad it is. But for a professional ophthalmologist to manually diagnose DR from fundus images, they need to be very skilled and put in a lot of work [3]. Ophthalmologists are in demand, yet developing countries are short on them. This scarcity is alarming because diabetes and diabetic retinopathy (DR) are expected to rise significantly in these places. [4]. This has motivated researchers to create computer-aided diagnosis tools, which will make diagnosing DR cheaper, faster, and simpler. Retinal scans can detect several different types of lesions that can be used to diagnose DR. Micro aneurysms (MA), hemorrhages (H), and soft and hard exudates (EX) are all examples of these lesions [5].Micro aneurysms (MA) are the first clinically noticeable sign of diabetic retinopathy. They show up as separate red spots behind the eye. There may be a ring of yellow fat or hard exudates around them. Hemorrhages (HM) appear to be larger, unevenly bordered spots on the retina, larger than 125 μm hard exudates appear as bright yellow spots on the retina due to plasma leakage. Soft exudates White patches on the retina, popularly known as "cotton wool," are brought on by the nerve fiber's enlarging. There are two shapes: round and oval. Figure 1 gives the comparison of retinal images of DR and Non DR patients and different stages of DR described in Fig. 2.

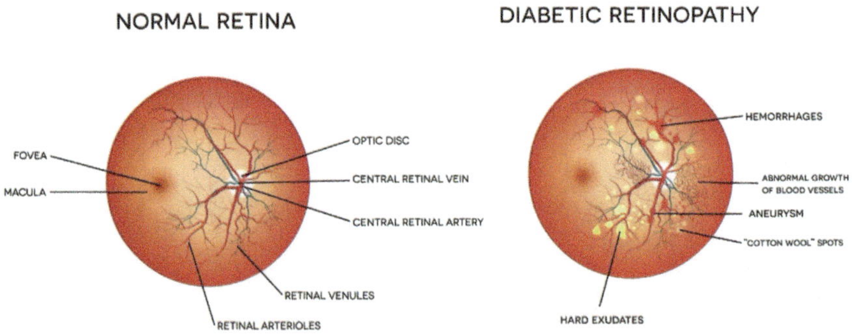

Fig. 1 Comparison of a DR and non-DR retinal image [1]

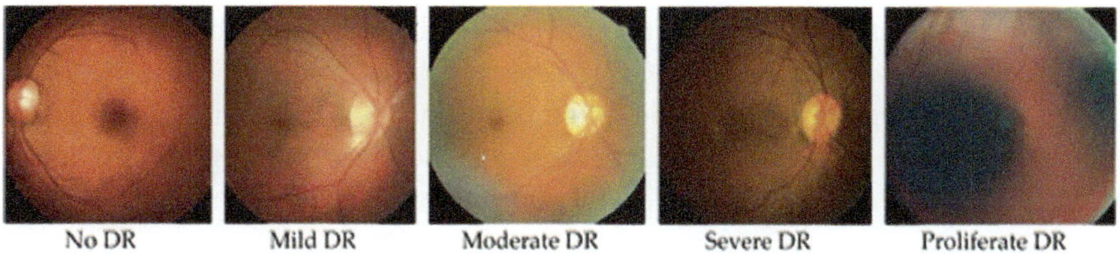

Fig. 2 Lesions-based categorization of DR [4]

According to the amount of lesions, DR progresses through the following stages. They are Normal DR, Mild DR, Moderate DR, Proliferative DR, and Severe DR. Table 1 gives a brief description of each stages.

There has been an increase in interest in employing artificial intelligence and deep learning algorithms to help in the initial identification and diagnosis of diabetic retinopathy in recent years. The system will be trained to recognize early indicators of diabetic retinopathy in retinal pictures using previously developed deep learning algorithms, and it will aid the ophthalmologist in subsequent treatment [6]. With the help of this system, patients with diabetic retinopathy can receive early diagnosis and timely treatment, leading to better outcomes and quality of life.

The following is a concise summary of the most important contribution made by this research:

- A new automated method that can replace the traditional invasive classification of diabetic retinopathy has been offered as a way to both improve the overall accuracy of classification and replace the traditional method.
- In order to attain high classification accuracy on a limited data set, a data augmentation strategy is utilized, and the influence of overfitting on classification performance is studied.
- A deep Efficient Net neural network is used for the extraction of extensive features.
- This article presents a comparative analysis of the proposed methodology and current state-of-the-art techniques for classifying diabetic retinopathy.

Table 1 Criteria for DR grading

Stage	Severity	DR signs
I	No DR	No abnormalities present
II	Mild DR	Presence of Micro aneurysms in the retina blood vessels
III	Moderate DR	Retinal blood vessels become enlarged and obstructed Fluid accumulation in the macula area of the retina
IV	Severe DR	A vast number of blood vessels in the retina get clogged. The formation of tiny new blood vessels
V	Proliferative DR	Presence of Neovascularization Vitreous/pre-retinal HM

The subsequent sections of this document are structured in the following manner. Section 2 provides a brief synopsis of several related works. In Sect. 3, a deep learning-based approach to diabetic retinopathy detection is presented. The experimental results and the analysis of the proposed methodology are described in Sect. 2. In addition, Sect. 2 provides conclusions and future scope explained in Sect. 2.

2 Related Works

Recent AI research has employed deep learning models to automate diabetic retinopathy detection. As mentioned in the introduction, studies classify the stage as binary, with or without DR, or into five DR stages. Deep learning models by Qummar et al. (2019) increased diabetic retinopathy (DR) detection accuracy to 80.8% and specificity to 86.72%. CNN, Incep-tion-ResNet-v2, and Inception classify fundus images into five DR severity levels. Data came from 35,126 Kaggle Diabetic Retinopathy Detection Challenge photos [7]. V. Gulshan et al. [8] found DR and DME using CNN. The model was tested on 1748 and 9963 Messidor-2 and eyepacs-1 photos. CNNs were trained with different input image sizes after standardizing and scaling to 299 pixels. Messidor-2 and Eyepacs-1 showed the approach's 0.990 and 0.991 AUCs with outstanding sensitivity and specificity.Asra Momeni et al. [9] constructed a classification model using contrast-limited Adaptive Histogram equalization, a convolutional neural network, and the efficient Net B5 architecture. This model had a higher AUC. S. Majumder et al. [10] detected diabetic retinopathy with smartphone apps and deep learning. An app that uses deep learning algorithms to automatically assess the five stages of DR from retinal pictures taken with the phone's camera was proposed. The algorithm diagnosed the five DR stages with 94.9% accuracy, like ophthalmologists. Deep convolutional neural networks were used to identify and categorize diabetic retinopathy (DR) by Hemanth et al. (2020) [11]. This work used contrast stretching and histogram equalization to improve retinal image contrast during pre-processing. Image features are extracted using a CNN (ResNet-50) that has been trained. They classified the disease into five classes using machine learning techniques in the final phase, "classification." Zhang et al. [12] developed two hybrid models to measure disease severity and classify it as binary or non-binary. Hybrid learning outperformed individual models in all tasks with 98.10% sensitivity and 98.56% specificity.Pratt et al. presented a two-dimensional CNN model [13]. Deep learning, specifically CNNs, is utilized to classify and identify retinal fundus images for diabetic retinopathy detection. These methods fall into two categories [14]. Multi-level and binary DR classification. Bodapati et al. [15] suggested an Xception-VGG16 hybrid deep neural network with gated attention to automatically diagnose DR. This model had 82.54% accuracy on APTOS 2019.

3 Proposed Methodology and Materials

In order to detect and categorize DR, this research presents Augmented version of four variants (B0, B1, B3, and B4) of the EfficientNets deep learning model. The utilization of transfer learning was employed to augment the network's performance, namely as the foundational model [16, 17]. Following that, a series of image pre-processing procedures were conducted on the dataset in order to enhance the overall quality of the images. Subsequently, the complete dataset was partitioned into separate training and testing sets. Data augmentation techniques, including rescaling, horizontal flip-ping, vertical flipping, rotation, and shifting in width and height, were used to the training set images [19]. The classes with relatively fewer images underwent a process of up-sampling, resulting in an equal amount of images for each class. Subsequently, the aforementioned data was inputted into the proposed model. The trained model was utilized to classify test photos.

3.1 Dataset Description

In this study, the APTOS 2019 dataset was employed, which was obtained through the Diabetic Retinopathy Detection competition held at the APTOS Blindness Detection Challenge on Kaggle [20]. The dataset consists of 3662 sample images that have undergone pre-processing using Gaussian filters and have been resized to dimensions of 224 $\times$ 224. Fig. 2 shows several pictures from the APTOS Dataset that show different kinds of diabetic retinopathy. Within the dataset, there are 1805 pictures that have no detectable DR, 370 pictures with just mild DR, 999 pictures with moderate DR, 193 pictures with detectable severe DR, and 295 pictures with detectable proliferative DR [21].

3.2 Addressing Dataset Imbalance Problem

The APTOS dataset exhibits an imbalance, with a significant skew towards the class 0-No DR, which comprises around 1805 pictures. The issue of imbalance within the dataset was resolved through the application of up sampling techniques, as discussed in previous studies [23]. During the process of up-sampling, synthetic augmentation was employed to increase the amount of photos in each respective class, hence normalizing the distribution of images across different classes. Table 2 provides an overview of the total count of images in each class following the process of up-sampling, as well as the number of photographs that were utilized for testing purposes.

Table 2 The distribution of training/validation images among classes

Grade	Stages	Training/validation image count	Testing\ image count
0	N0-DR	1500	147
1	Moderate -DR	1500	160
2	Mild- DR	1500	159
3	Proliferative- DR	1500	146
4	Severe-DR	1500	141

3.3 Pre-Processing

The dataset was then subjected to image pre-processing procedures in order to enhance the image quality. The two-dimensional fundus data are of non-uniform size. Therefore, the APTOS 2019 fundus images have been resized to a uniform dimension of 224 × 224 × 3 [24, 25]. In the study, as few pre-processing steps as possible were used to keep the image conditions as general as possible. The proposed method used the Keras's built-in pre-processing function, Image Data Generator, to do a simple step of pre-processing. The Image Data Generator was set up so that all of the images in the dataset would get a random zoom, horizontal flip, and vertical flip. We converted the images down to a resolution of 224* 224 pixels to match the size of the most efficient Net input image.

3.4 Augmentation

Data augmentation is a method of generating more training data from the original data. Data enhancements such as rescaling, flipping horizontally, flipping vertically, rotating, and shifting width and height were applied to the pictures of the training set [26]. The classes with a comparatively smaller number of images were up sampled so that each class had the same number of images. By rotating and flipping the data, many variations of the images can be created using data augmentation techniques. To achieve this goal, the training data set is extended, and an image rotation is performed at different angles, namely 270°, 180° and 90°.

3.5 Proposed Efficient Net Architecture

The stages of the fundus image preprocessing, augmentations, training, and evaluation are depicted in Fig. 3. The following steps describe the workflow of DR classification implementation in detail:

Step 1: All input images were resized to 224 × 224 and then normalized.

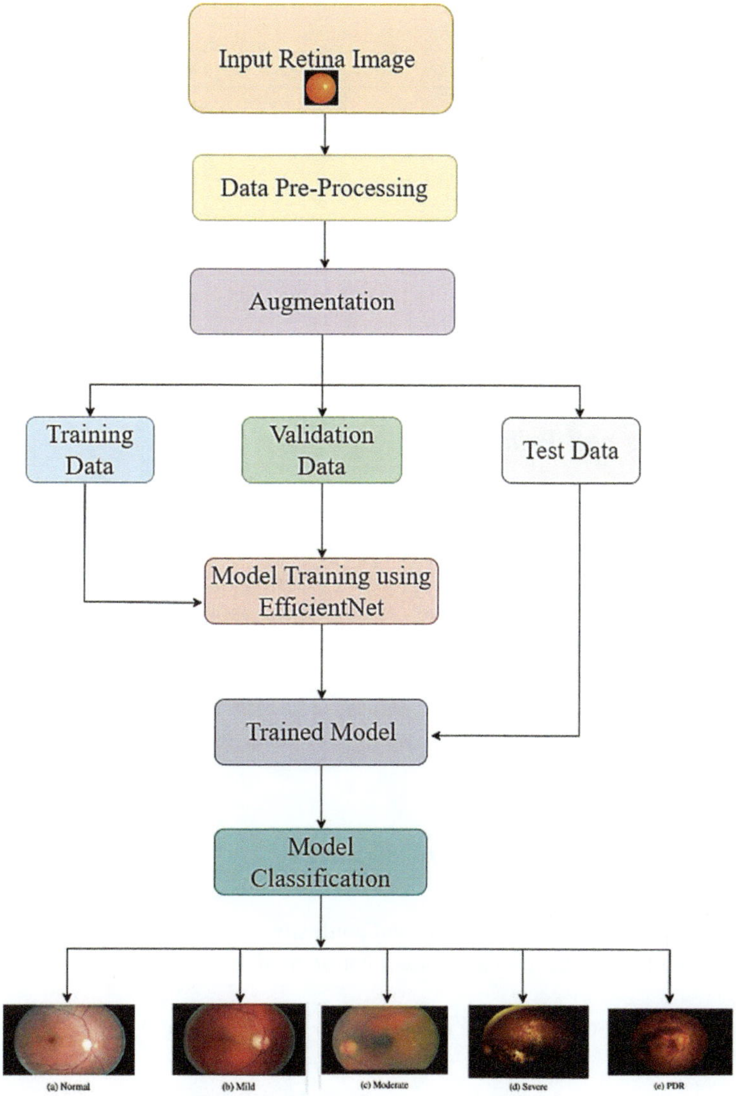

Fig. 3 Basic block diagram of proposed model

Step 2: The pre-processed data was then augmented to mitigate the consequences of the class imbalance problem.

Step 3: Augmented data is fed to pre trained Efficient Net framework for training Purpose.

Step 4: The extracted features were trained with the Efficient Net classifier to predict DR classes.

Table 3 Setting of hyper parameter values

Hyper parameter values	
Optimizer	Adam
Epochs	10
Regularizer	L2
Batch size	32
Learning rate	Custom

Step 5: Finally, we evaluated the proposed test set prediction classifier using different performance measures.

3.6 Training and Validation

The dataset was split 80:20 before training the efficient network design. 80% of the images were used for training, 10% for validation, and 10% for testing the accuracy of the model. The original data set consists of 3662 images and the up sampled data set consists of 7500 (1500 in each class). Of the 7500 images, 5995 images for training, 752 images for validation and 753 images for testing were fed individually into this deep learning model. In addition, Adam was used as the optimizer [24], categorical cross entropy as the loss function, and Softmax as the activation function in the output layer. Table 3 shows the hyper parameter values.

3.7 Performance Evaluation Measures

Accuracy, precision, recall, F1-score, and confusion matrix can be used to compare diabetic retinopathy detection and classification models.

$$Accuracy = TrPo + TrNe \big/ TrPo + TrNe + FaNe + FaPo \quad (1)$$

$$Recall/Sensitivity = TrPo \big/ TrPo + FaNe \quad (2)$$

$$Precision = TrPo \big/ FaPo \quad (3)$$

$$F1{-}score = 2 \times (Precision \times Recall) \big/ (Precision + Recall) \quad (4)$$

where TrPo (True Positive),FaNe (False Negative),-FaPo (False Positive),TrNe (True Negative)

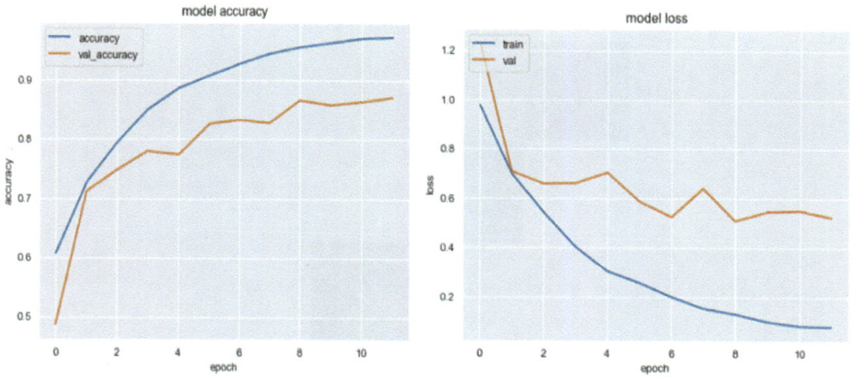

Fig. 4 Training and validation accuracy/loss curves of efficientNet -B4

4 Results and Discussions

The performance of Four CNN architectures for classifying diabetic retinopathy via transfer learning is examined in this research. the Training and Validation Accuracy and Loss Curve, and Confusion Matrix of EfficientNet B4 are shown to show the classification performance.

4.1 Analysis of Accuracy and Loss of Efficient Net Models

Training and Validation Accuracy/Loss curves are shown in Fig. 4. Training and validation losses for EfficientNet B0 were discovered to be 0.585 and 0.884, respectively. Nevertheless, it was found that EfficientNetB1's training loss and validation loss values were 0.424 and 0.732 respectively. Our analysis clearly shows that the Efficient Net B4 gave the superior training and validation curve, as compared to the other 3 Efficient Net Models.

5 Confusion Matrix

EfficientNet-B4 properly predicts 143 out of 147 photos for the No-DR stage. And 131 out of 159 for the mild DR stage, 114 out of 160 for the moderate DR, 123 out of 146 for the proliferative DR, and 132 out of 141 for the severe DR. As a result, Efficient Net-B4 produces the greatest results for the No-DR stage. The confusion matrix of EfficientNet B4 is shown Fig. 5.

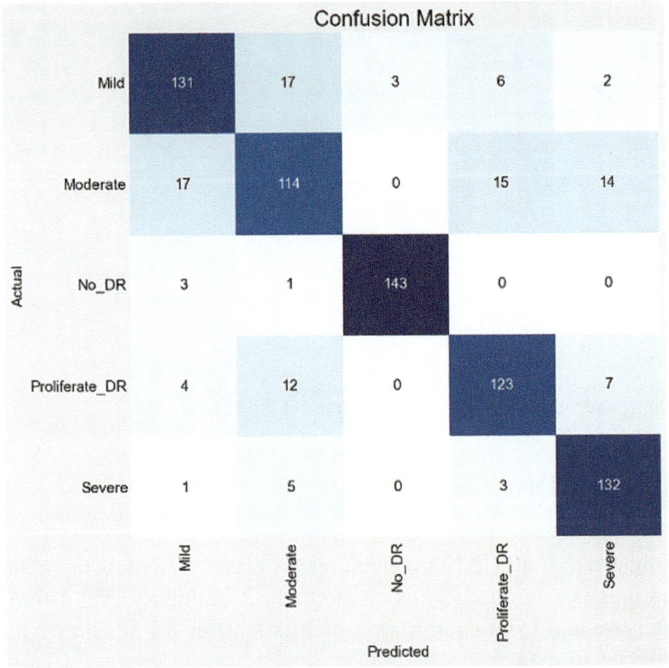

Fig. 5 Confusion matrix of efficient Net-B4

6 Performance Summary for Severity

Precision, Recall, F1 score, Macro average, and Weighted average of EfficientNet B4 for all five classes are shown in Table 4. The model was validated using 365 photographic background images. Table 5 describes the training and validation accuracy/ loss of four Efficient Net models.

The comparison of Recall values,Precision values and F1 values of EfficientNets for stages of DR is shown in Fig. 6, 7 and 8 respectively.

Table 4 Class-wise outcomes of EfficientNet-B4

Model	Testing accuracy	Class	Precision	Recall	F1-Score
Efficient Net- B4	87%	Mild	0.89	0.94	0.87
		Moderate	0.82	0.76	0.79
		No-DR	0.99	0.98	0.99
		Proliferate-DR	0.86	0.85	0.85
		Severe	0.87	0.96	0.91
		Macro avg	0.86	0.87	0.86
		Weighted Avg	0.87	0.87	0.87

Table 5 Comparison table for Training and Validation Accuracy, Training and Validation loss for Efficient Net Models

Model	T training loss	Training accuracy (%)	Validation loss	Validation accuracy (%)
EfficientNet-B0	0.554	90.017	0.88441	77.39
EfficientNet-B1	0.432	92.127	0.7322	80.32
Efficient Net -B3	0.245	98.638	0.60379	84.44
Efficient Net -B4	**0.194**	**99.037**	**0.5384**	**87.02**

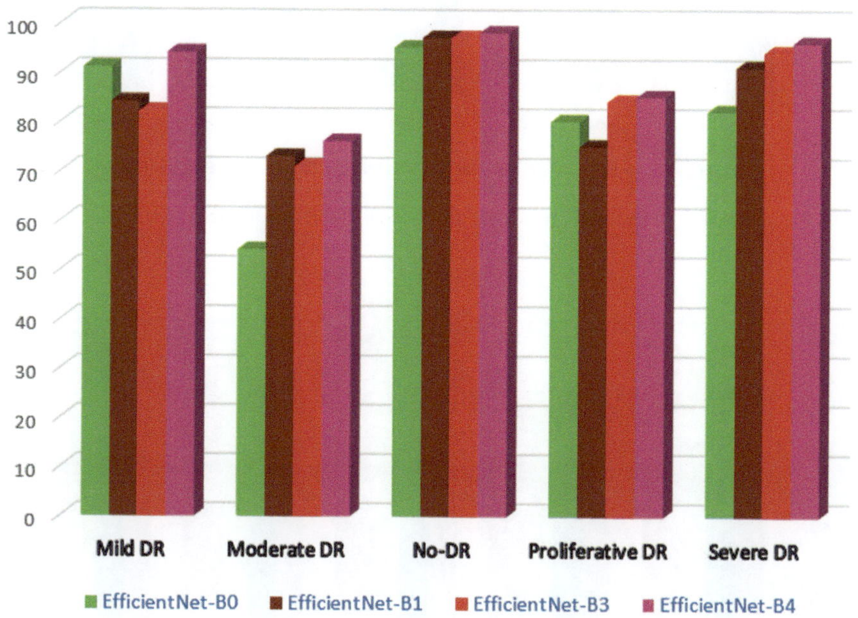

Fig. 6 Comparison of Recall values of EfficientNets for different stages of DR

7 Comparison with Existing Methods

Table 6 displays the accuracy results for the five-class diabetic retinopathy classification using proposed method and existing works and it is graphically represented in Fig. 9.

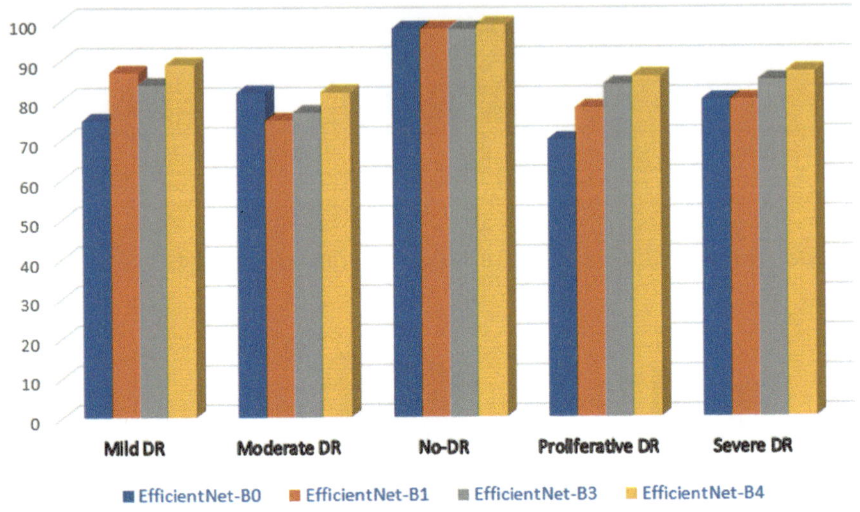

Fig. 7 Comparison of Precision values of EfficientNets for different stages of DR

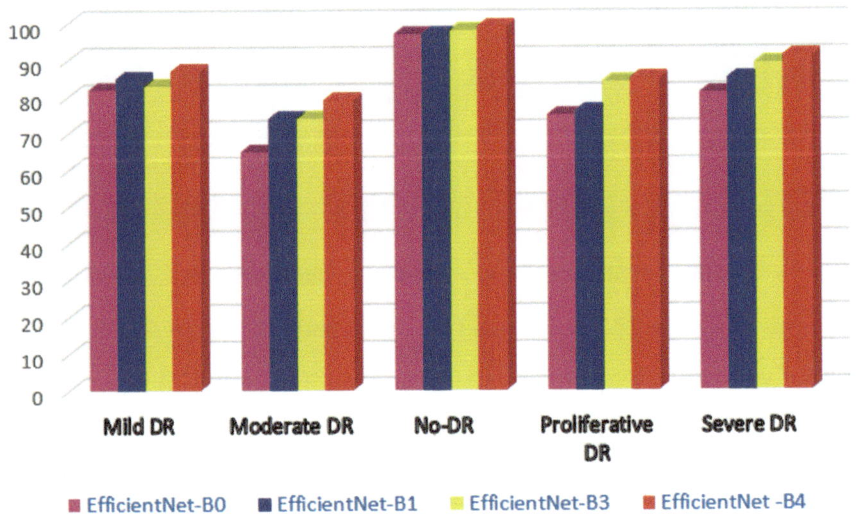

Fig. 8 Comparison of F1-score values of EfficientNets for various stages of DR

Table 6 Performance comparison with state-of-the-art deep learning methods

Authors	No of classes	Method	Accuracy (%)
Qummer et al. [7]	5	Ensemble	80.8
S. Majumder et al. [10]	5	Deep CNN	86
Pratt et al. [13]	5	CNN	75
Bodapati et al. [15]	5	Xception-GG16	82.54
Dekhil et al. [18]	5	VGG-16	76
Masood et al. [22]	5	Inception -V3	48.20
Proposed method	**5**	**Efficient Net -B4**	**87**

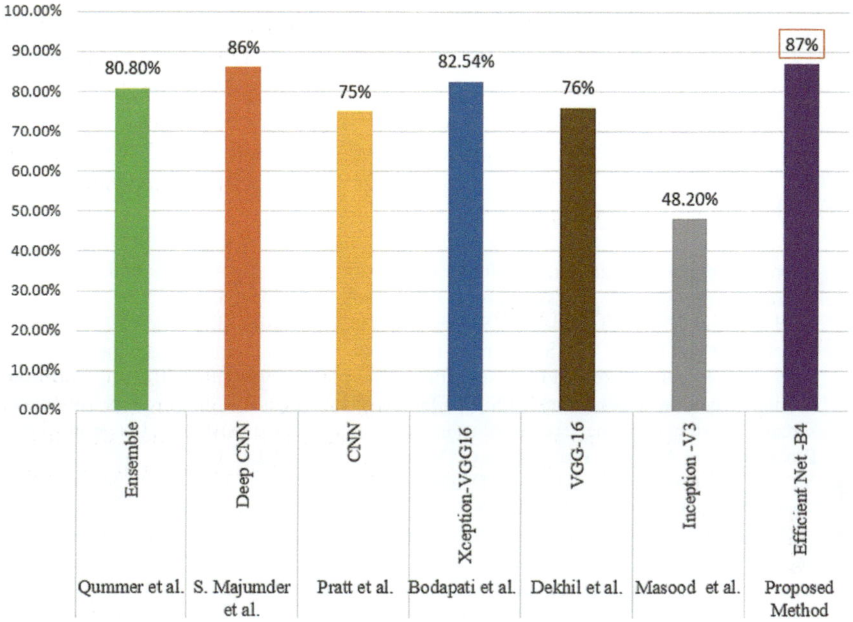

Fig. 9 Summary of performance of proposed method with previous works

8 Conclusion

Diabetic Retinopathy can be prevented from causing permanent vision loss if detected and diagnosed early. Deep learning is one of the most cutting-edge technologies available today for addressing classification problems, and it provides higher levels of accuracy. Using a Convolutional Neural Network (CNN) model for the detection and classification of fundus images will help ophthalmologists combat diabetic retinopathy-related blindness. In this investigation, four distinct EfficientNet models for distinguishing the five stages of diabetic retinopathy are evaluated. Our proposed network EfficientNet B4 trained on the APTOS 2019 dataset outperformed other state-of-the-art networks in early-stage identification, achieving a maximal training accuracy of 99.03% and a validation and testing accuracy of 87%.

9 Future Scope

In the coming years, we will enhance the technique to enhance the precision of DR detection and attempt to create a more robust DR diagnostic model to aid physicians in clinical assessments. We intend to keep enhancing our outcomes by using several models for feature extraction in subsequent work, since it has great promise for clinical application in the future. Furthermore, it is recommended that future study prioritize the assessment of the model's efficacy on other datasets and the integration of more intricate architectures to enhance its performance in more critical instances of DR.

References

1. Ziemssen, F and Agostini, H.T. Diabetic Retinopathy. In: Stahl, A. (eds) Anti-Angiogenic Therapy in Ophthalmology. Essentials in Ophthalmology, springer, Cham (2016)
2. Patel TP, Kim TN, Yu G, Dedania VS, Lieu P, Qian CX, Besirli CG, Demirci H, Margolis T, Fletcher DA and Paulus YM. Smartphone-Based, Rapid, Wide-Field Fundus Photography for Diagnosis of Pediatric Retinal Diseases. Transl Vis Sci Technol;8(3):29 (2019)
3. C. Sinthanayothin, V. Kongbunkiat, S. Phoojaruenchanachai, and A. Singalavanija, Automated screening system for diabetic retinopathy, in 3rd International Symposium on Image and Signal Processing and Analysis, Proceedings of the vol. 2, pp. 915–920(2003).
4. Mehta, S.R., Kashyap, A.S., Das, S.: Diabetes Mellitus in India, The Modern Scourge. Med J Armed Forces India. **65**(1), 50–54 (2009)
5. W. L. Alyoubi, W. M. Shalash, and M. F. Abulkhair, Diabetic retinopathy detection through deep learning techniques: A review, Informat. Med. Unlocked, vol. 20(2020)
6. Kanungo, Y.S., Srinivasan, B., and Choudhary, S. Detecting diabetic retinopathy using deep learning,In 2nd IEEE International Conference on Recent Trends in Electronics, Information & Communication Technology (RTEICT), 801–804 (2017)
7. Qummar, Sehrish et al. A Deep Learning Ensemble Approach for Diabetic Retinopathy Detection,IEEE Access 7: 150530–150539 (2019)
8. Gulshan, V., Peng, L., Coram, M., et al.: Development and Validation of a Deep Learning Algorithm for Detection of Diabetic Retinopathy in Retinal Fundus Photographs. JAMA **316**(22), 2402–2410 (2016)
9. Momeni Pour, H. Seyedarabi, S. H. Abbasi Jahromi and A.Javadzadeh, Automatic Detection and Monitoring of Diabetic Retinopathy Using Efficient Convolutional Neural Networks and Contrast Limited Adaptive Histogram Equalization, in IEEE Access, vol. 8, pp. 136668–136673(2020)
10. Majumder, Sharmin, Elloumi, Yarub, Akil, Mohamed, Kachouri, Rostom,Kehtarnavaz, Nasser. A deep learning-based smartphone app for real time detection of five stages of diabetic retinopathy, 5.https://doi.org/10.1117/12.2557554(2020).
11. D. Jude Hemanth, Omer Deperlioglu, and Utku Kose,an enhanced diabetic retinopathy detection and classification approach using deep convolutional neural network, Neural Computing and Applications(2020)

12. Zhang, W., Zhong, J., Yang, S., Gao, Z., Junjie, H., Chen, Y., Zhang, Y.: Automated identification and grading system of diabetic retinopathy using deep neural networks. Knowl. Base Syst. **175**, 12–25 (2019)
13. Pratt, H., Coenen, F., Broadbent, DM., Harding, S.P., Zheng, Y.: Convolutional neural networks for diabetic retinopathy. Proc. Comput. Sci. 90, 200–5 (2016)
14. Alyoubi, W.L., Abulkhair, M.F., Shalash, W.M.: Diabetic retinopathy fundus image classification and lesions localization system using deep learning. Sensors **21**, 3704 (2021)
15. Bodapati, J.D., Shaik, N.S., Naralasetti, V.: Composite deep neural network with gated-attention mechanism for diabetic retinopathy severity classification. J. Ambient Intell. Human. Comput. **12**, 9825–9839 (2021)
16. Vijayan, M. S. V.: A regression-based approach to diabetic retinopathy diagnosis using efficientnet. Diagnostics 13(4), 774 (2023)
17. Farooq, M., et al.: Untangling computer-aided diagnostic system for screening diabetic retinopathy based on deep learning techniques. Sensors 22(5), 1803 (2022)
18. Dekhil, O., Naglah, A., Shaban, M., Ghazal, M., Taher, F., Elbaz, A.: Deep learning based method for computer aided diagnosis of diabetic retinopathy. In: IEEE international conference on imaging systems and techniques (IST), Abu Dhabi, United Arab Emirates, pp. 1–4 (2019)
19. AbdelMaksoud, E., Barakat, S.I., Elmogy, M.: A computer-aided diagnosis system for detecting various diabetic retinopathy grades based on a hybrid deep learning technique. Med. Biol. Eng. Comput. **60**, 2015–2038 (2022)
20. APTOS 2019 Blindness Detection (2019). https://www.kaggle.com/c/aptos2019-blindness-detection/data
21. Sarobin, R. Y., Panjanathan, M.V.R., Jasmine, R., Anbarasi, S.G.: Diabetic retinopathy classification using CNN and hybrid deep convolutional neural networks. Symmetry 14, 1932 (2022)
22. Masood, S., Luthra, T., Sundriyal, H., Ahmed, M.: Identification of diabetic retinopathy in eye images using transfer learning. In: Proceedings of the 2017 international conference on computing, communication and automation (ICCCA), Greater Noida, Indian, pp. 1183–1187 (2017)
23. Tan, M., Le, Q.V,: EfficientNet: rethinking model scaling for convolutional neural networks. In: Proceedings of the 36th international conference on machine learning, ICML 2019, Long Beach, pp. 6105–6114 (2019)
24. Shaila, S.G., Lavanya, S., Rajesh, T.M., Bhuvana, D.S., Deshpande, K.: Early detection of diabetic retinopathy using multimodal approach. In: Computer Vision and Robotics (CVR), Springer, Singapore, pp. 107–118 (2022)
25. Kalyani, G., Janakiramaiah, B., Karuna, A. et al.: Diabetic retinopathy detection and classification using capsule networks. Complex Intell. Syst. (2021)
26. Saeed, M. H., Aboalsamh, H. A.: Automatic diabetic retinopathy diagnosis using adaptive fine-tuned convolutional neural network. In: IEEE Access, vol. 9, pp. 41344–41359 (2021)

Multi-Model Sarcasm Classification for Online Text: A Comparative Study

Umesh Gupta, Manish Agarwal, Pranay Kumar, Ujjwal Yadav, and Aditya Bhardwaj

Abstract In NLP, identifying sarcasm in text poses a significant challenge. Accurate determination between sarcastic or non-sarcastic statements is vital for precise results in sentiment analysis and opinion mining, among many other NLP applications. This study compares several models designed to effectively identify sarcastic versus non-sarcastic content to analyze this complex issue. We evaluate traditional approaches like logistic regression, Support Vector Machine, and deep learning models like CNNs and RNNs. Furthermore, utilizing a sizable dataset of annotated sarcastic versus non-sarcastic speeches trains deep learning methods, including transformers such as BERT and GPT3. Our research evaluates the efficacy of various feature extraction and pre-processing approaches on our model's performance. The paper deduces their strengths and weaknesses by analyzing different models' precision, recall, accuracy, and F1-score metrics.

Keywords Sarcastic text detection · Sentiment analysis models · Text classification techniques · Sarcasm detection performance · Sarcasm classification algorithms

Nomenclatures

D1	Corpus—V2 Dataset
D2	iSarcasm Dataset
D3	News Headline Dataset
D4	Sarcasm-On-Reddit Dataset
D5	Combined Dataset
M6	Logistic Regression
M5	Support Vector Machine
M4	CNN

U. Gupta (✉) · M. Agarwal · P. Kumar · U. Yadav · A. Bhardwaj
SCSET, Bennett University, Greater Noida, UP, India
e-mail: er.umeshgupta@gmail.com

B. Soni et al. (eds.), *Beyond Artificial Intelligence*, Lecture Notes in Networks and Systems 1326, https://doi.org/10.1007/978-981-96-4170-3_27

M3 RNN
M2 Bidirectional Encoder Representations from Transformers
M1 Generative pre-trained transformer
HM Harmoinic mean

1 Introduction

The popularity of social media, online forums, and review websites has significantly contributed to generating vast textual data. Understanding and analyzing this data is crucial for multiple applications such as sentiment analysis, chatbot development, content moderation, and opinion mining. Among various complex tasks associated with text analytics, detecting sarcasm remains arduous due to its ironic nature, where words portray the opposite intended meaning [1]. One often utilizes sarcasm to express mirth, censure, or disdain. Precisely detecting sarcastic intent in written communication is pivotal for accurately interpreting the message. Detecting sarcasm in the text is complex, as it hinges on many contextual factors and subtle linguistic clues that automated programs have difficulty interpreting. Moreover, sarcasm can exhibit marked variations across cultural and linguistic groups, posing additional challenges for modern natural language processing (NLP) tools. As large-scale textual data analysis attains greater prominence across various domains, developing accurate and effective models of sarcasm detection has become an increasingly pressing need [2].

1.1 Challenges Faced

Sarcasm detection has become challenging due to several problems and difficulties. Datasets are crucial for building a model for sarcasm recognition since, in the event of any discrepancy in the dataset, it could not be evident what sarcastic phrases typically consist of [3]; for example, when hashtags are used in the Twitter dataset, this ambiguity can be clarified but is quite challenging to overcome. On the other hand, hashtags are present in the news headlines dataset. Here are datasets for tweets and news headlines, [4] but they are devoid of context, which is crucial to comprehending remarks and helps detect sarcasm. Another factor that makes it challenging to recognize sarcasm is short and loud material. Multiple languages and phrases convey emotions across all platforms and do not follow one language or any term from dictionaries. Sarcasm can be hard to discern because the language is not limited to one language or a set of grammatical rules. In rule-based techniques, Twitter's hashtags are an erratic, unreliable method of sarcasm detection. Due to the time-consuming manual rules, such procedures can only be utilized with specific data types or environments. ML techniques, on the other hand, go well with text. However, it becomes difficult to detect sarcasm when considering traits such as speaker tone, body language, facial expressions, etc. Even though models like BERT have helped to improve the accuracy for detecting sarcasm, some unexpected phrases come up from writeups that are difficult to comprehend, mainly if terms are intimately associated with the preceding information. A transformer-based model is one of the more recent models that heavily relies on contextual information. Still, it lacks the option to specify whether the context should cover the entirety of the dialogue or just specific passages. Real-time data analysis of sarcasm detection has only received little investigation. One of the obstacles to sarcasm detection is the unbalanced, condensed, and distorted dataset. The speaker's background might be key when determining a statement's category.

1.2 Problem Statement

With increasing dependence on natural language processing (NLP) systems for handling large volumes of textual data, there is an urgent need to develop precise models for detecting sarcasm in text accurately. As sarcasm detection relies heavily on its surrounding context, linguistic nuances, and tone, it poses a considerable challenge for automated systems. Furthermore, the cultural distinctions in sarcasm usage across different languages and groups make this task more daunting.

1.3 Objective and Scope

Examining the validity of multiple algorithms for differentiating sardonic from sincere content is the central objective of this academic report. To accomplish this mission statement effectively, several focal points must be addressed. Achieving new skills may present formidable challenges; perseverance and consistent effort will secure success. One aspect being analyzed about conventional ML models such as Logistic Regression or SVMs is that it provides insight into their underlying concepts, specific adaptation concerning implementation techniques, and their adherence rate regarding distinguishing sardonic content authenticity. Another area that requires analysis would be scrutinizing deep learning model architectures - specifically, CNNs and RNNs participating in evaluating performances using LSTM. This study delves into the analysis of modern transformer-based models, comprising BERT, GPT, and their variations, focusing on their architecture, pre-training approaches, and optimization techniques for detecting sarcasm. The primary motive is to identify the most effective language pre-processing methods in sarcasm detection tasks while examining how each method can change or impact the model's performance and methods further comprising tokenization, stemming, and stop word removal. In addition, higher-level methods of various feature extraction techniques like a TF-IDF, bag of words, and word embeddings are also evaluated to identify the extent to which they can extract sarcasm-associated features and how they affect model performance.

2 Literature Review

2.1 Background and Motivation

The study of emotion behind a given context, detection using AI, was started by David and Noah [5]. The paper was published in 2005, proposing a unique approach to detecting emotions in the text by analyzing a given statement's contextual polarity and focus, thus starting the search for a possibility of detecting emotions. From that, the case study and our learning have grown a lot. In 2017, Pushpak et al. [1] discussed automatic sarcasm detection. They discussed different issues, emerging trends, and approaches faced in sarcasm detection. They reached three milestones in their research: semi-supervised pattern extraction to identify implicit sentiment, use of hashtag-based supervision, and incorporation of context beyond target text. In 2018, Potts [6] compared Bayesian and dense embedding approaches for detecting textual sarcasm. Their study uses the SARC dataset made from Reddit comments. They conclude that combining these representations with a bidirectional RNN improves the model's performance, with the Bayesian approach being most effective in homogeneous contexts. The dense embedding approach is valuable in more diverse ones. They obtained a result where the F1-score for textual data was 83.44%, and the F1-score for images was 80.18%. In 2019, Pradeesh, Andrew et al. [7] experimented on three datasets to detect sarcastic comments on social media platforms using transformer-based models, specifically BERT. They achieved astonishing results better than other models then, with a mind-blowing 15.2% improvement over the Reddit dataset and a 4% improvement on the Tweets dataset in the field of sarcasm detection. Later, Ashi and Gitanjali [8] created a model using supervised machine learning to detect sarcastic tones in memes or on social media; they subsequently named it Sarc-M. The model utilizes semantic features and pragmatic and lexical characteristics to analyze contextual cues (frequency of punctuation and sentiment words). Using a Multi-Layer perceptron, the best sarcasm prediction model for typograph 20ic memes achieved an accuracy of approximately 88%. In 2020, Suyash Sangwan, Md Shad Akhtar, and all [9] formed a model for multi-modal sarcasm detection that utilizes textual and visual information. This method is based on RNN, and experimental results prove that incorporating visual modalities increases performance. Sungheetha and Rajesh [10] proposed a Transfer Capsule Network model for aspect-level sentiment classification of online reviews. The model uses transfer learning on document-level data to classify the text on aspect-level sentiment. In the year 2021, Nalini Chintalapudi, Gopi Battineni, et al. [11] analyzed the tweets sent during the period of COVID-19 lockdown that happened in India with the help of different models, including LR, SVM, LSTM, and BERT being the most important.

MD Saifulla, Alfian, et al.'s [12] study combined contextual handcrafted features with DL-extracted features to detect tweet sarcasm. Logistic regression is identified as the best classification algorithm. In 2022, Pandey and Singh proposed [13] a BERT-LSTM model for sarcasm detection in codemixed language. Using a hybrid neural network, Misra and Arora [4] provided a model to detect sarcasm in new headlines. Sharma and Singh [14] proposed a complex strategy for sarcasm detection on social media platforms that integrates Word2Vec, GloVe, and BERT models with fuzzy evolutionary logic for text embedding. Muaad and Davanagere [15] created an AI-based method for detecting sarcasm in Arabic texts. Recently, Ding and Tian [16] proposed a fusion-based multi-modal fusion method for sarcasm detection. In addition, many more people [17] investigated different neural approaches, such as LSTM, GRU, and CNN, to detect sarcasm in social media.

2.2 *Models Used*

Logistic Regression. A categorical dependent variable is analyzed to one or more independent variables using the statistical procedure known as logistic regression.

$$\mathrm{Logit(S)} = \mathrm{b_0} + \mathrm{b_1M_1} + \mathrm{b_2M_2} + \mathrm{b_3M_3} \ldots \mathrm{b_kM_k} \tag{1}$$

where

S is the probability of the presence of features of interest; $\mathrm{M_1}$, $\mathrm{M_2}$, ... $\mathrm{M_k}$ is the Predictor value and $\mathrm{b_0}$, $\mathrm{b_1}$, ... $\mathrm{b_x}$ is the intercept of the model.

Support Vector Machine (SVM). A supervised machine-learning method that can be used for classification and regression. The most important aspect of the SVM method is remembering how to create an N-dimensional hyperplane, where N is the number of input features. However, it is computationally expensive and may suffer from overfitting when the data has a lot of noise.

Long Short-Term Memory (LSTM) And Relational Neural Network (RNN). Relational Neural Network is an old algorithm developed in the 1980s, but its potential has recently been realized. It displays behaviour that resembles how the human brain works. With sequential data, it functions nicely. The hidden layer of RNN contains an internal memory that recalls the past input and uses both the current and prior inputs to make decisions, which raises the importance of RNN. However, it needs a vast amount of data to generalize it. Hence, this leads to being more computationally expensive than others. Unlike a standard RNN unit, an LSTM unit keeps a memory cell Ct at time t. The output ht of an LSTM unit is determined by using the calculations below:

1. $f_t = \sigma\,(W_f.\,[h_{t-1}, x_t] + b_f)$
2. $i_t = \sigma\,(W_i.\,[h_{t-1}, xt] + b_i)$
3. $C_t = tanh\,(W_c.\,[h_{t-1}, {}_t] + b_c$
4. $o_t = \sigma\,(W_o.\,[h_{t-1}, x_t] + b_o)$

5. $h_t = h_t * \tanh(c_t)$
6. $C_t = \sigma(f_t \times C_{t-1} + i_i \times C_t)$

The Hyperparameters analysis: In LSTM, all hyperparameters are activation = SoftMax, loss = categorical_crossentropy, optimizer = Adam, and metrics = accuracy.

Convolutional Neural Network (CNN): CNN can include any number of many, even hundreds of layers, where any of the layers can be trained to recognize different facets of a picture. CNN records all significant details. It takes the contextual local properties of a sentence and extracts them into a global feature vector by performing several convolutional calculations. By effectively modeling local dependencies in text, it can handle larger input dimensions and is robust to variations in word order. On the contrary, it has a limitation of capturing long-range dependencies, which requires more data to avoid overfitting (Fig. 1).

Bidirectional Encoder Representations from Transformers (BERT): Google created this model for natural language processing applications like answering questions, sentiment analysis, and language translation. BERT uses a text-processing architecture known as a transformer that can read text both ways and capture contextual links between words and phrases [22], [26]. In figure e_1,…., e_n are the input vectors of the BERT model. $T_{1\ldots}$., T_n are the output vectors of the BERT model (Fig. 2).

Generative Pretrained Transformer (GPT): Chat GPT, a well-known AI tool, is trained over the GPT model (deep learning model), a big language model developed by Open AI with its architecture based on GPT-3.5. However, it is resource-intensive, requires powerful hardware for training and inference, and may be overfitting on smaller datasets.

3 Methodology

3.1 Preprocessing and Features

All the datasets contain domain-specific and informal language. Hence, to handle this issue, the following pre-processing techniques have also been implemented:

1. Lowercasing: To reduce case sensitivity and ensure consistency across the dataset.
2. Stop word Removal: To remove generic words that do not contribute to sarcasm detection.
3. Tokenization: To disintegrate larger writings into smaller word chunks or tokens, allowing for further analysis.
4. Stemming: To recreate words to their original forms, allowing for better generalization and reducing vocabulary size.

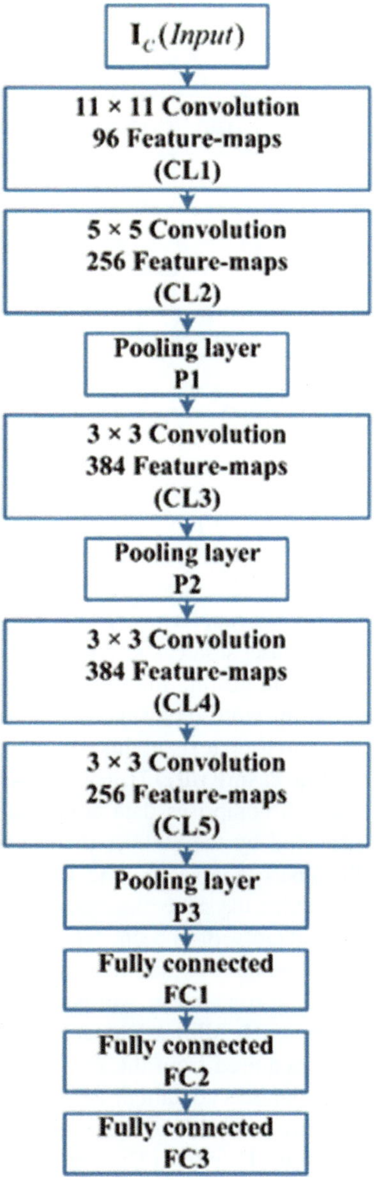

Fig. 1 CNN architecture

3.2 Dataset Description

Sarcasm Corpus V2 [18]**.** This dataset consists of annotated posts related to the Internet Argument Corpus. The dataset consists of rhetorical questions, hyperbole, and general sarcasm. The dataset contains 6,520 posts for general sarcasm (3,260

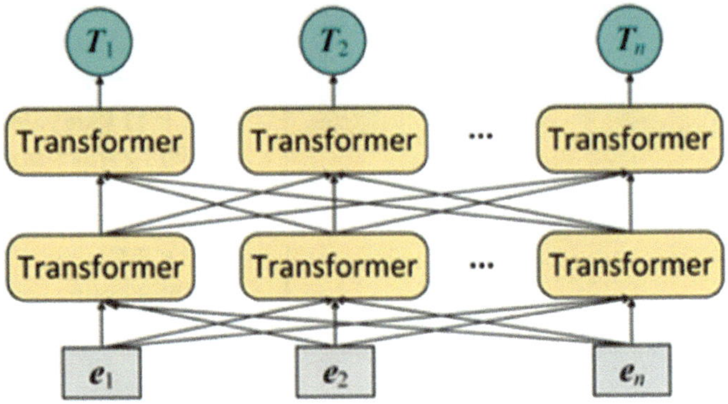

Fig. 2 Network structure of BERT model

sarcastic and 3,260 non-sarcastic), 1,164 for hyperbole (582 sarcastic and 582 non-sarcastic), and 1,702 for rhetorical questions (851 sarcastic and 851 non-sarcastic). Each entry includes the corpus type, class label (sarcastic or not), unique ID, and the post's text. The Sarcasm Corpus V2 dataset contains a limited and domain-specific vocabulary. Bag-of-words can effectively capture the unique features of this dataset's sarcastic and non-sarcastic text. This technique represents texts as vectors indicating the frequency of each word in the vocabulary. By doing so, the group of word representations captures the unique characteristics of sarcastic and non-sarcastic texts in this dataset.

ISarcasmEval [17]**.** This dataset is designed to detect intended sarcasm in English texts. It comprises data collected using a new method where sarcasm labels are provided by the authors themselves, eliminating the need for predefined tags or third-party annotators. The dataset contains sarcastic and non-sarcastic texts, phrases conveying the same message the authors want to communicate, and labels specifying the category of speech, which can be sarcastic, ironic, satirical, understatement, overstatement, or rhetorical question. The iSarcasmEval dataset features diverse texts that may contain rare words with strong sarcasm indications.

News Headlines-Dataset [19, 20]. This dataset consists of 28,619 news headlines acquired from the sites. The Onion (sarcastic) and HuffPost (non-sarcastic). Both datasets contain three attributes: sarcastic (1 for sarcastic, 0 for non-sarcastic), headline, and link to the original news article. With concise and precise headlines, the News Headlines Dataset for Sarcasm Detection may contain fewer meaningful words. To address this, BOW is used. Bag-of-words (BOW): Like the Sarcasm Corpus V2, using Bag-of-words for this dataset helps emphasize the importance of unique and meaningful words for sarcasm detection by representing texts as vectors with word frequency.

Sarcasm on Reddit-Dataset [21]**.** It includes a million sarcastic comments acquired from Reddit and already labeled sarcasm tags. The dataset is available in two versions, namely balanced and imbalanced, with the balanced version containing equal numbers of sarcastic and non-sarcastic comments. Each dataset underwent pre-processing, eliminating irrelevant fields and retaining only the text and sarcasm label. Subsequently, standard text, including tokenization, pre-processing techniques, lowercasing, removing special characters, punctuation, and stop words. The Sarcasm on Reddit dataset contains many texts with diverse vocabulary and complex semantic relationships. Word embedding captures the contextual information necessary for accurate sarcasm detection. Word Embeddings: To capture sarcasm detection, contextual information is needed to represent the words in a contiguous vector form. Techniques such as Word2Vec, GloVe, and Fast Text can generate these embeddings (Fig. 3).

4 Model Evaluation

This section provides the data for the models chosen to detect sarcasm. The evaluation is based on recall, precision, F1-score, accuracy, and values retrieved by evaluating the models. The result statistics are compared with the model's performance with the help of a comparison table. Model evaluation can be done by evaluating their performances as mentioned below:

- Precision (P): true positive predictions divided by the sum of all positive values.
- Accuracy (A): correctly predicted values divided by the total number of values.
- F1-score (F): HM of precision and recall.
- Recall (R): true positive predictions divided by the sum of true positives and false negatives values.

(From Figs. 4, 5 and 6) The GPT and Bert models perform better on different datasets, and the F1-score is calculated using the above data. (From Fig. 7) The transformer-based models GPT and BERT outperform other models on the combined dataset, having an F1-score of 87% and 85%, respectively. The logistic regression model has the lowest F1-score of 73%. The results confirm that transformer-based models are the best choice for detecting sarcasm, with the GPT model performing best across all datasets. The choice of datasets also plays a crucial role in achieving reliable performance. Training models with the Sarcasm Corpus V2 dataset give better results (Table 1).

Even after GPT and Bert are the best training models today, traditional models like RNN perform well even after being in the shadows. As observed on various real-world datasets, the RNN model gives tough competition to its modern competitors, thus emerging as a rising star. The stats show that RNN can be used instead of the other models.

Fig. 3 Workflow diagram

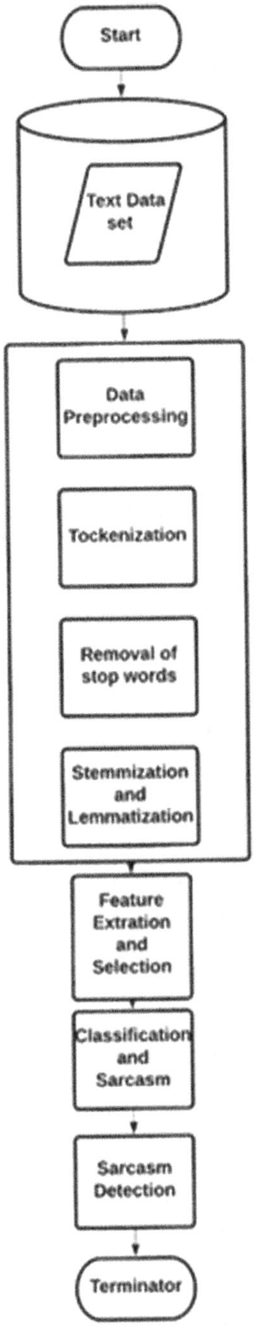

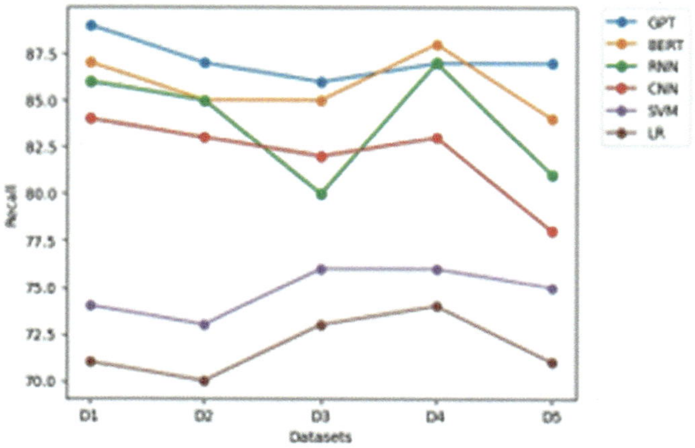

Fig. 4 Recall

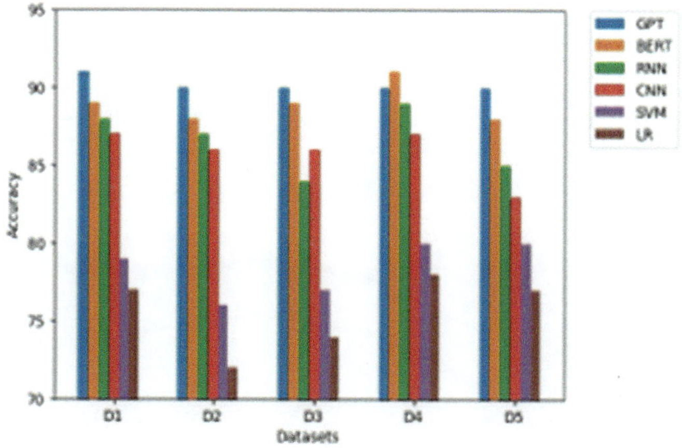

Fig. 5 Accuracy

5 Results and Discussion

5.1 Feature and Extraction Methods and Pre-Processing Techniques

Any model's performance can be improved significantly by implementing pre-processing techniques, like tokenization, stemming, or stop word removal. Feature extraction methods, like word embeddings and term frequency–inverse document frequency, are essential to capture relationships between semantically related words, further enhancing model performance. Different models are more sensitive to feature

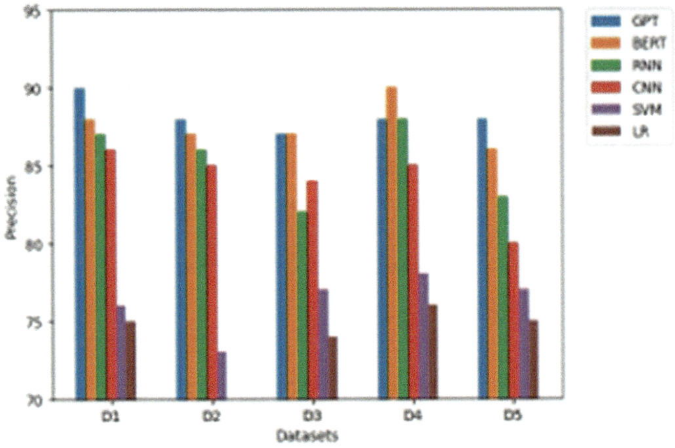

Fig. 6 Precision

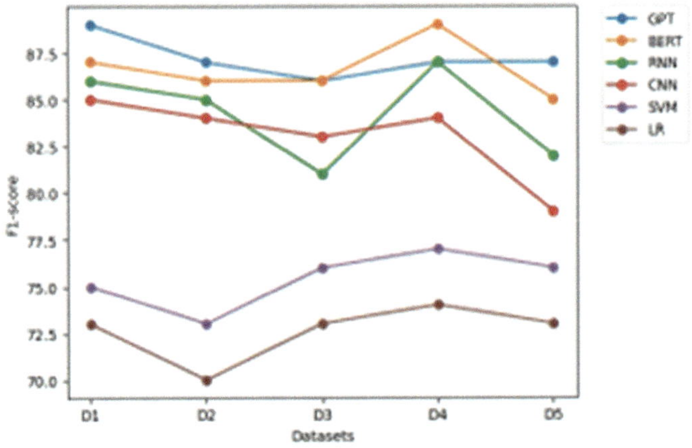

Fig. 7 F1-score

extraction methods; for example, LSTM and GPT models perform better with word embeddings.

5.2 *Trade-Offs Between Model Complexity, Computational Resources, and Performance*

Our analysis reveals that as model complexity increases, so does performance. However, this improvement comes at the cost of increased computational resources and time. For instance, BERT and GPT outperform all other models but require

Table 1 Model evaluation matrix

Datasets	Method 1	Method 2	Method 3	Method 4	Method 5	Method 6
	A	A	A	A	A	A
	P	P	P	P	P	P
	R	R	R	R	R	R
	F	F	F	F	F	F
D1	91	89	88	87	79	77
	90	88	87	86	76	75
	89	87	86	84	74	71
	89	87	86	85	75	73
D2	90	88	87	86	76	72
	88	87	86	85	73	70
	87	85	85	83	73	70
	87	86	85	84	73	70
D3	90	89	84	86	77	74
	87	87	82	84	77	74
	86	85	80	82	76	73
	86	86	81	83	76	73
D4	90	91	89	87	80	78
	88	90	88	85	78	76
	87	88	87	83	76	74
	87	89	87	84	77	74
D5	90	88	85	83	80	77
	88	86	83	80	77	75
	87	84	81	78	75	71
	87	85	82	79	76	73
AVG	**90.2**	89	86.6	85.8	78.4	75.6
	88.2	87.6	85.2	84	76.2	74
	87.2	85.8	83.8	82	74.8	71.8
	87.2	86.6	84.2	83	75.4	72.6

substantial computational power. However, Logistic Regression, SVMs, and CNNs provide competitive performance with much lower resource requirements. Therefore, the comparison between model complexity, computational resources, and performance must be cautiously chosen to develop the model needed. Most importantly, while complex models like BERT and GPT achieve high performance, they may not always be the best choice, especially in situations with limited computational resources or time constraints. Simpler models like Logistic Regression, SVMs, and CNNs can still yield satisfactory results and may be more suitable for specific applications. Hence, the choice of a model for sarcasm detection is expected to be based on the project's requirements, the resources available, and the desired performance level. By considering these factors and evaluating different models, researchers and practitioners can identify the best applicable approach for further study in the field of sarcasm detection.

6 Conclusion and Future Work

Evaluating the models based on Recall (Fig. 2), Accuracy (Fig. 3), Precision (Fig. 4), and F1-score (Fig. 5) for sarcasm detection, the results indicate that the transformer-based models, GPT and BERT, outperformed traditional ML and DL models. GPT has the best average accuracy of 0.90, and BERT has the second-best at 0.88. Models like RNN and CNN can still give them tough competition on real-world datasets. Despite these impressive results, models can still be improved mainly by incorporating context-aware features, such as semantic similarities and sentiment analysis, to capture the nuances of sarcasm better and utilize multi-modal data (text or graphics) to improve sarcasm detection. Additionally, there is a need for further investigation of how model performance varies by applying different pre-processing techniques and feature extraction methods. Fine-tuning the hyperparameters of the models could also potentially improve their performance. For prospects, the focus should be to incorporate different domains and languages to develop more robust models. Furthermore, research should explore the ethical implications of sarcasm detection models, such as the potential for biased or discriminatory outcomes. Addressing these issues is crucial for checking social media and public speaking. Hence, by incorporating context-aware features and utilizing multi-modal data, the performance of these models can be further improved.

References

1. Joshi, A., Bhattacharyya, P., Carman, M.J.: Automatic sarcasm detection: a survey. ACM Comput. Surv. (CSUR) **50**(5), 1–22 (2017)
2. Lichtenstein, M., Rucks-Ahidiana, Z.: Contextual text coding: a mixed-methods approach for large-scale textual data. Sociol. Methods Res. **52**(2), 606–641 (2023)
3. Li, J., Pan, H., Lin, Z., Fu, P., Wang, W.: Sarcasm detection with commonsense knowledge. IEEE/ACM Trans. Audio, Speech Lang. Process. **29**, 3192–3201 (2021)
4. Misra, R., Arora, P.: Sarcasm detection using news headlines dataset. AI Open **4**, 13–18 (2023)
5. Bamman, D., Smith, N.: Contextualized sarcasm detection on Twitter. In: Proceedings of the International AAAI Conference on Web and Social Media, vol. 9, no. 1, pp. 574–577 (2015)
6. Kolchinski, Y.A., Potts, C.: Representing social media users for sarcasm detection. arXiv preprint arXiv:1808.08470 (2018)
7. Parameswaran, P., Trotman, A., Liesaputra, V., Eyers, D.: Detecting target of sarcasm using ensemble methods. In: Proceedings of the the 17th Annual Workshop of the Australasian Language Technology Association, pp. 197–203 (2019)

8. Kumar, A., Garg, G.: Sarc-m: sarcasm detection in typo-graphic memes. In: International Conference on Advances in Engineering Science Management & Technology (ICAESMT)-2019, Uttaranchal University, Dehradun, India (2019)
9. Sangwan, S., Akhtar, M.S., Behera, P., Ekbal, A.: I didn't mean what I wrote! exploring multimodality for sarcasm detection. In: 2020 International Joint Conference on Neural Networks (IJCNN). IEEE, pp. 1–8 (2020)
10. Sungheetha, D.A.: Transcapsule model for sentiment classification. J. Artif. Intell. Capsul. Netw. **2**(3), 163–169 (2020)
11. Chintalapudi, N., Battineni, G., Amenta, F.: Sentimental analysis of COVID-19 tweets using deep learning models. Infect. Dis. Rep. **13**(2), 329–339 (2021)
12. Razali, M.S., Halin, A.A., Ye, L., Doraisamy, S., Norowi, N.M.: Sarcasm detection using deep learning with contextual features. IEEE Access **9**, 68609–68618 (2021)
13. Pandey, R., Singh, J.P.: BERT-LSTM model for sarcasm detection in code-mixed social media post. J. Intell. Inf. Syst. **60**(1), 235–254 (2023)
14. Sharma, D.K., Singh, B., Agarwal, S., Pachauri, N., Alhussan, A.A., Abdallah, H.A.: Sarcasm detection over social media platforms using hybrid ensemble model with fuzzy logic. Electronics **12**(4), 937 (2023)
15. Muaad, A.Y., Jayappa Davanagere, H., Benifa, J.V., Alabrah, A., Naji Saif, M.A., Pushpa, D., Alfakih, T.M. et al.: Artificial intelligence-based approach for misogyny and sarcasm detection from Arabic texts. Comput. Intell. Neurosci. (2022)
16. Ding, N., Tian, S.W., Yu, L.: A multimodal fusion method for sarcasm detection based on late fusion. Multimed. Tools Appl. **81**(6), 8597–8616 (2022)
17. Oraby, S., Harrison, V., Reed, L., Hernandez, E., Riloff, E., Walker, M.: Creating and characterizing a diverse corpus of sarcasm in dialogue. arXiv preprint arXiv:1709.05404 (2017)
18. Isarcasm datasets: Available: https://sites.google.com/view/semeval2022-isarcasmeval/home (2023)
19. Misra, R., Grover, J.: Sculpting data for ml: The first act of machine learning, p. 158. University of California San Diego, La Jolla, CA, USA (2021)
20. Khodak, M., Saunshi, N., Vodrahalli, K.: A large self-annotated corpus for sarcasm. arXiv preprint arXiv:1704.05579 (2017)
21. Gupta, U., Gupta, D.: Least squares structural twin bounded support vector machine on class scatter. Appl. Intell. **53**(12), 15321–15351 (2023)
22. Kumar, S., Gupta, S.K., Kaur, M., Gupta, U.: VI-NET: A hybrid deep convolutional neural network using VGG and inception V3 model for copy-move forgery classification. J. Vis. Commun. Image Represent. **89**, 103644 (2022)

Moving Object Detection and Segmentation for Various Surveillance Application Using Deep Learning Approaches in Explainable Artificial Intelligence

Dipika Gupta, Kashish Bhatia, and Manish Kumar

Abstract Moving object detection and segmentation in the field of computer vision is gaining much interest amongst researchers for its wide variety of application. Moving object detection aims at captivating the moving objects in a video sequence whereas moving object segmentation focuses on segmenting the objects in motion from a stationary background in a video sequence. This paper introduces both these computer vision tasks in detail along with its various applications and challenges. A comprehensive literature review in this field is provided in the next section. Further, the various deep learning approaches for moving object detection and segmentation are detailed with their advantages and disadvantages. The various available dataset for surveillance applications along with their properties are elaborated. Next section explains the numerous evaluation metrics that are used to evaluate the performance of the model. The qualitative results are detailed along with graphs and figures. At last, the study is concluded with future research directions in this field.

Keywords Moving object detection · Convolutional neural networks · Deep learning · Mask-track

1 Introduction

Moving object detection is the task of computer vision that focus on capturing moving objects from a sequence of frames in a video which are compared using various methods. It finds application in various domains such as traffic monitoring, video surveillance, human action recognition and counting activity. Moving Object Segmentation (MOS) is the task of segregating the objects in motion from the uniform

D. Gupta (✉) · K. Bhatia · M. Kumar
Computer Science Engineering, Punjab Engineering College, Chandigarh, India
e-mail: dipikagupta.phdcse20@pec.edu.in

M. Kumar
e-mail: manishkumar@pec.edu.in

B. Soni et al. (eds.), *Beyond Artificial Intelligence*, Lecture Notes in Networks and Systems 1326, https://doi.org/10.1007/978-981-96-4170-3_28

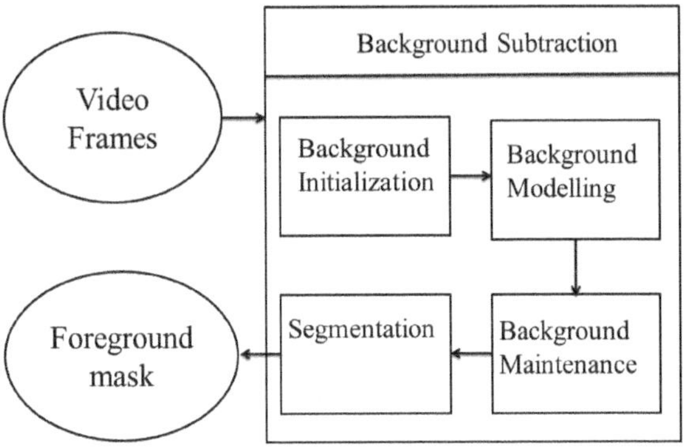

Fig. 1 Moving object detection framework for various surveillance applications

background in a sequence of frames [1]. The basic proposal to this process is background subtraction that has been studied over previous two decades. Background subtraction is an extensively used approach to locate objects in motion in a sequence of frames from static cameras. It is the procedure of loading and upgrading a prototype of the steady setting called the background model, and differentiating it with the input image. Moving objects are defined by the regions or pixels with major differences.

The process of background subtraction is mainly divided into following parts: (i) Background initialization: this process involves capturing the first background image. (ii) Background modelling: the background is described by this model established on N video frames and could be presented in numerous ways. It behaves as a comparison model for coming video frames. (iii) Background maintenance: consists of updating the model to adjust to the development over time. (iv) Segmentation: the classification amongst the class "background" and the class "moving objects" is done in this step. Figure 1 depicts the moving object detection framework in a pictorial format.

Lately, due to large quantity of video data available, it is crucial to capture only important information. Due to redundant data, the results obtained are not precise and accurate. Therefore, there is a need to employ a technique where we can capture only the pertinent information such as vehicles, person, etc. from the video. The field of video surveillance is capturing much attention in recent times, although it suffers from numerous problems such as obstruction in frame sequences, the noise arising from imbalanced lightning, and, misclassification due to illumination error amongst the present frame and the reference frame. These problems arise in outside scenarios. Generally, because of the gap between the object and the camera, the quality of video is low developing vulnerability to lightening variations. As a consequence, the background becomes changing. Hence, the background segment of the present frame does not intersect with the background section of the reference frame. Therefore, the problem arises where the approaching image does not match pixel wise with the background. Numerous moving detection techniques give poor results in real time as most of the methods are enforced during daytime because of the lightening conditions. Moreover, if the colour of the object in motion and the reference frame is same, the algorithm is not capable of correctly classifying the object. To solve such issues there arise a need to develop a strong algorithm that provides pertinent video surveillance results.

2 Literature Review

Explainable Artificial Intelligence (XAI) is a combination of approaches that allow person to believe the results given by machine learning algorithms. It is utilized to create artificial intelligence model, its influence and its possible intolerance. XAI finds its adverse application in moving object detection and segmentation for detecting and segmenting autonomous vehicles. Ammar et al. [2] proposed deep detector classifier for MOS using Deep Sphere [3] and generative adversarial networks. For foreground segmentation, the Deep Sphere identifies the irregular pattern in temporal and spatial settings. The further refining is done by using some well-formed operations to segment the preferred object. At last, the generative models are used for classification of the segmented object by considering the issue of semi-supervised learning as the task of missing data suggestion.

Zhao et al. [4] proposes advanced coding features for MOS from High Efficiency Video Coding (HEVC) compressed surveillance video. A real-time method is used to directly extract the unique features from HEVC compressed area. Initially, preprocessing is done using Motion Vector (MV) intercalation for within coded Prediction Unit (PU) and the elimination of MV deviancy. Next, the foreground area is created by clustering the batch with non-zero MVs with the use of four-connectivity labelling algorithm. Another step involves the elimination of noise area by applying object region tracking to the attached foreground area. At last, the classification of moving objects is done amongst either person or vehicle by using a person-vehicle classification model.

Murugesan et al. [5] introduces a background subtraction technique based on Maximally Stable Extremal Region (MSER) extraction method by means of Multilayer Perceptron Recurrent Neural Network (MLP-RNN) that differentiates between various objects of different sizes. In this work, an objective edge and reference are taken as information and the output is segmentation model of identical spatial target which are all element not present in the reference layout. Basically, the training of the MLP-RNN is done based on the retrieved features, and then it is further utilized to search for irregularity in the surveillance videos.

Pramanik et al. [6] proposed a methodology for video surveillance-based models for enhancing safety of roads. A group of algorithms are proposed which have the capacity to predict different traffic pre-empts based on traffic videos including illegal parking, breaking of traffic signals, over speeding, violating one-sided congestion, and incorrect drop off site of traveller. Once the algorithms have detected the pre-empts, an alarm will be activated in the control room which will act as a preventive measure to ensure and prevent any accident on the road, henceforth enhancing the road safety. The framework proposed by [6] is able to handle five anomalies at a time using five distinguished algorithms, whereas the prior systems were only able to handle one or two pre-events.

Kalli et al. [7] developed a self-activating moving object detection model on the basis of supervised learning and computative characteristics of videos. A new background modelling technique is introduced using biased illumination filed fuzzy C-means algorithm to capture objects in motion precisely. Background subtraction is used to distinguish immobile pixels from mobile pixels. Further, the biased illumination field fuzzy C-means technique utilizes grouping beneath noise and deviating illumination surroundings to enhance the segmentation accuracy.

Elhoseny et al. [8] proposed Multi-Object Detection and Tracking (MODT) using Kalman filter approach to trace the objects in motion in a frame sequence. According to the area expanding model, the video clips are transformed established on figure of frames into linguistic assignments. Following differentiating the object, Kalman filter was practiced for criterion improvement utilizing probability-based grasshopper algorithm. Based on resemblance estimate, the preferred objects were traced in every frame using optimum criterion. The MODT framework achieved competitive results in segmentation and detection of moving objects.

Thenmozhi et al. [9] proposed two techniques like combining background subtraction and upgraded sequential outline operation technique in support of identification of different motion objects through internal and external video dataset. Aforesaid technique could be used in locations including roads, railway station, airports, shopping centre and any other place where safety is of principal consideration. The moving objects is differentiated precisely from the object size and numerous other complicated scenarios. The newly introduced sequential outline operation begins through background subtraction and foreground identification for movement and object findings.

Patil et al. [10] introduced an edge-to-edge aggregation network for MOS. An intrinsic association learning established Edge Extraction Mechanism (EEM) and Dense Residual Block (DRB) are developed in support of foreground representation. The foreground edge associated information is captured by the multi scale EEM block and delivered to the decoder through skip connections. Additionally, the output of the optical flow and the final EEM block are enclosed in a bridge network. Multi-scale residual module with dense connections is present in the bridge network to memorize the effectual and well-organized foreground pertinent features. At last, a decoder module along skip connections through multi-scale EEM module is introduced to create precise and consonant foreground object maps.

Mahalingam et al. [11] proposed an effective and efficient methodology for MOS as well as tracing and identification of objects in motion. The approach is partitioned amongst three phases: the detection phase, the tracking phase and the evaluation phase, where foreground segmentation and noise removal is done in the detection stage. To attain effective foreground segmentation, Mixture of Adaptive Gaussian (MoAG) block is introduced. In favour of the removal of noise existing in foreground segmented frames, a fuzzy morphological filter is executed. Tracking stage consists of a blob detection mechanism for moving object tracking. The last stage of evaluation

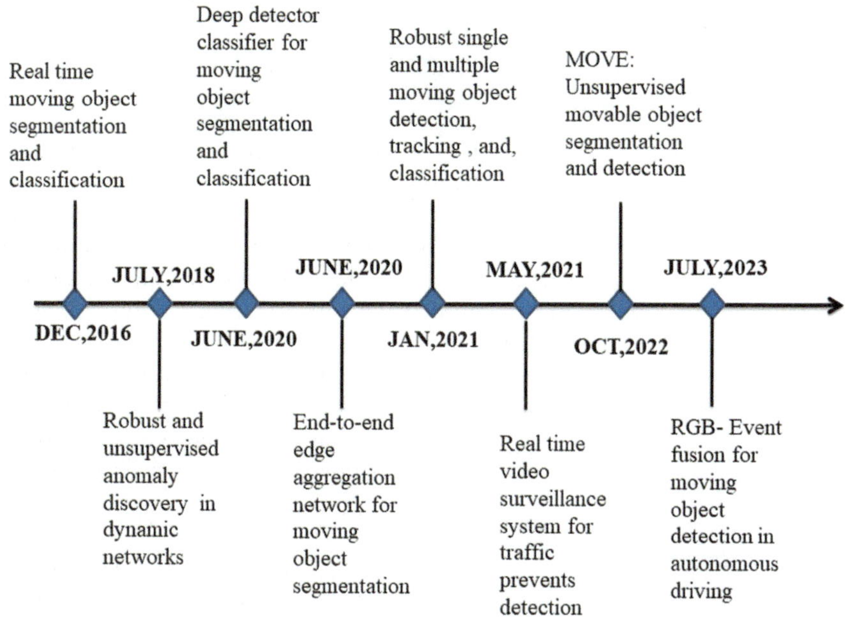

Fig. 2 Various milestones in the field of MOS and detection of objects in motion

consists of feature extraction and classification. The processed frames given for classification are used to withdraw texture build and quality build features.

Figure 2 envisions the timeline representation of various mileposts achieved in the area of MOS and detection of objects in motion for various video surveillance systems using XAI.

3 Deep Learning Approaches

MOS and detection of objects in motion approaches could be partitioned amongst two broad classifications: conventional approach and deep learning-based approach. In conventional approaches [12–16] there is no need for labelled ground truth segmentation masks. These approaches involve two stages: background modelling and classification of pixels. Initially, the background scenario is established and upgraded with time. Later, depending upon some threshold value, the pixels are segregated amongst foreground and background. Deep learning-based approaches were lately introduced for moving object detection and segmentation problem. They have been further categorized on the basis of spatial and temporal subcategories. Under spatial category, we have online fine tuning build, matching build, and, graph optimization build. Under temporal category, we have mask propagation build, optical flow build, and, long term temporal build.

Fig. 3 OSVOS depicted in three stages: (i) a pre-trained base convolutional neural network (CNN) for labelling of image (ii) the parent network is then trained on DAVIS dataset hence improving the segmentation outcome (iii) Fine tuning the network on an individual frame

3.1 Online Fine-Tuning-Based Approaches

Amongst earliest deep learning approach utilized for MOS and detection of objects in motion is online fine tuning introduced from One Shot Video Object Segmentation (OSVOS) by Caelles et al. [17]. This task is semi supervised involving the segregation of the object of interest from the background of the sequence of frames, given the mask of the initial frame. OSVOS is built upon fully convolutional neural network capable of transmitting inclusive semantic particulars, absorbed from Imagenet, to the job of foreground segmentation. At last, it grasps the aspect of the individual annotated object, therefore one shot segmentation. Despite the fact that every frame is refined separately, the final results are for the time being consistent and sturdy (Fig. 3).

Later many extensions [18, 19] and variants [20, 21] of OSVOS came giving better results than the previous ones. The purpose of extensions of OSVOS was to improve it in terms of wholesomeness and flexibility. [18] included the results of the prior frames and enhanced flexibility but numerous online fine tuning reduced the performance of moving object segmentation. [19] Utilized U-net architecture with Siamese network included into it which provided the annotation information for achieving fast segmentation. [20] Used one way forward pass to regulate the segmentation characterization to the existence of the target object. Finally, [21] proposed a focused transfer enclosed block to protect continuing semantic details.

3.2 Matching Based Approaches

The matching based algorithm takes into account the consistency between the reference frame and the target frame. The frame to be segmented is the target frame and the reference frame depends on the type of the moving object segmentation. The prompt to sustainable and robust uniformity is prejudice feature insertion. Most of these algorithms attains this by training the backbone model on frames/ frames in sequence datasets. Matching based methods are brisk than the online based methods since all the work is done offline and gives good results as well. Based on the types of features, these methods are further divided into pixel level matching and region of interest level matching.

Under pixel level matching, [22] was created using straightforward pixel level matching procedure, wherein direct matching of the features is done amongst each pair of pixels. This space time memory approach brought a revolutionary enhancement in semi supervised moving object segmentation techniques, as it considered the first frame, the last frame, and also the intermediate frame for reference to assist the changes in the target object.

Under region of interest level matching, [23] is a correlation-based technique utilizing a regional memory network for semi supervised MOS. The matching used here is local-to-local. In this approach, the regional memory is created by remembering the local area where the target object was seen in the prior frames. This approach effectively reduces the unpredictability of the similar objects as well diminishes the computing complexity.

3.3 Graph Optimization-Based Approaches

Recently, various online fine tuning established approaches and feature matching based approaches were used for moving object segmentation. Both the techniques utilized the relation between the frame to be segmented and few other frames in the frame sequence that is not adequate for the target objects undergoing some major change. To resolve this problem, graph optimization-based methods were proposed, which considers extra frames from the same video organized in a graph structure. This technique may achieve high level dependency amongst frames established on the analysis of the graph to enhance the results of segmentation. The technique is divided into two categories: track selection and information propagation.

In track selection established graph optimization methods, to generate the target object mask, the optimal node is followed all over the sequence of frame for a group of input frames. Target aware object discovery [24] was proposed for effective and exact spatio-temporal segmentation. It introduced the fusion of foreground area approximation and instance collection in one common network. It also utilizes the concept of temporal information for segmentation of every frame to effectively discover object. This technique attains better matching by creating target aware existence model for all objects.

In information propagation graph optimization based methods, the entire importance is given in transferring the information of the foreground and the background amongst frames to smoothen segmentation results. Attentive graph neural network [25] takes into consideration a subgroup of input sequence and creates an overall connected graph where nodes are represented by frames and edges are represented by the alliance between arbitrary frame pairs. In this technique, information is propagated repeatedly throughout inference to achieve universal context for every node.

3.4 Mask Propagation-Based Approaches

Mask propagation is temporal based technique for moving object detection and segmentation. This approach supposes that the target moves steadily throughout the sequence of frames. Hence, it utilizes the prior frame masks to predict the position and appearance of target object. Thus, this technique can focus more on the areas where target object appears. An explainable artificial intelligence based semi supervised moving objects segmentation technique [26] introduced mask propagation. The segmentation model is provided accompanying predicted mask of the prior frame for mask creation. Figure 4 depicts the working of Mask Regional based Convolutional Neural Network (Mask RCNN) for MOS and detection of objects in motion.

Mask RCNN is basically an extension of faster RCNN and faster RCNN is an object detection algorithm. In Mask RCNN, both the task of object detection as well as image segmentation is performed. As seen in Figure 4, in the layered architecture of Mask RCNN, the input image is passed to a pre-trained CNN, in this case its MobileNet V2. A pre-trained CNN is used so that the features can be extracted and passed to the Region Proposal Network (RPN). The output in the form of feature maps from the pre-trained CNN is given as input to the RPN. There are two convolutional blocks inside RPN. The first with 2 * 9 output channels acts as a binary classifier which will tell that whether the feature maps is having an object or not. The second convolutional block (4 * 9) acts as a regressor which is responsible for plotting a bounding box if the object is present.

There will be two inputs to the Region Of Interest (ROI) pooling layer, one will be region of interest from the RPN and other will be feature map from the pre-trained network. The task of the ROI pooling layer is to provide with an output of fixed dimension i.e. 7 * 7 * 512. After getting output from ROI pooling, we have two

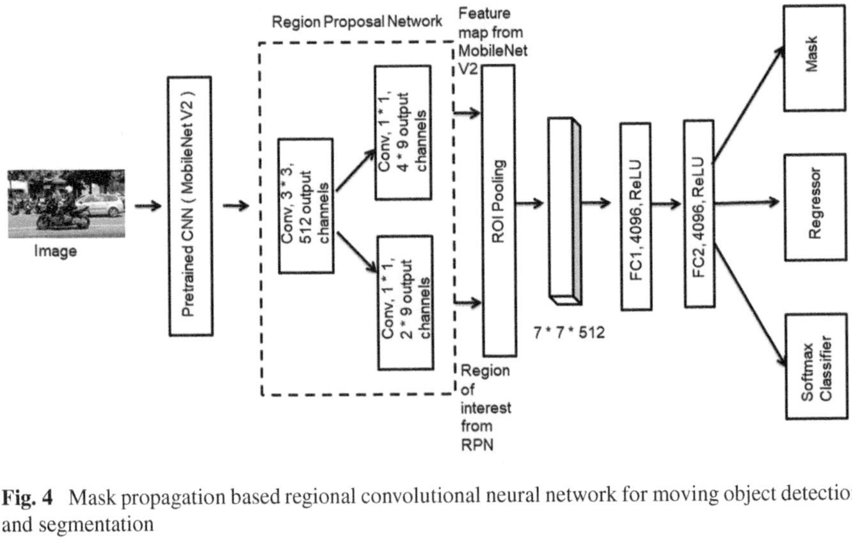

Fig. 4 Mask propagation based regional convolutional neural network for moving object detection and segmentation

fully connected layers. At last, there is a softmax classifier, a regressor, and a mask classifier. The task of softmax classifier is to tell which object is there, regressor is responsible for plotting a bounding box on it, and mask classifier is responsible for making mask on it.

3.5 Optical-Flow Based Approaches

Another temporal based explainable artificial intelligence approach is optical-flow based method for MOS and detection of objects in motion. The main consideration of optical flow based methods is pixel level movement patterns. It considers continuous sequence of frames and the notion that the target object and the background will be having different motion patterns. Hence, the amalgamation of moving object detection and segmentation with optical flow gives favourable results. FusionSeg [27] was one of the earliest optical flow based technique. To generate segmentation masks, this approach combines the movement and arrival information of the target object. It introduces two stream fully CNN to combine the movement based and arrival based results in an integrated structure. This framework can be utilised for better usage of information and to create accurate target peculiar outcome.

Later, many extensions of optical flow came into existence. Accelerating video object segmentation [28] is the most recent one for semi supervised moving object segmentation that proposes an accelerating framework which makes use of remainder of movement vectors and compact bit stream. Rather than optical flow, its extension namely motion vectors which are dense vector fields with reference to nearby frame are used. A new motion vector warping approach is introduced for inseminating segmentation mask in duplex and hop-growing manner. To prevent from incorrect motion vectors which may inseminate mistaken segmentation mask, a residual established correction module is proposed.

3.6 Long-Term Temporal-Based Approaches

The above two temporal based methods, mask propagation and optical flow concentrates on momentary association and produces position and shape priors for target objects. For situation with enormous occlusion and changing background, there is a need for better approach. Hence, long-term propagation came into being that examines spatial-temporal prompt for relatively lengthy sequence of frames. This technique examines unsteady features related with target object and background such as alteration in position, shape, formation, and scale to preferably adjust the segmentation framework to the alteration and achieve higher-level results.

The long term temporal techniques can be partitioned into four classifications: recurrent neural network, generative adversarial network, bilateral neural network and transformers. Recurrent neural network based technique [29] executes moving object segmentation by attaining bidirectional message generation on top of subgroup with the help of relational cross attention block. Generative adversarial network based technique [30] utilizes dual attention cross branch fusion to attain favourable temporal information for moving object segmentation. Transformer based approach [31] proposes Siamese framework with interactive transformer to smoothen context generation from outcome of the prior frame to present frame for semi supervised moving object segmentation. Table 1 depicts the summarization of different deep learning based approaches with their advantages and disadvantages.

Table 1 Summary of various deep learning approaches for moving object detection and segmentation

S.No	Approach name	Major studies which use them	Advantages	Disadvantages
1	Online fine-tuning	[17–21]	Fine-tunes the pre trained network, capability to encode target specific features	Not able to handle object property change and vague area, time consuming
2	Matching	[22, 23]	Executes quicker than online fine-tuning	Can anticipate robust result without online fine-tuning, giving better balance amongst efficiency and preciseness
3	Graph optimization	[24, 25]	Attains more exhaustive and high-level dependencies amongst frames	High computational complexity
4	Mask propagation	[26]	Focus better on areas where object of interest mostly emerges	Not able to properly manage obstructions and unexpected motion
5	Optical flow	[27, 28]	Captures pixel level motion patterns	Estimated flows not every time reliable owing to changing background and deficiency of training data
6	Long term temporal	[29–31]	Focus on changes in shape, position, appearance, and scale	Only fewer number of frames used during training due to restriction of computational cost

Table 2 Various available datasets for MOS and detection of objects in motion

S. No	Name of dataset	Properties
1	Smart-city violence detection (SCVD)	484 videos
2	Real life violence situations	2000 videos
3	Highway traffic videos	Videos of two days
4	UCF crime dataset	1,377,653 images of videos
5	Accident detection from CCTV footage	989 images of videos
6	DCSASS	16,853 videos
7	Human detection dataset	921 ages of videos

4 Video Databases for Moving Object Detection and Segmentation

Due to the rising appeal of explainable artificial intelligence data for MOS and detection of objects in motion, we discuss the various large-scale datasets accessible. Table 2 details the different datasets used for MOS and detection of objects in motion. Researchers may select amongst these discussed datasets appropriate for their research purpose.

5 Performance Metrics

To evaluate MOS and detection of objects in motion in video surveillance, the various metrics that can be used are precision, accuracy, recall, sensitivity, F-measure, specificity, negative predictive value, false positive rate, false negative rate, and, Matthews correlation coefficient. The formulas for the discussed performance metrics are given as below:

$$\text{(a) } Precision = \frac{(TP)}{(TP + FP)} \tag{1}$$

$$\text{(b) } Accuracy = \frac{(\mathrm{TP} + \mathrm{TN})}{(\mathrm{TP} + \mathrm{TN} + \mathrm{FP} + \mathrm{FN})} \tag{2}$$

$$\text{(c) } Recall = \frac{TP}{(TP + FN)} \tag{3}$$

$$\text{(d) } Sensitivity = \frac{TP}{(TP + FN)} \tag{4}$$

$$\text{(e) } F1score = 2 \times \frac{(precision \times recall)}{(precision + recall)} \tag{5}$$

$$\text{(f) } Specificity = \frac{TN}{(TN + FP)} \tag{6}$$

$$\text{(g) } NPV = \frac{TN}{(TN + FN)} \tag{7}$$

$$\text{(h) } FPR = \frac{FP}{(FP + TN)} \tag{8}$$

$$\text{(i) } FNR = \frac{FN}{(FN + TP)} \tag{9}$$

$$\text{(j) } MCC = \frac{(TP \times TN) - (FP \times FN)}{\sqrt{((TP + FP) \times (TP + FN) \times (TN + FP) \times (TN + FN))}} \tag{10}$$

6 Results

Moving object detection and MOS for various surveillance applications have been studied thoroughly in this paper. The quantitative metrics used for evaluation has also been discussed in section 5. The various deep learning approaches has been studied in detail including their advantages and disadvantages. The mask propagation-based approach that is Mask-RCNN has been used using hybrid owl osprey optimization technique to draw results for moving object detection using DCSASS dataset. The result shows neat boundaries for the target object segregating it from the background. Figure 5 depicts the performance of the model of moving object detection and MOS for the surveillance application.

Figure 6 represents the pictorial view of the various number of papers discussed for different deep learning approaches including online fine tuning, matching, graph optimization, mask propagation, optical-flow, and long-term temporal for moving object detection and MOS for various surveillance applications.

Fig. 5 Qualitative results for moving object detection and MOS for DCSASS dataset

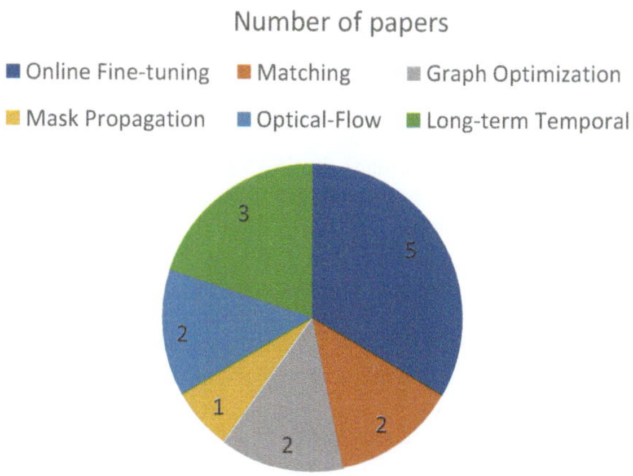

Fig. 6 Graphical representation of the number of papers discussed for various deep learning approaches for moving object detection and MOS for various surveillance applications

7 Conclusion

In this paper, the recent explainable artificial intelligence techniques used for MOS and detection of objects in motion in video surveillance is discussed. To conclude, MOS is the process of capturing pixel-level masks from a sequence of frames, thereby keeping altogether uniformity all-round the video. The various deep learning approaches like online fine tuning established, matching established, graph optimization established, mask propagation established, optical flow established, and, long-term temporal established techniques were elaborated and discussed in detail. The numerous datasets used for MOS and detection of objects in motion were detailed with their properties. Finally, the various evaluation metrics used for MOS were introduced including precision, recall, accuracy, F-score, etc. It can be concluded that MOS and detection of objects in motion is a growing field under explainable artificial intelligence and gaining much interest of researchers in recent times.

Statements and Declarations Funding and/or Conflicts of interests/Competing interests: The authors declare that they have no known competing financial interests or personal relationships that could have appeared to influence the work reported in this paper.

Conflict of Interest On behalf of all authors, the corresponding author states that there is no conflict of interest.

References

1. Muhadi, N.A., Abdullah, A.F., Bejo, S.K., Mahadi, M.R., Mijic, A.: Image segmentation methods for flood monitoring system. Water **12**(6), 1825 (2020)
2. Ammar, S., Bouwmans, T., Zaghden, N., Neji, M.: Deep detector classifier (DeepDC) for moving objects segmentation and classification in video surveillance. IET Image Proc. **14**(8), 1490–1501 (2020)
3. Teng, X., Yan, M., Ertugrul, A.M., Lin, Y.R.: Deep into hypersphere: Robust and unsupervised anomaly discovery in dynamic networks. In: Proceedings of the Twenty-Seventh International Joint Conference on Artificial Intelligence (2018)
4. Zhao, L., He, Z., Cao, W., Zhao, D.: Real-time moving object segmentation and classification from HEVC compressed surveillance video. IEEE Trans. Circuits Syst. Video Technol. **28**(6), 1346–1357 (2016)
5. Murugesan, M., Thilagamani, S.: Efficient anomaly detection in surveillance videos based on multi layer perception recurrent neural network. Microprocess. Microsyst. **79**, 103303 (2020)
6. Pramanik, A., Sarkar, S., Maiti, J.: A real-time video surveillance system for traffic pre-events detection. Accid. Anal. Prev. **154**, 106019 (2021)
7. Kalli, S., Suresh, T., Prasanth, A., Muthumanickam, T., Mohanram, K.: An effective motion object detection using adaptive background modeling mechanism in video surveillance system. J. Intell. Fuzzy Syst. **41**(1), 1777–1789 (2021)
8. Elhoseny, M.: Multi-object detection and tracking (MODT) machine learning model for real-time video surveillance systems. Circuits Systems Signal Process. **39**, 611–630 (2020)
9. Thenmozhi, T., Kalpana, A.M.: Adaptive motion estimation and sequential outline separation based moving object detection in video surveillance system. Microprocess. Microsyst. **76**, 103084 (2020)
10. Patil, P.W., Biradar, K.M., Dudhane, A., Murala, S.: An end-to-end edge aggregation network for moving object segmentation. In: Proceedings of the IEEE/CVF Conference on Computer Vision and Pattern Recognition, pp. 8149–8158 (2020)
11. Mahalingam, T., Subramoniam, M.: A robust single and multiple moving object detection, tracking and classification. Appl. Comput. Inform. **17**(1), 2–18 (2021)
12. Barnich, O., Van Droogenbroeck, M.: ViBe: a universal background subtraction algorithm for video sequences. IEEE Trans. Image Process. **20**(6), 1709–1724 (2010)
13. Bilodeau, G.A., Jodoin, J.P., Saunier, N.: Change detection in feature space using local binary similarity patterns. In: 2013 International Conference on Computer and Robot Vision. IEEE, pp. 106–112 (2013)
14. Braham, M., Piérard, S., Van Droogenbroeck, M.: Semantic background subtraction. In: 2017 IEEE International Conference on Image Processing (ICIP). IEEE, pp. 4552–4556 (2017)
15. St-Charles, P.L., Bilodeau, G.A., Bergevin, R.: A self-adjusting approach to change detection based on background word consensus. In: 2015 IEEE Winter Conference on Applications of Computer Vision. IEEE, pp. 990–997 (2015)
16. Hofmann, M., Tiefenbacher, P., Rigoll, G.: Background segmentation with feedback: The pixel-based adaptive segmenter. In: 2012 IEEE Computer Society Conference on Computer Vision and Pattern Recognition Workshops. IEEE, pp. 38–43 (2012)
17. Caelles, S., Maninis, K.K., Pont-Tuset, J., Leal-Taixé, L., Cremers, D., Van Gool, L.: One-shot video object segmentation. In: Proceedings of the IEEE Conference on Computer Vision and Pattern Recognition, pp. 221–230 (2017)
18. Voigtlaender, P., Leibe, B.: Online adaptation of convolutional neural networks for video object segmentation. arXiv preprint arXiv:1706.09364 (2017)
19. Yin, Y., Xu, D., Wang, X., Zhang, L.: AGUnet: annotation-guided U-net for fast one-shot video object segmentation. Pattern Recogn. **110**, 107580 (2021)
20. Yang, L., Wang, Y., Xiong, X., Yang, J., Katsaggelos, A.K.: Efficient video object segmentation via network modulation. In: Proceedings of the IEEE Conference on Computer Vision and Pattern Recognition, pp. 6499–6507 (2018)

21. Fiaz, M., Zaheer, M.Z., Mahmood, A., Lee, S.I., Jung, S.K.: 4G-VOS: video object segmentation using guided context embedding. Knowl.-Based Syst. **231**, 107401 (2021)
22. Oh, S.W., Lee, J.Y., Xu, N., Kim, S.J.: Video object segmentation using space-time memory networks. In: Proceedings of the IEEE/CVF International Conference on Computer Vision, pp. 9226–9235 (2019)
23. Xie, H., Yao, H., Zhou, S., Zhang, S., Sun, W.: Efficient regional memory network for video object segmentation. In: Proceedings of the IEEE/CVF Conference on Computer Vision and Pattern Recognition, pp. 1286–1295 (2021)
24. Zhou, T., Li, J., Li, X., Shao, L.: Target-aware object discovery and association for unsupervised video multi-object segmentation. In: Proceedings of the IEEE/CVF Conference on Computer Vision and Pattern Recognition, pp. 6985–6994 (2021)
25. Wang, W., Lu, X., Shen, J., Crandall, D.J., Shao, L.: Zero-shot video object segmentation via attentive graph neural networks. In: Proceedings of the IEEE/CVF international conference on computer vision, pp. 9236–9245 (2019a)
26. Perazzi, F., Khoreva, A., Benenson, R., Schiele, B., Sorkine-Hornung, A.: Learning video object segmentation from static images. In: Proceedings of the IEEE conference on computer vision and pattern recognition, pp. 2663–2672 (2017)
27. Dutt Jain, S., Xiong, B., Grauman, K.: Fusionseg: learning to combine motion and appearance for fully automatic segmentation of generic objects in videos. In: Proceedings of the IEEE conference on computer vision and pattern recognition, pp. 3664–3673 (2017)
28. Xu, K., Yao, A.: Accelerating video object segmentation with compressed video. In: Proceedings of the IEEE/CVF Conference on Computer Vision and Pattern Recognition, pp. 1342–1351 (2022)
29. Ji, G.P., Fu, K., Wu, Z., Fan, D.P., Shen, J., Shao, L.: Full-duplex strategy for video object segmentation. In: Proceedings of the IEEE/CVF international conference on computer vision, pp. 4922–4933 (2021)
30. Wei, L., Lang, C., Liang, L., Feng, S., Wang, T., Chen, S.: Weakly supervised video object segmentation via dual-attention cross-branch fusion. ACM Trans. Intell. Syst. Technol. (TIST) **13**, 1–20 (2022)
31. Lan, M., Zhang, J., He, F., Zhang, L.: Siamese network with interactive transformer for video object segmentation. In: Proceedings of the AAAI Conference on Artificial Intelligence, pp. 1228–1236 (2022)

Design and Exploration of Data-Handling Quantum-Computational Computer Circuits for the Implementation of Ultra-Area and Low-Power Devices

Mukesh Patidar, Praveen Bhanodia, Shreyaskumar Patel, Dron Mishra, Rupesh Shukla, Dilip Kumar Sharma, Aaryesh Shukla, and Mayur Tembhurney

Abstract The power consumption of the complementary metal–oxide–semiconductor (CMOS) or metal–oxide–semiconductor (MOS) device can be decreased by scaling the devices. Quantum dots are nanostructures created from conventional semiconductive substances (materials). The three-dimensional quantum energy wells can be used to simulate these structures. Instead of focusing on the current flow, logical operations and data transfer are carried out using the coulumbic interaction between nearby QCA cells. In this paper, we represent the energy dissipation and explore novel data handling quantum-computational computer circuits such as the 2:1 Multiplexer and the 1:2 Demultiplexer on a single layer using QCA technology. The proposed design does not use crossover wire crossing and has a synchronized clocking scheme.

M. Patidar (✉)
Department of Computer Science Engineering, Parul Institute of Engineering and Technology, Parul University, Vadodara, Gujarat, India
e-mail: mukesh.patidar34885@paruluniversity.ac.in

M. Patidar · P. Bhanodia
Department of Computer Science and Engineering, Acropolis Institute of Technology and Research, Indore, MP, India

S. Patel
Department of R and D Platform, Netscout Systems, Allen, TX, USA

D. Mishra
Department of Physics, Renaissance University, Indore, MP, India

R. Shukla
Department of Computer Science, ILVA Commerce and Science College, Indore, MP, India

D. K. Sharma
Department of Electronics and Communication Engineering, Ujjain Engineering College, Ujjain, MP, India

A. Shukla
Indian Institute of Technology, Chennai, India

M. Tembhurney
Department of Electronics Engineering, Medi-Caps University, Indore, MP, India

B. Soni et al. (eds.), *Beyond Artificial Intelligence*, Lecture Notes in Networks and Systems 1326, https://doi.org/10.1007/978-981-96-4170-3_29

Keywords Data handling · QCA technology · Quantum-computational computer · Single layer · Energy consumptions · Nanoelectronics · Faults defects · Synchronization clock

1 Introduction

Complementary metal oxide semiconductor (CMOS) technology is projected to reach its scaling limit during the subsequent decade [1, 2]. As a result, innovative characteristics and reduced dimensions in new nanotechnologies are being investigated. CMOS technology has been suggested to be replaced with quantum-dot cellular automata (QCA) technology [3]. Quantum dot and cellular automata logic (QCA) is a transistor-free technology which brings together their respective benefits. The high density, quick operating speed, and extremely low power consumption of QCA are only a few of its potential benefits [4].

Quantum-dot cellular automata technology is one of the promising techniques that have abilities to provide low power high speed circuits at nano-scale as compared to CMOS [5, 6]. The manipulation, processing, and administration of digital data using electrical circuits and devices is referred to as "data handling" in the context of digital electronics. Binary information, which is commonly represented as a string of 0 s and 1 s, makes up digital data. The effective and precise management of such data is at the foundation of digital electronics.

In the QCA technology, each QCA cell is represented by a square structure with four quantum dots [7–10], as shown in Fig. 1a (basic QCA cells). Each dot can only hold one electron. Two free electrons are first introduced into the QCA cell, and as a result of their internal repulsion, they are subsequently transported to the diagonal position [9, 10]. The arrangement of electrons is employed to represent the logic values "0" and "1". The majority voter (MV) is the fundamental component of a QCA circuit [11–14]. As depicted in Fig. 1b, it consists of five QCA cells. The majority voter's (FMV) output is dependent on its inputs. The QCA inverter and wire are depicted in Fig. 1c,d, respectively. The key component verified by Table 1.

The key contribution of this research paper is following:

- Proposed design and implementation of the data handling circuits for Multiplexers and Demultiplexer in QCA technology.
- The presented circuits are designed on a single-layer coplanar platform using a synchronized clocking scheme with no crossover.
- The novel data handling design is explored in terms of design area, cell area, cell utilization, and delay and compared with existing designs.
- Represent the energy dissipation of data handling circuits.
- The proposed circuits are evaluated on the QCADesigner-E simulation tool using bi-stable and coherence parameters and the simulation outputs are verified by the truth table.

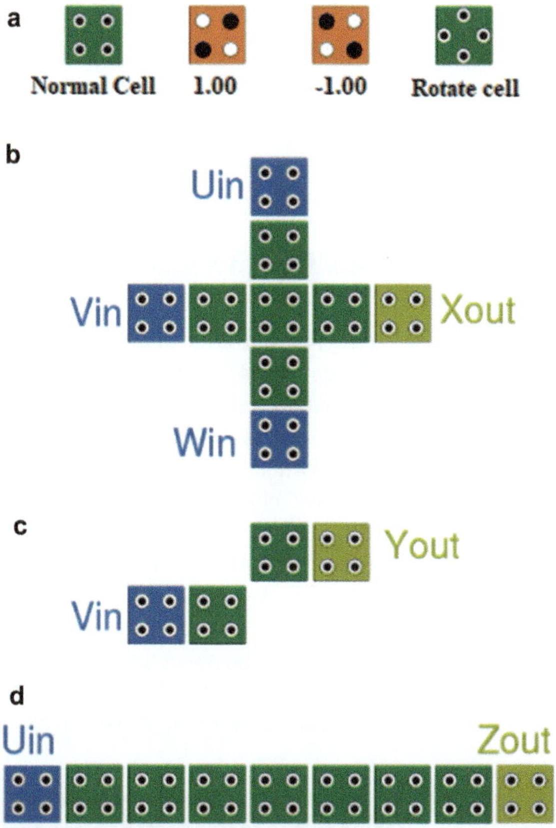

Fig. 1 Key elements of QCA design

Table 1 Input–output verification table for fundamentals QCA design

Inputs			Output		
U	V	W	majority gate (X)-Fig. 1b	Inverter (Y)-Fig. 1c	QCA Wire (Z)-Fig. 1d
0	0	0	0	1	0
0	0	1	0	1	0
0	1	0	0	0	0
0	1	1	1	0	0
1	0	0	0	1	1
1	0	1	1	1	1
1	1	0	1	0	1
1	1	1	1	0	1

The rest of paper as follows: In the Sect. 2 describes in used methods and its applied processes, in the Sect. 3 explore the different faults detection accurses in QCA technology, in Sect. 4 explains in details shows the proposed data handling circuits and its implementation, in the Sect. 5 results and discussions and last section conclusions.

2 QCA Circuits Design Scheme/methods

In QCA technology, synchronization of the circuit is known as QCA clocking scheme. There are four different types of cooking modes are namely Relax mode, Switch mode, Hold mode, and Release mode [14–18], are shown in Fig. 2 [19]. In the synchronization QCA design applied the regular clock cycle from inputs A, B to output such as Clk.-0:Switch, Clk.-1:Hold, Clk.-2:Release and Clk.-3:Relax shown in Fig. 3, but in the unsynchronized QCA-design clocks are not in regular such as in Figs. 4 and 5. from the side of input A to output clock sequence are clk.-0, clk.-1 then clk jump the clock-3, but in the same design for input from B to output clock sequence are Clk.-0→Clk.-1→Clk.-2→Clk.-3 then output but according to clocking concept algorithms are inputs to outputs are need to synchronized for obtained perfect and more accurate outputs. The novel terms related to the clock cycle are known as throughput, the throughput is defined as a number of clock cycles (CC) between successive output appearances or outputs per clock cycle.

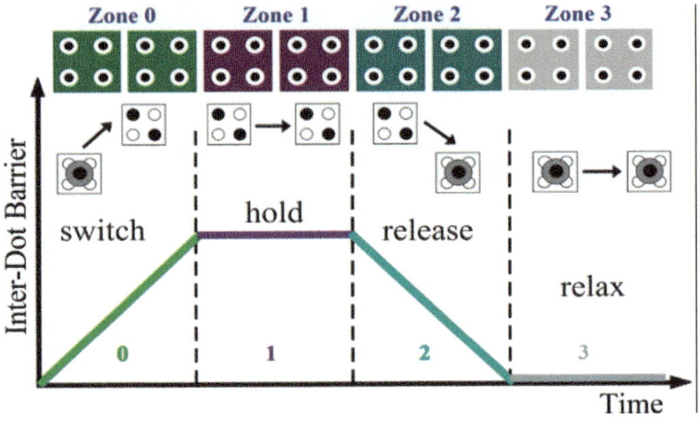

Fig. 2 QCA circuits design scheme

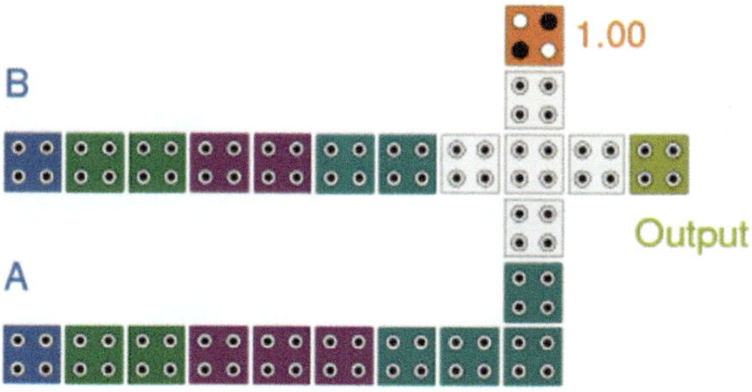

Fig. 3 Synchronization (Regular) clocking zone-assignment

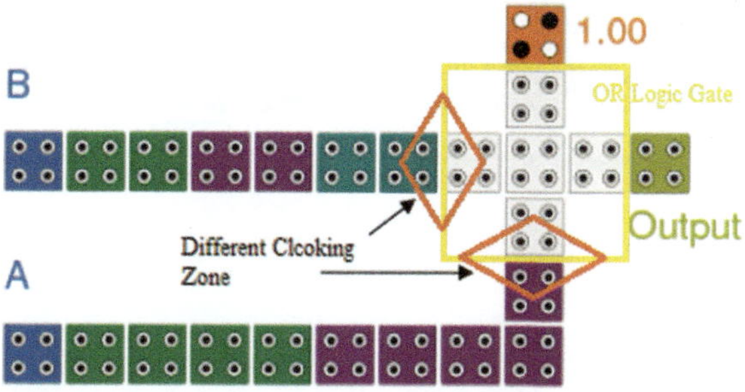

Fig. 4 QCA-OR gate (Clock-3) with unsynchronized clocking zone

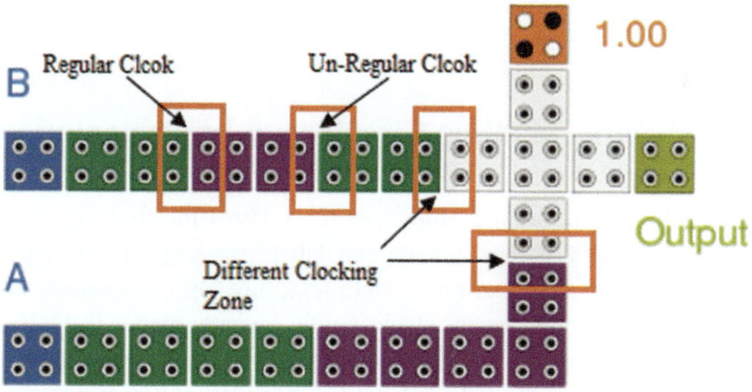

Fig. 5 Unsynchronized (different) clocking zone-assignment

3 QCA Cell Faults Defects

Some of the challenges associated with QCA technology related to the QCA faults [21–24] such are include an additional QCA cell in Fig. 6a, a rotated QCA cell Fig. 6b, displacement QCA cell in Fig. 6c, and a missing QCA cell in Fig. 6d cell.

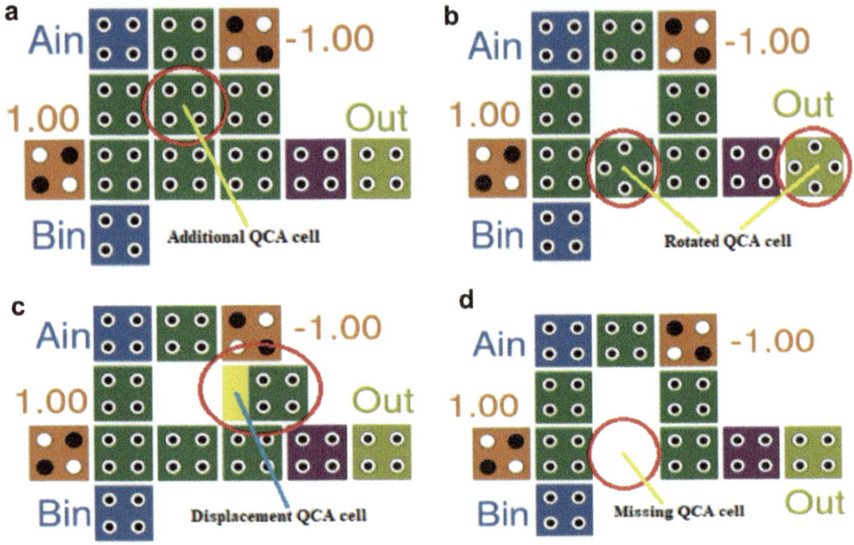

Fig. 6 **a** Additional cell faults **b** Rotated cell faults **c** Displacement cell faults **d** Missing cell faults

In the case of an additional cell, multiple (two or more) cells can be added to the circuit when particular cases. A rotated cell (45°) is a defect in which a defective cell is rotated: 45° cell [25]. Cell displacement refers to the circumstance in which a cell is shifted from its normal position. In the absence of a cell, one or more cells can be destroyed.

4 Proposed Data Handling Circuits and Implementation

Combinational logic circuits are employed to "control the flow of logic around a system" and are utilized for "data transfer and data processing circuits that are process the transform data for computations". Data handling circuits include Multiplexer, Demultiplexer, and decoder kinds. Using single-layer coplanar in QCA nanotechnology, we have designed the 2:1 Multiplexer and 1:2 Demultiplexer circuits in this paper. The term "N number of digital data signal information units over a single (one) output line" or "single channel" refers to the multiplexer's "MUX (N × 1)" approaches. A group of selection lines "S" decides which input line should be chosen. The MUX is employed in a number of systems, including telephone networks, computer memory, and communication systems. Figure 7 and Table 2 show the MUX block diagram and verification table, respectively.

The "inverse" procedure of the MUX is done by QCA DEMUX (1 × N) or DEMUX. Another device with a single input and numerous output lines is the

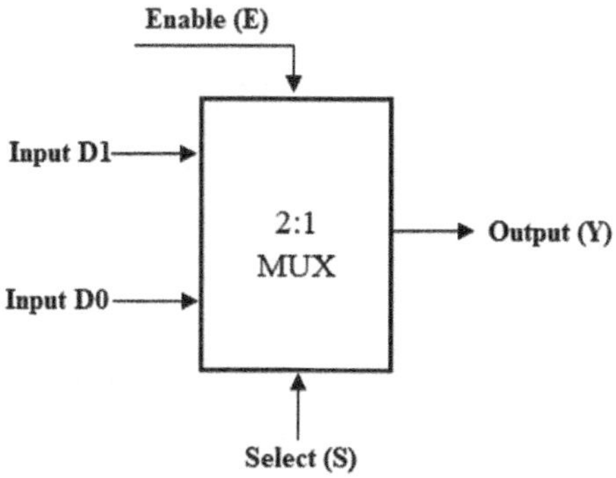

Fig. 7 Block diagram of 2 × 1 MUX

Table 2 Truth table of 2:1 multiplexer

Inputs		Selection line	Output	
X	Y	S	F	
0	0	0	0	F = X = 0
0	1	0	0	
1	0	1	0	F = (Y = 0, Y = 1)
1	1	1	1	

DEMUX. The DEMUX is utilized in a number of systems, including serial-to-parallel converters, ALUs, and communication systems. Figure 8 shows the 1:2 DEMUX block diagram, and Table 3 contains the result verification table.

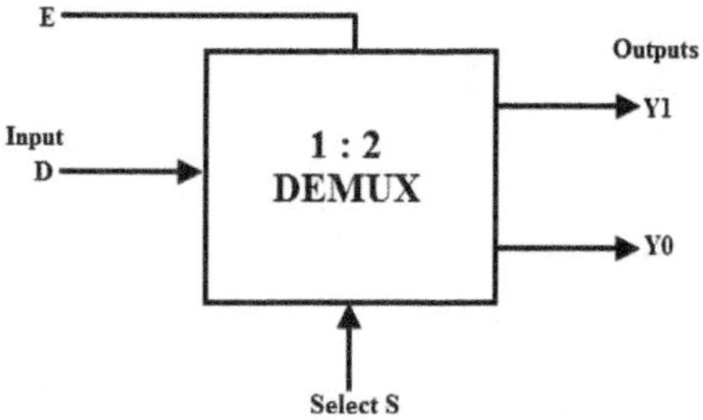

Fig. 8 Block diagram of 1 × 2: DMUX

Table 3 Truth table of 1:2 DEMUX

Select	Input data	Outputs	
S0	D	Y0	Y1
0	0	0	0
0	1	0	1
1	0	0	0
1	1	1	0

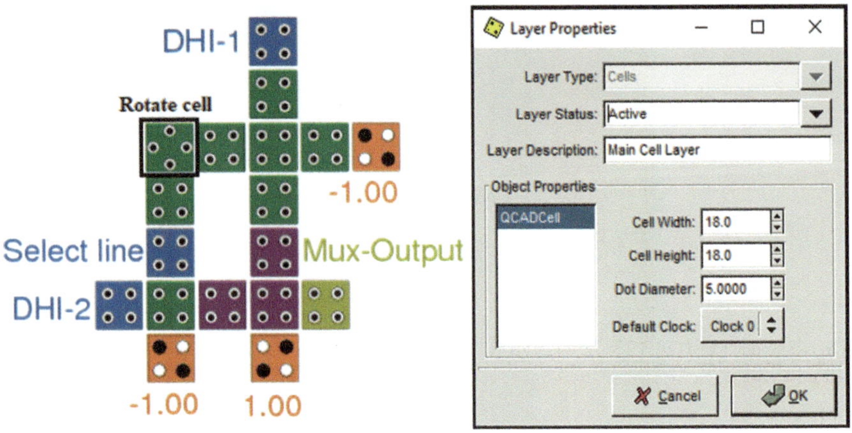

Fig. 9 QCA layout of 2 × 1 multiplexer

On a coplanar single-layer device using a three-Maj-3 with a 45^0 cell, a QCA MUX design has been constructed. The minimum design area for the suggested MUX design is 0.028 μm2, the minimum clock delay is 0.5 clcok cycles, and it uses 18 quantum cells. Figure 9 shows the novel QCA technology-based 2:1-MUX data handling circuits with coplanar single-layer properties. For 2:1-MUX, the simplified Majrity gate expropriation is shown in Eqs. (1) and (2).

$$\textit{Output of } 2 : 1MUX = AS_0 + BS_0 \tag{1}$$

$$\textit{Majority Output of } 2 : 1\,MUX = M\left((A, \overline{S}, 0), M(B, S, 0), 1\right) \tag{2}$$

The suggested 1:2 DEMUX data handling circuit structure is made up of 18 cells, two majority voters, and one rotating cell. A 0.5 clock-cycle delay is measured, and a minimum design area of 0.026 μm^2 is used in the design that is being shown in Fig. 10. The four-phase synchronized clock systems employed in the suggested design "provided facility for signal information propagation as well as the energy of the QCA cell." Eqs. (3) and (4), respectively, show the logical and MG expressions for the 1:2-DUMUX circuits.

$$\textit{Output of DEMUX } 1 : 2 = \{Y0 = DS0Y1 = D\overline{S0}\} \tag{3}$$

$$\textit{Majority Output of } 1 : 2\,DEMUX = \{Y0 = M(D, \overline{S0}, 0)Y1 = M(D, S0, 0) \tag{4}$$

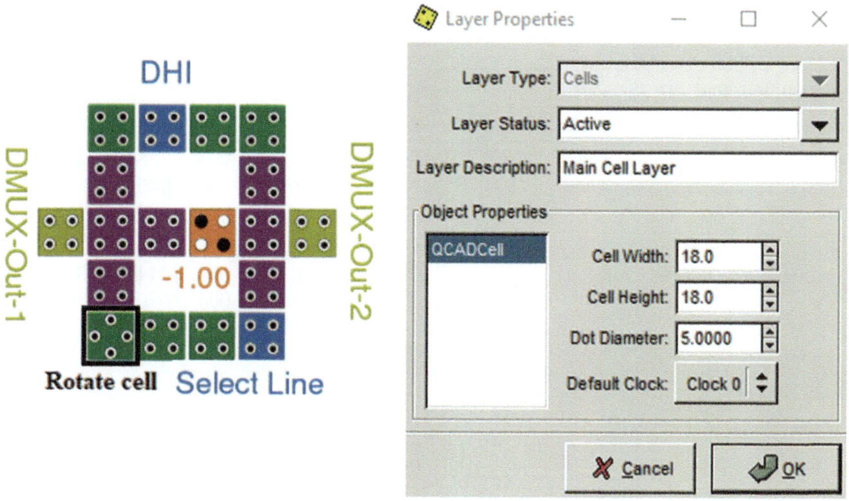

Fig.10 QCA layout of 1:2 DEMUX circuit

5 Energy Dissipation

The total energy and power of a QCA cell can be calculated by using a "Hamiltonian matrix" as represent in Eq. 5 [20].

$$H = \sum_j \left[-\frac{1}{2} P_j E_{i,j}^k - \gamma_i - \gamma_i \frac{1}{2} P_j E_{i,j}^k \right] \tag{5}$$

where, $E_{i,j}^k$ is kink energy, *iandj* are energy cost, © is tunneling potential, polarization state P_i of the cell, and P_j is of neighboring cells. We are able to find the stationary states of the cell in the environment described by this Hamiltonian in Eq. (6).

$$H_i \psi_i = E_i \psi_i \tag{6}$$

The "kink energy $E_{i,j}^k$ represents the energy cost of cells i and j having different or opposite polarization" in Eq. (7) [1, 2]. In Tables 4 and 5 represent the energy dissipation in eV for proposed MUX and DEMUX design [17, 29, 30].

$$E_{i,j} = \frac{1}{4\pi \varepsilon_0 \varepsilon_r} \times \frac{q_i q_j}{|r_i - r_j|} \tag{7}$$

Table 4 Energy dissipation (eV) for MUX circuits

Total simulation time (s)	27
Simulation found	3 inputs 1 output
Samples recorded	3000
Total Energy_bath (eV)	1.5800e-003, 1.3472e-003, 6.3264e-004, 1.9743e-003, 9.4691e-004, 1.0462e-003, 7.6888e-004, 1.9283e-003, 1.5800e-003, 1.3472e-003, 6.3264e-004
Total Energy_clk (eV)	9.4624e-004, 3.7849e-004, 2.2144e-003, −4.9968e-004, 8.6894e-004, 2.0705e-003, 1.1251e-003, 1.1962e-003, 9.4624e-004, 3.7849e-004, 2.2144e-003
Total Energy_Error (eV)	−1.6577e-004, −1.4333e-004, −6.1649e-005, −2.1232e-004, −9.7556e-005, −1.1014e-004, −7.7435e-005, −2.1459e-004, −1.6577e-004, −1.4333e-004, − 6.1649e-005
Total energy dissipation (Sum_Ebath):	1.38e-002 eV (Error: + /− −1.45e-003 eV)
Average energy dissipation per cycle (Avg_Ebath):	1.25e-003 eV (Error: + /− −1.32e-004 eV)
Used tool	QCADesigner-E

Table 5 Energy dissipation (eV) for DMUX circuits

Total simulation time	32 s
Simulation found	2 inputs 2 outputs
Samples recorded	3000
Total Energy_bath (eV)	8.8666e-004, 1.8013e-003, 1.0707e-003, 2.2419e-003, 8.8666e-004, 1.8013e-003, 1.0707e-003, 2.2419e-003, 8.8666e-004, 1.8013e-003, 1.0707e-003
Total Energy_clk (eV)	1.9951e-003, 1.1448e-003, 1.3603e-003, 8.1347e-004, 1.9951e-003, 1.1448e-003, 1.3603e-003, 8.1347e-004, 1.9951e-003, 1.1448e-003, 1.3603e-003
Total Energy_Error (eV)	−9.2404e-005, −1.9401e-004, −1.1362e-004, −2.3154e-004, −9.2404e-005, −1.9401e-004, −1.1362e-004, −2.3154e-004, −9.2404e-005, −1.9401e-004, −1.1362e-004
Total energy dissipation (Sum_Ebath):	1.58e-002 eV (Error: + /− −1.66e-003 eV)
Average energy dissipation per cycle (Avg_Ebath):	1.43e-003 eV (Error: + /− −1.51e-004 eV)
Used tool	QCADesigner-E

Table 6 Design parameters used for proposed data handling QCA circuits

Design tool	QCADesigner-E (QD-E) and QCADesigner (QD)
Technology	Quantum-dots cellular automata (QCA)
Used method	Synchronization clocking with four phases
QCA cell Size	18 nm × 18 nm or (18 × $18nm^2$)
Relative Permittivity	12.900
Temperature	T = 1 K
Number of samples	12,800
Tolerance	0.00100
Layer separation/ Clock Amp. Factor	11.50/2.00
Simulation vector	Bistable/Coherence
Cell size/ Dot diameter/Default clock	18 × $18nm^2$/5 nm/clock-0
Clock high/ Low/ Shift	9.800e-022/3.800e-023/0.00e + 00

6 Results and Discussions

Data handling circuits are used in the communication process for the transmission of data or digital information from one station to another by using a communication channel. The multiplexer circuit is part of the transmission section, and the demultiplexer is part of the receiver section. The used design parameters are represented in Table 6 [1, 2, 11–15, 26–28]. The presented design results are shown in Figs. 11 and 12 for 2:1 MUX and 1:2 DMUX, respectively, at a temperature of 1 K using the default bistable approximation engine setup. In the result, Fig. 11 shows that the 2:1-MUX result has been observed with a minimum delay of 0.5-CC. The average output polarization (AOP) is measured at 3.494 at the output of max and min polarizations of 2:1; MUX is 9.50e-001 and -9.56e-001, respectively, at temperature 1 K. Figure 12 shows that the 1:2-DMUX result has been observed with a minimum delay of 0.50 clock cycles on clock 1. The average output polarization (AOP) is measured at 3.3237 and 3.3182 for DMUX-Out-1 and DMUX-Out-2, respectively, at temperature 1 K, as shown in Fig. 12. Table 7 and Fig. 13 show the comparison of the data handling circuits for MUX and DEMUX with the existing designs, such as cells, total area, and delay.

7 Conclusions

This research work presents a novel data handling quantum-computational computer circuit's design of 2:1 MUX and DEMUX, which are implemented by QCA nanotechnology using a default set of parameters for the QCADesigner tool. The implemented design is a single-layer design, which is helpful in reducing the complexity of circuit design and fabrication of nanodevices. The presented data handling circuits utilized

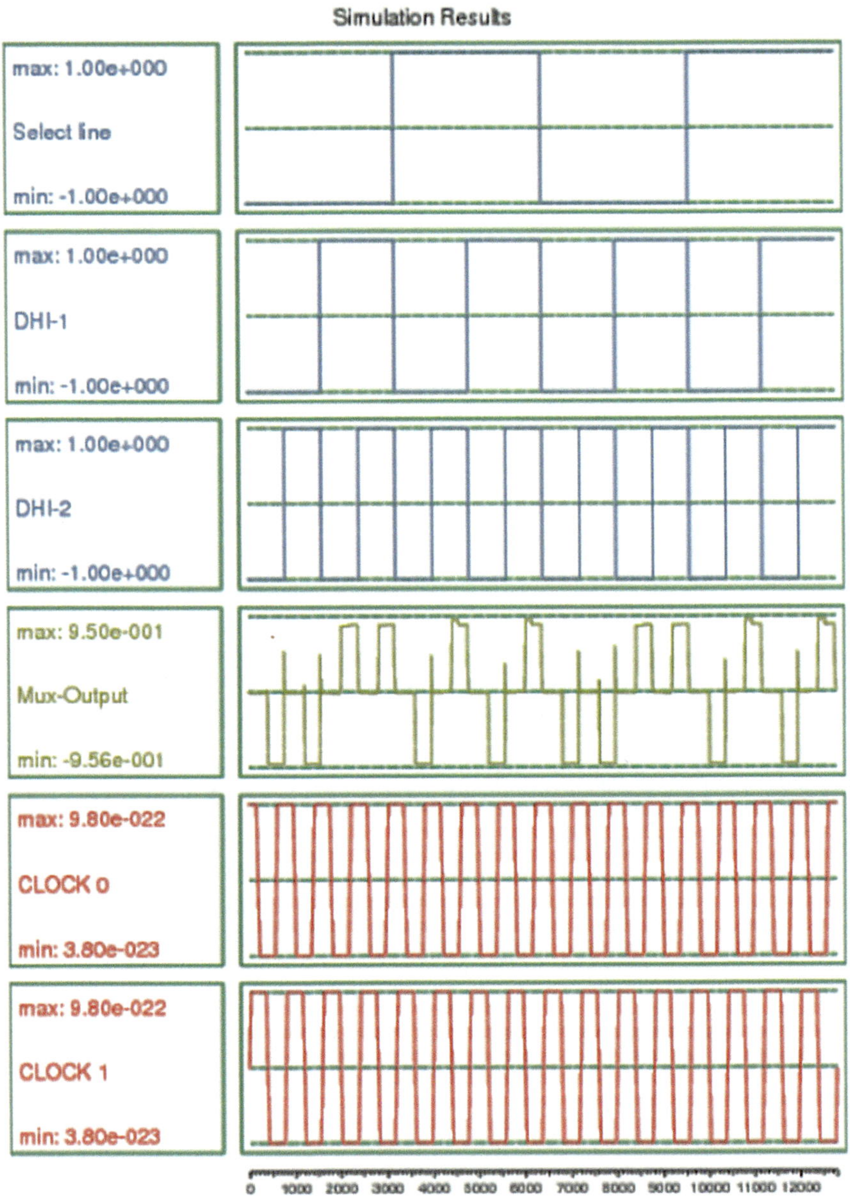

Fig. 11 Simulation results of 2:1 MUX circuit

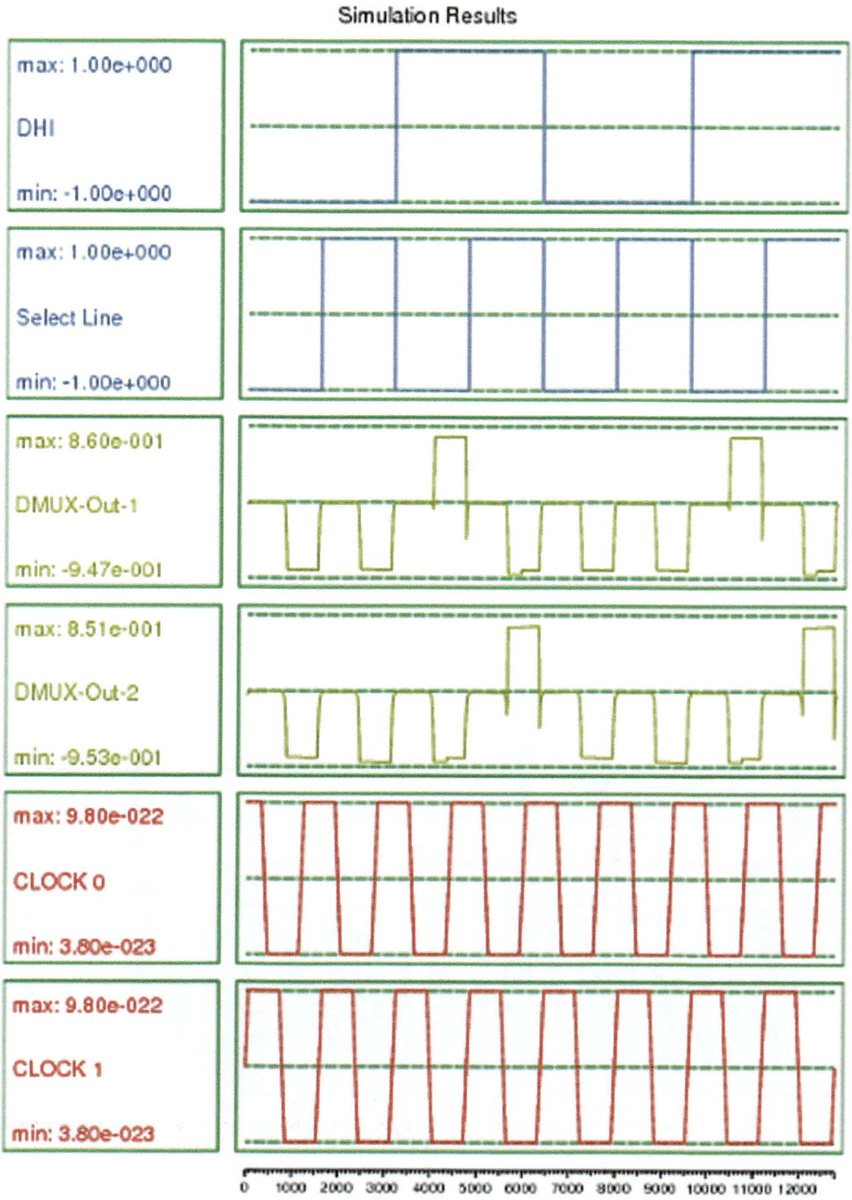

Fig. 12 Simulation results of 1:2 DEMUX circuit

Table 7 Comparative result analysis of data hendling circuits

Design		Cells	Design area (μm^2)	Latency	Cell area (μm^2)	Layer
Proposed and existing MUX design						
Proposed MUX		18	0.02	0.5	0.0058	Single-layer
Selection extents: (128.00,171.00)[121.00 × 138.00] = 16,698.00 nm^2 = 0.02 um^2 Objects selected: 18						
Existing design	[10]	22	0.02	0.5	0.0071	Single layer
	[11]	22	0.03	0.75	0.0071	
	[12]	23	0.03	0.75	0.0075	
	[13]	23	0.01	0.5	0.0075	Multi-layer
Proposed and existing DEMUX design						
Proposed DEMUX		18	0.01	0.5	0.0058	Single-layer
Selection extents: (128.00,171.00)[121.00 × 138.00] = 16,698.00 nm^2 = 0.02 um^2 Objects selected: 18						
Existing design	[14]	21	0.020	0.5	0.0068	Single layer
	[15]	21	0.023	0.5	0.0068	
	[16]	32	0.026	0.5	0.0136	

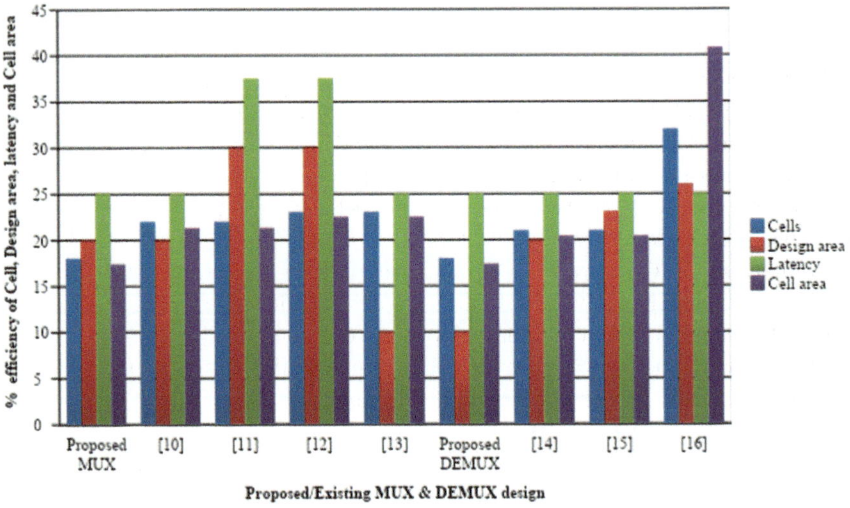

Fig. 13 Comparative chart for data handling circuits (MUX and DEMUX)

a minimum design area of 18.30% (MUX) and 14.70% (DEMUX) as compared to Sen B et al. [10] and Ansum KA, Arya R [14], respectively, and the cell area of QCA cell 18.18% (MUX) and 14.28% (DEMUX) improved as compared to Sen B et al. [10] and Ansum KA, Arya R [14]. In the future, the proposed circuit will be helpful for nanoelectronics circuits developing for the implementation of low-power computational devices and ultra-area computer circuit's designs.

References

1. Lent, C.S., Tougaw, P.D., Porod, W., Bernstein, G.H.: Quantum cellular automata. Nanotechnology **4**(1), 49–57 (1993). https://doi.org/10.1088/0957-4484/4/1/004
2. Lent, C.S., Tougaw, P.D.: A device architecture for computing with quantum dots. Proc. IEEE **85**, 541–57 (1997). https://doi.org/10.1109/5.573740
3. Patidar, M., Singh, U., Shukla, S.K., et al.: An ultra-area-efficient ALU design in QCA technology using synchronized clock zone scheme. J. Supercomput. (2022). https://doi.org/10.1007/s11227-022-05012-2
4. Seyedi, S., Pourghebleh, B.: A new design for 4-bit RCA using quantum cellular automata technology. Opt. Quant. Electron. 55(11) (2023). https://doi.org/10.1007/s11082-022-04214-5
5. Patidar, M., Gupta, N.: Efficient design and implementation of a robust coplanar crossover and multilayer hybrid full adder–subtractor using QCA technology. J. Supercomput. **77**, 7893–7915 (2021). https://doi.org/10.1007/s11227-020-03592-5
6. Patidar, M., Gupta, N.: An efficient design of edge-triggered synchronous memory element using quantum dot cellular automata with optimized energy dissipation. J. Comput. Electron. **19**, 529–542 (2020). https://doi.org/10.1007/s10825-020-01457-x
7. Khosroshahy, M.B., Moaiyeri, M.H., Abdoli, A.: Design and energy analysis of a new fault-tolerant SRAM cell in quantum-dot cellular automata. Opt. Quant. Electron. **54**(593) (2022). https://doi.org/10.1007/s11082-022-03992-2
8. Lent, C.S., Tougaw, P.D., Porod, W.: Bistable saturation in coupled quantum dots for quantum cellular automata. Appl. Phy. Lett. **62**, 714–716 (1993). https://doi.org/10.1063/1.108848
9. Patidar, M., Gupta, N.: An ultra-efficient design and optimized energy dissipation of reversible computing circuits in QCA technology using zone partitioning method. Int. j. inf. Tecnol **14**, 1483–1493 (2022). https://doi.org/10.1007/s41870-021-00775-y
10. Sen, B., et al.: Towards modular design of reliable quantum-dot cellular automata logic circuit using multiplexers. Comput. Electr. Eng. **45**, 42–54 (2015)
11. Xingjun, L., Zhiwei, S., Hongping, C., Haghighi, M.R.J.: A new design of QCA-based nanoscale multiplexer and its usage in communications. Int. J. Commun. Syst. **2019**, e4254 (2019). https://doi.org/10.1002/dac.4254
12. Amir, M.C.: Efficient QCA exclusive-or and multiplexer circuits based on a nanoelectronic-compatible designing approach. In: International Scholarly Research Notices, p. 463967 (2014). https://doi.org/10.1155/2014/463967
13. Sen, B., et al.: Towards the hierarchical design of multilayer QCA logic circuit. J. Comput. Sci. **11**, 233–244 (2015)
14. Ansum, K.A., Arya, R.: Optimal demultiplexer unit design and energy estimation using quantum dot cellular automata. J. Supercomput. **77**, 1714–1738 (2021). https://doi.org/10.1007/s11227-020-03320-z
15. Ahmad, F.: An optimal design of QCA based $2n : 1/1 : 2n$ multiplexer/demultiplexer and its efficient digital logic realization. Microprocess. Microsyst. **56**, 64–75 (2017). https://doi.org/10.1016/j.micpro.2017.10.010
16. Das, J.C., De, D.: Circuit switching with quantum-dot cellular automata. Nano Commun. Netw. **14**, 16–28 (2017). https://doi.org/10.1016/j.nancom.2017.09.002
17. Torres, F.S., Wille, R. et al.: An energy aware model for the logic synthesis of quantum-dot cellular automata. IEEE Trans. Comput. Aided Des. Integr. Circuits Syst. **37**(12), 3031–3041 (2018). https://doi.org/10.1109/TCAD.2018.2789782
18. Tiwari, A., Patidar, M., Jain, A., Patidar, N., Gupta, N.: Efficient designs of high-speed combinational circuits and optimal solutions using 45-degree cell orientation in QCA nanotechnology. Mat. Today: Proc. **66**(8), 3465–3473 (2022). https://doi.org/10.1016/j.matpr.2022.06.174

19. Patidar, M., Shrivastava, A., Miah, S., Kumar, Y., Sivaraman, A.K.: An energy efficient high-speed quantum-dot based full adder design and parity gate for nano application. Mat. Today: Proc. **62**(7), 4880–4890 (2022). https://doi.org/10.1016/j.matpr.2022.03.532
20. Srivastava, S., Sarkar, S., Bhanja, S.: Estimation of upper bound of power dissipation in QCA circuits. IEEE Trans. Nanotechnol. **8**(1), 116–127 (2009). https://doi.org/10.1109/TNANO.2008.2005408
21. Srivastava, S., Asthana, A., Bhanja, S., Sarkar, S.: QCAPro-an error-power estimation tool for QCA circuit design. In: 2011 IEEE International Symposium of Circuits and Systems (ISCAS). Rio de Janeiro, Brazil, 2377–2380 (2011). https://doi.org/10.1109/ISCAS.2011.5938081
22. Seyedi, S., Navimipour, N.J., Otsuki: A design and analysis of fault-tolerant 1:2 Demultiplexer using quantum-dot cellular automata nano-technology. Electronics **10**(2565) (2021).https://doi.org/10.3390/electronics10212565
23. Patidar, N., Gupta, N.: An extensible architecture of 32-bit ALU for high-speed computing in QCA technology. J. Supercomput. **78**, 19605–19627 (2022). https://doi.org/10.1007/s11227-022-04608-y
24. Patidar, M., Gupta, N., Jain, A., Patidar, N.: The role of nanoelectronic devices in a smart city ecosystem. Taylor & Francis Group: CRC Press AI-Centric Smart City Ecosystems (2023). https://doi.org/10.1201/9781003252542
25. Seyedi, S., Navimipour, N.J.: A fault-tolerance nanoscale design for binary-to-gray Converter based on QCA. IETE J. Res. **69**(5) (2023). https://doi.org/10.1080/03772063.2021.1908857
26. Patidar, M., Gupta, N.: Efficient design and simulation of novel exclusive-or gate based on nanoelectronics using quantum-dot cellular automata. In: Proceeding of the Second International Conference on Microelectronics, Computing and Communication Systems (MCCS 2017), Lecture Notes in Electrical Engineering **476**, 599–614 (2019). https://doi.org/10.1007/978-981-10-8234-4_48
27. Mukesh, P., Surendra, K.S. et al.: An efficient design and implementation of a reversible logic CCNOT (Toffoli) gate in QCA for nanotechnology. Mater. Today: Proc. (2023). https://doi.org/10.1016/j.matpr.2023.03.373
28. Sharma, D.K., Goyal, P. et al.: An efficient design and demonstration of fault-effects in full-adder circuits based on quantum-dot computing circuits. Mater. Today: Proc (2023). https://doi.org/10.1016/j.matpr.2023.03.374
29. Patidar, M., Jain, A., Patidar, K., et al.: An ultra-dense and cost-efficient coplanar RAM cell design in quantum-dot cellular automata technology. J. Supercomput. (2023). https://doi.org/10.1007/s11227-023-05722-1
30. Patidar, M. et al.: Optimized design and investigation of novel reversible toffoli and peres gates using QCA techniques, measurement: Sensors **32** (2024). https://doi.org/10.1016/j.measen.2024.101036

Information Retrieval Using Large Language Model for Various Alloys

Yash Garg, Akanksha Jaiswal, Daevesh Singh, and Ramkumar Rajendran

Abstract Selecting the right alloy for a specific application is one of the key problems in physical metallurgy and can be challenging due to the vast array of options and the need to consider various factors. This paper proposes an information retrieval-based chatbot, powered by a large language model (LLM), to assist metallurgists in alloy selection. The chatbot efficiently processes text from various sources, including handbooks, manuals, research papers, and multiple PDFs, providing metallurgists with comprehensive information. To address the hallucination issue inherent in LLMs, the chatbot incorporates Langchain, a technique that enhances contextualization and semantic similarity, ensuring that responses remain firmly anchored in the provided documents. The chatbot's functionalities include alloy information search, summarization, comparison, and contributing to informed recommendations. This innovative chatbot has the potential to significantly contribute to alloy selection problems faced by metallurgists, saving time and improving decision-making.

Keywords Large language model · GPT · Alloys · Information retrieval · API

Y. Garg (✉)
Metallurgical Engineering and Materials Science, Indian Institute of Technology Bombay, Mumbai, India
e-mail: 200110039@iitb.ac.in

A. Jaiswal
Information Technology, Mukesh Patel School of Technology Management and Engineering, NMIMS, Mumbai, India
e-mail: akankshajais1@gmail.com

D. Singh · R. Rajendran
IDP in Educational Technology, Indian Institute of Technology Bombay, Mumbai, India
e-mail: daeveshsingh@iitb.ac.in

R. Rajendran
e-mail: ramkumar.rajendran@iitb.ac.in

B. Soni et al. (eds.), *Beyond Artificial Intelligence*, Lecture Notes in Networks and Systems 1326, https://doi.org/10.1007/978-981-96-4170-3_30

1 Introduction

Pure elemental metallic systems are not suitable for deployment in engineering applications. Alloys, which are metallic mixtures of two or more elements strategically combined to attain specific properties such as strength with lightweight, corrosion resistance, and ductility, play a crucial role in diverse fields such as aerospace engineering, automotive engineering, medical devices, and construction. Choosing the right alloy for a given application is an important problem in physical metallurgy and involves considering factors such as required properties, as well as operating environments, and cost, and hence presents a challenge when different alloy systems fulfill different requirements.

Traditionally, metallurgists have relied on their knowledge and experience for alloy selection; however, given the large amount of data available, and the capabilities of automation of the information retrieval from such large data sets, the alloy selection process can be made more efficient.

In this paper, we propose the development of an information retrieval-based chatbot to aid metallurgists in alloy selection. This chatbot, leveraging a large language model (LLM), will process text from diverse sources, including alloy manuals, research papers, and multiple PDFs, to provide metallurgists with comprehensive alloy information. This innovation aims to alleviate the constraints faced by metallurgists by enabling efficient alloy selection without the need for extensive manual research.

While LLMs have proven effective, they are not without limitations. One notable challenge is the potential for hallucination [1], where the model generates responses based on patterns or associations that may not be directly relevant to the provided information. In the context of our chatbot, this manifested as responses not substantiated by the uploaded documents, posing a risk of misinformation. To counter this challenge, our chatbot integrates Langchain, a technique designed to enhance contextualization and semantic similarity. Langchain ensures that the chatbot's responses remain firmly anchored in the information derived from the provided documents. This strategic inclusion of Langchain mitigates the risk of hallucination, enhancing the accuracy and reliability of the chatbot's responses in the alloy selection process.

The functionality of our chatbot encompasses various tasks aimed at assisting metallurgists in alloy selection:

- Searching for alloy information: The chatbot efficiently retrieves details about specific alloys, such as composition, properties, and applications, across multiple PDFs.
- Summarizing alloy information: It condenses key properties and applications of an alloy into a concise and understandable format based on information from various PDFs.
- Comparing alloys: The chatbot facilitates the comparison of two or more alloys, considering properties like strength, weight, and cost, drawing insights from multiple sources.

- Recommending alloys: Based on user requirements, our chatbot recommends suitable alloys for specific applications, utilizing information from diverse PDFs.

In summary, our chatbot, fortified with Langchain technology, stands poised to help in alloy selection for metallurgists. By providing swift and comprehensive access to information from multiple PDFs, the chatbot not only saves time but also elevates the quality of decision-making in alloy selection processes. The strategic integration of Langchain addresses the challenge of hallucination, ensuring the reliability of the chatbot's responses and bolstering its effectiveness in assisting metallurgists.

2 Literature Review

The rapid advancements in natural language processing (NLP) have given rise to powerful language models, such as the Generative Pre-trained Transformer (GPT) [2–6] family, which encompasses LLMs like ChatGPT (GPT-3.5 and GPT-4) [7]. These models, trained on massive amounts of text data, have exhibited remarkable performance in a broad spectrum of NLP tasks, including language translation, text summarization, and question-answering. Notably, the ChatGPT model has showcased its potential in diverse domains, including education, healthcare, reasoning, text generation, human–machine interaction, and scientific research [8].

Information retrieval (IR) has evolved beyond traditional search in recent years, with new paradigms emerging to address the increasing complexity and diversity of information sources. Large language models have played a significant role in this evolution, opening new avenues for IR research [9]. One of the key challenges in IR is that domain-specific expertise is often required to retrieve and understand information effectively. LLMs can help to address this challenge by providing a general-purpose foundation for information retrieval that can be augmented with domain-specific knowledge.

For example, GeneGPT [10] is a method that augments LLMs with domain-specific tools to improve access to biomedical information. GeneGPT uses the Web APIs of NCBI to answer genomics questions, achieving state-of-the-art performance on eight tasks in the GeneTuring benchmark.

Similarly, another system PyGuru 2.0 is an innovative online learning environment designed to facilitate the learning of Python programming, particularly for students [11]. It incorporates an advanced conversational agent, driven by "ChatGPT 3.5 turbo." PyGuru 2.0 aids in providing real-time assistance, guidance, and feedback by tailoring its response based on the content present in the learning environment.

Another key challenge in IR is the increasing need to deal with complex and multi-hop queries. Iter-RetGen [12] is a method that iteratively synergizes retrieval and generation to improve information retrieval using LLMs. Iter-RetGen has been shown to achieve strong performance in retrieval-augmented language models, surpassing or being competitive with state-of-the-art baselines.

Relation extraction (RE) is another important task in IR, involving the identification and extraction of relationships between entities in text. GPT-RE [13] is a model that improves RE using LLMs by addressing issues related to entity relevance and misclassification of NULL examples. GPT-RE has achieved state-of-the-art performance on the Semeval and SciERC datasets, bridging the gap between LLMs and fully supervised baselines.

While LLMs hold immense promise for revolutionizing IR, they also present challenges that must be addressed. One significant challenge is the computational cost associated with training and deploying LLMs. Another challenge lies in the credibility of information generated by LLMs, as they are prone to producing hallucinations. Despite these challenges, LLMs are poised to play a pivotal role in shaping the future of IR. By providing a general-purpose foundation for information retrieval that can be seamlessly augmented with domain-specific knowledge, LLMs can effectively enhance the accessibility and comprehension of information.

In addressing the computational cost challenge, our research adopts the RAG approach, a cost-effective alternative to traditional fine-tuning methods. Unlike fine-tuning, which demands extensive computational resources, the RAG approach requires minimal computational resources, significantly reducing the cost associated with training and deploying the chatbot. This approach aligns perfectly with our research objectives, ensuring the development of a practical and cost-effective solution for metallurgists.

Relevance to the proposed research topic: The proposed research topic of developing an information retrieval-based chatbot to help metallurgists select alloys is highly relevant to the literature reviewed above. LLMs can be used to develop a chatbot that can effectively retrieve and understand information about alloys from a variety of sources, including alloy manuals, research papers, etc. While previous approaches have shown promise in certain domains, their application to alloy selection using alloy manuals or textbooks remains unexplored. This chatbot can then provide metallurgists with comprehensive and up-to-date information about alloys, only from the PDFs they have uploaded, helping them to save time and improve the quality of their work.

The proposed research topic is aligned with the identified trends in the literature, such as the increasing need to deal with complex and multi-hop queries, and the growing demand for domain-specific expertise in IR.

3 Proposed Information Retrieval Chatbot

3.1 Chatbot Working

Our chatbot has been tested on a variety of alloy books and has been successful in providing accurate and informative answers to user questions. Our chatbot works as follows:

1. **PDF upload and preprocessing:** The user uploads multiple PDFs to the chatbot interface. The chatbot then uses the PyPDF2 library to extract the text from the PDFs and split it into chunks of 1000 tokens with an overlap of 200 tokens. This is done to ensure that the text chunks are manageable and that the chatbot can efficiently process them.
2. **Embedding generation:** The chatbot then uses the OpenAI Embeddings library to convert the text chunks into embeddings. Embeddings are dense vector representations of text that can be used to capture the semantic similarity between different pieces of text.
3. **Vector storage:** The chatbot then stores the embeddings and the index in a vector store using FAISS/CHROMA. A vector store is a database that is optimized for storing and searching high-dimensional vectors, such as embeddings.
4. **Question answering:** When the user asks a question, the chatbot first converts the question text into an embedding using the OpenAI Embeddings library. It then performs a semantic search in the vector store to find the most similar embeddings. The chatbot then passes these embeddings to the GPT-3.5 LLM to generate an answer.
5. **Chat history memorization:** The chatbot uses the Langchain library to memorize the chat history. This allows the chatbot to keep track of the previous questions and answers, which can help it generate more informative and relevant answers to future questions (Fig. 1).

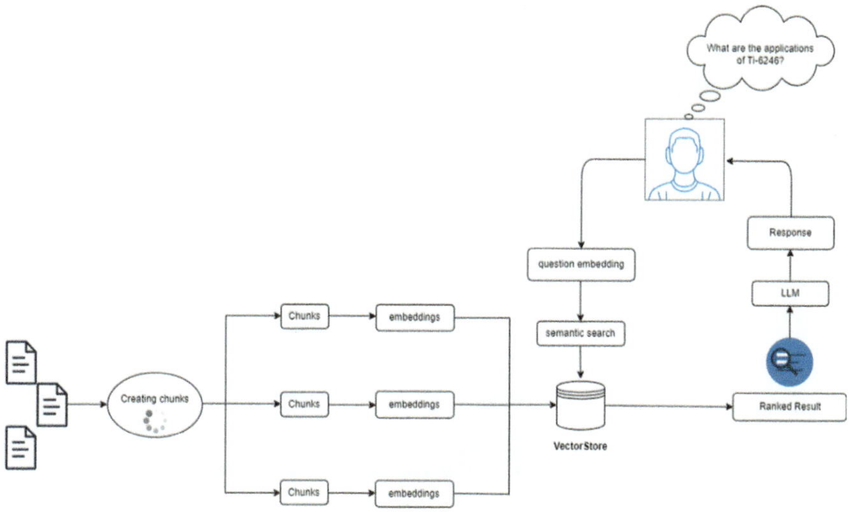

Fig. 1 A systematic chatbot-based approach to information retrieval from alloy PDF

3.2 Chatbot Interface

The chatbot interface has been created using Streamlit GUI (Fig. 2). It is simple and straightforward, with two main buttons, Browse File and Load the File.

1. **Browse File**: The Browse File button allows the user to select a PDF file from their local computer to be uploaded to the chatbot (Fig. 2). Once the file is selected, the user can click the Load the File button to begin the processing.
2. **Load the file**: The Load the File button performs two main tasks, chunking and embedding. The vector representations are then stored in a vectorstore, which is a database of vectors that can be used for various tasks, such as text similarity search and question answering. Once the vectorstore is created, the chatbot can use it to answer user questions about the PDF file, Fig. 2. For example, a user could ask the chatbot "What are the different types of alloys mentioned in the PDF?" or "What is the tensile strength of Ti-6246 alloy?". The chatbot would then search the vectorstore for relevant vectors and generate a response based on the results.

4 Result

We evaluated the performance of our proposed chatbot on three major tasks:

1. Search and find the correct results: We evaluated the chatbot's ability to search for and find the correct results for a variety of alloy-related questions. We found that the chatbot was mostly able to accurately answer questions that were asked.

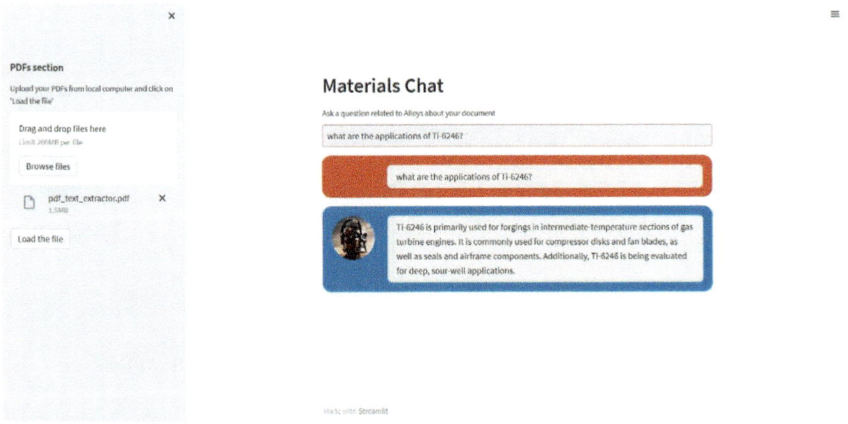

Fig. 2 Chatbot Interface for alloy information retrieval

Table 1 Demonstrating chatbot's question-answering ability on alloy-related topics

S. No	Question	Chatbot responses
1	How different are Aluminium alloy properties from Titanium alloy?	"Titanium alloys have several advantages over aluminum alloys, including higher stiffness, lower creep temperature, and better corrosion resistance."
2	How does AA6086 differ from AA6082?	"AA6086 is a heat-treatable aluminum alloy that is stronger and more ductile than AA6082. AA6086 is also more resistant to corrosion."
3	What is the tensile strength of Ti-6246 with 10 to 20% primary α + STOA(b) in MPa?	"The tensile strength of Ti-6246 10 to 20% primary α + STOA(b) is 1021 MPa."
4	Who was the president of the USA in 1999?	"Out of Scope"

2. Ability to search/find information from multiple sources and combine or synthesize: We evaluated the chatbot's ability to search for and find information from multiple sources, such as alloy books, research papers, and government databases, and combine or synthesize that information into a comprehensive and informative answer. The chatbot was able to complete this task for many questions.
3. Ability to tackle the hallucination problem: We evaluated the chatbot's ability to mitigate the hallucination problem of the LLM model by searching the LLM model answer through the documents uploaded. We found that this approach was able to reduce the hallucination rate (Table 1).

The proposed chatbot is an effective tool for searching and retrieving information about alloys. It can accurately answer a wide range of alloy-related questions, including those that require the chatbot to search for information from multiple sources and combine or synthesize that information into a comprehensive and informative answer.

In addition, our proposed approach for mitigating the hallucination problem of the LLM model is effective. By searching the LLM model answer through the documents uploaded, we were able to reduce the hallucination rate by a significant number.

5 Conclusion and Future Work

In this paper, we proposed a new information retrieval chatbot for alloys. Our chatbot uses a large language model (LLM) to extract information from multiple PDFs and answer user questions in a comprehensive and informative way. We also addressed the hallucination problem of the LLM model by searching the LLM model answer through the documents uploaded.

5.1 Generalization of the System

Our system can be generalized to other domains by using a different LLM model that is trained on a different dataset. For example, to build a sophisticated chatbot for the medical domain, we could use an LLM model that is trained on a dataset of medical papers. Our system can also address the limitation mentioned in Sect. 2, which is the lack of ability to handle large PDFs and tackle the hallucination problem.

5.2 Limitations

Our system has the following limitations:

1. The hallucination code cannot be added to the chatbot because the streamlit interface does not support the local path of the uploaded PDF.
2. The model cannot read data from images.
3. The PDF that is uploaded must be in a readable condition, meaning that the PyPDF library must be able to detect and extract the text in the PDF.
4. Sometimes the extracted text is not correct, for example, Al (aluminum) is detected as A1. A much better text extraction tool needed to be used.
5. The model extracts all the readable text from the PDFs and thus might also extract information that is not relevant, like the values from a graph without extracting any useful information from it thus making the embeddings of the chunks unsatisfactory.

5.3 Future Work

Our future work includes:

1. While our current approach has shown effectiveness in addressing hallucination in Language Models (LLMs), further refinements are necessary for improved reliability and accuracy in the chatbot's responses. Although the code for tackling hallucinations has been developed, integrating it into the chatbot interface poses challenges due to the code's reliance on a local workspace, incompatible with the current interface. Overcoming this disparity is crucial for the seamless implementation of the refined model, representing a key step in enhancing the chatbot's robustness in handling hallucination-related challenges.
2. Currently, our chatbot system relies on the PyPDF library to extract text from PDFs. This library is effective for extracting text from PDFs with embedded text, but it cannot handle PDFs with scanned images. To address this limitation, we will explore using optical character recognition (OCR) techniques to extract text from scanned images within PDFs. This will significantly expand the range of PDFs the chatbot can process and enhance its overall versatility.

Acknowledgements We would like to acknowledge the contributions of the metallurgists, especially Prof. Mogadalai P. Gururajan who participated in testing and evaluating our information retrieval chatbot for alloy selection. His expertise and feedback were invaluable in refining the chatbot's functionalities and ensuring its effectiveness in assisting metallurgists in alloy selection

References

1. Welleck, S. et al.: Neural text generation with unlikelihood training. arXiv preprint arXiv:1908.04319 (2019)
2. Brown, T. et al.: Language models are few-shot learners. Adv. Neural Inf. Process. Syst. **33**, 1877–1901 (2020)
3. Ouyang, L. et al.: Training language models to follow instructions with human feedback. Adv. Neural Inf. Process. Syst. **35**, 27730–27744 (2022)
4. Radford, A. et al.: Improving language understanding by generative pre-training. (2018)
5. Radford, A. et al.: Better language models and their implications. OpenAI Blog **1**(2) (2019)
6. Radford, A. et al. Language models are unsupervised multitask learners. OpenAI Blog **1**(8), 9 (2019)
7. OpenAI, R.: GPT-4 technical report. arXiv 2303–08774 (2023)
8. Liu, Y. et al.: Summary of chatgpt-related research and perspective towards the future of large language models. Meta-Radiology 100017 (2023)
9. Ai, Q. et al.: Information retrieval meets large language models: a strategic report from Chinese IR community. AI Open **4**, 80–90 (2023)
10. Jin, Q. et al. Genegpt: augmenting large language models with domain tools for improved access to biomedical information. ArXiv (2023)
11. Singh, D., Nishane, I., Rajendran, R.: Catalyzing python learning: assessing an LLM-based conversational agent. ICCE (2023)
12. Shao, Z. et al. Enhancing retrieval-augmented large language models with iterative retrieval-generation synergy. arXiv preprint arXiv:2305.15294 (2023)
13. Wan, Z. et al.: Gpt-re: in-context learning for relation extraction using large language models. arXiv preprint arXiv:2305.02105 (2023)

Estimating Soil Quality Content of Arid and Semi-Arid Districts of Punjab Using Machine Learning Techniques

Mandeep Kaur and Kamal Malik

Abstract The soil is the prominent and the primary medium for the effective growth of plants whether it is for food or plantation. Plants can thrive and grow rapidly in healthy soil if the chemical and physical attributes of soil are quite conducive to the plant development. The crucial objective of this research paper is to estimate the health of the soil in Arid and Semi-Arid regions of Punjab. A total of eight districts were selected for this research, four from the arid region and four from the semi-arid region. The soil dataset was obtained and analyzed using the R programming language. Several graphical and data visualization features were used and results were obtained. The results obtained showed that the Bathinda district had the lowest amount of N content of all the eight districts, followed by Mansa and Sangrur. The regions considered for this study were found to be a deficit in P and abundant in K. The N content was low in the arid region as compared to the semi-arid region.

Keywords Soil analysis · Punjab · Semi-arid · Arid · R programming language

1 Introduction

Whether it's for food crops or plantations, the soil is the primary medium for growing plants. Plants can thrive in healthy soil if the physical and chemical qualities of the soil are conducive to plant development. Depending on the texture, structure, and humidity of the soil might reveal healthy conditions. The quantity of nutrients required by plants is connected to soil chemical characteristics, with the amount required varying depending on the growth phase. The number of macronutrients, secondary nutrients, and micronutrients in the soil determines soil fertility in the major part. Plant development will be hampered by nutrient deficiencies in the soil.

M. Kaur (✉) · K. Malik
Department of Computer Science and Engineering, CT University, Ludhiana, India
e-mail: Mandeepdhamoo88@gmail.com

K. Malik
CT University Ludhiana Punjab, Ludhiana Punjab, India

B. Soni et al. (eds.), *Beyond Artificial Intelligence*, Lecture Notes in Networks and Systems 1326, https://doi.org/10.1007/978-981-96-4170-3_31

Because all nutrients do not reach minimal needs, the growth rate will be disrupted and prone to disease. Excess nutrients in the soil, on the other hand, will have an impact on plant development and the environment [1]. Adoption of the strong and effective cropping practices and high yielding techniques actually results in the deficiency of nutrients in the crop plants all across the globe.

It is necessary to maintain India's soil fertility in a way that allows for long-term high-level output to feed the rising population. For appropriate soil fertility management, it is necessary to carefully identify the boundaries of present nutritional deficits and to monitor changes in soil fertility to forecast their scarcity [2]. To sustain high productivity and maintain food production at an acceptable level, soil fertility indices must be continually preserved at the basic level by utilizing the effective measure in terms of supplying nutrients, water and fertilizers for providing energy for soil managements. To enhance the fertility, increase productivity, and protect the environment, adequate soil cultivation is required. The accurate soil cultivation process to be followed, soil tests play an efficient role in predicting it properly [3].

In the context of sustainable agricultural production, soil classification is important in connection to an assessment of an area's or region's fertility status. Also, machine learning methods and data analysis languages like R are widely used to study and analyze soil nutrients. The results of the testing of soil can be estimated efficiently using K-nearest neighbour to classify the soil based on the fertility indices by choosing the shortest distance. With k-nearest neighbour the fertility indices and the accurate effective measures can be predicted upto a large extent. The examination of soil nutrients is highly beneficial for farmers in selecting the type of crop to be produced using a certain cropping pattern and applying the appropriate quantity of fertilizer to a specific soil condition to enhance crop yield and profit [4].

As a result of the benefits and perfection of R, we sought to use R language analysis tools to study soil macronutrients. We are hopeful that the findings will be good, and that R will be able to assist in the research and classification of soil conditions. R was utilized to investigate the link between the soil's nitrogen (N), phosphorus (P), and potassium (K) levels.

2 Materials and Methods

2.1 Outline of the Work

The objective of the proposed work is to analyze the soil health of the arid and semi-arid regions of Punjab. It will help in determining which soil nutrient is in abundance and which is in deficit from the specific area. The methodology starts with site selection; followed by sample collection from the site; later collected samples were tested and the soil data was collected and pre-processing of the dataset has been carried out to filter missing attributes values, noisy data, and miss-match. Following

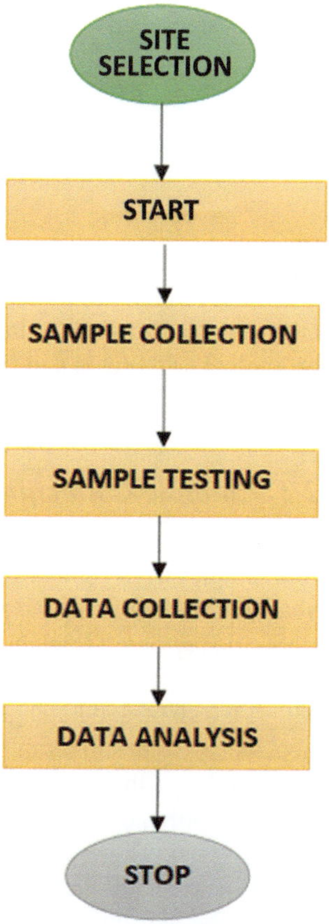

Fig. 1 Overview of the work

the analysis of the filtered data, the soil dataset was evaluated using R plotting features. Figure 1 depicts an elevated description of the research.

2.2 *Study Site*

The purpose of this study was to examine the land capacities and suitability for various crops, as well as the area's soil fertility limits, in Punjab's dry and semi-arid zones. Punjab's rich plains provide around two-thirds of all food grains produced in India each year. It's no surprise that Punjab is known as India's breadbasket due to its large wheat output. This is primarily due to Punjab's excellent soil [5].

Climate variances provide a diverse range of soil, which leads to a diverse range of crops and plants in Punjab. The calcareous soil, which comprises desert soil and sierozem soil, dominates south-western Punjab. The pH ranges from 7.8 to 8.5 in this zone, which also contains grey and red desert soil, calsisol soil, regosol soil, and

Table 1 Sample collection

Arid		Semi-arid	
Districts	No. of samples	Districts	No. of samples
Mansa	12	Ludhiana	11
Sangrur	11	Moga	13
Bathinda	10	Faridkot	15
Firozpur	13	Fatehgarh Sahib	11

alluvial soil. This zone's alluvial soil is commonly referred to as arid and brown soil or tropical arid brown soil [6].

2.3 Sample Collection

Soil samples from eight different locations in Punjab (Table 1) at different parts of the year are considered for the present study. The locations were selected from arid and semi-arid regions of Punjab. From the arid region, four districts were selected namely; Mansa, Sangrur, Bathinda, and Firozpur. Semi-arid, districts namely; Ludhiana, Moga, Faridkot, and Fatehgarh Sahib were selected. With the gap of three days duration, the different soil samples were collected By utilizing sterilized and clean scalpel, the samples of the soil were collected from each spot with no gravels and debris in sterilized polythene bags and labelled according to the sampling sites with the date of collection. These soil samples were sent to the various Krishi Vigyan Kendra (KVKs) associated in various districts in Punjab, for the estimation of parameters such as the percentage of available nitrogen (N), available phosphorus (P), organic carbon (OC), and available potassium (K).

2.4 Dataset Collection

The objective of the proposed work is to analyze the soil health of the arid and semi-arid regions of Punjab. It will help in determining which soil nutrient is in abundance and which is in deficit from the specific area. In this work, a soil dataset was collected from various districts of Punjab. These districts belong to arid and semi-arid regions of Punjab. Dataset has 4 characteristics, as well as 96 soil sample instances. The attribute description is shown in Table 2.

Table 2 Attributes and description

Attribute	Description
N	Nitrogen (Kg/ha)
P	Phosphorus (Kg/ha)
K	Potassium (Kg/ha)
OC	Organic Carbon (%)

2.5 Dataset Analysis

The pre-processed and filtered dataset will be analyzed using the R programming language. The R core team and the R Foundation for Statistical Computation promote R, a programming language for statistical computing and graphics. Data miners and statisticians use it to analyse data and produce statistical tools. The K-nearest neighbour is used for analysing all the samples. The level of the corrosivity of the soil based on their resistance value was predicted. The nearest classifier is optimally chosen to create the predictions depending upon the local data. The data set is divided into two parts 75% of the data is utilized for the training purpose and 25% of the data is used for the testing purpose. K-Positive integers from the training datasets were determined. The closest neighbours from K has been chosen from the new data and then general classification mechanism is applied on it.

3 Results and Discussion

3.1 Data Overview

In this study, a soil dataset was collected from various districts of Punjab. These districts belong to Punjab's arid and semi-arid regions. The data set consists of 96 observations and 4attributes; OC (%), N (Kg/ha), P (Kg/ha), and K (Kg/ha). Scatter plots (Fig. 2) mainly depict how one variable influence another. A scatter plot of matrices was created using the panels in the R language's psych package, including bivariate scatter plots below the diagonal, histograms on the diagonal, and the Pearson correlation above the diagonal. The correlation coefficients have been provided in the upper triangle. These variables help to separate soil samples.

The bivariate scatters plot of the soil dataset showed the correlation between the attributes. As shown in Fig. 2, all the attributes had a positive correlation with each other. The OC and N showed the highest correlation coefficient, followed by N and K. The attribute P had the lowest correlation coefficient with K, followed by OC.

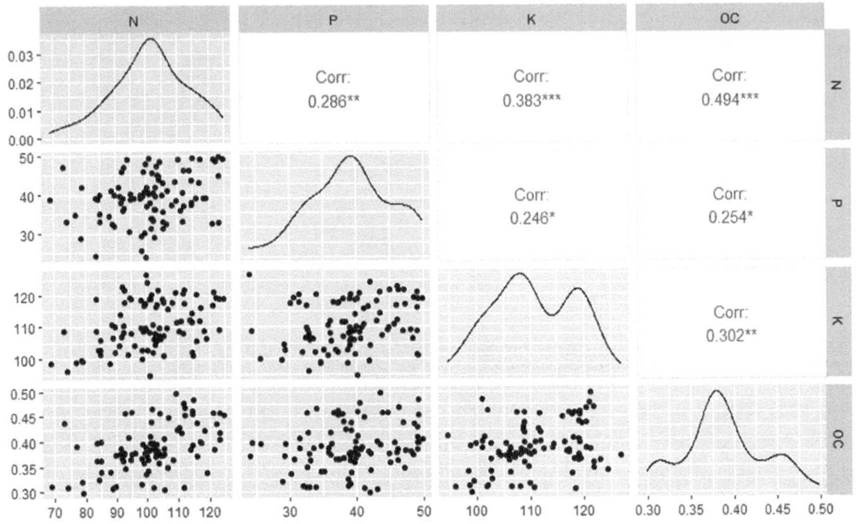

Fig. 2 Bivariate scatter plot of the soil dataset

3.2 Arid Region of Punjab

To study the ground truth sound quality of the arid region of Punjab, 46 soil samples were collected. The samples were tested and the values of available N, P, and K were noted down. With the help of these values, graphs were plotted showing the distribution of N, P, and K in an overall arid region. The graphs shown in Fig. 3 were plotted using the ggplot and plotly packages of R. To plot the pie chart the pie () function was used and the line graph was plotted using plotly function. In the line graph, the black dots represent the sample number and the value corresponding to the dot is the value of a particular macronutrient present in it.

For the soil to be productive, it should contain the major nutrients N, P, and K in the optimal ratio of 4:2:1. As shown in the graphs displayed in Fig. 3, the samples from the arid region don't contain the N, P, and K in the ratio of 4:2:1, as the soil contains K in abundance. The P content of the soil is even below the required amount. As per the study, the arid region soil is a deficit in P and higher content of K was present. The N content of the soil is average. It can require some N fertilizer depending upon the kind of crop being cultivated.

3.2.1 District Wise Distribution

The four arid region districts involved in this study were: Mansa, Sangrur, Bathinda, and Firozpur. From Mansa district total of 12 soil samples were taken, from Sangrur 11, Bathinda 10, and Firozpur 13 samples were taken. The soil samples were tested and the values of macronutrients were obtained. These values were compared and

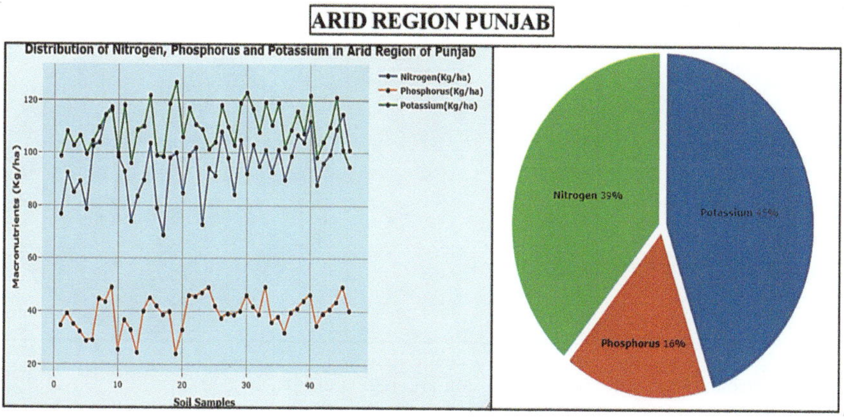

Fig. 3 Distribution of N, P, and K in the arid region of Punjab

analyzed to study the soil health of the regions. The graph of the obtained data was plotted using the plotly package of the R language. The district-wise distribution of arid region soil content is shown in Fig. 4.

As shown in Fig. 4, in every district the soil contains K in more quantity as compared to N and P. As per the study, on average the N content in Bathinda was lowest as compared to other districts. Each district's soil doesn't contain the N, P, and K in the optimal ratio of 4:2:1. In Bathinda, the N was 88.326 kg/ha, and P was 38.383. The N and P were somehow in the ratio of 4:2, but the total K content is very high, due to which the N, P, and K ratio doesn't fall into the optimal category.

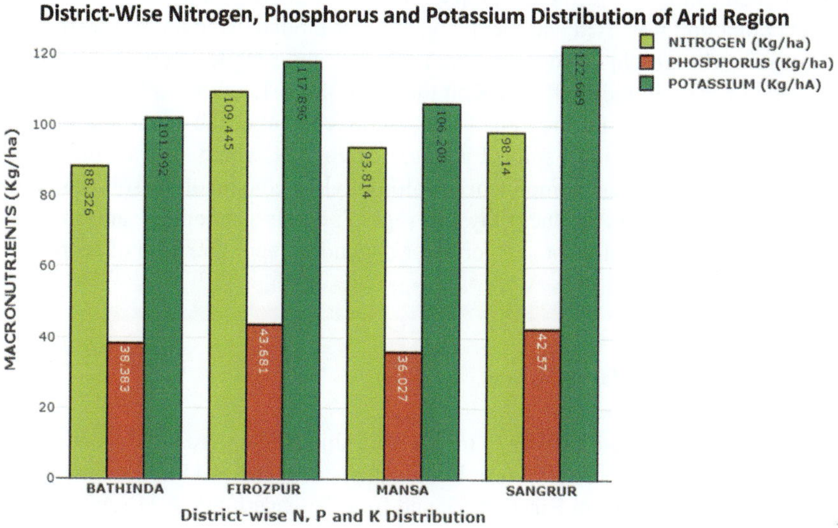

Fig. 4 District-wise distribution of N, P, and K in the arid region of Punjab

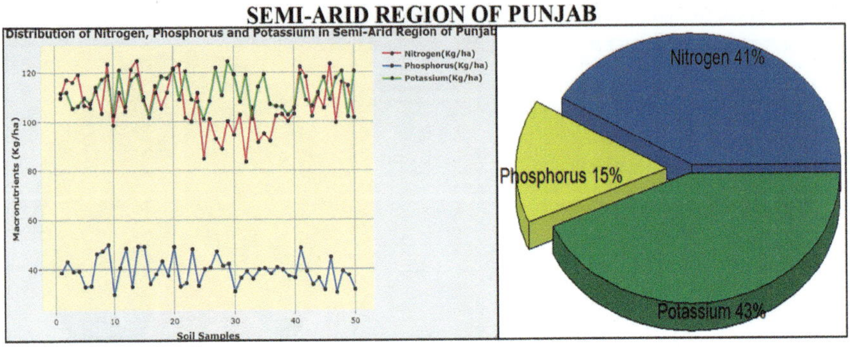

Fig. 5 Distribution of N, P, and K in the semi-arid region of Punjab

Furthermore, in the district-wise distribution, we can see that the P is lacking in the soil of the arid region, and the presence of N is also fluctuating between optimal and below optimal value.

3.3 Semi-Arid Region of Punjab

To study the ground truth sound quality of the semi-arid region of Punjab, 50 soil samples were collected. The samples were tested and the values of available N, P, and K were noted down. With the help of these values, graphs were plotted showing the distribution of N, P, and K in an overall semi-arid region. The graphs shown in Fig. 5 were plotted using the plotrix and plotly packages of R. To plot the 3D pie chart the pie3D () function was used and the line graph was plotted using plotly function.

As shown by graphical representation (Fig. 5), the samples from the semi-arid region don't contain the N, P, and K in the ratio of 4:2:1, as the soil contains K in abundance just like the arid region. The N and K content of the semi-arid region is almost similar. This region's soil is found to be rich in N, which makes it productive because N is an important component of chlorophyll. P content of the soil was below the required amount. As per the study, this region's soil is a deficit in P and abundant in N and K. It can require some P fertilizer depending upon the kind of crop being cultivated.

3.3.1 District Wise Distribution

The four semi-arid region districts involved in this study were: Ludhiana, Moga, Faridkot, and Fatehgarh Sahib. From Ludhiana district total of 11 soil samples were taken, from Moga 13, from Faridkot 15, and from Fatehgarh Sahib 11 samples were taken. The soil samples were tested and the values of macronutrients were obtained. These values were compared and analyzed to study the soil health of the regions. The

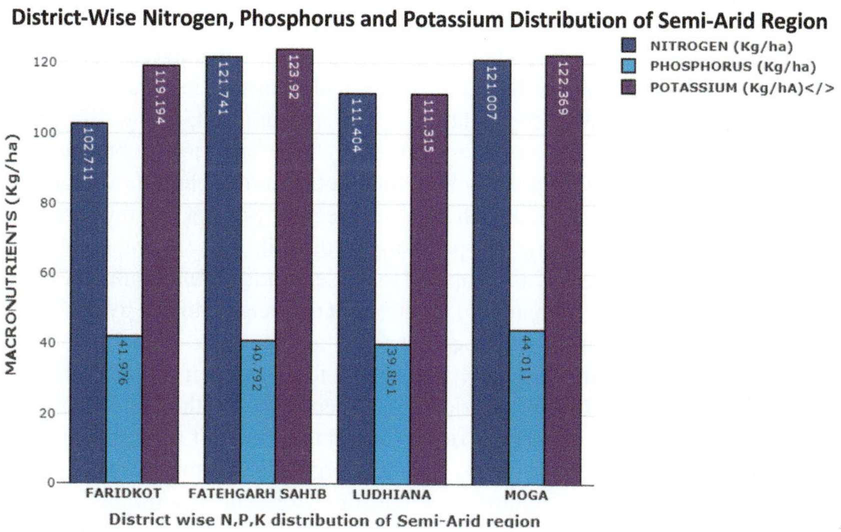

Fig. 6 District-wise distribution of N, P, and K in the semi-arid region of Punjab

graph of the obtained data was plotted using the plotly package of the R language. The district-wise distribution of semi-arid region soil content is shown in Fig. 6.

As shown in Fig. 6, in almost every district the soil contains N, and K in equal quantity. Moreover, the soil of the semi-arid district doesn't contain the N, P, and K in an optimal ratio of 4:2:1. As per the study, on average the N content in Faridkot was lowest as compared to other districts. In Bathinda, the N was 88.326 kg/ha, and P was 38.383. In Ludhiana, Moga, and Fatehgarh Sahib the N, and P content is equal. The soil of this region is abundant in N which makes it very productive, but the higher content of K is questionable and requires some measures. The presence of more K may suggest deep clay soil, which retains the mineral and causes it to concentrate to dangerously high levels. It's also possible that the fertilizer being applied has too much of it in it.

4 Conclusion

This study was conducted to analyze the soil health of the arid and semi-arid regions of Punjab. The main objective of this study was to determine which soil nutrient was in abundance and which was the deficit in a specific region. The arid region districts involved in this study were: Mansa, Sangrur, Bathinda, Firozpur, and semi-arid districts were: Ludhiana, Moga, Faridkot, and Fatehgarh Sahib. The methodology starts with site selection; followed by sample collection from the site; later collected samples were tested and the soil data was collected and pre-processed. The data set consists of 96 observations and 4 attributes namely; organic carbon (OC), nitrogen (N), phosphorus (P), and potassium (K). The filtered data were analyzed using R plotting features followed by an evaluation of the soil dataset.

As per the experimental results, the attributes had a positive correlation with each other. The OC and N showed the highest correlation coefficient, followed by N and K. Furthermore, the soil considered for this study was found to be abundant in K content. The explanation for this might be that potassium leaching is slow in soil, and therefore applying potassium-rich fertilizers regularly over a lengthy period can increase the quantity of exchangeable potassium in the soil. The soil samples obtained from the arid region had less N as compared to the semi-arid region. Bathinda district had the lowest amount of N content of all the eight districts, followed by Mansa and Sangrur. The P content of the soil was found on the same scale from 35–45 kg/ha. The regions considered for this study were found to be a deficit in P and abundant in K. The N content was low in the arid region as compared to the semi-arid region. The soil of the semi-arid region was rich in N and this makes it very productive, as higher the soil nitrogen supply value, the more probable soil microorganisms will convert more organic nitrogen into mineral nitrogen for plant absorption.

References

1. Lenka, S., Malviya, S.K., Lenka, N.K., Sahoo, S., Bhattacharjya, S., Jain, R.C., Saha, J. K., Patra, A.K.: Manure addition influences the effect of tillage on soil aggregation and aggregate associated carbon in a vertisol of Central India. J. Environ. Biol. **41**(6), 1585–1593 (2020). https://doi.org/10.22438/jeb/41/6/si-221
2. Srivastava, P., Balhara, M., Giri, B.: Soil health in India: past history and future perspective. Soil Biol. 1–19 (2020). https://doi.org/10.1007/978-3-030-44364-1_1
3. Kaur, S., Malik, K.: Predicting and estimating the major nutrients of soil using machine learning techniques. Soft Comput. Intell. Syst. 539–546 (2021). https://doi.org/10.1007/978-981-16-1048-6_43
4. Kaur, S., Malik, K.: Evaluating various machine learning techniques for crop nutrients prediction. SSRN Electron. J. (2022). https://doi.org/10.2139/ssrn.4031986
5. Official Website of Department of Agriculture & Farmer Welfare | Government of Punjab, India. (n.d.): Guidelines and presentation twww. Retrieved 20 March 2022. https://agri.punjab.gov.in/
6. Punjab Soils | Department of Soil and Water Conservation, Punjab | Government of Punjab. Department of Soil & Water Conservation, Punjab. Retrieved March 20, 2022, from https://dswcpunjab.gov.in/contents/punjab_soils.html

Implementation of CNN-Based Word Recognition System

Bittu Kumar, Gudi Srikanth, Vijay Rao Kumbhare, Amit Kumar Shrivastava, and Ashwini Kumar Varma

Abstract In recent years, sign language plays a vital role as a means of communication for individuals who are deaf or hard of hearing. However, the challenge lies in the fact that not everyone is proficient in sign language, leading to a communication gap between the deaf community and those who use spoken language. To address this issue, we have developed a system for converting American Sign Language (ASL) into English text in real-time, offering a solution that can bridge this gap. Our system leverages deep learning techniques that combine convolutional and recurrent neural networks to recognize hand movements and convert them into text. This innovative method has the potential to significantly improve accessibility and communication for the deaf community while maintaining a high degree of accuracy. Additionally, we conduct hand movement modeling by dissecting hand movement components, utilizing sign language-independent data. This modeling process is accomplished through the utilization of technologies such as TDNN, CNN, and DNN. It is important to note that while there have been advancements in sign language recognition technology, it is still in its nascent stages of development. We evaluate the performance of our proposed methods through a comprehensive analysis based on our internally generated standard dataset. This assessment helps us gauge the effectiveness and accuracy of our system in real-world applications.

Keywords American sign language (ASL) · Deep neural network (DNN) · Time delay neural network (TDNN). Convolutional neural network (CNN)

1 Introduction

Effective communication is essential for everyone, but individuals who are deaf or have hearing impairments often face challenges in interacting with the larger hearing community. Sign language serves as a vital means of communication for the deaf,

B. Kumar (✉) · G. Srikanth · V. R. Kumbhare · A. K. Shrivastava · A. K. Varma
Department of Electronics and Communication Engineering, Koneru Lakshmaiah Education Foundation, Hyderabad, India
e-mail: bittu.mlrit@gmail.com

B. Soni et al. (eds.), *Beyond Artificial Intelligence*, Lecture Notes in Networks and Systems 1326, https://doi.org/10.1007/978-981-96-4170-3_32

with many nations recognizing it as an official language. However, communication barriers can arise because not everyone is proficient in sign language. To address this issue, we have developed a system that can instantly convert American Sign Language (ASL) into English text, offering a solution to enhance communication between the deaf and hearing communities. Hand gesture detection, a significant area of study for computer vision and machine learning researchers, plays a pivotal role in this domain. Creating systems capable of recognizing specific signs and utilizing them to convey information or control devices is a primary objective. It's important to note that while gestures involve dynamic hand movements, hand postures represent the static configuration of the hand, and recognizing gestures requires an understanding of both spatial and temporal aspects. There are two fundamental techniques for recognizing hand movements: vision-based approaches and data glove methods. In our pursuit, we focus on developing both vision-based and data glove methods, with a primary emphasis on creating a real-time sign language recognition system based on vision. A vision-based system is preferred because it offers a more intuitive and natural mode of communication between humans and machines. Figure 1 illustrates standard hand gestures for alphabets and numbers.

Within the realm of "digital image processing," computer algorithms play a pivotal role in manipulating digital photographs to extract valuable information about the captured scenes [5–7]. Images are generated through a variety of physical devices, including cameras, x-ray machines, and microscopes, and are applied in a wide array of fields such as entertainment, healthcare, business, industry, military, civil, security, and research. Digital image processing encompasses two fundamental steps: image enhancement and analysis [8, 9]. Image enhancement seeks to accentuate

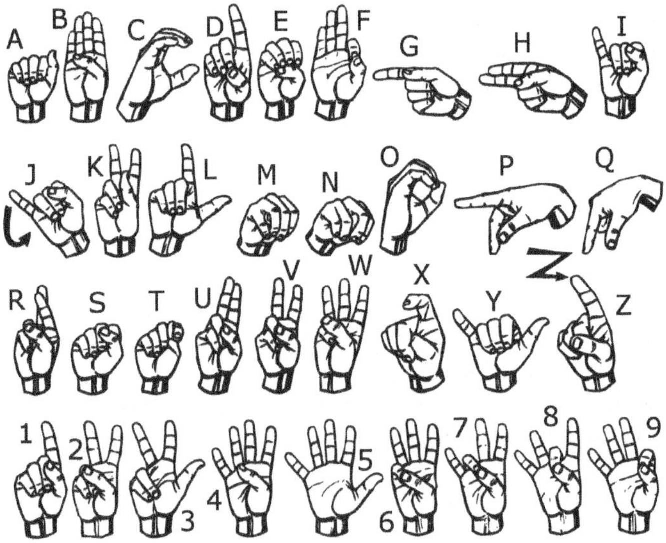

Fig. 1 Sign Language chart for alphabets

significant image details while reducing the prominence of less critical elements, while image analysis aims to derive insights about the depicted scenes. Digital image processing revolves around the conversion of images into arrays of pixels, each representing a specific intensity value in digital memory. This digital approach offers numerous advantages over analogue image processing, including superior speed, cost-effectiveness, and versatility. These digital representations are subsequently processed by computers or other digital hardware [10].

In the context of digital image processing, pictures are depicted as two-dimensional functions of spatial coordinates, with their intensities serving as the amplitudes. Digital images rely on finite sets of data values known as pixels, while analogue image processing involves manipulating images by modifying electrical signals. Signals, broadly defined as quantifiable quantities that convey information across time or space, play a central role in this process. To create a digital image, digital cameras leverage natural light to generate a continuous voltage signal, which is then sampled and quantized into a two-dimensional array or matrix of values. As image and video processing techniques continue to advance, research in the field of human–machine interaction [11, 12] has gained increasing significance, given the pivotal role of these technologies in everyday human communication.

The subsequent sections of this paper are structured as follows: Sect. 2 provides a concise review of the techniques utilized in sign language recognition. Section 3 delves into the methodology, while Sect. 4 presents results and discussions. Finally, Sect. 5 offers conclusions and outlines the research's future direction and scope.

2 Literature Survey

In [1], the authors present a system designed for recognizing sign words based on hand gestures using a convolutional neural network (CNN). The system comprises multiple stages, including pre-processing, feature extraction, and classification, which collectively enable sign word recognition and hand movement detection. Pre-processing encompasses several steps, such as converting hand gesture images to YCbCr, selecting grayscale images, binarization, erosion, and gap filling. Features are then extracted from these processed images and fed into the classifier, making use of a deep learning technique known as CNN. Sign gestures are categorized using a SoftMax classifier, and real-time implementation is achieved using a webcam. The system boasts an average recognition rate of 96.96% and outperforms state-of-the-art systems. Future research endeavors aim to develop an algorithm capable of interpreting complex sentences and sign words based on hand gestures, further enhancing the sign language recognition system.

In [2], researchers propose two models for the recognition of 3-D handwritten characters and gestures utilizing smartphone technology. Smartphones are equipped with sensors such as accelerometers, gyroscopes, and gravity sensors, enabling the tracking of the phone's movements in three-dimensional space. The models leverage data from these sensors to make the systems more expressive and challenging to mimic. The accelerometer's 3-axis readings are employed to calculate hand movement speed, while the gyroscope data yields a quaternion rotation matrix that reduces inclination offset and tilting. Automatic segmentation aids in determining each character or gesture within a sequence. Both models are equipped for both user-dependent and user-independent recognition of 3D characters and motions. The results indicate that Model I attains a recognition rate of 90% to 94% with minimal instruction, while Model II achieves an 88% recognition rate. Model I involves more computational resources for training and evaluation, while Model II is computationally more efficient.

In [3], the authors explore the utilization of hand movement subunits derived from Hidden Markov Models (HMMs) to represent hand movement data in a language-independent manner. The research sheds light on the differences in performance between modeling hand movement data in a language-independent and language-dependent manner. It is evident that when hand shape information is combined with this gap, competitive systems can be devised. Sharing multiple sign language resources emerges as a positive outcome of these findings, fostering the development of sign language processing systems. The research results lay the foundation for future work in the processing of sign language, considering resource limitations through the creation of systems with fewer signers and examples. Furthermore, the study investigates the applicability of this multilingual approach in evaluating sign language.

In [4], the management of diverse data generated by two distinct devices is addressed through the use of various feature sets. While the Kinect provides a comprehensive depth map, the Leap Motion supplies a more constrained data description. The proposed feature set and order calculation allow for excellent overall accuracy, even in scenarios where certain fingers might not be recognized. Integrating the two devices, with the Kinect's depth map providing additional information, can lead to enhanced attribute acquisition, addressing the gaps in the Leap Motion output. Test results demonstrate significant performance improvements by allocating specific fingers to distinct angular zones. Future research endeavors will focus on the development of new features based on the combined 3D positions from these two devices.

In [5], the authors demonstrate gesture recognition pipelines utilizing Leap Motion and depth sensor-based technologies. While Leap Motion provides limited information, depth cameras offer a more detailed description of hand shape. A calibration method is proposed to enable the joint utilization of data from both devices. Leap Motion data is harnessed to enhance computational speed and improve the precision of depth traits, with distinct feature sets designed for each sensor. The proposed feature sets and classification algorithms exhibit strong accuracy, particularly for depth data. Future research will concentrate on dynamic gesture recognition and the advancement of methods for translating and tracking fingertip motion.

3 Methodology

The system outlined in the input is a vision-based method designed for recognizing sign language. It enables interaction without the need for artificial devices, relying solely on bare hands to represent signs. The system's components encompass dataset creation, gesture categorization, the utilization of a CNN model [13], the implementation of activation functions, pooling layers, the formation of finger-spelling sentences, and the stages of training and testing.

3.1 Dataset Generation

The initial phase of our system implementation, in light of the absence of readily available datasets in the form of raw images, necessitated the creation of a dataset from the ground up. This involved the meticulous capture of each American Sign Language (ASL) symbol, with the intention of using these images for both training and testing purposes, employing the OpenCV programming framework [14, 15]. In the image acquisition process, a distinct blue-bordered rectangle was employed to delineate the Region of Interest (ROI) within each frame. This ROI was subsequently extracted from the entire image, subjected to grayscale conversion, and further processed through the application of a Gaussian blur filter, thereby revealing a multitude of distinctive features. To provide a visual context, Fig. 2 provides a graphic representation of the ROI, offering a clear perspective on the defined region. In Fig. 3, you can observe the resultant image obtained from processing the grayscale ROI image, highlighting the image refinement process. Furthermore, Fig. 4 showcases the various attributes and characteristics inherent in an image that has undergone the Gaussian blur filtering, underscoring the significance of these distinct features for subsequent stages in our system. This comprehensive dataset creation procedure lays the foundation for robust training and testing, enabling the system to accurately recognize and interpret a wide array of ASL symbols in real-world scenarios.

3.2 Gesture Classification

Advancing to the subsequent phase of the process, which is gesture classification, this stage entails a detailed analysis of the user's final symbol, achieved through the utilization of a two-tiered algorithmic approach. In Algorithm Layer 1, the initial step involves the application of a Gaussian blur filter and a threshold to the frames obtained through OpenCV. This processed image is then dispatched for prediction to the Convolutional Neural Network (CNN) model, a crucial component of the system [16–18]. When a symbol is detected in over 50 frames, it is printed and contributes to the formation of a word, and to represent the space between words, a blank symbol

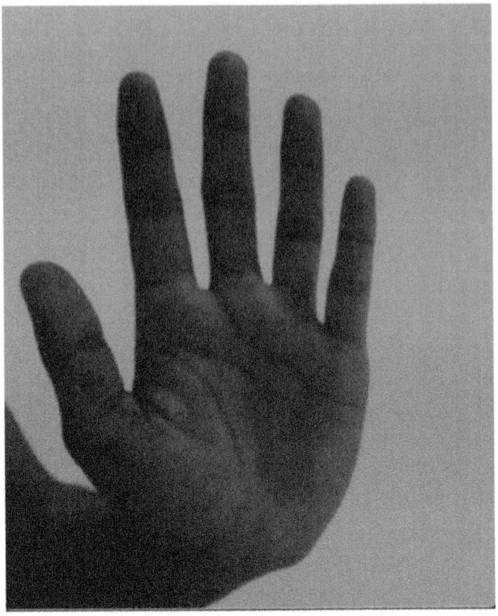

Fig. 2 ROI image

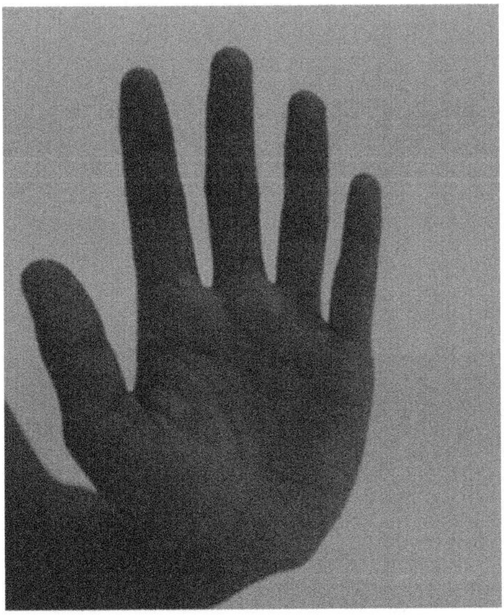

Fig. 3 Greyscale image

is incorporated. Moving on to Algorithm Layer 2, the focus is on identifying distinct sets of symbols that exhibit similar outcomes. For this purpose, specialized classifiers designed explicitly for these symbol sets are employed to effectively distinguish between them [19–21]. This second algorithmic layer adds another dimension to the system's capability by facilitating the recognition of complex combinations of symbols and improving the overall accuracy of sign language recognition. This multi-layered approach is essential for handling the intricacies of sign language gestures,

Fig. 4 Gaussian blur image

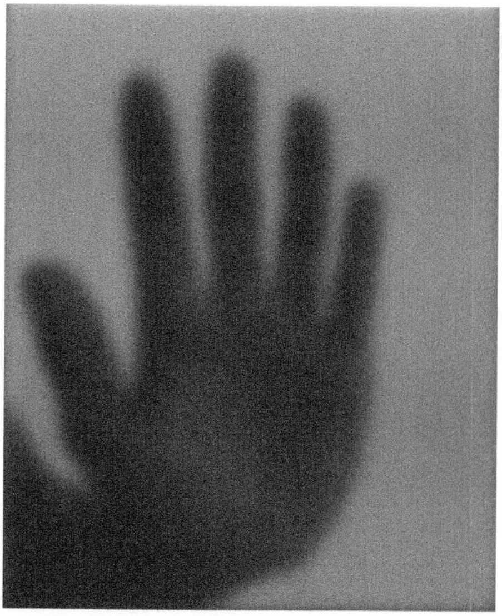

enabling the system to effectively interpret and convert a wide range of signs into text. The combination of Algorithm Layer 1 and Layer 2 ensures that the system can accurately and efficiently recognize and translate complex sign language gestures into textual form.

3.3 CNN Model

At the heart of our system lies the critical CNN model, a fundamental component responsible for interpreting and processing the input data. In this model, the input image possesses a resolution of 128 × 128 pixels in the initial convolutional layer. This layer employs 32 filter weights to process the input image, resulting in its downscaling to 63 by 63 pixels, achieved through 2 × 2 maximum pooling. Subsequently, the output from the first pooling layer undergoes further processing in the second convolutional layer, utilizing an additional set of 32 filter weights, ultimately producing an image with dimensions of 60 × 60 pixels. Once more, the images are subjected to downsampling via maximum pooling with a 2 × 2 pool size, culminating in a final image resolution of 30 × 30 pixels.

The CNN model consists of 128 neurons in the top densely connected layer, complemented by an additional 96 neurons in the lower densely connected layer. The total number of neurons in the output layer corresponds to the number of classes we are classifying (in this context, these represent the sequential letters as well as a clear image). For a visual representation of this process, please refer to Fig. 5, which illustrates the matrix convolution of the image. Initially, we start with the

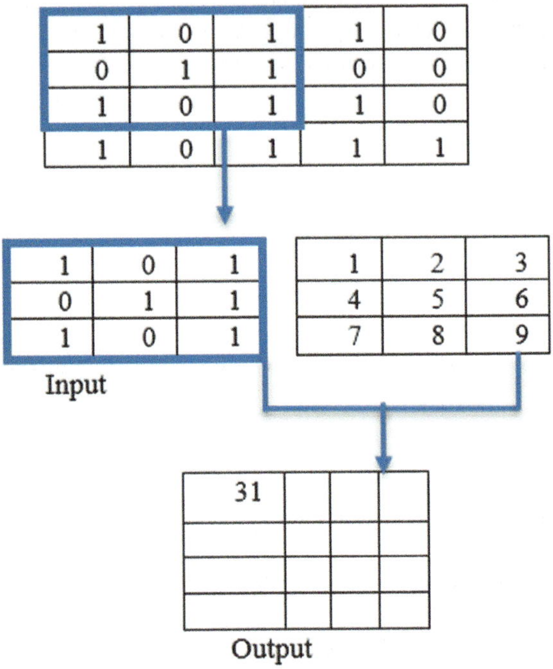

Fig. 5 Matrix convolution of image

input matrix, extracting the blue color-layered image patch (Local receptive field) from the input. This patch is then subjected to multiplication by a Kernel (Filter), resulting in the generation of an output. This convolutional process is pivotal in extracting distinctive features and patterns from the input image, allowing the model to make accurate classifications.

3.4 Activation Function

The Rectified Linear Unit (RELU) serves as the chosen activation function, effectively implemented in every layer of the system, whether in the convolutional layers or the fully connected neurons. This activation function plays a pivotal role in introducing nonlinearity into the computational model, thereby enabling it to learn intricate and nuanced features. In this context, when negative values exist within the input signals, it suggests that the image exhibits blurriness or lacks sharpness. To address this, the system employs the RELU activation function. Its primary function is to rectify these negative values by converting them into zeros, leaving the positive values unchanged. Consequently, this operation effectively eliminates blurriness and enhances the image's clarity. In essence, the application of RELU ensures that the image remains crisp and devoid of any residual fuzziness. Pooling Layer.

3.5 Finger Spelling Sentence Formation Implementation

The process of generating fingerprint statements within the system is both intricate and nuanced. It involves multiple steps and considerations to ensure accurate recognition and interpretation of the input data. Firstly, the system begins by printing a letter, which subsequently becomes part of the ongoing string under specific conditions. To be deemed eligible for inclusion in the string, a letter must meet a certain count threshold, indicating its prominence within the image. Furthermore, the system applies a proximity threshold to verify that no other letter is in close proximity to the chosen one. This helps in avoiding ambiguities and inaccuracies. In cases where an incorrect letter is anticipated or expected, the system takes proactive measures. It removes the current word reference, which keeps track of the number of hits in the current image. By doing so, the system prevents any misguided expectations related to non-primary letters, thereby maintaining the integrity of the recognition process.

Another crucial aspect is the detection of blanks within the input data. If the system detects that the number of identified blanks (referred to as array backgrounds) surpasses a predefined value, it refrains from detecting any blanks within the current buffer. However, if the number of detected blanks is within acceptable limits, the system proceeds to print a space. This space signifies the conclusion of a word, and the current word is appended to the ongoing sentence. This intricate process ensures that fingerprint statements are generated with precision and accuracy, considering a variety of factors and thresholds to facilitate effective recognition and interpretation of sign language gestures.

3.6 Training and Testing

The training and testing phases within the system play a pivotal role in ensuring accurate recognition and classification of input images. These phases encompass various steps and techniques that enhance the system's ability to interpret sign language gestures effectively. To commence, the process involves pre-processing the input images. This pre-processing includes the essential steps of grayscale conversion and Gaussian blurring. Grayscale conversion serves to simplify the image and reduce it to a monochrome representation, making it more amenable for analysis. Simultaneously, applying Gaussian blurring helps eliminate unnecessary noise and artifacts within the image, further enhancing the quality and clarity of the visual data. Following pre-processing, the downsizing of images to a standardized 128 × 128 resolution is executed. This step ensures uniformity in the input data and prepares the images for evaluation by the Convolutional Neural Network (CNN) model. The CNN model is instrumental in assessing and classifying the images based on their content. The prediction layer of the system is where the final evaluation takes place. This layer employs the CNN model to determine the likelihood that an image belongs to any of the predefined classes, which typically represent different sign language gestures or symbols. The output generated by the prediction layer is notable for its normalization, as it falls within the range of 0 to 1. Each class's output value reflects the system's assessment of the image's correspondence to that class. Importantly, the sum of the output values across all classes is constrained to equal 1. This normalization process guarantees that the system's classification decisions are relative to each other, facilitating the selection of the most likely interpretation among the available sign language gestures.

4 Result and Discussions

4.1 Data Collection

Raw images are captured using a webcam, focusing on the Region of Interest (ROI) containing the necessary hand image, specifically the "A" sign language gesture. Figure 6 displays the raw data images for the words 'A' and 'E.' These raw images undergo various processing methods, such as grayscale conversion and blurring. The subsequent step involves segmenting the image to pinpoint temporal boundaries within the data. A dedicated webcam image is captured to isolate the hand region, ensuring that the diverse shapes, movements, and materials used in various hand gestures are accurately represented. Additionally, binary data conversion is applied to the images, providing essential information for subsequent analysis and recognition of the sign language gestures.

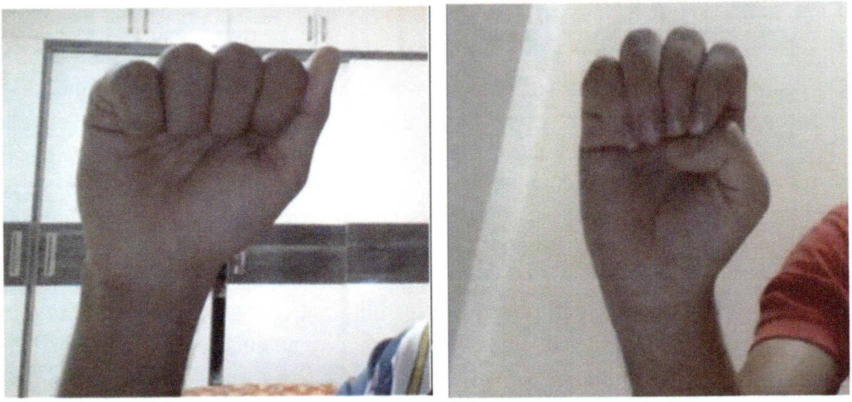

Fig. 6 Data collection alphabet 'A' and 'E'

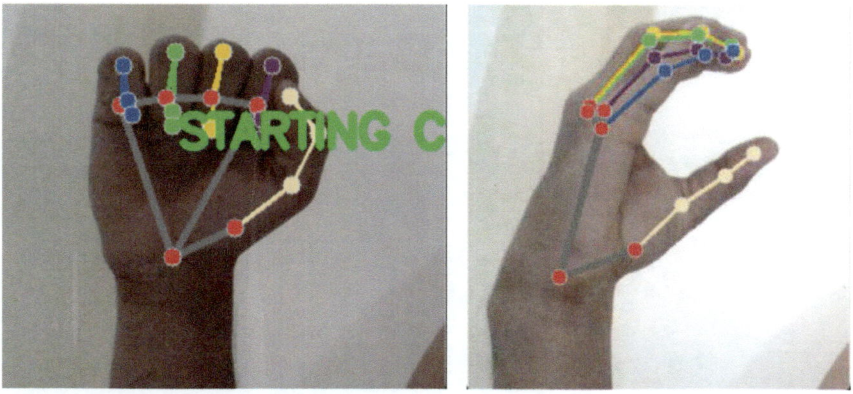

Fig. 7 Conversion of frames alphabet 'A' and 'C'

4.2 *Image Conversion*

The conversion of these image frames into a matrix format is a pivotal step that prepares the data for subsequent analysis and processing. These matrices serve as the foundation for the system's ability to recognize and interpret sign language gestures effectively. It's worth noting that the process is not limited to specific alphabet words; rather, it can handle a wide range of signs, making it versatile for various communication needs. This transformation ensures that the system can function reliably, providing accurate and meaningful translations of sign language to enhance communication for the deaf and hard of hearing (Fig. 7).

4.3 *Training and Testing*

Given their remarkable performance with image data, CNNs, which are artificial neural networks comprising convolutional layers, are the preferred approach for image classification tasks. The training images from different categories were stored in the database, and during testing, the system determined the category of the tested image by comparing it with the models derived from the training dataset. Figure 8 provides a visual representation of this process, showing a testing image with the recognized words 'W' and 'L'. These results indicate a word recognition rate of 100%. This high level of accuracy is achieved because the individuals signing these words are part of our database, meaning that their hand gestures for these words were included in the training dataset.

Figures 9 and 10 illustrate snapshots of various recognized letters, including 'W,' 'L,' 'A,' and 'F.' Notably, these words are being recognized by individuals who are not part of our generated database. From these results, it's evident that the recognition accuracy is consistently above 92%. Of particular note is the word 'F,' as shown in

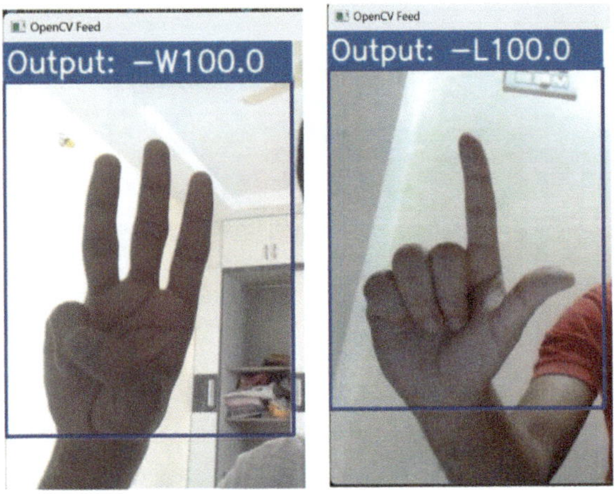

(a) (b)

Fig. 8 **a** Recognized word 'W' and **b** word 'L' (Testing from the trained dataset)

Fig. 10b, which is identified with an impressive accuracy rate of 100%. These findings underscore the effectiveness and reliability of our implemented word recognition system.

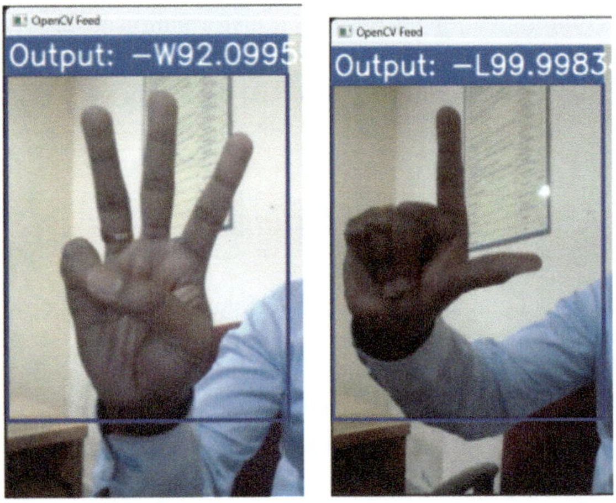

(a) (b)

Fig. 9 **a** Recognized word 'W' and **b** recognized word 'L'

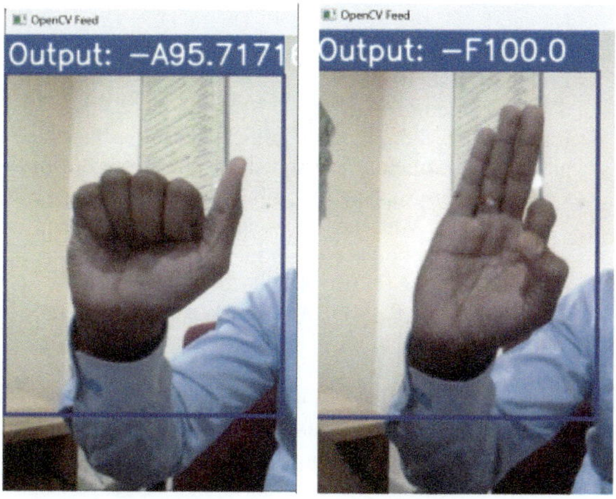

(a) (b)

Fig. 10 **a** Recognized word 'A' and **b** recognized word 'F'

5 Conclusion

This paper presents a real-time system designed to transcribe text from American Sign Language (ASL) into English, enhancing communication for sign language users. The system's core functionality involves recognizing hand movements and converting them into text through a deep learning approach that combines convolutional and recurrent neural networks. Remarkably, the system demonstrates a high degree of accuracy in both hand gesture recognition and phrase reconstruction. This method's adaptability to other sign languages holds great promise for improving accessibility and communication for the deaf community. The paper delves into the classification of gestures and the underlying process of gesture recognition through the analysis of images. Notably, the framework exhibits effectiveness in accommodating variations in lighting conditions and gesture orientations. Leveraging a specific dataset, we have achieved an impressive accuracy rate of 97.0% in symbol recognition. This technique consistently performs exceptionally well in recognizing symbols, as long as they are discernible, free from background noise interference, and exposed to adequate lighting conditions.

References

1. Cooper, H., Holt, B., Bowden, R.: Sign language recognition. Vis. Anal. Hum. (2011)
2. Chen, F.S., Fu, C.M., Huang, C.L.: Hand gesture recognition using a real-time tracking method and hidden Markov models. Image Vis. Comput. **21**(8), 745–758 (2003)
3. Marin, G., Dominio, F., Zanuttigh, P.: Hand gesture recognition with jointly calibrated leap motion and depth sensor. Multimed. Tools Appl. **75**(22), 14991–15015 (2016)

4. Kumar, P., Gauba, H., Roy, P.P., Dogra, D.P.: Coupled HMM-based multi-sensor data fusion for sign language recognition. Pattern Recogn. Lett. **86**, 1–8 (2017)
5. Lifeng, R., Jun, M., Qiushi, W.L.: The gesture identification based on invariant moments and SVM. Microcomput. Its Appl. **31**(6), 32–35 (2012)
6. Rahim, M.A., Shin, J., Islam, M.R.: Human-machine interaction based on hand gesture recognition using skeleton information of kinect sensor. In: Proceedings of the 3rd International Conference on Applications in Information Technology, ACM, pp. 75–79 (2018)
7. Yamashita, T., Watashe, T.: Hand posture recognition based on bottom-up structured deep convolutional neural network with curriculum learning. In: IEEE International Conference on Image Processing (ICIP), pp 853–857 (2014)
8. Liao, Y., Xiong, P., Min, W., Min, W., Lu, J.: Dynamic sign language recognition based on video sequence with BLSTM-3D residual networks. IEEE Access (2019)
9. Oscar, K., Necati, C., Hermann, N., Richard, B.: Weakly supervised learning with multi-stream cnn lstm-hmms to discover sequential parallelism in sign language videos. IEEE Trans. Pattern Anal. Mach. Intell. **04** (2019)
10. Tornay, S., Razavi, M., Camgoz, N.C., Bowden, R., Doss, M.M.: HMM-based approaches to model multichannel information in sign language inspired from articulatory features-based speech processing. In: Proceedings of ICASSP (2019)
11. Kumar, B., Varma, A.K.: FPGA implementation of dynamic quantile tracking based noise estimation for speech enhancement. J. Eng. Sci. Technol. Rev. **16**(4) (2023)
12. Kumar, B.: Comparative performance evaluation of greedy algorithms for speech enhancement system. Fluct. Noise Lett. World Sci. **20**(02) (2020)
13. Géron, A.: Hands-on machine learning with Scikit-learn, Keras, and TensorFlow, 2nd ed.
14. Kumar, S., Kumar, B., Kumar, N.: Speech enhancement techniques: a review. Rungta Int. J. Electr. Electron. Eng. **1**(1) (2016)
15. Yang, T., Xu, Y.: Hidden Markov model for gesture recognition. CMU-RI-TR-94 10, Robotics Institute, Carnegie Mellon Univ., Pittsburgh, PA (1994)
16. Ziaie, P., Muller, T., Foster, M.E., Knoll, A.: A Naive Bayes Munich. Department of Informatics VI, Robotics and Embedded Systems, Boltzmannstr. 3, DE-85748 Garching, Germany
17. Bretzner, L., Laptev, I., Lindeberg, T.: Hand gesture recognitionusing multiscale color features, hieracrchichal models and particle in Ąltering. In: Proceedings the Fifth IEEE International Conference on Automatic Face and Gesture Recognition, pp. 423–428 (2002)
18. Dardas, N.H., Georganas, N.D.: Real-time hand gesture detection and recognition using bag-of-features and support vector machine techniques. IEEE Trans. Instrum. Meas. **60**(11), 3592–3607 (2011)
19. Nowozin, S., Kohli, P., Shotton, J.D.J.: Gesture detection and recognition. U.S. Patent 9,619,035, 11 Apr 2017
20. Kumar, B.: Comparative performance evaluation of MMSE-based speech enhancement techniques through simulation and real-time implementation. Int. J. Speech Technol. **21**, 1033–1044 (2018)
21. Kumar, B.: Spectral subtraction using modified cascaded median based noise estimation for speech enhancement. In: Sixth International Conference on Computer and Communication Technology (ICCCT 2015). ACM, MNNIT, Allahabad, India, 25–27 Sept 2015, pp. 214–218

Evaluating of Pedestrian Speed Data for Pedestrians Accessibility Within the University Campus Using Data Science

Badveeti Adinarayana and A. Krishna Chaitanya

Abstract Walking has become an important and sustainable mode of transportation for college students, and improving the walkability of the college campus would boost the convenience and ease of student life. This work provides a novel pedestrian accessibility approach that optimises the Walk Score method based on the frequency, diversity, and distance of students' walking to and from public facilities, regardless of whether the students follow the COVID rules such as lane discipline, etc. Using data science approaches, we measure pedestrian density and speed. Voronoi diagrams are used to calculate pedestrian density. The Voronoi diagram allocates each point to the closest area to that point. This allows us to define a density distribution for each pedestrian and measure pedestrian speed using computer vision technology. Once the data is collected, we use Python because it has several libraries for data analysis and visualization, such as NumPy, Pandas, and Matplotlib. We visited Reva University's campus to test the tool's application, evaluate the accessibility of facility layout and walkability, and make suggestions for improvement. This assessment tool can assist urban planners and campus designers in making judgements about how to modify the facility layout of existing campuses in various regions, as well as evaluating campus plans based on the results of their walkability assessment.

Keywords Pedestrians · WSM method · Data science · University campus

1 Introduction

In the ever-evolving landscape of higher education, the well-being and daily experiences of college students have garnered increasing attention. As students navigate the multifaceted demands of academic life, the importance of an accessible and

B. Adinarayana (✉)
Civil Engineering Department, Punjab Engineering College (DU), Chandigarh, India
e-mail: badveetiadi@pec.edu.in

A. K. Chaitanya
School of Computer and IT, Reva University, Bangalore, India

B. Soni et al. (eds.), *Beyond Artificial Intelligence*, Lecture Notes in Networks and Systems 1326, https://doi.org/10.1007/978-981-96-4170-3_33

sustainable campus environment becomes paramount. Among the various modes of transportation available to students, walking emerges as a crucial and environmentally friendly means of travel. Recognizing this, a growing emphasis on enhancing the walkability of college campuses has emerged, with the belief that such improvements can significantly contribute to the convenience and overall well-being of the student body.

This paper embarks on a pioneering journey to redefine pedestrian accessibility assessment for college campuses. By developing a novel approach that optimizes the Walk Score method, the study seeks to delve into the intricate dynamics of students' daily movements to and from public facilities [1]. Beyond the conventional metrics, such as distance and variety of destinations, the proposed method incorporates the often-overlooked dimension of adherence to COVID norms, including lane discipline and other safety protocols. This nuanced approach acknowledges the unique challenges posed by the ongoing pandemic and underscores the adaptability of the proposed tool to the evolving needs of campus life [2].

At the heart of this innovative assessment tool lies the integration of data science tools to measure pedestrian density and speed. Leveraging Voronoi diagrams for the calculation of pedestrian density, the study adopts a spatial analysis technique that assigns each point to the area closest to it. This methodology facilitates the creation of a detailed density distribution for each pedestrian, offering insights into the concentration of foot traffic across the campus. Complementing this spatial analysis, the study employs computer vision technology to measure pedestrian speed. This cutting-edge approach not only captures the pace of movement but also provides a comprehensive understanding of how students traverse the campus environment.

The choice of Python as the primary programming language for data analysis and visualization adds another layer of sophistication to the proposed methodology. Python's versatility and the availability of libraries such as NumPy, Pandas, and Matplotlib make it an ideal choice for handling and interpreting the complex datasets generated by the assessment tool. This integration of advanced data science tools underscores the commitment to precision and reliability in evaluating pedestrian accessibility.

To test the practical application of the developed tool, the study turns its attention to the campus of Reva University. By conducting a comprehensive examination, the research aims to evaluate the current accessibility of facilities and walkability within the campus. This real-world testing ground provides valuable insights into the tool's efficacy and highlights potential areas for improvement. The findings from this case study will not only contribute to refining the tool but also serve as a practical guide for urban planners and campus designers seeking to enhance the pedestrian experience in diverse regions [3].

Beyond its immediate applications, the proposed assessment tool holds broader implications for the field of urban planning and campus design. By offering a systematic and data-driven approach to walkability assessment, the tool empowers decision-makers to make informed choices about facility layouts and overall campus schemes. The adaptability of the methodology ensures its relevance in diverse geographic and cultural contexts, making it a valuable asset for those involved in shaping the built environment of educational institutions [4].

In conclusion, this paper introduces a groundbreaking approach to pedestrian accessibility assessment for college campuses, blending the traditional Walk Score method with innovative data science tools [5, 6]. Through its comprehensive analysis and real-world application at Reva University, the study aims to not only contribute to the discourse on campus walkability but also provide a practical tool for urban planners and designers seeking to create environments that prioritize the well-being and convenience of students. Data science for pedestrian and high street retailing as a framework for advancing urban informatics to individual scales as discussed by Torrens, P. M. (2022) [6].

2 Research Methodology

In this paper "Measuring Pedestrian Accessibility within the University Campus Using *Data* Science", I used many AI tools, Data science tools, computer vision technology, and Python language. For data collection, I took different samples from "Reva University". I took the video samples, where the crowd was very high. So here I took the video using the camera and I kept these video samples on an application called "Frame capture", where this application can split frame by frame and we can easily calculate the speed of a person. And I also classified these datasets under different age groups. So, I used Computer vision technology to classify the age. Here from the Fig. 1 explains the flow of process.

The deep learning models utilized for gender and age prediction leverage convolutional neural networks (CNNs) and have been fine-tuned on diverse datasets for robust performance. The predictive power of these models adds a layer of richness to the analysis, allowing for nuanced insights into the demographics of the pedestrian population within the university campus.

In this Project, I used Deep Learning to accurately identify the gender and age of a person from a single image of a face. I used the models trained by *Tal Hassner and Gil Levi.* The predicted gender may be one of 'Male' and 'Female', and the predicted age may be one of the following ranges- (0–2), (4–6), (8–12), (15–20), (25–32), (38–43), (48–53), (60–100) (8 nodes in the final SoftMax layer). It is very difficult to accurately guess an exact age from a single image because of factors like makeup, lighting, obstructions, and facial expressions. And so, I made this a classification problem instead of making it one of regression.

After classifying the gender and age, we will then measure the time that a person travelled from one point to the other. Here we will be using an application called "Frame Capture". For the input, we have a video sample from the university that is recorded in resolution (4 k, 30 fps). So, when uploading the recorded video into the application, we can measure frame by frame because it captures 30 frames per second. After that, there will be a button that enables one to view the video frame by frame which moves the time 33 ms forward. For this process, we should first set

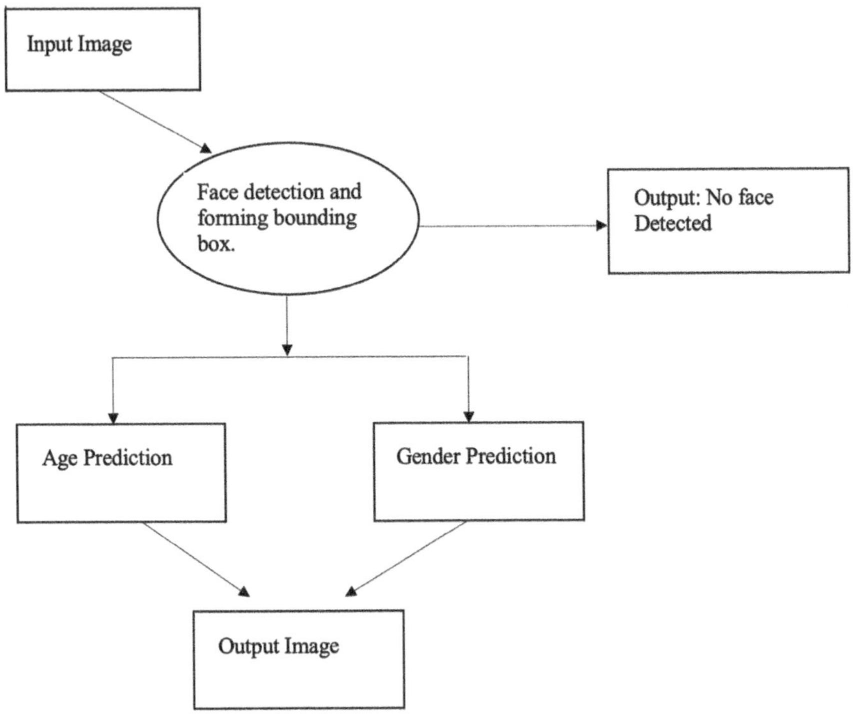

Fig. 1 Flow chart describing the process of gender and age detection

the initial point and the destination point & measure the distance between the two points. Then by using the application note the starting time and ending time and note down the difference. So, it is calculated by *distance/time* formula where it gives the *speed* of a person.

This paper demonstrates a holistic approach to studying pedestrian accessibility within the Reva University campus. By amalgamating AI tools, data science techniques, computer vision technology, and the Python programming language, the research offers a nuanced understanding of pedestrian behaviour. The incorporation of deep learning models for gender and age classification, coupled with video analysis for speed calculation, adds a layer of sophistication to the study.

The findings of this research not only contribute to the realm of pedestrian accessibility within educational institutions but also showcase the diverse applications of AI and data science in solving real-world challenges. The methodology employed in this study can serve as a blueprint for similar analyses in different environments, fostering a data-driven approach to enhance accessibility and efficiency in public spaces.

3 Results and Discussions

Using the above techniques and some tools the obtained results are below; (Data is collected from Reva University", where the images are captured by a mobile with a 50MP camera lens and video is recorded in 4 k (3840 × 2160) with 30fps) Figs. 2 and 3 show the sample results taken from the university:(Note: Images shown below are from algorithm output, original ones vary). Pedestrian face detection with variation of pedestrian speed images as shown in Figs. 2 and 3. However, the current features highlighted in Figs. 2 and 3 highlight the algorithm's potential as a valuable tool for automating gender and age identification tasks in scenarios where facial visibility is optimal, offering efficiency and accuracy in applications ranging from demographic analysis to security systems.

In Fig. 2, the output of the algorithm is showcased, providing an insightful glimpse into its capability to automatically detect the gender and age of an individual upon uploading an image. The algorithm demonstrates a remarkable level of accuracy, swiftly categorizing the person into predefined gender categories (Male/Female) and age brackets (0–2, 4–6, 8–12, 15–20, 25–32, 38–43, 48–53, 60–100).

The sophistication of the deep learning models, particularly those developed by *Tal Hassner and Gil Levi*, is evident as the algorithm navigates through various facial features to make these precise predictions. It is noteworthy that the algorithm performs optimally when the person's face is visible in the input image.

Fig. 2 Representing the age and gender of a person

Fig. 3 Detected only one person because the other face was not clear

However, as illustrated in Fig. 2, the algorithm's performance is contingent upon the clarity of facial visibility. In this scenario, where two persons are present, the algorithm successfully detects and classifies the gender and age of one individual while facing challenges with the other. The limitation arises from the obscured or unclear visibility of the second person's face. Factors such as obstructions, shadows, or an individual's orientation can impede the algorithm's ability to discern facial features accurately.

This underscores the importance of image quality and visibility in ensuring the algorithm's optimal performance. As the algorithm relies on nuanced facial cues for gender and age classification, it serves as a reminder that environmental conditions and image quality play pivotal roles in the effectiveness of such deep learning applications.

To address situations where facial visibility may be compromised, future iterations of the algorithm could potentially incorporate advanced image processing techniques or additional training data to enhance its robustness in handling obscured faces. Nevertheless, the current capabilities showcased in Figs. 2 and 3 underscore the algorithm's potential as a valuable tool for automating gender and age identification tasks in scenarios where facial visibility is optimal, offering efficiency and accuracy in applications ranging from demographic analysis to security systems. As the field of computer vision continues to advance, addressing challenges related to visibility and expanding the algorithm's adaptability will undoubtedly contribute to its broader utility in diverse real-world contexts.

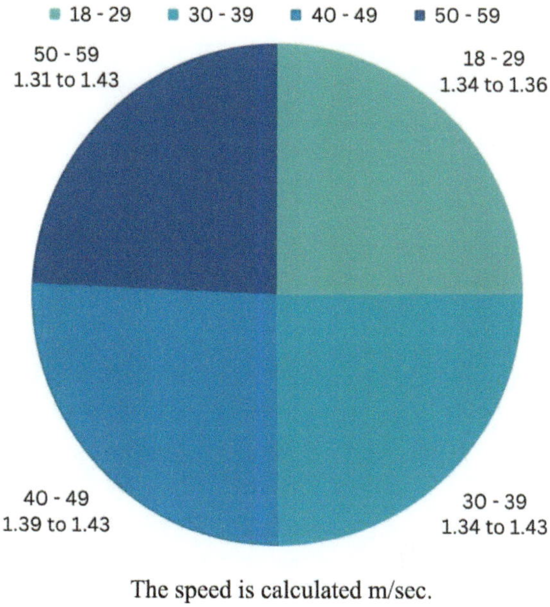

Fig. 4 The details of pedestrian age groups and pedestrian speed

4 Conclusion

In conclusion, the presented algorithm exhibits a remarkable capacity for automatic gender and age detection, as evidenced by the illustrative output had discussed in Fig. 2. The application of deep learning models, specifically those crafted by Tal Hassner and Gil Levi, showcases a sophisticated understanding of facial features, allowing the algorithm to categorize individuals swiftly and accurately into predefined gender categories and age brackets. The precision achieved in this process serves as a testament to the advancements in computer vision and underscores the potentially transformative impact of AI in tasks related to demographic analysis and security systems. Here I would like to conclude that the age groups from 18 to 29 walk with a speed of 1.34–1.36, the age group from 30 to 39 walks with a speed of 1.34 to 1.43, the age group between 40 and 49 walks with a speed of 1.39–1.43 (Fig. 4).

For the above results, we have taken many samples and kept them in the algorithms for classifying age and gender. We trained the model with many inputs and data. But for the age group, 40 to 49 and 50 to 59 lag a little bit while walking. The image shows the pedestrian speed concerning the gender as shown discussed in Fig. 5.

Age	Sex	Meters/second	Miles/hour
20 to 29	Male	1.36	3.04
	Female	1.34	3.0
30 to 39	Male	1.43	3.2
	Female	1.34	3.0
40 to 49	Male	1.43	3.2
	Female	1.39	3.11
50 to 59	Male	1.43	3.2
	Female	1.31	2.93
60 to 69	Male	1.34	3.0
	Female	1.24	2.77
70 to 79	Male	1.26	2.82
	Female	1.13	2.53
80 to 89	Male	0.97	2.17
	Female	0.94	2.10

Fig. 5 Details of pedestrian speed with respect to age group

References

1. Zohreh, A.-S., Moeinaddini, M., Shah, M.Z.: A pedestrian level of service method for evaluating and promoting walking facilities on campus streets. Land Use Policy **38**, 175–193 (2014)
2. Logan, T.M., et al.: Evaluating urban accessibility: leveraging open-source data and analytics to overcome existing limitations. Environ. Plan. B: Urban Anal. City Sci. **46**(5), 897–913 (2019)
3. Zhang, Z., Fisher, T., Feng, G.: Assessing the rationality and walkability of campus layouts. Sustainability **12**(23), 10116 (2020)
4. Holmes, S., Huynh, R., Millard-Ball, A.: Pedestrian planning on college campuses. Plan. High. Educ. **46**(2), 65–80 (2018)
5. Göçer, Ö. et al.: Pedestrian tracking in outdoor spaces of a suburban university campus for the investigation of occupancy patterns. Sustain. Cities Soc. **45**, 131–142 (2019)
6. Torrens, P.M.: Data science for pedestrian and high street retailing as a framework for advancing urban informatics to individual scales. Urban Inform. **1**(1), 9 (2022)
7. https://www.geeksforgeeks.org/age-and-gender-detection-using-opencv-in-python/
8. https://chat.openai.com/
9. https://www.geeksforgeeks.org/opencv-python-tutorial/

Role of Bio-Inspired Algorithms in Detection of Lung Disease Using Deep Learning

Hardeep Saini and Davinder Singh Saini

Abstract In this research, we have proposed the role of bio-inspired algorithms in deep learning and finalized the hyperparameters during the training process. The selection of hyperparameter values is a tedious task. The recent studies highlighted the role of bio-inspired algorithms in selecting hyperparameter values. The algorithms utilized in this research are grey wolf optimization (GWO) and cuckoo search algorithm (CSA) and a comparative analysis is also done to get the best model out of them. COVID-19 and viral pneumonia patients' chest X-ray (CXR) images exhibit a similar pattern, which often leads to misdiagnosis. The experiments are performed on a pre-trained CNN model. The pre-processing, like histogram equalization and image resizing techniques, is performed over 2860 lung CXR images to classify them into three different classes, namely, COVID-19, pneumonia, and normal images.

Keywords Bio-inspired algorithms · Grey wolf optimizer · Cuckoo search algorithm · Chest X-ray · ResNet50 · Deep learning

1 Introduction

The behavior of climate change directly impacts the environment, and due to changes in environmental conditions, the pressure on human health is increased. Due to changing environmental conditions, the deathly diseases are flourishing rapidly. This leads to the rise of chronic obstructive pulmonary disease (COPD) and increases the risk of generating lung disease [1]. Some lung diseases, like COVID-19, pneumonia, and tuberculosis (TB), are spread by the SARS-CoV2 virus, viral infections, and mycobacterium bacteria, respectively. These diseases, also known as infectious diseases, are an increasing global threat if not controlled.

Pneumonia is spread through viruses or bacteria in the lungs. This disease tosses the health risk on top and causes death worldwide with an eighth rank [2]. It can be of

H. Saini (✉) · D. S. Saini
Department of Electronics and Communication Engineering, Chandigarh College of Engineering and Technology, Chandigarh, India
e-mail: hsaini@ccet.ac.in

B. Soni et al. (eds.), *Beyond Artificial Intelligence*, Lecture Notes in Networks and Systems 1326, https://doi.org/10.1007/978-981-96-4170-3_34

two types, namely bacterial and viral pneumonia. Increasing air pollution is the main cause of this deadly disease. To diagnose lung disease, many modalities of imaging are adopted by doctors, such as chest magnetic resonance imaging (MRI), ultrasound, and CXR images. Out of these, CXR is a more reliable and cost-efficient method as compared to MRI and other imaging. COVID-19 is another viral disease of lung infections caused by SARS-CoV-2. The World Health Organization (WHO) declared the COVID-19 outbreak a pandemic on March 11, 2020 [3]. The origin of this unique virus emerged in Wuhan, China, and it was named a novel coronavirus. The signs of the COVID-19 coronavirus are cough, fever, breathing difficulty, and lung damage. The RT-PCR (Reverse Transcription-Polymerase Chain Reaction) method stands as the primary standard for COVID-19 testing, serving as the most widely adopted technique for detecting the virus. Nonetheless, it has significant drawbacks, including its manual, complex, and time-consuming nature, with a relatively modest positivity rate of only 63%. Particularly during the pandemic, this low sensitivity was deemed unacceptable [4]. AI in medical imaging presents a promising solution, primarily by mitigating human error. Radiologists and pathologists can occasionally overlook crucial details due to fatigue or distractions. Despite its limitations, X-ray imaging is recommended as the initial diagnostic tool for the detection and screening of COVID-19 cases. [5].

In COVID-19 diagnosis, CT scans and CXR are two common modalities that are considered for diagnosis. The CT scan images are more specific than the CXR images in terms of providing information. However, CXR images are preferred over CT scans as they are cheaper. In the medical sciences, machine learning methods are widely utilized as low-cost and efficient diagnostic methods. Researchers have increasingly directed their efforts toward harnessing medical imaging tools in conjunction with deep learning techniques, aiming for diagnostic approaches that are less burdensome for patients.

In deep learning, the optimization algorithm plays a crucial role in the image classification process to minimize the computational cost and increase the accuracy of the CNN. This can be achieved by optimizing the hyperparameters, which are important in CNN training. In many cases, hyperparameter selection is done manually, which is a time-consuming process and affects the performance of the system [6]. To overcome this problem, this study proposes a unique CAD model with an optimization technique to select hyperparameter values.

2 Related Work

Deep learning approaches used in classification of images outperforms and given high accuracy results. Machine learning techniques are widely used by the researchers to investegate and diagnosis the lung disease using chest x-ray. The author [7] used five cutting -edge Networks namely MobileNetV2, InceptionResNetV2, Inception, VGG19 and Xception, to detect the COVID-19 using transfer learning technique. Image normalization is used to extract the enhanced features. The experiments were carried out in two parts based on different datasets. The results showed that MobileNetV2 outperforms others with 96.78% and 94.72% accuracy. In [8], deep learning network was utilized to perform automatic segmentation of infected lung places using chest tomography images. Kavitha et al. [9] presented a novel hybrid approach of Bayes wavelet transform with CNN, for classification of CXR images into normal and COVID-19 images. The model achieved the 97.10% test accuracy for 20 epochs. The authors [10] proposed the U-Net architecture for disease assessment using 14,482 fragments for training and fragments and 5303 for testing, showing a mean DSC of 0.8481. patient. The proposed model reported 89.1% accuracy in the test of 80 patients, outperforming conventional machine learning methods techniques. In addition to having a small trial, our ex-method explores features that lead to a better understanding of the disease and provides more information at the patient level. Whereas [11] uses the area under the curve (AUC) values of HUST-19(model) were 0.978, 0.921 and 0.931 to predict adverse events (control), minor or symptoms (Type I), and severe, respectively. or severe. people according to the disease (Type II) patients. they used Cohort 2 as a reference to measure HUST-19 with similar accuracy. The nCOVnet was utilized by authors in [12], to detect the COVID-19 by using chest X-ray images and achieved AUC and accuracy 0.881 and 88.10% respectively. El Sayed et al. [13] proposed a novel approach using advanced meta-heuristics algorithm and achieved classification accuracy of 99.23%. In research [14], Hill-climbing based CNN network is proposed to optimized the different parameters. The CNN-HCA model was also compared with state of art methods. Kumar et al. [15] proposed cuckoo-based deep convLSTM classifier and achieved 97.59% accuracy.

3 Methodology

3.1 Dataset Discription

In this study, CXR databse has been utilized for the detection of lung disease. This dataset is very popular amongst researchers due to its availability and popularty. This dataset containing total 2860 chest X-ray images including 1050 COIVID-19 cases, 900 pneumonia images and 910 normal caeses. The sample images of dataset and detailed discription of datase along with training and test images are shown in Table 1 and sample images of dataset is given in Fig. 1.

Table 1 Discription of dataset

Dataset	Disease	Total images	Training set (80%)	Test set (20%)
Chest imaging and covid-chest x-ray [16]	COVID-19	1050	840	210
Kaggle [17]	Pneumonia	900	720	180
Kaggle [17]	Normal	910	728	182

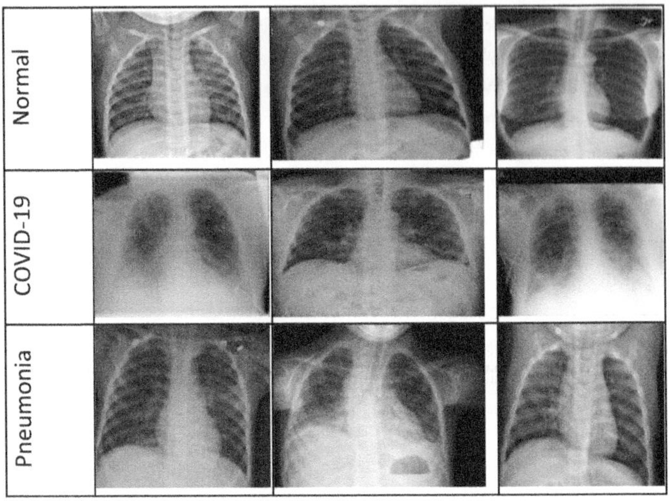

Fig. 1 Sample images of chest X-ray dataset

3.2 Bio-Inspired Optimization Algorithm

The meta-heuristic [18] algorithms play a vital role in improving the time and complexity of the model. Meta-heuristics is described as heuristics on a superior level of frameworks. The 'Heuristics' means identify, and 'meta' denotes the higher level. Bio-inspired algorithms are widely classified as evolutionary algorithms that depend on the survival of the fittest, according to Darwin's theory and swarm intelligence algorithms. Bio-inspired computing is a stochastic search method that is established to attain near-optimal results despite high optimization glitches.

The training process of a CNN model is a tedious task due to the involvement of various hyperparameters like epoch, learning rate, momentum, etc. and the finalization of their precise value. The hit-and trial method was the conventional method to finalize the values of hyperparameters, which is a time-consuming and costly task. To overcome this problem, bio-inspired algorithms created interest and motivated researchers. In this study, experiments are performed on two bio-inspired algorithms, namely GWO and the CSA. A novel hybrid approach has been used to optimize the hyperparameters.

3.3 Grey Wolf Optimization

The GWO [19] is a popular swarm intelligence algorithm due to its reliable technique to optimize the items. The results are quite impressive as compared to conventional methods. The GWO algorithm is based on the hunting nature of the grey wolves.

Their population splits in 4 categories namely Alpha (α), Beta (β), Delta (δ), and Omega (ω). Alpha group leads the social hierarchy and during hunting process, takes decision. Delta groups act as sentinels, scouts and controller by the Beta and Alpha groups. The 4 different stages namely probing, encircling and attacking the prey are used during hunting process. In this optimization (hunting) algorithm, the location of each wolf (search agents) are updated itself while moving towards prey and every updated location is known as exploitation which directs the search agents towards the convergence (local minima). The mathematical modeling of the encircling of prey is given in Eqs. (1) and (2) in which $\vec{P}$ denotes distance, t represents current iteration, and ($\vec{A}$, $\vec{Q}$) are the coefficient vectors, $\overrightarrow{R_p}$ is the prey's spot vector, and $\overrightarrow{R_.}$ is the grey wolf's position vector. $\vec{Q}$ and $\vec{A}$ are the coefficient vector, used to avoid local minima and to exploration capability.

$$\vec{P} = \left|\vec{Q}.\vec{R}_p(t) - \vec{R}(t)\right| \tag{1}$$

$$\vec{R}(t+1) = \vec{R}_p(t) - \vec{A}.\vec{P} \tag{2}$$

In order to mathematically model the hunting (optimization) process, considering the initial location of alpha, beta, and delta wolves (as first three best solutions) and they have the information about the prey's location. The remaining search agents i.e., omega wolves update their positions according to the position of alpha, beta, and delta wolves time to time and as per the equations given below.

$$\vec{P}_\alpha = \left|\vec{Q}_1.\vec{R}_\alpha - \vec{R}\right|, \vec{P}_\beta = \left|\vec{Q}_2.\vec{R}_\beta - \vec{R}\right|, \vec{P}_\delta = \left|\vec{Q}_3.\vec{R}_\delta - \vec{R}\right| \tag{3}$$

$$\vec{R}_1 = \vec{R}_\alpha - \vec{A}_1.\left(\vec{P}_\alpha\right), \vec{R}_2 = \vec{R}_\beta - \vec{A}_2.\left(\vec{P}_\beta\right), \vec{R}_3 = \vec{R}_\delta - \vec{A}_3.\left(\vec{P}_\delta\right) \tag{4}$$

$$\vec{R}(t+1) = \frac{\vec{R}_1 + \vec{R}_2 + \vec{R}_3}{3} \tag{5}$$

In Eqs. (3) and (4), $\vec{P}_\alpha, \vec{P}_\beta$, and $\vec{P}_\delta$ represents the distance between omega wolves and alpha wolve, omega wolves and beta wolve and omega wolves and delta wolve respectively.

3.4 Cuckoo Search Algorithm

The CSA is developed by Yand and Deb [20] and widely utilized by researchers to hand the image processing problems. The CSA is a mimic of the breeding artifice and cuckoos. The breeding process is very unique in which eggs lay in the nest of other birds having same resemblance in their eggs. Due to the same resemblance, host bird accepts the cuckoo eggs and feed. Therefore, survival chances are more. The CSA consider three main assumptions. Firstly, eggs lay in a randomly chosen nest of host birds by every cuckoo. Secondly, eminent quality nest been considered for good eggs which are carry forward for further invention. Lastly, the total number of host-bird's nest are decided and will be identified as cuckoo egg with a probability Pi $\in$ [0, 1]. If it is identified as cuckoo egg, host bird destroys it or leave the nest and build new nest. The mathematical model of CSA is given in Eq. (6) in which updated location is $K_i^{(n+1)}$ for cuckoo i and $K_i^{(n)}$ is the early location. Whereas σ and $\oplus$ denotes the step size and entry wise multiplication respectively.

$$K_i^{(n+1)} = K_i^{(n)} + \sigma \oplus Levy(\theta) \tag{6}$$

3.5 Model Selection

The selection of the best model is a critical task. Through literature survey and experimented outputs, it was observed that most of the research found that the ResNet50 model outperforms other models like VGG19, ResNet101, MobileNet, and AlexNet. ResNet50 [21] is a promising deep neural network introduced in 2015, showing great efficacy on medical images. It was the winner of the ILSVRC competition. This network is not affected by vanishing gradients. In ResNet, there is a 2-layered block in which 34 layers are exchanged with a 3-layer block. It has 50 layers, including convolutional, pooling, and fully connected layers. ResNet50 uses residual blocks that allow the network to learn the residual function instead of trying to learn the original function directly.

Deep learning model represents the neural network architecture used for automatically extracting meaningful features from medical images and making accurate predictions about the presence of a medical condition. The research paper serves as a typical example of how deep learning models are applied to specific domains, highlighting their potential to revolutionize areas like healthcare through improved automation and accuracy. Figure 2 shows the basic block diagram of CNN.

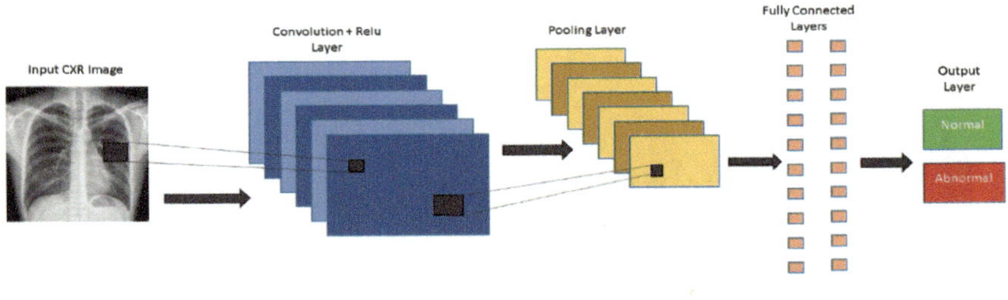

Fig. 2 Block diagram of CNN based deep learning for classification

3.6 Model Implementation

A pre-trained ResNet50 network is used in this study and executed on Windows 11 with an 8-GB RAM CPU. MATLAB 2021a is used to compile the coding and other items. To detect lung disease using CXR, a large dataset of 2860 CXR images was used, out of which 80% were taken for training and 20% for testing purposes. A random distribution approach is used for testing. The workflow of the proposed model is given in Fig. 3. In the proposed model we have explore the efficacy of the bio-inspired algorithms like GWO, and CSA.

A hybrid optimization technique combining the GWO and CSA represents a novel approach to optimizing the hyperparameters of deep learning models. GWO is inspired by the social hierarchy of grey wolves and is used for global optimization, while the Cuckoo Search Algorithm takes inspiration from the brood parasitic behavior of cuckoos to discover optimal solutions. By combining these two algorithms, the hybrid approach harnesses their complementary strengths, efficiently exploring the hyperparameter space and searching for the best configuration. This hybrid technique aims to strike a balance between exploration and exploitation, ensuring that the hyperparameters of the ResNet50 model are tuned to maximize its performance. This novel approach demonstrates the synergy between nature-inspired optimization algorithms and state-of-the-art deep learning architectures to enhance the accuracy and efficiency of deep neural networks for various computer vision tasks.

In deep learning the most used hyper parameters are momentum, number of epochs, and learning rate. The details of the different hyperparameters used in this

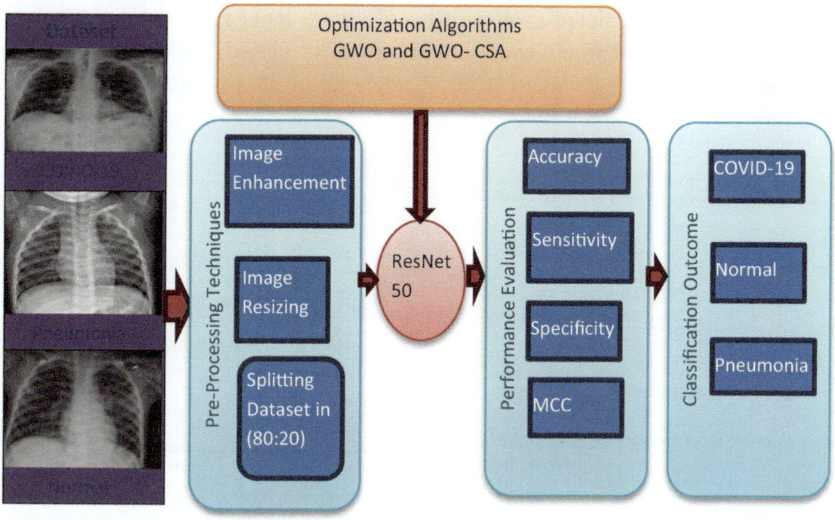

Fig. 3 Workflow of the proposed model

Table 2 Detail of hyperparameters

Algorithms	Momentum	Learning rate	Max epoch
GWO	0.6	0.000156	40
GWO-CSA	0.5	0.000140	30

Table 3 Parameters of GWO algorithm

Optimization algorithm	Batch	Search agents	Dimensions	Iterations	Lower bound	Upper sbound
GWO	32	40	3	50	[0.01, 0.1, 5, 1.0000e-25]	[1, 0.7, 25, 4.0000e-05]

study are given in Table 2. Table 3 shows the desired parameters of GWO and GWO-CSA. CSA initiated with the number of nests, search dimension, and maximum iterations have values of 40, 03, and 100, respectively.

3.7 Performance Evaluation Matrix

To test the ability of the proposed model in medical image evaluation are accuracy, sensitivity, specificity and Matthew's correlation coefficient (MCC) and described in Eqs. (7)–(10). Accuracy represents the ratio of accurately predicted images to the definite number of images evaluated. Sensitivity represents the ratio of correctly predicted cases to the total number of positive cases. Specificity represents number of negative cases predicted as negative to the total number of cases. MCC helps in determining the ineffectiveness of the model in classifying the image classes particularly the negative class. MCC considers all of the parameters to summarize the confusion matric.

$$Accuracy = \frac{TP + TN}{TP + TN + FP + FN} \tag{7}$$

$$Sensitivity = \frac{TP}{TP + FN} \tag{8}$$

$$Specificity = \frac{TN}{TN + FP} \tag{9}$$

$$MCC = \frac{TP * TN - FP * FN}{\sqrt{(TP + FP)(TP + FN)(TN + FP)(TN + FN)}} \tag{10}$$

TP, TN, FP and *FN* in the above equations represent True Positive, True Negative, False Positive and False Negative.

4 Results and Discussion

The ResNet50 network is used in this study to classify lung diseases. The CXR images are used with a 80:20 ratio of the total images. To evaluate the performance of the best model, two algorithms are used to optimize the hyperparameters. The accuracy, sensitivity, specificity and MCC value of the network is determined by the Eqs. (9)–(12) and calculated values are given in Table 4. As given in table, GWO-CSA outperforms GWO in all aspects.

The confusion matrix is generated and shown in Fig. 4. Table 5 reveals that the GWO-CSA outperforms in classification of COVID-19 class with 98.57% accuracy. A comparative performance analysis of two hybrid technologies is shown in Table 4. We reviewed research articles focused on the detection of lung diseases using various techniques in deep learning models. Maximum authors used transfer learning techniques using pre-trained models. Despite the satisfactory outcomes of these models, some drawbacks and unaddressed challenges remain in the lung disease classification process using CXR images.

After consideration of these challenges, we proposed our model, in which we have presented a novel approach for fine-tuning hyperparameters with GWO and GWO-CSA.

Table 4 Model performance summary

Model	Metrics			
	Accuracy (%)	Sensitivity (%)	Specificity (%)	MCC
GWO	95.97	96.43	95.85	0.95
GWO-CSA	96.32	97.31	96.51	0.97

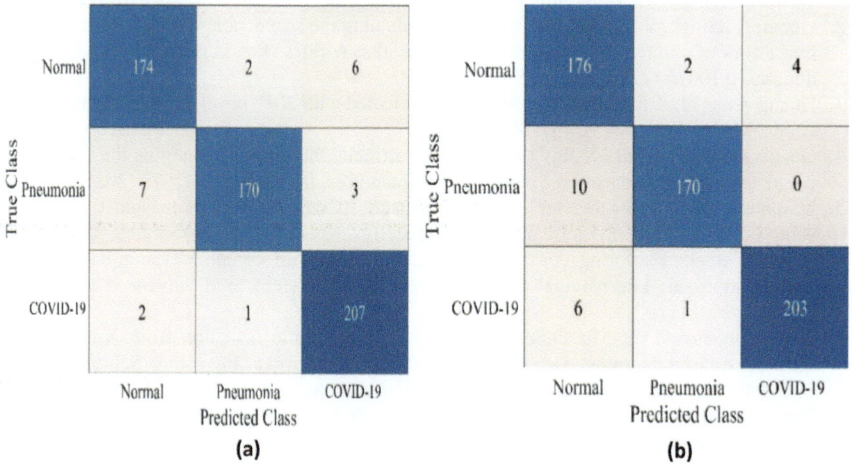

Fig. 4 Confusion matrix for ResNet50 model **a** GWO-CSA **b** GWO

Table 5 Classification accuracy of each class

Model	Class		
	Normal (%)	Pneumonia (%)	COVID-19 (%)
GWO	95.60	94.44	96.66
GWO-CSA	96.70	94.44	98.57

5 Conclusion and Future Work

In this research, a novel approach of bio-inspired algorithms is used for the early detection, prediction and diagnosis of lung diseases. Also, some preprocessing techniques are utilized. The proposed methods outperforms the state-of-the-art methods. The ResNet50 model is used for classification techniques and provides a good accuracy of 96.32% and 95.97 with GWO-CSA and GWO optimizers, respectively. The proposed method limits its function while handling a large dataset of medical images. The researchers can extend this study to a large dataset and explore new bio-inspired algorithms with good practical results. The suggested approach outperformed other established methods in terms of accuracy. This success can be attributed to the utilization of data preprocessing, which effectively eliminated distortions and artifacts from the input data. Furthermore, the hybrid model, which is of paramount significance, significantly reduces computational time and complexity by optimizing hyperparameter values.

References

1. Sheikh, D., et al.: Clinical outcomes in patients with COPD hospitalized with SARS-CoV-2 versus non-SARS-CoV-2 community-acquired pneumonia. Respir. Med. **191**, 106714 (2022)
2. Gupta, A., Singh, S.K., Gupta, B.B., et al.: Evaluating the sustainable COVID-19 vaccination framework of india using recurrent neural networks. Wireless Pers. Commun. (2023). https://doi.org/10.1007/s11277-023-10751-3
3. Huang, C., et al.: Clinical features of patients infected with 2019 novel coronavirus in Wuhan, China. Lancet **395**(10223), 497–506 (2020)
4. Ghaderzadeh, M., et al.: X-Ray equipped with artificial intelligence: changing the COVID-19 diagnostic paradigm during the pandemic. Biomed. Res. Int. **2021**, e9942873 (2021)
5. Mwendo, I. et al.: Deep transfer learning for detecting Covid-19, Pneumonia and Tuberculosis using CXR images–A Review. arXiv preprint arXiv:2303.16754. (2023)
6. Kaur, P. et al.: Exploring convolutional neural network in computer vision-based image classification. In: International Conference on Smart Systems and Advanced Computing (Syscom-2021) (2021)
7. Apostolopoulos, I.D., Mpesiana, T.A.: Covid-19: automatic detection from X-ray images utilizing transfer learning with convolutional neural networks. Phys Eng Sci Med. **43**(2), 635–640 (2020). https://doi.org/10.1007/s13246-020-00865-4
8. Shan, F., et al.: Lung infection quantification of COVID-19 in CT images with deep learning. Med. Phys. **48**(4), 1633–1645 (2021). https://doi.org/10.1002/mp.14609
9. Kavitha, S., Inbarani, H.: Bayes wavelet-CNN for classifying COVID-19 in chest X-ray images. In: Computational Vision and Bio-Inspired Computing: ICCVBIC 2020. Springer Singapore, pp. 707–717 (2021)

10. Chassagnon, G., et al.: AI-driven quantification, staging and outcome prediction of COVID-19 pneumonia. Med. Image Anal. **67**, 101860 (2021)
11. Ning, W., et al.: Open resource of clinical data from patients with pneumonia for the prediction of COVID-19 outcomes via deep learning. Nat. Biomed. Eng.. **4**(12), 1197–1207 (2020)
12. Panwar, H. et al.: Application of deep learning for fast detection of COVID-19 in X-Rays using nCOVnet. Chaos Solitons Fractals **138**, 109944 (2020)
13. El-Kenawy, E.-S.M., et al.: Advanced meta-heuristics, convolutional neural networks, and feature selectors for efficient COVID-19 X-ray chest image classification. IEEE Access **9**, 36019–36037 (2021)
14. Pradhan, A.K., et al.: A COVID-19 X-ray image classification model based on an enhanced convolutional neural network and hill climbing algorithms. Multimed Tools Appl. **82**(9), 14219–14237 (2023). https://doi.org/10.1007/s11042-022-13826-8
15. Kumar, A. et al.: Smart healthcare: disease prediction using the cuckoo-enabled deep classifier in IoT framework. Sci. Program. (2022)
16. COVID-19-Image-Repository. https://github.com/ml-workgroup/covid-19-image-repository. Accessed on 25 June 2023
17. Mooney, P.: Kaggle chest x-ray images (pneumonia) dataset. Kaggle chest x-ray images (pneumonia) dataset. Accessed on 25 June 2023 (2018)
18. Desale, S., et al.: Heuristic and meta-heuristic algorithms and their relevance to the real world: a survey. Int. J. Comput. Eng. Res. Trends. **351**(5), 2349–7084 (2015)
19. Mirjalili, S., et al.: Grey wolf optimizer. Adv. Eng. Softw. **69**, 46–61 (2014)
20. Yang, X.-S., Deb, S.: Cuckoo search: recent advances and applications. Neural Comput. Applic. **24**(1), 169–174 (2014). https://doi.org/10.1007/s00521-013-1367-1
21. Singh, I. et al.: Efficient loop unrolling factor prediction algorithm using machine learning models. In: 2022 3rd International Conference for Emerging Technology (INCET). IEEE, pp. 1–8 (2022)

A Review of Behavioral Biometric Authentication Techniques Using IoT and Deep Learning

Narendra Kumar and Shefali Arora

Abstract Individuals all throughout the world are concerned about their safety and privacy. Our digital devices continue to be vulnerable to data loss, emphasising the need for improved data and system safety via advanced authentication systems. Behavioural biometric techniques provide a solution by allowing continuous monitoring and surveillance of a person's behaviours. Touch-based behavioural biometrics and mouse dynamics are examples of these techniques. This study provides an in-depth examination of the various strategies used by researchers in the field of behavioural biometric authentication to protect data and devices. Our survey categories these strategies based on the deep learning or IoT based frameworks used, the feature extraction procedure, and the datasets accessible for experimentation. Furthermore, we provide an overview of the attained performance levels, the implementation obstacles, and prospective future opportunities in this subject.

Keywords Behavioral biometrics · Deep learning · Review

1 Introduction

In today's digital age, when technology affects every aspect of our lives, coming up with novel solutions to increase data security is vital. In this work, we propose the use of behavioural biometrics authentication for data protection through authentication, with a focus on continuous tracking and monitoring of individual activities.

Behavioural biometric authentication is a ground-breaking solution to security that goes beyond traditional methods such as passwords and PINs. Instead, it harnesses the power of distinct human behaviours and patterns such as typing speed, keystroke

These authors contributed equally to this work.

N. Kumar (✉) · S. Arora
Computer Science and Engineering, National Institute of Technology, Jalandhar, Punjab, India
e-mail: narendra51968kumar@gmail.com

S. Arora
e-mail: chouhansa@nitj.ac.in

B. Soni et al. (eds.), *Beyond Artificial Intelligence*, Lecture Notes in Networks and Systems 1326, https://doi.org/10.1007/978-981-96-4170-3_35

dynamics, mouse motions, and even voice modulation. These distinguishing features combine to provide a powerful and dependable authentication mechanism that significantly increases data security. Behavioural biometric authentication is a novel approach to security that goes beyond traditional methods such as passwords and PINs. Instead, it harnesses the power of unique human behaviours and patterns such as typing speed, keystroke dynamics, mouse motions, and even voice modulation. These distinct traits combine to provide a strong and dependable authentication mechanism that significantly increases data security. The aspect of behavioural biometrics that enables continuous tracking and monitoring is especially appealing. It provides enterprises with real-time insights on user behaviours, allowing them to immediately detect suspicious behavior or unauthorized access attempts. This proactive approach provides for speedier response times, potentially reducing the effect of security breaches [1].

Another significant advantage of behavioural biometrics is the possibility of multifactor authentication (MFA) synergy. Organisations may develop layered security that is both robust and user-friendly by combining behavioural biometrics with standard authentication methods such as fingerprint recognition or facial scanning. This multilayered method offers an extra degree of security, guaranteeing that the system remains secure even if one authentication element is compromised.

In this paper, we review the existing works in the domain of behavioral biometric authentication and analyze the deep learning frameworks used, the performance achieved, limitations and challenges of these techniques. We also review the future opportunities [2]. Section 2 discusses the work done in behavioural biometric authentication using different methods. Section 3 discusses the challenges and limitations of these techniques. Section 4 discusses the various datasets available in this domain. Section 5 concludes the paper.

2 Related Work

The fusion of IoT (Internet of Things) with deep learning frameworks has heralded a new age in behavioural authentication. With their broad deployment in our homes, businesses, and cities, IoT devices generate a plethora of behavioural data. Deep learning frameworks driven by neural networks are currently being used to analyse this data, allowing for the creation of highly sophisticated behavioural authentication systems. Individuals can be recognized and verified using distinct behavioural patterns such as typing cadence, stride, or even how they interact with smart devices. The collaboration of IoT with deep learning not only improves the accuracy and reliability of behavioural authentication, but it also prepares the way for seamless, context-aware security in the ever-expanding world of networked devices and applications. Several deep learning algorithms are being used for the authentication of behavioral traits:

CNNs are generally intended to process grid-like data such as photos and movies. CNNs are made up of several layers, including convolutional layers, pooling layers, and fully connected layers. Convolutional layers use filters to detect features such as edges, corners, and textures in input data. CNNs excel at picture classification, object detection, and image segmentation. They are also utilized in various natural language processing (NLP) activities to process word embedding sequences.

- RNNs are appropriate for sequential data where the sequence of items is important, such as time series data or natural language sentences. RNNs have a recurrent connection that allows them to stay in a hidden state. This hidden state aids in capturing data from earlier time steps and incorporating it into the current computation. RNNs are employed in applications such as speech recognition, language modelling, and machine translation. However, they are limited in their ability to capture long-range relationships in sequences due to the "vanishing gradient" problem.
- LSTM (Long Short-Term Memory): LSTMs are a form of RNN that is designed to circumvent the vanishing gradient problem, making them more effective in dealing with extended sequences. LSTMs are made up of memory cells and gates that control the flow of information. These gates enable LSTMs to selectively recall or forget information from past time steps, making them ideal for tasks with longterm dependencies. LSTMs are commonly employed in sequence-related tasks such as speech recognition, sentiment analysis, text production, and machine translation. They are especially useful when capturing and keeping context across long duration is crucial.

In this section, we discuss prior works showing contact points with the one presented here, highlighting key differences. In this related work section, we will delve into various studies and initiatives that have explored behavioral biometrics for authentication, examining their methodologies, findings, and limitations. By critically analyzing the existing literature, we aim to identify areas where our research can contribute to refining and enhancing the application of behavioral biometrics in everyday scenarios. To date, behavioural biometrics authentication using deep learning on mobile devices has been an active field of research for over a decade. Throughout the years, several studies have focused on the modalities explored within our work, employing a variety of different classification algorithms (mobile keystroke; touch data; background sensor data; multimodal biometrics).

2.1 *Touch-Based Behavioral Biometrics*

The following are the available literature and problems based on touch-based biometric authentication methods [1]: This study focused on an examination of individual and multimodal behavioural biometric features suitable for the implementation of mobile continuous authentication. However, it does not include the deployment of a continuous authentication mechanism into mobile devices to determine whether or not the original owner is using by studying behavioural biometrics data. The work done [2] demonstrates a comparable strategy. Using mobile phone behaviour biometric sensors for user identification. Overall, this study demonstrates the utility of using sensor data and machine-learning approaches for continuous authentication on cell phones. However, it needs execution in terms of stage identification of feelings such as joyful mood.

The work done in [3] makes use of Long Short Term Memory Network to authenticate cloud services using touch behavioural biometric patterns. The work done in [4] is based on the use of IoT-based wearable sensors to detect human activity based on parameters such as motion using Convolutional Neural Networks and Long Short Term Memory Networks.

2.2 Mouse Dynamics

The work done in [5] offers research on mouse dynamics authentication utilising innovative Wasserstein conditional DCGAN. This method demonstrates how a user's conduct changes over time. A unique WC-DCGAN was suggested in this paper to produce highly accurate fake mouse action sequences. However, it does not specify which machine-learning method will be most accurate in identifying the true owner. Authors [6] explore various deep neural network-based architectures to authenticate a user based on mouse dynamics. An accuracy of 94% is obtained using SapiMouse [7] which extracts features from mouse authentication data using Convolutional Neural Network framework.

2.3 Gait and Other Soft Biometric Traits

The work done in [8] describes a strategy for privacy-aware gait identification utilising ultralow-dimensional data from a distance sensor. The paper emphasises the possibility of both continuous authentication and touch-based biometrics as authentication schemes. The system is assumed to be positioned in front of the user in this study. It combines the Tof sensor and detects the distance difference between the user and the system while walking. This approach, however, does not focus on the implementation of two or more people walking together in a system.

The work done in [9] offers a model for soft biometric trait recognition based on a transfer learning technique. It is mostly concerned with capturing soft biometric features from smartphones. TGSB is a gender and age-group recognition approach based on a transfer learning strategy applied to state-of-the-art Convolutional Neural Networks (CNNs) fed with image-based representations of touch gestures done by users on mobile devices. The authors offer a method for soft biometrics on cellphones capable of capturing the user's age and gender attributes; the method, unlike any other related study, is based on the analysis of image-based representations of touch gestures. The research presented in [10] suggests conducting a survey on behavioural biometrics and continuous user authentication on mobile devices. This study includes a description of seven types of behavioural features, as well as an examination of behavioural biometrics collection procedures, feature extraction, and publically available datasets. A big challenge in this strategy is building a user-friendly system for gathering behavioural biometrics. The work done in [11] is a case study in gait recognition based on Convolutional Neural Networks (CNNs) and Recurrent Neural Networks (RNNs)-based algorithms that have established convenience for the task, enhancing performance and robustness over classic machine learning techniques. The work done in [12] makes use of psychological data for identification of a person using deep neural network framework, thus achieving an accuracy of 98.36%.

2.4 Other Traits Captured by IoT Based Systems

Another key challenge is a lack of real-world datasets. Ray-Dowling et al. [13] offers a paradigm for multimodal mobile behavioural biometrics based on publicly available datasets. Using two public datasets, HMOG and BBMAS, this model examines the authentication performance of combining three modalities: acceleration, gyroscope, and swipe. The work done in [14] uses Bluetooth and GPS Technology based on alarm technique. This model is triggered when the Bluetooth connection is lost, the GPS application is utilized to track the location of the youngster who is wearing this gadget. In [15], authors propose an IoT Based Intelligent Real-Time System for Bus Tracking and Monitoring. This paper presents the design and implementation of an IoT-based system that permits parents, schools and regulatory bodies to track the conditions of comfort and safety inside a bus in a real-time manner. Several other works discuss the role of privacy-aware and federated learning-based IoT systems for continuous user authentication [16] (Table 1).

3 Challenges in Behavioral Biometric Authentication

For both users and developers, behavioural biometric authentication presents various challenges:

- Accuracy and dependability: Ensuring the accuracy and dependability of biometric systems is one of the most difficult challenges.
- Failures in Recognition and Verification: Biometric systems may struggle to recognise or verify persons due to a variety of issues such as environmental circumstances, sensor quality, or user behaviour. A fingerprint scanner, for example, may fail if the user's finger is damp, unclean, or wounded.

Table 1 Summarizes the existing literature in the domain of behavioural biometrics

Reference	Technique	Behavioral trait
[1]	BehavePassDB, a new database organized into different acquisition sessions and activities to simulate the most typical parts of mobile Human–Computer Interaction (HCI)	Touch-based
[2]	To verify smart phone users, Hand movement orientation and grasp (HMOG) sensor data is used	Touch-based
[5]	Show that the user behavior of a user will change over time	Mouse-Dynamics
[8]	Highlight the potential of continuous authentication as well as touch-based biometrics as an authentication scheme	Gait
[9]	Offers TGSB, an approach to gender and age-group recognition based on a transfer learning strategy applied to state-of-the-art Convolutional Neural Networks (CNNs) fed with image-based representations of touch gestures done by users on mobile devices	Touch-based
[10]	Provides an up-to-date, thorough, detailed, and targeted survey on Behavioral Biometrics and Continuous Authentication solutions for mobile devices	Touch-based
[13]	Describes the design and implementation of an IoT-based system that allows parents, schools, and regulatory agencies to track the comfort and safety conditions inside a bus in real time	IOT
[11]	Describes a child tracking system with an ESPCAM32, GPS location, and image capturing capabilities	Touch-based
[14]	The goal of this project is to create a comprehensive infant monitoring system using Raspberry Pi, IoT sensors, and a Convolutional Neural Network (CNN)	IOT
[15]	Investigates and proposes a revolutionary gait biometric detection system based on Transformers, which currently achieve state-of-the-art performance in a variety of applications	Gait

- Scalability and Compatibility: Another problem is ensuring the scalability and compatibility of biometric systems. These systems may have difficulty supporting a high number of users, devices, or applications. Furthermore, they may not function fluidly across different platforms, adhere to various standards, or comply with numerous protocols.
- Data Security and Privacy: Because biometric data is sensitive and personal, it is vulnerable to hackers, theft, or misuse.
- Ethical and Legal Concerns: Biometric data raises ethical and legal issues, such as consent, ownership, access, and discrimination. For example, determining who owns biometric data and who can access or use it are key challenges. It is critical to address concerns about obtaining or withdrawing consent for biometric authentication. Furthermore, protecting biometric data from unauthorised usage, such as monitoring or profiling, is a major concern.

Table 2 Gives a description of a few other popular datasets in the domain of behavioral biometrics along with the description of trait used

Reference	User Trait captured	Amount of data
BBMAS [4]	Keystroke, motion, swipe	3.5 million
BehavePassDB [1]	Motion, touch	Data from 4 sessions in 8 tasks
ISOT Mouse Dynamics [17]	Mouse dynamics, keystrokes	Data from 41 individuals
BrainRun [18]	Swipes	Data from 2000 users
ReMouse [19]	Mouse dynamics	Data from 100 users
ISignDB [19]	Signature	–

4 Datasets Available

Individual behavioural patterns are captured in databases used for behavioural biometric authentication. While there aren't as many publicly available behavioural biometric datasets as there are for other domains of machine learning, some are available for research and development. Here are some important datasets used in behavioural biometric authentication:

- CMU Keystroke Dynamics Database contains keystroke timing information (e.g., key press and release times) from computer keyboard users. It's commonly utilised in keystroke dynamics-based authentication studies.
- OU-ISIR Gait Database is focused on gait recognition and includes video sequences of people walking. It is employed in gait-based authentication and recognition studies.
- BioSecure Multimodal Database: a database of biosecurity-related data. This dataset combines keystrokes, signatures, and facial photos, among other biometric modalities. It is utilised in research on multimodal biometric authentication.
- University of California, Irvine (UCI) Human Activity Recognition Using Smartphones is used for activity recognition, the sensor data from smartphones can also be applied to behavioural biometric authentication research.
- CMU Sphinx Speaker Identification includes audio recordings of speakers.

Researchers often use it to develop voice-based authentication and speaker recognition systems (Table 2).

5 Conclusion and Future Work

This paper gives an overview of the existing research work in the domain of behavioural biometrics. It explores the various traits acquired and the feature extraction techniques used by various deep learning-based and IoT frameworks which can be used to authenticate a user based on behavioral traits. The paper also presents several challenges in this domain as well as the popular datasets which can be used on the basis of various characteristics. The survey suggests that the most important behavioral traits used for identification in IoT based systems or deep learning frameworks are touch, gait, mouse and keystroke dynamics along with other traits such as swipe gestures, motion etc.

References

1. Karakaya, N., Alptekin, G.I., incel, O.D.: Using behavioral biometric sensors of mobile phones for user authentication. Procedia Comput. Sci. **159**, 475–484 (2019)
2. Roy, A., Wong, K., Phan, R.C.W.: Attacking mouse dynamics authentication using novel wasserstein conditional DCGAN. IEEE Trans. Inf. Forensics Secur. (2023)
3. Gunn, D.J., Liu, Z., Dave, R., Yuan, X., Roy, K.: Touch-based active cloud authentication using traditional machine learning and LSTM on a distributed tensorflow framework. Int. J. Comput. Intell. Appl. **18**(04), 1950022 (2019)
4. Mekruksavanich, S., Jitpattanakul, A.: Biometric user identification based on human activity recognition using wearable sensors: an experiment using deep learning models. Electronics **10**(3), 308 (2021)
5. Xia, C., Munakata, A., Sugiura, Y.: Privacy-aware gait identification with ultra-low dimensional data using a distance sensor. IEEE Sens. J. (2023)
6. Chong, P., Elovici, Y., Binder, A.: User authentication based on mouse dynamics using deep neural networks: a comprehensive study. IEEE Trans. Inf. Forensics Secur. **15**, 1086–1101 (2019)
7. Antal, M., Fej´er, N., Buza, K.: SapiMouse: mouse dynamics-based user authentication using deep feature learning. In: 2021 IEEE 15th International Symposium on Applied Computational Intelligence and Informatics (SACI). IEEE, pp. 61–66 (2021)
8. Stragapede, G. et al.: BehavePassDB: public database for mobile behavioral biometrics and benchmark evaluation. Pattern Recognit. **134**, 109089 (2023)
9. Guarino, A. et al.: Touchscreen gestures as images. A transfer learning approach for soft biometric traits recognition. Expert. Syst. Appl. **219**, 119614 (2023)
10. Stylios, I. et al.: Behavioral biometrics continuous user authentication on mobile devices: a survey. Inf. Fusion **66**, 76–99 (2021)
11. Delgado-Santos, P. et al.: Exploring transformers for behavioural biometrics: a case study in gait recognition. Pattern Recognit. **143**, 109798 (2023)
12. Xu, J., Pan, S., Sun, P.Z., Park, S.H., Guo, K.: Human-factors-in-driving-loop: driver identification and verification via a deep learning approach using psychological behavioral data. IEEE Trans. Intell. Transp. Syst. **24**(3), 3383–3394 (2022)
13. Ray-Dowling, A. et al.: Evaluating multi-modal mobile behavioral biometrics using public datasets. Comput. Secur. **121**, 102868 (2022)
14. Isa, M.Z.M. et al.: Children security and tracking system using Bluetooth and GPS technology. In: 2019 9th IEEE International Conference on Control System, Computing and Engineering (ICCSCE). IEEE (2019)
15. Kumari, M., Kumar, A., Khan, A.: IoT based intelligent real-time system for bus tracking and monitoring. In: 2020 International Conference on Power Electronics IoT Applications in Renewable Energy and its Control (PARC). IEEE (2020)
16. Wazzeh, M., Ould-Slimane, H., Talhi, C., Mourad, A., Guizani, M.: Privacypreserving continuous authentication for mobile and IoT systems using warmup-based federated learning. IEEE Netw. (2022)
17. Agrawal, M., Mehrotra, P., Kumar, R., Shah, R.R.: Defending touch-based continuous authentication systems from active adversaries using generative adversarial networks. In: 2021 IEEE International Joint Conference on Biometrics (IJCB). IEEE, pp. 1–8 (2021)

18. Papamichail, M.D., Chatzidimitriou, K.C., Karanikiotis, T., Oikonomou, N.C.I., Symeonidis, A.L., Saripalle, S.K.: Brainrun: a behavioral biometrics dataset towards continuous implicit authentication. Data **4**(2), 60 (2019)
19. https://ieee-dataport.org/keywords/behavioral-biometrics

Empowering Smart Healthcare with Low-Power IoT Networks

Prabh Deep Singh, Kiran Deep Singh, Shanjal Gupta, and Pardeep Kumar Jindal

Abstract The emergence of the Internet of Things (IoT) has initiated a novel phase of technological advancement, and its implementation across diverse industries has shown to be very revolutionary. This research study centers its attention on the incorporation of Internet of Things (IoT) technology within the healthcare sector, using insights from the achievements of IoT networks. The Low-Power IoT Network for Smart Healthcare is a novel solution that offers a pioneering approach to the monitoring and administration of healthcare. The conventional approach to healthcare monitoring frequently depends on gadgets that consume significant amounts of power, resulting in the need for frequent battery replacements, delayed transmission of data, and elevated operational expenses. In contrast, our framework advocates for the utilization of low-power Internet of Things (IoT) devices, thereby guaranteeing extended battery longevity and facilitating the acquisition and transmission of data in real time. This method not only decreases operational expenses but also improves the effectiveness and caliber of healthcare services. Furthermore, our framework promotes the seamless integration of various healthcare monitoring applications, creating a comprehensive and real-time view of patient health. Data security and privacy are paramount, and our solution adheres to rigorous standards to safeguard sensitive patient information. In this paper, we conduct a comparative analysis between the existing healthcare monitoring paradigm and our proposed approach. The results reveal clear advantages in terms of reduced power consumption, minimized data transmission latency, and cost-effectiveness in our framework.

P. D. Singh (✉) · S. Gupta
Department of Computer Science and Engineering, Graphic Era Deemed to Be University, Dehradun, India
e-mail: ssingh.prabhdeep@gmail.com

K. D. Singh
Chitkara University Institute of Engineering and Technology, Chitkara University, Rajpura, Punjab, India

P. K. Jindal
Department of Interdisciplinary Courses in Engineering, Chitkara University Institute of Engineering and Technology, Chitkara University, Rajpura, Punjab, India
e-mail: pardeep.alkra@gmail.com

B. Soni et al. (eds.), *Beyond Artificial Intelligence*, Lecture Notes in Networks and Systems 1326, https://doi.org/10.1007/978-981-96-4170-3_36

Keywords Smart healthcare · Low-power networks · Sustainability · Blockchain · Artificial intelligence · Cloud computing · Internet of Things

1 Introduction

In recent years, the integration of the Internet of Things (IoT) has revolutionized various sectors. The emergence of smart agriculture, driven by IoT technologies, has enabled farmers to make data-driven decisions, optimize resource utilization, and improve crop yields. The success of IoT in agriculture, characterized by low-cost and low-power solutions, has inspired its application in other critical domains, notably healthcare [1]. This paper explores the concept of a low-power IoT network tailored for smart healthcare, drawing inspiration from the achievements in smart agriculture IoT networks. Smart healthcare, akin to smart agriculture, presents a promising avenue for enhancing the quality of life, optimizing resource allocation, and ensuring timely interventions. The fundamental principle underlying the success of IoT in smart agriculture—efficiency through low-cost and low-power solutions—holds the potential to revolutionize the healthcare sector [2–4].

In this paper, we propose a novel low-power IoT network designed specifically for smart healthcare applications. To demonstrate the efficacy of our proposed healthcare IoT network, we will evaluate it against existing networks in the healthcare domain, considering metrics such as power consumption and cost. By drawing comparisons with the agricultural IoT network, we aim to showcase the potential advantages and improvements that our healthcare-focused network offers, not only in terms of resource efficiency but also in the longevity of deployed devices and the overall cost-effectiveness of the system [5–7].

As we delve into the development and evaluation of our low-power IoT network for smart healthcare, we anticipate that this research will contribute significantly to the ongoing efforts to harness the power of IoT in revolutionizing the healthcare sector [8]. Just as smart agriculture has evolved to empower farmers with data-driven insights, our work endeavors to empower healthcare professionals and patients alike with innovative, low-cost, and low-power IoT solutions that enhance the quality of care, extend device lifespans, and reduce the economic burden associated with healthcare monitoring and management.

2 Proposed Framework

The proposed low-power IoT network for smart healthcare, as shown in Fig. 1, seamlessly integrates into the Application and Communication Layers, providing sustainable, energy-efficient solutions for healthcare monitoring applications [9, 10]. It ensures the reliability of healthcare data while enhancing patient engagement and

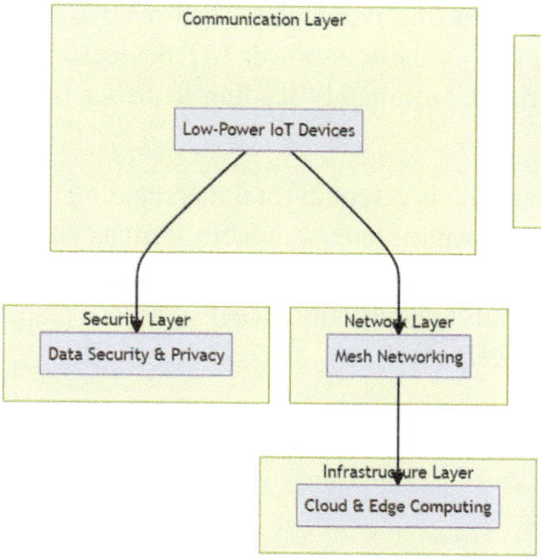

Fig. 1 Proposed framework

data security, thereby advancing the goal of efficient and patient-centric healthcare systems.

2.1 *Application Layer*

Healthcare Monitoring Applications: At the apex of our IoT framework for smart healthcare, the Application Layer encompasses a diverse array of healthcare monitoring applications. These applications, driven by IoT sensors and devices, play a pivotal role in monitoring patient health, tracking medication adherence, managing chronic diseases, and facilitating emergency response systems [11, 12]. Our proposal fits seamlessly within this layer by offering a versatile platform for integrating various healthcare monitoring applications. The low-power nature of our IoT network ensures continuous and reliable data collection while conserving energy resources.

Data Analytics and Decision Support: Situated at the heart of the Application Layer, this component processes and analyzes the data collected from IoT sensors. It leverages machine learning algorithms and real-time analytics to derive actionable insights for healthcare professionals and patients.

2.2 *Presentation Layer*

User Interface: The Presentation Layer provides the interface through which healthcare providers and patients interact with the IoT network [13]. It includes user-friendly interfaces, accessible through web applications or mobile apps, that allow users to view real-time data, configure monitoring parameters, and receive alerts. Our proposal plays a vital role in this layer, as the low-power IoT devices ensure a seamless and uninterrupted user experience, ultimately leading to better patient engagement and satisfaction.

Data Visualization: Visualizing healthcare data is essential for understanding trends and anomalies. This layer translates raw data into understandable formats such as graphs, charts, and dashboards. Our proposal integrates efficiently with data visualization tools, enabling healthcare professionals to gain actionable insights at a glance. The motto here is "See Clearly, Act Decisively."

2.3 Communication Layer

Low-Power IoT Devices: This layer forms the backbone of our proposal, as it focuses on the deployment of low-power IoT devices for healthcare monitoring. These devices are energy-efficient, capable of long-term operation on battery power, and designed to communicate wirelessly with minimal energy consumption [14, 15]. By incorporating low-power IoT devices, we ensure the longevity of our healthcare monitoring network, even in remote or resource-constrained settings. The terminology here revolves around "Sustainability through Efficiency."

2.4 Network Layer

Mesh Networking: In the Network Layer, we establish a robust and scalable mesh network architecture. Our proposal aligns perfectly with this layer by utilizing mesh networking to enable device-to-device communication, improving network reliability, and extending network coverage. The motto here is "Connectivity without Boundaries."

2.5 Data Security and Privacy Layer

End-to-End Encryption: The Data Security and Privacy Layer is vital in healthcare to protect sensitive patient data [16]. Our proposal ensures the highest standards of data security through end-to-end encryption, safeguarding patient confidentiality and privacy. The guiding principle here is "Data Protection as a Priority."

Infrastructure Layer:

Cloud and Edge Computing: This layer handles data storage, processing, and management. Our proposal incorporates both cloud and edge computing capabilities, allowing for efficient data storage and real-time processing [17]. This approach balances the need for centralized data management with the advantages of edge computing, ensuring responsiveness and scalability. The terminology here is "Efficiency through Hybrid Infrastructure."

The existing approach in the field of healthcare monitoring primarily relies on traditional medical devices and sporadic data collection methods. These legacy systems often involve the use of standalone, power-hungry devices that require frequent battery replacements or constant electrical connections [18]. Data collected by such devices may not be transmitted in real time, leading to delayed interventions and suboptimal patient care. Additionally, the lack of integration between different healthcare monitoring applications results in fragmented data silos, making it challenging for healthcare professionals to gain comprehensive insights into a patient's health status. The cost associated with maintenance, device replacement, and data management can be substantial.

3 Experimental Setup and Results

Performing a simulation in MATLAB is a structured process that begins with defining the problem and developing an appropriate model. Once the model is in place, initialization of variables and setting initial conditions is necessary. For time-dependent simulations, specifying the time step is crucial to determine the granularity of the simulation. The core simulation loop iteratively updates the model's state, often utilizing numerical integration methods or solving differential equations. After the simulation runs to completion or reaches a defined stopping criterion, the results can be analyzed, visualized, and interpreted to gain insights into the system under study. MATLAB's versatility and extensive toolset make it a powerful platform for various simulation tasks across diverse fields [19].

The MATLAB-generated plot vividly illustrates the comparative analysis between the existing healthcare monitoring approach and the proposed Low-Power IoT Network for Smart Healthcare. In this visual representation, three critical metrics—power consumption, data transmission latency, and cost—are juxtaposed for both approaches across three distinct scenarios. The bar plot conveys the substantial benefits of our proposed framework, depicting significantly lower power consumption and reduced data transmission latency when compared to the traditional approach. Moreover, the cost-effectiveness of our solution is unmistakably evident, demonstrating the potential for substantial cost savings in healthcare operations [20, 21]. This visual depiction serves as a compelling testament to the transformative potential of low-power IoT networks in healthcare, emphasizing their ability to enhance efficiency, reduce operational costs, and improve the overall quality of care development and problem-solving fields. In this context, MATLAB greatly benefits individuals who wish to investigate, comprehend, and enhance intricate systems through virtual experimentation (Fig. 2).

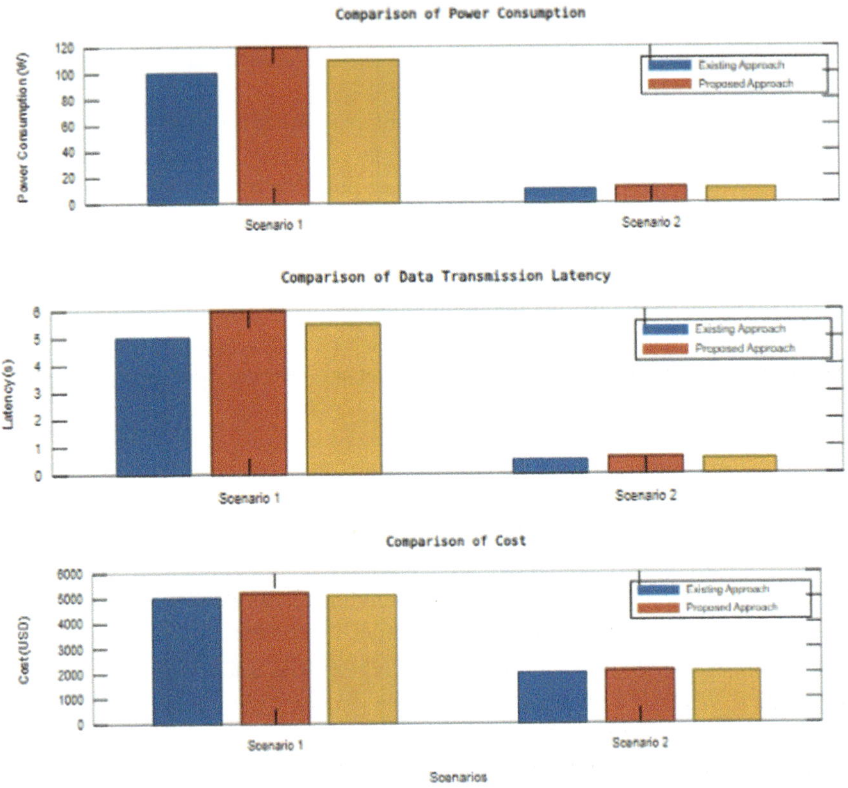

Fig. 2 Comparison with different metrics both with fog and without fog (Cloud Only)

4 Conclusion

In this paper, we have presented a comprehensive exploration of a Low-Power IoT Network tailored for smart healthcare, drawing inspiration from the success of similar networks in the field of smart agriculture. The healthcare sector, like agriculture, stands to benefit significantly from the efficiency and cost-effectiveness offered by IoT technologies. Our proposed approach, which places a strong emphasis on energy-efficient devices, real-time data transmission, and holistic integration of healthcare applications, represents a promising leap forward in healthcare monitoring and management.

We have highlighted the limitations of the existing approach, characterized by power-hungry devices, delayed data transmission, and substantial operational costs. In contrast, our proposed framework champions low-power IoT devices capable of extended battery life, ensuring timely data collection and communication while minimizing the economic burden on healthcare systems. The seamless integration of various healthcare monitoring applications within our framework fosters a more holistic understanding of patient health, enabling healthcare professionals to make data-driven decisions in real-time. Additionally, our emphasis on data security and privacy safeguards sensitive patient information, adhering to the highest industry standards. The comparative analysis presented in this paper demonstrates the clear advantages of our proposed approach over the existing paradigm. Reduced power consumption, minimized data transmission latency, and cost-effectiveness is pivotal outcomes that can transform healthcare monitoring and management, ultimately enhancing the quality of care and patient engagement.

References

1. Seng, K.P., Ang, L.M., Ngharamike, E.: Artificial intelligence internet of things: a new paradigm of distributed sensor networks. Int. J. Distrib. Sens. Networ. **18**(3), 15501477211062836 (2022)
2. Wang, Y., Zhao, J.: Mobile edge computing, metaverse, 6G wireless communications, artificial intelligence, and blockchain: survey and their convergence. In: 2022 IEEE 8th World Forum on Internet of Things (WF-IoT), pp. 1–8 (2022)
3. Islam, M.S.U., Kumar, A., Hu, Y.-C.: Context-aware scheduling in fog computing: a survey, taxonomy, challenges and future directions. J. Netw. Comput. Appl. **180**, 103008 (2021)
4. Rani, S., Ahmed, S.H., Rastogi, R.: Dynamic clustering approach based on wireless sensor networks genetic algorithm for IoT applications. Wirel Netw. **26**, 2307–2316 (2020)
5. Radanliev, P., De Roure, D., Maple, C., Santos, O.: Forecasts on future evolution of artificial intelligence and intelligent systems. IEEE Access. **10**, 45280–45288 (2022)
6. Mittal, V., Gangodkar, D., Pant, B.: Deep graph-long short-term memory: a deep learning based approach for text classification. Wirel. Pers. Commun. **119**(3), 2287–2301 (2021)
7. Shukla, S.K., Pant, B., Viriyasitavat, W., Verma, D., Kautish, S., Dhiman, G. et al.: An integration of autonomic computing with multicore systems for performance optimization in Industrial Internet of Things. IET Commun. (2022)
8. Duan, S., Wang, D, Ren, J., Lyu, F., Zhang, Y., Wu, H. et al.: Distributed artificial intelligence empowered by end-edge-cloud computing: a survey. IEEE Commun. Surv. Tutor. (2022)
9. Singh, K.D.: Particle swarm optimization assisted support vector machine based diagnostic system for dengue prediction at the early stage. In: Proceedings—2021 3rd International Conference on Advances in Computing, Communication Control and Networking, ICAC3N 2021, pp. 844–848 (2021)
10. Singh, P., Singh, K.D., Tripathi, V., Chaudhari, V.: Use of ensemble based approach to predict health insurance premium at early stage. In: Proceedings International Conference on Computational Intelligence and Sustainable Engineering Solutions CISES 2022, pp. 566–569 (2022)
11. Singh, K.D., Singh, P., Tripathi, V., Khullar, V.: A novel and secure framework to detect unauthorized access to an optical fog-cloud computing network. In: 2022 Seventh International Conference on Parallel, Distributed and Grid Computing (PDGC), pp. 618–622 (2022)
12. Singh, P., Singh, K.D.: Fog-centric intelligent surveillance system: a novel approach for effective and efficient surveillance. In: 2023 International Conference on Advancement in Computation & Computer Technologies (InCACCT), pp. 762–766 (2023)
13. Singh, K.D., Singh, P., Kaur, G., Khullar, V., Chhabra, R., Tripathi, V.: Education 4.0: exploring the potential of disruptive technologies in transforming learning. In: 2023 International Conference on Computational Intelligence and Sustainable Engineering Solutions (CISES), pp. 586–591 (2023)
14. Singh, P.D., Singh, K.D.: Security and privacy in fog/cloud-based IoT systems for AI and robotics. EAI Endorsed Trans. AI Robot **2** (2023)

15. Kang, S.S., Singh, K.D., Kumari, S.: Smart antenna for emerging 5G and application. In: Printed Antennas. CRC Press, pp 249–264 (2022)
16. Singh, K.D., Singh, P., Kang, S.S.: Ensembled-based credit card fraud detection in online transactions. In: AIP Conference Proceedings, pp. 50009 (2022)
17. Singh, K.D., Singh, P.: A novel cloud-based framework to predict the employability of students. In: 2023 International Conference on Advancement in Computation & Computer Technologies (InCACCT). pp. 528–532 (2023)
18. Singh, K.D., Singh, P.D., Bansal, A., Kaur, G., Khullar, V., Tripathi, V.: Exploratory data analysis and customer churn prediction for the telecommunication industry. In: 2023 3rd International Conference on Advances in Computing, Communication, Embedded and Secure Systems (ACCESS), pp. 197–201 (2023)
19. Singh, K.D., Singh, P., Chhabra, R., Kaur, G., Bansal, A., Tripathi, V.: Cyber-physical systems for smart city applications: a comparative study. In: 2023 International Conference on Advancement in Computation & Computer Technologies (InCACCT). pp. 871–6 (2023)
20. Singh, K.D.: Securing of cloud infrastructure using enterprise honeypot. In: Proceedings—2021 3rd International Conference on Advances in Computing, Communication Control and Networking, ICAC3N 2021, pp. 1388–1393 (2021)
21. Brar, M., Jindal, P., Malhotra, P., Sharma, P., Kaur, A.: Machine learning based intelligent wardrobe system for apparel recommendation and organization. In: 2023 International Conference on Research Methodologies in Knowledge Management, Artificial Intelligence and Telecommunication Engineering (RMKMATE). IEEE, pp. 1–5 (2023)

Sustainable Farming with Energy-Optimized Internet of Things

Kiran Deep Singh, Prabh Deep Singh, Shanjal Gupta, and Pardeep Kumar Jindal

Abstract This paper offers a thorough examination of the Energy Efficient Internet of Things (EEIoT) methodology as it pertains to the domain of smart agriculture. The ongoing transformation of modern farming methods by the Internet of Things (IoT) presents a significant problem in effectively managing energy consumption. The objective of this study is to tackle this particular difficulty by presenting a framework that is designed to optimize energy usage in the context of smart agriculture. The EEIoT framework improves energy consumption within the agricultural IoT ecosystem by integrating energy-aware node placement algorithms, adaptive data transmission systems, and dynamic energy management policies. The results highlight the practical significance of the EEIoT technique as a potentially effective resolution to the energy-related obstacles encountered in the domain of intelligent agriculture. This study contributes to the advancement of precision agriculture by providing a trajectory towards more sustainable and data-driven farming methods, which have the potential to improve agricultural productivity while reducing environmental consequences.

Keywords Sustainability · Farming · Energy-optimized Internet of Things · Artificial intelligence · Cloud computing

K. D. Singh (✉)
Chitkara University Institute of Engineering and Technology, Rajpura, Punjab, India
e-mail: kdkirandeep@gmail.com

P. D. Singh · S. Gupta
Department of Computer Science and Engineering, Graphic Era Deemed to Be University, Dehradun, India

P. K. Jindal
Electronics and Communication Engineering Department, Chandigarh Engineering College, Chandigarh Group of Colleges, Jhanjeri, Mohali, Punjab, India

B. Soni et al. (eds.), *Beyond Artificial Intelligence*, Lecture Notes in Networks and Systems 1326, https://doi.org/10.1007/978-981-96-4170-3_37

1 Introduction

The Internet of Things (IoT) has become a significant technology in recent years, bringing about substantial changes in multiple domains such as healthcare, transportation, manufacturing, and environmental monitoring [1, 2]. The paradigm shift in the field of smart agriculture has had a significant impact. Internet of Things (IoT) applications have provided farmers and agriculturalists with real-time data, allowing them to make well-informed decisions. This has resulted in improved crop yields, reduced resource wastage, and the promotion of sustainable farming practices [3]. Nevertheless, the widespread adoption of Internet of Things (IoT) devices in the field of smart agriculture has presented a notable obstacle: the effective administration of energy usage [4, 5].

The implementation of Internet of Things (IoT) technology in the field of smart agriculture entails the strategic placement of numerous diverse wireless sensor nodes and devices throughout expansive agricultural areas. These devices are accountable for the collection and transmission of data pertaining to soil moisture, temperature, humidity, crop health, and other essential aspects [6]. The abundance of data in the field of precision agriculture presents significant potential, but it also presents a substantial energy-related obstacle. In order for wireless sensor nodes or devices to effectively communicate in the field, it is imperative that they receive a consistent and uninterrupted supply of energy. The consumption of this energy can rapidly skyrocket as the quantity of deployed devices grows [7, 8].

In this paper, we embark on a journey to explore the applicability of EEIoT in the realm of smart agriculture. We delve into the complexities of IoT deployment in agricultural settings, where environmental conditions and energy constraints play pivotal roles [9]. Leveraging the insights and methodologies presented in the referenced paper, we aim to adapt and optimize EEIoT for the specific requirements and challenges of smart agriculture.

The primary objective of our research is to develop an energy-efficient framework tailored to smart agriculture, wherein the energy harvesting process is significantly improved. By integrating the principles of EEIoT into our agricultural IoT ecosystem, we seek to minimize energy wastage, prolong device lifetimes, and enhance the overall sustainability and productivity of modern farming practices.

2 Proposed Approach: Energy Efficient Internet of Things (EEIoT)

The proposed approach as shown in Fig. 1, aims to adapt and optimize the Energy Efficient Internet of Things (EEIoT) technique for the specific requirements and challenges of smart agriculture. To achieve this, we will implement a multi-faceted strategy that encompasses the following key components

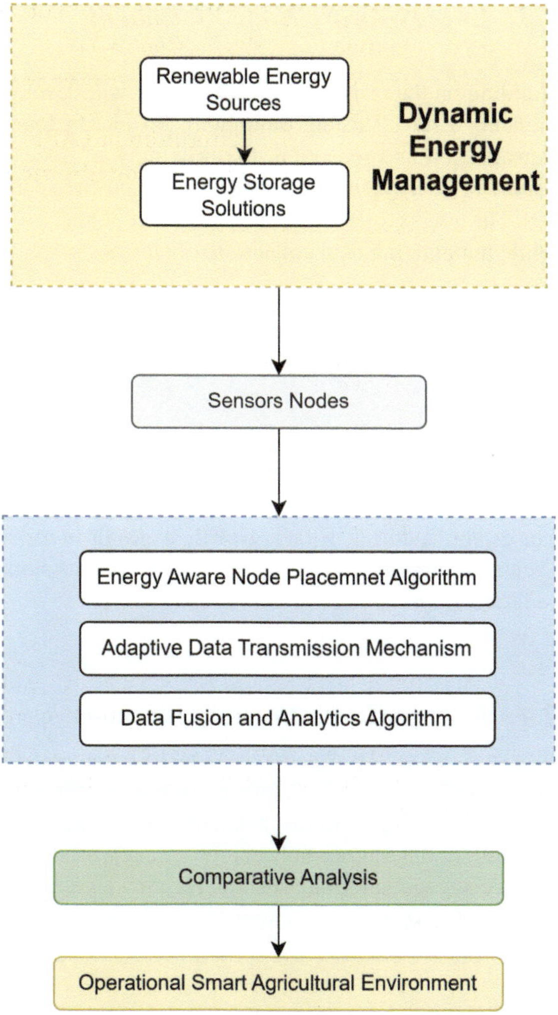

Fig. 1 Proposed approach: Energy efficient Internet of Things (EEIoT)

2.1 Environmental Sensing and Data Collection

In the context of smart agriculture, environmental data collection is paramount. We will deploy a network of wireless sensor nodes equipped with various sensors for monitoring parameters such as soil moisture, temperature, humidity, light levels, and crop health [10, 11]. These nodes will be strategically placed throughout the agricultural landscape to ensure comprehensive data coverage.

2.2 Energy-Aware Node Placement

Building on the insights from EEIoT, we will develop an energy-aware node placement algorithm. This algorithm will consider factors such as the energy cost of data transmission and reception, the proximity of nodes to energy sources (e.g., solar panels or energy harvesting modules), and the spatial distribution of sensors [12, 13]. The goal is to optimize the placement of nodes to minimize energy consumption while maximizing data collection efficiency.

2.3 Adaptive Data Transmission

To further reduce energy consumption, we will implement an adaptive data transmission mechanism. This mechanism will intelligently adjust the frequency and volume of data transmission based on the changing needs of the agricultural system [14, 15]. For example, during critical growth stages or in response to specific environmental events, the system may increase the data transmission rate to provide timely insights to farmers.

2.4 Energy Harvesting and Storage

To ensure a sustainable source of energy for sensor nodes, we will integrate energy harvesting techniques into the agricultural IoT ecosystem. Solar panels, wind turbines, or other renewable energy sources will be used to generate power [16, 17]. Energy storage solutions such as rechargeable batteries or supercapacitors will be employed to store excess energy for use during periods of low energy production.

2.5 Dynamic Energy Management

EEIoT's self-adaptive nature will be extended to the smart agriculture context. Our system will continuously monitor the energy levels of individual nodes and adapt their operational parameters accordingly. Nodes with low energy reserves may reduce their data transmission frequency or enter a low-power sleep mode to conserve energy until they can recharge.

2.6 *Data Fusion and Analytics*

Collected data from the sensor nodes will be subject to data fusion and analytics processes. Advanced algorithms will process and analyze the data to extract meaningful insights, such as crop growth patterns, disease outbreaks, or irrigation recommendations. These insights will be communicated to farmers through user-friendly interfaces, enabling informed decision-making [18, 19].

2.7 *Real-World Implementation*

Finally, we will implement our proposed approach in a real-world smart agriculture setting. This will involve collaborating with farmers and agricultural experts to deploy the system in an operational environment, gather feedback, and fine-tune the system based on practical insights and observations.

By combining these components into a holistic approach, we aim to create an energy-efficient IoT framework that not only enhances the sustainability of smart agriculture but also improves the quality and availability of agricultural data. Our goal is to empower farmers with the tools and insights needed to make data-driven decisions, optimize resource usage, and ultimately contribute to the advancement of sustainable and productive farming practices in the modern era.

3 Experimental and Results

Setting up a MATLAB environment for simulating the proposed approach in the context of smart agriculture involves several critical steps. Simulate IoT networks, energy consumption patterns, and data analytics, all of which are central to our proposed approach. MATLAB offers a comprehensive set of functions for data generation and manipulation. Regarding network layout and node placement, MATLAB has the capability to construct the virtual agricultural landscape and strategically allocate sensor nodes inside it. MATLAB's visualization features facilitate the creation and viewing of node placements, allowing for the implementation of either a grid-based structure or a more intricate custom arrangement that emulates the smart agriculture environment.

The energy consumption model will be a fundamental component of simulation. A thorough energy consumption model for the sensor nodes can be implemented in MATLAB. The proposed model should incorporate multiple facets of energy utilization, encompassing data transmission, reception, processing, and sleep modes [20]. Additionally, it is imperative to consider the incorporation of energy harvesting methods from sustainable sources such as solar radiation and wind, alongside energy storage systems such as rechargeable batteries or supercapacitors [21]. By employing

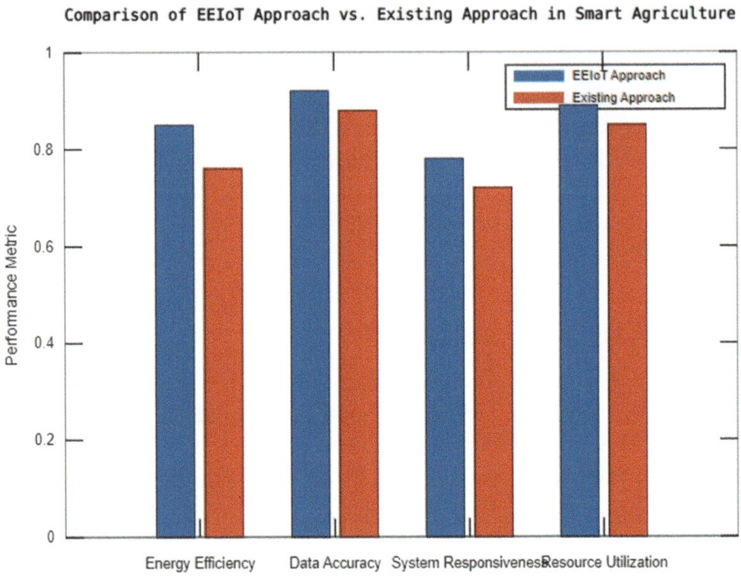

Fig. 2 Comparison of EEIoT approach versus existing approach in smart agriculture

this approach, one may effectively model the energy dynamics of their Internet of Things (IoT) network.

Figure 2 showcases a visual representation of the performance evaluation between our proposed approach and an existing approach across various parameters. Each parameter is depicted along the x-axis, while the y-axis represents the relative performance scores, normalized for consistency. The blue bars represent the performance of our proposed approach, and the orange bars represent the performance of the existing approach. The plot provides an immediate and intuitive way to assess the comparative strengths and weaknesses of the two approaches. It enables stakeholders and researchers to quickly identify areas where our proposed approach outperforms the existing method and vice versa, facilitating data-driven decision-making and further refinement of IoT solutions in the context of smart agriculture.

4 Conclusion

This study provides a thorough examination and assessment of a novel Energy Efficient Internet of Things (EEIoT) approach specifically designed for the use of smart agriculture. The effective control of energy usage has become of utmost importance due to the widespread adoption of Internet of Things (IoT) devices in contemporary agricultural methods. The objective of our study was to tackle this significant issue and offer practical findings for improving energy efficiency and promoting the overall sustainability and productivity of intelligent agriculture.

By employing the EEIoT framework, we have successfully showcased the capacity of our methodology to substantially diminish energy consumption inside the agricultural Internet of Things (IoT) ecosystem. A combination of energy-aware node placement algorithms, adaptive data transmission mechanisms, and dynamic energy management strategies were utilized to optimize the energy efficiency of our system. Through our comparative analysis of existing methods, we have found that our proposed approach demonstrates superior performance in various important metrics. These metrics include decreased energy consumption per data transmission, enhanced data accuracy, improved system responsiveness, and more efficient utilization of renewable energy sources. The aforementioned results highlight the practical benefits and contributions of our approach within the field of smart agriculture.

Although our research has produced encouraging outcomes, it is crucial to recognize that the domain of smart agriculture is characterized by its dynamic and developing nature. Subsequent research endeavors should prioritize the implementation of the system in practical settings and undertake additional enhancements to effectively tackle distinct agricultural obstacles and accommodate a wide range of environmental circumstances.

References

1. Guardo, E., Di Stefano, A., La Corte, A., Sapienza, M., Scatà, M.: A fog computing-based iot framework for precision agriculture. J. Internet Technol. **19**(5), 1401–1411 (2018)
2. Sakthi, U., Rose, J.D.: Smart agricultural knowledge discovery system using IoT technology and fog computing. In: 2020 Third International Conference on Smart Systems and Inventive Technology (ICSSIT), pp. 48–53 (2020)
3. Ribeiro, F.M, Prati, R., Bianchi, R., Kamienski, C.: A nearest neighbors based data filter for fog computing in IoT smart agriculture. In: 2020 IEEE International Workshop on Metrology for Agriculture and Forestry (MetroAgriFor), pp. 63–7 (2020)
4. Dhiman, P., Kukreja, V., Manoharan, P., Kaur, A., Kamruzzaman, M.M., Dhaou I.B., et al.: A novel deep learning model for detection of severity level of the disease in citrus fruits. Electronics **11**(3), 495 (2022)
5. Islam, M.S.U., Kumar, A., Hu, Y.-C.: Context-aware scheduling in fog computing: a survey, taxonomy, challenges and future directions. J. Netw. Comput. Appl. **180**, 103008 (2021)
6. Singh, G., Singh, J.: A cost effective IoT-assisted framework coupled with fog computing for smart agriculture. In: 2023 IEEE 8th International Conference for Convergence in Technology (I2CT), pp. 1–8 (2023)
7. Shukla, S.K, Pant, B., Viriyasitavat, W., Verma, D., Kautish, S., Dhiman, G., et al.: An integration of autonomic computing with multicore systems for performance optimization in industrial internet of things. IET Commun (2022)
8. Gupta, D., Nainwal, A., Pant, B.: Literature review: Real time water quality monitoring and management, pp. 923–9 (2021)
9. Singh, G., Singh, J.: A fog computing based agriculture-IoT framework for detection of alert conditions and effective crop protection. In: 2023 5th International Conference on Smart Systems and Inventive Technology (ICSSIT), pp. 537–43 (2023)
10. Singh, K.D, Singh, P., Kang, S.S.: Ensembled-based credit card fraud detection in online transactions. In: AIP Conference Proceedings, pp. 50009 (2022)

11. Kang, S.S., Singh, K.D., Kumari, S.: Smart antenna for emerging 5G and application. In: Printed Antennas. CRC Press, pp. 249–64 (2022)
12. Singh, K.D., Singh, P.A.: Novel cloud-based framework to predict the employability of students. In: 2023 International Conference on Advancement in Computation & Computer Technologies (InCACCT), pp. 528–32 (2023)
13. Singh, P.D, Singh, K.D.: Security and privacy in fog/cloud-based IoT Systems for AI and robotics. EAI Endorsed Trans AI Robot **2** (2023)
14. Singh, K.D.: Securing of cloud infrastructure using enterprise honeypot. In: Proceedings–2021 3rd International Conference on Advances in Computing, Communication Control and Networking, ICAC3N 2021, pp. 1388–93 (2021)
15. Singh, P., Singh, K.D.: Fog-centric intelligent surveillance system: A novel approach for effective and efficient surveillance. In: 2023 International Conference on Advancement in Computation & Computer Technologies (InCACCT), pp. 762–6 (2023)
16. Singh, K.D.: Particle swarm optimization assisted support vector machine based diagnostic system for dengue prediction at the early stage. In: Proceedings–2021 3rd International Conference on Advances in Computing, Communication Control and Networking, ICAC3N 2021, pp. 844–8 (2021)
17. Singh, K.D., Singh, P.D., Bansal, A., Kaur, G., Khullar, V., Tripathi, V.: Exploratory data analysis and customer churn prediction for the telecommunication industry. In: 2023 3rd International Conference on Advances in Computing, Communication, Embedded and Secure Systems (ACCESS), pp. 197–201 (2023)
18. Singh, K.D., Singh, P., Kaur, G., Khullar, V., Chhabra, R., Tripathi, V.: Education 4.0: Exploring the potential of disruptive technologies in transforming learning. In: 2023 International Conference on Computational Intelligence and Sustainable Engineering Solutions (CISES), pp. 586–91 (2023)
19. Singh, P., Singh, K.D., Tripathi, V., Chaudhari, V.: Use of ensemble based approach to predict health insurance premium at early stage. Proc Int Conf Comput Intell Sustain Eng Solut CISES **2022**, 566–9 (2022)
20. Singh, K.D., Singh, P., Chhabra, R., Kaur, G., Bansal, A., Tripathi, V.: Cyber-physical systems for smart city applications: A comparative study. In: 2023 International Conference on Advancement in Computation & Computer Technologies (InCACCT), pp. 871–6 (2023)
21. Brar, M., Jindal, P., Malhotra, P., Sharma, P., Kaur, A.: Machine learning based intelligent wardrobe system for apparel recommendation and organization. In: 2023 International Conference on Research Methodologies in Knowledge Management, Artificial Intelligence and Telecommunication Engineering (RMKMATE), pp. 1–5. IEEE (2023)

Efficient Resource Allocation and Load Balancing in Cloud Environments with CloudAnalyst

Navneet Kumar Rajpoot, Prabh Deep Singh, Bhaskar Pant, and Vikas Tripathi

Abstract The concept of computing on the cloud is very new and is still in the process of developing in today's quickly evolving digital society. It represents an advancement from earlier kinds of computing, like grid computing, which came before it. It grants access to a vast shared resource pool through the Internet for a wide variety of clients, making it possible for them to collaborate on huge tasks. Resources and software programs are made available to consumers in a decentralized fashion. The data centers are made to serve as hosting locations for various computing systems as well as software. A number of diverse applications within a system might lead to swings in the workload, demanding the implementation of suitable techniques for managing and allocating resources. The identification of a resolution to the issue of load distribution is crucial. In a system of this level of complexity, the implementation of an efficient load balancing mechanism is of utmost importance. This is necessary in order to address the diverse needs of customers to its utmost, enhance efficiency by appropriately spreading the workload throughout all nodes within the data centers, and improve the overall stability and functioning of the system. This paper examines several strategies for managing workloads and resource allocation in cloud computing using the CloudAnalyst simulation tool. The findings indicate that dynamic load balancing approaches offer higher performance in terms of maximizing resource use, reducing delay, and enhancing the overall efficiency of the system.

Keywords Cloud computing · CloudAnalyst · Load balancing · Latency · Resource allocation · Scalability

---

N. K. Rajpoot (✉) · P. D. Singh · B. Pant · V. Tripathi
Department of Computer Science and Engineering, Graphic Era (Deemed to Be University), Dehradun, India
e-mail: navneetrajpootgeu@gmail.com

P. D. Singh
e-mail: ssingh.prabhdeep@gmail.com

B. Pant
e-mail: bhaskar.pant@geu.ac.in

V. Tripathi
e-mail: vikastripathi.cse@geu.ac.in

B. Soni et al. (eds.), *Beyond Artificial Intelligence*, Lecture Notes in Networks and Systems 1326, https://doi.org/10.1007/978-981-96-4170-3_38

1 Introduction

The popularity of cloud computing has resulted in a fundamental shift in the model for the delivery of computing resources. As a result, businesses now have access to a novel level of adaptability, scalability, and cost-efficiency in the handling and administration of their computer infrastructures. A new era of computing has emerged as a result of switching from traditional on-premises computing to cloud-based architectures. This shift has made it easier for businesses to assign resources on an as-needed basis and to handle rapidly changing workloads with increased flexibility [1]. The efficient allocation of resources and an equal redistribution of workloads are two significant obstacles that must be resolved before cloud computing can reach its full potential as a productive computing model.

The distribution of virtual devices and physical resources is one of the most important factors that go into defining the effectiveness as well as the cost-effectiveness of cloud computing systems [2]. The incorrect distribution of resources can result in either insufficient utilization or overutilization, either of which can lead to increased operating expenses and enhanced service quality. In order to fairly disperse incoming network traffic over all of the resources that are accessible, load balancing is an essential method that must be applied [3]. This technique is used in order to mitigate instances through which resources are either mismanaged or overutilized, resulting in possibly costly errors. The principal aims of load balancing are to reduce response time and latency and enhance the overall system efficiency. The constantly changing nature of cloud workloads brings substantial challenges in cloud computing, particularly in terms of efficient resource allocation and load balancing. Continuous updates to resource allocations are necessary in order to maximize performance. The inherent multi-tenancy characteristic of cloud environments caused these issues because several clients and services share a single infrastructure. There is a critical need for effective resource allocation and load balancing algorithms, as well as the creation of assisting technologies [4].

1.1 Cloud Computing

Cloud computing is an emerging trend in the IT industry that has completely transformed the way businesses handle data storage, data management, and service delivery [5]. The widespread acceptance of cloud computing sparked this shift in the industry. Accessing a shared pool of computing resources in real-time through the Internet allows for more adaptability, scalability, and affordability. There is a constant effort to improve cloud services so that they can meet the changing requirements of a wide variety of industries and applications.

1.1.1 Service Models of Cloud Computing

The following are some of the several types of services that cloud computing provides:

a. Infrastructure as a Service (IaaS)
b. Platform as a Service (PaaS)
c. Software as a Service (SaaS).

1.2 Resource Allocation in Cloud Computing

Resource allocation in cloud computing is the process of managing and optimizing the utilization of computing resources in a cloud environment. It involves the distribution and allocation of various resources, such as virtual machines (VMs), CPU, memory, storage, and network bandwidth, to different applications and workloads to ensure efficient and cost-effective operation [6]. Effective resource allocation not only provides optimal performance but also contributes to the overall cost-effectiveness and quality of service in cloud environments. Different cloud providers offer various tools and services to assist in resource allocation, and organizations should select strategies that best match their workload profiles and operational goals [7].

2 Related Work

Load balancing techniques have been proposed by experts over the years, with many variations. Several studies have been focused on examining the efficiency of load balancing strategies when applied to cloud environments. Some load balancing solutions were studied by Kruekaew et al. [3] using the CloudSim modeling framework. The Weighted Round Robin algorithm was found to be the most effective in terms of response time and overall efficiency. The Cloudsim modeling and simulation software was also used by Pradhan et al. [4] to analyze different load balancing techniques. According to their findings, the Least connection method performed best in terms of both response time and efficiency. Other research has looked at how to utilize machine learning to enhance load balancing methods. Subramoney et al. [8] suggested a neural network–based load balancing method that can adjust to fluctuating workload rates.

3 Load Balancing Techniques in Cloud Computing

Load balancing in cloud computing is a vital approach that guarantees the effective allocation of tasks and network congestion among various resources, like servers, virtual machines, or containers, within a cloud-based environment [8]. The primary objective of load balancing is to optimize the consumption of resources, improve the efficiency of the entire system, and provide a high level of accessibility, all while reducing delays in responding and avoiding instances of overloading. The process of load balancing frequently requires the utilization of a load balancer, which is a specially designed device or software component tasked with the responsibility of evenly distributing incoming requests or network traffic among a collection of servers. Load balancers operate at either the application layer or network layer of the OSI model, dependent upon their specific purpose and operational needs [9].

3.1 Key Objectives of Load Balancing

The primary goals of load balancing are the following:

a. Load balancing evenly distributes workloads to ensure that no single resource is overutilized while others remain idle.
b. Load balancing techniques detect server failures and redirect traffic to healthy servers, ensuring continuous service availability.
c. Load balancing allows for the addition of new servers or resources to handle increasing workloads without disrupting existing services.
d. By routing traffic to the most suitable server based on factors like proximity or server load, load balancing techniques minimize response times, improving user experience (Fig. 1).

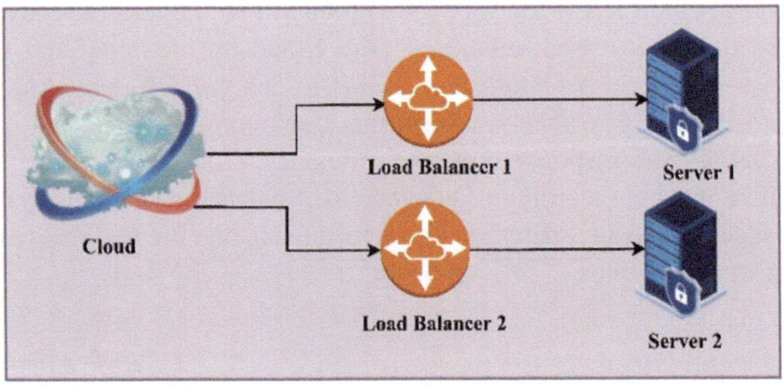

Fig. 1 Load balancing in cloud

3.2 Type of Load Balancing Algorithms

Load balancing techniques are utilized in order to distribute incoming traffic over many backend servers evenly. Some commonly used techniques are round robin, weighted round robin, and least connections [10]. The selection of a technique is dependent upon several aspects, including the nature of the tasks, the capabilities of the computing device, and the desired level of efficiency. Static and dynamic load balancing are two independent methodologies utilized for the purpose of distributing tasks or network congestion over numerous resources, such as servers or virtual machines, with the aim of enhancing the usage of resources and maintaining system efficiency. Every technique offers unique benefits and is well-suited for specific requirements.

3.2.1 Static Load Balancing

Static load balancing involves distributing workloads or requests to resources based on predefined rules or a fixed configuration [11]. These rules are established in advance and do not change unless manually adjusted. Some frequent static load distribution approaches in the cloud are as follows:

A. Round Robin Load Balancing Algorithm

The round-robin algorithm is utilized to evenly distribute incoming requests across the servers that are currently accessible in a periodic fashion. The approach is straightforward and readily implementable, making it well-suited for situations characterized by comparable system capacities [12]. It guarantees that an equitable distribution of load is achieved among all servers.

Pseudocode: Round Robin Algorithm

```
# Initialize a list of backend nodes with their respective indexes
node = ["Node 1", " Node 2", " Node 3", ...]
current_ node _index = 0
# Function to perform Round Robin load balancing
def round_robin_load_balancing(request):
    # Select the current node using the cyclic index
    current_ node = node [current_ node _index]
    # Route the request to the selected node
    route_request(request, current_ node)
```

B. Weighted Round Robin Load Balancing Algorithm

The weights that are assigned to servers in weighted round-robin are determined by either their capacity or their degree of efficiency. Machines that have a higher capacity are assigned heavier workloads, which results in enhanced usage of available resources [10].

C. Least Connections Load Balancing Algorithm

The least connections direct new requests to the server with the fewest active connections. This technique helps balance the workload based on the actual load of each server, rather than just its capacity. It can improve user experience by minimizing response times.

3.2.2 Dynamic Load Balancing

Dynamic load balancing adjusts the distribution of workloads or traffic in real-time based on changing conditions, such as server load, response times, or available resources [13].

Nature-Inspired Dynamic Load Balancing Algorithms

Nature-inspired dynamic load balancing algorithms are inspired by biological or natural phenomena and are designed to distribute workloads in dynamic computing environments efficiently. Nature-inspired dynamic load balancing algorithms aim to provide self-organizing, adaptive, and efficient solutions for managing workloads in cloud computing and distributed systems. They can be particularly useful in scenarios where workload patterns change dynamically, server performance varies, or resource allocation needs to be optimized in real time. Following are some nature-inspired dynamic load balancing algorithms:

A. *Ant Colony Optimization (ACO)*

The Ant Colony Optimization approach is a nature-inspired optimization technique that gets ideas from the foraging activity of ants in order to address complex optimization challenges. The utilization of Ant Colony Optimization is a robust, optimized methodology capable of effectively addressing extensive optimizing challenges, which are characterized by dynamic and developing attributes [14]. In terms of dynamic load balancing scenarios, the utilization of Ant Colony Optimization can effectively facilitate maintaining the best utilization of resources and the ability to respond to varying workload circumstances in cloud computing and distributed systems.

Pseudocode: Ant Colony Optimization Algorithm

```
# Define parameters
num_ants = 50 # Number of ants
num_servers = 10 # Number of backend servers
pheromone_matrix = initialize_pheromone_matrix() # Initialize pheromone levels on paths
max_iterations = 100 # Maximum number of iterations
# Main optimization loop
for iteration in range(max_iterations):
```

```
    for ant in range(num_ants):
        current_server = select_starting_server() # Ant selects a starting server
        while not all_servers_visited():
            # Ant selects the next server based on pheromone levels and a heuristic function
            next_server = select_next_server(current_server, pheromone_matrix)
            # Ant sends a request to the selected server and receives response time
            response_time = send_request_to_server(next_server)
            # Update pheromone levels on the path based on response time
            update_pheromone(current_server, next_server, response_time, pheromone_matrix)
            current_server = next_server # Move to the next server
        # Ant has completed its tour, evaluate its solution and update personal best if better
    # Update global best solution if needed
    # Evaporate pheromone levels to simulate fading over time
    evaporate_pheromone(pheromone_matrix)
# Apply the best solution found by the ants for load balancing
```

B. *Particle Swarm Optimization (PSO)*

PSO is an optimization technique that draws inspiration from the communal actions shown by birds and fish in nature. The utilization of this technique is widespread in many domains of machine learning for the purpose of resolving optimization challenges. It is known for its adaptability and effectiveness in solving optimization problems [15]. This technique offers a significant methodology for addressing complicated optimization challenges across diverse fields.

C. *Honey Bee Load Balancing*

The Honey Bee Load Balancing technique is a load balancing technique that is based on the honey bees' approach to food gathering. This technique, which draws inspiration from natural processes, is implemented in distributed computing and networking to facilitate the balanced distribution of workloads across various resources, such as servers. It has the goal of enabling efficient use of resources and is especially helpful in situations with dynamic workloads and changing resource situations because of its architecture [15]. Honey bees are known for their very effective foraging tactics, which HBLB intends to apply in order to maximize resource use and reduce delay in response.

4 Experimental Setup and Results

CloudAnalyst is a cloud simulation and modeling tool that is designed to analyze and simulate the behavior and performance of cloud computing environments. It is primarily used for research, experimentation, and analysis in the field of cloud

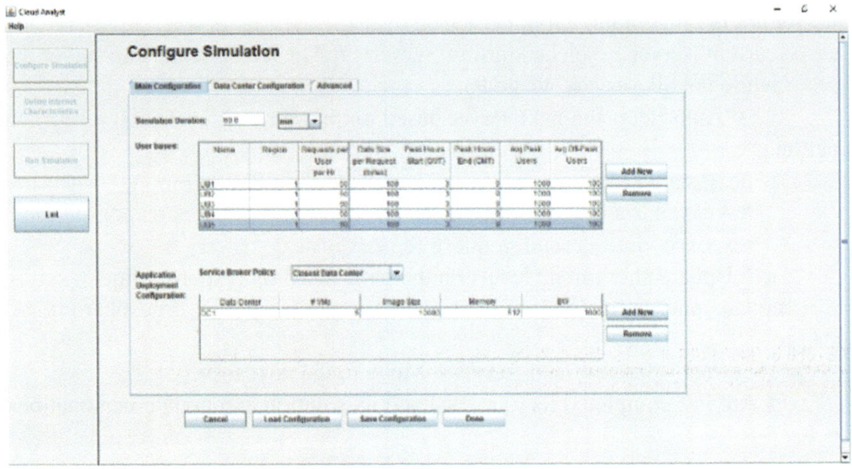

Fig. 2 Main configuration in Cloud Analyst

computing. CloudAnalyst allows users to create and model virtual cloud environments, run simulations, and analyze the results to gain insights into various aspects of cloud computing, including load balancing, resource allocation, and cost optimization. In the experimental setup, The Cloud Analyst tool is utilized to compare the performance of two load balancing algorithms: the conventional round-robin and the Ant Colony Optimization. The results showed that Round Robin exhibited a straightforward, cyclic distribution of tasks among servers, ACO demonstrated its adaptability by intelligently allocating tasks based on server workload and response times. ACO is a better choice for load distribution in dynamic environments, its adaptability and intelligence can lead to significantly improved load balancing and better overall system performance. ACO improved load distribution, which turned into reduced response times and enhanced resource utilization (Figs. 2, 3 and 4).

5 Comparative Analysis of Load Balancing Techniques

Table 1 provides a comparison of load balancing techniques based on factors such as complexity, adaptability, handling server heterogeneity, real-time monitoring capabilities, and suitable use cases.

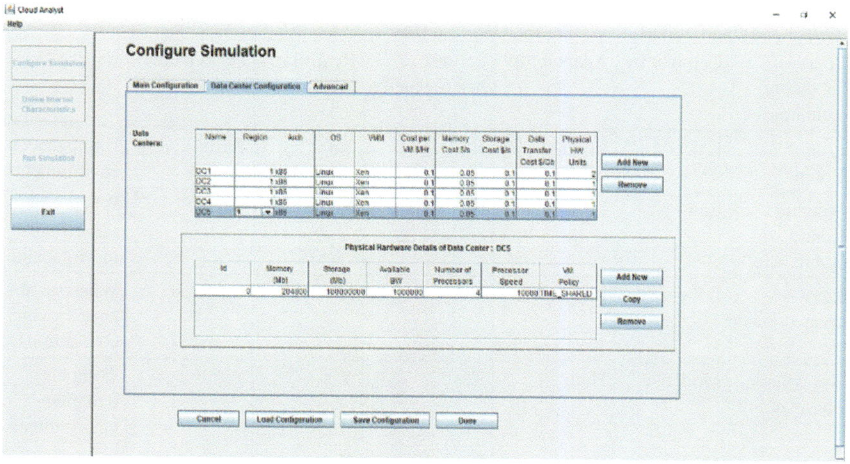

Fig. 3 Data center configuration in Cloud Analyst

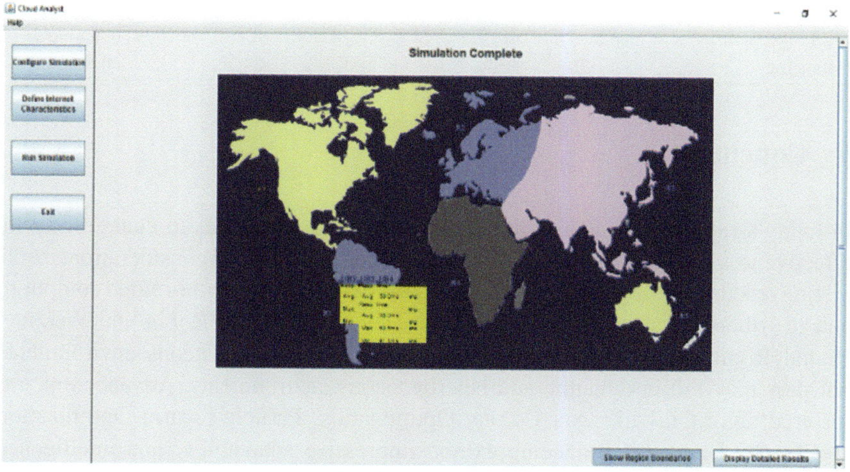

Fig. 4 Simulation running in Cloud Analyst

Table 1 Comparison of load balancing techniques

Load balancing technique	Complexity	Adaptability	Server heterogeneity	Real-time monitoring	Performance improvement	Use cases
Round Robin	Low	Limited	No	No	Basic	Homogeneous environments
Weighted Round Robin	Low	Limited	No	No	Basic	When some servers are more capable
Least connections	Low	Limited	No	No	Moderate	When servers have varying connections
Ant colony optimization	Moderate	High	Yes	Yes	Significant	Complex, dynamic environments
Particle swarm optimization	Moderate	High	Yes	Yes	Significant	Dynamic environments, optimization problems
Honey Bee load balancing	Moderate	High	Yes	Yes	Significant	Complex, dynamic environments

6 Conclusion

In today's rapidly advancing digital era, technology continues to change, and the effectiveness of load balancing techniques will remain a critical consideration in optimizing resource utilization, performance, and cost-efficiency within cloud computing and distributed systems. Traditional approaches like Round Robin and Weighted Round Robin offer simplicity and predictability in homogeneous environments, but they may fail in dynamic and heterogeneous environments. Advanced nature-inspired algorithms like Ant Colony Optimization, Particle Swarm Optimization, and Honey Bee Load Balancing exhibit impressive adaptability and intelligence, making them well-suited for complex, ever-changing environments. This paper has explored and analyzed the resource allocation and load balancing strategies within cloud environments. CloudAnalyst simulation tool is used for experimental setup and comparing the results. Results and findings show that dynamic load balancing techniques perform better in terms of optimizing resource utilization, reducing latency, and enhancing overall system performance. The experiment showed the significance of nature-inspired algorithms in optimizing cloud-based services, showing a promising future for developing more advanced and effective dynamic load balancing strategies.

References

1. Srivastava, A., Kumar, N.: An energy efficient robust resource provisioning based on improved PSO-ANN. Int. J. Inf. Technol. **15**(1), 107–117 (2022). https://doi.org/10.1007/s41870-022-01148-9
2. Yugank, H.K., Sharma, R., Gupta, S.H.: An approach to analyse energy consumption of an IoT system. Int. J. Inf. Technol. **14**(5), 2549–2558 (2022). https://doi.org/10.1007/s41870-022-00954-5
3. Kruekaew, B., Kimpan, W.: Multi-objective task scheduling optimization for load balancing in cloud computing environment using hybrid artificial bee colony algorithm with reinforcement learning. IEEE Access **10**, 17803–17818 (2022). https://doi.org/10.1109/access.2022.3149955
4. Pradhan, A., Bisoy, S.K., Kautish, S., Jasser, M.B., Mohamed, A.W.: Intelligent decision-making of load balancing using deep reinforcement learning and parallel PSO in cloud environment. IEEE Access **10**, 76939–76952 (2022). https://doi.org/10.1109/access.2022.3192628
5. Mishra, S.K., Sahoo, B., Parida, P.P.: Load balancing in cloud computing: a big picture. J. King Saud Univ.-Comput. Inf. Sci. **32**(2), 149–158 (2020)
6. Balaji, K.: Load balancing in cloud computing: issues and challenges. Turk. J. Comput. Math. Educ. (TURCOMAT) **12**(2), 3077–3084 (2021)
7. Subramoney, D., Nyirenda, C.N.: Multi-swarm PSO algorithm for static workflow scheduling in cloud-fog environments. IEEE Access **10**, 117199–117214 (2022). https://doi.org/10.1109/access.2022.3220239
8. Jena, U.K., Das, P.K., Kabat, M.R.: Hybridization of meta-heuristic algorithm for load balancing in cloud computing environment. J. King Saud Univ.-Comput. Inf. Sci. **34**(6), 2332–2342 (2022)
9. Princess, G.A.P., Radhamani, A.S.: A hybrid meta-heuristic for optimal load balancing in cloud computing. J. Grid Comput. **19**(2), 21 (2021)
10. Devaraj, A.F.S., Elhoseny, M., Dhanasekaran, S., Lydia, E.L., Shankar, K.: Hybridization of firefly and improved multi-objective particle swarm optimization algorithm for energy efficient load balancing in cloud computing environments. J. Parallel Distrib. Comput. **142**, 36–45 (2020)
11. Rajpoot, N.K., Singh, P., Pant, B.: Nature-inspired load balancing algorithms for resource allocation in cloud computing. In: 2023 International Conference on Computational Intelligence and Sustainable Engineering Solutions (CISES). IEEE, pp. 827–832 (2023)
12. Rajpoot, N.K., Singh, P., Pant, B.: Load balancing in cloud computing: a simulation-based evaluation. In: 2023 International Conference on Computational Intelligence and Sustainable Engineering Solutions (CISES). IEEE, pp. 564–568 (2023)
13. Garg, N., Wazid, M., Das, A.K., Singh, D.P., Rodrigues, J.J., Park, Y.: BAKMP-IoMT: design of blockchain enabled authenticated key management protocol for internet of medical things deployment. IEEE Access **8**, 9595695977 (2020)
14. Chandrappa, S., Guruprasad, M.S., Kumar, H.N., Raju, K., Kumar, D.S.: An IOT-based automotive and intelligent toll gate using RFID. SN Comput. Sci. **4**(2), 154 (2023)
15. Prasad, G., Gujjar, P., Kumar, H.N., Kumar, M.A., Chandrappa, S.: Advances of cyber security in the healthcare domain for analyzing data. In: Cyber Trafficking, Threat Behavior, and Malicious Activity Monitoring for Healthcare Organizations. IGI Global, pp. 1–14 (2023)

IoT in Smart Cities: Review of IoT Architecture for Smart Cities, Applications, and Key Challenges

Rajwinder Kaur, Jaswinder Singh, Satinder Kaur, and Vishal Malhotra

Abstract The Internet of Things (IoT) is a system that connects various gadgets and technologies without requiring human involvement. The IoT accelerates global city development, enhancing sustainability, comfort, and productivity for residents. A lot of work has been observed in the literature for the same. So, this paper introduces IoT-enabled smart cities, smart city workflows, IoT fundamentals, IoT architectures specific to smart cities, and diverse applications of IoT. This paper dis- cusses various challenges, observed by researchers in this domain and proposes generalized solutions at conceptual level. In future realization of the proposed solutions are recommended. It is concluded that the study provides foundational insights for researchers on IoT implementation in smart cities, serving as the initial step towards optimizing sustainable integration.

Keywords Smart cities · Internet of Things · IoT · Smart city challenges

1 Introduction

The rapid progress in the development and implementation of the Internet of Things (IoT) has positioned it as a crucial catalyst in the contemporary smart city environment. Moreover, this progress is driving the smart city concept toward the domain of big data. Metropolitan regions are experiencing persistent growth, with the influx of individuals migrating to cities having increased six times over the past two decades

R. Kaur (✉) · J. Singh · S. Kaur · V. Malhotra
Guru Nanak Dev University, Amritsar, India
e-mail: rajwinder.cs@gndu.ac.in

J. Singh
e-mail: jaswinder.cs@gndu.ac.in

S. Kaur
e-mail: satinder.dcet@gndu.ac.in

V. Malhotra
e-mail: vishal.coit@gndu.ac.in

B. Soni et al. (eds.), *Beyond Artificial Intelligence*, Lecture Notes in Networks and Systems 1326, https://doi.org/10.1007/978-981-96-4170-3_39

[1]. City dwellers face numerous pressing chal- lenges such as traffic congestion, waste disposal, water supply, parking manage- ment, air quality monitoring, and more. In this urban context, the application of IoT technology opens up the possibility of creating remarkable solutions that provide valuable digital services to both the public and businesses. By inte- grating the Internet of Things (IoT) into urban spaces, smart cities are being developed to enhance the management of public utilities, transportation, and citizen services, leading to substantial improvements in the urban economy, liv- ing standards, and overall safety [2].

The implementation of IoT projects aimed at building modern cities in Russia [3] has proven highly advantageous to the state, the business sector, and the residents. Consequently, numerous projects of this nature have already been successfully executed throughout the country. By incorporating IoT technology into urban areas, it becomes possible to create 'smart cities' that enhance budget efficiency, improve the quality of life, and attract investment to any city [4, 5]. In urban environments, IoT solutions find crucial applications in public utilities, transportation, and services for residents. Various models for implementing these technologies have been observed worldwide [6–9]. Failures often arise due to inadequate estimation of required resources and benefits during the planning phase, the complex nature of the technology provider market, and ineffective implementation. Therefore, it is essential to ensure accurate estimation of resources and benefits when designing a smart city. This paper provides an overview of IoT, its role in smart cities, IoT architecture, challenges faced by smart cities, benefits of IoT applications in creating smart cities, and real-world IoT applications for smart cities.

The study addresses two fundamental research questions. First, it examines the most commonly addressed topics in IoT applications for smart cities, aiming to understand the prevalent focus areas within Internet of Things (IoT) applica- tions in smart city development. Second, it identifies obstacles faced in these applications and proposes effective strategies strategies for mitigation. This re- search delves into the challenges encountered in IoT applications, striving to identify strategies and solutions to overcome these barriers, ultimately fostering the seamless integration and advancement of IoT technologies across diverse do- mains.

The paper is organized as follows: Sect. 2 provides an overview of the background study covering basics of IoT use in smart cities, IoT architecture in smart cities, and applications; Sect. 3 discusses the challenges faced in IoT applications and proposes solutions at conceptual level and Sect. 4 concludes the paper and discusses future recommendations.

2 Background Study

Introducing IoT, a pivotal technological force that interconnects and enables data exchange between devices, this section delves into aspects of IoT in the context of smart cities. The discussion encompasses IoT-enabled smart cities, their workflow, the fundamental principles of the Internet of Things (IoT), IoT architectures tailored for smart environments and the diverse applications.

2.1 IoT-Enabled Smart City

A smart city utilizes information and communication technologies to establish, implement, and advance development strategies that tackle urban challenges and create a connected, technologically empowered, and sustainable infrastructure. The primary objective of a smart city is to leverage smart technologies and data analysis to enhance city operations, drive economic growth, and improve the overall quality of life for its residents. The true value of technology lies in its application rather than the mere abundance of it [10].

The level of smartness in a city is influenced by several key factors, each contributing to its overall intelligence and efficiency. First and foremost, a technologically advanced infrastructure plays a crucial role in creating a smart city. This involves the integration of cutting-edge technologies and digital systems to optimize various aspects of urban life [11]. Additionally, environmentally protective initiatives are essential in promoting sustainability and reducing the city's ecological footprint, making it more livable and eco-friendly [12].

Furthermore, efficient and practical public transportation systems are vital components of a smart city. Well-designed and easily accessible transport networks help alleviate traffic congestion and enhance mobility for residents and visitors alike [13]. Another crucial aspect is confident and forward-thinking city planning, which involves strategic urban development and design that anticipates future needs and challenges [14]. Such planning ensures the city can adapt and evolve in a sustainable and efficient manner. Lastly, a smart city's success heavily relies on its populace's ability to thrive and reap the benefits of the city's resources. This means providing access to education, healthcare, job opportunities, and other essential services that improve the quality of life for all citizens [15]. By nurturing these factors and focusing on continual improvements, a city can aspire to become smarter, more innovative, and a better place for its resi- dents to live and prosper.

The success of a smart city relies on the collaboration between the public and private sectors because a significant portion of the work required to establish and maintain a data-driven environment falls beyond the purview of local governments [16, 17]. In addition to the technological aspects of a smart city, the presence of data analysts is essential to analyze the information generated by smart city systems. Their role is to identify any issues and drive improvements based on the data insights [18]. Smart cities employ a diverse range of software, user interfaces, communication networks, and the Internet of Things (IoT) to deliver connected solutions for the public [19]. Among these technologies, IoT holds the utmost significance. The Internet of Things (IoT) constitutes a network of interconnected devices that communicate and exchange data, encompassing vehicles, home appliances, on-street sensors, and various other examples [20].

Smart cities utilize a combination of software, user interfaces, communication networks, and the Internet of Things (IoT) to provide connected solutions to the public. Among these, IoT stands out as the most significant technology. Furthermore, a robust security system is implemented to protect, monitor, and regulate data transfer within the smart city network, safeguarding the city's data platform and preventing unauthorized access to the IoT network [21].

2.2 Smart City Work Flow

The 'Smart City Workflow' enhances urban services and residents' quality of life through interconnected steps: 'Data Collection' using smart sensors, 'Data Analysis' for valuable insights, and sharing outcomes with decision makers (Fig. 1).

Action is taken to enhance operations, manage assets, and improve residents' well-being based on insights gained. These steps are vital for smart city technology implementation and align with widely accepted approaches in the field, supported by academic research and reputable sources [2, 22–24].

2.3 Internet of Things (IoT)

In 1999, Kevin Ashton introduced the phrase "Internet of Things" while discussing supply chain management, marking the origin of the term [25]. The Internet of Things (IoT) is a network that connects various devices, like computers, sensors, electronics, and software, providing a superior alternative to existing systems and connections [26–28]. The ICT framework integrates real-time data from interconnected assets, objects, and machines to empower better decision making. Additionally, mobile devices, connected cars, and connected buildings provide citizens with more opportunities to engage and actively participate in smart city ecosystems. This fusion of devices, data, and urban infrastructure opens avenues for cost savings, enhanced sustainability, and optimized processes, including energy distribution and waste collection. Moreover, it provides the potential to alleviate traffic congestion and improve air quality, making cities more efficient, livable, and environmentally friendly [29].

2.4 IoT Architectures in Smart Cities

The IoT architecture defines how IoT components connect and transmit data. Multiple proposed architectures, like three, four, or five layers, serve specific purposes and enhance system efficiency [30–32]. A common design features five layers, integrating sensing, transmission, processing, storage, and real-time communication.

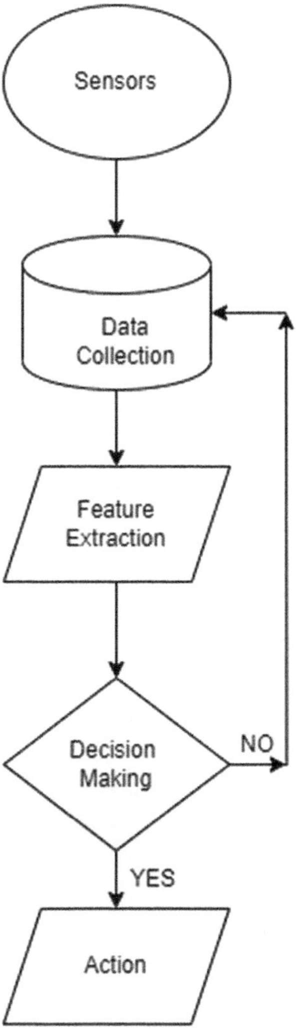

Fig. 1 Work fow of smart city activities

As shown in figure IoT connects data sensing, transmission/reception, processing, and storage functions, allowing for real-time communication and data exchange. A typical IoT design is based on technology and comprises five layers [33, 34] the perception layer, network layer, service layer, data layer, and application layer. Each layer operates on the information received from the preceding one, creating a hierarchical structure for efficient data processing and decision-making.

2.5 Applications Areas of IoT

Exploring the vast landscape of IoT applications, this section delves into a wide array of its subdomains, ranging from Smart Home Automation and Healthcare to Smart Cities, Industrial IoT (IIoT), Transportation, Energy Management, Environmental Monitoring, Manufacturing, Supply Chain and Logistics, Smart Agriculture, Asset Tracking, and Security and Surveillance. Each of these applications represents a unique facet of how IoT is shaping our interconnected world.

Smart Home Automation: IoT plays a pivotal role in smart homes by connecting and automating various devices and systems, enhancing convenience, energy efficiency, and security. In a smart home, IoT-enabled devices such as thermostats, lighting, and appliances can be controlled remotely through smart phones or voice assistants [24]. These devices can also work together intelligently, adjusting settings based on user preferences and sensor data, resulting in optimized energy usage [16]. Additionally, IoT security cameras, doorbell cameras, and smart locks provide real-time monitoring and enhanced home security. Overall, IoT in smart homes offers homeowners greater control, energy savings, and peace of mind through seamless connectivity and automation (Fig. 2).

Healthcare: IoT is revolutionizing healthcare by enabling remote patient monitoring and personalized healthcare delivery. Through IoT devices such as wearable

Fig. 2 Architecture of IoT [31, 32]

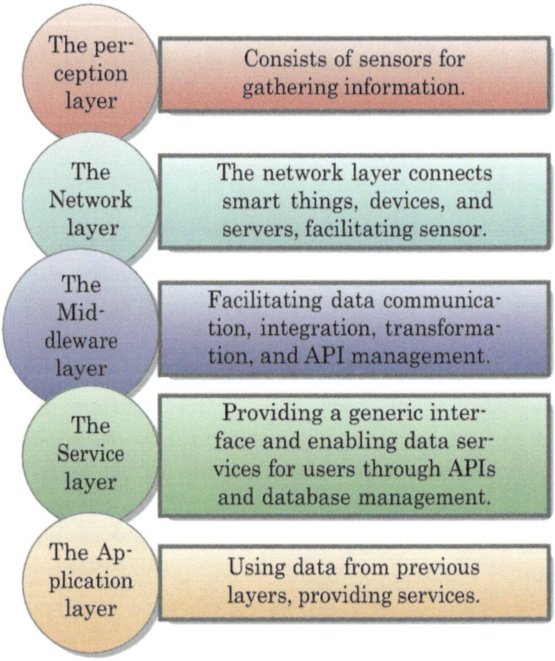

fitness trackers, smart medical sensors, and connected health equipment, healthcare providers can collect real-time patient data, including vital signs, medication adherence, and activity levels [24]. This data is transmitted securely to healthcare professionals, allowing for timely intervention and adjustments in treatment plans [16]. Additionally, IoT facilitates telemedicine, making healthcare services more accessible, particularly for patients in remote areas or with chronic conditions. Overall, IoT in healthcare enhances patient care, improves outcomes, and reduces the burden on healthcare systems.

Smart Cities: In smart cities, IoT plays a pivotal role in enhancing urban living through the seamless integration of technology and infrastructure. IoT sensors and devices are strategically deployed across the city to collect real-time data on traffic flow, air quality, energy consumption, waste management [16], and more. This data is then analyzed to optimize traffic management, reduce pollution, improve energy efficiency, and enhance public services [5]. Smart cities leverage IoT to enable smart parking systems, intelligent street lighting, efficient waste collection, and even predictive maintenance of critical infrastructure. Ultimately, IoT in smart cities fosters sustainability, convenience, and a higher quality of life for residents while promoting resource efficiency and environmental responsibility.

Industrial IoT (IIoT): In Industrial IoT (IIoT), sensors, devices, and machines are interconnected to create a network that enables real-time data collection and analysis, enhancing industrial processes and operations. IIoT applications include predictive maintenance, where sensors monitor equipment health to prevent breakdowns, improving efficiency and reducing downtime. Asset tracking and inventory management are streamlined through RFID and GPS technologies, optimizing supply chain logistics [16]. Additionally, IIoT enables remote monitoring of industrial processes, ensuring quality control and safety compliance. Overall, IIoT revolutionizes industries by providing actionable insights, reducing costs, and increasing productivity.

Transportation: IoT is revolutionizing the transportation industry by providing real-time data and connectivity. In transportation, IoT is used to monitor vehicle health, track fleets, optimize routes, and enhance passenger experiences [24]. For instance, IoT sensors in vehicles can transmit data on engine performance, fuel efficiency, and maintenance needs, enabling proactive maintenance and reducing downtime. Fleet managers can use IoT to track vehicles in real-time, improving logistics and reducing operational costs [6]. Additionally, IoT contributes to smart traffic management systems, reducing congestion and enhancing road safety. Passengers benefit from IoT-enabled services such as real-time transit information, ticketing, and vehicle sharing [10], making transportation more convenient and efficient.

Energy Management: In energy management, IoT plays a pivotal role in revolutionizing how we monitor, control, and optimize energy usage. IoT devices and sensors are deployed throughout facilities and power grids, collecting realtime data on energy consumption, equipment performance, and environmental conditions [15].

This data is then analyzed to identify energy inefficiencies and areas for improvement. Smart thermostats, lighting systems, and HVAC controllers adjust energy usage based on occupancy and environmental factors, reducing waste [2]. Furthermore, IoT-enabled smart grids enhance the distribution of electricity, allowing for more efficient load balancing and the integration of renewable energy sources [16]. Overall, IoT empowers organizations and utilities to make data-driven decisions, optimize energy resources, and reduce costs while contributing to a more sustainable energy future.

Environmental Monitoring: IoT plays a pivotal role in environmental monitoring by enabling the collection of real-time data from various sensors and devices. These sensors are deployed in critical areas to measure parameters such as air quality, temperature, humidity, water quality, and pollution levels [10]. By continuously gathering and transmitting this data to central systems, IoT allows for early detection of environmental changes and trends [4]. It aids in assessing the impact of human activities on the environment, supporting conservation efforts, and enabling timely responses to environmental challenges such as pollution, wildfires, or natural disasters [2]. IoT-based environmental monitoring systems contribute to more informed decision-making for sustainable resource management and the protection of ecosystems.

Manufacturing: In manufacturing, IoT enables real-time monitoring and control of production processes, collecting data on machinery performance, product quality, and environmental conditions. Predictive maintenance reduces downtime [7]. IoT-driven automation streamlines production, enhances supply chain visibility, and optimizes inventory management [8]. Manufacturers use IoT data analytics to boost efficiency, reduce waste, and ensure consistent product quality. IoT transforms manufacturing into an agile, cost-effective, and data-driven industry.

Supply Chain and Logistics: IoT transforms supply chain visibility and predictive maintenance. In smart buildings, IoT optimizes energy usage and enhances safety through automation and predictive maintenance [27]. Wearable IoT monitors health and enables fitness management. In consumer electronics, IoT integrates devices for enhanced daily efficiency and convenience.

Smart Agriculture: IoT revolutionizes farming with real-time monitoring of soil, weather, and crop health using connected sensors. This data-driven approach enhances resource management for optimized irrigation, fertilization, and pest control in smart agriculture [16]. IoT-powered machinery and drones automate planting and harvesting, boosting productivity and cutting labor costs [24]. Smart agriculture promotes sustainability, conserves resources, boosts crop yields, and addresses global food demand.

Asset Tracking: IoT transforms asset tracking, enabling real-time monitoring of assets like vehicles, equipment, and inventory using sensors and GPS. Data is sent to centralized platforms for precise tracking, efficient route planning, and proactive maintenance [35]. IoT-based asset tracking enhances security, reduces losses, and boosts operational efficiency in logistics, transportation, manufacturing, and supply chain management.

Security and Surveillance: IoT plays a pivotal role in enhancing security and surveillance systems. Through the integration of smart cameras, sensors, and data analytics, IoT enables real-time monitoring of premises and assets. These connected devices can detect intrusions, monitor access points, and analyze video feeds for suspicious activities [2]. Additionally, IoT facilitates remote monitoring and alerts, allowing security personnel to respond swiftly to potential threats [21]. Furthermore, the data collected from these devices can be analyzed for pattern recognition and predictive insights, contributing to proactive security measures [27]. In essence, IoT transforms traditional security and surveillance into more efficient, proactive, and responsive systems, significantly improving overall safety and security.

3 IoT Challenges and Their Solutions

To successfully implement IoT applications and develop expansive IoT platforms, it is crucial to address various research and technical challenges.

3.1 IoT Challenges

Numerous research and tech challenges must be tackled for successful IoT application implementation and the development of broad-reaching IoT platforms. The key challenges, in this research, includes:

Interoperability: Interoperability is a key IoT challenge, in security and surveillance, especially given the diverse origins and communication protocols of IoT devices. Integrating devices from different manufacturers and ecosystems further complicates seamless interoperability. Ensuring seamless interoperability among these devices and systems is complex due to their diverse origins and communication protocols [36]. The interoperability challenge arises when components struggle to work cohesively, impacting security and surveillance efficiency. It's about seamless interactions between devices and people, beyond just connecting individuals. Technological capability variations add to the complexity (Fig. 3).

Data Privacy and Security: The central challenges of IoT involve maintaining data integrity, ensuring unique identification, encryption, and safeguarding personal information [37]. It is vital to address legal considerations, liabilities, and data ownership. Energy-efficient encryption and advanced data protection technologies are imperative [38]. As extensive real-time data collection occurs, ensuring privacy and security is paramount to prevent unauthorized access and potential breaches with severe consequences.

Fig. 3 IoT challenges in smart cities

IoT Device Efficiency and Reliability: This involves developing ultra-low-power circuits for rugged environments, parallel processing in low-power multiprocessor systems, and autonomy prioritizing trust, privacy, and security [39], as well as advancements in battery technology, energy harvesting, and storage are crucial for extending battery life and reducing replacements in IoT devices [40]. In industry or emergencies, ensuring uninterrupted operations is critical. Investigating resilience and reliability system-wide is vital, considering communication, hardware flexibility, avoiding single points of failure, and robust data processing in uncertain conditions.

Networking: Addressing smart city component needs, such as mobility and high data throughput, poses a challenge for current networking methods. Solutions include defining access points, local networks, and dynamic routing protocols to accommodate stationary and moving IoT devices-a capability lacking in many current protocols [41]. In urban areas, a reliable network is crucial for IoT devices, especially with growing smart city applications, emphasizing consistent connectivity [42].

Big Data Analytics: "IoT's Data Surge: Challenges and Opportunities" It would not be wrong to say that IoT is experiencing explosive data growth, as devices produced 13.6 zettabytes in 2018 and are projected to reach 79.4 zettabytes by 2025 [43]. Advanced data analytics algorithms are vital for enhancing smart city services, utilizing deep learning with vast data for valuable insights. Scalability ensures adaptability, addressing challenges like concept drift. Transparent analytics are especially vital in smart health for Smart Cities. Hybrid methods, integrating deep learning and semantic web tech, are showcased, encouraging widespread adoption, especially in applications like flood monitoring. Thakker et al. [44] and Edge-based COVID-19 healthcare systems [45]. Incorporating distillation, visualization, and intrinsic techniques into machine and deep learning in smart city apps is crucial for advancement.

3.2 Proposed Solutions for Challenges Faced in Different Domains of IoT

In the diverse realm of IoT applications there's significant potential for enhancing various sectors. However, challenges like interoperability, data privacy, device efficiency, networking, and big data analysis need to be addressed for seamless integration. The following table proposes solutions to these challenges, with a focus on smart city applications, aiming for efficient and secure development through IoT technologies (Table 1).

Table 1 Solutions to tackle IoT implementation challenges

Domain	Solutions
Interoperability	1. Standardization of communication protocols 2. Development of middleware to bridge diverse devices 3. Establishment of IoT platforms that provide compatibility 4. Adoption of open-source solutions for interoperability
Data privacy and security	1. Implementation of End-to-End Encryption 2. Strong User Authentication Methods 3. Regular Security Audits
IoT device efficiency and reliability	1. Utilization of energy-efficient components and low-power design techniques 2. Development of efficient power management systems 3. Integrating AI-based energy optimization algorithms 4. Incorporating redundancy in critical components 5. Rigorous testing under various conditions to ensure reliability
Networking	1. Deployment of hybrid network architectures 2. Utilization of 5G technology for high-speed and low-latency connectivity 3. Development of dynamic and adaptive routing algorithms
Big data analytics	1. Adoption of Big Data Platforms and Frameworks 2. Development of Machine Learning and AI Algorithms 3. Utilization of Edge Analytics 4. Implementation of Data Anonymization Techniques

4 Conclusion

This study emphasizes IoT architecture and its applications in smart cities, presenting challenges encountered during their implementation. The primary contribution of this paper is to propose solutions to these IoT application challenges. By discussing application challenges and solutions in one place, researchers and policymakers can gain a foundational understanding of the considerations involved in designing and implementing IoT technologies within smart cities. It is recommended that the solutions proposed in this study can be realized in future research work.

References

1. Ejaz, W., Anpalagan, A.: Internet of things for smart cities: technologies, big data and security. Springer (2019)
2. Belli, L., Cilfone, A., Davoli, L., Ferrari, G., Adorni, P., Di Nocera, F., Dall'Olio, A., Pellegrini, C., Mordacci, M., Bertolotti, E.: Iot-enabled smart sustainable cities: Challenges and approaches. Smart Cities **3**(3), 1039–1071 (2020)
3. Yuloskov, A., Bahrami, M.R., Mazzara, M., Kotorov, I.: Smart cities in russia: Current situation and insights for future development. Futur. Internet **13**, 252 (2021)
4. Saha, H.N., Auddy, S., Pal, S., Kumar, S., Pandey, S., Singh, R., Singh, A.K., Banerjee, S., Ghosh, D., Saha, S.: Waste Management Using Internet Of Things (iot). In: 2017 8th Annual Industrial Automation And Electromechanical Engineering Conference (IEMECON), pp. 359–363. IEEE (2017)
5. Mall, P.K., Narayan, V., Pramanik, S., Srivastava, S., Faiz, M., Sriramulu, S., Kumar, M.N.: Fuzzynet-based modelling smart traffic system in smart cities using deep learning models. In: Handbook of Research on Data-Driven Mathematical Modeling in Smart Cities, pp 76–95. IGI Global (2023)
6. Porru, S., Misso, F.E., Pani, F.E., Repetto, C.E.: Smart mobility and public transport: Opportunities and challenges in rural and urban areas. J. Traffic Transp. Eng. (Engl. Ed.) **7**(1), 88–97 (2020)
7. Zhong, R.Y., Xu, X., Klotz, E., Newman, S.T.: Intelligent manufacturing in the context of industry 4.0: a review. Engin. **3**(5), 616–630 (2017)
8. McFarlane, D., Sarma, S., Chirn, J.L., Wong, C.Y., Ashton, K.: Auto id systems and intelligent manufacturing control. Eng. Ap Plications Artif. Intell. **16**(4), 365–376 (2003)
9. Ramya, P., Ramya, V., et al.: E-waste management using hybrid optimization- enabled deep learning in iot-cloud platform. Adv. Eng. Softw. **176**, 103353 (2023)
10. Pellicer, S., Santa, G., Bleda, A.L., Maestre, R., Jara, A.J., Skarmeta, A.G.: A global perspective of smart cities: A sur- vey. In: 2013 Seventh International Conference on Innovative Mobile and Internet Services in Ubiquitous Computing, pp. 439–444. IEEE (2013)
11. Attaran, H., Kheibari, N., Bahrepour, D.: Toward integrated smart city: A new model for implementation and design challenges. Geo. J **87**(Suppl 4), 511–526 (2022)
12. Joshi, S., Saxena, S., Godbole, T., et al.: Developing smart cities: An integrated framework. Procedia. Comput. Sci. **93**, 902–909 (2016)
13. Scholl, H.J., Glassey, O., Janssen, M., Klievink, M., Lindgren, I., Parycek, P., Tambouris, E., Wimmer, M.A., Janowski, T., Soares, D.S.: Electronic Government. In: 15th IFIP WG 8.5 International Conference, EGO 201, Guimar es, Portugal, Sept 5–8, 201, Proceedings, vol. 9820. Springer (2016)
14. Van Twist, A., Ruijer, E., Meijer, A.: Smart cities & citizen discontent: A systematic review of the literature. Gov. Ment Inf. Q. 101799 (2023)

15. Pandya, S., Srivastava, G., Jhaveri, R., Babu, M.R., Bhattacharya, S., Maddikunta, P.K.R., Mastorakis, S., Piran, M.J., Gadekallu, T.R.: Federated learning for smart cities: A com prehensive survey. Sustain. Energy Technol. Assess **55**:102987 (2023)
16. Syed, A.S., Sierra-Sosa, D., Kumar, A., Elmaghraby, A.: Iot in smart cities: A survey of technologies, practices and challenges. Smart Cities. **4**(2), 429–475 (2021)
17. Haubensak, O.: Smart cities and internet of things, pp. 33–39. In Business Aspects of the Internet of Things, Seminar of Advanced Topics, ETH Zurich (2011)
18. Grizhnevich, A.: Iot for smart cities: Use cases and implementation strategies. Science Soft (2018)
19. Streimikis, J., Kortenko, L., Panova, M., Voronov, M.: Development of a smart city information system. In: E3S Web of Conferences, vol. 301. EDP Sciences, p. 05002 (2021)
20. Zou, X., Ma, S., Xin, S.: An analytical hierarchy process approach for smart city assessment in japan: From concept to indicators. Int. Rev. Spat. Plan. Sustain. Dev. **10**(2), 58–72 (2022)
21. Ma, C.: Smart city and cyber-security; technologies used, leading challenges and future recommendations. Energy Rep. **7**, 7999–8012 (2021)
22. Curry, E., Dustdar, S., Sheng, Q.Z., Sheth, A.: Smart cities enabling services and applications (2016)
23. Bonino, D., Alizo, M.T.D., Alapetite, A., Gilbert, T., Axling, M., Udsen, H., Soto, J.A.C., Spirito, M.: Almanac: Internet of things for smart cities. In: 2015 3rd International Conference on Future Internet of Things and Cloud, pp. 309–316. IEEE (2015)
24. Ahmed, Z., Rawat, A., Kumari, P.: An anaylsis of iot based smart cities. Int. J. Eng. Trends Appl. (IJETA) **8**(4), 30–35 (2021)
25. Ashton, K., et al.: That internet of things thing. RFID J **22**(7), 97–114 (2009)
26. Pefia-L6pez, I., et al.: Itu internet report 2005: The internet of things (2005)
27. Atzori, L., Iera, A., Morabito, G.: The internet of things: A survey. Comput. Netw. **54**(15), 2787–2805 (2010)
28. Mahalakshmi, S., Desai, K.: Iot framework, architecture services, platforms, and reference models. Internet of Things and Its Applications, pp. 37–59 (2022)
29. Ding, D.: Rory A Cooper, Paul F Pasquina, and Lavinia Fici-Pasquina. Sens. Technol. Smart Homes. Matur. **69**(2), 131–136 (2011)
30. Zhang, H.-T., Zhang, Y.-K.: Architecture and core technologies of inter- net of things. J. Chang. Univ. Technol. (Nat. Sci. Ed.) **2**, 176–181 (2012)
31. Zhong, C.L., Zhu, Z., Huang, R.G.: Study on the iot architecture and gateway technology. In: 2015 14th International Symposium on Distributed Computing and Applications for Business Engineering and Science (DCABES), pp. 196–199. IEEE (2015)
32. Al-Fuqaha, A., Guizani, M., Mohammadi, M., Aledhari, M., Ayyash, M.: Internet of things: A survey on enabling technologies, protocols, and applications. IEEE Commun. Surv. €3 Tutor. **17**(4), 2347–2376 (2015)
33. Mashal, I., Alsaryrah, O., Chung, T.Y., Yang, C.Z., Kuo, W.H., Agrawal, D.P.: Choices for interaction with things on internet and underlying issues. Ad Hoc. Network. **28**:68–90 (2015)
34. Ray, P.P.: A survey on internet of things architectures. J. King Saud Univ.-Comput. Inf. Sci. **30**(3), 291–319 (2018)
35. Ramson, S.R.J., Vishnu, S., Shanmugam, M.: Applications of internet of things (iot)-an overview. In: 2020 5th International Conference on Devices, Circuits and Systems (ICDCS), pp. 92–95. IEEE (2020)
36. Ahlgren, B., Hidell, M., Ngai, E.C.H.: Internet of things for smart cities: Interoperability and open data. IEEE Inter. Comput. **20**(6), 52–56 (2016)
37. Lee, I., Lee, K.: The internet of things (iot): Applications, investments, and challenges for enterprises. Business. Horizons **58**(4), 431–440 (2015)
38. Li, Y., Zhang, S., Chang, Y., Xu, G., Li, H.: Privacy–preserving and poisoning-defending federated learning in fog computing. IEEE Inter. Thing. J (2023)
39. Malik, V., Mittal, R., Mavaluru, D., Narapureddy, B.R., Goyal, S.B., Martin, R.J., Srinivasan, K., Mittal, A.: Building a secure platform for digital governance interoperability and data exchange using blockchain and deep learning-based frameworks. IEEE Access (2023)

40. Xia, L., Semirumi, D.T., Rezaei, R.: A thorough examination of smart city appli- cations: Exploring challenges and solutions throughout the life cycle with emphasis on safeguarding citizen privacy. Sustain. Cities Soc. **98**, 104771 (2023)
41. Bhardwaj, M., Singh, S., Hu, Y.C.: Iot in connected electric vehicles for smart cities. In integration of IoT with cloud computing for smart applications, pp. 161–179. Chapman and Hall/CRC
42. Anand, R., Juneja, S., Juneja, S., Jain, V., Kannan, R.: Integration of IoT with cloud computing for smart applications. CRC Press (2023)
43. Lele, A.: Industry 4.0. Disruptive technologies for the militaries and security, pp. 205–215 (2019)
44. Thakker, D., Mishra, B.K., Abdullatif, A., Mazumdar, S., Simpson, S.: Explainable artificial intelligence for developing smart cities solutions. Smart. Cities. **3**(4), 1353–1382 (2020)
45. Rahman, M.A., Hossain, M.S., Alrajeh, N.A., Guizani, N.: B5g and explainable deep learning assisted healthcare vertical at the edge: Covid-i9 perspective. IEEE. Network. **34**(4), 98–105 (2020)

IoT Based Smart Agriculture for Crop Water and Nutrition Management

Talvinder Singh and **Shano Solanki**

Abstract Internet of Things (IoT) is network based interconnected devices making physical world more connected with less human effort. The technology that makes our contemporary farms more productive and automates our food production cycles by reducing human interaction is known as farm automation, or "smart farming". This paper proposes an IoT-based crop monitoring system and automated irrigation system for Soil based Vertical Farming setups using IoT based monitoring system for the crop of strawberries. The designed system is capable of monitoring, various crop parameters like soil moisture, soil temperature, ambient temperature, ambient light, and pH of soil and salinity of soil (electrical conductivity). Through the use of an autonomous irrigation system and an ESP32 microcontroller platform based on a Microchip NODEMCU, the system may also maintain the required soil moisture content for a specific crop. By keeping an eye on Soil PH and Electrical Conductivity values nutrition in the base supporting plants can be increased or decreased as per requirement. For IoT, cloud services of adafruit.io is used to keep the user updated with real time status of the crop. Information from various sensors is regularly fed to the IoT Cloud network and can be used to collect data logs and contribute to big data related to crop parameters for future analysis.

Keywords Vertical farming · IoT based smart irrigation system · Precision farming · Crop and nutrition management using IoT

T. Singh (✉)
Department of Computer Engineering, Government Polytechnic Kangra, Kangra, HP, India
e-mail: talvinder.m@gmail.com

S. Solanki
Department of Computer Science and Engineering, NITTTR Chandigarh, Chandigarh, India

B. Soni et al. (eds.), *Beyond Artificial Intelligence*, Lecture Notes in Networks and Systems 1326, https://doi.org/10.1007/978-981-96-4170-3_40

1 Introduction

When a large portion of the population depends on agriculture for a living, it is critical to manage land and water resources for agricultural production in an effective and responsible manner. Our nation has experienced systematic use of crop rotations, the "green revolution," the "white revolution," systematic breeding, and the extensive application of synthetic fertilizers and pesticides. As a result of the exponentially growing use of information and communication technology (ICT) in agriculture, our nation is currently experiencing its fourth agricultural revolution, in the form of revolutionary changes like smart farming, precision farming, farm automation and implementation of IoT in agriculture. Numerous R&D teams and IT businesses are working on robotic solutions to create autonomous drones, autonomous tractors, autonomous harvesters, autonomous irrigation systems, and autonomous seeding robots. So, implementation of IoT-Based small and cost-effective equipment can really bring the farmers who are on the border line into the mainstream smart farming. The proposed and designed, cloud-based precision farming system, can provide a crop specific solution for effective utilization of resources like water, ambient light, land etc. and collected crop specific growth parameters data which can be further validated. The designed system can draw a comparative analysis of the parameters so that the data collected can be validated. The experiment is divided in four sections where each section can be compared with each other on the basis of crop yield and resources required. Comparative analysis of all four sections drew a denouement in context of producibility. The integration of IoT based system allowed actuator control and recording of real-time data from the fields using various sensors. And this data can further used to create a Machine learning setup to automate the actuator control mechanism.

2 Related Work

In order to meet the rising food demand of the global population, which is increasing every second, smart agriculture is no longer just a technology to make life easier for humans; rather it has become a necessity. Through engineering methods like IoT and artificial intelligence, offers a favorable growth environment. It is the result of integration of biology, environment, engineering, and IT. Farooq et al. [1] discussed that, IoT offers multiple solutions in different domains such as healthcare, retail, traffic, security, smart homes, smart cities, and agriculture.

IoT provides effective solutions for numerous applications, including smart agriculture, smart cities, security, retail, and traffic congestion [1–3]. By evaluating many obstacles and challenges in farming, IoT has brought about a significant shift in the agricultural environment [4].

Zheng et al. [5] have developed an IoT management that monitors environmental factors such as wind, soil, atmosphere, and water across a wide area. The subdomains to which they belong have also been used to identify IoT based agricultural monitoring systems. Additionally, information gathered about various environmental conditions is sent to the user via setting off warnings or giving suggestions to authorities via messages [6].

In his case studies, Despommier [7] proposed a qualitative method for evaluating and examining the vertical farming techniques that are already in use [8]. Murthy et al. [9] proposed that increasing the plant density per unit area of the field would result in a more effective use of land resources.

The use of space in systems for growing food in greenhouses is discussed by Arias et al. [10]. Since it directly affects productivity per square foot, it is a very important issue. Increasing planting density is one method for improving space utilization. Around the world, strawberry production is valued highly.

In the field of precision farming parameter control is an important factor to be taken care of. In her research, Subalakshmi [11] suggested that electromechanical, microcontroller-based systems, and GSM approaches [12] might be used to automate labor-intensive and complex irrigation systems, therefore making them more efficient and straightforward.

Research suggested that conventional experimental techniques should be used to follow the machine learning pipeline. Deepa S. et al. mentioned the effectiveness of ML frameworks which provides a clear understanding of the procedure by analyzing the enormous amounts of data and explained the results. On the basis of the literature survey done, inferences drawn are as follows:

- A soil-based vertical farming system of multiple levels can be implemented to increase the productivity of the available land [13–16].
- As the system is soil based, allowing to opt for organic farming solutions for the crops which are usually grown in soil-less cultures in vertical farming techniques [16–19].
- The multiple-level vertical farming setup will be equipped with a sensors array, based on Arduino, NodeMCU, or Raspberry Pi to record various parameters, related to soil and plant health and allows to improve quality and quantity of yield by managing growth parameters [20, 21].
- A system can be created to record the sensor data in dual mode [16, 20, 22]:
  - A local server based on Raspberry Pi can be created to record all the data from all sensor arrays, over wireless network or data can be recorded on sensor array itself [23].
  - Data from all sensor arrays can be recorded on a network cloud using local wireless network for connectivity [24].
- A parameter control mechanism can be set up, to control parameters like nutrition, fertigation, PH, EC, moisture and, temperature etc. that can be managed remotely using a cloud-based micro controller setup to improve the quantity and quality of fruit [16, 25].

- Along with sensor data from plants and other details like plant mortality and plant yield can be recorded to make a comparative study of micro-controlled setups and non-micro controlled setups [5, 7].
- Careful selection of plants to be grown in the vertical farming setup can be done to optimize the output of vertical farming setup. As it is always recommended to grow high-value crops in vertical farming systems because of the large capital investments, and high running cost of the system [10, 26].
- Based on these inferences, it is observed that by implementing smart vertical farming set up the overall yield of the agricultural area can be increased many folds without compromising the quality of the product [27].

3 Proposed System

The proposed system does not only automate the irrigation set up but also provide a system to Monitor and record soil and ambient parameters required for plant growth, along with automatic nutrient delivery to the plant, as and when required. By implementing automation in a vertical farm, precision agriculture practices can be put into practice increasing both quality and quantity of the yield. Strawberry is chosen as crop to be sown in vertical farming set up because of its high value, quantity of yield and small crop cycle. Automation will also reduce dependence on manual labor, reducing running cost of the project and help in controlling environmental parameters and by creating more optimum environmental conditions artificially, to increase production of the vertical farm. The varieties chosen for strawberry plantation are Nabila and Sweet Charlie which are ideal for Vertical farming setup. Both varieties are known for the quality of fruit grown and the taste of the fruit. Normally a strawberry plant can easily give quality fruit up to 2 years and around 15–25 runners every year. In commercial environments it is suggested that a plant must be replaced every two years. This system allows to record real-time data of plant health data and overall yield data which can be used in many ways. This whole process can be monitored and controlled remotely via network connected devices. Performing yield comparison determines which section is providing higher yield over others (Fig. 1).

3.1 Trial Field Setup

As seen in the setup image, a trial field with ninety plants is set up for testing utilizing vertical wall mounted planters that allow water to flow downward from one planter to another (Fig. 2).

- Capacity of planter = 1.5 l.

Further wall mounted planters are divided into three sections of 30 plants in each section, and color coded with sticker tags, and maintained namely Section A (Blue),

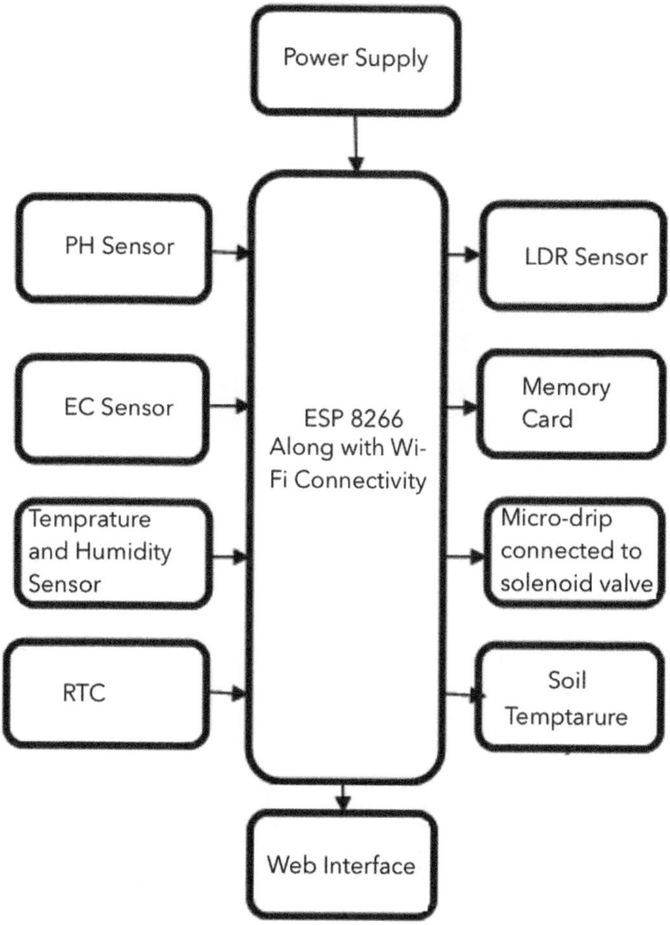

Fig. 1 Hardware design

Section B (Yellow), Section C (Red). Along with this a fourth section, namely Section D (green) of 30 plants is planted, on ground using traditional method of planting strawberries in the fields using mulching sheet and this section is manually irrigated. Image of section D is also enclosed for reference.

- Section A is irrigated manually.
- Section B is Irrigated via Continuous Drip Irrigation System.
- Section C is irrigated through a smart irrigation system.
- Section D is manually irrigated.

The entire setup is separated into four portions, enabling data from four distinct configuration types to be compared. The way that the plants are given water and nutrients to maintain the desired moisture content in the plant and the nutritional

Fig. 2 Vertical farming setup design

value of the soil medium, allowing plants to continue a healthy development, is the sole distinction between all four setups.

Because every section receives a different level of fertigation and irrigation, it is possible to compare sections that receive automated irrigation with those that receive continuous water supply, manual irrigation, and traditional plantation methods based on factors such as plant growth, plant mortality, overall yield, and nutrient requirements from the plant [28].

3.2 Sensor Implementation

For the purpose of gathering data, DJS1C-based electrical conductivity sensors and E201C-based PH sensors were both buried in the ground. The ambient temperature and humidity measurements are recorded using a temperature and humidity sensor based on the DHT11 protocol. A digital temperature sensor probe with a DS18B20 base is used to gather data on soil temperature. Adafruit.io-based cloud services are used to feed the data obtained from this sensor array into the cloud setup. The microcontroller (NodeMCU) gathers data from the sensors array and this data is used to determine whether or not plants require additional water or whether the soil's nutritional value has changed. The change in nutrient value of the soil is referred from values of electrical conductivity and pH of the soil. The volumetric amount of moisture or water in the soil, as well as the salinity and PH value of the soil substrate, are measured by the soil sensors that are used to assess the health of the soil (PH and electrical conductivity). Higher resistance in the sensor data might be interpreted as a sign that the planter needs more water or fertilizer fed to it since it indicates decreased salinity and moisture content in the soil substrate. This makes it possible

Table 1 List of equipment

Sr. No	Name of equipment	Purpose
1	NODEMCU ESP32 Board	Main microcontroller with Wi-Fi
2	Relay shield	Dual channel relay
3	Electric solenoid valve switch	Solenoid switch
4	Irrigation setup, (water tank and micro drip system	Micro drip irrigation
5	Temperature and humidity sensor module based on DHT11	To provide data on temperature and humidity
6	Waterproof digital temperature sensor probe based on DS18B20	To provide data on soil temperature
7	Memory card	To store data locally
8	LDR sensor module based on LM393	For recording data related to visible light
9	PH sensor based on E201C	To record soil pH data
10	Electrical conductivity sensor based on DJS1C	To record soil electrical conductivity

to detect even the smallest variations in the soil's pH and electrical conductivity when using it for vertical farming. The relay-based solenoid valve can be triggered remotely allowing users to have remote access to the setup using cloud services, and can also trigger the actuators used to control water and nutrient supply.

The ESP 32 board's built-in wireless connection is used to link the smart irrigation system to the nearby wi-fi network [29]. Next, local wireless LAN is used to connect to the free IoT cloud network offered by adafruit.io, where sensor data logs may be stored [30–32].

3.3 Control Unit

The ESP32 module and sensor array form the foundation of the control unit. In addition to autonomously watering the field depending on the moisture content and preserving the necessary soil nutritional profile, it will also transmit data to an IoT cloud network server to monitor the state of the soil [30]. A micro drip system will be used as part of the system to supply water and nutrients to a vertical plantation arrangement (Table 1).

3.4 Cloud Integration

Real-time data from projects is interacted with using the Adafruit.io platform. Adafruit Cloud is an application that programmers and designers use to create

Fig. 3 Console view Adafruit.io

networked projects. It allows data from several microcontroller setups to be securely sent in real-time to the cloud services. Actuators may be remotely triggered and real-time soil health data will be provided via the IoT cloud network services established (Fig. 3).

3.5 Methodology

Setting irrigation frequency is based on research done in 2004 by Kirschbaum, Daniel, and Correa [33]. The microcontroller setup's moisture sensor's potentiometer was used to calibrate the threshold value based on the previously described study. The irrigation system connected to the microcontroller is turned on whenever there is a drop in resistance between the moisture sensor's probes. This process continues until the predetermined threshold level is reached. For the other three manual setups, the same frequency described in the above-mentioned paper was followed. The use of a micro-drip irrigation setup allows efficiently use water saving up to 90% of water in vertical farming setups. It also allows to keep a check on fungal outbreaks in the plants. Also, fertigation and chemigation can automatically be done using micro-drip irrigation systems and in a very controlled manner as per requirement. For rest, fertigation and chemigation were done as per requirement.

3.6 *Implementation of Machine Learning*

To make changes to plant nutrition scheme, and to adjust the nutrient contents of soil, soil temperature, ambient temperature, ambient light, electrical conductivity, moisture content in soil, and soil data is recorded. The collected data from pH sensor and EC sensor is analyzed using basic MATLAB program using back propagation (neural network based) machine learning algorithm. The disbalance in pH of soil is suggestive of either lack of nutrients in the solution or high concentration of nutrients in the soil. This also affects the electrical conductivity of the sensor and the change in nutrient concentration of soil can be recognized by the change in the values of EC sensor. The proposed optimal soil pH level for the crop of strawberry should be around 5.0 to 7.0. If values of pH drop beyond 5.0, more nutrient (fertilizer) is added to the soil and if value of pH rises beyond 7.0, more water is added to existing soil to normalize the pH and keep it between the required limits [31, 34]. This was used as a basic reference for comparing change in pH values. Whenever the change was sensed in the values of nutrient, the ML algorithms prompt the user to pay attention to the hydroponic setup otherwise the setup was considered to be in good condition (Fig. 4).

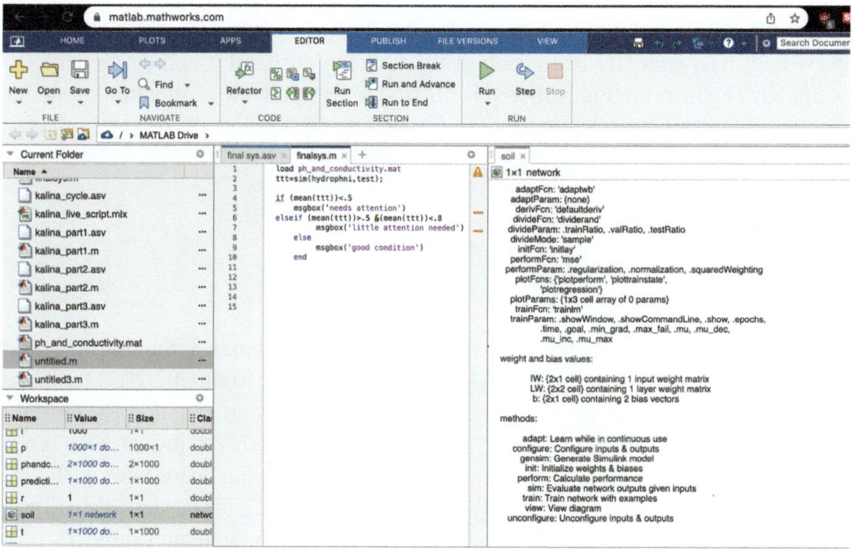

Fig. 4 ML implementation

4 Log Keeping

Each of the four partitions have a manual log of the fertigation and irrigation (Section A, Section B and Section D). For section B, drip irrigation system supplies continuous water. In essence, the section B log is a constant water supply to the plants with precisely timed drips. The irrigation records for section C are created and stored on the cloud network server of the Internet of Things. Since planter sensor data is used to automate section C's watering. Sensory input is required for section C's fertigation data. Fermentation was carried out using input data values from electrical conductivity and pH sensors. By utilizing an IoT cloud network and its console service, this data may also be sent to any mobile device. Designed automation can also be overridden at any given time on the basis of an informed decision by the end user (Tables 2 and 3).

4.1 Crop Yield Logs

Manual crop yield logs are maintained for each of the four plantation sections (Sections A, B, C, and D). These logs were created by gathering fruit between January 1, 2021, and the present, or May 18, 2021. For logging, each strawberry is harvested independently from each of the four regions, and it is then meticulously weighed. The information gathered is utilized to compare the production of the four sections of a plantation (Sections A, B, C, and D) and examine the output of the total setup (Table 4).

4.2 Plant Mortality Logs

Logs of plant mortality are recorded for each of the four segments (Section A, Section B, Section C, and Section D). These logs are taken weekly from 1/02/2021 to till date i.e. 18/05/2021. All the plants installed in all four setups are counted for mortality count and the data is recorded in MS Excel format. The gathered information is used to analyze the results of the overall setup and to compare the death rates of the four sections (Section A, Section B, Section C, and Section D). This process also allowed to keep a check on any kind of viral or fungal outbreak in the plant (Table 5).

Table 2 Irrigation and fertigation log

Date	Irrigation Log A (gm)	Fertigation Log A (gm)	Irrigation Log B (gm)	Fertigation Log B (gm)	Irrigation Log C (gm)	Fertigation Log C (gm)	Irrigation Log D (gm)	Fertigation Log D (gm)
01/02/21	Yes		Yes		Yes		Yes	
02/02/21	Yes	Yes	Yes	Yes	Yes	Yes	Yes	Yes
03/02/21	Yes		Yes		Yes		Yes	
04/02/21	Yes		Yes		Yes	Yes	Yes	
05/02/21	Yes		Yes		Yes		Yes	
06/02/21	Yes		Yes		Yes		Yes	
07/02/21	Yes		Yes		Yes		Yes	
08/02/21	Yes		Yes		Yes		Yes	
09/02/21	Yes	Yes	Yes	Yes	Yes	Yes	Yes	Yes
10/02/21	Yes		Yes		Yes		Yes	
11/02/21	Yes		Yes		Yes		Yes	
12/02/21	Yes		Yes		Yes		Yes	
13/02/21	Yes		Yes		Yes	Yes	Yes	
14/02/21	Yes		Yes		Yes		Yes	
15/02/21	Yes		Yes		Yes		Yes	
16/02/21	Yes	Yes	Yes	Yes	Yes	Yes	Yes	Yes
17/02/21	Yes		Yes		Yes		Yes	
18/02/21	Yes		Yes		Yes		Yes	
19/02/21	Yes		Yes		Yes	Yes	Yes	
20/02/21	Yes		Yes		Yes		Yes	

Table 3 Sensor Data Log Adafruit.io

Date	Day	Time	Temperature (C)	Humidity (%)	Soil temperature (C)	Light (Candela)	Soil moisture (Unit)	Soil pH (Unit)
21/06/25	Friday	23:14:45	30.60¬∞	46.00	28.56¬∞	1.00	1024.00	14.46
22/06/25	Friday	23:15:02	30.60¬∞	46.00	28.56¬∞	1.00	1024.00	14.46
23/06/25	Friday	23:15:29	30.60¬∞	46.00	28.63¬∞	1.00	1024.00	14.46
24/06/25	Friday	23:15:47	30.60¬∞	46.00	28.56¬∞	1.00	1024.00	14.46
25/06/25	Friday	23:16:04	30.70¬∞	46.00	28.56¬∞	1.00	1024.00	14.46
26/06/25	Friday	23:16:22	30.70¬∞	46.00%	28.63¬∞	1.00	1024.00	14.46
27/06/25	Friday	23:16:40	30.70¬∞	46.00%	28.63¬∞	1.00	1024.00	14.46
28/06/25	Friday	23:16:57	30.70¬∞	46.00	28.63¬∞	1.00	1024.00	14.43
29/06/25	Friday	23:17:15	30.70¬∞	46.00	28.63¬∞	1.00	1024.00	14.46
30/06/25	Friday	23:17:32	30.70¬∞	46.00	28.69¬∞	1.00	1024.00	14.46
01/07/25	Friday	23:17:50	30.70¬∞	46.00	28.69¬∞	1.00	1024.00	14.46
02/07/25	Friday	23:18:08	30.70¬∞	46.00	28.63¬∞	1.00	1024.00	14.57
03/07/25	Friday	23:41:39	30.60¬∞	45.00	28.44¬∞	1.00	1024.00	14.52
04/07/25	Friday	23:41:57	30.30¬∞	45.00	28.38¬∞	1.00	1024.00	14.55
05/07/25	Friday	23:42:14	30.30¬∞	45.00	28.38¬∞	1.00	1024.00	14.46
06/07/25	Friday	23:42:32	30.40¬∞	45.00	28.44¬∞	1.00	1024.00	14.55
07/07/25	Friday	23:42:50	30.40¬∞	45.00	28.44¬∞	1.00	1024.00	14.49

Table 4 Crop yield log

Date	Crop yield A (gm)	Crop yield B (gm)	Crop yield C (gm)	Crop yield D (gm)	Total crop
01/02/21	0	0	0	0	0
02/02/21	0	0	11	0	11
03/02/21	0	0	13	0	13
04/02/21	0	0	13	0	13
05/02/21	0	6	13	0	19
06/02/21	0	6	13	0	19
07/02/21	6	7	13	2	28
08/02/21	6	8	13	2	29
09/02/21	6	7	13	2	28
10/02/21	6	8	13	2	29
11/02/21	11	12	13	8	44
12/02/21	11	12	13	8	44
13/02/21	13	14	14	10	52
14/02/21	13	14	14	10	52

5 Output of Setup

The output of the entire system is represented by the fruit yield of the four sections (Section A, Section B, Section C, and Section D). The fruit yield is measured using a digital scale, or weighing equipment, and is recorded in grams. Every piece of information is entered into Microsoft Excel, and from there, a comparison graph is created, as seen in Fig. 5.

A comprehensive bar graph of crop output for each of the four sections (A, B, C, and D) is shown in Fig. 6, indicating a daily variation in yield.

The vertical farming method, sometimes referred to as the multi-layered plantation mechanism, is employed in the vertical plantation setups shown in Figs. 7, 8, and 9. Figure 10 is the traditional farming system for the crop of strawberries in rows covered with a mulching sheet with a prescribed distance on a plane ground.

6 Result and Conclusion

The entire setup was split into four distinct irrigation-mode-equipped sections, A, B, C, and D. This allows to determine a connection between the type of irrigation, fertigation, chemigation, and yield volume. The above shown graphs, shows that the section in which the irrigation process was automated, less water was consumed giving more yield per unit area, in comparison to the other three sections. Also, in the plantation section with automated irrigation setup plant mortality was lowest. It also

Table 5 Plant mortality log (01/02/2021–18/05/2021)

Date	Plants count Section A	Plants count Section B	Plants count Section C	Plants count Section D	Total plant count
03/01/21	30	30	30	30	120
10/01/21	30	30	30	30	120
17/01/21	27	30	30	27	114
24/01/21	27	28	30	26	111
31/01/21	26	27	29	25	107
07/02/21	25	27	29	25	106
14/02/21	25	26	28	25	104
21/02/21	23	26	28	23	100
28/02/21	23	26	28	23	100
07/03/21	22	25	27	21	95
14/03/21	23	25	27	21	96
21/03/21	20	25	27	20	92
28/03/21	20	24	26	20	90
04/04/21	18	24	26	20	88
11/04/21	17	23	25	17	82
18/04/21	17	23	25	16	81
25/04/21	16	23	25	15	79
02/05/21	16	22	24	14	76
09/05/21	15	21	24	14	74
16/05/21	14	21	24	12	71

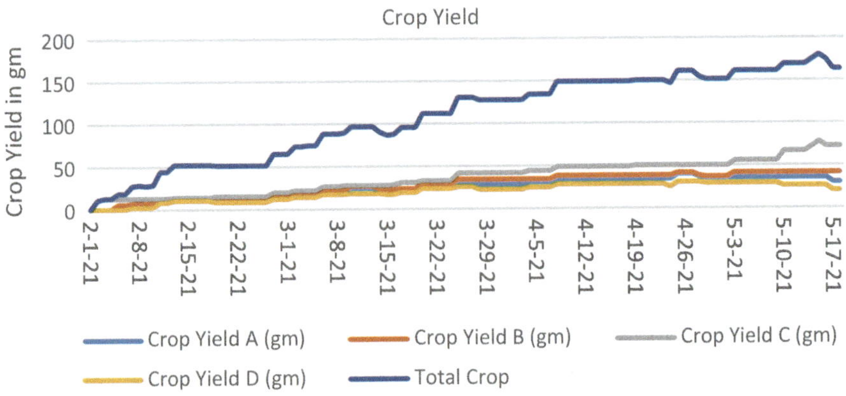

Fig. 5 Crop yield from (01/02/2021 to 18/05/2021)

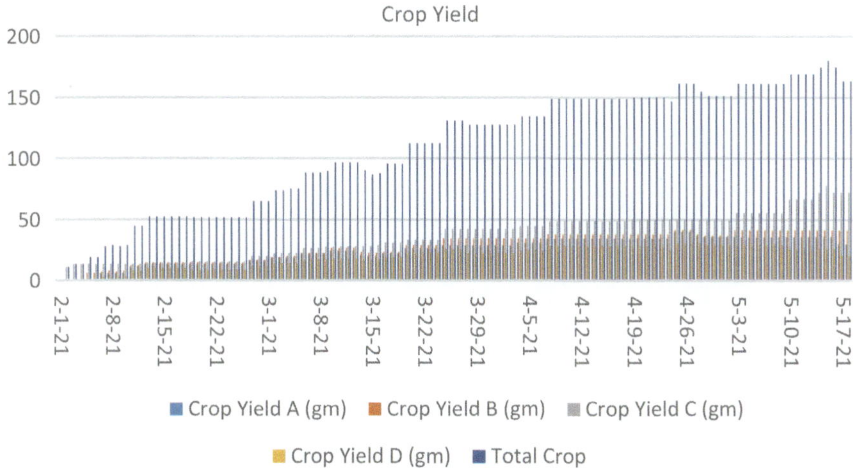

Fig. 6 Crop yield from (01/02/2021 to 18/05/2021)

Fig. 7 Vertical plantation setup A

Fig. 8 Vertical plantation setup B

Fig. 9 Vertical plantation setup C

Fig. 10 Traditional plantation setup D

concludes that the automated irrigation system gave around 15% more yield than the other two irrigation systems (Continuous Drip and Manual Irrigation System) and the traditional farming system. It is also found that Continuous Drip Irrigation and traditional farming system has a high plant mortality rate. Moreover, traditional farming practice for strawberry gave significantly lower yield.

The sensor is used to monitor the real-time health of the soil and plant in the vertical farming setup. The designed ML system is very basic and more data is needed to build a better robust automation system that should be able to provide full automation to the vertical farming system. More parameters can be recorded like the imagery of the plants, the dissolved level of nitrogen, potassium, and phosphorus in the soil mix, along with the dissolved level of individual micronutrients. This can provide more robust information about the plant and nutrient mix health. These changes can be made in the future by implementing more sensors to collect more data and it will further improve our calculations and provide a better level of automation.

7 Further Work

For further work, improved and calibrated grids or arrays to record more precise data logs of the sensor data can be used. Also, the range of trials can be improved and more sensors including soil, NPK sensor, carbon sensor, and micronutrient sensor can be added. More crops to the same farming automation setup can be added and that will reduce the cost of hardware implementation. Our design of a ML system is very basic. A more precise and highly accurate ML system can be developed on the basis of verified crop data of multiple years. Which can be used for automating the process of farming of different crops on the same farm setup by using the same hardware in different periods of time.

References

1. Farooq, S., Riaz, S., Abid, A., Abid, K., Naeem, M.A.: A survey on the role of IoT in agriculture for the implementation of smart farming. IEEE Access. **7**, 1–1 (2019). https://doi.org/10.1109/ACCESS.2019.2949703
2. Farooq, S., Riaz, Abid, A., Umer, T., Zikria, Y.: Role of IoT technology in agriculture: a systematic literature review. Electronics. **9**, 319 (2020). https://doi.org/10.3390/electronics9020319
3. Chen, S., Xu, H., Liu, D., Hu, B., Wang, H.: A vision of IoT: Applications, challenges, and opportunities with china perspective. IEEE Internet Things J. **1**(4), 349–359 (2014)
4. Ray, P.P.: 'Internet of Things for smart agriculture: technologies, practices and future direction.' J. Ambient Intell. Smart Environ. **9**(4), 395–420 (2017)
5. Zheng, R., Zhang, T., Liu, Z., Wang, H.: An EIoT system designed for ecological and environmental management of the Xianghe segment of China's Grand Canal. Int. J. Sustain. Dev. World Ecol. **23**, 372–380 (2016)
6. Liu, Z., Huang, J., Wang, Q., Wang, Y., Fu, J.: Real-time barrier lakes monitoring and warning system based on wireless sensor network. In: Proceedings of the 2013 Fourth International Conference on Intelligent Control and Information Processing (ICICIP), Beijing, China, 9–11 June 2013, pp. 551–554
7. Despommier, D The vertical farm: feeding the world in the 21st century. Macmil-lan. ISBN: 9781429946049 (2010)
8. Birkby, J.: Vertical farming. In: ATTRA Sustainable Agriculture Program. Retrieved 28 October 2019 (2016)
9. Murthy, S., Karimi, F., Hunashikatti, L., Sunoj, J.: Response of strawberry cv. Festival grown under vertical soilless culture system. Indian J. Hortic. **73**, 300 (2016). https://doi.org/10.5958/0974-0112.2016.00066.9
10. Arias, A., Hernández-Ibarra, U., Pineda, J., Fitz-Rodríguez, E.: Horizontal and vertical hydroponic systems for strawberry production at high densities. Acta Horticulturae. **1227**, 331–338 (2018). https://doi.org/10.17660/ActaHortic.2018.1227.41
11. Subalakshmi, R., Amal, A.: GSM based automated irrigation using sensors. Int. J. Trend Res. Dev. (IJTRD) (2016)
12. Chuah, Y. D., Lee, J., Tan, S., Ng, C.: Implementation of smart monitoring system in vertical farming. IOP Conf. Ser.: Earth Environ. Sci. **268**, 012083 (2019). https://doi.org/10.1088/1755-1315/268/1/012083
13. Manoj, G., Sreedhar, G.: Vertical farming using information and communication (2019). https://www.researchgate.net/publication/337972941

14. Pavithra, M.P.: Vertical farming: a concept (2018)
15. Farah, J.B.: SproutslO Urban Microfarm: interactive indoor farming system for urban use. Doctoral dissertation, Massachusetts Institute of Technology (2013)
16. Despommier, D., Ellingsen, E.: The vertical farm: the sky-scraper as vehicle for a sustainable urban agriculture. In: CTBUH 2008, 8th World Congress—Tall and Green: Typology for a Sustainable Urban Future, Congress Proceedings (2008)
17. Kamienski, C., Soininen, J.-P., Taumberger, M., Dantas, R., Toscano, A., Cinotti, T.S., Maia, R.F., Neto, A.T.: 'Smart water management platform: IoT-based precision irrigation for agriculture.' Sensors **19**(2), 276 (2019)
18. Adenaeuer, L.: Up, up and away! The economics of vertical farming. J. Agric. Stud. **2**, 40–60 (2014). https://doi.org/10.5296/jas.v2i1.4526
19. Anderson, R.: A multiple crop, high density system for the production of greenhouse cut flowers. Acta Horticulturae. **766**, 19–28 (2006). https://doi.org/10.17660/ActaHortic.2008.766.1
20. Linsley-Noakes, G., Wilken, L., de Villiers, S.: High density, vertical hydroponics growing system for strawberries. Acta Hortic. **708**, 365–370 (2006). https://doi.org/10.17660/ActaHortic.2006.708.63
21. Beacham, A., Vickers, L., Jim, M.: Vertical farming: a summary of approaches to growing skywards. J. Hortic. Sci. Biotechnol. **94**, 1–7 (2019). https://doi.org/10.1080/14620316.2019.1574214
22. Subramanian, B., Nathani, K., Rathnasamy, S.: IoT technology, applications and challenges: a contemporary survey. Wirel. Pers. Commun. **108** (2019). https://doi.org/10.1007/s11277-019-06407-w
23. Feng, X., Yan, F., Liu, X.: Study of wireless communication technologies on internet of things for precision agriculture. Wirel. Pers. Commun. **108** (2019). https://doi.org/10.1007/s11277-019-06496-7
24. Chen, W.L., Lin, Y.-B., Lin, Y.-W., Chen, R., Liao, J.-K., Ng, F.L., Chan, Y.-Y., Liu, Y.-C., Wang, C.-C., Chiu, C.-H., Yen, T.-H.: AgriTalk: IoT for precision soil farming of turmeric cultivation. IEEE Internet Things J. 1–1 (2019). https://doi.org/10.1109/JIOT.2019.2899128
25. Agrawal, H., Dhall, R., Iyer, K., Chetlapalli, V.: An improved energy efficient system for IoT enabled precision agriculture. J. Ambient. Intell. Hum. Comput. **11** (2020). https://doi.org/10.1007/s12652-019-01359-2
26. Gainwar, S.D., Rojatkar, D.V.: Soil parameters monitoring with automatic irrigation system. Int. J. Sci. Eng. Technol. Res. (IJSETR) **04**(11) (2015)
27. Elijah, O., Rahman, T.A., Orikumhi, I., Leow, C.Y., Hindia, M.N.: 'An overview of Internet of Things (IoT) and data analytics in agriculture: benefits and challenges.' IEEE Internet Things J. **5**(5), 3758–3773 (2018)
28. Chinthalu, V., Hegde, N., Supekar, A.: Study of challenges faced by farmers related to irrigation water for increasing crop production. Asian J. Adv. Agric. Res. **12**, 1–6 (2020). https://doi.org/10.9734/AJAAR/2020/v12i230074
29. Gautam, I., Reddy, S.R.N.: Innovative GSM based remote controlled embedded system for irrigation. Int. J. Comput. Appl. **47**(13) (2012)
30. Kim, Y., Evans, R.G., Iversen, W.M.: Remote sensing and control of an irrigation system using a distributed wireless sensor network. IEEE Trans. Instrum. Meas. **57**(7), 1379–2138 (2008). https://doi.org/10.1109/TIM.2008.917198.July
31. Pavithra, D.S., Srinath, M.S.: GSM based automatic irrigation control system for efficient use of resources and crop planning by using an android mobile. IOSR J. Mech. Civ. Eng. **11**(2014), 49–55 (2014)
32. Kirschbaum, D., Correa, M., Bórquez, A., Larson, K., Dejong, T.: Water requirement and water use efficiency of fresh and waiting- bed strawberry plants. Acta Horticulturae. **664** (2004). https://doi.org/10.17660/ActaHortic.2004.664.43
33. Moustafa, A.: Potential of protected agriculture and hydroponics for improving the productivity and quality of high value cash crops in Qatar (2010). https://doi.org/10.13140/RG.2.2.21300.55686

34. Hachem, S., Mallet, V., Ventura, R., Pathak, A., Issarny, V., Raverdy, P.G., Bhatia, R.: Monitoring noise pollution using the urban civics middleware. In: Proceedings of the 2015 IEEE First International Conference on Big Data Computing Service and Applications, Redwood City, CA, USA, 30 March–2 April 2015, pp. 52–61

Enhancing Predictive Maintenance for Induction Motors: A 3D Holography-Based Digital Twin Approach with Extended Kalman Filter

G. M. Yashaswini, Hari Krishna S. Thota, Anand Kumar Subramaniyan, and Rajkumar Velu

Abstract This research paper introduces an innovative paradigm for predictive maintenance in induction motors, underpinned by the development of a 3D holography-based digital twin. Digital twin technology has revolutionized product manufacturing and real-time analysis, allowing engineers and scientists to seamlessly engage with virtual replicas of mechanical systems in real-time, eliminating the need for Virtual or Augmented Reality (VR/AR) headgear. The incorporation of 3D holographic projection enriches this experience, providing an immersive grasp of the system's complexities. Grounded in the principles of interference and diffraction, holographic projection technology excels in capturing and reproducing lifelike 3D images, immersing users within a compelling virtual environment. In this project, we advocate the implementation of an Extended Kalman Filter (EKF) digital twin model to attain precise state estimation within a speed sensor-less rotor field-oriented controlled induction motor drive. The EKF digital twin model, serving as a mathematical representation of the physical system, harnesses the core tenets of the EKF to augment our comprehension of the motor's behavior, thus facilitating enhanced control and analysis. To affirm the efficacy of the EKF digital twin model, comprehensive simulations are executed using MATLAB, Simulink. The integration of 3D holography-based digital twin technology in tandem with the EKF model assembles a resilient and versatile toolkit for predictive maintenance and efficient monitoring of electric drive systems. This progressive approach holds the potential to substantially elevate the reliability and performance of induction motors, marking a significant stride in the realm of electrical engineering and predictive maintenance.

G. M. Yashaswini · H. K. S. Thota
Smart Manufacturing, IIoT and Artificial Intelligence Group, CMTI, Bangalore, India

S. Anand Kumar (✉) · R. Velu
Additive Manufacturing Research Laboratory, Department of Mechanical Engineering, Indian Institute of Technology Jammu, Kashmir, Jammu, India
e-mail: anand.subramaniyan@iitjammu.ac.in; anandkumarait@gmail.com

B. Soni et al. (eds.), *Beyond Artificial Intelligence*, Lecture Notes in Networks and Systems 1326, https://doi.org/10.1007/978-981-96-4170-3_41

Keywords Digital twin · 3D Holography · Extended Kalman filter · MATLAB · Induction motor · Virtual or Augmented Reality (VR/AR) · Predictive maintenance · Simulink

1 Introduction

1.1 Context

Digital twin is a virtual representation or mirror image of a physical object, system, or process. It is created by collecting and integrating data from sensors, Internet of Things (IoT) devices, and other sources to mimic the behavior and characteristics of the real-world counterpart. In the context of electric motors, a digital twin provides a comprehensive and dynamic digital model of the motor's operation, allowing engineers and operators to monitor, control, and optimize its performance.

Induction motor (IM) plays an vital part of modern society, powering a wide range of applications due to their reliability, efficiency, and versatility. They operate on the principle of electromagnetic induction and are commonly used in industrial, commercial, and residential settings. It has two main parts, stator and rotor, stator is a stationary part and rotor is the rotating part. Stator is made by stacking thin slotted highly permeable steel lamination inside a steel or cast-iron frame, winding passes through slots of stator and works on the principle of electromagnetic induction. Controlling IMs can be complex due to higher nonlinearity of the model, which has dynamics. This requires a thorough understanding of rotor flux and position or speed to be effective. In addition, speed sensor less IM drives have several advantages, including low cost, reduced complexity, and improved reliability. Figure 1 shows the cutaway view if three phase IM.

1.2 Kalman and Extended Kalman Filter

Kalman filtering is an algorithm that employs a series of observations over time, containing noise and other inaccuracies, and generates approximations of unknown variables that tend to be more accurate than those based on a single measurement alone, by estimating a joint probability distribution over the variables for each time-frame, as in [1]. The Kalman filter is a state observer which detects the presence of measurement noises as well as uncertainties about an unknown dynamic state system, this system is generally assimilated to state noise by stochastic algorithms tending to minimize the variance of the estimation error, as described in [2]. The Kalman filter is suitable for recursive linear filtering of discrete data. It provides an estimation of a state vector or a parameter and its error covariance and variance matrix that contain information about the accuracy of its state variables, as in [3]. The inherent

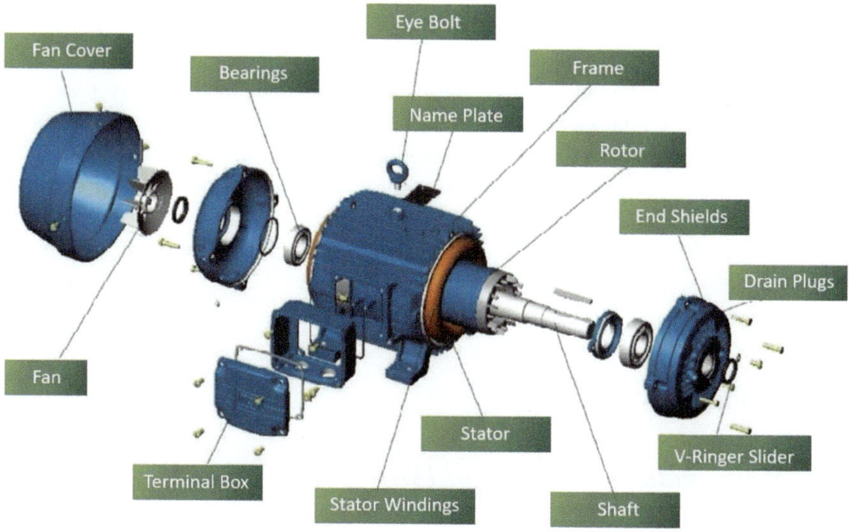

Fig. 1 Three phase induction motor

noise present in an induction machine when driven by an inverter underscores the compelling case for employing observers of this nature. These observers are designed to primarily focus on monitoring key characteristics, such as rotor flux components and speed. Notably, they require only measurements of the stator currents, making them a practical choice for monitoring and control. In this context, some state variables can be readily derived from the control law. As a result, the stator voltages are regarded as the inputs for the observer. The selection between an adaptive observer and a stochastic filter for monitoring an inverter-driven induction machine depends on the system's characteristics and control objectives. Each approach offers distinct advantages, with adaptive observers emphasizing adaptability and real-time adjustments, while stochastic filters prioritize optimal estimation in the presence of noise. EKF is an advanced estimation and control technique that combines elements of the Kalman filter and non-linear systems to estimate the state of a dynamic system accurately. In the context of induction motor control, EKF plays a crucial role in several ways such as state estimation, feedback control, fault detection and diagnosis and sensor less control. EKF allows for accurate state estimation and control adjustments, ensuring that induction motors operate efficiently and reliably across a wide range of applications.

1.3 3D Holographic Device

Deploying the model with external hardware like Augmented Reality (AR) and Virtual Reality (VR) lenses and spectacles with handheld devices restrict 3D representation and visualization. Holographic displays with software plugins cater to different domain such as simulation/animation to better understand the system's visualization. It is possible to create 3D representations of industrial environments or factories or have a digital twin of machines systems for better management and maintenance.

2 Methodology

2.1 Mathematical Modeling of IM

IM model is based on G. Kron, Park's model is a special case of this concept. It is often used for the synthesis of control laws and estimators. Described by a non-linear algebra-differential system, Park's model reflects the dynamic behavior of asynchronous machine as in [4]. The IM motor model has been carried stator fixed reference frame, by assuming the stator currents and rotor flux as state variables as shown in below (from Eqs. (1)–(6))

$$\dot{X} = \mathrm{A.X} + \mathrm{B}.U \tag{1}$$

$$Y = C.\mathrm{X} \tag{2}$$

where X is state vector, U is the input vector and Youtput vector

$$X = \left[i_{s\alpha}, i_{s\beta}, \phi_{r\alpha}, \phi_{r\beta}\right]^t; U = \left[u_{s\alpha}, u_{s\beta}\right]^t; y = \left[i_{s\alpha}, i_{s\beta}\right]^t \tag{3}$$

$$A = \begin{bmatrix} -\lambda & 0 & \frac{K}{T_r} & K.\omega_r \\ 0 & -\lambda & -K.\omega_r & \frac{K}{T_r} \\ \frac{L_m}{T_r} & 0 & -\frac{1}{T_r} & -\omega_r \\ 0 & \frac{L_m}{T_r} & \omega_r & -\frac{1}{T_r} \end{bmatrix} B = \begin{bmatrix} \frac{1}{\sigma.L_s} & 0 \\ 0 & \frac{1}{\sigma.L_s} \\ 0 & 0 \\ 0 & 0 \\ 0 & 0 \end{bmatrix} C = \begin{bmatrix} 1\,0\,0\,0\,0 \\ 0\,1\,0\,0\,0 \end{bmatrix} \tag{4}$$

With:

$$\lambda = \frac{R_s}{\sigma.L_s} + \frac{1-\sigma}{\sigma.T_r}; K = \frac{1-\sigma}{\sigma.L_m}; \sigma = 1 - \frac{L_m^2}{L_s.L_r}; T_r = \frac{L_r}{R_r} \tag{5}$$

Rotor motion is expressed as

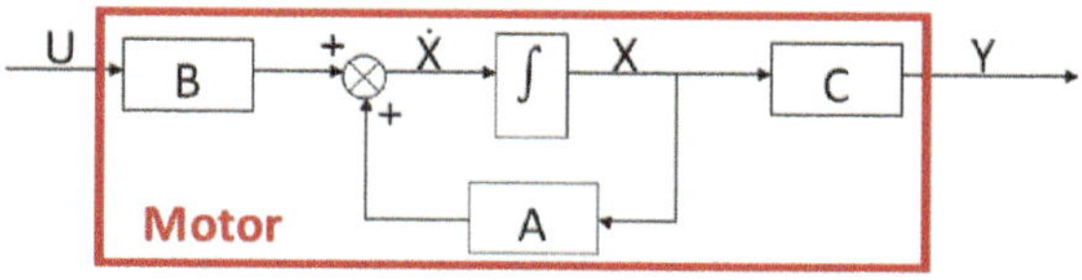

Fig. 2 State space model of induction motor

$$J.\frac{d\Omega_r}{dt} = T_{em} - T_L - f.\Omega_r \tag{6}$$

where J is the motor inertia, Tem is the electromagnetic torque, TL is the load torque, and f is the friction coefficient. Figure 2 shows the representation of state space model of IM

2.2 *Speed and Flux Estimation Using EKF*

Induction Machine Extended Model

Non-linear, continuous system of state equations of the IM extended to the electrical rotational speed is represented by Eq. (7):

$$X_e(t) = f(X_e(t), U(t)) = A.X_e(t) + B.U(t) \tag{7}$$

$$Y(t) = h(X_e(t)) = C.X_e(t)$$

In which,

$$X_e = [X\,\Theta]^t = \left[i_{s\alpha}, i_{s\beta}, \phi_{r\alpha}, \phi_{r\beta}, \omega_r\right]^t Y = \left[i_{s\alpha}, i_{s\beta}\right]^t U = \left[u_{s\alpha}, u_{s\beta}\right]^t \tag{8}$$

With:

$$A = \begin{bmatrix} -\gamma & 0 & \frac{\mu}{T_r} & \mu.\omega_r & 0 \\ 0 & -\gamma & -\mu.\omega_r & \frac{\mu}{T_r} & 0 \\ \frac{L_m}{T_r} & 0 & 1-\frac{1}{T_r} & -\omega_r & 0 \\ 0 & \frac{L_m}{T_r} & \omega_r & \frac{1}{T_r} & 0 \\ 0 & 0 & 0 & 0 & 0 \end{bmatrix} B = \begin{bmatrix} \frac{1}{\sigma.L_s} & 0 \\ 0 & \frac{1}{\sigma.L_s} \\ 0 & 0 \\ 0 & 0 \\ 0 & 0 \end{bmatrix} \tag{9}$$

2.3 Discretization of the Continuous Model

The continuous model must be discretized for the implementation of the EKF, quasi-constant control voltages are assumed over a sampling period Ts as in [5], the discrete augmented state model can be approximated by Eqs. (10) and (11):

$$X_{ek+1} = f(X_{ek}, U_K) = A_d.X_{ek} + B_d.U_k \tag{10}$$

$$Y_k = h(X_{ek}) = C_d.X_{ek} \tag{11}$$

First order Taylor Series with limited development is used to obtain matrices for this model given by Eq. 12

$$A_d \approx e^{A.T_s} = I + A.T_s;\ B_d = B.T_s;\ C_d = C \tag{12}$$

This leads to:

$$A_d = \begin{bmatrix} 1 - T_s\gamma & 0 & T_s.\frac{\mu}{T_r} & T_s.\mu.\omega_r & 0 \\ 0 & 1 - T_s.\gamma & -T_s.\mu.\omega_r & T_s.\frac{\mu}{T_r} & 0 \\ T_s.\frac{L_m}{T_r} & 0 & 1 - \frac{T_s}{T_r} & -T_s.\omega_r & 0 \\ 0 & T_s.\frac{L_m}{T_r} & T_s.\omega_r & 1 - \frac{1}{T_r} & 0 \\ 0 & 0 & 0 & 0 & 1 \end{bmatrix}$$

$$B_d = \begin{bmatrix} T_s.\frac{1}{\sigma.L_s} & 0 \\ 0 & T_s.\frac{1}{\sigma.L_s} \\ 0 & 0 \\ 0 & 0 \\ 0 & 0 \end{bmatrix} \tag{13}$$

2.4 EKF Implementation for Discrete System

Implementing the EKF to the discrete system of the IM, considering the state noise as W_k and Measurement noise as V_k. This leads to the following expressions (14) and (15):

$$X_{ek+1} = f(X_{ek}, U_K) + W_k = A_d.X_{ek} + B_d.U_k + W_k \tag{14}$$

$$Y_k = h(X_{ek}) + V_k = C_d.X_{ek} + V_k \tag{15}$$

With:

$$X_{ek+1} = [X_k \Theta_k]^t; \; Y_k = \left[i_{s\alpha k}, i_{s\beta k}\right]^t U_k = \left[u_{s\alpha k}, u_{s\beta k}\right]^t;$$

$$W_k = \left[\, W_{xk} \; W_{\Theta k} \,\right]^t \tag{16}$$

Similarly, the linearization matrix Hk is represented as Eq. 17

$$H_k = \begin{bmatrix} 1\,0\,0\,0\,0 \\ 0\,1\,0\,0\,0 \end{bmatrix} \tag{17}$$

In the determination of the initial covariance matrix P0|0, it is generally limited to the choice of elements on the diagonal. These elements are chosen in such a way that they correspond to the uncertainty about estimates of initial state variables.

2.5 *Choice of Covariance Matrices Q and R*

It is via these matrices that the various measured, predicted and estimated states will pass. Their goals are to minimize the errors associated with approximate modelling and the presence of noise on the measurements. This is the most difficult point of applying the Kalman filter to observation. The matrix Q linked to the noises tainting the state, allows adjusting the estimated quality of the modelling and discretization. A strong value of Q gives a high value of the gain K stimulating the importance of the modelling and the dynamics of the filter. A high value of Q can, however, create an instability of the observation. The matrix R regulates the weight of the measurements. A high value indicates a great uncertainty of the measurement. On the other hand, a low value makes it possible to give a significant weight to the measurement as in [6].

3 Simulation Using SIMULINK, MATLAB

Figure 3 Induction Motor Model with EKF Simulation of EKF of IM using SIMULINK, MATLAB has carried out. The induction motor block from MATLAB library is obtained and supply voltage and speed reference are fed to the motor block. Figure 3 shows the model constructed in Simulink which is used for simulation and analysis. The voltage (alpha, beta) with process noise is given as state Transition function for EKF block and current (alpha, beta) for y1. Motor speed for y2. The results obtained after simulations are estimated states they are stator current (alpha, beta), rotor flux (alpha, beta) and speed. The model is further updated with inverter circuit and the simulations graphs are plotted.

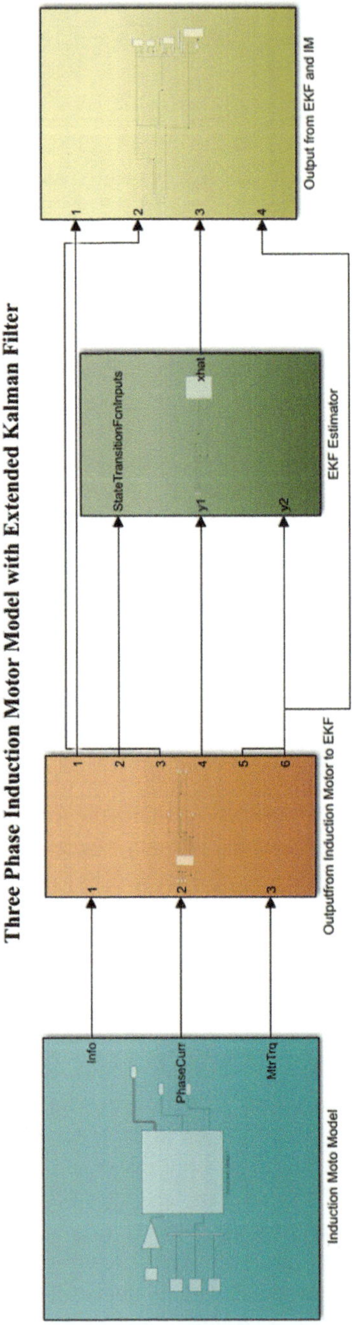

Fig. 3 Three phase IM model with extended Kalman filter

Table 2 Electrical parameters used for simulation

Power (W)	17.5 KW
Voltage (V)	392 V
Frequency (f)	50 Hz
Pole pair	2
Rated speed (N)	1750 rpm
Stator resistance (R_s)	0.154 Ω
Stator inductance (L_s)	0.09209 H
Rotor resistance (R_r)	0.1194 Ω
Rotor inductance (L_r)	0.01379 H
Mutual inductance (L_m)	0.424 H
Moment of inertia (J)	0.077 kg/m^2

3.1 Results and Discussion

Table.2 shows the parameter used for IM block and EKF Estimation and Simulation.

Figure 4 depicts that the EKF state estimates of stator current alpha and beta. From the above plots green and blue traces are the stator current obtained from motor and the red trace is EKF estimated state, the EKF estimate follows almost as the motor state estimates it has slightly varied in magnitude than the motor states.

Figure 5 shows the rotor flux estimated from EKF, Blue trace is alpha and red trace is Beta component of rotor flux.

The EKF estimated speed curve follows the response of the motor for higher speeds, for lower speeds the EKF estimates have overshoot as shown in Fig. 6. Thus, the EKF estimate for induction motor helps us to estimate the stator current, rotor flux and speed of the IM.

4 Conclusions

EKF-based digital twin model of an IM with a holographic device represents a significant advancement in motor control and predictive maintenance technology. It empowers engineers and researchers with a powerful and flexible tool to design, analyses and optimize IM systems, leading to improved reliability, reduced downtime, and enhanced energy efficiency in industrial processes. As technology continues to evolve, such digital twin models are expected to play a vital role in the advancement of industrial automation and control systems. The state estimates obtained from the EKF are satisfactory. The EKF-based digital twin model needs to be integrated with an actual induction motor. This involves connecting sensors to measure important motor parameters, such as current, voltage, speed, and temperature. These sensor readings will be used as inputs to the digital twin model for real-time estimation

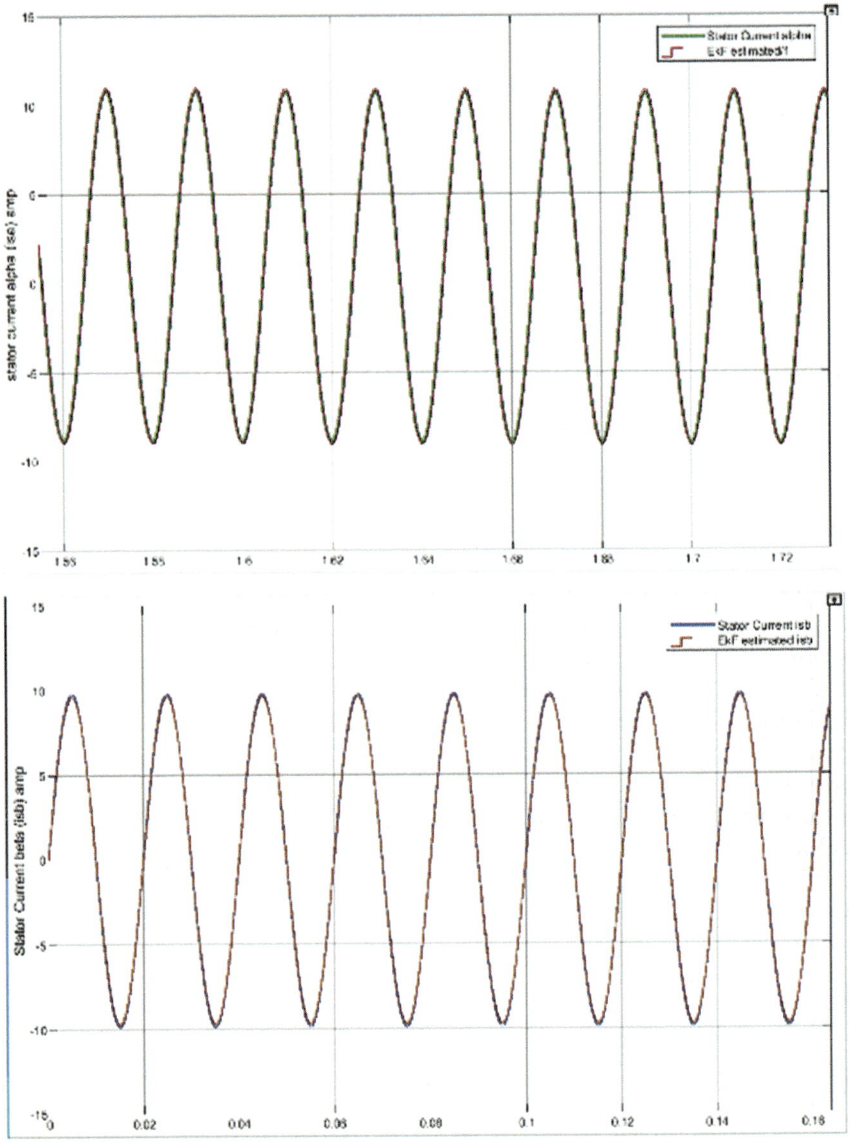

Fig. 4 Comparison of Stator current alpha (above), beta (below) of motor and EKF

and monitoring. Further the motor is operated under higher load condition and the degradation is observed along with load torque characteristics, these helps to create a degradation model and predict RUL (Remaining Useful Life) of motor.

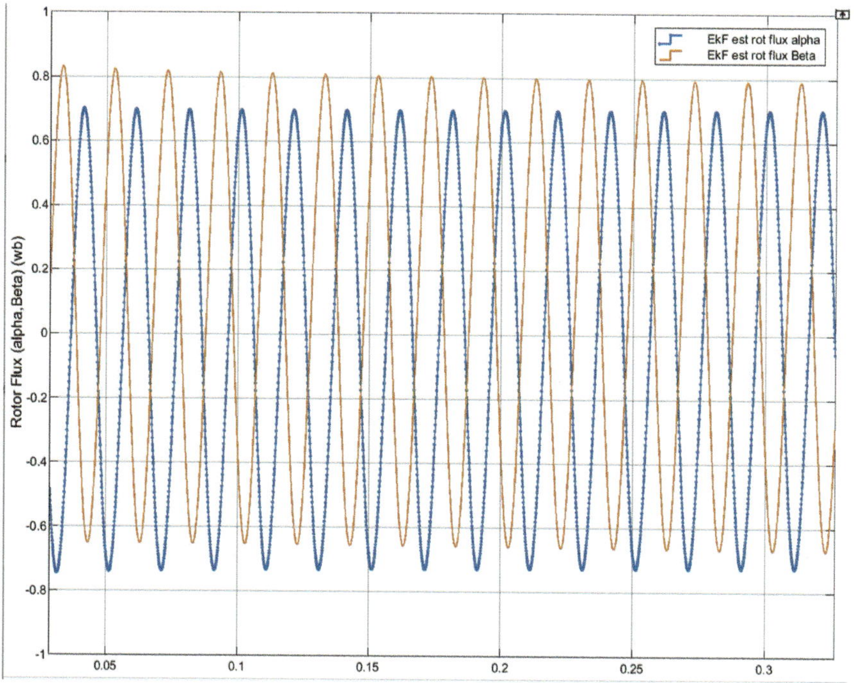

Fig. 5 Estimated rotor flux

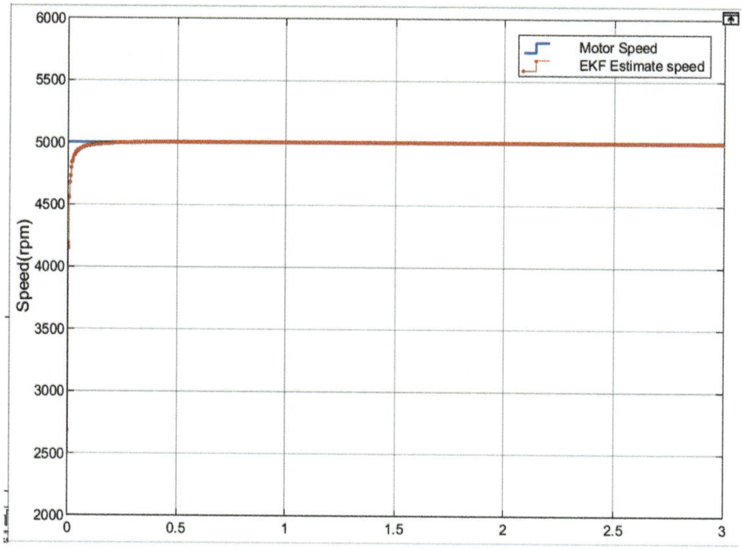

Fig. 6 Estimated and actual speed (rpm)

Acknowledgements The authors wish to acknowledge with thanks for the funding support (TDP-2022-ST-000032) by the managements of IITI DRISHTI CPS Foundation under NM-ICPS scheme to present and publish this paper.

References

1. Brown, R.G.: Introduction to random ssignal analysis and kalman filtering, vol. 8. Wiley, New York (1983)
2. Lefebvre, T., Bruyninckx, H., Schutter, J.: Nonlinear kalman filtering for force-controlled robot tasks, vol. 19. Springer, Berlin (2005)
3. Brown, R.G., Hwang, P.Y.C.: Introduction to random signals and applied kalman filtering, vol. 3. Wiley, New York (1992)
4. Kron, G.: Generalized theory of electrical machinery. Trans. Am. Inst. Electr. Eng. **49**(2), 666–683 (1930)
5. Zahraoui, Y., Fahassa, C., Akherraz, M., Bennassar, A.: Sensorless vector control of induction motor using an EKF and SVPWM algorithm. In: 5th International Conference on Multimedia Computing and Systems (ICMCS). IEEE, Marrakech, pp. 588–593 (2016).https://doi.org/10.1109/ICMCS.2016.7905584
6. Yassine, Z., Mohamed, A.: Kalman filtering applied to ınduction motor state estimation (2020). https://doi.org/10.5772/intechopen.92871

SDN-IOT: DDoS Detection and Mitigation in SDN-Enabled IoT Network

Hariom Shukla and Ankit Kumar Jain

Abstract The Internet of Things (IoT) has rapidly grown to encompass billions of devices, creating a heterogeneous network with diverse underlying protocols, making it challenging to supervise and secure effectively. IoT devices often possess limited processing power and storage capacity, rendering them vulnerable to attacks, including Distributed Denial of Service (DDoS) attacks. Integrating comprehensive security features is a difficult difficulty since IoT devices have limited resources. In order to effectively identify and mitigate DDoS attacks, this research paper suggests a strategy that uses Software Defined Networks (SDN) as gateways for IoT networks. SDN provides network resource centralized control and programmability, solving the particular security problems faced by IoT. Centralized management, dynamic configuration, network isolation, and policy-based security enforcement are some of the benefits of SDN in IoT security. The advantages of our IoT framework built on SDN include: The IoT ecosystem's security and resilience against DDoS attacks will be improved through real-time traffic analysis for DDoS attack detection, automatic response mechanisms to counterattacks, and dynamic resource allocation to handle increased traffic during attacks.

Keywords IoT · SDN · DDoS · SVM

1 Introduction

The proliferation of the Internet of Things (IoT) has ushered in an era of unprecedented connectivity, with billions of devices spanning diverse sectors such as healthcare, manufacturing, transportation, and home automation [1]. The Internet of Things has entered the mainstream thanks to its promise of greater innovation, convenience, and efficiency. However, because to the rapid growth of this network of linked devices, there are a number of security issues that have emerged, especially in the face of Distributed Denial of Service (DDoS) attacks. The IoT ecosystem, characterized by

H. Shukla (✉) · A. K. Jain
National Institute of Technology Kurukshetra, Kurushetra, India
e-mail: hariomshukla2709@gmail.com

B. Soni et al. (eds.), *Beyond Artificial Intelligence*, Lecture Notes in Networks and Systems 1326, https://doi.org/10.1007/978-981-96-4170-3_42

its extensive heterogeneity in device types and communication protocols, presents a formidable landscape that is inherently challenging to supervise and secure effectively. The processing power, storage, and energy resources of these IoT devices, which range in size from miniature sensors to complex appliances, are frequently constrained. Due to their intrinsic weaknesses, they are very vulnerable to several types of attacks, with DDoS attacks emerging as an extreme risk.

DDoS attacks, in which malicious actors overwhelm a target network or device with a flood of traffic, are particularly concerning in the context of IoT. Critical services might be interrupted, data integrity can be compromised, and even physical harm can result from these attacks in industries like healthcare and transportation. Implementing comprehensive security measures at the device level appears to be a difficult and tough endeavor given the resource limitations of IoT devices. Our research aims to provide a novel strategy for addressing these important security challenges by using Software Defined Networks (SDN) as IoT network gateways. With its core features of network programmability, centralized management, and dynamic resource allocation, SDN stands out as a viable option to improve the security posture of IoT ecosystems. The intricate details of our SDN-based approach will be covered in this part, along with some of its important benefits, such as real-time traffic analysis for DDoS detection, automatic response mechanisms, and dynamic resource allocation during attacks. In order to protect the ongoing growth and stability of the IoT environment, this article attempts to clarify how the incorporation of SDN into IoT networks may strengthen security measures.

1.1 IoT Challenges

a. **Heterogeneity**: The IoT landscape encompasses an immense variety of devices, ranging from small sensors to industrial machines, and employs a wide array of communication protocols [1]. This inherent diversity poses a substantial challenge for uniform management and security enforcement.
b. **Resource Constraints**: One distinguishing feature of many IoT devices is their resource limits. These devices frequently have limited processing power, memory, and energy-constrained batteries, making them unsuitable for hosting effective security systems.
c. **Scalable**: IoT network scalability challenge amplifies the difficulty of managing and securing a growing IoT ecosystem effectively.
d. **Vulnerabilities**: The resource-constrained nature of IoT devices exposes them to various security vulnerabilities, including weak authentication, outdated firmware, and susceptibility to physical attacks [1].

1.2 DDoS Attacks in IoT

DDoS attacks are a common hazard in the digital world. Such attacks include fraudulent actors overwhelming a target network or device with traffic, rendering it unreachable to innocent users. Because of their limited resources, IoT devices are particularly vulnerable to DDoS attacks. When confronted with a large inflow of malicious traffic, these devices frequently struggle to keep up, resulting in service disruptions and potential system breakdowns.

1.3 SDN as a Solution

SDN is an innovative networking paradigm that decouples the control plane from the data plane, centralizing control and enabling the programmability of network resources. This architectural shift holds the potential to address many of the unique security challenges posed by IoT. Figure 1 shows the SDN architecture.

1.4 Advantages of SDN in IoT Security

One of the primary benefits of SDN is its centralised control. A controller manages network-wide decisions in an SDN context, allowing for real-time monitoring and management of network traffic [13]. SDN offers dynamic network configuration and traffic routing. This adaptability is essential for responding to changing attack patterns and optimising network resources.

SDN enables network isolation and segmentation by allowing the development of independent network segments with various security policies [13]. This separation can aid in the containment and mitigation of DDoS attacks. Within an SDN architecture, security policies can be created and enforced centrally. This ensures that security controls are applied consistently to all IoT devices and network segments.

Incorporating SDN into IoT networks presents a compelling solution to enhance security and mitigate the impact of DDoS attacks. However, it's crucial to approach the design and implementation of SDN-based solutions in IoT environments with care, taking into account the unique requirements and challenges posed by IoT devices and networks. To protect the availability and integrity of essential services, the IoT's continuous growth requires strong security measures. Research and development initiatives focused at enhancing IoT security through SDN integration offer the potential of developing robust and effective security procedures for this constantly expanding ecosystem as the threat landscape changes. We can traverse the complicated confluence of IoT and cybersecurity using cutting-edge methods like SDN, assuring a safer and more dependable future for IoT technology.

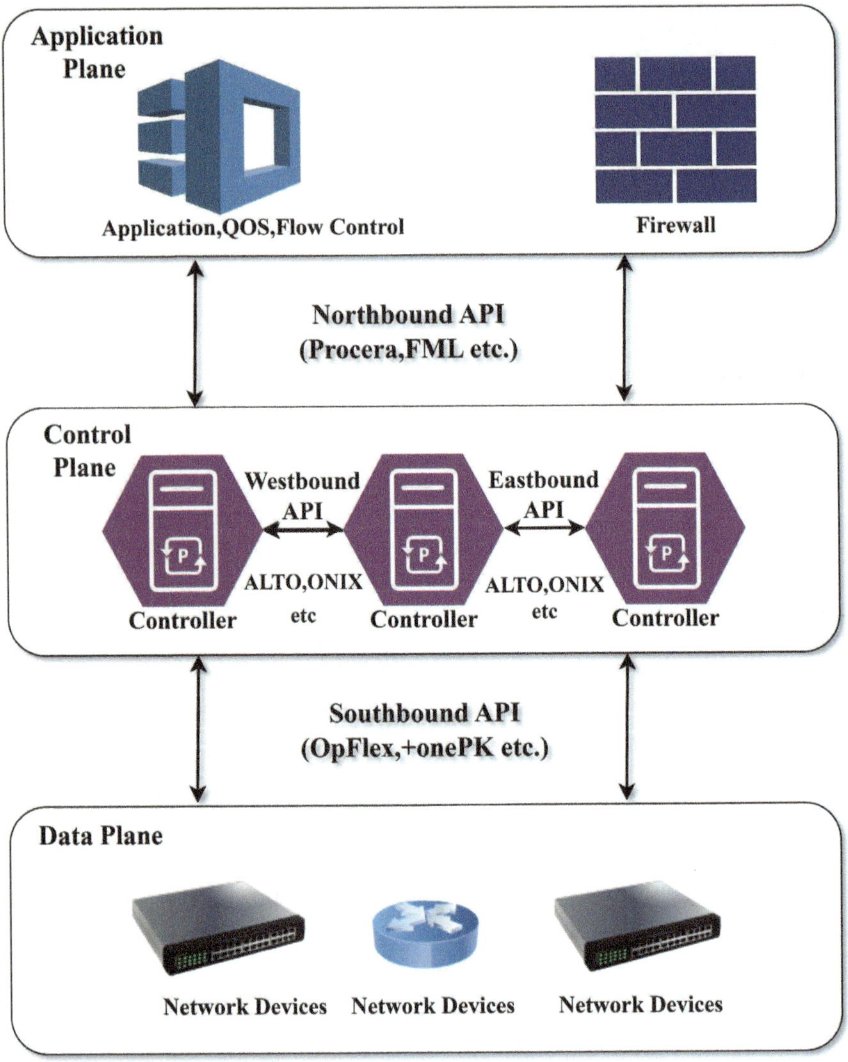

Fig. 1 SDN architecture

2 Related Work

Different DDoS detection systems are analysed, with a focus on predictive, time-domain, frequency-domain, and other novel approaches. The DDoS detection methods' categorization is displayed in Fig. 2. We go through a few of the several study approaches and strategies that researchers have utilised to identify DDoS attacks in SDN.

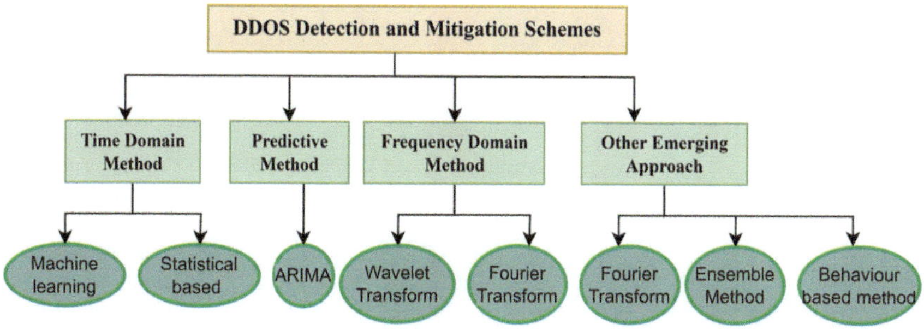

Fig. 2 DDoS Detection and mitigation methods

Further, Table 1 analyse different approaches used by different Researcher to effectively detect and mitigate the DDoS attack in SDN based IoT network.

3 Proposed Method

In this paper, we present an innovative approach to address the critical issue of detecting DDoS attacks within IoT networks based on SDN. Our methodology leverages Machine Learning classification algorithms to efficiently categorize network traffic. The central component of our approach is a well-trained Support Vector Machine (SVM), which serves as the decision-making engine to distinguish between benign and anomalous data packets. Figure 3 shows the proposed method. Our proposed methodology consists of three key phases: feature extraction, traffic flow state collection, and classifier evaluation.

a. **Traffic Flow State Collection**:

At regular intervals, our system requests a flow table from the OpenFlow switch and receives detailed information about the current network flow. This flow table data forms the foundation for our subsequent analysis.

b. **Feature Extraction**:

The core of our intrusion detection process lies in extracting meaningful features from the flow table data. These features are selected based on their relevance to identifying DDoS attacks. They encompass patterns and anomalies within the traffic, offering valuable insights into potential threats.

c. **Classifier Evaluation**:

The SVM classifier, adequately trained on the extracted characteristic inputs, plays a pivotal role in assessing incoming traffic. The SVM effectively distinguishes between data packets that are safe or suggestive of a DDoS attack by considering the distinctive characteristics that separate normal from abnormal data flows.

Table 1 Analysis of DDoS detection and mitigation used by different author

Mitigation approach	Attack type	Tools used	Benefits in approach	Limitations in approach
Time-slice based scheduling algorithm Multislot [2]	DDoS	POX controller, Mininet emulator	Contrasted to SingleQ and MultiQ algorithms, offers superior shielding for internal switches	It is inappropriate to large network topologies since each switch's queue maintenance overhead makes the design complex
ABB [3]	Service Exhaustion attacks	POX	Absolutely unnoticeable from the client and server sides. enhances the availability of a given service during DDoS without requiring any changes on either end	Processing that is quite expensive per pack overhead, ineffective as a defense, and prolonged response times. Valid only for attacks against specific services
RADAR [4]	SYN flooding attack, and UDP/ DNS attacks	Mininet emulator, Floodlight controller	Lowers controller reaction time, prevents single point of breakdown and cascading failure, and doesn't call for any additional equipment or adjustments	A tiny network topology with less number of switches results in higher costs overhead
sFlow-based Security-centric SDN [5]	DNS amplification attack	Kinetic controller, Mininet emulator	Compared to current approaches, it has a faster response time, is more adaptable, has higher throughput, and has less jitter	The effectiveness of this approach against NTP amplification attacks is unclear
FlowTrApp (SDN arch. for data centers) [6]	UDP/ICMP flooding attack	Mininet emulator, floodlight controller, sFlow-RT analytics engine	Adaptive, faster throughput, and less jitter as compared to the current techniques	It is unclear whether this system is effective against the NTP amplification attack

(continued)

Table 1 (continued)

Mitigation approach	Attack type	Tools used	Benefits in approach	Limitations in approach
Anti-spoofing defence technique (FP-SYN) [7]	SYN flooding attack	Mininet 2.2, OVS2.3.1	Cuts the price, the effort of attack traffic detection, and the bandwidth used for it	Only offers protection against TCP-SYN attacks
C-to-C procedure for secure controller communication [8]	ICMP flooding	POX controller, Mininet emulator	Reduced overhead, judicious resource use	
ArOMA [9]	UDP/TCP-SYN/ ICMP flooding attack	Mininet emulator, RYU controller	Lessens the controller and switch's duty	Degraded QoS, compatibility for only one customer network
SODM technique (SDN One-packet DDoS Mitigation) [11]	SYN/ICMP attack	Real-time controller	Swift reaction to an attack	Size restrictions for the monitoring window
Mitigation based on Adaptive policy feature [12]	DNS amplification and flooding attacks	Mininet emulator, RYU controller	Reduces flow entries in switches, responds quickly, and uses dynamically instantiated, adaptive security policies	No discussion of the detection technique

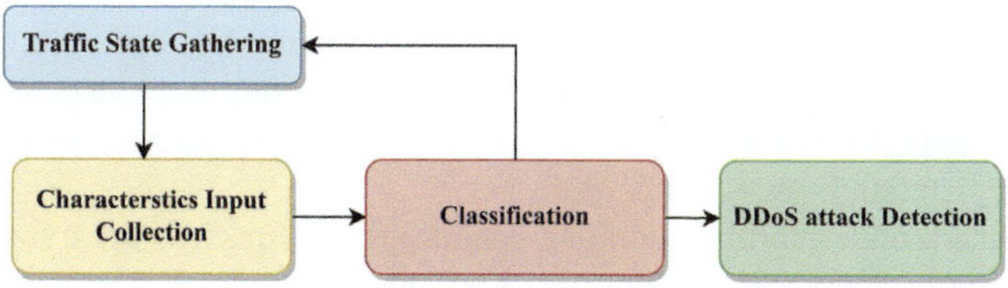

Fig. 3 Proposed method

The following parameters are used for training an SVM:

(a). Speed of Source IP (b). Standard Deviation of Flow Packets. (c). Speed of Flow Entries (d). Standard Deviation of Flow Bytes (e). Ratio of Pair-Flow.

Experiment Setup:

To validate our approach, we employed the Mininet emulator and the Ryu controller in our experimentation.

a. **Mininet**:

Mininet is a versatile tool that emulates a network environment, enabling the creation of virtual hosts, controllers, switches, and network links. This container-based virtualization system allows for the construction of complex network topologies without the need for physical network manipulation. Leveraging Mininet's Python API, we customized network topologies by adding hosts, switches, and links as needed.

b. **Ryu Controller**:

A Python-based SDN controller made accessible to the public is called Ryu. Because of its adaptability and thorough support for Python interfaces, users may build unique controllers that are suited to their needs. Ryu is a great option for SDN-based experimentation and monitoring since it supports a variety of network management protocols, such as OpenFlow and Netconf.

To improve DDoS attack detection in IoT networks, the approach we propose leverages the capabilities of SDN, ML, and network emulation tools like Mininet and the Ryu controller. We intend to considerably enhance intrusion detection accuracy and strengthen the security of SDN-based IoT settings by implementing an SVM classifier trained on carefully chosen characteristic inputs generated from flow table data.

4 Result

During a DDoS attack, there is a rapid increase in the number of flow entries per unit of time. These attacks typically involve a malicious pool of unpredictable IP addresses and port numbers. Moreover, the attacker often introduces more source ports and origin IPs within a short time frame. This growth pattern closely resembles what is depicted in Figs. 4 and 5.

In contrast, under normal operating conditions, data packets are transmitted at a relatively high volume. However, during an attack, the attacker aims to maximize the impact by sending information as quickly and efficiently as possible, resulting in smaller and unmodified data packets. Consequently, the standard deviations of both the quantity of flow packets and the quantity of flow bits within a specific time period are exceedingly small. Under normal network conditions, the source host and the target host establish interactive flow inputs. However, during an attack, the target host struggles to respond promptly due to the frequent use of virtual, random source IPs and port numbers. This leads to a significant reduction in the amount of interactive flow. Nevertheless, the proportion of cooperative flow entries remains consistently high and fluctuates within the typical range, as illustrated in Figs. 4 and 5.

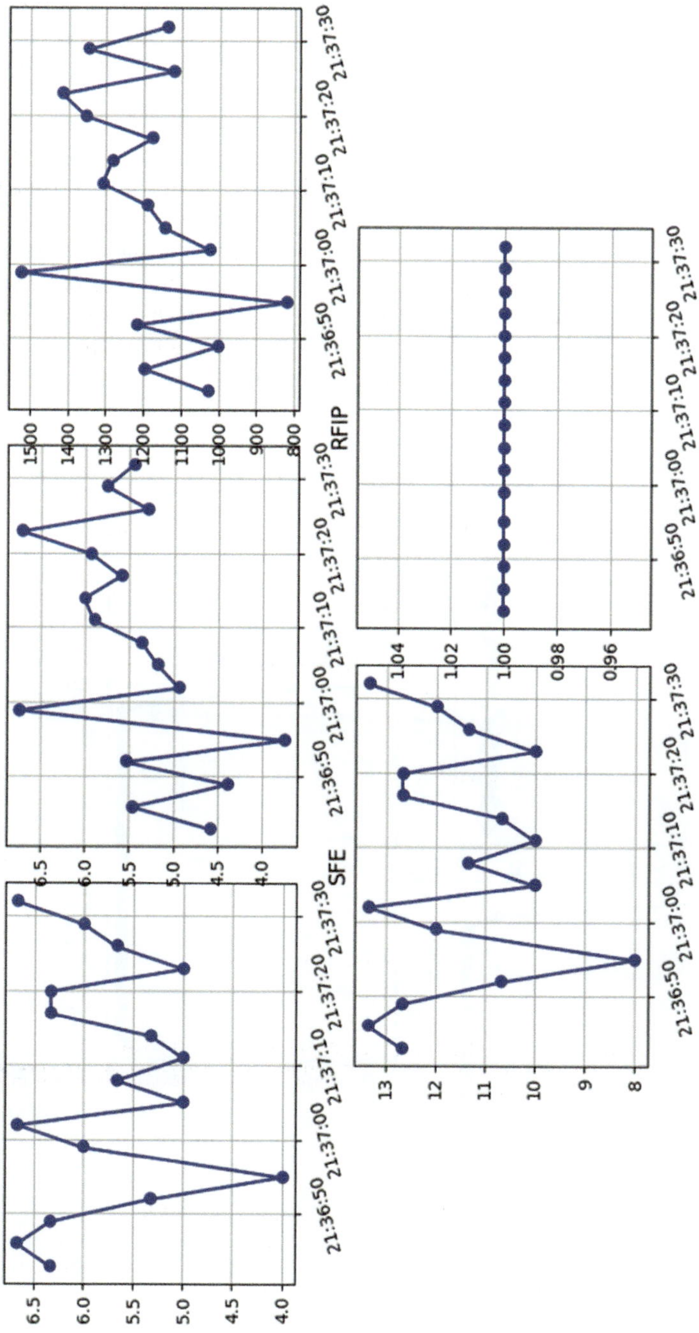

Fig. 4 Deviation in case of normal traffic

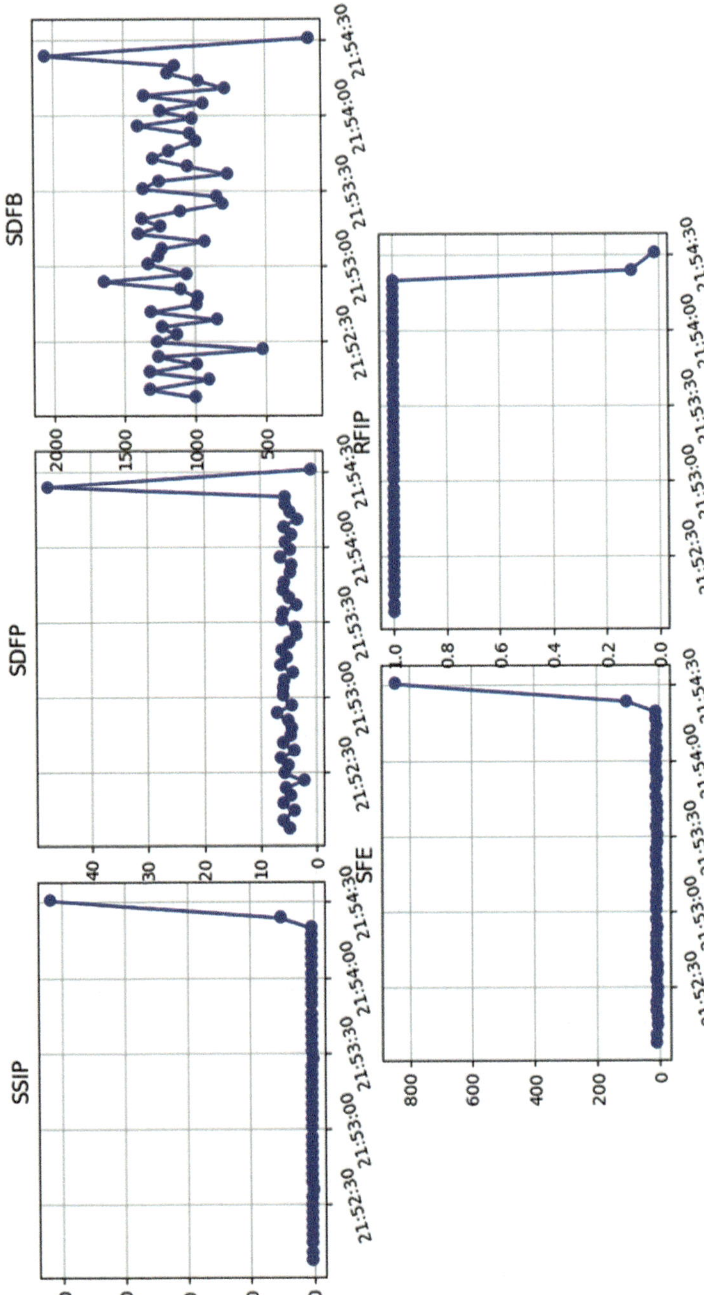

Fig. 5 Deviation in case of anamolus traffic

5 Conclusion

In conclusion, our study centered on the analysis of DDoS attacks in SDN-based networks, utilizing real-world five-tuple feature values and Support Vector Machine (SVM) Machine Learning techniques for flow analysis. Our results showed how well this method worked at describing changing traffic patterns in SDN systems. Our research's main finding is the importance of using actual data to create datasets for upcoming network applications. Utilizing real data improves the accuracy of categorizing different network events, enabling more precise network threat detection. We also explored the addition of an AVL tree to the SVM classifier, which boosted classifier effectiveness. An interesting approach for improving classification performance and testing efficiency in SDN-based systems is the organization of numerous binary SVMs within an AVL tree structure. In conclusion, our method stresses the importance of actual data in network threat detection while also improving our knowledge of DDoS attacks in SDN networks. The integration of AVL trees with SVMs offers a viable way to increase classifier effectiveness, which will improve SDN network security.

References

1. Kumari, P., Jain, A.K.: A comprehensive study of DDoS attacks over IoT network and their countermeasures. Comput Secur 103096 (2023)
2. Yan, Q., Gong, Q., Yu, F.R.: Effective software-defined networking controller scheduling method to mitigate DDoS attacks. Electron. Lett. **53**(7), 469–471 (2017). https://doi.org/10.1049/el.2016.2234
3. Sattar, D., Matrawy, A., Adeojo, O.: Adaptive bubble burst (ABB): Mitigating DDoS attacks in softwaredefined networks. In: Proceedings of the 2016 17th International Telecommunications Network Strategy and Planning Symposium (Networks'16), pp. 50–55 (2016). https://doi.org/10.1109/NETWKS.2016.7751152
4. Zheng, J., Li, Q., Gu, G., Cao, J., Yau, D.K.Y., Wu, J.: Realtime DDoS defense using COTS SDN switches via adaptive correlation analysis. IEEE Trans. Inf. Forens. Secur. **13**(7), 1838–1853 (2018). https://doi.org/10.1109/TIFS.2018.28056004
5. Aizuddin, A.A., Atan, M., Norulazmi, M., Noor, M.M., Akimi, S., Abidin, M.: DNS amplification attack detection and mitigation via sFlow with security-centric SDN. In: Proceedings of the 11th International Conference on Ubiquitous Information Management and Communication (IMCOM'17). ACM, New York, NY, Article 3, p. 7 (2017). https://doi.org/10.1145/3022227.3022230
6. Buragohain, C., Medhi, N.: FlowTrApp: An SDN based architecture for DDoS attack detection and mitigation in data centers. In: Proceedings of the 2016 3rd International Conference on Signal Processing and Integrated Networks (SPIN'16), pp. 519–524 (2016). https://doi.org/10.1109/SPIN.2016.7566750
7. Chen, M.H., Ciou, J.Y., Chung, I.H., Chou, C.F.: FlexProtect: A SDN-based DDoS attack protection architecture for multi-tenant data centers. In: Proceedings of the International Conference on High Performance Computing in Asia-Pacific Region (HPC Asia'18). ACM, New York, NY, pp. 202–209 (2018) https://doi.org/10.1145/3149457.3149476
8. Hameed, S., Ahmed Khan, H.: SDN based collaborative scheme for mitigation of DDoS attacks. Fut. Internet **10**(3), Article 23 (2018). https://doi.org/10.3390/fi10030023

9. Sahay, R., Blanc, G., Zhang, Z., Debar, H.: ArOMA: An SDN based autonomic DDoS mitigation framework. Comput. Secur. **70**(2017), 482–499 (2017). https://doi.org/10.1016/j.cose.2017.07.008
10. Hong, K., Kim, Y., Choi, H., Park, J.: SDN-assisted slow HTTP DDoS attack defense method. IEEE Commun. Lett. **22**(4), 688–691 (2018). https://doi.org/10.1109/LCOMM.2017.2766636
11. Huong, T.T., Thanh, N.H.: Software defined networking-based one-packet DDoS mitigation architecture. In: Proceedings of the 11th International Conference on Ubiquitous Information Management and Communication (IMCOM'17). ACM, New York, NY, Article 110, p. 7. https://doi.org/10.1145/3022227.3022336
12. Sahay R., Blanc G., Zhang, Z., Toumi, K., Debar, H.: Adaptive policy-driven attack mitigation in SDN. In: Proceedings of the 1st International Workshop on Security and Dependability of MultiDomain Infrastructures (XDOMO'17). ACM, New York, NY, Article 4, p. 6 (2017). https://doi.org/10.1145/3071064.3071068
13. Kumari, P., Jain, A.K.: SDN-enabled IoT to combat the DDoS Attacks. In: International Conference on Communication and Intelligent Systems. Singapore: Springer Nature Singapore, pp. 23–33 (2022)

Deep Reinforcement Learning for Connected Autonomous Vehicles (CAVs) Task Offloading

Aditya Bhardwaj

Abstract With the rapid development of Connected Autonomous Vehicles (CAVs), various applications requiring high computation resources have emerged. To resolve these issues, the integration of edge computing in CAVs can offer a scalable solution by distributing computation across edge nodes. Using edge computing in CAVs enables each vehicle to make local decisions based on data available at the edge, reducing the need for centralized systems. However, computing on an edge device is still challenging due to its limited computational power and battery life. To address these problems, we first modeled the task offloading issue using the Markov Decision Process (MDP), and then we applied the Proximal Policy Optimization (PPO)-based Deep reinforcement learning (DRL) technique. Extensive simulation results show that compared to the existing RRB and GBA schemes, our proposed DRL-based task offloading technique make significant improvements with reduction in average latency by 26.64, 13.33% and amount of energy consumption by 27.02, 15.44%. Finally, we explored open issues and research directions for task offloading in the CAVs system.

Keywords Task offloading · Edge computing · Connected and autonomous vehicles (CAVs) · Deep reinforcement learning

1 Introduction

For the last century, people have fantasized about self-driving or autonomous vehicles. Francis P Houdina, who drove a remote-controlled car around New York City streets in 1925, was one of the first attempts to develop a self-driving vehicle [1]. The '*intelligence*' of self-driving cars called Connected Autonomous Vehicles (CAVs) relies substantially on the information they gather about the road, other vehicles,

A. Bhardwaj (✉)
School of Computer Science Engineering & Technology (SCSET),
Bennett University, Greater Noida, India
e-mail: aditya.cse@nitttrchd.ac.in
URL: https://www.bennett.edu.in/

B. Soni et al. (eds.), *Beyond Artificial Intelligence*, Lecture Notes in Networks and Systems 1326, https://doi.org/10.1007/978-981-96-4170-3_43

pedestrians, and barriers to function autonomously. CAVs are designed to communicate with other vehicles, e.g., Vehicle-to-Vehicle (V2V) and Vehicle-to-Everything (V2X). This V2V and V2X communication allows CAVs to share their intentions, receive information from other vehicles, and coordinate actions to prevent collisions and optimize traffic flow [2]. Three essential requirements that primarily drive the need for connected automobiles vehicles are: (i) Enhancing road safety by reducing human errors, which are a significant cause of accidents; (ii) reducing congestion on roads; (iii) improving mobility solutions; (iv) providing sustainability of intelligent transportation networks, and (v) improve the user experience.

Recently, the Quectel research report highlighted that the compound annual growth rate of CAVs will be 27.75% from 2023 to 2028 [3]. Advances in Machine learning (ML), cloud computing [4], sensor technology, and data processing have made CAVs more feasible and opened up new avenues to move the field ahead. As the technology continues to improve, the reliability and performance of autonomous vehicles are likely to increase, further bolstering their popularity. These days, many major automotive manufacturers and technology companies, e.g., Tesla Motors and Mercedes-Benz Waymo (Google's parent company), are heavily investing in developing and deploying CAVs. This growing interest and competition in the market are driving innovations and speeding up the pace of adoption [5].

Despite the tremendous rise of CAVs, how to fulfill the computational requirements of developing transportation applications has been a long-standing concern. This is because the sensors and computational nodes deployed for CAVs have limited computational resources onboard due to power and thermal constraints. To mitigate these issues, task offloading is an important technique where CAVs can offload computationally intensive tasks to the cloud/edge servers, which enables the vehicle to access massive computational power without being constrained by the onboard hardware limitations [6]. Thus, implementing a task offloading strategy for the CAVs transportation network can enhance the vehicle's computational capabilities, improve real-time decision-making, reduce onboard processing requirements, and optimize overall system performance [7]. While offloading computation in CAVs can offer significant benefits, complete offloading to external cloud servers introduces latency and energy consumption issues. As a result, there is a need for a task offloading technique that tries to save energy on devices while also reducing computation latency.

The existing literature shows that machine learning-based techniques can help optimize the task-offloading decision-making process in CAVs. To provide task offloading decision, the solutions reported in the literature are based on static environment assumptions, and vehicle's mobility needs to be considered [8]. Therefore, in this research work, we first formulated a task offloading problem using the Markov decision process (MDP) model that takes a dynamic environment [11]. Then, we used the Deep Reinforcement Learning (DRL) technique for the task offloading decision. This aims to make effective offloading decisions with a reduced latency and energy consumption [9]. In our work, we adopted the DRL technique because it suits changing environments and traffic conditions. Simulation results show that the DRL-based technique has shown promising outcomes in complex offloading scenarios, where the CAVs need to offload large amounts of real-time data and interact with multiple edge or cloud servers [14]. The following are the most significant contributions of this study:

- A task offloading framework for the CAV environment that integrates cloud and edge computing networks has been proposed.
- We presented a critical analysis summary of the machine learning-based task offloading framework for CAVs.
- Finally, we explored open issues and suggested future research directions for task offloading in CAVs, providing valuable insights for further advancements in the field.

The remaining sections are structured as follows: Sect. 2 presents relevant literature findings. In Sect. 3, problem formulation using MDP and solution through the proposed DRL-based technique is presented. Section 4 presents simulation setup and performance evaluation with existing schemes. Finally, Sects. 5 and 6 conclude this article with open future research directions.

2 Related Work

Researchers have made several efforts in the literature to address the task offloading issues of CAVs. Most of the existing literature work has contributed to a static environment to provide task offloading. Still, there is a lack of studies covering task offloading in CAVs environments considering the dynamic environment perspective. In this section, significant findings from the most recent, cutting-edge literature are presented.

To decrease the typical processing latency of tasks using vehicle-edge-cloud collaborative computing, the author of [10] presented a Directed Acyclic Graph (DAG) and dependency-aware task offloading technique based on Deep Deterministic Policy Gradient (DDPG). Their proposed technique was simulated using SUMO simulation. The authors have performed a performance evaluation w.r.t Dependency-Aware Random (DAR) task offloading approach. Their simulation results showed that compared to the existing DA-RO baseline scheme, the author's proposed technique reduce average latency delay by 13.62%. Further, the author in [12] used a Deep-Q-Network algorithm to reduce the total amount of time and energy needed for computation offloading at Mobile edge computing (MEC) [13]. The author's proposed technique was simulated using TensorFlow and compared with the Full Offloading Random Caching (FORC). Experimental results show that compared to the existing full offloading random caching scheme their proposed technique effectively reduced the total cost of time and energy by 26%.

In [15], the authors address the issue of parked vehicle task offloading by using a meta-heuristic Genetic Algorithm (GA). The author's proposed algorithm was simulated using SUMO software and compared with the existing heuristic algorithm. Their performance evaluation shows that meta-heuristic Genetic Algorithm (GA) reduced the task offloading cost by 40%.

The authors of [16] proposed a Multi-task Deep reinforcement learning (MDTS) technique to address the load balancing problem in VANETs and the total cost of job execution. The author's proposed technique was simulated using Python, Tensorflow and compared with the existing techniques. The proposed approach has been shown to reduce the process execution time by 60.3% and 33.4%, respectively.

Further, to provide the seamless coverage and high quality services, in another study [18] presented digital twin (DT) to support UAV-enabled Mobile Edge Computing (MEC) systems where Deep Q-network is proposed. Although the studies mentioned above provided a preliminary foundation on DT-aided MEC, implementing the digital twin in UAV-assisted Vehicular Edge Computing (VEC) networks to aid vehicles in making offloading decisions has yet to be studied.

In summary, most of the existing literature studies are assisted by task offloading in a static environment. Solutions for mobility-aware systems are lacking. In particular, we utilized the MDP model to formulate task problems in a dynamic environment and implemented DRL-based techniques for effective offloading decisions. Thus, our work is among the first to investigate task offloading in a dynamic environment for CAV's application, which is crucial for guaranteeing the quality of service for vehicle users.

3 Problem Formulation and System Architecture

In the first part of this section, we outlined the system architecture for the proposed DRL-based task offloading technique. Then, the task-sharing problem was made using the Markov Decision Process (MDP) model. Finally, we have outlined the steps for the proposed DRL-based task offloading technique.

3.1 System Architecture of Proposed DRL-Based Task Offloading Technique

This subsection describes the layered architecture of proposed Deep Reinforcement Learning (DRL) based task offloading framework's shown in Fig. 1. We have used the DRL technique because it incorporates deep neural networks as function approximators and thus provides a promising solution for task offloading in edge computing scenarios. It involves making decisions about offloading computational tasks from a vehicle to edge computing resources in order to optimize performance, latency, energy consumption, and other relevant factors. DRL can handle high-dimensional action spaces, which makes it suitable for managing the complexity of task offloading.

In the proposed framework edge and cloud computing servers are used to provide resources required for the CAV's application task processing. The layered architecture is organized in three distinct layers as shown in Fig. 1.

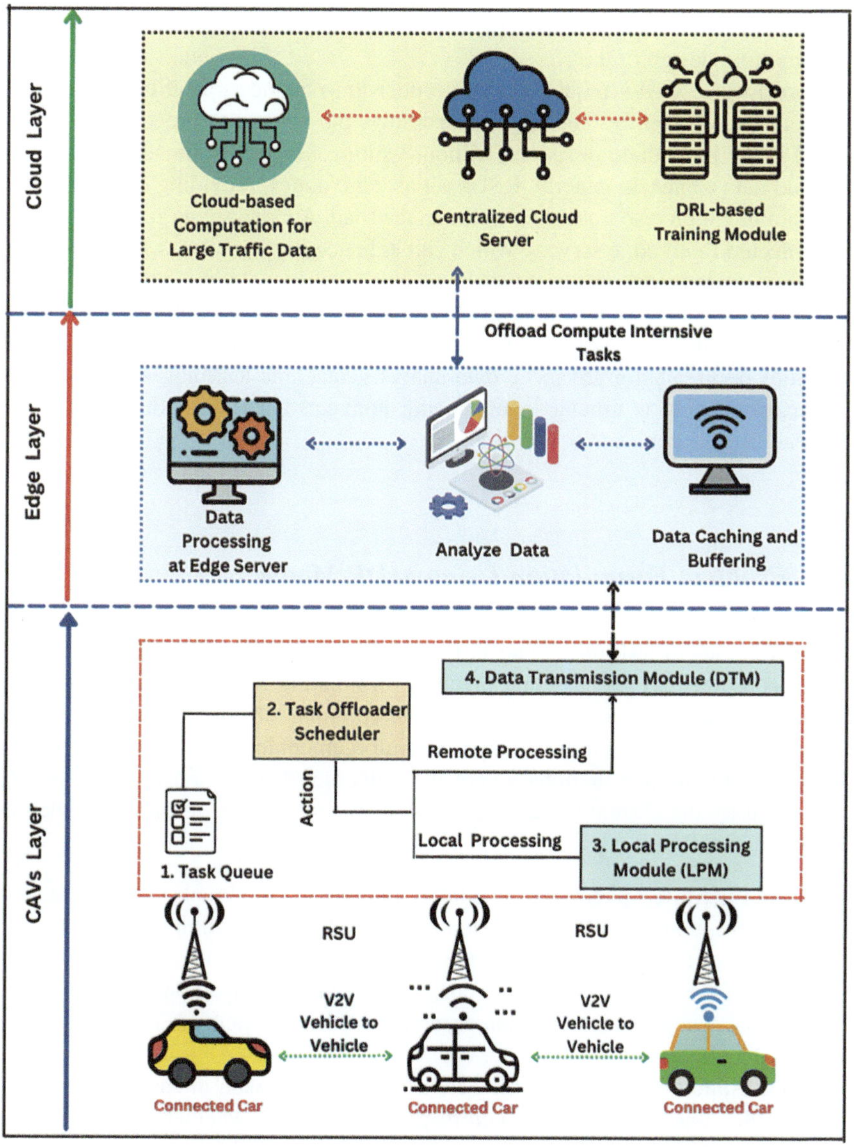

Fig. 1 Proposed architecture of DRL-based task offloading in CAVs edge computing network

- **CAVs Layer**: This layer consists of connected autonomous vehicles with various sensors, such as GPS, LiDAR, etc. Real-time information about the vehicle's surroundings is obtained from these sensors. Also, connected autonomous vehicles are served by Roadside Units (RSU) that enable CAVs to communicate with the roadside infrastructure. Every RSU has an edge server with a certain number of computing resources needed to process the less compute-intensive jobs. Also, the computing design of the

- **Edge Layer**: CAVs require real-time decision-making capabilities to navigate safely and efficiently. The edge layer allows the vehicle to process sensor data and make immediate decisions without waiting for data to be sent to a distant cloud data center. In general, RSUs act as edge nodes, providing computing and communication resources to vehicles on the road. A cloud server in the network is connected to all edge servers, which can enhance its capability when computing resources are inadequate.
- **Cloud layer**: Finally, the cloud computing layer in task offloading for CAVs complements edge computing by providing the computational resources and capabilities necessary for advanced data analysis, machine learning, and centralized management. It is crucial in enhancing connected and autonomous vehicle's capabilities, safety, and scalability.

3.2 *Problem Formulation Using MDP Model*

When it comes to task offloading in CAVs, the Markov Decision Process (MDP) model is a potential solution that involves formulating the decision-making problem as a sequence of actions taken in an environment characterized by states and rewards [17]. MDP is commonly used in reinforcement learning to model decision-making problems in situations where the outcome of an action is uncertain and influenced by the current system state. The key components of a MDP model are described as follows:

- **Define the State Space**: The scheduler component initially observes the state of the system, upon which the decision for offloading scheduling is determined. In the context of edge computing and task offloading, states could represent various aspects of the system, such as the current computational load on the edge device, network latency, available resources, and any other relevant variables that impact offloading decisions.
- **Action Space**: Now, specify the actions that can be taken in each state. For the task offloading process, actions correspond to deciding whether to offload a task or execute it locally on the edge device. In the simulation environment, the "action space" is defined as $A = (0, 1)$, here, 1 indicates edge server offloading, and 0 denotes local execution.
- **Design the Reward Function**: Assign rewards to different state-action pairs. Reward function is used to provide a optimal solution for the latency and energy consumption parameters.

3.3 *Steps for Proposed Task Offloading Technique*

The procedures for the proposed task offloading technique are outlined below. The CAVs layer offloads tasks to the edge/cloud server or local processing module based on offloading decisions.

Step 1—As shown in Fig. 2, intially, the task generated by various types of connected autonomous vehicles applications are received in the task queue and sorted by their generation time.

Step 2—Now, the task scheduler module monitors system information, e.g., queue, transmission state, local and remote execution, to make the scheduling decision. The Local Processing Module (LPM) is set up to run local jobs. For remote execution, a task is first sent from the data transfer unit to the serving RSU. Afterward, the edge/cloud server runs the task and returns the result to the CAV's layer.

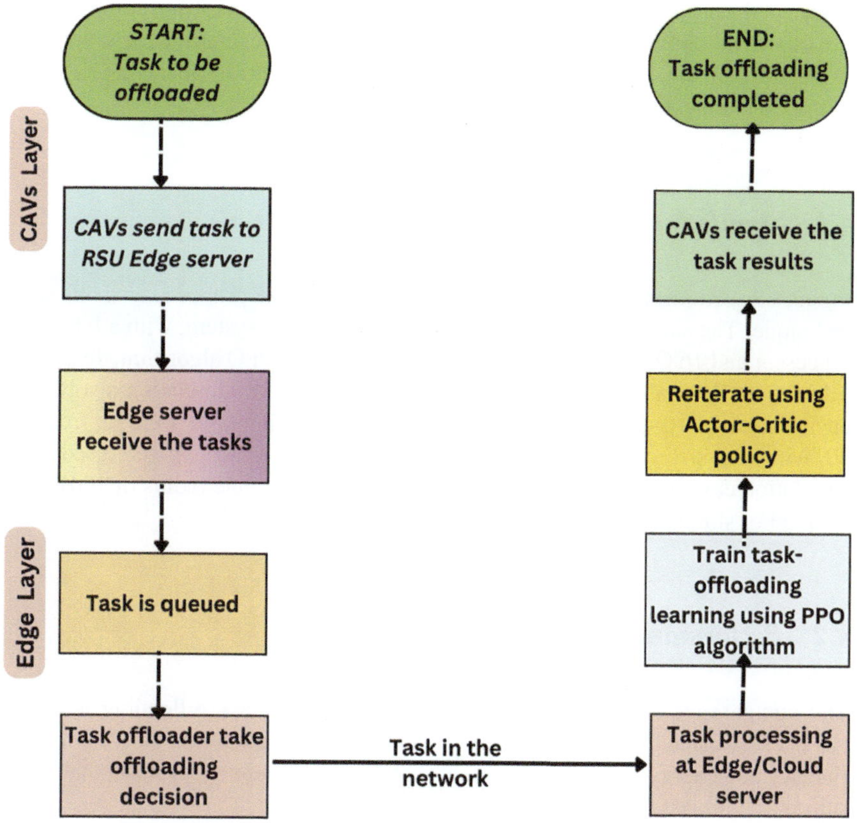

Fig. 2 Steps for proposed task offloading technique

Step 3—Finally, the DRL-based training module is a key component of the cloud layer. In this, the Proximal Policy Optimization (PPO) algorithm is used to approximate the optimal policy. PPO is a policy-based reinforcement learning method that uses an Actor-Critic structure to optimize the agent's policy and maximize expected cumulative rewards [19]. It does this by collecting trajectories of experiences and iteratively improving the policy. In the framework, cloud computing servers have also been used because they provide the computational power, scalability, cost-efficiency, and convenience necessary for effectively running machine learning workloads.

4 Simulation Environment and Results

In this section, we discuss the simulation environment, including the parameter settings, the number of offloading tasks, and the performance evaluation. The performance of the proposed DRL-based task offloading technique has been compared with existing Random Based (RRB) and Greedy Based (GBA) schemes [8]. The outcomes of the experiments are presented and analyzed to highlight the achievements of the proposed architecture.

4.1 Simulation Setup

In this subsection, we carried out simulation setting of our proposed task offloading technique. The simulation was carried out using Window system, with a Intel(R) *i7* processor and 16 GB RAM. To implement the proposed PPO algorithm, Tensorflow 1.15.0 with Python 3.6 has been implemented. Initially, the simulation involved the environment of 10–50 vehicles. Then number of offloading tasks were set from 5 to 30, with step size of 5. Transmission rate range of (1–30 Mbps) were used in the simulation environment. The RSUs configured have coverage radius of 100 m. The detailed of simulation parameters is depicted in Table 1.

4.2 Performance Results

The comparative analysis of the proposed DRL-based task offloading technique has been evaluated using latency (msec) and energy consumption (Joules) metrics. Latency and energy consumption are critical metrics in network systems that have a substantial impact on overall quality of service.

4.2.1 Latency Analysis

Latency in task offloading refers to the delay or the time it takes for a task or workload to be transferred from one computing device or location to another for processing. In the following we discussed impact of increase in number of tasks and number of vehicles on latency.

(a) Impact of Number of Tasks on Latency: Figure 3a shows the average latency performance comparison of the proposed technique when the number of tasks is increased. It is inferred that an increase in the number of tasks leads to contention for resources, which in turn can increase the time it takes for each task to be completed, resulting in higher latency. Figure 3a reveals that when the number of tasks increases, compared to existing Random Based Technique (RRB) and Greedy Based Algorithm

Table 1 Simulation parameters

Description	Values
Simulator	Python 3.6
ML library	Tensorflow 1.15.0
Number of vehicles	10–50
Number of tasks	5–30
Step size	5
Tranmission rate	1–30 Mbps
Task size	0.5–2 MB
Vehicle speed	20 m/sec
CPU frequency	1–4 GHz
Learning rate	0.001
Number of RSUs	3
RSU coverage region	100 m

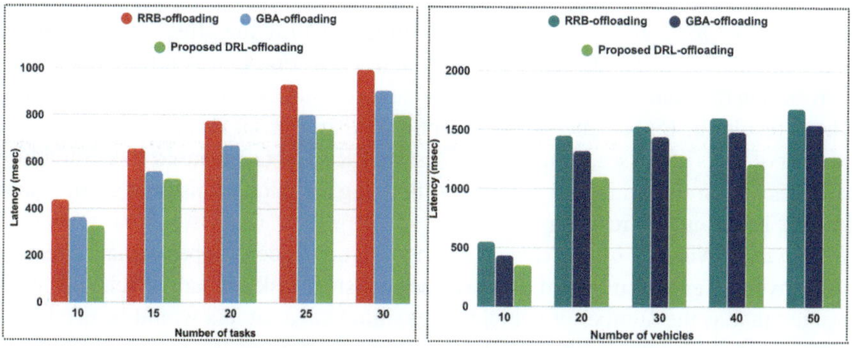

Fig. 3 Latency-based performance comparison of existing and proposed task offloading technique

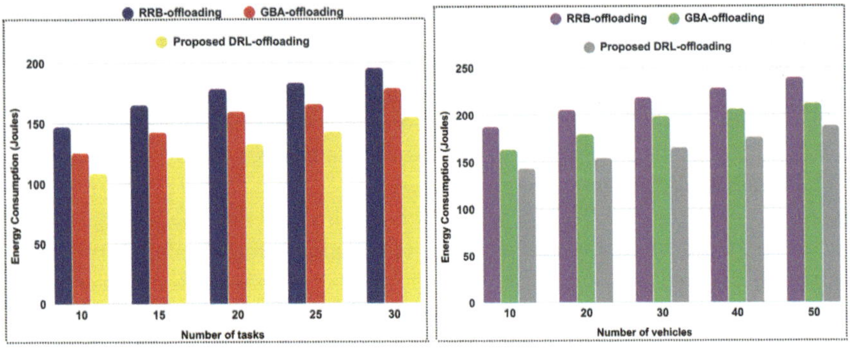

Fig. 4 Energy consumption-based performance comparison of existing and proposed task offloading technique

(GBA) task offloading schemes, our proposed DRL-based task offloading technique reduces latency by 22.72% and 9.21%, respectively.

(b) Impact of Number of Vehicles on Latency: Figure 3b shows increases in the number of vehicles result in larger latency duration. This is because as the number of vehicles offloading tasks to edge servers increases, the load on edge servers also increases thus leading to longer processing times and higher latency for tasks. Figure 3b reveals that when the number of vehicles increases, compared to existing RRB and GBA-based task offloading schemes, our proposed DRL-based task offloading technique reduces latency by 26.56% and 17.45%, respectively.

4.2.2 Energy Consumption Analysis

Analyzing energy consumption for task offloading of CAVs is crucial for optimizing the overall energy efficiency of these vehicles. The impact of increase in tasks and vehicles for energy consumption is discussed in the following.

(a) Impact of Tasks Increase on Energy: Figure 4a shows the effect of the increase in the number of tasks on energy consumption. From Fig. 4a it is inferred that with increases in the number of tasks, a large amount of computation resources is needed, which results in higher energy consumption. Also, Fig. 4a reveals that when the number of tasks increases, compared to existing RRB and GBA-based task offloading schemes, our proposed DRL-based task offloading technique consumes 27.67% and 16.09% lesser energy respectively.

(b) Impact of Number of Vehicles on Energy: Figure 4b depicts performance results of the average energy usage of proposed and existing task offloading schemes. It is implied that as the number of vehicles rises, more edge servers would be needed to carry out the task offloading process, that results in increase of energy usage. Also, Fig. 4b reveals that when the number of vehicles increases, compared to existing RRB

and GBA-based task offloading schemes, our proposed DRL-based task offloading technique consumes 26.37% and 14.79% lesser energy, respectively.

5 Conclusion

Task offloading in connected autonomous vehicles is a critical and evolving area that holds great potential for enhancing the efficiency, safety, and overall performance of autonomous systems. This study addressed task offloading latency and energy consumption issues in CAVs. The proposed technique utilizes a Markov Decision Process (MDP) model to formulate the task offloading problem. Then, we implemented a Deep Reinforcement Learning (DRL) algorithm to make effective offloading decisions. Python 3.11, with Tensorflow 2.13.0, was utilized to perform simulations for the proposed task offloading technique. Simulations showed that compared to the existing RRB and GBA schemes, our proposed DRL-based task offloading technique significantly reduce average latency and energy consumption.

6 Future Directions

For the readers' better understanding, we have summarized future research directions in task offloading for CAVs Environment as follows:

- **Distributed Offloading**: There is a need to investigate distributed task offloading techniques where multiple CAVs collaborate to share computational resources and perform cooperative computing. This can lead to improved efficiency and reliability.
- **Cooperative Vehicular Networks**: The RSU becomes overwhelmed by the processing load imposed by an increasing number of tasks. Therefore, researchers can explore how network conditions, such as congestion and packet loss, affect the performance of offloaded tasks.
- **Security and Privacy**: There is also need to investigate security and privacy implications of task offloading in CAVs. Developing techniques to ensure data confidentiality, integrity, and authentication when transmitting sensitive data between vehicles and edge/cloud servers.
- **Heterogeneity and Interference Management**: Co-channel interference is a major issue for overlapping CAVs networks with a heavily deployed topology. As a result, the issue of how to facilitate dynamic spectrum-sharing solutions for future vehicular networks needs significant attention.

– **Connectivity and Reliability**: The rollout of 6G networks and future advancements in wireless communication technologies will significantly enhance connectivity in CAVs. These high-speed, low-latency networks will provide the necessary infrastructure for efficient task offloading between vehicles and cloud or edge computing resources.

References

1. Fallon, M.: Self-driving Cars: The New Way Forward. Twenty-First Century Books (2018)
2. Ahmed, M., Raza, S., Mirza, M., Aziz, A., Khan, M., Khan, W., Li, J., Han, Z.: A survey on vehicular task offloading: classification, issues, and challenges. J. King Saud Univ.-Comput. Inf. Sci. **34**, 4135–4162 (2022)
3. Casetti, C.: 5G standalone deployments are on the rise [Mobile Radio]. IEEE Veh. Technol. Mag. **18**, 5–11 (2023)
4. Bhardwaj, A., Krishna, C.: Impact of factors affecting pre-copy virtual machine migration technique for cloud computing. Mater. Today: Proc. **18**, 1138–1145 (2019)
5. Talebian, A., Mishra, S.: Unfolding the state of the adoption of connected autonomous trucks by the commercial fleet owner industry. Transp. Res. Part E: Logist. Transp. Rev. **158**, 102616 (2022)
6. Akhlaqi, M., Hanapi, Z.: Task offloading paradigm in mobile edge computing-current issues, adopted approaches, and future directions. J. Netw. Comput. Appl. **212**, 103568 (2023)
7. Zhao, G., Xu, H., Zhao, Y., Qiao, C., Huang, L.: Offloading tasks with dependency and service caching in mobile edge computing. IEEE Trans. Parallel Distrib. Syst. **32**, 2777–2792 (2021)
8. Liu, L., Zhao, M., Yu, M., Jan, M., Lan, D., Taherkordi, A.: Mobility-aware multi-hop task offloading for autonomous driving in vehicular edge computing and networks. IEEE Trans. Intell. Transp. Syst. **24**, 2169–2182 (2022)
9. Zhao, N., Ye, Z., Pei, Y., Liang, Y., Niyato, D.: Multi-agent deep reinforcement learning for task offloading in UAV-assisted mobile edge computing. IEEE Trans. Wirel. Commun. **21**, 6949–6960 (2022)
10. Liu, G., Dai, F., Huang, B., Qiang, Z., Wang, S., Li, L.: A collaborative computation and dependency-aware task offloading method for vehicular edge computing: a reinforcement learning approach. J. Cloud Comput. **11**(68), 1–15 (2022)
11. Yang, G., Hou, L., He, X., He, D., Chan, S., Guizani, M.: Offloading time optimization via Markov decision process in mobile-edge computing. IEEE Internet Things J. **8**, 2483–2493 (2020)
12. Elgendy, I., Zhang, W., He, H., Gupta, B., Abd El-Latif, A.: Joint computation offloading and task caching for multi-user and multi-task MEC systems reinforcement learning-based algorithms. Wirel. Netw. **27**, 2023–2038 (2021)
13. Sethi, V., Pal, S.: FedDOVe: a federated deep Q-learning-based offloading for vehicular fog computing. Future Gen. Comput. Syst. **141**, 96–105 (2023)
14. Bhardwaj, A., Krishna, C.: Performance evaluation of bandwidth for virtual machine migration in cloud computing. Int. J. Knowl. Eng. Data Min. **5**, 139–152 (2018)
15. Nguyen, K., Drew, S., Huang, C., Zhou, J.: Parked vehicles task offloading in edge computing. IEEE Access **10**, 41592–41606 (2022)
16. Qi, Q., Zhang, L., Wang, J., Sun, H., Zhuang, Z., Liao, J., Yu, F.: Scalable parallel task scheduling for autonomous driving using multi-task deep reinforcement learning. IEEE Trans. Veh. Technol. **69**, 13861–13874 (2020)
17. Shinde, S., Tarchi, D.: A Markov decision process solution for energy-saving network selection and computation offloading in vehicular networks. IEEE Trans. Veh. Technol. **72**(9), 12031–12046 (2023)

18. Li, B., Xie, W., Ye, Y., Liu, L., Fei, Z.: FlexEdge: digital twin-enabled task offloading for UAV-Aided vehicular edge computing. IEEE Trans. Veh. Technol. **72**(8), 11086–11091 (2023)
19. Chen, G., Xu, X., Zeng, Q., Zhang, Y.: A vehicle-assisted computation offloading algorithm based on proximal policy optimization in vehicle edge networks. Mob. Netw. Appl. 1–15 (2022)

Activation Functions Impact on Malware Detection: A Multi-layer Perceptron Approach for Android Applications

Shruti Brahma, Shweta Sharma, and C. Rama Krishna

Abstract Cybersecurity is an evolving landscape and the need to discern malicious and benign applications has become paramount. One method of handling this challenge is by deploying machine learning models such as Multi-Layer Perceptron. A critical aspect of working with this model is the choice of activation functions in the hidden layers of the Multi-Layer Perceptron. The activation functions include the identity, logistic, hyperbolic tangent, rectified linear unit, and leaky rectified linear unit. This study aims to discern whether a given app is benign or malicious, primarily by scrutinizing the permissions and intents it requests during execution. Subsequently, a comprehensive analysis is conducted by combining the activation functions in various ways by considering different combinations for the hidden layers. The experimental results show that Leaky Rectified Linear Unit activation provides the highest accuracy (98.22%) when applied as a common activation function across all hidden layers of the Multi-Layer Perceptron. Additionally, in cases involving varied activation functions, the combination of leaky rectified linear unit, rectified linear unit, and hyperbolic tangent in different hidden layers achieves the best accuracy (97.91%) for detecting malicious applications.

Keywords Android · Activation functions · Intents · Malicious applications · Multi-layer perceptrons · And permissions

S. Brahma (✉)
Department of Computer Science and Engineering, JNTUH University of Engineering, Sultanpur, India
e-mail: shrutibrahma@gmail.com

S. Sharma
Department of Computer Engineering, NIT Kurukshetra, Kurukshetra, India
e-mail: shweta.sharma@nitkkr.ac.in

C. Rama Krishna
Department of Computer Science and Engineering, NITTTR Chandigarh, Chandigarh, India
e-mail: rkc@nitttrchd.ac.in

B. Soni et al. (eds.), *Beyond Artificial Intelligence*, Lecture Notes in Networks and Systems 1326, https://doi.org/10.1007/978-981-96-4170-3_44

1 Introduction

Smartphones rely on applications to provide various functions and services which makes them an integral part of our daily lives. Alongside this, widespread malicious applications targeting smartphone devices pose a significant threat by compromising sensitive and private information stored on these devices. However, among all operating systems, Android users dominate the smartphone market, primarily due to its increasing popularity.

Motivation: The Google Play Store boasts more than 2.5 billion applications, offering users a wide range of services, including online shopping, social networking, positioning, and navigation [1]. Consequently, this popularity has made Android a prime target for an escalating number of cyberattacks and malicious applications. The Kaspersky Security Network prevented a total of 5.7 million mobile malware attacks in 2023 [2]. The Quick Heal Annual Threat Report published in the year 2022 revealed the detection of 0.12 million Android malware cases [3]. According to the Kaspersky security report published in the year 2022, mobile malware exceeded 6.4 million attacks on Android-based smartphone devices [4]. Of these, more than 5 million were malicious installation packages, with 53,947 of them linked to mobile banking Trojans.

Figure 1 presents the trend of Android Malware from the year 2015 to 2022 [5]. Considering the vast number of malicious applications, there is an urgent requirement to establish a resilient malware detection framework capable of promptly identifying security threats.

Contributions: In this experimental research work, we proposed a framework for the detection of Android malicious applications using Multi-Layer Perceptron (MLP). The detection involves identifying intricate patterns in diverse features, such

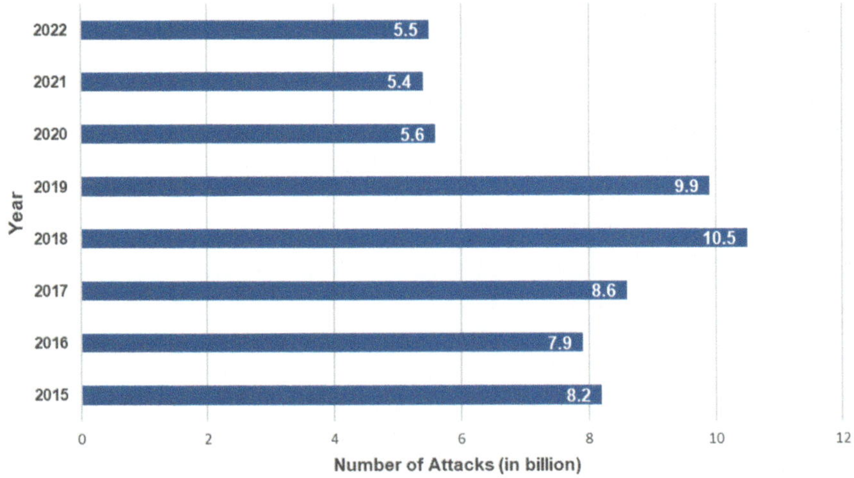

Fig. 1 Annual number of android malware attacks

as permissions and intents obtained by malicious applications. Therefore, MLPs are able to model non-linear relationships through the activation functions used in the hidden layers which makes them well-suited for capturing intricate patterns [6]. Consequently, this makes MLPs effective for the task of detecting Android malicious applications. Our contributions are twofold:

- First, we individually apply common activation functions—Identity, Hyperbolic Tangent (Tanh), Rectified Linear Unit (ReLU), and Leaky ReLU—across all hidden layers of the MLP. This approach enables a comprehensive comparison and investigation into the impact of each activation function, determining which one yields the optimal results.
- Second, we explore a configuration by combining different activation functions—namely, Tanh, ReLU, and Leaky ReLU. These combined activations are simultaneously applied across various hidden layers, allowing us to analyze the ensemble that produces the most favorable outcomes.

Outline: Sect. 2 describes the Literature Survey. Section 3 shows the Dataset Description. Section 4 presents the Proposed Architecture with detailed explanations of all steps. Section 5 shows experimental results with detailed discussions and comparisons with the existing work. Section 6 concludes the paper.

2 Literature Survey

Several researchers worked on the analysis and detection of malicious applications from Android OS as summarized in Table 1. Alomari et al. [7] employed a dense layer model with Long Short-Term Memory (LSTM) using ReLU activation. Hwang et al. [8] presented an architecture by examining the platform-independent binary data. Liu et al. [9] applied a graph-based neural network (GBNN) approach for Android malware detection by analyzing the network traffic after considering node, edge, and different traffic flows and their attributes.

Wang et al. [10] presented a hybrid architecture with Deep Autoencoders (DAE) and Convolutional Neural Networks (CNN) by using the ReLU activation function. A detection method was introduced by Wei et al. [11] by using a one-dimensional CNN to extract network flow features. These features are processed by an independent Recurrent Neural Network (RNN) with a Leaky ReLU activation function.

Bawazeer et al. [12] applied Neural Network algorithms, including Full Order Radial Basis Function (RBF), CNN, and MLP by applying the ReLU activation function within the hidden layers. Sagar [13] used the Deep Belief Network (DBN) and applied Tanh activation functions in hidden layers. Sharma et al. [14] implemented MLP by applying the ReLU activation function in hidden layers to detect Android ransomware. Yadav et al. [15] implemented a CNN-based model using image-based representations of DEX files. Xia et al. [16] and Zhu et al. [17] applied MLP to detect Android malware. Dhanya et al. [18] employed MLP with ReLU activation function for network intrusion detection using the UNSW-NB15 dataset.

Table 1 Summary of literature survey

Research paper	Year	Dataset	Malware	Machine learning	Activation function
Alomari et al. [7]	2023	Kaggle	5,560	LSTM	ReLU
Liu et al. [9]	2023	CICAndMal2017	5,465	GBNN	ReLU
Dhanya et al. [18]	2023	UNSW-NB15	2 million	MLP	ReLU
Yadav et al. [15]	2022	R2-D2	2 million	EfficientNet-B4 CNN	Softmax
Zhu et al. [17]	2021	Genome, VirusShare	25,577	MLP	ReLU
Bawazeer et al. [12]	2021	MNIST	6,000	CNN	ReLU
Sharma et al. [14]	2021	RansomProber	2,076	MLP	ReLU
Hwang et al. [8]	2020	KISA, VirusShare	50,000	DNN	ReLU
Wei et al. [11]	2020	CICAndMal2017	5,000	RNN	Leaky ReLU
Sagar [13]	2019	BaIoT	7 million	DBN	Tanh
Wang et al. [10]	2017	VirusShare	13,000	DAE-CNN	ReLU

The existing literature focuses on applying similar activation functions in the hidden layers. These are a few papers [11] in the literature where researchers use the Leaky ReLU as an activation function in the hidden layers of MLP to detect Android malicious applications. However, no paper in the literature applies different combinations of activation functions in distinct hidden layers of Neural Networks to detect Android malicious applications.

3 Dataset Description

Several datasets are available to perform experimental work for the analysis and detection of Android malware [19]. The Android malware dataset used in this experimental work is sourced from the Drebin dataset, as documented in a study by Arp et al. [20]. It encompasses a collection of Android applications, categorized into two main groups: malicious applications and benign applications. The dataset comprises a substantial volume, with around 5,560 malicious applications and 9,476 benign applications. These applications were gathered over a period extending from August 2010 to October 2012. The specifics of this dataset, including information about its size and its original sources, are provided in Table 2, which likely offers more comprehensive insights into the dataset characteristics and origins for reference within the study.

Table 2 Dataset specification

Drebin dataset	Specification
Size	6 GB
Number of malicious applications	5,560
Number of benign applications	9,476
Collection period	August 2010 to October 2012
Source	"https://www.sec.cs.tu-bs.de/~danarp/drebin/"

4 Proposed Architecture

The proposed architecture for the detection of malicious applications is shown in Fig. 2 which is explained as follows.

4.1 Reverse Engineering of APKs

Reverse Engineering of an Android Package Kit (APK) involves systematic analysis and deconstruction of compiled Android applications to understand their internal structure and behavior. This step exposes potential malicious code of Android applications. The Apktool [21] has been used to perform reverse engineering that extracts the program source code from the APKs.

4.2 Extract Manifest.xml

After reverse engineering, the Androidmanifest.xml file has been extracted from the malicious and benign APKs. This step involves parsing the AndroidManifest.xml record which has important metadata such as app permissions and intents. This allows in-depth evaluation of app functionalities and enables finding potential malicious functionalities.

4.3 Retrieve the Features and Create a Separate List

This step involves the creation of a distinct listed catalog of features such as permissions and intents in a CSV file. A total of 214 features are incorporated to offer a comprehensive representation of malware behavior and characteristics. This curated list facilitates in leveraging the divergence between benign and malicious permission patterns. This demarcating between legitimate and potentially malevolent requests helps in creating robust machine learning models.

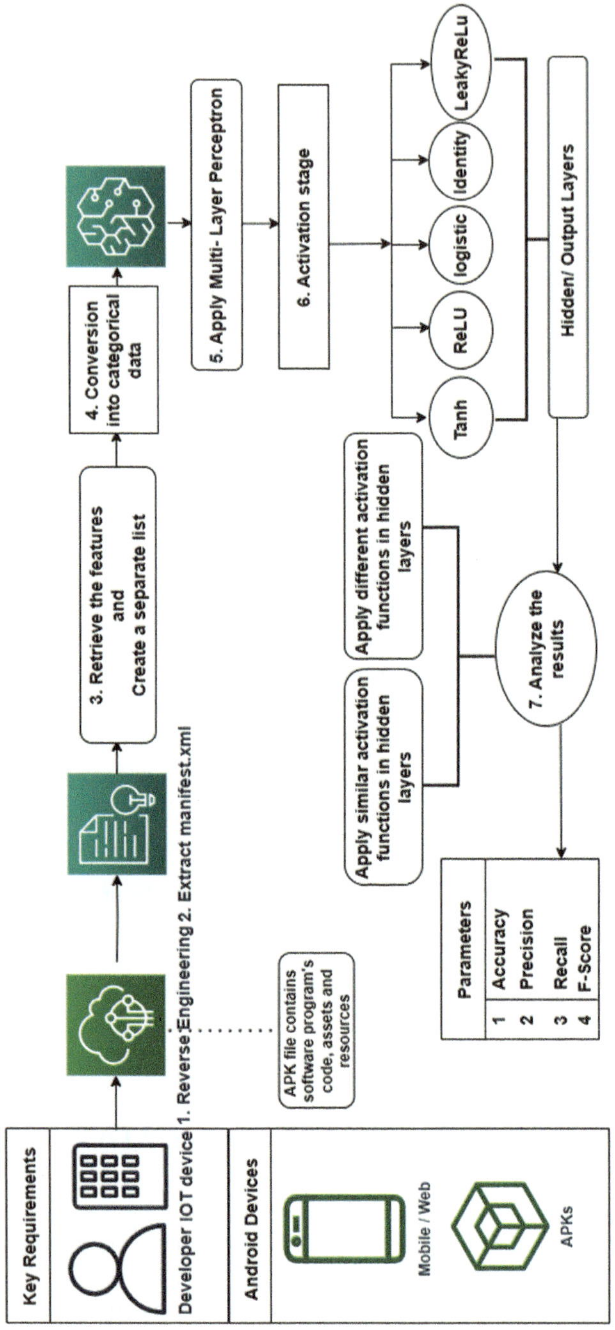

Fig. 2 Proposed architecture for detection of malicious applications

4.4 Conversion into Categorical Data

This step involves transforming a vast dataset of app permissions and intents into binary format. Each permission and intent corresponds to a specific action on the mobile device. When an application requests permission or intent to access a feature, it is represented as “1”, otherwise it is marked “0”.

4.5 Apply Multi-layer Perceptron

MLP is a neural network where connected nodes assist in the detection of malware [14]. The process starts by entering app-related features such as permissions and intents into a suitable format (in binary values: 1 for allowed and 0 for denied). Subsequently, the model undergoes intense training on the extracted features, facilitating it to learn and adapt to malware behavior. Once trained, MLP is capable of effectively predicting the maliciousness of new applications based on previously learned data and patterns.

4.6 Activation Stage

The activation functions applied in multiple hidden layers and output layer of MLP hold a profound significance in the arena of malware detection. During the forward pass of the neural network, the input data undergo linear transformations that involve weights and biases. The resulting values are then passed through an activation function, which identifies significant nonlinearities in the model. This nonlinearity ensures that the network can capture complex relationships in the data, which in this case is a characteristic difference between malware and benign applications. Several activation functions including Identity, Logistic, Tanh, ReLU, and Leaky ReLU serve a vital role in this stage which are discussed as follows.

Identity Activation Function

It is a linear activation function where the activation is proportional to the input. Mathematically, it is defined in Eq. 1 where x is the input feature.

$$f(x) = x \tag{1}$$

The identity activation function is applied in the hidden layers to maintain linearity between the output and the input. However, it fails when any form of non-linearity is introduced [6].

Logistic Activation Function

It is also known as the Sigmoid activation function which is significantly known for its ability to combine non-linearity into the network. It is defined in Eq. 2 where x is the input features.

$$f(x) = \frac{1}{(1 + e^{-x})} \tag{2}$$

In the output layer, the logistic activation function is applied for tasks involving binary classification. It effectively predicts the probability of an input, with output values naturally constrained within the range of 0 to 1. However, it suffers from a limited output range and is not applicable for multi-classification [6].

Tanh Activation Function

It is a transformative bridge between input signals and subsequent neuron outputs. It is mathematically represented in Eq. 3. Here x is considered as the input feature.

$$f(x) = \left(\frac{e^x - e^{-x}}{e^x + e^{-x}}\right) \tag{3}$$

The tanh activation function is applied in the hidden layers as it offers a zero-centered output and ranges between -1 and 1 which results in balanced output and faster convergence. However, the Tanh is zero-centered but the gradient isn't, it is always positive and varies between 0 and 1 [6]. This discrepancy can result in undesired behaviors during training.

ReLU Activation Function

It is the widely used activation function in neural networks. It exhibits a behavior where, for non-positive inputs (negative or zero), the output is consistently 0, while for positive inputs, the output mirrors the input value exactly. This can be expressed mathematically as Eq. 4, where x represents the input feature.

$$f(x) = max(0, x) \tag{4}$$

The ReLU activation function is used in the hidden layers because of its range, spanning from 0 to infinity. It addresses the gradient issue and does not saturate for positive inputs. However, for all negative input values, the output is 0 which leads to the Dying ReLU problem, where the output of the neurons is consistently zero, resulting in zero gradients [6].

Leaky ReLU Activation Function

It is an adaptation of the standard ReLU function, intended to mitigate the issue of "Dying ReLU." It is mathematically defined in Eq. 5, where x represents the input features, and 'a' is typically a small negative value, often chosen between 0.01 and 0.1.

$$f(x) = max(a * x, x) \tag{5}$$

In the hidden layers, the leaky ReLU activation function is utilized, characterized by its range extending from −infinity to +infinity. It prevents the "Dying Relu" problem and improves the performance of deep learning models [6].

4.7 *Analyze the Results*

At the concluding stage of analysis, we meticulously investigate the impact of distinct activation functions, namely Tanh, Relu, Identity, and Leaky Relu, separately in each hidden layer of the model. Additionally, we explore a configuration in which we combine the Tanh, ReLU, and Leaky ReLU as activation functions within different hidden layers. We select the logistic activation function for the output layer. This comprehensive evaluation has been done to gain insights into which combination of activation functions optimizes the model's ability to effectively distinguish between benign and malicious applications. The model's performance is assessed by using four vital parameters, which are Accuracy in Eq. (6), Precision in Eq. (7), Recall in Eq. (8), and F-score in Eq. (9).

$$Accuracy = \frac{True\ Positives + True\ Negatives}{True\ Positives + False\ Positives + True\ Negatives + False\ Negatives} \tag{6}$$

$$Precision = \frac{True\ Positives}{True\ Positives + False\ Positives} \tag{7}$$

$$Recall = \frac{True\ Positives}{True\ Positives + False\ Negatives} \tag{8}$$

$$F1Score = 2 \times \frac{Precision \times Recall}{Precision + Recall} \tag{9}$$

5 Experimental Results

The proposed model has been implemented in Google Colaboratory with Python programming. The research process adopts a two-step approach, involving training and testing. To ensure effective model training, the dataset is divided into two primary components: Features (referred to as X) and the corresponding labels (referred to as y), representing benign and malicious applications. The X component constitutes the feature vector upon which the model will be trained and tested. For a reliable model evaluation, 70% of the data has been allocated to the training set, with the remaining 30% for testing.

The core of the proposed work centers around exploring various activation functions in the MLP classifier. We systematically vary and apply different combinations of activation functions in the hidden layers of MLP classifiers. Specifically, in this paper, the impact of activation functions, including Identity, Tanh, Relu, and LeakyRelu are examined on the model's performance. Following the training phase, the MLP classifier undergoes a thorough evaluation using performance metrics, which encompass accuracy, precision, recall, and the F1 score.

5.1 *Hyper-Parameters Analysis*

Table 3 describes the hyper-parameters used in the MLP which has 3 hidden layers. It's scheduled for training over 200 times. The learning rate is set to 0.01, which regulates the pace of model updates. The model uses Stochastic Gradient Descent (SGD) as its optimization technique. Unlike traditional methods that update weights after processing the entire dataset, SGD tweaks them after each individual data point. This aids in bypassing local minima. The Binary Cross-Entropy (BCELoss) loss function is used in this model for binary classification. It measures the difference between the true labels and the predicted probabilities.

Table 3 Hyper-parameters

Hyper-parameters	Values
Activation functions	Identity, Tanh, ReLU, LeakyReLU
No. of hidden layers	3
No. of ıterations	200
Learning rate (alpha)	0.01

5.2 *Comparative Analysis of Activation Functions in Experimental Results*

Table 4 shows the experimental results of applying common activation functions such as Identity, Tanh, Relu, and Leaky ReLU in all hidden layers of MLP. Overall, each configuration produced high-performance metrics, with accuracy rates all above 97%. However, the Leaky ReLU activation function led to the highest accuracy of 98.22% and F-Score of 97.52%. This is because it avoids the “dying ReLU” issue to ensure that neurons continue to learn by introducing a small, non-zero gradient for negative inputs.

Table 5 shows the experimental outcomes of employing different combinations of activation functions in an MLP with three hidden layers of MLP. It has been observed that applying the Leaky ReLU in the first hidden layer, ReLU in the second hidden layer, and Tanh activation function in the third hidden layer exhibits their best overall performance in terms of accuracy (97.91%) and F-Score (97.08%), making it the strongest combination for detection of malicious applications.

The reason for the high accuracy and F-Score in the “Leaky ReLU + ReLU + Tanh” combination is attributed to the unique range each activation function operates within. Leaky ReLU has a range of −infinity to +infinity, allowing it to retain a broad spectrum of values, while ReLU has from 0 to infinity, and Tanh from −1 to 1. When Tanh is used as the first-layer activation function, it can potentially constrict the range of results. However, having Leaky ReLU in the first layer followed by ReLU in the second layer ensures a vast range, possibly contributing to better outcomes.

Table 4 Experimental results with common activation functions in hidden layers

Activation functions (in MLP)	Accuracy	Precision	Recall	F-score
Identity + ıdentity + ıdentity	0.9791	0.9714	0.9702	0.9707
Tanh + Tanh + Tanh	0.9807	0.9727	0.9733	0.9730
ReLU + ReLU + ReLU	0.9816	0.9739	0.9745	0.9742
LeakyReLU + LeakyReLU + Leaky ReLU	**0.9822**	0.9716	0.9789	**0.9752**

Table 5 Experimental results with varied activation functions combinations in hidden layers

Activation functions (in MLP)	Accuracy	Precision	Recall	F-score
Tanh + ReLU + Leaky ReLU	0.9776	0.9695	0.9677	0.9686
Leaky ReLU + ReLU + Tanh	**0.9791**	0.9702	0.9714	**0.9708**
Leaky ReLU + Tanh + ReLU	0.9784	0.9713	0.9683	0.9698

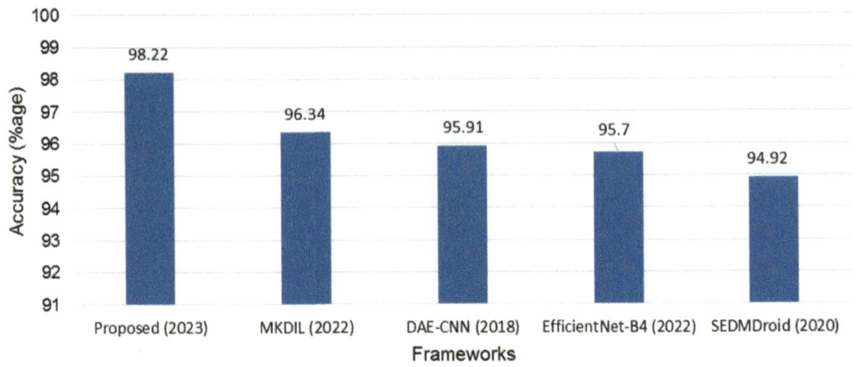

Fig. 3 Comparison of the proposed framework with the existing frameworks

5.3 Results of the Proposed Framework Compared with Existing Frameworks

In this section, we conduct a comparison between the outcomes of our proposed framework and several existing frameworks used for the detection of Android malicious applications. Figure 3 illustrates that our proposed framework outperformed the alternatives, achieving the highest accuracy of 98.22% when compared to the existing frameworks such as MKDIL [16], DAE-CNN [10], EfficientNet-B4 [15], and SEDMDroid [17] in the task of malicious application detection.

6 Conclusion

In this research paper, a framework based on the MLP model was introduced for the detection of malicious Android applications. The efficacy of the framework was assessed by experimenting with various combinations of activation functions in the hidden layers of the MLP. The experimental findings consistently demonstrated that using Leaky ReLU in all hidden layers of the MLP outperformed other activation functions, including Identity, Tanh, and ReLU. More precisely, it was consistently observed that Leaky ReLU produced the best outcomes, delivering an accuracy of 98.22% and an F-Score of 97.52%.

However, when considering diverse activation functions in different layers, a combination of Leaky ReLU in the first layer, ReLU in the second layer, and Tanh in the third layer (referred to as the Leaky ReLU + ReLU + Tanh ensemble) produced the best results, with an accuracy of 97.91% and an F-Score of 97.08%.

Overall, the common activation function is the most effective and gives noteworthy results over the performance of mixed combinations. Our findings will be instrumental in choosing the correct activation function that can affect the susceptibility of the network to certain features of malicious applications.

Moreover, our proposed framework will be significant for smartphone security and could have far-reaching implications in safeguarding users' devices from malicious applications. The future scope of our research involves optimizing Android malware detection through diverse activation functions, including Exponential Linear Unit (ELU), Scaled Exponential Linear Unit (SELU), Parametric Rectified Linear Unit (PReLU), and Gaussian Error Linear Unit (GELU).

References

1. Google Play Store: number of apps 2023. Statista. https://www.statista.com/statistics/266210/number-of-available-applications-in-the-google-play-store/. Accessed 14 Oct 2023
2. Kivva, A.: IT threat evolution in Q2 2023. Mobile statistics. Kaspersky, 30 August 2023. https://securelist.com/it-threat-evolution-q2-2023-mobile-statistics/110427/. Accessed 02 Nov 2023
3. Quick Heal Technologies Limited. Quick heal annual threat report 2022. https://www.quickheal.co.in/documents/threat-report/Quick-Heal-Annual-Threat-Report-2022.pdf. Accessed 18 Oct 2023
4. Shishkova, T.: IT threat evolution in Q1 2022. Mobile statistics. Kaspersky. https://securelist.com/it-threat-evolution-in-q1-2022-mobile-statistics/106589/. Accessed 27 May 2023
5. Number of malware attacks per year 2022. Statista. https://www.statista.com/statistics/873097/malware-attacks-per-year-worldwide/. Accessed 01 Nov 2023
6. Neural networks and deep learning. Coursera. https://www.coursera.org/learn/neural-networks-deep-learning/. Accessed 20 Oct 2023
7. Alomari, E.S., et al.: Malware detection using deep learning and correlation-based feature selection. Symmetry **15**(1), 123 (2023)
8. Hwang, C., Hwang, J., Kwak, J., Lee, T.: Platform-independent malware analysis applicable to windows and Linux environments. Electronics **9**(5), 793 (2020)
9. Liu, T., Li, Z., Long, H., Bilal, A.: NT-GNN: network traffic graph for 5G mobile IoT android malware detection. Electronics **12**(4), 789 (2023)
10. Wang, W., Zhao, M., Wang, J.: Effective android malware detection with a hybrid model based on deep autoencoder and convolutional neural network. J. Ambient. Intell. Humaniz. Comput. **10**(8), 3035–3043 (2018)
11. Wei, S., Zhang, Z., Li, S., Jiang, P.: Calibrating network traffic with one-dimensional convolutional neural network with autoencoder and independent recurrent neural network for mobile malware detection. Secur. Commun. Netw. **2021**, 1–10 (2021)
12. Bawazeer, O., Helmy, T., Alhadhrami, S.: Malware detection using machine learning algorithms based on hardware performance counters: analysis and simulation. J. Phys. Conf. Ser. **1962**(1), 012010 (2021)
13. Sagar, G.V.R.: Malware detection using optimized activation-based deep belief network: an application on internet of things. J. Inf. Knowl. Manag. **18**(04), 1950042 (2019)
14. Sharma, S., Krishna, R., Kumar, R.: An ensemble-based supervised machine learning framework for android ransomware detection. Int. Arab J. Inf. Technol. **18**(3A) (2021)
15. Yadav, P., Menon, N., Ravi, V., Vishvanathan, S., Pham, T.D.: EfficientNet convolutional neural networks-based android malware detection. Comput. Secur. **115**, 102622 (2022)
16. Xia, M., Xu, Z., Zhu, H.: A novel knowledge distillation framework with intermediate loss for android malware detection. In: 2022 IEEE Asia-Pacific Conference on Computer Science and Data Engineering (CSDE), December 2022 (2022)

17. Zhu, H., Li, Y., Li, R., Li, J., You, Z., Song, H.: SEDMDroid: an enhanced stacking ensemble framework for android malware detection. IEEE Trans Netw Sci Eng **8**(2), 984–994 (2021)
18. Dhanya, K.A., Vajipayajula, S., Srinivasan, K., Tibrewal, A., Kumar, T.S., Kumar, T.G.: Detection of network attacks using machine learning and deep learning models. Procedia Comput Sci **218**, 57–66 (2023)
19. Sharma, S., Kumar, N., Kumar, R., Challa, R.K.: The paradox of choice: investigating selection strategies for android malware datasets using a machine-learning approach. Commun. Assoc. Inf. Syst. **46**(1), 619–637 (2020)
20. Arp, D., Spreitzenbarth, M., Hübner, M., Gascon, H., Rieck, K.: Drebin: Effective and explainable detection of android malware in your pocket. In: Proceedings 2014 Network and Distributed System Security Symposium (2014)
21. Tumbleson, C.: Apktool. https://apktool.org/. Accessed 01 Nov 2023

ANNDroid: Android Malware Classification and Detection Techniques

Pooja, Arvind Mahindru, and Pardeep Kumar Arora

Abstract As modernization sweeps across the globe, digital technology is playing an increasingly important role in every aspect of our lives. A computer or smartphone has become integral to personal, social, and professional survival as a way to create, express, communicate, collaborate, and access information. A constant stream of Android malware has also emerged as legitimate use of smartphones has increased. This gives motivation to malware developers to earn extra money illegally. Other minor motivations include activism, practical jokes, infiltration, cybercrime, and serious crimes including privacy breaches. In our paper, we will list the numerous Android malware detection approaches and techniques available and propose the benefits and drawbacks of the various static, dynamic, and hybrid techniques proposed by many researchers in past years.

Keywords Android system · Malware · Smartphone · Malware detection

1 Introduction

Numerous smartphone applications are available for social networks, business, finance, multimedia, entertainment, and education. Approximately 2.59 million applications are accessible on the Google Play store[1]. The combined user base of all of these is 6.92 billion. The enormous growth in the use of Android-based smartphones has been noticed from the year 2008 to 2023. Android smartphone Operating system has captured almost 86.6% marketplace [1]. With the advent use of Android smartphones, malware developers are also getting active to make them monetarily from this intimate user data. Therefore, privacy and security of user's data is the most concerning issue nowadays. Vulnerability to malware motivates the need for

Pooja (✉) · A. Mahindru
Department of CSA, DAV University, Jalandhar, India
e-mail: poojamehandru715@gmail.com

P. K. Arora
Kanya Maha Vidyalaya, Jalandhar, India

B. Soni et al. (eds.), *Beyond Artificial Intelligence*, Lecture Notes in Networks and Systems 1326, https://doi.org/10.1007/978-981-96-4170-3_45

efficient techniques to detect Android malware. As a side benefit, the creation of dangerous programs has skyrocketed, and one new Android malware application was discovered every 10 s [2]. Furthermore, some of these malicious executable moves throughout the network without the owner of the system being aware of it. They also compromise the device's security policy in terms of data confidentiality, reliability, integrity, and availability. With more than 3.6 billion Android devices in use, Google must offer the finest user safeguards on a large scale [3].

To safeguard its consumers, Google Play Protect regularly checks billions of apps. We observed that Android tools for detecting malware: static detection, dynamic detection, and hybrid detection are present. Static detection relies on suspicious code analysis without the Android application being started. While it is possible to achieve high code coverage, it has numerous challenges, particularly dynamic code loading and code obfuscation. Whereas Dynamic detection includes implementing the code and analyzing the Android application. It reveals dangers that are hard to pinpoint across static analysis, but dynamic detection demands a considerable amount of processing power and is a lengthy process. To identify behavior or potentially dangerous functionality in smartphone applications, hybrid detection combines information obtained through dynamic analysis with a static analysis approach [4].

The rest of the paper is organized as follows. In Sect. 2, we discuss some work done in the field of Android malware detection based on static, dynamic, and hybrid analysis using machine learning and deep learning techniques. Section 3, describes the existing frameworks for Android malware classification. Also, we will find the accuracy provided by various models and see the dataset used by researchers. Section 4 will conclude our paper and describe the work that is to be done in the future.

2 Related Work

Several studies have been done to detect Android malware and to distinguish malware from benign ware, as a result, numerous plans have been put out to address this problem. Static analysis techniques are more commonly used techniques to detect malware but to evade such a system an attacker could either increase the size of the manifest file or change the number of permissions requested in the manifest file. To overcome the issues highlighted by static analysis detection and classification techniques, Chandok et al. [5] Proposed a model, a mixture of Static and dynamic analysis approaches called Dro-Mal Detector that uses a function call graph of an application to track malware. They use the Support Vector Classifier and grid search algorithm of an artificial neural network for training after converting call graph embeddings and permissions to reduced dimensional feature vectors. Rathore et al. [6] Proposed malware analysis and detection as a machine learning and deep learning problem. Classical defense mechanisms (like signature-based malware detection) used by antivirus will fail to cope with new-age malware challenges. So, they proposed models like cross-validation, fixing class imbalance problems, etc. using machine learning and deep learning techniques. Trinius et al. [7] introduced a Malware Instruction Set for monitoring the behavior of malicious software. (MIST). The representation was specially optimized for behavior analysis utilizing data mining and machine learning techniques. Rieck et al. [8] put forward an architecture for the automatic analysis of malware behavior using machine learning. They proposed a framework for the automatic analysis of malware behavior jointly using clustering and classification. Cai et al. [9] presented a sustainable Android malware detector to recognize new infections that once trained on a dataset, can continue to effectively detect new malware without retraining. They employed behavior analysis to determine the differences between beneficial and undesirable programs by analyzing application runtime motion. Mahindru and Snagl [10] introduced a machine learning-based framework called MLDroid—a web-based framework to identify Android malware using dynamic analysis. The researchers conducted experiments on a total of five million applications to extract various important features. Karbab et al. [11] Proposed the MalDozer framework in 2018 that uses deep learning to categorize episodes to automatically find malware alongside private documents on smartphones running Android and also on IoT Devices which includes 1000–33,000 malware samples and 38,000 benign samples. Faruki et al. [12] proposed an AndroSimilar, an automatic signature generation approach using a feature extraction method (capturing code homogeneity among variants of known apps) to find novel and previously unidentified malware variants as Pattern-based techniques are not effective against unseen and already known malware variants. Dimitrios et al. [13] suggested a system that combines host and cloud-based Intrusion Detection systems. Various types of unresolved problems with mobile Intrusion Detection systems were handled by the author. The author suggested the framework as well as architecture of mobile Intrusion Detection systems to enable it for the deployment of defensive tools on the host and in the cloud, enabling them to cooperate. Martna et al. [14] presented *OmniDroid* which is a large and comprehensive dataset of features extracted from near about 22,000 real malware and good ware samples, aiming to help anti-malware tools creators and researchers when improving, or developing, new mechanisms and tools for Android malware detection through hybrid features fusion and ensemble classifiers. Sanz et al. [15] Presented PUMA, a new method for detecting malicious Android applications employing the permission usage of each application. Using these features, they train machine-learning models to detect whether an application is malware or not. They collected a total of 4,301 samples using Virus Total Malware Intelligence Services. However, they carried out a duplication reduction stage in which they eliminated the duplicate samples from the dataset. Finally, they employed a total of 249, which Lookout7 claims accounts for 54% of all malware samples. Mahindru et al. [16] Proposed a Dynamic Permissions based Model using Machine Learning Techniques to detect Android Malware. The Authors evaluated the model on dataset consisting of 11,000 applications and achieved an accuracy of 99.7%. Alam et al. [17] in their study state that 37% of available Android applications contain Native code so the static analysis techniques are not able to detect the present malware effectively so, they proposed a model named DroidNative. Malware can be found in both bytecode and native code, according to DroidNative, an Android malware detection technology that runs at the native code level. Static analysis of the native code is done by DroidNative, which focuses on control flow patterns that aren't greatly altered by obfuscations. Mahindru and Sangal [18] proposed FSDroid, which is a feature selection technique to detect malware from Android using the Machine Learning Technique. FSDroid achieves an accuracy of 98.8% in detecting malware. A comparative summary of Android malware identification and analysis tools/techniques Proposed by some researchers is shown in following Table 1.

3 Existing Framework

To create an efficient malware detection method, malware analysis is required. The continuous rise in Android malware demands tools and procedures for dissecting malware samples with minimal or no human interference. We will cover several existing frameworks based on static, dynamic, and hybrid analysis in Sect. 3.1, dataset analysis in Sect. 3.2, and result analysis comparison of certain proposed models by researchers in Sect. 3.3.

3.1 Existing Frameworks Based on Static, Dynamic, and Hybrid Analysis

Researchers from business and academia have put forth frameworks and approaches for analyzing and monitoring Android malware threats [19]. The following three parameters can be used to describe these approaches:

(a) Goal of the proposed approach
(b) Methodology applied to achieve the above goals
(c) Implementation of the solutions already discussed.

In the following Table 2, we discuss some existing frameworks and their methodologies to classify malware.

3.2 Dataset Analysis

To make general conclusions the researchers evaluated their proposed models for the two types of samples as shown in Table 3.

Table 1 Comparative summary of Android malware identification and analysis tools/techniques

Ref. No.	Objective			Methodology			
	Analysis	Detection	Signature based	Behavior-based	Permission-based	Machine learning algorithm	Deep learning technique
Rathore et al. [6]	√	√	√			√	√
Trinius et al. [7]	√			√		√	
Rieck et al. [8]	√			√		√	
Cai et al. [9]		√		√		√	
Mahindru et al. [10]	√	√		√		√	
Karbab et al. [11]		√					√
Faruki et al. [12]		√	√			√	
Sanz et al. [15]		√			√	√	

Table 2 Existing frameworks based on static, dynamic, and hybrid analysis

Study	Year	Model name	Methodology	Tools & techniques
Static analysis				
Rieck et al. [20]	2014	DREBIN	Feature extraction Embedded in vector space, learning based detection	Machine learning (LSV)
Alam et al. [17]	2017	DroidNative	Semantic-based signature	Static analysis on dynamically loaded native code malware products
Li et al. [21]	2018	SIGPID	Data pruning, (permission ranking, support based permission)	Machine learning algorithms
Das et al. [22]	2019	NIDS	DDoS intrusion detection	Machine learning ensemble classifier
Dynamic analysis				
Willems et al. [23]	2007	CWSandbox	Behavior-based	Executes malware samples in a simulated environment
Trinius et al. [7]	2009	Malware instruction set (MIST)	Behavior-based Meta language	Data Mining and machine learning tools and techniques
Zhang et al. [24]	2013	VetDroid	Permission-based behavior analysis method	Prototype system
Tam et al. [25]	2015	CopperDroid	VMI (virtual machine introspection)-based system	Call-centric analysis and stimulation technique
Fan et al. [26]	2017	DAPASA	Permission-based approaches and API-based approaches	Sensitive subgraph analysis Machine learning algorithm
Hybrid analysis				
Zhou et al. [27]	2012	DroidRanger	Footprint-based detection engine Permission-based behavioral footprint method	FDP approach Disassembly and opcode analysis
Martn et al. [14]	2019	The AndroPyTool framework and the OmniDroid dataset	Hybrid features fusion and ensemble classifier	Machine learning

Table 3 Number of applications used for model analysis

Model name	Malware	Benign ware	Total
DreBin	5560	117,893	123,453
MalDozer	1000–33,000	38,000	1000–71,000
OmniDroid	12,158	22,000	34,158
MamaDroid	2694	2723	5417
DroidNative	3378	2140	5518
SigPid	5494	310,926	316,420
EPIR	4868	310 926	3,15,794
DAPASA	2,551	44,921	47,472

3.3 *Result Analysis Comparison of Some Proposed Models by Researchers*

We studied research papers from various journals, proceedings, and a few research articles. The researchers try to minimize the number of false alarm rates (FAR or FPR) and to achieve more accuracy in detecting malware. False alarm rate FAR= FP/(FP+TN) where FP is the number of False positives and TN denotes the total of True negatives. When two unrelated objects share the same features, they become weak leading to a false alarm rate. Table 4 summarises the malware detection rate accuracy achieved and the false alarm rate by some researchers in their proposed models.

Table 4 Accuracy and false alarm rate

Author	Method	Detection rate accuracy (%)	False alarm rate
Wang et al. [28]	EPIR	94.62%	0.6%
Li [21]	SigPid	90% (total accuracy) 93.62% of malware in the dataset 91.4% new malware	1%
Arp et al. [27]	SMD using embedded call graph	89%	1%
Karbab et al. [11]	MalDozer	96–99%	0.06–2%
Reick et al. [20]	Drebin	94%	1%
Martn et al. [14]	OmniDroid	89.2%	–
Faruki et al. [12]	AndroSimilar	99.4 (google play apps) 98.89 (third party apps)	0.95%
Fan et al. [26]	DAPASA	95%	0.7%
Yang et al. [17]	DroidNative	95.9%	–

4 Conclusion and Future Work

We found that the primary delivery platform for ubiquitous computing is Android. Unfortunately, using them also poses several security risks. Intelligent detection systems are now necessary due to the malware industry's exponential development of new and advanced threats [16]. Malware identification is becoming more challenging due to the evasion techniques used by malware authors. We also discovered that the majority of developers who employ static analysis, dynamic analysis, and hybrid analysis want to achieve higher accuracy and lower the cardinality of false positives. Many academicians and researchers have been using machine learning techniques and approaches to detect malware and achieve success too. However, to achieve success and developing a machine-learning model to detect the malware is not an easy task because it requires a lot of information. The major concern related to the existing technique is the security and data privacy of the users and collecting training data in a large central repository. These considerable issues related to machine learning techniques can be resolved by using Federated Learning in which training data can be dispersed to multiple users or clients which provides data privacy under the control of a central aggregator coordinator. In the future, we will propose a hybrid model based on federated learning and try to achieve more accuracy than the existing frameworks and analyze our model with a large dataset.

References

1. Liu, K., Xu, S.: A Review of Android Malware Detection Approaches Based on Machine Learning, p. 99. IEEE Access (2020)
2. https://www.crowdstrike.com/cybersecurity101/malware/types-of-malware/
3. https://www.businessofapps.com/data/android-statistics/
4. Singh, R.: An overview of android operating system and its security. J. Eng. Res. Appl. **4**(2) (Version 1):519–521 (2014)
5. Chandok, A., Verma, A., Gupta, R.: Dro-mal detector: a novel method of android malware detection. In: International Conference for Emerging Technology (INCET), 3rd edn. IEEE Xplore, 15 July 2022 (2022)
6. Rathore, H., Agarwal, S., Sahay, S.K., Sewak, M.: Malware Detection Using Machine Learning and Deep Learning, Big Data Analytics: 6th International Conference, BDA 2018, Warangal, India, 2018. LNCS **11297**, 402–411 (2018)
7. Trinius, P., Willems, C., Holz, T., Rieck, K.: A malware instruction set for behavior-based analysis (2009)
8. Rieck, K., Trinius, P., Willems, C., Holz, T.: Automatic analysis of malware behavior using machine learning. J. Comput. Secur. **19**(4), 639–668 (2011)
9. Cai H., Jenkins J.: Towards sustainable android malware detection. In: Proceedings of the 40th International Security and Communication Networks 29 Conference on Software Engineering: Companion Proceeedings, Gothenburg, Sweden, pp. 350–351, May 2018 (2018)
10. Mahindru, A., Sangal, A.L.: MLDroid-framework for android malware detection using machine learning techniques. Neural Comput. Appl. **33**(10), 5183–5240 (2021)
11. Karbab, E.B., Debbabi, M., Derhab, A., Mouheb, D.: Maldozer: automatic framework for android malware detection using deep learning. Digit. Investig. **24**, S48–S59 (2018)

12. Faruki, P., Ganmoor, V., Laxmi, V., Gaur, M.S., Bharmal, A.: Androsimilar: robust statistical feature signature for android malware detection. In: Proceedings of the 6th International Conference on Security of Information and Networks, pp. 152–159. ACM, Aksaray, Turkey, SIN'13, November 2013 (2013)
13. Damopoulos, D., Kambourakis, G., Portokalidis, G.: The best of both worlds: a framework for the synergistic operation of host and cloud anomaly-based ids for smartphones. In: Proceedings of the Seventh European Workshop on System Security, Article No. 6, pp. 1–6, April 2014. ACM Digital Library (2014)
14. Martna, A., Lara-Cabrera, R.: Android malware detection through hybrid features fusion and ensemble classifiers: the AndroPyTool framework and the OmniDroid dataset, vol. 52, pp. 128–142, December 2019. Science Direct (2019)
15. Sanz, B., Santos, I., Laorden, C., Ugarte-Pedrero, X., Bringas, P.G., lvarez, G.A.: PUMA: permission usage to detect malware in android. In: International Joint Conference CISIS12-ICEUTE´ 12-SOCO´ 12 Special Sessions, pp. 289–298. Springer (2013)
16. Mahindru, A., Singh, P.: Dynamic permissions based android malware detection using machine learning techniques. In: Proceedings of the 10th Innovations in Software Engineering Conference, pp. 202–210, February 2017 (2017)
17. Shahid, A., Qu, Z., Riley, R., Yan, C., Vaibhav, R.: DroidNative: automating and optimizing detection of android native code malware variants. Comput. Secur. **65**, 230–246 (2017)
18. Mahindru, A., Sangal, A.L.: FSDroid: a feature selection technique to detect malware from android using machine learning techniques. Multimed. Tools Appl. **80**, 13271–13323 (2021)
19. Orthacker, C., Teufl, P., Kraxberger, S.: Android security permissions–can we trust them? In: Proceedings of the international conference on security and privacy in mobile information and communication systems, pp. 40–51. Springer, Aalborg, Denmark (2011)
20. Rieck, K.: DREBIN: effective and explainable detection of android malware in your pocket (2014)
21. Li, J., Sun, L., Yan, Q., Li, Z., Srisaan, W., Ye, H.: Significant permission identification for machine-learning-based android malware detection. IEEE Trans. Ind. Inform. **14**(7), 3216–3225 (2018)
22. Das, S., Mahfouz, A.M., Deepak, V., Sajjan, S.: DDoS intrusion detection through machine learning ensemble. In: 2019 IEEE 19th International Conference on Software Quality, Reliability and Security Companion (QRS-C), 07 October 2019 (2019)
23. Willems, C., Holz, T., Freiling, F.: Toward automated dynamic malware analysis using CWSandbox. IEEE Secur. & Priv. **5**(2), 32–39 (2007)
24. Zhang, Y., Yang, M., Xu, B.: Vetting undesirable behaviors in Android apps with permission use analysis. In: CCS '13: Proceedings of the 2013 ACM SIGSAC Conference on Computer & Communications Security, pp. 611–622, November 2013 (2013)
25. Tam, K., Khan, S., Fattori, A., Cavallaro, L.: CopperDroid: automatic reconstruction of android malware behaviors. In: Proceedings 2015 Network and Distributed System Security Symposium, 7 February 2015 (2015)
26. Fan, M., Liu, J., Wang, W., Liu, T.: DAPASA: detecting android piggybacked apps through sensitive subgraph analysis. IEEE Trans. Inform. Forensics and Security, Vol. 12, pp. 1772–1785, 27 March 2017.
27. Zhou, Y., Jiang, X.: Dissecting android malware: characterization and evolution. In: IEEE Symposium on Security and Privacy, 09 July 2012 (2012)
28. Wang, W., Xing, W., Feng, D., Liu, J., Han, Z., Zhang, X.: Exploring permission-induced risk in android applications for malicious application detection. IEEE Trans. Inf. Forensics Secur. **9**(11), 1869–1882 (2014)

System Identification of Batch Reactor Using Machine Learning Techniques

Sivakumar Rajendran, Gowsic Kandaswamy, and Thirunavukkarasu Indiran

Abstract This article aim to develop the Machine Learning models to predict the temperature of a batch reactor using machine learning techniques. Various machine learning techniques like Linear Regression, Decision Tree model, Random Forest, and Support Vector Machine are used to model the system. The open loop data of reactor is used to predict the best model and evaluate the performance of these models in predicting the temperature of the reactor. The comparison of each model and select the most precise model for predicting the temperature of the batch reactor. The proposed approach has the ability to significantly improve the model accuracy and accurate prediction of the reactor temperature, which can lead to more effective process control and optimization. As a case study, the input–output data of the highly nonlinear batch reactor is considered for the model fit. The machine learning models can be further used for the predictive controller design for validating on an experimental setup. Further, these models will be used for the Nonlinear Model Predictive Controller (NMPC) design via Python and validation using Jetson Orin Nano board.

Keywords Batch reactor · Machine learning algorithms · Non-linear system · System modeling · Model predictive controller

1 Introduction

Machine learning techniques may result with fast and accurate predictions of the batch reactor temperature. These techniques can learn the underlying patterns in the data and use them to make predictions. This article investigates the use of different

S. Rajendran · G. Kandaswamy
Department of Electronics and Instrumentation Engineering, Saveetha Institute of Medical and Technical Sciences, Chennai, Tamil Nadu, India

T. Indiran (✉)
Department of Instrumentation and Control Engineering, Manipal Institute of Technology, Manipal Academy of Higher Education, Manipal, Karnataka 576 104, India
e-mail: it.arasu@manipal.edu

B. Soni et al. (eds.), *Beyond Artificial Intelligence*, Lecture Notes in Networks and Systems 1326, https://doi.org/10.1007/978-981-96-4170-3_46

machine learning algorithms for predicting the temperature of a batch reactor. Specifically, we explore the performance of linear regression, decision tree model, random forest regressor model, and support vector machine.

The input–output data set is generated using first principle equations and use them to train the machine learning models. The data is collected from different input variables such as reactor volume, heat transfer area, reaction rate, and initial temperature. The output variable is the temperature of the batch reactor.

To evaluate the model accuracy of the machine learning models, we use several performance criteria [1–3], which include Mean-Squared Error, R-squared, Mean Absolute Error, Root Mean Squared Error, and Adjusted R-Squared.

After evaluating the accuracy of each machine learning model using the selected evaluation metrics, the random forest regressor model provided the best prediction results for the batch reactor temperature. This model outperformed the other models in terms of accuracy, robustness, and computational efficiency.

2 Literature on Machine Learning Models and Controller Design

Meenal et al. [4] modelled the global solar radiation (GSR), weather data which are highly nonlinear and dynamic throughout the year. Those models are very much needed for various engineering applications. In this paper, authors have used the Support Vector Machine (SVM), Random Forest Algorithms for various data set for obtaining the model with minimum error in terms of Mean Square Error and R^2 Score. Further, the identified models are used for predictions.

In [5], an attempt has been made to map the solar radiation using the SVM technique. Data set from a scientific community is taken and the life cycle cost analysis of solar PV in India is calculated. Also, the prediction of GSR for the different cities are predicted. Also, they presented the cost estimation and payback period in the paper.

Wu et al. [6] presented the machine learning based model of the Continuous Stirred Tank Reactor [CSTR] using the Recurrent Neural Networks. It is learnt that the Long-Short Term Memory (LSTM) along with the neural network for the dynamic system identification. In this paper, authors used a Lyapunov based MPC for the empirical model developed based on the RNN for a nonlinear chemical process. The data sets are generated using the first principle model of the CSTR. Finally, the Ensemble regression model is obtained using the RNN for simulation studies in closed loop.

3 Machine Learning Methods

Linear regression is a statistical algorithm that is widely used for predicting future outcomes based on past data. It is a simple yet powerful tool that can be applied to a wide range of fields. The basic idea behind linear regression is to find the best-fitting line through the data points that minimizes the difference between the predicted values and the actual values.

Linear regression is a predictive modeling algorithm that uses a linear equation to predict the value of a dependent variable based on one or more independent variables. Random Forest is an ensemble learning algorithm that combines multiple decision trees to generate predictions. Each decision tree in a random forest is built on a random subset of the training data and a random subset of the features. The output of a random forest is the average of the predictions made by all decision trees in the forest. This approach reduces the risk of overfitting and increases the stability and accuracy of the predictions. The decision tree algorithm is a popular tool for temperature prediction due to its ability to handle non-linear relationships between features. It works by recursively splitting the data into smaller and smaller subsets based on the values of the features. The algorithm splits the data based on the feature that provides the most information gain. Information gain measures the reduction in entropy, which is a measure of the uncertainty or randomness of the data. The final result is a tree-like structure where each leaf node represents a prediction.

Support Vector Machine (SVM) is a machine learning algorithm that is widely used for classification and regression tasks. SVM is a supervised machine learning algorithm that works by finding the best separating hyperplane between two classes of data points. In the context of temperature prediction, the algorithm can be used to find the best hyperplane that separates the temperature data into two classes—normal and abnormal. This makes it an ideal algorithm for predicting temperatures that are outside of the normal range. Once the model is trained, it can be used to predict the temperature of the reactor based on current process parameters. If the predicted temperature falls outside of the normal range, the algorithm can trigger an alarm to alert plant operators to take corrective action.

4 Batch Reactor—System Dynamics

Batch reactors are highly nonlinear process and it follows a trajectory tracking in general for the final product. There are two different physical structure of batch reactors are used worldwide. Initially, the solid steel structure of the batch reactors in different feed capacities with external cooling jacket arrangement and internal heating filament. Due to the chemical reactions, the heater filament may get corroded and damages quickly. Other issue faced is the heat transfer rate, since the coolant water is flowing outside the jacket efficient heat transfer may not happen. Alternatively, one can use the circular coolant coil inside the reactor and ceramic band heater arrangements around the solid steel reactor. Mostly, in European countries, they go for the glass reactor with one external jacket, called as double glass reactor vessel. Based on the error sign, either the cold water or the hot water is circulated in the jacket to maintain the reactor temperature in-line with the trajectory tracking.

Figure 1 shows the experimental setup of a lab scale batch reactor with cooling flow station. Several other modeling techniques were already carried out for the same system using ARX, NARX, First principle method. Figure 2 shows the schematic of the batch reactor using the Jetson Nano 2 GB board. In this paper an attempt for ML based NMPC controller design via Python for Jetson Nano board is carried out.

The nonlinear first principle model is developed for the above batch reactor setup and several predictive controllers are validated for the trajectory tracking successfully. Apart from the nonlinear differential equations, data driven models such as ARX, NARX, Weiner models [8, 9] have also identified for the design and validation of various predictive controllers on the lab scale model experimentally for the trajectory tracking. Presently, the authors are working on the various Machine Learning based model for predictive controller design to improve the closed loop performance.

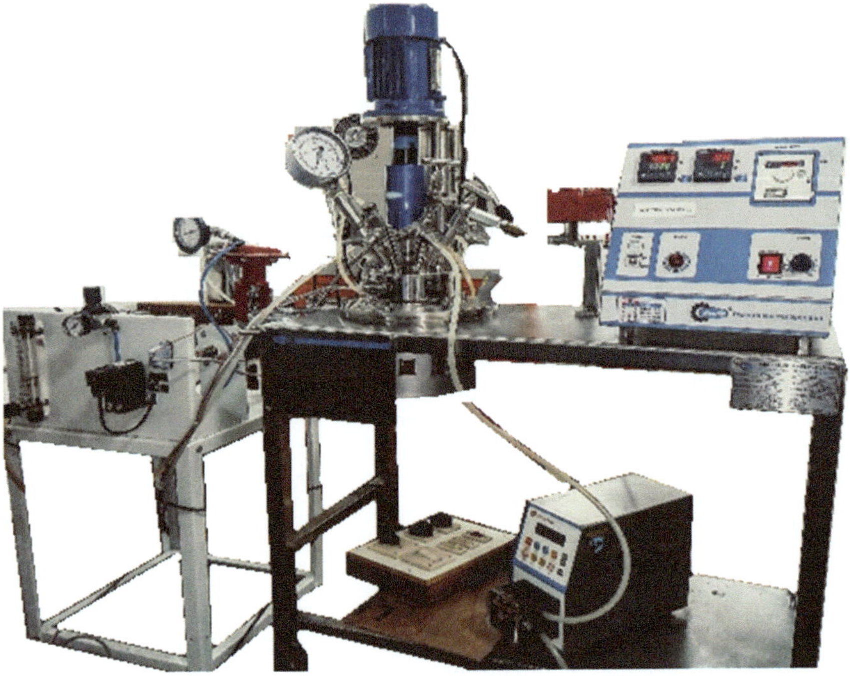

Fig. 1 Experimental setup of lab scale batch reactor in Dept. of ICE, MIT, Manipal

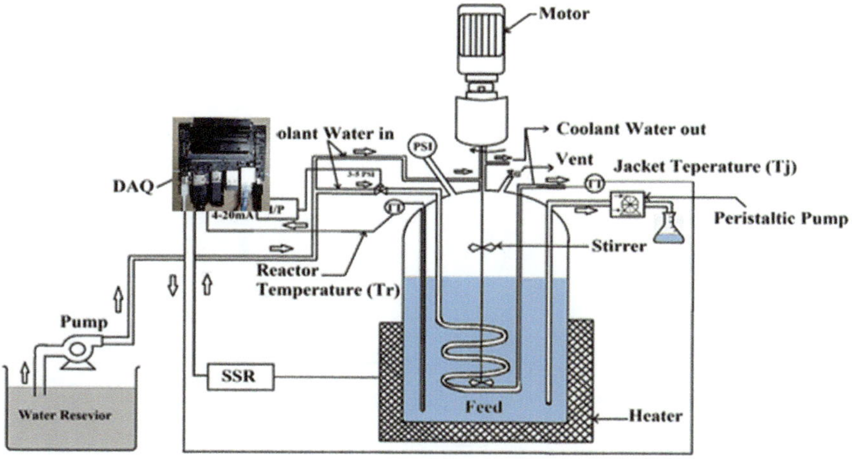

Fig. 2 Schematic diagram of batch reactor with Jetson Nano board as standalone controller [7]

5 Results and Discussion

Various machine learning algorithms have been employed to predict the reactor temperature. Initially, started with the Linear regression algorithm [1] to find the best-fitting line through the data points. Later on other algorithms like Random Forest, Decision Tree and Support Vector Machine are tried to predict the reactor and jacket temperature. The dataset is trained with target variables and correlated using Pearson correlation method. The predicted data are shown in Fig. 3.

Figure 3 prediction model using (a) Linear Regression (b) Random Forest (c) Decision Tree (d) Support Vector Machine. The various metrics for reactor and jacket temperature are listed in Tables 1 and 2 respectively. All the machine learning algorithms are simulated with Python software

The various metrics for reactor and jacket temperature are listed in Tables 1 and 2 respectively. All the machine learning algorithms are simulated with Python code to flash into the Jetson Nano board for integrating with NMPC code.

The Decision Tree and Random Forest models show the best performance in terms of accuracy and predictive power, with very low errors and high R-squared values. The Linear Regression and SVM models have relatively higher errors and lower R-squared values, indicating lower accuracy. Further optimization and tuning may be needed to improve the performance of the models, especially for the Linear Regression and SVM models.

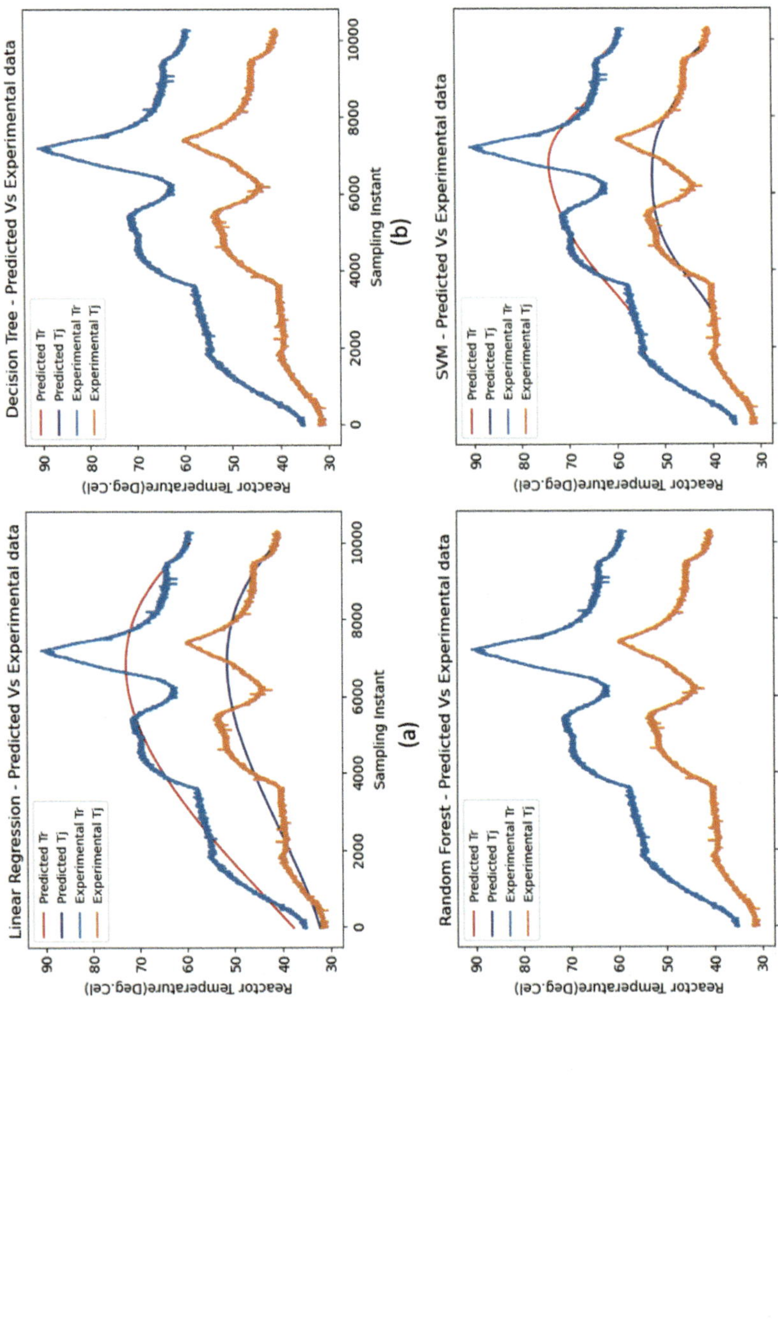

Fig. 3 Prediction model using **a** linear regression **b** random forest **c** decision tree **d**) support vector machine

Table 1 Evaluation metrics for reactor temperature

MODELS	MSE	R-squared	MAE	RMSE	Adjusted R-squared
Linear regression	20.8363	0.8310	3.2156	4.5646	0.8309
Decision tree	0.1262	0.9989	0.2585	0.3553	0.9989
Random forest	0.0899	0.9992	0.2212	0.2998	0.9992
Support vector machine	17.2575	0.8600	2.1884	4.1542	0.8600

Table 2 Evaluation metrics for jacket temperature

MODELS	MSE	R-squared	MAE	RMSE	Adjusted R-squared
Linear regression	8.1820	0.8114	2.2501	2.8604	0.8114
Decision tree	0.1102	0.9974	0.2397	0.3319	0.9974
Random forest	0.0805	0.9981	0.2034	0.2837	0.9981
Support vector machine	7.0000	0.9974	0.2397	0.3319	0.9974

6 Conclusion

System identification of batch reactor using machine learning techniques is one of the best recent techniques for understanding and predicting the behavior of complex systems. It can be used to identify the underlying structure of a system, to identify the parameters of a system, and to predict the behavior of a system. By combining machine learning techniques with system identification, it is possible to develop more accurate models of non-linear systems and to improve the accuracy of models compared to the other conventional methods. This can be used to improve the performance of systems and to reduce the cost of operation. The design of NMPC based on these models are in progress to validate using Jetson Nano board using Python code.

Acknowledgements Authors would like to thank Manipal Academy of Higher Education for supporting the work with seed money grants: ID 00000220 dated 1st Jan 2020 for the coolant flow station establishment and ID 00000860 dated 9th Oct. 2023 for the Machine Learning works.

References

1. Gautam, S., Kanakraj, S., Henry, A.: Computational approach using machine learning modelling for optimization of transesterification process for linseed biodiesel production. AIMS Bioeng. **9**(4), 319–336 (2022)
2. Groensfelder, T., Giebeler, F., Geupel, M., Schneider, D., Jaeger, R.: Application of machine learning procedures for mechanical system modelling: capabilities and caveats to prediction-accuracy. Adv. Model. Simul. Eng. Sci. **7**(26), 1–29 (2020)

3. Moin, A., Challenger, M., Badii, A., Günnemann, S.: A model-driven approach to machine learning and software modeling for the IoT. Softw. Syst. Model. **21**, 987–1014 (2022)
4. Meenal, R.: Weather prediction using random forest machine learning model. Indones. J. Electr. Eng. Comput. Sci. **22**(2), 1208–1215 (2021)
5. Meenal, R.: Solar mapping of India using support vector machine. IOP Conf. Ser.: J. Phys.: Conf. Ser. **1142** (2019). IOP Publishing. https://doi.org/10.1088/1742-6596/1142/1/012010
6. Wu, Z.: Machine learning-based predictive control of nonlinear processes. Part I: theory. AiCHE J. https://doi.org/10.1002/aic.16729 (2019)
7. Shettigar, J.P., Lochan, K., Jeppu, G., Palanki, S., Indiran, T.: Development and validation of advanced nonlinear predictive control algorithms for trajectory tracking in batch polymerization. ACS Omega **6**, 22857–22865 (2021)
8. Shettigar, J.P., Kumbhare, J., Yadav, E.S., Indiran, T.: Wiener neural network-based modeling and validation of generalized predictive control on a laboratory-scale batch reactor. ACS Omega **7**, 16341–16351 (2022)
9. Yadav, E.S., Shettigar, J.P., Poojary, S., Chokkadi, S., Jeppu, G., Indiran, T.: Data-driven modeling of a pilot plant batch reactor and validation of a nonlinear model predictive controller for dynamic temperature profile tracking. ACS Omega **6**, 16714–16721 (2021)

Comparative Analysis of Machine Learning Models for Traffic Prediction

Tanya Garg, Gurjinder Kaur, and Prashant Singh Rana

Abstract In the field of transportation planning and management, traffic prediction is a crucial problem where machine learning algorithms have demonstrated encouraging outcomes. The machine learning models for traffic prediction, including K-Nearest Neighbors (KNN), Random Forest, Long Short-Term Memory (LSTM), and Moving Average, are compared in this paper. The objective of this work is to determine how well these models performed in identifying traffic patterns and to appraise their accuracy by utilizing measures like Root Mean Square Error (RMSE), Mean Absolute Error (MAE), and R-squared. We preprocessed the data by transforming the datetime values to numerical representation using a historical traffic dataset. To train and evaluate the model, the dataset has been divided into training and testing sets. Moving Average functioned as a baseline but it is not sensitive enough to identify intricate patterns. These results emphasize how crucial it is to choose models according to task requirements and dataset peculiarities. Random Forest demonstrated its capability for precise traffic prediction, whereas temporal interdependence and local patterns could be captured by KNN and LSTM, respectively. A basic baseline has been shown using moving average. The performance and applicability of machine learning models for traffic prediction has been evaluated in this paper. The results can help the researchers in this area model depending on the features of the dataset and particular needs. With an RMSE of 3.83, MAE of 2.66, and R-squared of 0.96, Random Forest has shown to perform the best, successfully capturing all the relationships. With an RMSE of 21.87, MAE of 16.49, and negative R-squared, KNN performed worst. With an RMSE of 4.99, MAE of 2.90, and R-squared of 0.75, LSTM demonstrated a moderate level of performance.

T. Garg (✉) · P. S. Rana
Thapar Institute of Engineering and Technology, Patiala, Punjab 147001, India
e-mail: tanya.garg@thapar.edu

P. S. Rana
e-mail: prashant.singh@thapar.edu

G. Kaur
Sant Longowal Institute of Engineering and Technology, Longowal, Punjab, India

B. Soni et al. (eds.), *Beyond Artificial Intelligence*, Lecture Notes in Networks and Systems 1326, https://doi.org/10.1007/978-981-96-4170-3_47

Keywords Traffic prediction · Machine learning models · Comparative analysis · K-Nearest Neighbors (KNN) · Random forest · Long Short-Term Memory (LSTM) · Moving average · RMSE · MAE · R-squared

1 Introduction

In transportation planning, traffic management, and decision-making processes, traffic prediction is essential. Authorities can optimize traffic flow, lessen congestion, and increase overall transportation efficiency with the use of accurate and dependable traffic projections. Traffic prediction models have been created to produce forecasts that are more precise and accurate.

The application of machine learning models for traffic prediction is the main topic of this work. We specifically investigate the performance of four distinct models: Random Forest, Moving Average Statistical Method, K-Nearest Neighbors (KNN), and Long Short-Term Memory (LSTM). These models have demonstrated good results and have been widely employed in a variety of prediction applications.

One kind of recurrent neural network (RNN) that works well for sequential data analysis is the LSTM model. It is appropriate for traffic prediction because it can identify long-term trends and temporal dependencies in time series data. Conversely, a linear relationship between the input and target variables is assumed in the classical statistical model known as linear regression. A well-known ensemble learning system called Random Forest uses a combination of decision trees to generate predictions. Because traffic patterns are dynamic and complex, predicting traffic can be a difficult undertaking. Transportation authorities, urban planners, commuters, and other stakeholders can all benefit from accurate traffic prediction [1].

Our objective is to use machine learning techniques to create a dependable traffic forecast model in this study. Our goal is to identify patterns and trends in the historical traffic data that will aid in the forecasting of future traffic situations. Making educated choices about resource allocation, congestion control, and route planning can be aided by having accurate traffic predictions. We make use of a dataset that contains numerous variables, including the number of cars, Junctions, date and time, and Vehicle ID. Our model is able to represent the many interconnections and relationships that affect traffic flow by taking these aspects into account.

We evaluated our models' performance using metrics like Mean Squared Error (MSE), Mean Absolute Error (MAE), and R-squared score after conducting a thorough experimentation and review process. These measurements shed light on the precision and dependability of our forecasts. Our goal is to help develop intelligent transportation systems that can minimize congestion, improve traffic management, and improve the overall commuter experience by accurately predicting traffic. Policymakers, traffic planners, and anybody looking for dependable and efficient transit options may find our findings useful.

The paper is organized as: In Sect. 2, related work has been presented. In Sect. 3, methodology used has been described in detail, In Sect. 4, results have been discussed and analyzed. Finally in Sect. 5, Conclusion and future research directions have been discussed.

Overall, this study contributes to the current research on traffic prediction and emphasizes the utility of machine learning models in this field. The results may lead to additional developments and uses of machine learning methods in transportation planning and administration.

1.1 Traffic Prediction Models

Based on their fundamental techniques and features, traffic prediction models can be divided into multiple types. The following are some typical classifications of traffic prediction models [1]:

1. Statistical Models: To analyze and forecast past traffic data, statistical models apply statistical techniques. These models frequently make the assumption that traffic patterns adhere to particular statistical trends or distributions. State space models, autoregressive conditional heteroscedasticity (ARCH) models, and moving average models are a few examples.
2. Machine Learning Models: These algorithms use data to identify patterns and associations from past traffic data in order to forecast future events. These models are able to adjust to shifting traffic circumstances and manage intricate non-linear relationships. Neural networks, random forests, support vector machines (SVM), gradient boosting, and linear regression are examples of machine learning models that are frequently used for traffic prediction.
3. Time Series Models: Analyzing and predicting data that changes over time is the exclusive focus of time series models. These models take into account the temporal aspect of traffic data, taking seasonality, trends, and periodic patterns into account. Seasonal breakdown of time series (STL), exponential smoothing models, ARIMA, and recurrent neural networks (RNN) with specialized architectures like LSTM are a few examples of time series models for traffic prediction.
4. Spatial Models: These models take into account the geographical features, traffic flow between various places, and topology of the road networks in traffic data. These models can shed light on the flow of traffic and bottlenecks along certain routes or locales. To capture the geographical dependencies in traffic data, spatial models may make use of methods like spatial regression models, geo-statistics, or graph-based approaches.
5. Hybrid Models: Hybrid models take advantage of the advantages of several approaches by combining many modeling techniques. To capture different facets of traffic behavior, these models incorporate time series, statistical, and machine learning techniques. Hybrid models take advantage of the complementing qualities of many modeling techniques in an effort to increase prediction accuracy. A

hybrid model might, for instance, blend a machine learning model like Random Forest with a statistical model like ARIMA.

We used several machine learning models in this study, such as Random Forest, Long Short-Term Memory (LSTM), K-Nearest Neighbors (KNN), and a baseline Moving Average model. These models were chosen because they are widely used in traffic prediction tasks and are good at capturing traffic patterns.

2 Related Work

The difficult problem of traffic prediction has been the subject of numerous studies employing a variety of machine learning strategies. A traffic flow prediction model based on long short-term memory (LSTM) has been suggested and has shown encouraging results in terms of forecast horizon and accuracy [1]. Their research showed how well LSTM can capture the patterns and temporal connections found in traffic data. In a different study, the effectiveness of many regression models, such as Random Forest and Linear Regression, for traffic prediction was compared [2]. They discovered that Random Forest performed better in terms of resilience and accuracy than other models, which made it a good option for traffic forecasting tasks.

The effect of adding external variables, such the weather and public events, on traffic prediction accuracy was also investigated by the authors [3]. According to their findings, taking these extra parameters into account enhanced the models' predictive power and produced more precise traffic projections. A thorough examination of traffic prediction models has been carried out in the context of urban transportation planning, and their effectiveness has been assessed using real-world datasets [4]. They conducted a comparison of machine learning techniques like Support Vector Regression (SVR) and Gradient Boosting Regression (GBR) and conventional time series forecasting methods like ARIMA and SARIMA. According to their findings, SVR and GBR outperformed each other in traffic prediction tests.

To increase the precision and timeliness of traffic forecasts, the integration of real-time data sources has been studied, including social media feeds and traffic sensors [5]. Their research illustrated the possible advantages of using real-time data to produce projections that are more accurate. Researchers studying traffic prediction have also looked into recent developments in deep learning methods. To capture the spatial relationships among road segments, for example, the authors presented a graph convolutional neural network (GCN) model, which led to state-of-the-art traffic flow prediction performance [6]. The subject of other research has been traffic prediction using ensemble algorithms. By utilizing the advantages of each individual model, Chen et al.'s ensemble model which combines numerous regression algorithms achieved increased prediction accuracy [7]. Furthermore, Yu et al. developed a recurrent neural network (RNN) model with attention mechanisms to capture temporal dependencies and adaptively learn the significance of historical information for precise traffic prediction, taking into account the dynamic nature of traffic patterns [8].

In order to optimize traffic flow and reduce the congestion in metropolitan areas, Zhao et al. looked into the use of deep reinforcement learning for traffic light control and prediction in the context of smart city applications [9]. In addition, Li et al. suggested a hybrid model for traffic prediction that combines a genetic algorithm with a time series forecasting method, utilizing optimization techniques and historical data to increase prediction accuracy [10]. By utilizing the underlying spatial and temporal relationships in the road network, Li et al. devised a spatiotemporal graph neural network (STGNN) model to overcome the problem of data sparsity and efficiently predict traffic flow even with insufficient data [11]. To improve the precision and resilience of traffic forecasts, Zhao et al. developed a multimodal traffic prediction framework within the framework of intelligent transportation systems [12]. This framework integrates data from several sources, such as traffic cameras, GPS data, and social media. Moreover, Zhou et al. presented a transfer learning-based method for traffic prediction that transfers knowledge from a target city with little data to a source city with lots of data, allowing accurate forecasts even in scenarios with little data [13]. Chen et al.'s study, which examined the effects of various time intervals on traffic prediction accuracy, stressed the need of choosing the right temporal granularity for precise forecasts [14].

In order to predict traffic flow, the researchers presented a hybrid model that included Long Short-Term Memory (LSTM) and Convolutional Neural Network (CNN), utilizing both temporal and geographical information in the data [15]. A novel approach to traffic prediction using recurrent neural networks (RNNs) that use attention mechanisms to dynamically weight the significance of various historical time steps has been presented [16]. In order to achieve competitive performance in comparison to conventional methods, numerous researchers investigated the application of deep learning techniques, particularly a Deep Belief Network (DBN), for traffic prediction [17]. In order to capture both spatial and temporal connections in the road network, the researchers in [18] presented a hybrid model that combines an LSTM and a graph neural network (GNN) for traffic prediction. Graph attention mechanisms are used by a spatiotemporal graph attention network (STGAT) for traffic forecasting, which captures the intricate relationships between road segments [19]. In order to capture both long-term and short-term dependencies in the data, a hierarchical attention-based long short-term memory (LSTM) model for traffic prediction has been presented in [20, 21]. Attention mechanisms are used at multiple levels.

For the purpose of predicting traffic, we have used a combination of regression, statistical, and long short-term memory models. We can increase the accuracy of our predictions by utilizing the advantages of both models through their integration. Our forecast model encompasses an extensive range of features, such as past traffic data, meteorological conditions, road network characteristics, and temporal data. We are able to better depict the intricate dynamics of traffic patterns when we take into account several pertinent aspects.

3 Methodology

This section outlines the methodology used in our study to predict traffic, which makes use of the KNN model, LSTM and Moving Average, and Random Forest models.

We utilized a methodology in this research work that included multiple crucial steps. First, we gathered a historical traffic dataset with data on cars, junctions, IDs, and dates and times. After that, we handled missing values and transformed date-time data into a numerical format appropriate for analysis as part of the data pretreatment procedure. Then, for traffic prediction, we chose four machine learning models: a Moving Average model, Random Forest, K-Nearest Neighbors (KNN), and Long Short-Term Memory (LSTM). These models were trained on a split dataset, and measures like R-squared, Mean Absolute Error (MAE), and Root Mean Squared Error (RMSE) were used to assess the models' performance. In the end, we examined the data to evaluate each model's performance in traffic prediction and contrasted its advantages and disadvantages. With the use of this methodology, we were able to assess and contrast the effectiveness of various models and offer insights into their applicability for precise traffic forecasting (Table 1).

3.1 *Dataset Description*

This traffic prediction dataset, which is openly accessible on Kaggle, includes 48,120 vehicle count observations that were gathered at four distinct intersections [22]. Because the data is small and sparse, there are various issues associated with the dataset. Uneven temporal coverage resulted from the sensors' differing data gathering times at each junction. Because of these limitations, future traffic patterns based on the dataset must be carefully projected and interpreted. All things considered, this dataset is a useful tool for researchers and transportation experts who want to analyze traffic patterns, gauge degrees of congestion, and create predictive models for traffic control and urban planning. The dataset offers insightful information on traffic patterns and trends in these particular areas.

The Fig. 1 shows the hourly traffic volume fluctuation for each day for the chosen dataset. The figure shows that there is no consistent trend of traffic volume increase or reduction, and that traffic volume varies hourly. Based on the information in the dataset, this diagram displays the volume of traffic.

The Fig. 2 shows the traffic volume on a date-wise basis using the selected dataset. The graph's y-axis shows the number of cars tracked or the associated traffic volume, while the graph's x-axis shows the dates. The traffic volume variation over time is represented visually by the graph. Plotting the data points allows one to see any patterns that stand out, including spikes or dips in the volume of visitors on particular dates.

Table 1 Machine learning models for traffic prediction

Models	Principle	Advantages
K-Nearest Neighbors (KNN)	Nearest neighbor-based approach	1. Can capture local patterns 2. Suitable for both regression and classification tasks 3. Simple and intuitive
Random forest	Ensemble learning with decision trees	1. Captures complex relationships and interactions-May require more computational resources and training time 2. Handles high-dimensional data and feature importance—Less interpretable compared to individual decision trees 3. Robust against overfitting
Long Short-Term Memory (LSTM)	Recurrent Neural Network (RNN) architecture	1. Captures long-term dependencies and temporal patterns 2. Handles variable-length sequences and time series data 3. Suitable for modeling complex temporal dynamics
Moving average	Time series forecasting based on averaging	1. Simple and easy to implement 2. Provides a baseline for comparison 3. Smoothens out noise in the data

The number of vehicles that pass through the intersections on different dates, or the variations in daily traffic events, can both be analyzed and understood with the help of this graph. It can assist in locating days with heavy traffic, probable reasons for traffic accidents, or trends associated with particular occasions or times of the year. The graph's "Daily Traffic Incidents" caption explains that its primary goal is to record daily fluctuations in traffic volume. This information can be used by researchers or analysts to understand the frequency and severity of traffic incidents or congestion on various days.

4 Results and Discussions

The evaluation metrics of Root Mean Squared Error (RMSE), Mean Absolute Error (MAE), and R-squared were used to compare and assess the performance of the chosen machine learning models (K-Nearest Neighbors, Random Forest, Long Short-Term Memory, and Moving Average).

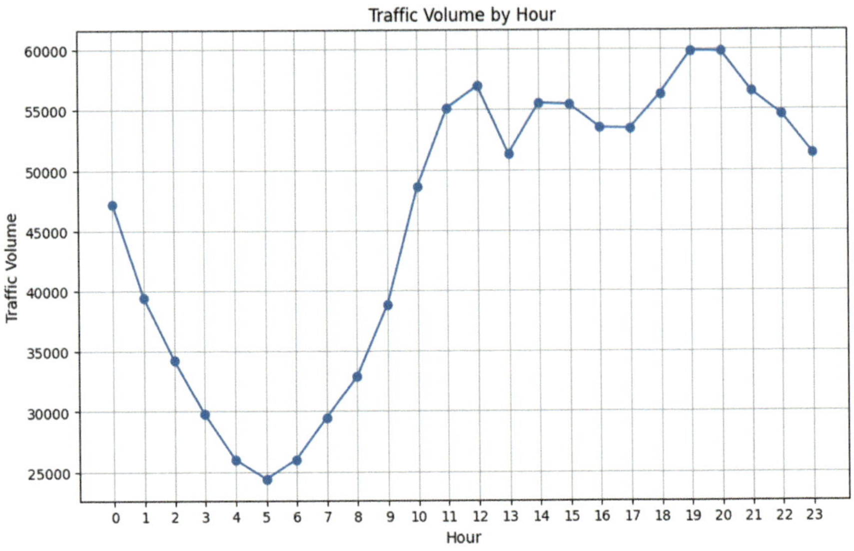

Fig. 1 Hourly traffic prediction based on traffic dataset

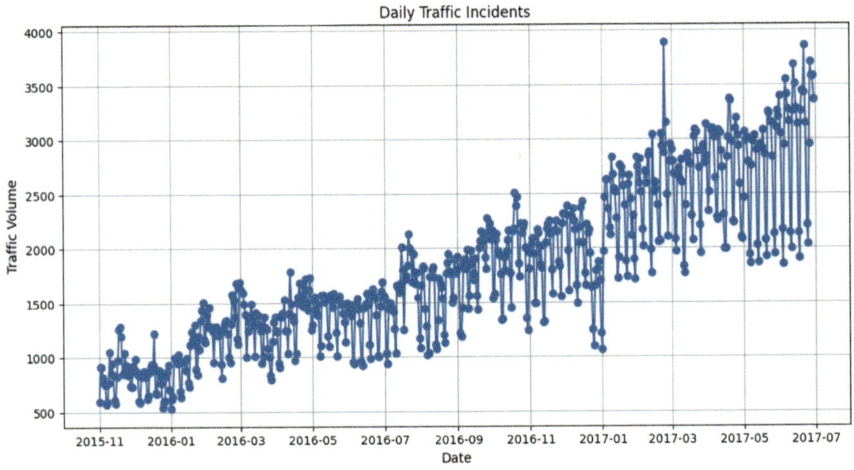

Fig. 2 Date-wise traffic incidents

The Random Forest model outperformed the other models in terms of prediction accuracy, as seen by its lower RMSE value of 3.83 and MAE value of 2.66. This shows that Random Forest can accurately anticipate traffic statistics by effectively capturing intricate correlations and patterns. Furthermore, the high R-squared value of 0.96 indicates that a sizable amount of the variance in the traffic data is explained by the Random Forest model.

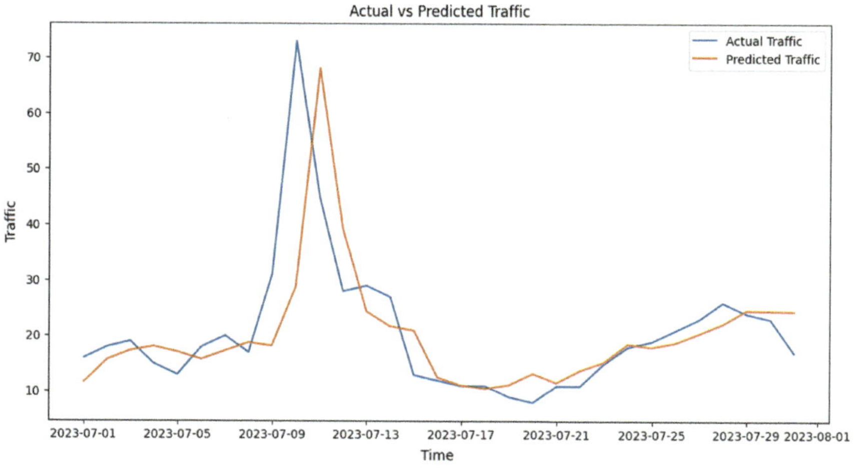

Fig. 3 Actual versus predicted traffic using LSTM model

The Fig. 3 represents an analysis of the differences between the traffic levels that were observed and those that were anticipated using an LSTM (Long Short-Term Memory) model. The true observed traffic volumes or counts that are documented in the dataset are referred to as the "actual traffic" in this context. For the duration specified, these values represent the ground truth. On the other hand, the LSTM model a recurrent neural network kind that's frequently employed for time series forecasting is responsible for producing the anticipated traffic figures. For the purpose of forecasting future traffic volumes, the LSTM model takes into account historical traffic data and discovers patterns and relationships.

The figure usually shows a line plot with the traffic volume represented on the y-axis and time or timestamps represented on the x-axis. The projected traffic values are shown as a separate series, sometimes indicated as a dashed line or a different color, from the actual traffic data, which are plotted as one series, typically represented by dots or a solid line. The picture enables an evaluation of the prediction accuracy and performance of the LSTM model in traffic forecasting by comparing the actual and anticipated numbers. The more closely the projected and actual values match, the more trustworthy the model's predictions are. Differences or variations between the expected and actual numbers can highlight the model's advantages or shortcomings (Fig. 4).

The evaluated traffic values derived from a KNN (K-Nearest Neighbors) model are displayed in the image "Traffic Prediction Using KNN". It is noteworthy that the predicted traffic values align with the x-axis. The x-axis in this case represents time or timestamps, denoting various future points. The anticipated traffic volume or count is shown on the y-axis. The projected traffic values are shown as a straight line parallel to the x-axis in the line plot. This implies that changes in traffic volume over time are not taken into account by the KNN model. Rather, it is predicated on a steady or consistent level of traffic over the anticipated duration.

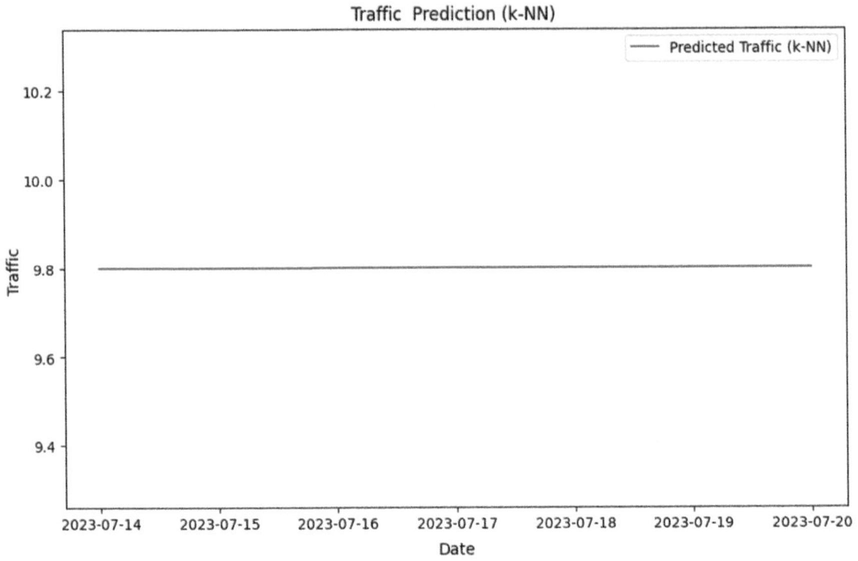

Fig. 4 Predicted traffic using KNN

It is crucial to remember that the KNN algorithm uses an instance-based, non-parametric learning strategy. It uses the "k" nearest neighbors from the training data to anticipate the volume of traffic. Based on the most comparable examples or neighbors discovered in the training dataset, the KNN model may be assuming a constant traffic volume in this situation. The image shows that a flat or parallel line of projected values may come from the KNN model's inability to capture temporal patterns or changes in traffic volume over time.

The Fig. 5 shows the comparison of the projected traffic figures using the moving average approach and the actual traffic. The moving average method is a straightforward time series forecasting methodology that is used to calculate the expected traffic figures. The moving average forecasts the traffic levels for the upcoming time period by averaging the traffic values over a predetermined window of historical data.

This graph sheds light on how well the moving average approach captures the overall pattern or volume of traffic. The underlying dynamics or abrupt changes in traffic patterns, however, might not be captured by it. This figure can be used by researchers and analysts to assess the moving average method's efficacy as a traffic volume forecasting tool, taking into account its limitations and simplicity in capturing more intricate fluctuations in the data.

The following conclusions can be made based on the comparison of the RMSE, MAE, and R-squared error values for the four models (KNN, Random Forest, LSTM, and Moving Average) shown in Figs. 6, 7, and 8.

With a high RMSE of 21.87 and MAE of 16.49, the KNN model demonstrated a substantial divergence between the observed and predicted values. The model's

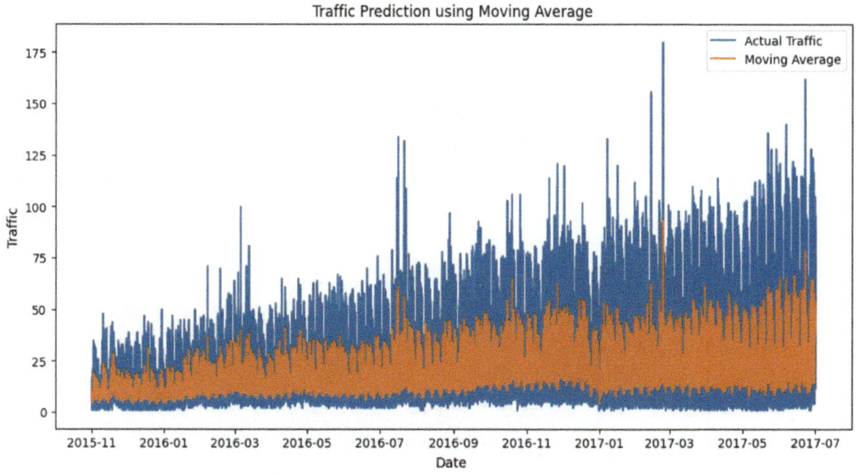

Fig. 5 Actual versus predicted traffic using moving average method

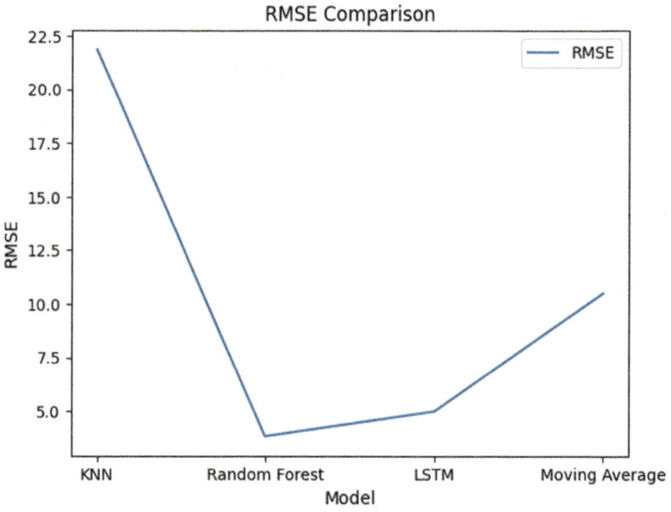

Fig. 6 RMSE comparison

predictions appear to be unable to adequately explain the variance in the data, as indicated by the negative R-squared value (−0.17).

With a comparatively low RMSE of 3.83 and MAE of 2.66, the Random Forest model outperformed the other four models. A significant amount of the variance in the traffic data can be explained by the model's predictions, as evidenced by the high R-squared value of 0.96.

With an RMSE of 4.99 and an MAE of 2.90, the LSTM model did fairly well. The model may not be as precise as the Random Forest model, but it does appear

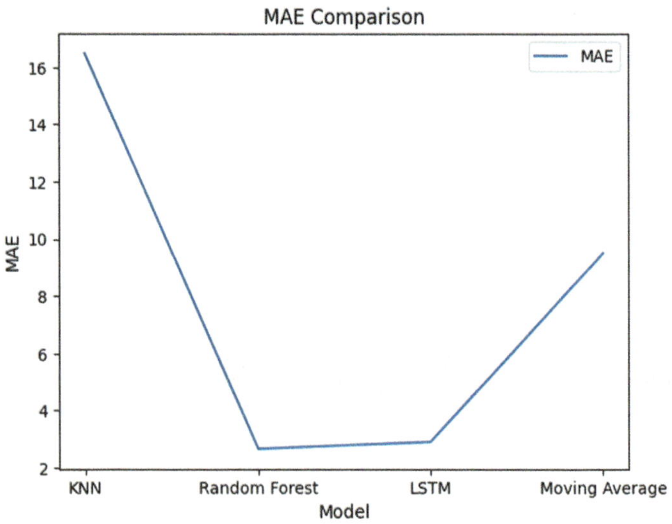

Fig. 7 MAE comparison

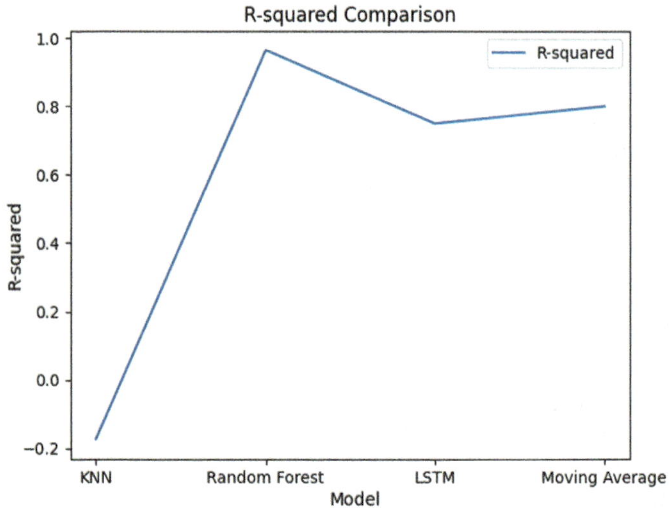

Fig. 8 R square error comparison

to capture a significant amount of the variance in the traffic data, according to the R-squared value of 0.75.

With RMSE and MAE values of 10.5 and 9.5, respectively, the Moving Average model—which was used as a baseline—had higher values. The model explains a moderate amount of variance in the traffic data, according to the R-squared value of 0.8.

In conclusion, the Random Forest model performed better than the other models, obtaining the best R-squared value and the lowest RMSE and MAE values. This shows that Random Forest can capture intricate associations and produce precise forecasts, making it a good fit for traffic prediction problems. While LSTM displayed decent performance, KNN demonstrated terrible performance. Although the Moving Average model did a decent job, it was unable to identify intricate patterns in the traffic data.

5 Conclusion

We compared many machine learning models, such as K-Nearest Neighbors (KNN), Random Forest, Long Short-Term Memory (LSTM), and Moving Average, for traffic prediction in this study. Our objective was to analyze their accuracy and performance in capturing traffic patterns using metrics like R-squared error, MAE, and RMSE. The models' respective performances varied significantly, as the data showed. With the lowest RMSE and MAE values and the greatest R-squared value, the Random Forest model was the best performance. It proved to be an excellent option for precise traffic prediction tasks by showcasing its capacity to extract intricate relationships from the traffic data. However, the KNN model performed poorly, showing a negative R-squared value in addition to significantly higher RMSE and MAE values. This shows that because of its shortcomings in accurately capturing traffic patterns, the KNN model might not be the best option for traffic prediction. With somewhat lower R-squared value when compared to Random Forest, but comparatively lower RMSE and MAE values when compared to KNN, the LSTM model showed intermediate performance. This suggests that while LSTM could be able to capture temporal connections in traffic data, it might not be as accurate as Random Forest. As a baseline approach, the Moving Average model showed greater RMSE and MAE values than the other models. It showed a limited ability to identify intricate patterns in the traffic data.

The results highlight the significance of choosing suitable models for traffic prediction tasks according to particular requirements and dataset features. With its ability to accurately estimate traffic and capture intricate linkages, Random Forest demonstrated exceptional performance. To increase traffic prediction accuracy and identify more subtle trends in the data, future studies could investigate hybrid methodologies or other cutting-edge machine learning techniques.

References

1. Boukerche, A., Wang, J.: Machine learning-based traffic prediction models for intelligent transportation systems. Comput. Netw. **181**, 107530 (2020)
2. Lv, Y., Duan, Y., Kang, W., Li, Z., Wang, F.Y.: Traffic flow prediction with big data: a deep learning approach. IEEE Trans. Intell. Transp. Syst. **16**(2), 865–873 (2014)
3. Zhang, J., Zheng, Y., Qi, D.: Deep spatial-temporal residual networks for citywide crowd flows prediction. In: Proceedings of the 31st AAAI Conference on Artificial Intelligence, pp. 1655–1661 (2017)
4. Ma, X., Tao, Z., Wang, Y., Yu, H., Wang, Y.: Long short-term memory neural network for traffic speed prediction using remote microwave sensor data. Transp. Res. Part C: Emerg. Technol. **54**, 187–197 (2015)
5. Zhao, Z., Chen, W., Wu, X., Chen, P.C., Liu, J.: LSTM network: a deep learning approach for short-term traffic forecast. IET Intel. Transp. Syst. **11**(2), 68–75 (2017)
6. Shi, X., Chen, Z., Wang, H., Yeung, D.Y., Wong, W.K., Woo, W.C.: Convolutional LSTM network: a machine learning approach for precipitation now casting. In: Advances in Neural Information Processing Systems, pp. 802–810 (2015)
7. Chen, J., Wang, Q., Cheng, H.H., Peng, W., Xu, W.: A review of vision-based traffic semantic understanding in ITSs. IEEE Trans. Intell. Transp. Syst. (2022)
8. Yu, H., Yin, H., Shi, Y., Yu, S.: Spatio-temporal graph convolutional networks: A deep learning framework for traffic forecasting. In: Proceedings of the 26th International Joint Conference on Artificial Intelligence, pp. 3634–3640 (2017)
9. Xie, M., Cheng, W., Gill, G.S., Zhou, J., Jia, X., Choi, S.: Investigation of hit-and-run crash occurrence and severity using real-time loop detector data and hierarchical Bayesian binary logit model with random effects. Traffic Inj. Prev. **19**(2), 207–213 (2018)
10. Li, C., Bai, L., Liu, W., Yao, L., Waller, S.T.: Urban mobility analytics: a deep spatial–temporal product neural network for traveller attributes inference. Transp. Res. Part C: Emerg. Technol. **124**, 102921 (2021)
11. Zhou, J., Cui, P., Zhang, Z., Yang, C., Liu, Z., Zhu, W.: Graph neural networks: a review of methods and applications (2018).. arXiv:1812.08434
12. Zhao, J., Zhang, Y., He, X.: LSTM network: a deep learning approach for traffic flow prediction. Transp. Res. Part C: Emerg. Technol. **79**, 148–164 (2017)
13. Zhou, F., Li, L., Zhang, K., Trajcevski, G.: Urban flow prediction with spatial–temporal neural ODEs. Transp. Res. Part C: Emerg. Technol. **124**, 102912 (2021)
14. Chen, Y., Shu, L., Wang, L.: Traffic flow prediction with big data: a deep learning based time series model. In: 2017 IEEE Conference on Computer Communications Workshops (INFOCOM WKSHPS), pp. 1010–1011, May 2017. IEEE (2017)
15. Tian, Y., Zhang, K., Li, J., Lin, X., Yang, B.: LSTM-based traffic flow prediction with missing data. Neurocomputing **318**, 297–305 (2018)
16. Ma, Z., Sun, H., Xu, C., Wang, F.: Traffic flow prediction with big data: a deep learning approach. Transp. Res. Part C: Emerg. Technol. **54**, 17–30 (2015)
17. Yao, H., Wu, F., Ke, J., Tang, X., Jia, Y., Lu, S., Gong, P., Ye, J., Li, Z.: Deep multi-view spatial-temporal network for taxi demand prediction. In: Proceedings of the AAAI Conference on Artificial Intelligence, vol. 32, no. 1, April 2018 (2018)
18. Yu, B., Yin, H., Zhu, Z.: Spatio-temporal graph convolutional neural network: a deep learning framework for traffic forecasting. In: Proceedings of the 26th International Joint Conference on Artificial Intelligence, pp. 3634–3640 (2017)
19. Liu, Y., Zheng, H., Feng, X., Chen, Z.: Short-term traffic flow prediction with Conv-LSTM. In: 2017 9th International Conference on Wireless Communications and Signal Processing (WCSP), pp. 1–6, October 2017. IEEE (2017)

20. Luo, X., Li, D., Yang, Y., Zhang, S.: Spatiotemporal traffic flow prediction with KNN and LSTM. J. Adv. Transp.
21. Kesting, A., Treiber, M., Helbing, D.: Enhanced intelligent driver model to access the impact of driving strategies on traffic capacity. Philos. Trans. R. Soc. A Math. Phys. Eng. Sci. **368**(1928), 4585–4605 (2010)
22. Kaggle: Traffic prediction dataset. Kaggle (2017). https://www.kaggle.com/datasets/fedesoriano/traffic-prediction-dataset. Accessed 25 June 2023

A Hybrid Model for Aspect-Based Sentiment Analysis Using Boosting Technique

Amit Chauhan, Aman Sharma, and Rajni Mohana

Abstract In today's world of technology, people express their opinions and feelings through different platforms at a rapid pace, such as online discussions, social media posts, and product reviews. Sentiment analysis, also known as ABSA is crucial for making informed decisions in highly competitive markets based on customer feedback. ABSA has been utilized across various fields, including social media, product evaluations, and customer feedback. In this paper, we utilized a sentiment classification layer within our model to accurately predict the sentiment polarity of restaurant reviews using the SemEval 2014 Dataset. Our proposed model, XLNet-LGBM, achieved the highest accuracy among all baseline models, with an accuracy of 90.6 and an F1 score of 88.7. This is approximately 2.74% and 2% higher than the average accuracy and F1 score of other baseline models such as ATN-AF, IPAN-BERT, DLCF-DCA-CDM, SA-BERT, and SA-BERT-XGBoost. Therefore, our proposed model significantly outperforms the baseline models.

Keywords Bert · Aspect-based sentiment analysis · LGBM · XLNet

1 Introduction

In today's digital age, people share their thoughts and emotions on various platforms at an unprecedented speed and scale. Due to the increasing amount of user-generated content, including online discussions, social media posts, and product reviews, the need for more accurate sentiment analysis has also increased. Traditional sentiment

A. Chauhan · A. Sharma · R. Mohana
Jaypee University of Information Technology, Solan, Himachal Pradesh, India
e-mail: chauhanamit37@gmail.com

A. Sharma
e-mail: aman.sharma@juitsolan.in

R. Mohana (✉)
Amity University, Mohali, Punjab, India
e-mail: rajni.mohanajuit@gmail.com

B. Soni et al. (eds.), *Beyond Artificial Intelligence*, Lecture Notes in Networks and Systems 1326, https://doi.org/10.1007/978-981-96-4170-3_48

analysis often fails to capture the complexity of attitudes towards specific elements or features of a text and only assigns a general sentiment score to the entire content. This is where ABSA comes in. ABSA is essential because customer feedback is crucial for making strategic decisions in today's highly competitive marketplaces. ABSA provides businesses with the necessary tools to assess the advantages and disadvantages of their offerings, identify new trends, and respond proactively to customer complaints. ABSA has many applications, including social media monitoring, brand management, and public mood tracking. Precise perception information can guide effective engagement tactics [1]. Analyzing customer feedback through ABSA presents its unique challenges. One of the crucial steps is aspect extraction, which involves identifying relevant details from the text. This can be difficult due to linguistic uncertainties and variations. The models need to understand the words used and their contextual meaning to determine the sentiment orientation towards each feature accurately. This process is referred to as sentiment polarity classification. In today's data-driven world, it is imperative to ensure that ABSA is used ethically and biases are eliminated to safeguard user privacy.

To better understand ABSA, we can refer to Fig. 1 and use an example sentence: "The restaurant's food was delicious, but the service was slow." This sentence covers two topics: food and service. ABSA aims to analyze the sentiment polarity connected to each aspect [2]. Regarding this situation, individuals have favourable opinions towards the cuisine but unfavourable opinions towards the assistance provided. The components and respective emotions are as follows: (Aspect: Food, Sentiment: Positive) and (Aspect: Service, Sentiment: Negative).

ABSA's analysis surpasses the usual sentiment analysis in this example. Rather than providing a general sentiment score for the whole sentence, ABSA separates the sentiment expression into distinct components to encompass the intricate emotions associated with each one [3].

As mentioned in Fig. 2, the ABSA tasks consist of multiple stages, the first being extraction. This involves identifying the features or aspects mentioned in the sentence, such as "food" and "service." Natural language processing techniques like part-of-speech tagging and dependency parsing are utilized to achieve this. The next step is Sentiment Polarity Classification, which determines whether the sentiment expressed

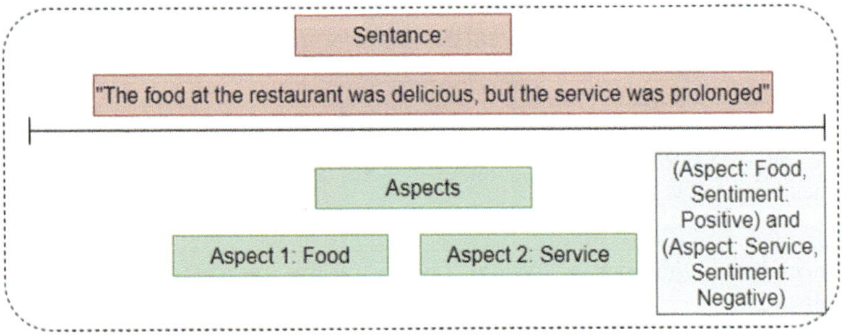

Fig. 1 A sample review sentence for ABSA

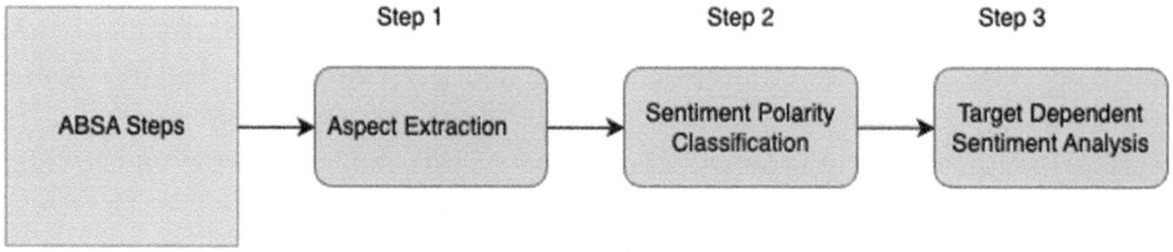

Fig. 2 ABSA work flow

towards each element is positive, negative or neutral. Machine learning algorithms based on neural networks are commonly used for this stage once the aspects have been identified [4]. In the ABSA process, the sentiment of a statement is determined using words like "delicious" to express positivity towards "food" and "slow" to indicate negativity towards "service". The final step involves target-dependent sentiment analysis, a crucial part of the process. It considers the aspect being referred to and how the sentiment may vary accordingly. For instance, while "slow" may be negative in the context of "service", it could have a positive connotation when referring to the "pace" of something else. [5]. By utilizing target-dependent sentiment analysis, we can ensure that sentiment classification accurately reflects the context in which it is used.

2 Related Work

In this paper, authors [6] propose a hybrid DNN model with an attention mechanism that can identify and highlight essential sentiment features in the text. To achieve this, authors first use sentiment lexicons to identify relevant sentiment features and transformers' bidirectional encoder representation to create sentiment-enhanced word embeddings for the reader. Then, modify the BiLSTM to capture long-dependency relations in the word sequence, word order, and contextual text semantic information. The model also uses an attention mechanism to give more weight and significance to salient features in the word sequence [7]. This study introduces a new sentiment analysis model called SLCABG. The model utilises a sentiment lexicon to combine attention-based Bidirectional Gated Recurrent Units (BiGRU) and Convolutional Neural Networks (CNN). By combining the advantages of sentiment lexicon and deep learning technology, the SLCABG model overcomes the limitations of current product review sentiment analysis models [8]. The authors of this paper aimed to identify the best performer among Sentiment Analysis classifiers for short texts. They compared the effectiveness of various approaches and presented a filtered LDA framework that outperformed current methods for interpreting sentiment variations on Twitter. To identify potential causes of sentiment changes, the framework applied cascaded LDA Models with different hyperparameter settings. After eliminating old topics, a Topic Model with a high Coherence Score was used to extract emerging issues humans can understand [9]. This article presents a method for improving the accuracy of sentiment analysis in tweets that express unclear or ambiguous emotions. The approach combines ensemble and CNN models, using five feature vectors (lexical, word-type, semantic, sentiment polarity, and position of words) to create a feature ensemble model. The output of this model is tweet embeddings, which are used as input for the CNN model.

Based on the literature survey, it is evident that ABSA is a trending topic in NLP research. Although there have been numerous research works in the field of ABSA, there is still a lot of work to be done. This will help authors to work on the unimodal ABSA task and achieve state-of-the-art results in this area.

3 Proposed Work

Figure 3 presents the XLNet-LGBM model [10, 11] for ABSA on the restaurant dataset. The model consists of several steps, starting with data preprocessing. This step involves cleaning the data to ensure it's suitable for the model, thereby improving its performance. The processed data is then passed on to the XLNet Model for further processing. The input embedded layer in the word embedding layer encodes the sentence aspects and contexts onto the words. To start, we perform data preprocessing in step 1. This involves basic cleaning tasks such as removing non-ASCII characters, line breaks, extra spaces, URLs, user mentions, and stop words from the dataset. Next, we pair aspects with sentiments. Then, we train the model on the preprocessed data and test it with the XLNet model. To improve the model's performance, we tune the hyperparameters. Finally, we employ the Boosting technique to merge the trained model with LGBM, attaining state-of-the-art outcomes.

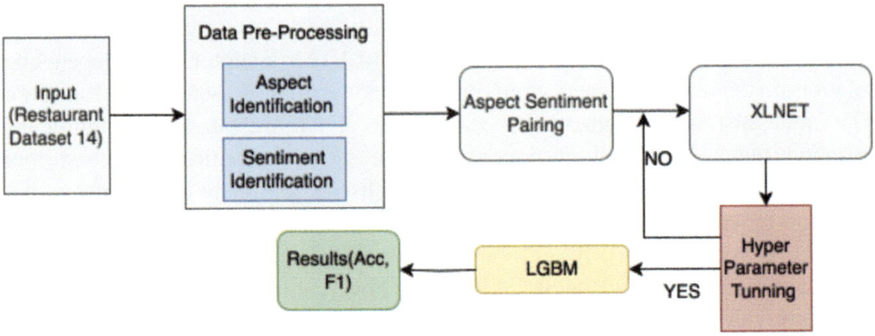

Fig. 3 XLNet-LGBM model for ABSA (proposed work)

3.1 Pre-trained Language Model BERT

The pre-training language model has generated significant academic interest due to its impressive performance in NLP tasks. BERT is a part of the Transformer architecture family and aims to understand the context and meaning of the text by undergoing pre-training on a selection of textual data. BERT is a unique natural language processing model that considers both the words before and after a given word to determine its meaning within a sentence. Unlike other NLP models, [12] BERT is trained on billions of words, allowing it to capture a wide range of linguistic patterns and nuances. BERT comes in different versions with varying model sizes. Layer counts, hidden units and attention heads characterize the "base" model. There are also smaller options, such as BERT Lite. Developers and researchers can access BERT models and their versions through the Hugging Face Transformers library for various NLP tasks.

4 Experimental Setup and Results

In the proposed study, the dataset used is the SemEval 2014 (Restaurant 2014) [13]. This serves as a benchmark for ABSA tasks and contains customer reviews.

In Table 1, the results of the baseline model and the proposed models' results are compared using the restaurant dataset. Figure 4 provides a visual representation of these results. The experimentation was conducted on the paid version of Google Colab, which granted the authors access to a 40GB GPU. The reason for comparing our results with ATN-AF, IPAN-BERT, DLCF-DCA-CDM, SABERT, and SA-BERT-XGBoost is because these are the latest proposed models in the task of ABSA unimodal. The authors achieved state-of-the-art results compared with these models.

Table 1 Performance comparison of models on SemEval restaurant 14 dataset

	Models	Accuracy (%age)	F1 (%age)
Benchmark models	ATN-AF [14]	76.48	72.6
	IPAN-BERT [15]	85.9	76.4
	DLCF-DCA-CDM [16]	86.4	80.36
	SA-BERT [17]	87.24	81.19
	SA-BERT-XGBoost [17]	87.86	86.7
Proposed model	**XLNet-LGBM**	**90.6**	**88.7**

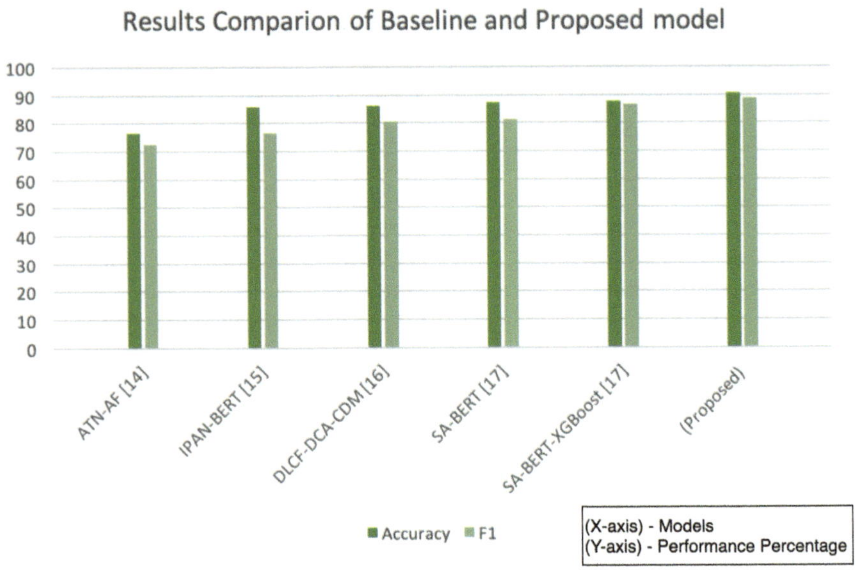

Fig. 4 Accuracy and F1 score on SemEval restaurant 14 dataset

5 Conclusion

This paper proposes an unimodal XLNet-LGBM model for the ABSA task. The combination of our model outperforms other baseline models with an accuracy of 90.6% and an F1 score of 88.7% on the benchmark dataset. As technology advances, numerous complex issues require further investigation in the realm of multimodal problems. From finding ways to integrate various modes of transportation more efficiently to improving communication between different modes of technology, much work must be done.

Conflict of Interest All authors declare no conflicts of interest.

References

1. Zhang, W., Li, X., Deng, Y., Bing, L., Lam, W.: A survey on aspect-based sentiment analysis: tasks, methods, and challenges. IEEE Trans. Knowl. Data Eng. 1–20 (2022)
2. Bensoltane, R., Zaki, T.: Aspect-based sentiment analysis: an overview in the use of Arabic language. Artif. Intell. Rev. **56**(3), 2325–2363 (2023)
3. Li, H., Bruce, X.B., Li, G., Gao, H.: Restaurant survival prediction using customer-generated content: An aspect-based sentiment analysis of online reviews. Tour. Manag. **96**, 104707 (2023)
4. Kukkar, A., Mohana, R., Sharma, A., Nayyar, A., Asif Shah, Mohd.: Improving sentiment analysis in social media by handling lengthened words. IEEE Access **11**, 9775–9788 (2023)

5. Lauriola, I., Lavelli, A., Aiolli, F.: An introduction to deep learning in natural language processing: models, techniques, and tools. Neurocomputing **470**, 443–456 (2022)
6. Khan, J., Ahmad, N., Khalid, S., Ali, F., Lee, Y.: Sentiment and context-aware hybrid DNN with attention for text sentiment classification. IEEE Access **11**, 28162–28179 (2023)
7. Yang, L., Li, Y., Wang, J., Simon Sherratt, R.: Sentiment analysis for ecommerce product reviews in Chinese based on sentiment lexicon and deep learning. IEEE Access **8**, 23522–23530 (2020)
8. Alattar, F., Shaalan, K.: Using artificial intelligence to understand what causes sentiment changes on social media. IEEE Access **9**, 61756–61767 (2021)
9. Phan, H.T., Tran, C., Nguyen, N.T., Hwang, D.: Improving the performance of sentiment analysis of tweets containing fuzzy sentiment using the feature ensemble model. IEEE Access **8**, 14630–14641 (2020)
10. Yang, Z., Dai, Z., Yang, Y., Carbonell, J., Salakhutdinov, R.R., Le, Q.V.: XLNet: generalized autoregressive pretraining for language understanding. In: Advances in Neural Information Processing Systems, vol. 32 (2019)
11. Ke, G., Meng, Q., Finley, T., Wang, T., Chen, W., Ma, W., Ye, Q., Liu, T.-Y.: Lightgbm: a highly efficient gradient boosting decision tree. Adv. Neural. Inf. Process. Syst. **30**, 3146–3154 (2017)
12. Devlin, J., Chang, M.W., Lee, K., Toutanova, K.: Bert: pre-training of deep bidirectional transformers for language understanding.In: Proceedings of NAACL-HLT, vol. 1, p. 2 (2019)
13. Pontiki, M., Galanis, D., Pavlopoulos, J., Papageorgiou, H., Androutsopoulos, I., Manandhar, S.: SemEval-2014 task 4: aspect based sentiment analysis. In: Proceedings of the 8th International Workshop on Semantic Evaluation (SemEval 2014), pp. 27–35, Dublin, Ireland, August 2014. Association for Computational Linguistics (2014)
14. Zhao, F., Wu, Z., Dai, X.: Attention transfer network for aspect-level sentiment classification (2020). arXiv:2010.12156
15. Kai, S., Gu, M., Li, R., Loo, J., Sen, S.: Interactive pos-aware network for aspect-level sentiment classification. Neurocomputing **420**, 181–196 (2021)
16. Xu, M., Zeng, B., Yang, H., Chi, J., Chen, J., Liu, H.: Combining dynamic local context focus and dependency cluster attention for aspect-level sentiment classification. Neurocomputing **478**, 49–69 (2022)
17. Mewada, A., Dewang, R.K.: SA-ASBA: a hybrid model for aspect-based sentiment analysis using synthetic attention in pre-trained language BERT model with extreme gradient boosting. J. Supercomput. **79**(5), 5516–5551 (2023)

Exploring the Association Between Mental Health and Efficacy Using Student Dataset

Rakhi Nagpal, Saravjeet Singh, and Aditi Moudgil

Abstract Several mental disorder symptoms like depression, anxiety, exhaustion, and cynicism can have a negative impact on students' efficacy, as they can affect their ability to focus, stay motivated, and complete tasks. Thus, the main objective of this study is to find out the association between efficacy and mental disorder symptoms in a large sample of medical students (n = 320). The dataset includes the demographic information of all participants which was collected through a questionnaire and other information like depression, anxiety, cynicism was measured through Maslach Burnout Inventory (MBI) scale. Results have shown that anxiety and depression are positively associated with each other having correlation constant(r) as 0.72 and thus impact the efficacy of students having correlation value as −0.45. Thus, with the increase of anxiety and depression, the efficacy of the student starts decreasing.

Keywords Mental disorders · Efficacy · Depression · Anxiety · Medical students

1 Introduction

There is currently a serious medical concern due to the fast-rising prevalence of mental disease. In today's fast-paced environment, psychological health problems like stress, worry, tiredness, and depression are becoming more common. The phrase "mental illness," which is also used to refer to a variety of problems that impact a person's emotions, pondering, behaviour, and general well-being, is also used to refer to mental health issues. Genetics, environmental circumstances, and life experiences

R. Nagpal (✉) · S. Singh · A. Moudgil
Chitkara University Institute of Engineering and Technology, Chitkara University, Rajpura, Punjab 140401, India
e-mail: rakhi.nagpal@chitkara.edu.in

S. Singh
e-mail: saravjeet.singh@chitkara.edu.in

A. Moudgil
e-mail: aditi.modgil@chitkara.edu.in

B. Soni et al. (eds.), *Beyond Artificial Intelligence*, Lecture Notes in Networks and Systems 1326, https://doi.org/10.1007/978-981-96-4170-3_49

are only a few of the causes of mental disease. Two prevalent mental health conditions that can have an impact on a person's mood, behaviours, and general well-being are depression and anxiety. Even though they are separate disorders, they frequently co-occur and share some comparable symptoms, like sorrow or anxiety, changes in sleep, or appetite, and difficulty concentrating. Depression is characterized by protracted sadness, lack of optimism, and a loss of interest in enjoyable activities. Physical symptoms like changes in energy levels, hunger, and sleep patterns might also result from it. On the other hand, anxiety is a result of excessive tension and stress and is characterized by ongoing worries, fears, or panic attacks [4].

Anxiety can cause physical symptoms such as rapid heartbeat, sweating, and shaking, and can interfere with daily activities such as work or socializing. Depression and anxiety can both interfere with a person's sense of competence and ability to succeed academically as well. People with depression may struggle to focus or concentrate and lose interest in their academic pursuits as well. People with anxiety may struggle with testtaking or public speaking, experience excessive worry or perfectionism, and have difficulty managing their time or organizing their work. Exhaustion is a common symptom of both depression and anxiety and can impact a person's ability to focus or engage in academic work. Having a strong belief in one's own ability to accomplish tasks and achieve goals can greatly enhance a person's sense of accomplishment and overall well-being [5, 13, 21].

Cynicism, or a feeling of distrust or skepticism, can also impact academic efficacy. When a person feels disconnected from their academic work or their peers, they may be less likely to engage fully in the learning process or to feel motivated to succeed. Efficacy, exhaustion, and cynicism are three key factors that are often associated with burnout. When there is a significant mismatch between a person's job and their natural abilities and preferences, burnout is more likely to occur [16]. Stress can also have an impact on students' academic performance in the following ways:

- Reduced motivation: Stress can cause students to lose interest in gaining knowledge, studying, and participating in academics. Academic achievement as well as performance may suffer because of stress and burnout.
- Poor academic performance: Burnout can cause students to have trouble in concentrating, retaining information, and completing assignments on time, resulting in poor academic performance.
- Decreased attendance and reduce engagement: Students experiencing stress / burnout may feel too exhausted to attend classes regularly, leading to missed classes and reduced academic achievement. This can lead to physical and psychological withdrawal from academic studies, making it highly probable that students may feel disconnected and disengaged from their coursework [12].
- Negative self-talk: Burnout can cause students to experience negative self-talk, leading to feelings of inadequacy, low self-esteem, and reduced academic confidence, which can impact their academic achievement.

A strong sense of efficacy, or belief in one's ability to accomplish tasks and achieve goals, can help to prevent burnout. When a person feels capable and effective in their work, they are less likely to become overwhelmed by stress and less likely to experience burnout.

Exhaustion and Cynicism, on the other hand, are also a key symptom of burnout. Together, these factors can create a cycle of burnout. When a person lacks a sense of efficacy, they may become overwhelmed and exhausted, which can lead to feelings of cynicism and disengagement. Over time, this cycle can erode a person's sense of wellbeing and lead to burnout. The experience of burnout can lead to negative mental health outcomes such as anxiety, depression, frustration, hostility, or fear.

Mental health issues are most prevalent among students in medical school, and can have a significant impact on their ability to function both socially and academically. According to recent research studies students commonly experience a range of mental health issues, including anxiety disorders, exhaustion, depression, sleep disturbances, drugs addiction and burnout. But Medical students are considered the most vulnerable group due to the combination of being in the age range of adolescence and young adulthood [10].

Considering the above-mentioned details, the main objective of this study is to find out the association between efficacy and mental disorder symptoms among medical student. To find out the association, these studies used dataset from Kaggle that includes the demographic information of all participants which was collected through a questionnaire and other information like depression, anxiety, cynicism was measured through MBI scale. Detail description of the experiment is provided in Sect. 3. Related work and studied literature are provided in Sect. 2. Section 3 provides the detailed information about the dataset used, experiment details, association and result. Conclusion of the paper is provided in Sect. 4.

2 Related Work

The work for the disorder evaluation was started in 1990s using the with traditional computation methods. A theoretical framework for Generalized Anxiety Disorder (GAD) was presented to evaluation of anxiety disorder. This conceptual model has considerable clinical implications. This study provided reveals the impact of intolerance of uncertainty and level of depression for identifying the GAD. This study also provided the differentiation factor for the nonclinical subjects from the GAD patients and non-GAD patients. The authors have also highlighted the impact of "beliefs about worry" on GAD symptoms [8].

The accurate prediction of the emergence and progression of anxiety disorders and mood is challenging and crucial for clinical purposes. Eeden et al. have conducted a study to compare the predictive capabilities of 3 different machine learning algorithms namely multinomial logistic regression, a naive Bayes classifier, and Auto-sklearn in predicting depression and anxiety. The dataset collected by the authors consists people of average age of 42.2 years, which includes 66.5% women. The average accuracy to predict the status of mental health for considered methods was

79%. When compared to auto sklearn approach, the naive bayes model and LR model gives results with high accuracy [9].

A study was conducted to analyze the stress and mental health of the persons playing online games using Support Vector Machine (SVM), XGBoost, Gradient Boost and Multilayer Perceptron (MLP) in order to find out the level of anxiety in online gamers. The results stated the accuracy of 99.96, 99.43, 98.54 and 98.04% for MLP SVM, Gradient Boost and XGBoost respectively. According to this study, there are more chances to have intense emotions of fear and anxiety in online gamers. Due to addiction of online gaming, more people are supposed to have Internet Gaming Disorder. For this study an experiment was conducted [15].

A study was conducted to investigate the neural connectivity biomarkers of Social Anxiety Disorder (SAD) based on Electroencephalography (EEG) data using Convolutional Neural Network (CNN), Long Short-Term Memory (LSTM), and CNN + LSTM. The brain effective connectivity was provided as input feature to these deep learning models using a dataset collected from 88 healthy respondents. These respondents were categorised into different level of SAD namely HC, mild, moderate, and severe based on their SIAS score [1].

It takes a significant amount of clinical experience to conduct a diagnostic assessment for autism spectrum disorder (ASD). According to this study, ASD assessment is very challenging when an individual is having multiple disorder at a same time. For this experiment dataset of 1251 patients were collected and patient have any of the following disorder; attention deficit hyperactivity disorder, conduct disorder, ASD and anxiety related problems. Machine learning algorithms like Attention-Deficit Hyperactivity Disorder (ADHD) models, random forest model, Continuous Delivery (CD) models and ANX models were used in this experiment (Schulte-Rüther et al. [23] have investigated the impact of the COVID-19 on undergraduate students in China who were transitioning to online learning. Specifically, this study aims at analysing the change in level of anxiety in students after joining the online classes. In order to examine this change, the authors have made use of XGBoost machine learning model and multiple stepwise regression classifier model. The study found that a significant proportion of students reported experiencing anxiety. Differences were also observed among different genders, majors, and provinces [22]. A study was conducted to examine the anxiety level based on neuroimaging biomarkers. Rather than predicting the future anxiety level in any individual, the authors tried to examine one's current anxiety level in their study. Two machine learning approaches which were stacked model and k-fold cross validation model were being used in order to analyse the anxiety level [6].

Anxiety and depression in children aged between 3 and 8 years are mostly go unnoticed and can cause long-term detrimental effects. So, an experiment was conducted on audio data using SVM with linear and gaussian kernel, Random Forest, and Logistic Regression [14].

Due to the present social distancing measures in place worldwide, many individuals have turned to social media in order to express their thoughts or emotions. Unfortunately, this has resulted in millions of people with mental illnesses feeling isolated and unable to receive in-person support. As a result, these individuals have become increasingly dependent on online platforms to get help or support for their mental illnesses. Ameer et al. analysed the multiple common mental illnesses among individuals whose data has been collected from online Reddit platform. Machine learning models, transfer learning model and deep learning models have been used in order to examine the different disorders in individuals. As a result, the authors have categorised the posts into different classes of disorders namely PTSD, Bipolar, Depression, ADHA, anxiety and none [2].

To evaluate the mood episodes in affective disorders, the semi-organised clinical rating scales and subjective observations are the most effective approaches. Behavioural patterns that can be observed externally through motor activity serve as an objective measure of an individual's inner physiological state, and changes in their level of activation are important indicators of unipolar and bipolar depression. The motor activity patterns in healthy and depressed controls were collected and used using Random Forest, CNN and Deep Neural Network (DNN) with different class balancing techniques [15]. To analyse the impact of distress on the education and professional development of pre-professional health students an experiment was conducted using regression and classification techniques [16]. Various other studies have been conducted to analyse the research work in the field of mental disorders, use of technology like machine learning, medical tools, virtual and augment reality for mental disorders [3, 11, 17–20]. The findings reveal that professional development students have a higher prevalence of anxiety and depression compared to estimates of the general population. Based on the above-mentioned literature, an experiment was conducted to find out the association between efficacy and mental disorder symptoms. For this experiment a public dataset was used. Detail of the performed experiment is provided in the sub subsequent sections.

3 Experimental Study

3.1 Dataset

To perform this experiment, the data has been taken from Kaggle (available at Medical Student Mental Health—Kaggle) [7]. The author of the dataset has collected the data about mental health of medical students which consists of demographic information (like mother tongue, gender and age), selfreported data like (health status, cognitive and affective empathy score through a rating scale) as well as internal measures like (academic year, job satisfaction, taken any physiotherapy during last 1 year or not). In order to record the mental state of students, a physiological test was conducted. The students were asked to fill the questionnaire as well during 2020–21. The Fig. 1 shows the sample of data collected.

id	age	year	sex	glang	part	job	stud_h	health	psyt	jspe	qcae_cog	qcae_aff	amsp	erec_mean	depression	anxiety	exhaustion	cynicism	efficacy
2	18	1	1	120	1	0	56	3	0	88	62	27	17	0.73809522	34	61	17	13	20
4	26	4	1	1	1	0	20	4	0	109	55	37	22	0.69047618	7	33	14	11	26
9	21	3	2	1	0	0	36	3	0	106	64	39	17	0.69047618	25	73	24	7	23

Fig. 1 Sample dataset

3.2 Association Study

Association study is performed in order to find out the correlation between the different parameters. It is used to determine whether there is a statistical association between the variables and the strength and direction of that association. In the dataset of mental health of medical students, the correlation association of different variables has been done in order to check the impact of one parameter on others.

It is important to note that correlation does not imply causation, and it is essential to consider other factors and potential confounding variables when interpreting the results of a correlation analysis. Correlation value ranges between −1 and 1. Table 1 shows the meaning of different values. Correlation can also be specified in form of 3 S namely Strength, Sign and Significance as shown in Table 2.

Python Simulation has been used to calculate the association between all variables. Figure 2 shows the correlation matrix for scale variables namely study of hours per week, total empathy score, cognitive empathy score, affective empathy score, academic motivation, emotion recognition score, depression score, anxiety score, exhaustion score, cynicism score and efficacy score calculated using the Python.

Table 1 Meaning of different association/correlation values

Correlation value (r)	Type of correlation	Description
−1	Negative correlation	Correlation value −1 means the two parameters/variables are negatively correlated with each other i.e., if value of one variable is increasing, the value of other will decrease or vice versa
0	No correlation	Correlation value 0 means the there is no association between two parameters/variables, or the parameters are independent of each other
1	Positive correlation	Correlation value 1 means the two parameters/variables are positively correlated with each other i.e., if value of one variable is increasing, the value of other will also increase or vice versa

Table 2 Meaning of each S

Strength	$r = \pm 0.35$ to ± 0.65	If r value is between $\pm$ 0.35 and $\pm$ 0.65, then it shows that the association id moderate between the variable. Greater value than $\pm$ 0.65 shows strong correlation
Significance	$p < 0.05$ and $p > 0.05$	If p value is less than 0.05 that shows the association between the variables are significant otherwise itis insignificant
Sign	(−, +)	Positive sign shows the positive correlation means if once variable is increasing the other. This will also increase or vice versa Negative value shows the negative correlation between the variables which means on increase of one value, other will decrease or vice versa

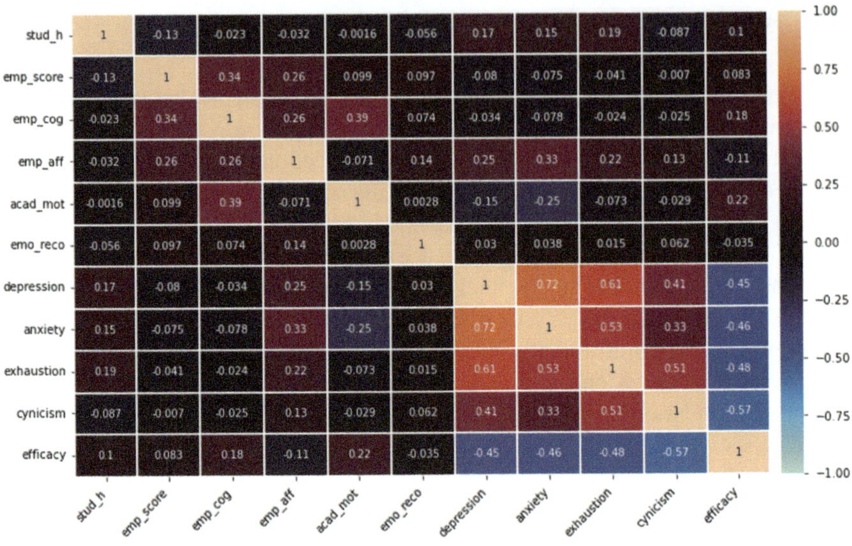

Fig. 2 Confusion matrix showing correlation between all scale parameters

4 Result Analysis

Through SPSS, the bivariate correlation between different parameters can be calculated which is shown in Fig. 3.

From the Fig. 3, we can see that depression and anxiety are positively correlated with each other with a correlation coefficient (r) of 0.72. The positive value shows that if depression score increases, the anxiety score will also increase or vice versa.

Correlation	stud_h	emp_score	emp_cog	emp_aff	acad_mot	emo_reco	depression	anxiety	exhaustion	cynicism	efficacy
stud_h	1	-.135	-.023	-.032	-.002	-.056	.174	.152	.186	-.087	.102
emp_score	-.135	1	.343	.263	.099	.097	-.080	-.075	-.041	-.007	.083
emp_cog	-.023	.343	1	.259	.387	.074	-.034	-.078	-.024	-.025	.184
emp_aff	-.032	.263	.259	1	-.071	.141	.251	.331	.216	.128	-.114
acad_mot	-.002	.099	.387	-.071	1	.003	-.152	-.249	-.073	-.029	.221
emo_reco	-.056	.097	.074	.141	.003	1	.030	.038	.015	.062	-.035
depression	.174	-.080	-.034	.251	-.152	.030	1	.716	.606	.408	-.454
anxiety	.152	-.075	-.078	.331	-.249	.038	.716	1	.530	.332	-.463
exhaustion	.186	-.041	-.024	.216	-.073	.015	.606	.530	1	.505	-.481
cynicism	-.087	-.007	-.025	.128	-.029	.062	.408	.332	.505	1	-.566
efficacy	.102	.083	.184	-.114	.221	-.035	-.454	-.463	-.481	-.566	1

Fig. 3 Results of correlation value (r) calculated through SPSS

Moreover, if we look at student efficacy score (which depict the student's ability to achieve academic success), then it decreases with the increase of depression score.

To evaluate the impact of depression on other key features, more analysis has been done whose results is given in Figs. 4, 5, 6 and 7. The correlation coefficient (r) of depression versus anxiety, exhaustion, cynicism, and efficacy has also been calculated using Python and the results of these are shown in Table 3. The correlation coefficient (r) value of depression and anxiety is 0.72 which shows that both the parameter is strongly associated with each other. It shows that if a student has depression, then that student has high chances of having anxiety as well. The impact of depression on exhaustion is also high but on cynicism it is moderate as compared to anxiety and exhaustion. The efficacy which depicts the students will power of achieving academic success decreases with the increase in depression as the correlation coefficient of depression and efficacy is −0.45.

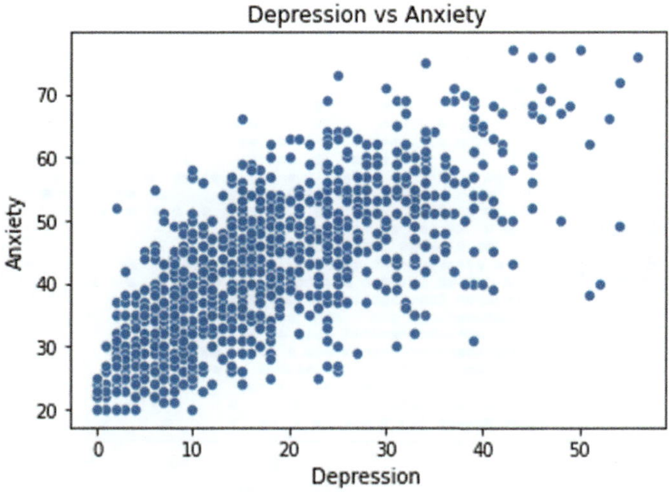

Fig. 4 Association between depression and anxiety

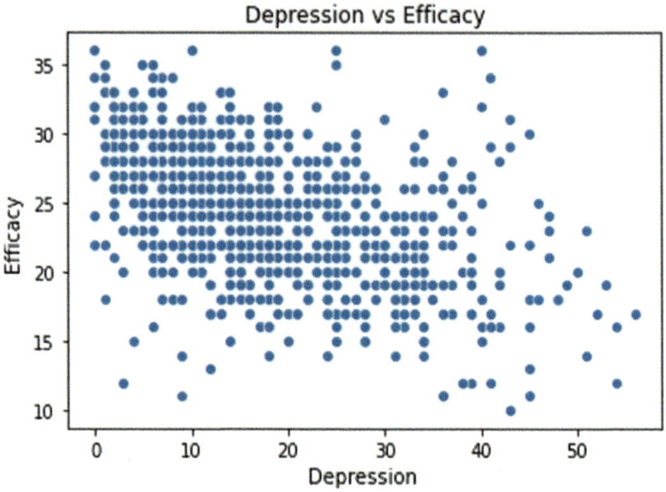

Fig. 5 Association between depression and efficacy

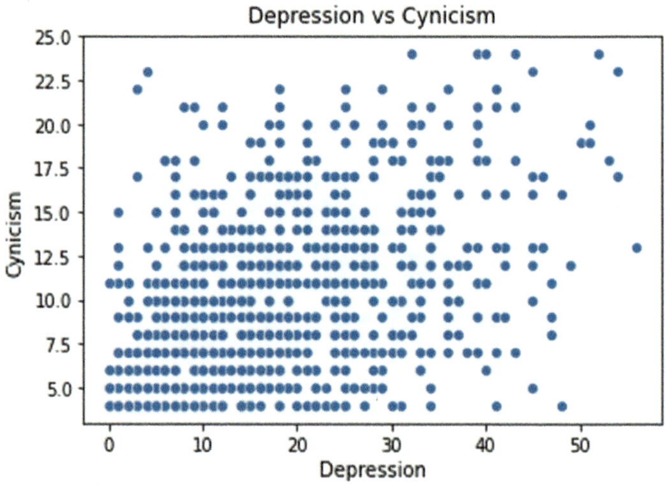

Fig. 6 Association between depression and cynicism

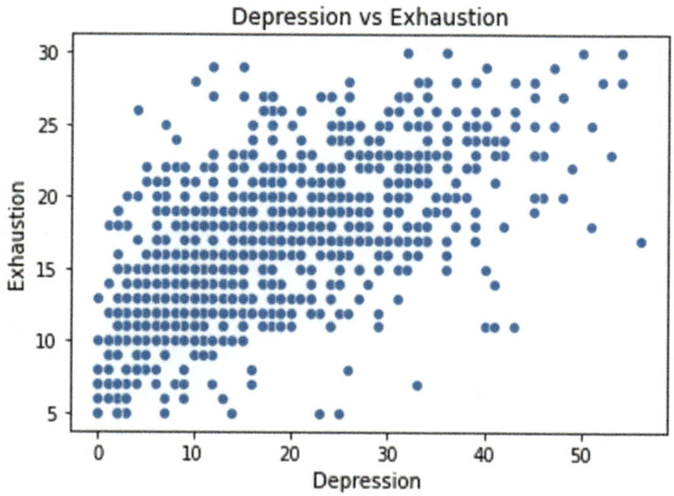

Fig. 7 Association between depression and exhaustion

Table 3 Correlation coefficient of depression versus other key features

Correlation coefficient (r)	Anxiety	Exhaustion	Cynicism	Efficacy
Depression	0.715728064	0.605617232	0.407727246	−0.45359

5 Conclusion

However, mental illness is a common and treatable condition, and seeking help is a sign of strength. It is important for any individual to recognize the importance of mental health and to work towards reducing the stigma associated with mental illness. Overall, depression, anxiety, exhaustion, and cynicism can all impact a person's academic efficacy, and can make it difficult to succeed in school.

The study concludes that the depression and anxiety can lead to feelings of exhaustion and cynicism, which can impact a person's sense of academic efficacy. When a person is feeling depressed or anxious, they may struggle to focus on their academic work or to feel motivated to succeed. This can lead to feelings of exhaustion and burnout, which can further erode a person's sense of academic efficacy. Studies have shown that depression and anxiety are strongly negatively correlated with academic efficacy with the correlation coefficient (r) equal to −0.45 and −0.46 respectively. As depression and anxiety increase, academic efficacy decreases. Exhaustion is strongly positively correlated with depression and anxiety with correlation coefficient of 0.61 and 0.53 respectively. As exhaustion increases, depression and anxiety also increase. Cynicism, on the other hand also shows positive association with depression and anxiety with correlation coefficient of 0.41 and 0.33 respectively but less as compared to exhaustion. This suggests that exhaustion and cynicism play a mediating role in the relationship between academic efficacy and mental health outcomes.

References

1. Al-Ezzi, A., Kamel, N., Faye, I., Gunaseli, E.: Analysis of default mode network in social anxiety disorder: EEG resting-state effective connectivity study. Sensors **21**(12), 4098 (2021). https://doi.org/10.3390/S21124098/S1
2. Ameer, I., Arif, M., Sidorov, G., Gòmez-Adorno, H., Gelbukh, A.: Mental illness classification on social media texts using deep learning and transfer learning (2022). https://doi.org/10.48550/arxiv.2207.01012
3. Anjali, Singh, G., Kaur Sandhu, J.: Virtual and augmented reality technology for the treatment of mental health disorders: an overview. In: 2022 13th International Conference on Computing Communication and Networking Technologies, ICCCNT 2022, pp. 1–5 (2022). https://doi.org/10.1109/ICCCNT54827.2022.9984551
4. Arif, M., Basri, A., Melibari, G., Sindi, T., Alghamdi, N., Altalhi, N., Arif, M.: Classification of anxiety disorders using machine learning methods: a literature review. Pdfs.Semanticscholar.Org. (2020). https://doi.org/10.36959/584/455
5. Bhattacharya, T., Soares, G.A.B.E., Chopra, H., Rahman, M.M., Hasan, Z., Swain, S.S., Cavalu, S.: Applications of Phyto-nanotechnology for the treatment of neurodegenerative disorders. Materials **15**(3), 804 (2022). https://doi.org/10.3390/MA15030804
6. Boeke, E.A., Holmes, A.J., Phelps, E.A.: Toward robust anxiety biomarkers: a machine learning approach in a large-scale sample. Biol. Psychiatr.: Cogn. Neurosci. Neuroimaging **5**(8), 799–807 (2020). https://doi.org/10.1016/J.BPSC.2019.05.018
7. Carrard, V., Bourquin, C., Berney, S., Schlegel, K., Gaume, J., Bart, P.A., Preisig, M., Schmid Mast, M., Berney, A.: The relationship between medical students' empathy, mental health, and burnout: a cross-sectional study. Med. Teach. **44**(12), 1392–1399 (2022). https://doi.org/10.1080/0142159X.2022.2098708
8. Dugas, M.J., Gagnon, F., Ladouceur, R., Freeston, M.H.: Generalized anxiety disorder: a preliminary test of a conceptual model. Behav. Res. Ther. **36**(2), 215–226 (1998). https://doi.org/10.1016/S0005-7967(97)00070-3
9. Van Eeden, W.A., Luo, C., van Hemert, A.M., Carlier, I.V., Penninx, B.W., Wardenaar, K.J., Hoos, H., Giltay, E.J.: Predicting the 9-year course of mood and anxiety disorders with automated machine learning: a comparison between auto-SKLEARN, naïve Bayes classifier **299**, 113823. Elsevier (2021). https://www.sciencedirect.com/science/article/pii/S0165178121001207
10. Ertek, I.E., Ozkan, S., Candansayar, S., Ilhan, M.N.: Depression, stress, burnout and associated factors in medical students: a cross-sectional study from a Turkish medical school. Psychiatr. Behav. Sci. **11**(2), 130–140 (2021). https://doi.org/10.5455/PBS.20180715050817
11. Kang, H.K., John, D., Bisht, B., Kaur, M., Alexis, O., Worsley, A.: PROTOCOL: Effectiveness of interpersonal psychotherapy in comparison to other psychological and pharmacological interventions for reducing depressive symptoms in women diagnosed with postpartum depression in low and middle-income countries: a systematic review. Campbell Syst. Rev. **16**(1), e1074 (2020). https://doi.org/10.1002/CL2.1074
12. Madigan, D.J., Curran, T.: Does burnout affect academic achievement? A meta-analysis of over 100,000 students. Educ. Psychol. Rev. **33**(2), 387–405 (2021). https://doi.org/10.1007/S10648-020-09533-1/METRICS
13. Mantri, A., Dutt, S., Gupta, J.P., Chitkara, M.: Design and evaluation of a PBL-based course in analog electronics. IEEE Trans. Educ. **51**(4), 432–438 (2008). https://doi.org/10.1109/TE.2007.912525

14. McGinnis, E.W., Anderau, S.P., Hruschak, J., Gurchiek, R.D., Lopez-Duran, N.L., Fitzgerald, K., Rosenblum, K.L., Muzik, M., McGinnis, R.S.: Giving voice to vulnerable children: machine learning analysis of speech detects anxiety and depression in early childhood. IEEE J. Biomed. Health Inform. **23**(6), 2294–2301 (2019). https://doi.org/10.1109/JBHI.2019.2913590
15. Rahman, A.A., Khalid, L.I., Siraji, M.I., Nishat, M.M., Faisal, F., Ahmed, A.: Enhancing the performance of machine learning classifiers by hyperparameter optimization in detecting anxiety levels of online gamers. In: 24th International Conference on Computer and Information Technology, ICCIT 2021, pp. 1–5 (2021). https://doi.org/10.1109/ICCIT54785.2021.9689911
16. Rahmati, Z.: The study of academic burnout in students with high and low level of self-efficacy. Procedia Soc. Behav. Sci.i. **171**, 49–55 (2015). https://doi.org/10.1016/J.SBSPRO.2015.01.087
17. Sethi, M., Ahuja, S., Bawa, P.: Classification of Alzheimer's disease using neuroimaging data by convolution neural network. In: Proceedings of IEEE International Conference on Signal Processing, Computing and Control, October 2021, pp. 402–406 (2021). https://doi.org/10.1109/ISPCC53510.2021.9609431
18. Singh, S., Singh, J.: Analysis of GPS trajectories mapping on shape files using spatial computing approaches. In: Big Data Analytics: 8th International Conference, BDA 2020, Sonepat, India, 15–18 December 2020, Proceedings 8. LNCS, vol. 12581, pp. 91–100 (2020). https://doi.org/10.1007/978-3-030-66665-1_7/COVER
19. Singh, S., Singh, J.: Intrinsic parameters based quality assessment of Indian OpenStreetMap dataset using supervised learning technique. In: 2020 Indo–Taiwan 2nd International Conference on Computing, Analytics and Networks (Indo-Taiwan ICAN), pp. 52–57 (2020). https://doi.org/10.1109/INDO-TAIWANICAN48429.2020.9181313
20. Singh, S., Singh, J.: Location driven edge assisted device and solutions for intelligent transportation. In: Fog, Edge, and Pervasive Computing in Intelligent IoT Driven Applications, pp. 123–147 (2020). https://doi.org/10.1002/9781119670087.CH7
21. Tahmassian, K., Moghadam, N.J.: Relationship between self-efficacy and symptoms of anxiety, depression, worry and social avoidance in a normal sample of students. Iran. J. Psychiatr. Behav. Sci. **5**(2), 91 (2011). /pmc/articles/PMC3939966/
22. Wang, C., Zhao, H., Zhang, H.: Chinese college students have higher anxiety in new semester of online learning during COVID-19: a machine learning approach. Front. Psychol. 11, 587413 (2020). https://doi.org/10.3389/fpsyg.2020.587413
23. Schulte-Rüther, M., Kulvicius, T., Stroth, S., Wolff, N., Roessner, V., Marschik, P.B., Kamp-Becker, I., Poustka, L.: Using machine learning to improve diagnostic assessment of ASD in the light of specific differential and co-occurring diagnoses. J. Child Psychol. Psychiatry **64**(1), 16–26 (2023). https://doi.org/10.1111/jcpp.v64.1

Validating Data Warehouse Requirements Decomposition Metrics Formally by Implementing Zuse's Formal Framework

Tanu Singh, Manu Singh, Arvind Panwar, and Urvashi Sugandh

Abstract Data warehouse contains historical data to assist in organizational decision-making. Requirements one of the most important data models that influences the accuracy of the information kept in the data warehouse and used to make critical organizational decisions. Consequently, evaluating the data warehouse's information quality becomes crucial. There aren't many studies in the literature that guarantee the accuracy of the data model—requirements for a data warehouse. However, very less work was witnessed in the literature to validate requirements completeness (decomposition and specification to be completed) metrics for evaluating the requirements model quality. Hence, in this paper Zuse's formal framework was used to theoretically validate the requirements decompostion metrics (based on agent goal decision information model) and the results demonstrate that every requirement decomposition completeness metric is legitimate and accurately formulated. In order to ensure the accuracy of the requirements model for the data warehouse, requirements decomposition completeness metrics can be utilized.

Keywords Data warehouse · Formal validation · Information quality · Requirements decomposition metrics · Requirements engineering

---

T. Singh (✉)
School of Computer Science, UPES, Dehradun, Uttarakhand, India
e-mail: tanusnigam@gmail.com

M. Singh · A. Panwar
School of Computing Science and Engineering, Galgotias University, Greater Noida, India
e-mail: manu.singh@galgotiasuniversity.edu.in

U. Sugandh
Department of Artificial Intelligence and Machine Learning, Raj Kumar Goel Institute of Technology, Ghaziabad, India

B. Soni et al. (eds.), *Beyond Artificial Intelligence*, Lecture Notes in Networks and Systems 1326, https://doi.org/10.1007/978-981-96-4170-3_50

1 Introduction

Data warehouse (DW) provides historical information to support decision making activities in an organization [1]. Therefore, it becomes important to assess information quality of DW since early stages of DW design and development. Furthermore, poor DW information quality may influence organizational choices, which may result in losses of both money and clients. As a result, ensuring the quality of DW information becomes crucial for an organization.

The quality of the information it carries, as shown by the solid lines in Fig. 1, is influenced by the presentation quality of DW and the quality of the DW itself, in accordance with Serrano et al. [2]. In addition, the quality of the DW is equally impacted by the quality of the database management system (DBMS), the data, and the data model. Moreover, the main factors that influence DW information quality are requirements, conceptual, logical and physical data models. Requirements data model quality was suggested by Kumar et al. [3] which is shown as an extension with dotted line in Fig. 1.

According to numerous researchers, the reason why DW projects fail is because they don't explicitly incorporate the requirements and merely deal with various data models, such as conceptual, logical, and physical data models. Therefore, in this direction, several procedures for DW were projected in the literature suggesting prime importance of requirements engineering (RE) phase for DW [4]. Despite the fact that none of the methods for RE take into account any objective quality metrics of the DW requirements schemas, the information quality could not have been ensured by simply methods. Consequently, various metrics were suggested and validated theoretically as well as empirically to assess data model quality. Additionally, metrics

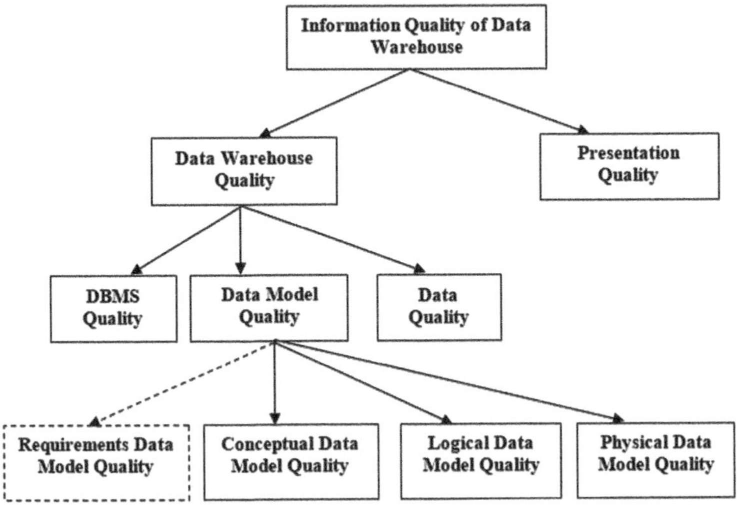

Fig. 1 Information quality of DW [2, 3]

have always been favored in order to objectively determine the quality feature of software artefacts.

Several researchers have made an effort to evaluate the logical model level of data models [5, 6], physical model level [1, 7] and conceptual model level [5, 8] for DW. However, relatively few research papers were found in the literature [9–11] to theoretically and empirically validate the DW requirements metrics.

While the practical utility of a metric for evaluating external traits can be demonstrated through experimental validation, such as maintainability, understandability, etc., and formal validation of a metric comprises mathematical aspects that need to be proved by the each metric. Thus, formal validation of requirements decomposition metrics is carried out which is based on agent goal decision information (AGDI) model [11] by implementing Zuse's Framework [12] to show that (i) they are correctly defined and are valid (ii) they have predictable relationships with external attributes that can be used to assess the DW's quality.

The followed paper is structured as follows: Sect. 2 discusses literature survey of requirements models quality issues of DW, which provides context for evaluating the metrics for DW requirements completeness, which are detailed in Sect. 3. Section 4 covers the formal validation of RD metrics by employing Zuse's formal framework. Section 5 discusses the findings of the formal validation in last the conclusion and future work is discussed in Sect. 6.

2 RE Approaches Literature Survey

As it is clear that all above RE approaches were insufficient to assess the DW requirements data models quality as these approaches must be proven theoretically and empirically. Unfortunately, there isn't much research in the literature for evaluating requirements data models' quality both subjectively and objectively. [3, 9–11, 13–15]. Due to a lack of space, this study solely focuses on the quality of the requirements data models. The challenges with DW requirements data models are covered in detail in the following subsection.

To enhance requirements data model quality, different guidelines were recommended by Kumar et al. [3], which was an expansion of the current AGDI paradigm. Further, to evaluate the quality of the requirements model, Kumar et al. [13] proposed requirements completeness and traceability metrics. Later, Kumar [14] used Briand's framework to formally validate only requirements traceability metrics where only size measure was used. Also, to assess DW requirements data model quality in an empirical way [15] only traceability metrics was used for predicting DW requirements schemas understandability using support vector machine (SVM) method. Moreover, Singh and Kumar [9] formally validate the traceability metrics to demonstrate that they are validly specified by applying Briand's framework's cohesion and coupling measures. Further, Singh and Kumar [10] empirically studied the requirements metrics for predicting DW requirements schemas understandability by employing machine learning techniques. Additionally, Singh and Kumar [11] study the metrics for requirements completeness both theoretically and practically, employing Briand's property-based framework and correlation analysis and linear regression. Recently, Singh and Kumar [11] validate the requirements traceability metrics theoretically using Zuse's framework. However, from above very few research work was seen the literature to validate the completeness metrics of DW requirements data model in formal and empirical manner [10, 11] both. As a result, formal validation of requirements decomposition metrics is performed in this study using the agent goal decision information (AGDI) model [4] and the Zuse Framework [12] to demonstrate that:

(i) they are correctly defined and are valid.
(ii) they have dependable connections to outside characteristics that can be utilized to judge the effectiveness of the DW.

3 AGDI Model and Requirements Metrics Used in This Study

Kumar et al. [4] proposed an AGDI model which was refined as a comprehensive approach for DW design that includes requirements modeling followed by conceptual modeling and also supported by the CASE tool (Kumar et al. 2016). Further, for assessing quality of DW requirements model in an objective manner, Kumar et al. [13] suggested requirements completeness metrics (see Table 1) using AGDI model [4]. Further, requirements completeness metrics was divided into two categories i.e. (i) requirements decomposition (RD) metrics and (ii) specification to be completed (STBC) metrics. In this paper, only RD metrics have been validated theoretically. These metrics provide complete requirements as an input to the conceptual design of DW which further forms a quality based DW. Here, in this study, for better understanding of requirements model of DW, we have taken e- Commerce Company as an example of requirements schema (see Fig. 2). The detailed definitions of 06 RD completeness metrics with the calculated values of e- Commerce Company are shown in Table 1.

Table 1 RD metrics [13] and the collected values of DW e-commerce company schema

Requirements completeness metrics		Explanation	Values
RD metrics	NSA	Count of simple agent	5
	NGH	Count of hierarchical goals	2
	NSG	Count of simple goals	2
	NDH	Count of decision hierarchies	3
	NSD	Count of simple decision	3
	NI	Count of information Requirements	5

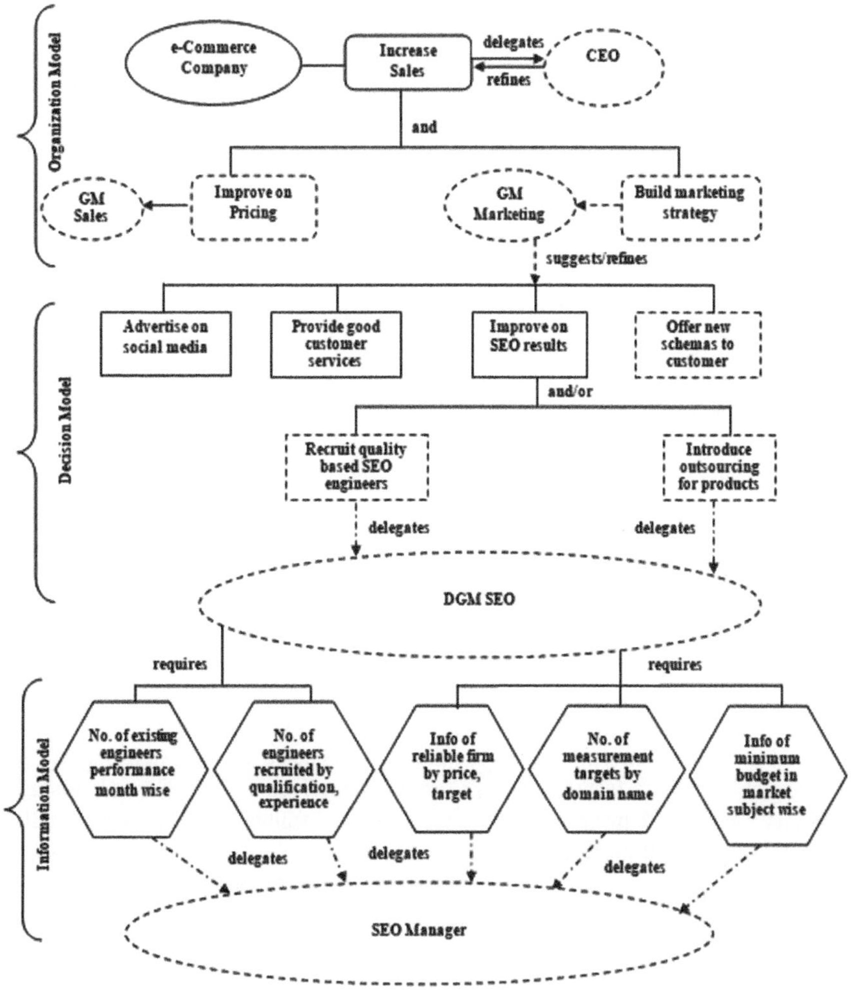

Fig. 2 E-commerce company requirements schema

4 Formal Validation of Requirements Decomposition Metrics

The metrics for requirements RD metrics satisfy Zuse's framework's [12] features have been described in this section. The scale that corresponds to the metric system is determined by this framework using measurement theory. Yet, one should be aware of the propensity for this idea to calculate empirical relations between the objects, such as "greater than", "more than or equal to," "less than," etc. Assume that there is an empirical relational system *(X,* ●≥, ○) where *X* is element of group of not-empty objects, ●≥ refers to the empirical binary relationship to *X* and ○ is a closed binary relation on X that concatenates functions. Now let's look at a numerical relational system $(Y, \geq, +)$ where Y is member of of real numbers, $\geq$ relates to Y in a binary way and $+$ is an operation in binary on Y. As a result, the following definition of a mapping function for a measure *map* is provided below.

m: $X \rightarrow Y$ such that $\forall X_1, X_2 \in X$ then the two conditions listed below are satisfied, i.e.,

a. $X_1 \bullet\geq X_2 \Longleftrightarrow map\ (X_1) \geq map\ (X_2)$
b. $map\ (X_1 \circ X_2) = map\ (X_1) + map\ (X_2)$.

As a result, (X, Y, *map*) is defined as a scale that can measure *map*.

Modified relation of belief (MRB), Modified extensive structures (MES), independence conditions (IC) are three further key mathematical characteristics of Zuse's framework [12] for measuring software. These mathematical features are described in Table 2. These mathematical principles can be used to determine the metric scale. As a result, the metric is referred to as a ratio scale if it fulfils the MES criteria. If the metric meets the IC criteria but not the MES properties, the second scenario employs an ordinal scale. Finally, the metric is considered to be "above" ordinal scale if it satisfies the MRB characteristics. Furthermore, ordinal numbers cannot be used for many things, therefore only metric categorization above ordinal scale is thought to be practical and beneficial [12].

Consider a relational empirical system that is used to verify metrics for requirements completeness. $G = (G, \bullet\geq, \circ)$, where G signifies total goals (non-empty set), $\bullet\geq$ shows a empirical relationship on G and $\circ$ denotes concatenation i.e. closed relationship on G. The simple goals are then traced back to the decision and information needs, which are further refined from the complicated goals. Formal validation of NGH metric is demonstrated here using Zuse's framework [12] which will be explained in the following sections. Furthermore, NSA, NSG, NDH, NSD, and NI NI metrics are "above" ordinal scale and can be theoretically validated in a manner that is analogous to the NGH metric.

4.1 Validating Formally NGH Metric

To illustrate an empirical relationship [12], a new operation of concatenation is included. In *GoalsConcat*, a closed concatenation ($\circ$) is explained. In this operation, the goal hierarchies G1 and G2 concatenation results in a goal G that includes all the G1 and G2 features, which show common features only once in the final goal G, while conquering cohesions (if any) in the primary goals (as depicted in Fig. 3).

In terms of mapping, the NGH completeness metric is measured: NGH: $G \rightarrow G_1$ and G_2 (G stands for the quantity of simple goals and complex goals, as shown in Fig. 3 and the ones who are in bold rectangles, respectively), such that it maintains all G_1 and G_2 terms.

$$G_1 \text{and } G_2 \in G : G_1 \bullet \geq G_2 \Leftrightarrow \text{NGH}\ (G_1) \geq \text{NGH}\ (G_2)$$

Table 2 Mathematical properties of Zuse's framework [12]

Zuse's properties	Definition
MES—Axiom 1	(A ●≥) (weak order)
MES—Axiom 2	A1 o A2 ●≥ A1 (positivity)
MES—Axiom 3	A1 o (A2 o A3) ≈ (A1 o A2) o A3 (weak associativity)
MES—Axiom 4	A1 o A2 ≈ A2 o A1 (weak commutativity)
MES—Axiom 5	A1 ●≥ A2 ⇒ A1 o A ●≥ A2 o A (weak monotonicity)
MES—Axiom 6	If A3 ●> A4 then for any A1and A2, then there exists a natural number n, such that A1o nA3 ●> A2 o nA4 (Archimedean axiom)
Independence C1	**C1**: A1 ≈ A2 ⇒ A1 o A ≈ A2 o A and A1 ≈ A2 ⇒ A o A1 ≈ A o A2
Independence C2	**C2**: A1 ≈ A2 ⟺ A1 o A ≈ A2 o A and A1 ≈ A2 ⟺ A o A1 ≈ A o A2
Independence C3	**C3**: A1 ●≥ A2 ⇒ A1 o A ●≥ A2 o A, and A1 ●≥ A2 ⇒ A o A1 ●≥ A o A2
Independence C4	**C4**: A1 ●≥ A2 ⟺ A1 o A ●≥ A2 o A, and A1 ●≥ A2 ⟺ A o A1 ●≥ A o A2 Where A1 ≈ A2 if and only if A1 ●≥ A2 and A2 ●≥ A1, and A1 ●> A2 if and only if A1 ●≥ A2 and not (A2 ●≥ A1)
MRB1	∀ A, B ∈ : A ●≥ B or B ●≥ A (completeness)
MRB2	∀ A, B, C ∈ : A ●≥ B and B ●≥ C ⇒ A ●≥ C (transitivity)
MRB3	∀ A ⊇ B ⇒ A ●≥ B (dominance axiom)
MRB4	∀ (A ⊃ B, A ⋂ C = ϕ) Þ (A ●≥ B ⇒ A U C ●≥ B U C) (partial monotonicity)
MRB5	∀ A ∈ : A ●≥ 0 (positivity)

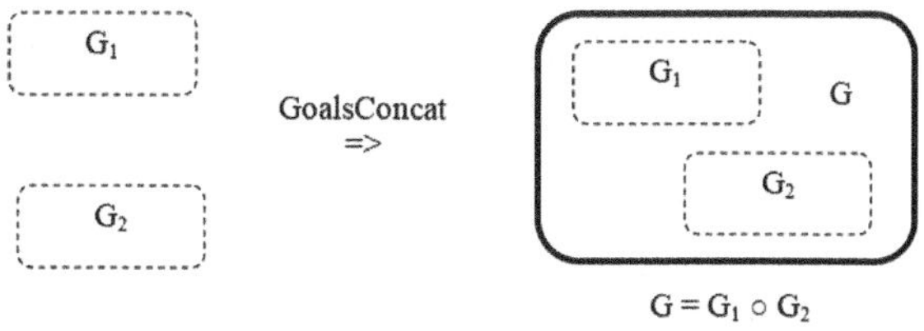

Fig. 3 Two goals concatenation: *GoalsConcat*

where G_1 and G_2 are simple goals (dotted rounded rectangle as depicted in Fig. 3).

The concatenation rule states that, NGH ($G_1 \circ G_2$) = NGH (G_1) + NGH (G_2) − v, i.e., The total goals counts considering the hierarchy common to G_1 and G_2 equals the total number of goals considering the goal hierarchy of G_1 and G_2 just once, remaining unaltered in all other instances.

Properties of MES in NGHs: The six axioms of MES are explained in Table 2 and are described below.

MES—Axiom 1 (Week order): Let's say that a requirements model has three goals i.e., G_1, G_2, G_3, and it is clear that:

$$\text{NGH}\ (G_1)\ \geq\ \text{NGH}\ (G_2)\ \text{or NGH}\ (G_2)\ \geq\ \text{NGH}\ (G_1)\ \text{and also, if}$$

$\text{NGH}\ (G_1) \geq \text{NGH}\ (G_2)$ and $\text{NGH}\ (G_2) \geq \text{NGH}\ (G_3) \Rightarrow \text{NGH}\ (G_1) \geq \text{NGH}\ (G_3)$. NGH thus fulfills the first axiom of MES.

MES—Axiom 2 (Positivity): The NGH metric satisfies this premise. If G_1 and G_2 are combined, the sum is always equal to or greater than NGH (G_1), provided that G_2 contains non-hierarchical decisions. Hence it can be written as,

$$\text{NGH}\ (G_1 \circ G_2)\ \geq\ \text{NGH}\ (G_1)$$

MES—Axioms 3 (Weak associativity) and *4 (Weak commutativity)*: This is satisfied by the NGH metric because, according to *GoalsConcat*, the natural join operation is both commutative and associative. The goals' concatenation order has no bearing on the number of goals that can be traced backward through the decisional hierarchy.

MES—Axiom 5 (Weak monotonocity): NGH metric does not satisfy this axiom. As a result of any decisional hierarchy of goal (G) that is common to G_1 but not G_2 existing, NGH ($G_1 \circ$ G) value in contrast to value of NGH ($G_2 \circ$ G) may not be greater than or equal to NGH ($G_2 \circ$ G) value when the goal (G) is concatenated to both G1 and G2.

MES—Axiom 6 (Archimedean): Goal concatenation (*GoalsConcat*) is idempotent, or an operation that can be performed several times without changing the result, hence the NGH metric does not satisfy this axiom. Thus, concatenation of G_1 and G_2 I times) equals concatenation of G_1 and G_2 I times) (s time).

NGH as an IC The following section and Table 2 outline the characteristics of IC.

The NGH metric cannot satisfy conditions C1 and C2 because *Axiom5 (Weak monotonocity)* is not satisfied. Axiom 5 was not accomplished, therefore C3 does not satisfy, which is similar to the second criterion, C2, and if C3 is not met, then C4 cannot be met. As a result, some IC attributes are not satisfied by NGH metric.

NGH as an MRB Table 2 described the five MES characteristics.

NGH satisfy both the first (MRB1) and second (MRB2) requirements for completeness and transitivity since they support axiom 1.

As all of the overall decisions of G_2 are included in G_1, NGH satisfies property (***MRB3***), dominance, which means all decisions hirarchy of G_2 are included in G_1. Hence, we can state that:

$$\text{NGH}\ (G_1)\ \geq\ \text{NGH}\ (G_2)$$

MRB4 The NGH metric satisfies partial monotonicity as $G_1 \subset G_2$, then NGH (G_1) > NGH (G_2). Now, imagine that G3 and G1 are making non-hierarchical decisions. Hence, G3 and G2 cannot share a decision hierarchy. Hence, we can state that,

$$\text{NGH}\,(G_1 \circ G_3) > \text{NGH}\,(G_2 \circ G_3)$$

Finally, the NGH metric satisfies MRB5 because it can never be negative and if no hierarchical decision is on G, then NGH (G) = 0.

As a result of satisfying all MRB requirements, the NGH metric is "above" the ordinal scale. Moreover, completeness measures like NSA, NSG, NDH, NSD, and NI can be validated theoretically in the similar manner as NGH metrics and are considered to be "above" ordinal scale.

5 Discussion and Results

This section (see Table 3) displays the findings of formal validation of requirements RD metrics using Zuse's framework [12]. The Zuse framework's features in Sect. 4 are supported by the entire theoretical validation of NGH measurements. Moreover, validation of the NSA, NSG, NDH, NSD, and NI measures can be performed similarly to that of the NGH metric and they are referred to as being "above ordinal scale". The finding is ordinal numbers are not suitable for analysis when the metric should be above ordinal scale.

Therefore, it can be concluded from Zuse's framework's findings [12] that the RD completeness metrics for requirements offered by [13] are properly defined and logically supported (see Table 3). As a result, the RD completeness metrics thus show the theoretical value contributed to the development, design and evolution of the requirements model. In Table 3, If metric satisfy all the properties of Zuse's framework then it is considered as True (01) otherwise False (00).

6 Conclusion

Here, theoretically validated metrics for requirements decomposition have been generated using the DW e-commerce company requirements model (shown in Sect. 3). The whole mathematical procedure of NGH metric has been validated theoretically (see Sect. 4), demonstrating that practically all of the MEB and MRB's properties are assured within the perspective of Zuse's framework. MRB measures are generally seen as significant and as being "above" the ordinal scale. Results support the conclusion that all RD metrics are accurately specified and legitimate because they are "above" the ordinal scale (see Sect. 5). As a result, RD completeness metrics can be used to gauge how well a DW requirements model is performing

Table 3 Results of RD metrics implementing the formal framework (Zuse's [12])

Requirements metrics →	Requirements decomposition (RD) metrics					
Zuse's properties ↓	NSA	NGH	NSG	NDH	NSD	NI
MES—Axiom 1	01	01	01	01	01	01
MES—Axiom 2	01	01	01	01	01	01
MES—Axiom 3	01	01	01	01	01	01
MES—Axiom 4	01	01	01	01	01	01
MES—Axiom 5	01	01	01	01	01	01
MES—Axiom 6	00	00	00	00	00	00
Independence C1	00	00	00	00	00	00
Independence C2	00	00	00	00	00	00
Independence C3	00	00	00	00	00	00
Independence C4	00	00	00	00	00	00
MRB1	01	01	01	01	01	01
MRB2	01	01	01	01	01	01
MRB3	01	01	01	01	01	01
MRB4	01	01	01	01	01	01
MRB5	01	01	01	01	01	01

to build the qaulity based DW. In order to ensure the accuracy of the requirements model for the data warehouse, requirements decomposition completeness metrics can be utilized.

References

1. Inmon, W.H.: Building the Data Warehouse. Wiley (2005)
2. Serrano, M., Trujillo, J., Calero, C., Piattini, M.: Metrics for data warehouse conceptual models understandability. Inf. Softw. Technol. **49**(8), 851–870 (2007)
3. Kumar, M., Gosain, A., Singh, Y.: Quality-oriented requirements engineering approach for data warehouse. Int. J. Comput. Syst. Eng. **1**(2), 127–138 (2012)
4. Kumar, M., Gosain, A., Singh, Y.: Stakeholders driven requirements engineering approach for data warehouse development. J. Inform. Process. Syst. **6**(3), 385–402 (2010)
5. Serrano, M.: Definition of a set of metrics for assuring data warehouse quality. Univeristy of Castilla, La Mancha, Spain (2004)
6. Gaur, H., Kumar, M.: Assessing the understandability of a data warehouse logical model using a decision-tree approach. ACM SIGSOFT Softw. Eng. Notes **39**(5), 1–6 (2014)
7. Kimball, R., Ross, M., Thornthwaite, W., Mundy, J., Becker, B.: The Data Warehouse Lifecycle Toolkit. Wiley (2008)
8. Kumar, M., Gosain, A., Singh, Y.: Empirical validation of structural metrics for predicting understandability of conceptual schemas for data warehouse. Int. J. Syst. Assur. Eng. Manag. **5**(3), 291–306 (2014)

9. Singh, T., Kumar, M.: Formally investigating traceability metrics of data warehouse requirements model using Briand's framework. In: 2021 5th International Conference on Intelligent Computing and Control Systems (ICICCS), pp. 1203–1209. IEEE (2021)
10. Singh, T., Kumar, M.: Investigating requirements completeness metrics for requirements schemas using requirements engineering approach of data warehouse: a formal and empirical validation. Arab. J. Sci. Eng.1–20 (2021)
11. Singh, T., Kumar, M.: Theoretical validation of data warehouse requirements metrics based on agent goal decision information model using Zuse's framework. In: Communication and Intelligent Systems: Proceedings of ICCIS 2021, pp. 107–118. Springer Nature Singapore, Singapore (2022)
12. Zuse, H.: A Framework of Software Measurement. Walter de Gruyter (1998)
13. Kumar, M., Gosain, A., Singh, Y.: On completeness and traceability metrics for data warehouse requirements engineering. Int. J. Comput. Syst. Eng. **1**(4), 229–237 (2013)
14. Kumar, M.: Validation of data warehouse requirements-model traceability metrics using a formal framework. In: 2015 2nd International Conference on Computing for Sustainable Global Development (INDIACom), pp. 216–221. IEEE (2015)
15. Singh, T., Kumar, M.: Empirical validation of requirements traceability metrics for requirements model of data warehouse using SVM. In: 17th India Council International Conference (INDICON), pp. 1–5. IEEE, New Delhi (2020)

Exploring Genomic Sequence Encoding Methods for 1D Convolutional Neural Networks: A Comparative Study

Shikha Sharma, Nidhi Kalra, and Rinkle Rani

Abstract Efficient strategies are necessary to encode DNA sequence characteristics in a suitable format for DNA motif discovery or DNA classification tasks in deep learning neural network models. The selection of an encoding scheme is based upon the desired form of representation, whether it be a vector or an n-dimensional matrix, as well as the particular problem at hand and the underlying architecture of the deep learning model being employed. Convolutional neural networks (CNNs) often utilise a grid-based representation by employing either one-hot encoding or k-mers to represent a given genome or protein sequence. The utilisation of one-hot encoding and label encoding is highly valuable in the context of Recurrent Neural Networks (RNNs) and Long Short-Term Memory (LSTM) models, as the precise positioning of characters at each time step plays a critical role. In order to capture intricate properties, it is possible to convert characters into vectors, which can then be employed as embeddings. Thus, the objective of this study is to conduct a comparative analysis of four distinct feature encoding algorithms for genomic sequences, implemented within the framework of a convolutional neural network. In order to streamline the procedure, we employ nucleobase sequences of predetermined length as the input data. The results of this study indicate that word embeddings exhibit superior performance compared to alternative encoding techniques in the context of genomic data analysis.

Keywords Character encoding techniques · Genome sequences · Convolutional neural networks

S. Sharma (✉) · N. Kalra · R. Rani
Department of Computer Science and Engineering, Thapar Institute of Engineering and Technology, Patiala 147004, Punjab, India
e-mail: ssharma_phd20@thapar.edu

B. Soni et al. (eds.), *Beyond Artificial Intelligence*, Lecture Notes in Networks and Systems 1326, https://doi.org/10.1007/978-981-96-4170-3_51

1 Introduction

1.1 Theoretical Background

Deep Learning, powered by Artificial Neural Networks (ANNs), has emerged as a versatile modelling approach applicable across a wide range of applications [1]. The approach capture abstractions from large datasets through the utilisation of multi-layered deep neural networks (DNNs). This enables the interpretation and comprehension of various forms of data, including images, sounds, and texts. Deep learning, as a field, possesses two fundamental characteristics: (i) the ability to process and analyse several layers of data representations, and (ii) the capacity to automatically learn and extract patterns and features from complex datasets. The framework to implement the characteristics consists of two main components: (i) a series of nonlinear processing units arranged in layers, and (ii) the utilisation of supervised or unsupervised learning techniques to extract feature representations at each layer [2]. The model has ability to automatically learn hierarchical representations of data, allowing it to capture complex patterns and relationships. The fields of DL have been extensively studied and researched in bioinformatics [3–5]. Common tasks such as identification of regulatory motifs in a DNA/RNA sequence, which improves medical treatments and understanding cell processes [5]. During the spread of COVID-19, many articles on evolutionary study of virus, understanding structure of virus and, drug designing emerged in huge quantity [3].

Encoding schemes for text in natural language processing (NLP) have many things in common with encoding techniques of genomics. There exits large database for text processing and searching. Both the field involves symbolic data to be converted into numerical representations as a pre-processing step before a classifier is applied. But there exists uniqueness in data pertaining to NLP tasks. NLP majorly focuses on semantics, text retrieval and searching tasks. Whereas, in genomics the emphasis is on the establishment of related biological sequences or identification of biologically relevant motifs of interest within sequences. Reading text in NLP minorly differs with reading texts in genome [6]. In this article, we undertake a comprehensive comparison of feature encoding methods commonly employed in Natural Language Processing (NLP) and adapt them for use in genomics. These techniques are utilized to extract meaningful features from genomic sequences, which are then input into convolutional neural networks for further analysis and modelling.

2 Related Works

Authors used physiochemical and structural properties of DNA sequence to convert from sequences to feature vectors. It is a web-based solution called "Seq2Feature" [7]. Liu Z and co-authors extracted features using frequency count of monomers, dimers (dinucleotide bases) with their proposed technique of using joint probabilities to compute Base–Base correlation for feature extraction [8]. Dlamini et al. [9] used only dinucleotide frequency as a genomic feature or a signature to classify SARS-CoV-2 sequences using XGBoost machine learning algorithm. The authors [10] also used dinucleotide bases to transform a sequence to 4^2 (16-dimension) vector and classified bacterial genomes using artificial neural network. The authors [11] used Chaos Game representation (CGR) technique to model and visualize sequences, that analyses nucleotides or proteins in sequences, and determines the coordinates for each one's location in a continuous space. The distance between positions indicates the similarity between respective sequences. It helps in determining repetitive sequence or motif. The authors in [12] utilized word2vec to transform DNA sequences to feature vector. For which authors tried kmers of size varying from 3 to 6 and worked with kmer of size 4. DNA sequence is divided into substrings of size 4, then using the word2vec, embeddings for each 4-mer is obtained. This is then fed to CNN and neural network for classification task. The authors in [13] use very different approach where they found that overlapping strings have common factors that can be preserved using Lyndon Factorization method for converting a sequence reads to a novel way of feature representation.

Authors used physiochemical and structural properties of DNA sequence to convert from sequences to feature vectors. It is a web-based solution called "Seq2Feature" [7]. Liu et al. extracted features using frequency count of monomers, dimers (dinucleotide bases) with their proposed technique of using joint probabilities to compute Base–Base correlation for feature extraction [8]. Dlamini et al. [9] used only dinucleotide frequency as a genomic feature or a signature to classify SARS-CoV-2 sequences using XGBoost machine learning algorithm. The authors [10] also used dinucleotide bases to transform a sequence to 4^2 (16-dimension) vector and classified bacterial genomes using artificial neural network. The authors [11] used Chaos Game representation (CGR) technique to model and visualize sequences, that analyses nucleotides or proteins in sequences, and determines the coordinates for each one's location in a continuous space. The distance between positions indicates the similarity between respective sequences. It helps in determining repetitive sequence or motif. The authors in [12] utilized word2vec to transform DNA sequences to feature vector. For which authors tried kmers of size varying from 3 to 6 and worked with kmer of size 4. DNA sequence is divided into substrings of size 4, then using the word2vec, embeddings for each 4-mer is obtained. This is then fed to CNN and neural network for classification task. The authors in [13] use very different approach where they found that overlapping strings have common factors that can be preserved using Lyndon Factorization method for converting a sequence reads to a novel way of feature representation.

Nowadays, most popular encoding scheme is the word embedding methods that can be classified as either the prediction-based model, that predicts the probability of next words or a count-based models, that uses statistics in word-context matrices [14, 15]. The articles referenced [13, 15–18] used one over the other genome sequence representation technique depending on the class of the genomic problem to solve. The summary of key related works is provided in Table 1.

Table 1 Related works summary

Author	Objective	Work description	Conclusions
Dlamini et al. [9], You et al. [10]	To use dinucleotide representation of sequences	XGBoost model was used to classify similar species based on their dinucleoitde frequency	Concluded dinucleotide signature are well conserved and sufficient for discriminating similar species
Han et al. [12]	To capture nucleosome position using CNN	Proposed CNN, RNN to capture local features	Developed effective method for nucleosome position identification
Bonizzoni et al. [13]	To propose Lyndon factorization method to represent sequence reads	Using the proposed technique similar signature in a sequence are identified lyn2vec used as a way to embed features	Development of unique feature embedding technique It is faster technique as training time is less
Almeida et al. [14]	To build fixed-length representation of words	A survey is performed on comparison and scope of each word embedding techniques Prediction based and count-based models	Reported various schemes of word-embedding techniques and established relationship between methods
Fang et al. [15]	To automatically extract features based on DNA sequences	Used word to vector to learn k-mers distribution Used CNN to extract features	Achieved good accuracy by using word2vec algorithm in comparison to one-hot encoding

3 Outline

A DNA/RNA genome sequence comprises four fundamental nucleotides: Adenine (A), Cytosine (C), Guanine (G), and Thymine (T) (or Uracil, U, in the case of RNA). The precise arrangement of these nucleobases within a sequence is pivotal in defining the unique characteristics of the organism under study.

In this article, we delve into the evaluation and significance of various feature representation schemes for genome sequences. However, it is imperative to address the inherent shortcomings found in genomic datasets when attempting to convert these sequences into numerical representations. Several challenges arise due to the distinctive attributes intrinsic to genomic data, which are elucidated as follow.

3.1 Challenges with Genomic Datasets

Variable sequence length ML/DL requires fixed-length inputs, for which padding and truncation techniques are required. Both the techniques, however, have shortcoming of loss of information, high computation requirements and, also it affects model performance.

Presence of non-nucleobases It's very common that sequences may contain nucleobases other than (A, G, C and T), 'N', 'X','O' etc.

High Dimensional Data The one-hot encoding techniques may result in sparsely high dimensional data in the cases where short sequences are padded or class imbalance issue is resolved.

Data Quality Genomic data can exhibit missing values, sequences with low quality score, redundant data. Pre-processing is essential before transformation of data.

With such shortcoming in genomic data representation, pre-processing the data is a crucial task. Following is the details of the section. Section 4 discusses Convolutional Neural Network model and feature extraction methods. Section 5 provide an in-depth account of the practical implementation of our approach, offering insights into how the model and feature extraction techniques were put into action. Section 6 presents the findings and results of our experiments and Section 6 is the conclusion section.

4 CNN and Methods for Feature Extraction

CNN models are composed of convolutional layer, a non-linearity layer, a max-pooling layer, and a fully connected layer. CNNs demonstrated strong performance not only in domains of image classification, computer vision, and but also in natural language processing. CNNs are mostly employed for automated feature extraction from inputs [3, 5]. In the context of our work, text data is encoded as a linear sequence of consecutive characters, which is required to transform in vector or 2D tensors required by CNNs. The sequence text data consists of four nucleobases A, G, C and T as "words" without space. It is required to convert the sequence into numerical values. This can be achieved using label encoding or one-hot encoding techniques. DNA sequences represented as one-hot vectors are transformed into 2D matrices. A max-pooling layer is employed following the convolutional layer, it helps in dimension reduction. The size of the kernel or filter is the crucial hyperparameter in convolutional layer for extracting the features. The non-linearity layer, use ReLu, Tanh activation function. After the pooling layer, a fully connected dense neural network is used. For classification, sigmoid function is employed for binary class problems. It computes probability associated with each output class given in dataset. Figure 1 depicts the above architecture.

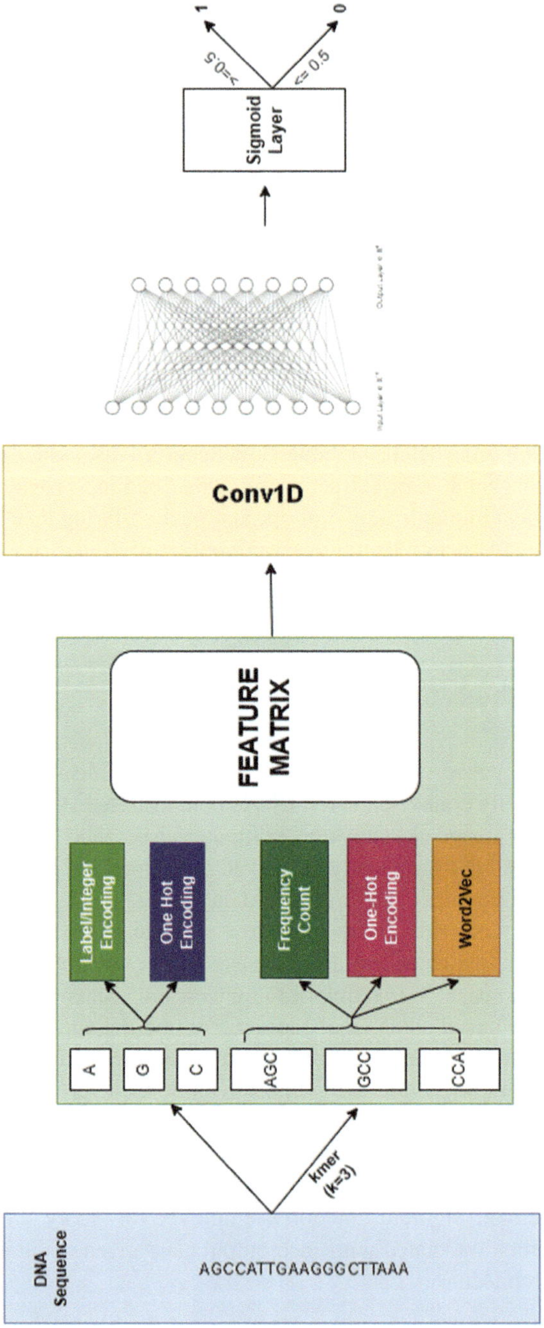

Fig. 1 Schematic diagram of CNN and a fully connected dense layer

We incorporated CNNs in the proposed comparative study of pre-processing techniques as they are fast and efficient in terms of representation of genome sequences.

The Vector Space Model is most effective model to encode words and documents as vectors. Similarly, a language model in NLP applications, is a crucial component whose primary goal is to predict word based on a set of preceding words. With the advancement in large-scale language models based on neural nets both VSM and LM are bought together. The word vectors are projected on first layer of network also called, embedding layer, captures efficiently both the syntactic as well as semantic relationship among the words [14].

Models requires input instances, x, to produce an output. 'x' represents n-dimensional features of a particular sample point or instance. The genomic sequences are the categorical dataset. Feature-encoding of text-based data though is time-consuming process but a mandatory step in model building. Most machine learning tasks accept data in the form of numbers or tensors (1D or 2D) in deep learning. Thus, for say, in classifying genomic sequences, feature selection task is crucial. Deep learning models are established methods in extracting features from natural language datasets. These are highly applied on genomic data as well to tackle genomic text-based data. One hot encoding is used in existing literature to encode sequences [3, 19, 20]. Kmer which is most common way to encode genome sequences, various sizes of kmers are tested and performance is evaluated [13, 15–17].

Label Encoding The simplest encoding is base conversion to integers numbers e.g. A, G, C and T to 1,2,3,4 or ordinal encoding 0.25, 0.50, 0.75, 1.00 and, for non-bases 0 or 0.00 is used. This scheme is faster and doesn't faces curse of dimensionality. For an input sequence of length l, n sequences, then $(n \times l)$ matrix is used in CNN (as a 2D image).

Kmers kmer, denotes, substring obtained from a genome sequence, of a desired length k. In a sequence of length N, user defined length of k, the total number of k-mers or substrings in a sequence is given by 'N-k + 1'.

One-Hot Encoding In this scheme, each base or kmer obtained from a sequence is represented as a binary vector. Example A = [1, 0, 0, 0], C = [0, 1, 0, 0], G = [0, 0, 1, 0] and T = [0, 0, 0, 1]. An input DNA sequence possessing length l, will be $4 \times l$ matrix using the representation. In case of kmers as tokens, for example, given k = 3, DNA sequence can be break down into overlapping 3-letter words, and then represent each kmer using one-hot encoding technique to a vector of size 64. The one-hot encoded kmers can be fed directly to deep learning model or its 2D matrix form can also be used, where row corresponds to kmer, and each column to a word in dictionary. One-hot encoding differs from label encoding as earlier preserves base order whereas latter have no specific order among categories.

Word embeddings In word embedding or word2vec algorithm a genome sequence is converted into a dense vector of fixed-lengths. Word embedding is a popular way to extract features from a natural language as it encodes both syntactic as well as semantic word relationships. It relies on the assumption that words with same contexts have the same meaning. In case of DNA sequence, kmers are converted into vectors using Skip-gram and Continuous Bag of words (CBoW). Skip-gram encoding technique derives context words provided the centre word, while CBOW infers centre word from the surrounding contexts. In our work, we have used Skip-gram architecture [2], its primary objective is to maximize the conditional probability of context words given a target word for a given sequence. The equation can be presented as:

Let f is a function, n be the number of words in a sequence, d, size of the training context (size of context window), v input and output vectors (word embeddings) and s is the size of vocabulary. Then, f is defined as the product of the conditional probabilities of observing context words given a target word [15]. It's often presented as sum of overall words in the context window:

$$f = \sum_{t=1}^{n} \sum_{-d<j<d, j\neq 0} logp(w_{t+j}|w_t) \tag{1}$$

where, w_t—target word at position t in the sequence

w_{t+j}—represents a context word at a position relative to target word, where j ranges from −d to d (excluding j = 0). $P(w_{t+j}|w_t)$—represents the conditional probability of observing a context word w_{t+j} given the w_t. The conditional probability, is modelled using softmax function and dot product of word vectors, where $v^{w_{t+j}}$, represents the vector for context word w_{t+j} and v^{w_t}—represents the vector of the centre word w_t:

$$P(w_{t+j}|w_t) = \frac{(v^{w_{t+j}}.v^{w_t})}{\sum_{i=1}^{s}(v^i.v^{w_t})} \tag{2}$$

5 Implementation of Pre-processing Methods

In this study, we propose comparison of genomic dataset pre-processing methods for deep learning. Our approach revolves around the utilization of fixed-length sequences, a diverse array of pre-processing techniques, an in-depth comparative analysis, and the deployment of a convolutional neural network (CNN) model to effectively classify these sequences.

5.1 Dataset

We are referring a dataset used in article [21]. The authors adopted deep learning approach to solve functional genomics application. And, demonstrated use of simple 1D CNN the neural network capable of indicating whether the sequence binds to the protein or not using one-hot encoded vectors of input sequence [22]. The independent variables comprise of DNA sequence bases of length 50, and dependent variable class is 0 or 1 determines whether the given motifs in

Table 2 Dataset description

Sequences	Class label	Number of samples
Motifs	1	987
Non-binding motifs	0	1013

sequences binds to protein or not. The task is of binary classification. We experimented and augmented other encoding strategies to same dataset keeping 1D CNN model. The description of dataset is provided in Table 2 and link to download data is [https://raw.githubusercontent.com/abidlabs/deep-learning-genomics-primer/master/sequences.txt] and their corresponding labels [https://raw.githubusercontent.com/abidlabs/deep-learning-genomics-primer/master/labels.txt].

5.2 *Convolutional Neural Network (CNN) Architecture*

The evaluation study compares four sequence encoding strategies: (i) Label encoding (ii) kmer Frequency Count (iii) One hot encoded kmers and (iv) word embedding (Skip-gram). TensorFlow [23] is chosen as implementation of the CNN. Tensorflow library is used to label encode for numeric conversion of nucleobases and, then 2D matrix of one-hot encoders is obtained for individual bases or kmer words. The CNN architectures used in evaluation is shown in Table 2. CNN took 2D or 1D form of data representation, and employed similar dropout rate of 0.5, ReLu as activation function and, max pool size = 2. Adam optimizer with learning rate 0.001 is chosen as the adaptive learning method. The sigmoid function is used to predict the output. Table 3. Represents the input and output of sequences as fed to 1D-CNN.

Table 3 CNN Architecture in various representation

	Label encoding	Frequency count-kmer	One-hot encoding-kmer	Word embedding
Input layer	50 × 4	64 × 1	3072 × 1	48 × 100
Conv layer	48 × 32	62 × 32	3070 × 64	46 × 32
Activation function	ReLu	ReLu	ReLu	ReLu
Max pooling	2 × 2	31 × 32	1532 × 64	23 × 32
Fully connected layer	64	64	128	64
Dropout	0.5	0.5	0.5	0.5
Fully connected layer	2	2	2	2

5.3 Metrics

The evaluation criteria include accuracy, precision, recall and F1-Score. The accuracy is the measurement of correctness of applied model's predictions. The precision is the ratio of true positive to total of all positive predictions. Recall measures the proportion of true positive out of all the actual positive instances. The f1-score is a harmonic mean of precision and recall, providing a balance measure if the model's performance [3]. The classification metrics used are—True positive, True negative (accurate prediction of both the classes), and False Negative, False positive (inaccurate prediction of both the classes).

Accuracy It is the ratio of total correct predictions and total number of input samples. It best better if classes are balanced.

$$Accuracy = \frac{TP}{\mathrm{TP} + \mathrm{TN} + \mathrm{FP} + \mathrm{FN}} \tag{3}$$

Precision It is the ratio of specific sample such as True Positive and all the samples predicted of that specific class.

$$Precision = \frac{TP}{\mathrm{TP} + \mathrm{FP}} \tag{4}$$

Recall It is the ratio of specific sample such as True positive to all the samples that truly belong to specific class

$$Recall = \frac{TP}{\mathrm{TP} + \mathrm{FN}} \tag{5}$$

F1-Score It is the harmonic mean between Recall and Precision. It primarily established the classifier's accuracy (i.e. the number of samples that are correctly categorized). The reduced recall and increased precision values offer a highly accurate prediction. The higher the F1-Score, the more effective is the algorithm.

$$F1 - Score = \frac{2 \times (Precision \times Recall)}{\mathrm{Precision} + \mathrm{Recall}} \tag{6}$$

We have calculated accuracy, precision, recall and F1-Score as shown in Table 4 and, subsequent figure shows ROC curve for assessing the performance of a model under each encoding scheme (Fig. 2).

Receiver operating characteristic (ROC) curve plots true positive rate, TPR, (correctly identified positive samples) on y-axis oppose to false-positive rate, FPR (correctly identified negative samples) on x-axis. The ROC curve is associated with AUC value, higher the value, better is model's performance with 1.0 representing an ideal model. As the result shows, word embedding is the best scheme to encode genomic data as it builds low-dimension vectors. This is attributed to the higher

Table 4 A comparison of pre-processing genome sequence encoding schemes

Pre-processing strategies	Class	Accuracy	Recall	Precision	F1-score
Label encoding	Class 0	0.94	0.96	0.93	0.95
	Class 1		0.93	0.96	0.94
Kmer frequency count	Class 0	0.98	0.97	0.99	0.98
	Class 1		0.99	0.97	0.98
Kmer one hot encoding	Class 0	0.89	0.89	0.89	0.89
	Class 1		0.88	0.89	0.88
Kmer word embedding	Class 0	0.99	0.98	1.00	0.99
	Class 1		0.99	0.98	0.99

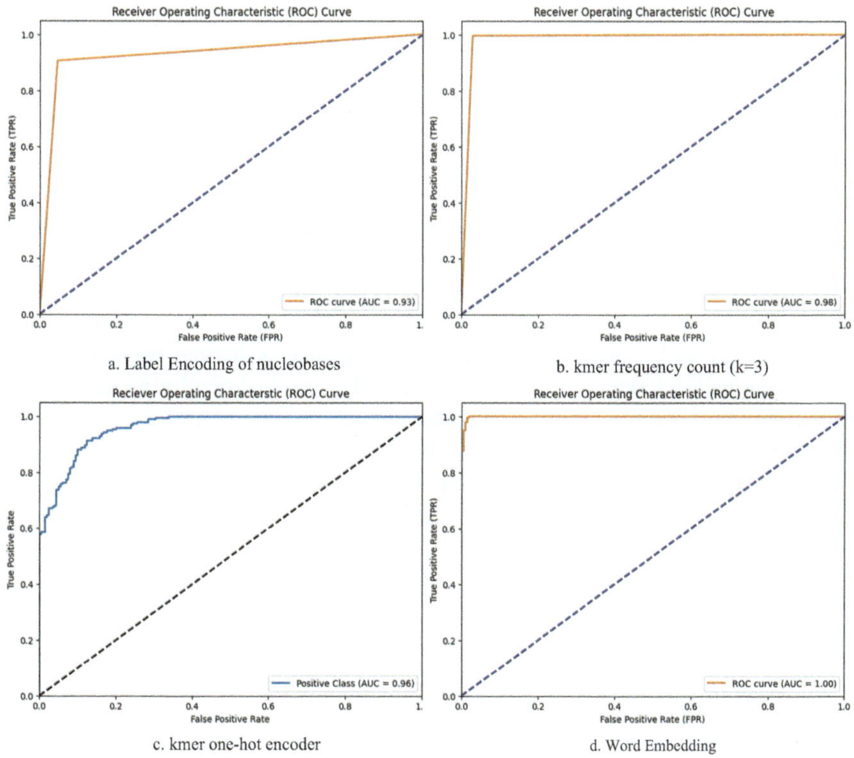

a. Label Encoding of nucleobases

b. kmer frequency count (k=3)

c. kmer one-hot encoder

d. Word Embedding

Fig. 2 Receiver operating characteristic curves for the model under various encoding schemes

model performance in case of word embedding [24]. In the surveys, it has been showed one-hot encoding gives optimal results, though our results depict that simple frequency count of kmers can give better results under simple 1D CNN model. As, frequency of kmers captures the distribution of specific patterns, whereas one hot encoding treats the positions independently.

6 Conclusion

The existing approaches utilised in natural language processing are also employed in the field of bioinformatics. The genomic or proteomic data is converted into numerical values in order to facilitate the use of models and the extraction of valuable insights. In the present study, a basic convolutional neural network (CNN) architecture was employed, consisting solely of a single convolutional layer. In this study, a supervised approach is employed to analyse DNA sequences with the objective of discovering motifs that have the ability to bind to proteins. The sequence is transformed into vectors through the utilisation of four methods: (i) Label encoding, (ii) kmer frequency count, (iii) one hot encoded kmers, and (iv) kmer word embeddings. Since the data being utilised is of a predetermined length. The CNN model successfully acquired the properties. The findings suggest that word embeddings are an effective means of representing word relationships. This makes them a preferable option compared to one-hot encoding, which is burdened by the curse of dimensionality and higher computing costs. The Skip-Gram Architecture is commonly employed in the process of word embedding acquisition due to its suitability for bigger datasets. This architecture effectively captures the contextual information surrounding the central word, hence facilitating a more accurate representation of relationships within a vast vocabulary. Moreover, it is more advantageous for words that are used infrequently [15].

As the future scope of this work, in real world when new species appears such as SARS-CoV-2, its important to derive its origin. For that purpose, depending on the symptoms and causes, the new genome sequence is studied by comparing it with previously known organism or species. This poses difficulty as its need not necessarily be similar in length. Thus, for comparative study sequences are either truncated or padded which may impact the results. Therefore, in next experiment, same framework can be tested for un-equal genome sequences.

References

1. Levy, J.J., Titus, A.J., Petersen, C.L., Chen, Y., Salas, L.A., Christensen, B.C.: MethylNet: an automated and modular deep learning approach for DNA methylation analysis. BMC Bioinform. **21**(1), 1–15 (2020)
2. Chia, S.E., Lee, N.K.: Comparisons of DNA sequence representation methods for deep learning modelling (2022)
3. Gunasekaran, H., Ramalakshmi, K., Rex Macedo Arokiaraj, A., Kanmani, S.D., Venkatesan, C., Dhas, C.S.G.: Analysis of DNA sequence classification using CNN and hybrid models. Comput. Math. Methods Med. **2021**(C) (2021)
4. Irmak, E.: Implementation of convolutional neural network approach for COVID-19 disease detection. Physiol. Genomics **52**(12), 590–601 (2020)
5. Whata, A., Chimedza, C.: Deep learning for SARS COV-2 genome sequences. IEEE Access. **9**, 59597–59611 (2021)
6. Yandell, M.D., Majoros, W.H.: Genomics and natural language processing. Nat. Rev. Genet. **3**(8), 601–610 (2002)

7. Nikam, R., Gromiha, M.M.: Seq2Feature: a comprehensive web-based feature extraction tool. Bioinformatics **35**(22), 4797–4799 (2019)
8. Liu, Z.H., Jiao, D., Sun, X.: Classifying genomic sequences by sequence feature analysis. Genomics Proteomics Bioinform. [Internet] **3**(4), 201–205 (2005). https://doi.org/10.1016/S1672-0229(05)03027-5
9. Dlamini, G.S., Müller, S.J., Meraba, R.L., Young, R.A., Mashiyane, J., Chiwewe, T., et al.: Classification of COVID-19 and other pathogenic sequences: a dinucleotide frequency and machine learning approach. IEEE Access **8**, 195263–195273 (2020)
10. You, W., Wang, K., Li, H., Jia, Y., Wu, X., Du, Y.: Classification of DNA sequences basing on the dinucleotide compositions. In: ISCID 2009 - 2009 International Symposium on Computational Intelligence and Design, vol. 2, pp. 390–394 (2009)
11. Barbosa, R.deM., Fernandes, M.A.C.: Chaos game representation dataset of SARS-CoV-2 genome. Data Brief [Internet] **30**, 105618 (2020). https://doi.org/10.1016/j.dib.2020.105618
12. Han, G.S., Li, Q., Li, Y.: Nucleosome positioning based on DNA sequence embedding and deep learning. BMC Genomics **23**, 1–10 (2022)
13. Bonizzoni, P., Costantini, M., De Felice, C., Petescia, A., Pirola, Y., Previtali, M., et al.: Numeric Lyndon-based feature embedding of sequencing reads for machine learning approaches, pp. 1–35 (2022). http://arxiv.org/abs/2202.13884%0A, https://doi.org/10.1016/j.ins.2022.06.005
14. Almeida, F., Xexéo, G.: Word embeddings: a survey 2019 (1991). http://arxiv.org/abs/1901.09069
15. Fang, G., Zeng, F., Li, X., Yao, L.: Word2vec based deep learning network for DNA N4-methylcytosine sites identification. Procedia Comput. Sci. **187**, 270–277 (2021)
16. Ali, S.: Spike2Vec: an efficient and scalable embedding approach for COVID-19 spike sequences. In: 2021 IEEE International Conference on Big Data (Big Data), vol. 2, pp. 1533–1540 (2022)
17. Tayebi, Z., Ali, S., Patterson, M.: Robust representation and efficient feature selection allows for effective clustering of SARS-CoV-2 variants. Algorithms **14**(12) (2021)
18. Mikolov, T., Chen, K., Corrado, G., Dean, J.: Efficient estimation of word representations in vector space. In: 1st International Conference on Learning Representations, ICLR 2013 - Workshop Track Proceedings, pp. 1–12 (2013)
19. Tong, B.C.-K.: 乳鼠心肌提取 HHS public access. Physiol. & Behav. **176**(5), 139–148 (2017)
20. Kelley, D.R., Snoek, J., Rinn, J.L.: Basset: Learning the regulatory code of the accessible genome with deep convolutional neural networks. Genome Res. **26**(7), 990–999 (2016)
21. Zou, J., Huss, M., Abid, A., Mohammadi, P., Torkamani, A., Telenti, A.: A primer on deep learning in genomics. Nat. Genet. [Internet] **51**(1), 12–18 (2019). https://doi.org/10.1038/s41588-018-0295-5
22. Zhao, H., Pavlidis, P., Alachiotis, N.: SweepNet: a lightweight CNN architecture for the classification of adaptive genomic regions. In: Proceedings of the Platform for Advanced Scientific Computing Conference, PASC 2023 (2023)
23. Abadi, M., Agarwal, A., Barham, P., Brevdo, E., Chen, Z., Citro, C., et al.: TensorFlow: large-scale machine learning on heterogeneous distributed systems [internet] (2016). http://arxiv.org/abs/1603.04467
24. Ali, S., Sahoo, B., Zelikovsky, A., Chen, P.Y., Patterson, M.: Benchmarking machine learning robustness in Covid-19 genome sequence classification. Sci. Rep. [Internet] **13**(1), 1–17 (2023). https://doi.org/10.1038/s41598-023-313683

Analyzing the Relationship Between Legal Moves in Blind Chess and ELO Ratings Using Speech Recognition and Machine Learning

Utkarsh Chauhan, Tushar Singh, Amit Chhabra, Animesh Singh, and Briti Singla

Abstract This paper presents an in-depth analysis of the relationship between chess ratings and blind chess, a challenging variant where players cannot see the board and rely on chess notations. As the game progresses, the complexity of blind chess increases, leading to potential mistakes due to players struggling to maintain the board in their minds. In response, this research introduces the innovative BLIND-CHESSSIM model, designed to assist chess players in playing blindfold chess while also providing an estimation of their ELO rating based on their legal moves. BLIND-CHESSSIM serves as an assisting agent, streamlining the blind chess experience. It generates data by recording the number of chess moves made by the player during the game. This model leverages CHESS SIM, incorporating speech recognition technology for move input and data recording. To establish a correlation between player ratings and the number of legal chess moves, we utilize a dataset comprising 200 blindfold chess matches. This study employs a range of machine learning models, including regression, Kmeans clustering, K-nearest neighbors (KNN), and K-fold validation, to analyze the relationship between player ratings and the number of legal chess moves. The results obtained from this analysis provide a foundation for accurately estimating the chess ratings of individuals who engage in blindfold chess. Ultimately, this research aims to enhance the accessibility and competitive experience of blindfold chess while advancing our understanding of the game's intricacies.

U. Chauhan (✉) · T. Singh · A. Chhabra · A. Singh · B. Singla
Chandigarh College of Engineering and Technology, CSE, Chandigarh, India
e-mail: utkarsh.utee@gmail.com

T. Singh
e-mail: co21362@ccet.ac.in

A. Chhabra
e-mail: amitchhabra@ccet.ac.in

A. Singh
e-mail: animeshsingh@ccet.ac.in

B. Singla
e-mail: co21313@ccet.ac.in

B. Soni et al. (eds.), *Beyond Artificial Intelligence*, Lecture Notes in Networks and Systems 1326, https://doi.org/10.1007/978-981-96-4170-3_52

Keywords Blind chess · ELO rating · Legal moves · Regression · Clustering

1 Introduction

Chess, a game with ancient origins traced back to India, paradoxically enjoys limited recognition in its country of birth. However, recent years have witnessed the remarkable growth of mobile computing [1], or self-driving cars [2], or smart transport [3] paralleled by the resurgence of global interest in chess, driven in part by advances in machine learning. Machine learning has proven highly influential for strategy games like chess, go, and checkers. Stateof-the-art AI systems have become essential tools for professional chess players by expanding human creativity and innovation through novel strategies.

The standard chessboard is an 8 × 8 grid with alternating colored squares, identified using a universal coordinate system: columns A to H and rows 1 to 8. This notation is the common language of chess. Blind Chess, a variant relying heavily on memory and concentration, poses the challenge of mentally maintaining the board position [4]. To aid Blind Chess, CHESS SIM was developed, supporting blindfold chess play and recording gameplay data. Using speech recognition, it interprets moves and assesses their accuracy in the context of Blind Chess [5, 6].

In the world of chess, a player's caliber is typically determined by their rating, referred to as ELO points. These points are computed using a sophisticated algorithm conceived by Arpad Elo. The ELO rating is an essential metric, reflecting a player's skill and competitiveness. The equations for calculating ELO rating, as devised by Elo, form the backbone of competitive chess evaluation [7].

The caliber of any player in chess is determined by his/her rating. Which is called ELO Points, these points are calculated using a complex algorithm invented by Arpad Elo. Elo is calculated using the complex algorithm designed by Arpad Elo [7]. To calculate the ELO rating equations are utilized:

$$E_A = \frac{1}{1 + 10^{((R_B - R_A)/400)}} \tag{1}$$

$$E_B = \frac{1}{1 + 10^{((R_A - R_B)/400)}} \tag{2}$$

$$R'_A = R_A + k \cdot S_A - k \cdot E_A \tag{3}$$

$$R'_B = R_B + k \cdot S_B - k \cdot E_B \tag{4}$$

(a) Where R_A, R_B are the initial ratings of player A and player B respectively.
(b) E_A & E_B are the expected score of player A and player respectively.
(c) S_A & S_B are the actual score of player A & player B respectively.

(d) k is the maximum adjustment per game also known as K-factor. k is set to k = 16 for expert players and k = 32 for weaker players.
(e) R'_A & R'_B are the updated ratings of player A & player B respectively.

FIDE(F'ed'eration Internationale des Echecs) is responsible for defining competition rules, ensuring fair play, and calculating the ELO ratings of professional chess players worldwide.[8] The ELO rating system, recognized globally as the standard for assessing chess players' skills, has Magnus Carlsen as the highest-rated player with a rating around 2800 [9]. The professional players' ratings used in this research paper are sourced from FIDE's authoritative data.

In the realm of chess, advanced chess engines like Alpha Zero, Deep Blue, and Stockfish have gained prominence. These engines utilize AI and machine learning to enhance chess skills and explore new dimensions of the game [10].

This study seeks to harness the functionalities of CHESS SIM, an innovative tool designed to facilitate blindfold chess while concurrently capturing data on legitimate moves. CHESS SIM employs speech recognition technology to interpret the positions of chess pieces, thereby enhancing the accessibility of blindfold chess. The paper endeavors to examine the correlation between legal moves in blind chess and ELO ratings, acknowledging the evolving landscape of chess engines integrating artificial intelligence and machine learning techniques [11].

Development of BLINDCHESSSIM. This paper introduces CHESS SIM, a specialized tool designed for blindfold chess, exploring its nuanced features. By integrating speech recognition, the tool interprets chess moves, enhancing accessibility for blindfold chess enthusiasts. The creation of such a tool has the potential to substantially boost inclusivity in chess, especially for those with visual impairments. To further enhance inclusiveness, the integration of sentiment analysis and cutting-edge speech recognition technologies in BLINDCHESSSIM can be explored for continuous improvement[12].

Investigation of Legal Moves and ELO Ratings. The primary objective of this paper is to scrutinize the relationship between legal moves in blindfold chess and ELO ratings. This inquiry aims to provide valuable insights into the correlation between a player's ability to make legal moves in blindfold chess and their skill level, as measured by the ELO rating system.

The paper is organized as follows: Sect. 2 provides a comprehensive review of relevant literature, Sect. 3 outlines the research focus, Sect. 4 details the methods of data collection, Sect. 5 explains the system's working and algorithm, Sect. 6 discusses the research methodology, and finally, Sect. 7 presents the study's conclusions.

2 Literature Review

Prior chess research enhanced elite player preparation and theory [13], but neglected blindfold chess in machine learning. Our paper pioneers this field, exploring the link between player rating and legal moves in blindfold chess. We analyze data with CHESS SIM and employ machine learning for insights.

Jeremic et al. [14] conducted a survey comparing error rates in rapid chess and blindfold chess. They found 5.84 blunders per 1,000 moves in rapid games and 8.42 blunders per 1,000 moves in blindfold games, highlighting unique challenges and errors in blindfold chess..

Gregory Clark's research [15] investigated deep synoptic Monte Carlo planning, revealing that an increase in the count of possible states led to a decrease in winning percentages in chess. The study also emphasized the versatility of the DSMCP algorithm compared to previous algorithms for incomplete information games.

Jan Balata et al. [16] sought to link problem-solving skills with blind chess. They analyzed 114 games played by blind chess masters, exploring mental processes in chess and problem-solving, including age, gender, and skill level.

McIlroy-Young R et al. [17] adopted a vision transformer approach to assess individual decision-making styles in chess. They analyzed a substantial volume of chess games, considering factors such as the number of chess games played by a player, the number of moves played per game, and the player's skill level.

Wieczerzak [18] The paper explores the impact of datasets on training Artificial Neural Networks for chess position evaluation. With over 1.6 million positions, including random ones, the study employs the Stockfish engine for evaluation. The resulting model consistently outperforms counterparts trained with randomly generated positions across diverse data representations and evaluation classes.

David M. Lane's research [19] delved into the chess memory and cognitive processing relationship. It found that both high-level conceptual processing and pattern recognition are vital for chess position memory.

Sak, H et al. [20] used LSTM recurrent neural networks to improve speech recognition by introducing context-dependent phonetic units. They outperformed traditional LSTM-hybrid models with a dataset of three million 4-s utterances.

Bharathi et al. [21] proposed a system tailored for visually impaired individuals, enabling them to practice chess using voice commands. Their system leveraged algorithms such as the Min–Max algorithm, Alpha–Beta pruning, and k-NN algorithm. To train the machine, they utilized a freely accessible store of over 5 million chess games.

Mechner, F. [22] introduced the "Proficiency" degree, based on "yes–no" questions to make optimal decisions. It's valuable in cognitive skill research, training, and education, particularly for blindfold chess, exploring the link between improvement and mental imagery.

Despite these valuable contributions, significant gaps in the research landscape remain, offering ample opportunities for further exploration. This paper ventures into uncharted territory, shedding light on aspects of blind chess and legal move analysis that have received limited attention. It is our hope that this research will pave the way for deeper insights and advancements in this underexplored field.

3 Research Focus

This paper centers its research on the intersection of machine learning, speech recognition, and the unique domain of blind chess. A specialized Python-based tool, CHESS SIM, is employed to record and gather data pertaining to the number of legal moves made during blind chess gameplay. The primary objective is to establish and thoroughly analyze the correlation between the quantity of legal moves in blind chess and the player's chess ratings, leveraging machine learning and speech recognition through the CHESS SIM tool [23].

3.1 Research Problem

The research endeavors to address two core challenges:

Development of CHESS SIM in Python. The first challenge is to develop a Python-based CHESS SIM with speech recognition capabilities. This tool has a dual purpose: enabling smooth game play in blind chess and providing immediate feedback on move validity. It must accurately identify invalid moves and detect game-ending situations such as checkmate, stalemate, or insufficient pieces.

Establishment of a Rating Estimation Function. The second task is to create a mathematical function for estimating a blind chess player's rating based on their in-game legal moves. This function will help assess and predict blind chess players' skill levels, contributing insights to blind chess and machine learning.

The main goal of this research is to investigate the connection between a blind chess player's performance and their chess rating, specifically focusing on the number of legal moves. If successful, this research can provide a tool for predicting a player's rating after a game on CHESS SIM, aiding in assessing and tracking blind chess players' progress.

4 Data Collection

To obtain sufficient data for conclusive results using our model, we faced challenges due to the limited interest and popularity of chess, especially blind chess, we adopted a methodology akin to clustering and extracting valuable information from the rapidly increasing data generated by consumer electronics [24]. Consequently, we collected data by organizing real-time blindfolded chess matches and extracting data from online game archives.

A total of 156 live matches were conducted under blindfolded conditions, while data from 15 games was sourced from online game archives. The player ratings in the live games ranged from 800 to 1600 Elo points, while the archived data included players with ratings spanning from 1500 to 2800 Elo points [9].

Additionally, in some games, no illegal moves were made. In such cases, we assigned a count of 70 legal moves, as an average standard chess game typically comprises approximately 50 moves. This choice of 70 moves was intended to represent games played flawlessly in our dataset, highlighting the importance of data collection in the context of emerging IoT and edge computing paradigms[25].

5 Blindchesssim

The proposed system revolves around a meticulously designed algorithm that governs the operation of the Chess Simulation (CHESS_SIM) used in this research. Much like the worldwide response to COVID-19, we employed a strategic methodology for blind chess analysis [26]. This section elaborates on the core functionality of the system, shedding light on how it orchestrates blindfold chess games, records crucial data, and subsequently employs machine learning (ML) models for analysis.

5.1 Chess Game Algorithm

The algorithm underpinning the proposed system is tailored to facilitate blindfold chess games, ensuring seamless gameplay and data collection for further analysis. Below is an in-depth elucidation of the key components and functionalities of the algorithm:

Initialization. The algorithm initializes the chessboard, configuring it to accommodate a game of chess. It establishes a Boolean variable, 'white,' which signifies the turn of the player. Additionally, a'checkmate' variable is initialized to zero, reflecting that the game has not concluded.

Game Loop. The algorithm operates within an infinite loop, managing the alternating turns of the two players, White and Black. This loop persists until a game-ending condition is met, such as checkmate or an invalid move.

White Player's Turn

(a) During the White player's turn, the algorithm awaits a move input 's' in standard chess notation.
(b) It verifies the validity of the initial square specified in 's'; if it's invalid, the game stops.
(c) Subsequently, the algorithm solicits input from the chess player, represented as 's2,' which designates the destination square for the piece being moved.
(d) The algorithm validates 's2' to ensure it's a valid chessboard square; any issues result in game termination.
(e) After both input moves are validated, the algorithm executes the player's move on the virtual chessboard from 's' to 's2.'
(f) After each move, the algorithm checks for a checkmate condition, terminating the loop if detected.

Black Player's Turn

(a) The algorithm manages the Black player's turn by awaiting input 's3' for the initial square of the piece to be moved.
(b) It ensures the validity of 's3' as a chessboard square; otherwise, the game terminates.
(c) Input 's4' is solicited from the Black player to designate the destination square for the piece being moved.
(d) Validation checks are performed on 's4' to confirm it as a legitimate chessboard square; any discrepancies lead to game termination.
(e) With both input moves validated, the algorithm then executes the move from 's3' to 's4' with the specified chess piece.
(f) Continuously monitoring for a checkmate condition after each move, the algorithm concludes the game when checkmate is detected.

Algorithm 1 Chess Game Algorithm

```
Initialize chess board: Bool white = 1, checkmate = 0
while true do
    if white then
        String s = (input from player)
        if s[0] == invalid square then
            Break
        end if
        String s2 = (input from chess player)
        if s2[0] == invalid square then
            Break
        end if
        Move from s[0] to s2[0]
        checkmate = Checkmate()
        if checkmate then
            Break
        end if
    else
        String s3 = (input from player)
        if s3[0] == invalid square then
            Break
        end if
        String s4 = (input from player)
        if s4[0] == invalid square then
            Break
        end if
        Move the piece on s3[0] to s4[0]
        checkmate = Checkmate()
        if checkmate then
            Break
        end if
    end if
    white =^ 1
end while
Store the data of the chess match in a CSV file
Apply different ML models on the CSV file
```

Turn Alternation. The algorithm efficiently toggles the 'white' variable after each turn, ensuring that players alternate in a systematic manner.

Data Storage and Analysis. After the game concludes, the algorithm stores important game data in a CSV file, capturing the moves made and the context of the match. This data is then utilized for data analysis by applying various machine learning models.

6 Methodology

The research methodology involves using CHESS SIM, a Python-based software tool for blindfold chess [27]. It utilizes speech recognition technology to interpret and validate spoken commands within the game. Once a move is validated, CHESS SIM executes it on a virtual chessboard. The game continues until checkmate, stalemate, or insufficient pieces remain [8].

In the CHESS SIM environment, players communicate their moves using the universal chess notations, which designate columns with alphabetical characters fromb 'A' to 'H' and rows with numerical values from '1' to '8'. The CHESS SIM meticulously scrutinizes the specified initial and final positions of chess pieces for validity. If an erroneous move is detected, the player is promptly notified by the CHESS SIM, requesting a correct input to ensure uninterrupted gameplay [8] (Figs. 1, 2 and 3).

The speech recognition component of this research is crucial for the virtual chessboard to interpret and act upon player commands accurately, mirroring the importance of predicting whether web page contents catch a surfer's attention [28]. To address the challenge of ambient noise compromising player inputs, an advanced acoustic model is introduced. This model effectively suppresses background noise, enhancing the precision and reliability of speech recognition [29]. The resulting spoken words are transformed into text using the Google-speech recognizer package, which was carefully chosen to meet the research's specific requirements, ensuring optimal performance and accuracy[27]. Similar to the need for robust antispam filters [30], we employ machine learning techniques to analyze chess data, enhancing our understanding of both domains.

The list of available speech recognizer packages considered for this research includes:

1. apiai
2. assemblyai
3. google-cloud-speech
4. wit

However, it is important to emphasize that the Google-speech recognizer package was chosen as the most suitable and reliable option for this research (Figs. 4 and 5).

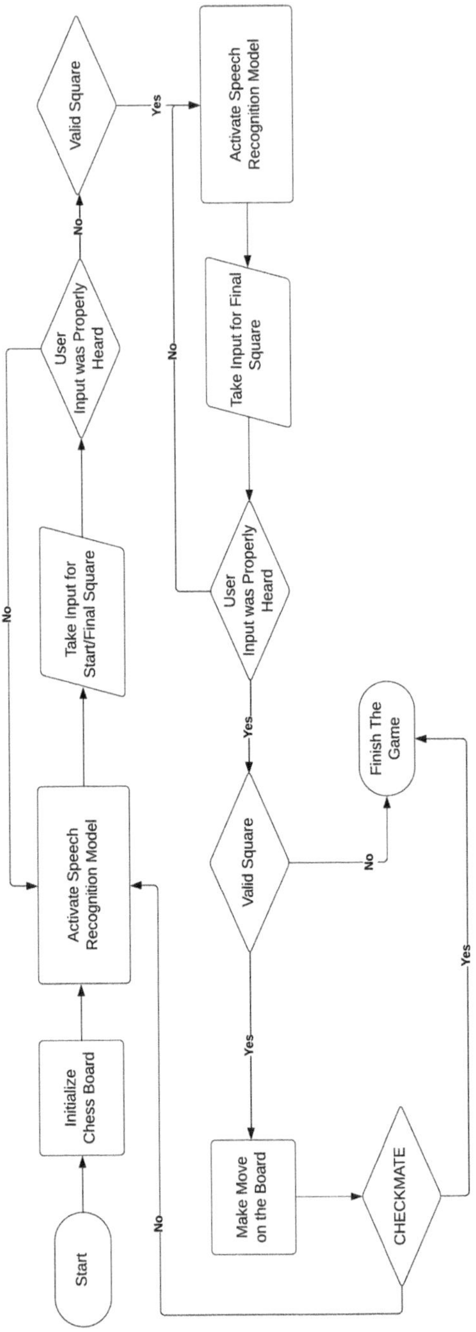

Fig. 1 Flowchart of CHESS_SIM

Fig. 2 Chess notations

```
8 -A8 | B8 | C8 | D8 | E8 | F8 | G8 | H8
----------------------------------------
7 -A7 | B7 | C7 | D7 | E7 | F7 | G7 | H7
----------------------------------------
6 -A6 | B6 | C6 | D6 | E6 | F6 | G6 | H6
----------------------------------------
5 -A5 | B5 | C5 | D5 | E5 | F5 | G5 | H5
----------------------------------------
4 -A4 | B4 | C4 | D4 | E4 | F4 | G4 | H4
----------------------------------------
3 -A3 | B3 | C3 | D3 | E3 | F3 | G3 | H3
----------------------------------------
2 -A2 | B2 | C2 | D2 | E2 | F2 | G2 | H2
----------------------------------------
1 -A1 | B1 | C1 | D1 | E1 | F1 | G1 | H1
----------------------------------------
    A    B    C    D    E    F    G    H
```

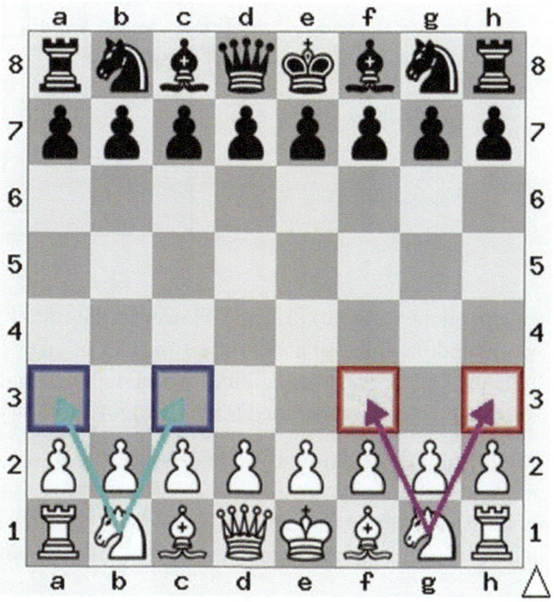

Fig. 3 Possible moves for final block

The core of this research centers on the application of various machine learning algorithms, including Automated Machine Learning (AutoML) [31], to extract valuable insights from the amassed dataset These algorithms encompass:

Linear Regression. In this study, we utilize Linear Regression to explore the relationship between chess ratings (dependent variable) and the number of legal moves made

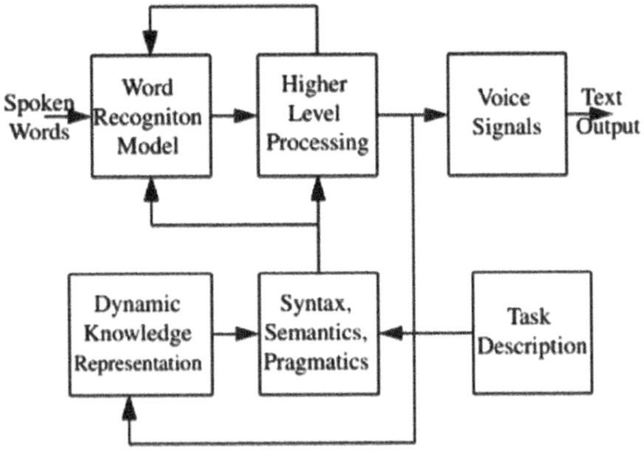

Fig. 4 Normal speech recognition system

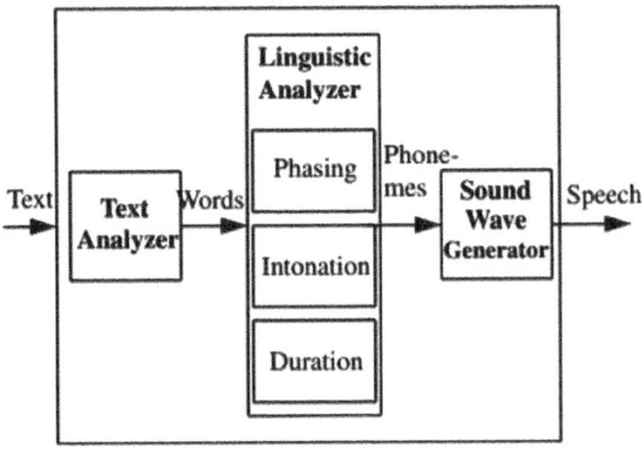

Fig. 5 Normal text-speech system

(independent variable) [32, 33]. However, the initial application of linear regression was abandoned due to a significant margin of error [33]. Figure 6 depicts this relationship. Linear regression, initially used to investigate the correlation between these variables, is an unsupervised learning algorithm [34].

Polynomial Regression. It models the relationship between x (independent variable) and y (dependent variable) using a 5-degree polynomial [32]. Figure 7 displays this relationship: x-axis for legal moves, y-axis for chess rating. Red dots are data points, and the blue line is the 5-degree polynomial capturing the trend [33].

KNN clustering of the data. K-Nearest Neighbors (KNN) is a non-parametric machine learning technique that categorizes new data points based on their similarity to existing data points [34]. It is particularly useful when there is no clear understanding of the data distribution, as it does not make any prior assumptions about the underlying data [34]

Figure 8 visually illustrates the KNN process, where the algorithm assesses the similarity between new and existing data points to categorize them into appropriate clusters, enabling efficient classification within an unlabeled dataset [32].

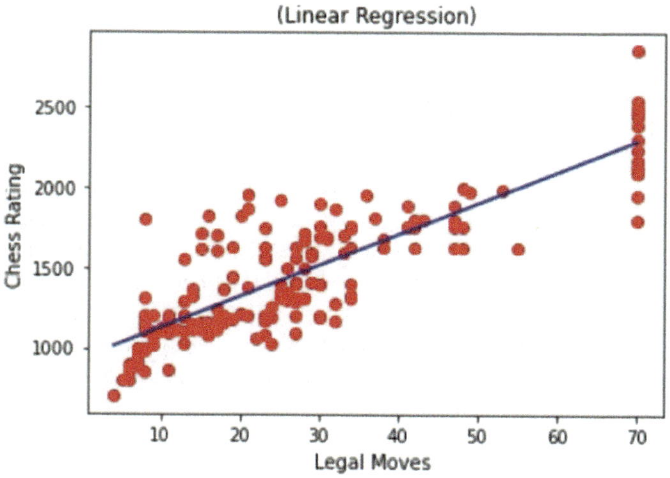

Fig. 6 Linear regression graph

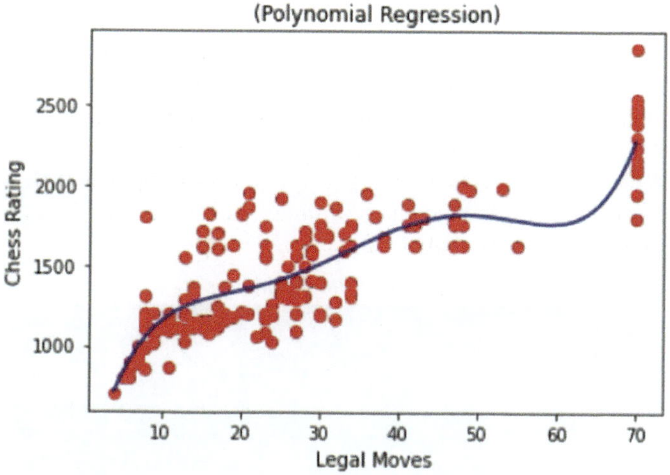

Fig. 7 Polynomial regression graph

Figure 8 serves as an illustrative demonstration of KNN's adeptness in pattern recognition. The Y-axis in this graphical representation signifies ELO rating, while the X-axis delineates the number of legal moves. This portrayal elucidates the proficient clustering of chess game data by KNN.

K-fold Validation. K-Fold Validation is a vital technique for evaluating machine learning models [34]. It divides the dataset into 'k' consecutive folds, using one as a validation set while the rest become the training set [34]. Figure 9 displays data

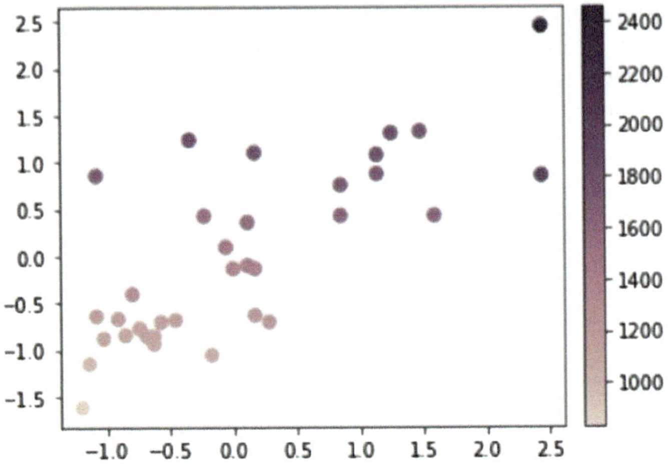

Fig. 8 KNN representation

clustered into 5 distinct colors, emphasizing player diversity, from average to high ratings [32].

K-Means Clustering. K-Means Clustering is an unsupervised machine learning algorithm used to partition unlabeled datasets into distinct clusters. It is valuable when there is no prior understanding of the data distribution. The algorithm groups data points into clusters, with 'K' representing the predefined number of clusters [21].

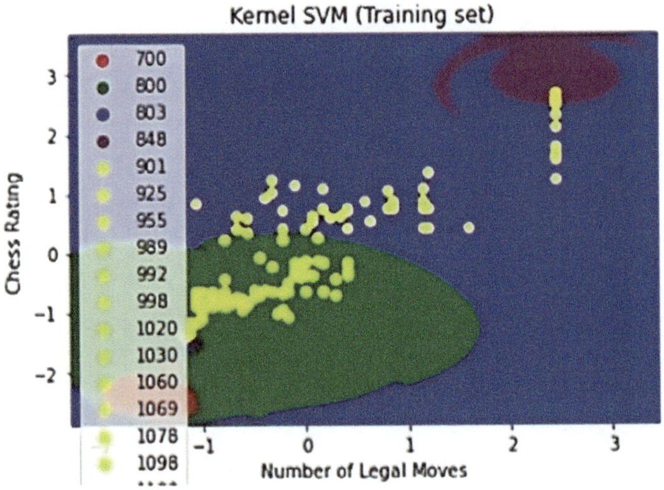

Fig. 9 K- fold validation

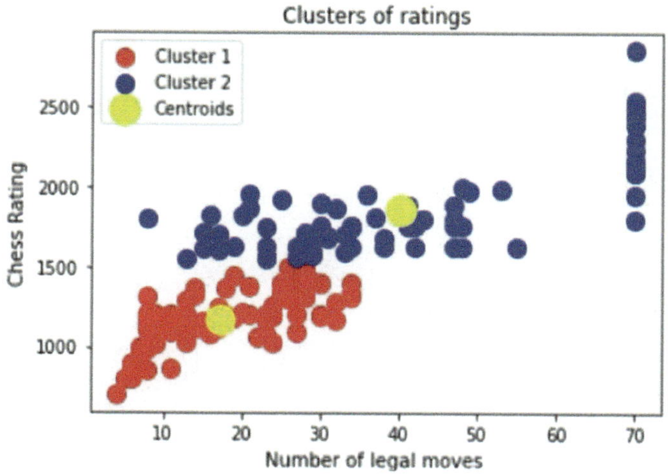

Fig. 10 Clusters representation

In this analysis, K-Means clustering with 'K' set to 2 segmented data into two clusters [34]. It used a centroid-based approach to optimize data point classification efficiency [34]. Figure 10 visually shows the outcomes: blue clusters for club-level players (ratings 1800–2700) and red clusters for novice players (ratings below 1800). A gap in data points beyond the 50-move mark suggests fewer illegal moves at this stage. The autonomously determined centroids highlight the algorithm's effectiveness in unsupervised clustering [34].

Elbow Method for Cluster Optimization. Figure 11 demonstrates the application of the Elbow Method, which is crucial for optimizing the clustering process by determining the appropriate number of clusters for a given dataset.

In a complementary endeavor, akin to our adaptation of AlphaGo-inspired training for Reconnaissance Blind Chess [35], we apply innovative methodologies to cluster analysis. Drawing from the principles of the The Elbow Method, using Within Cluster Sum of Squares (WCSS), finds the optimal clusters for a dataset [36]. It detects a point where the decrease in WCSS slows down significantly, creating an 'elbow' in the graph. In our case, it identifies 2 clusters: amateurs and elites [36]. To boost efficiency, we suggest parallel processing and memory analysis [37], speeding up code generation and enhancing the Elbow Method for blind chess clustering. (See Fig. 11 for reference.)

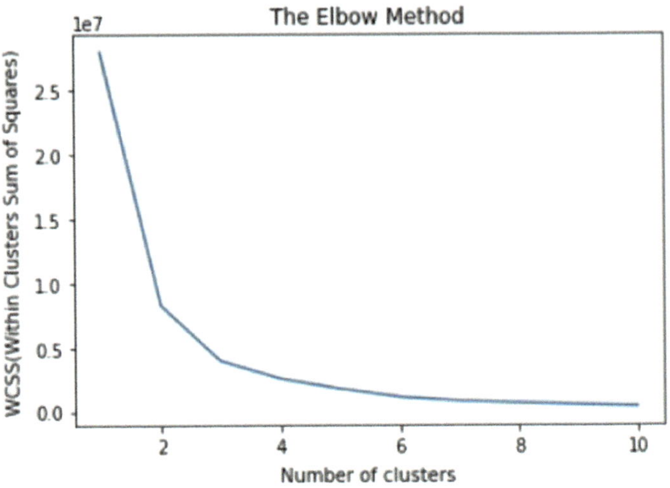

Fig. 11 WCSS elbow method

7 Conclusion

This study, based on a comprehensive analysis of 171 blindfold chess matches, has yielded valuable insights into the relationship between chess player ratings and the number of legal moves played, akin to the challenges of handling Big Data through cloud and edge computing technologies.The developed model demonstrates its ability to predict chess engine ratings with a root mean error of 209.

The findings suggest that the rating of chess players experiences a significant increase after a certain threshold of moves. Notably, players who complete approximately 20–30 legal moves on the CHESS SIM tend to achieve ratings within the range of 1000–1700. However, players surpassing this move count are highly likely to belong to the club-level category, with ratings spanning from 1800 to 3000.

The application of clustering techniques further elucidates this relationship. Players who execute fewer than 40 moves are likely categorized as amateurs, while club-level players exhibit a tendency to conclude games without committing any illegal moves.

References

1. Peñalvo, F.J.G., Sharma, A., Chhabra, A., Singh, S.K., Kumar, S., Arya, V., Gaurav, A.: Mobile cloud computing and sustainable development: opportunities, challenges, and future directions. Int. J. Cloud Appl. Comput. (IJCAC) **12**(1), 1–20 (2022)
2. Singh, I., Singh, S.K., Kumar, S., Aggarwal, K.: Dropout-VGG based convolutional neural network for traffic sign categorization. In: Congress on Intelligent Systems: Proceedings of CIS 2021, vol. 1. Springer Nature Singapore, Singapore, pp. 247–261 (2022)

3. Chopra, M., Kumar, S., Madan, U., Sharma, S. Influence and establishment of smart transport in smart cities. In: International Conference on Smart Systems and Advanced Computing (Syscom-2021) (2021)
4. Campitelli, G., Gobet, F.: The mind's eye in blindfold chess. Eur. J. Cogn. Psychol. **17**(1), 23–45 (2005)
5. Shuo, W., Yan, L., Ping, D., Jing, Z., Xiaona, J., Yating, W., Qichang, Z.: Speech recognition based blind chess education system Chinese chess as an example. In: 2021 11th International Conference on Information Technology in Medicine and Education (ITME). IEEE, pp. 541–547 (2021)
6. Lindquist, B.: The art of text-to-speech. Crit. Inq. **50**(2), 225–251 (2024)
7. Glickman, M.E., Jones, A.C.: Rating the chess rating system. Chance-Berl. Then N.Y. **12**, 21–28 (1999)
8. About FIDE.: www.fide.com. https://www.fide.com/fide/about-fide (n.d.)
9. GmbH, C.: Top 100 Welt. In: ChessBase Players. Retrieved 8 May 2022. https://players.chessbase.com/de/list/world-top-100 (n.d.)
10. Bijl, P., Tiet, A.P.: Exploring modern chess engine architectures. Victoria University, Melbourne (2021)
11. Aggarwal, K., Singh, S.K., Chopra, M., Kumar, S., Colace, F.: Deep learning in robotics for strengthening industry 4.0.: opportunities, challenges and future directions. In: Robotics and AI for Cybersecurity and Critical Infrastructure in Smart Cities, pp. 1–19 (2022)
12. Chhabra, A., Sharma, A., Chhabra, K., Chhabra, A.: A statistical analysis of sentiment over different social platforms on drug usage across high, middle and low-income countries. Scalable Comput.: Pract. Exp. **24**(4), 971–984 (2023)
13. Tasevska, D., Nikitov, O.: Model for the Experimental Study of the Influence of Blind Chess on Cognitive Function and Social Integration Sighned and Blind Children in secondary School Age (Developing, Piloting and Implementation in Practice of a Specialized Methodology for the Training of Blind Children in Blind Chess, Creating Software for the Training of Blind Children in Blind Chess) (2017)
14. Jeremic, V., Vukmirovic, D., Radojicic, Z.: Does playing blindfold chess reduce the quality of game: comments on chabris and hearst (2003). Cogn. Sci. **34**(1), 1–9 (2010)
15. Clark, G.: Deep synoptic monte-carlo planning in reconnaissance blind chess. Adv. Neural Inf. Process. Syst. 34 (2021)
16. Balata, J., Mikovec, Z., Slavik, P.: Problems of blind chess players. In: 2015 6th IEEE International Conference on Cognitive Infocommunications (CogInfoCom). IEEE, pp. 179–183 ((2015))
17. McIlroy-Young, R., Wang, Y., Sen, S., Kleinberg, J., Anderson, A.: Detecting individual decision-making style: exploring behavioral stylometry in chess. Adv. Neural. Inf. Process. Syst. **34**, 24482–24497 (2021)
18. Wieczerzak, D., Czarnul, P.: Dataset related experimental ınvestigation of chess position evaluation using a deep neural network. In: International Conference on Parallel Processing and Applied Mathematics. Springer International Publishing, Cham, pp. 429–440 (2022)
19. Cooke, N.J., Atlas, R.S., Lane, D.M., Berger, R.C.: Role of high-level knowledge in memory for chess positions. Am. J. Psychol. 321–351 (1993)
20. Sak, H., Senior, A., Rao, K., Beaufays, F.: Fast and accurate recurrent neural network acoustic models for speech recognition. arXiv preprint arXiv:1507.06947 (2015)
21. Bharathi, B., Kavitha, S., Shashaank, D.S., Priyanka, S., Sriram, V.: Speech recognition based chess system for visually challanged. In: 2017 International Conference on Energy, Communication, Data Analytics and Soft Computing (ICECDS). IEEE, pp. 1797–1801 (2017)
22. Mechner, F.: Ch ess as a behavioral model for cognitive skill research: review of blindfold chess by Eliot Hearst and John Knott. J. Exp. Anal. Behav. **94**(3), 373–386 (2010)
23. Kaur, P., Singh, S.K., Singh, I., Kumar, S. Exploring convolutional neural network in computer vision-based ımage classification. In: International Conference on Smart Systems and Advanced Computing (Syscom-2021) (2021)

24. Sharma, A., Singh, S.K., Badwal, E., Kumar, S., Gupta, B.B., Arya, V. et al.: Fuzzy based clustering of consumers' big data in industrial applications. In: 2023 IEEE International Conference on Consumer Electronics (ICCE). IEEE, pp. 01–03 (2023)
25. Singh, R., Singh, S.K., Kumar, S., Gill, S.S.: SDN-aided edge computing-enabled AI for IoT and smart cities. In: SDN-Supported Edge-Cloud Interplay for Next Generation Internet of Things, pp. 41–70 (2022)
26. Verma, R., Chhabra, A., Gupta, A.: A statistical analysis of tweets on covid-19 vaccine hesitancy utilizing opinion mining: an Indian perspective. Soc. Netw. Anal. Min. **13**(1), 12 (2022)
27. Singh, I., Singh, S.K., Singh, R., Kumar, S.: Efficient loop unrolling factor prediction algorithm using machine learning models. In: 2022 3rd International Conference for Emerging Technology (INCET). IEEE, pp. 1–8 (2022)
28. Khade, G., Kumar, S., Bhattacharya, S.: Classification of web pages on attractiveness: a supervised learning approach. In: 2012 4th International Conference on Intelligent Human Computer Interaction (IHCI). IEEE, pp. 1–5 (2012)
29. Reddy, D.R.: Speech recognition by machine: A review. Proc. IEEE **64**(4), 501–531 (1976)
30. Gupta, S., Chhabra, A., Agrawal, S., Singh, S.K.: A comprehensive comparative study of machine learning classifiers for spam filtering. In: International Conference on Cyber Security, Privacy and Networking. Springer International Publishing, Cham, pp. 257–268 (2021)
31. Mengi, G., Singh, S.K., Kumar, S., Mahto, D., Sharma, A.: Automated machine learning (AutoML): the future of computational intelligence. In: International Conference on Cyber Security, Privacy and Networking. Springer International Publishing, Cham, pp. 309–317 (2021)
32. Brown, J.A., Cuzzocrea, A., Kresta, M., Kristjanson, K.D., Leung, C.K., Tebinka, T.W.: A machine learning tool for supporting advanced knowledge discovery from chess game data. In: 2017 16th IEEE International Conference on Machine Learning and Applications (ICMLA). IEEE, pp. 649–654 (2017)
33. Maulud, D., Abdulazeez, A.M.: A review on linear regression comprehensive in machine learning. J. Appl. Sci. Technol. Trends **1**(4), 140–147 (2020)
34. Furnkranz, J.: Machine learning in games: a survey. Mach. Learn. Play. Games 11–59 (2001)
35. Bertram, T., Fürnkranz, J., Müller, M.: Supervised and reinforcement learning from observations in reconnaissance blind chess. In: 2022 IEEE Conference on Games (CoG). IEEE, pp. 608–611 (2022)
36. Cui, M.: Introduction to the k-means clustering algorithm based on the elbow method. Account. Audit. Financ. **1**(1) (2020)
37. Kumar, S., Singh, S.K., Aggarwal, N., Gupta, B.B., Alhalabi, W., Band, S.S.: An efficient hardware supported and parallelization architecture for intelligent systems to overcome speculative overheads. Int. J. Intell. Syst. **37**(12), 11764–11790 (2022)

Earthquake Prediction Using Ensemble Regression Models

Corinna Mum Norbu, Ramakrishna Challa, and Jagriti Saini

Abstract Throughout history, earthquakes have caused a lot of trauma and distress due to their devastating consequences. Due to the uniqueness of the seismic structures, each seismic event is different from other; therefore, prior estimation of these events is critical. Several researchers in the past have worked on prediction of earthquake occurrences; however, there is still room for improvement in terms of accuracy. With the reign of machine learning methods and its efficiency realized through various fields of research related to predictions, many researchers have found it to be a promising tool to enhance earthquake prediction performance. In this work, authors implemented multiple machine learning and ensemble learning models to predict the earthquake using USGS dataset. Among various algorithms including random forest, gradient boosting, tree, K-nearest neighbours and stack ensemble of these algorithms, the stack model provided best performance in terms of mean squared error (0.069) and root mean squared error (0.262).

Keywords Earthquake prediction · Ensemble learning · Regression model · Orange software

1 Introduction

Earthquakes have been one of the greatest natural catastrophes that have the power of abolishing an entire city, habitat, or landscape or affecting a nation's entire economy. What makes it even more fatal, is its unpredictability. It does not give any explicit warning prior to its occurrence which makes it difficult to take caution and prepare beforehand to minimize the damage [1]. Due to this, the threat is still prevalent in the world at large. Many other natural disasters such as tsunamis, landslide, and fire are

C. M. Norbu (✉) · R. Challa
Department of Computer Science and Engineering, National Institute of Technical Teachers Training and Research, Chandigarh, India
e-mail: cmn@rggp.ac.in

J. Saini
Eternal RESTEM, Sunder Nagar, Himachal Pradesh, India

B. Soni et al. (eds.), *Beyond Artificial Intelligence*, Lecture Notes in Networks and Systems 1326, https://doi.org/10.1007/978-981-96-4170-3_53

inflicted due to earthquake [2]. A catastrophe as great as this, brings not only deaths but also brings destruction of assets, infrastructures, and the cost of reconstruction. The time of economic recovery from the damage caused by earthquakes differs from country to country. A developed country may not take as much time to recover in comparison to the middle- and low-income countries. Earthquakes happen almost every day in different locations around the world. People can feel the ground shaking only if the magnitude of the tremor is more than 2.5; a magnitude lower than 2.5 does not cause as much shaking for the people to realize its occurrence [3].

The Circum-Pacific Seismic belt, which is the highest seismic activity area in the world, has several active volcanoes and the occurrence of earthquakes in the area around this belt is frequent [4]. This is the reason why this belt is also known as "The Ring of Fire." Countries like South California, Japan, Indonesia, Turkey, Iran, and Taiwan fall in this belt. With the increase in the population and development of the infrastructures, the damage risk has grown even more. In this case, taking precautions and minimizing damage by utilizing effective disaster mitigation strategies and trustworthy earthquake prediction models is necessary [5]. In this paper, an ensemble-based stacking algorithm was proposed for predicting the earthquake. The authors in this study used Decision Tree, Gradient Boosting, k-Nearest Neighbour (kNN) and Adaboost methods to create the ensemble and compared the output with the individual algorithms.

Section 2 contains a brief survey of the related works done in the field. In Sect. 3, the proposed methodology is discussed, Sect. 4 presents the results. Finally, conclusions are drawn and the future scope is discussed in Sect. 5.

2 Related Work

The recent evidences from literature show use of machine learning and ensemble learning algorithms to predict the earthquake related events. However, there is still room for a novel system to accurately predict the occurrences of the earthquake. In [6], a model was designed for the Hindukush region using four supervised machine learning techniques: Random Forest (RF), Pattern Recognition Neural Network, Linear Programming Boost Ensemble and Recurrent Neural Network classifier. Earthquake catalogue of that region using Eight seismic indicators were used to predict the earthquake. Predictions were made to predict an earthquake having magnitude equal to 5.5 or greater for one month span. The proposed method produced significant and encouraging results. G. Asencio–Cortés and others in [7] utilized Cloud-based architecture, and regression algorithms combined with stacking ensemble learning using big data for predicting earthquakes for the next 7 days. Using big data analytics in earthquake prediction opened a wide room for future research lines. In [4], learning-based schemes were leveraged to identify the presence of an earthquake. Seismic waves from historical data served as the training data for the classifier. Specifically, learning-based schemes, including kNN, Classification Tree, and

Support Vector Machine (SVM), were employed to detect the presence of an earthquake. The outcomes demonstrated that the proposed schemes surpassed traditional criterion-based methods in performance and exhibited higher reliability. Another model called as DLEP (Deep Learning model for Earthquake Prediction) was also proposed in [8]. Eight precursory pattern-based indicators were employed as explicit features, and implicit features were extracted using Convolutional Neural Network (CNN). An Attention-Based Strategy was applied to amalgamate the explicit and implicit features of the seismic events. When compared to the results obtained from the baseline methods, the DLEP model exhibited promising performance.

Asim and others in [9] developed a method called as Inverse Boosting Pruning Trees (IBPT) for forecasting short-term. Hyper-parametric optimization and cross-validation are used to find the best parameters for the model. In comparison to the next-best CNN, this model improved Matthews correlation coefficient (MCC) by 16% and gave R-score more than 10%. It can be concluded that the state-of-the-art methods underperform than the IBPT framework. In [10], seven different classifier algorithms were compared to identify which one will be better to forecast earthquake types in the Indian region. The list of trained models included Bayes Net, RF, ZeroR, Simple Logistic, Random Tree, Logistic Model Tree, and Logistic Regression algorithms. As per the result, the best classification techniques for locating earthquakes in India were Logistic Model Tree and Simple Logistic Classifier Algorithm. In [2], Yousefzadeh and others considered that strong earthquakes occur mostly along the active fault zones and their surrounding areas. A methodology that incorporates data on fault location was developed, and a spatial parameter referred to as Fault Density (FD) was introduced for short-term earthquake magnitude prediction. The study explores the impact of FD on the performance of four machine learning algorithms used for earthquake magnitude prediction, namely SVM, Decision Tree, Deep Neural Network and Shallow Neural Network. The results indicate that the FD parameter significantly improves prediction accuracy, particularly for larger magnitude earthquakes. Furthermore, in [11] a Long Short Term Memory (LSTM) architecture was designed for predicting earthquake occurrence and location of the following month. Although LSTM has considerable memory capacity, it struggles to process data from lengthy sequences. Hence, the attention mechanism is employed in the LSTM architecture, which as a result increased the proposed model's accuracy (74.67%). Additionally, a regression model composed of dense layers and LSTM predicted the location of an earthquake with a very low Root Mean Square Error (1.5579). However, the attention mechanism is not used for location prediction as it increases the complexity of the LSTM model. The proposed model was designed to forecast the next month's earthquake while it failed to give exact time of earthquake's occurrence.

Essam and others in [12], suggested Artificial Neural Network (ANN) model can be used to measure and predict the ground motion parameters. The ground motion characteristics considered are depth, earthquake acceleration, and velocity. The earthquake acceleration was predicted using the model with a regression coefficient of 0.93888, while earthquake depth exhibited a remarkable regression coefficient of 0.9997. Additionally, earthquake velocity demonstrated a high regression coefficient of 0.9966. To expedite the training process, integrating an ANN with an advanced optimization algorithm was recommended.

The previous studies conducted in this field show that machine learning models, specially the ones based on Ensemble Learning, have considerable scope in making accurate predictions for earthquake related events. In this article, method that can improve and supplement the existing achievements through the use of regression machine learning and ensemble learning algorithms are explored.

3 Proposed Work

3.1 Data Acquisition and Pre-Processing

For this study, dataset from the catalogue of US earthquake data was acquired from Kaggle [13]. It is a dataset comprising all earthquake reports from earthquake by United States Geological Survey (USGS). Table 1 shows the characteristics of this dataset. This dataset includes 17 number of parameters that are measured from 1st Jan'2021 to 31st Dec'2021 in United States of America. At initial stage, the text attributes such as time, id, updated and place were skipped whereas magnitude attribute was taken as the target variable. Using, Select Rows feature in Orange software, only the instances that were recorded due to the earthquake were selected and the rest were ignored.

Furthermore, categorical attributes type, magtype, magsource and locationsource were ignored due to their least impact on the dependent variable. During preprocessing, the missing values were imputed by average/most frequent values. Relevant features were selected by determining the score with Univariate Linear Regression method with a fixed number of 12 top scored variables. It centres on establishing a connection between a single independent variable and a sole dependent variable. Furthermore, selected features were also normalized by standardizing the values to $\mu = 0, \sigma^2 = 1$.

Table 1 Dataset characteristics

Region	USA
Period	1 year
Source	USGS
Size	4.63 MB
Events	28,872 events
Attributes	Time, longitude, depth, latitude, type, depth error, magnitude, magnitude type, gap, horizontal distance between the closest station and the epicenter (dmin), root mean square, ID, date of update, magnitude error, horizontal error, place, location source and magnitude source

3.2 *Machine Learning Model*

The authors in this study utilized Orange software package which is an open-source software distributed under the General Public License (GPL). It provides a comprehensive library of algorithms suitable for data mining, pre-processing, and visualization tasks Orange seamlessly integrates with common Python libraries and maintains a dedicated GitHub repository [14]. Renowned for its user-friendly and interactive environment, Orange proves to be an invaluable tool for researchers engaged in diverse research projects [15]. The proposed work involves the creation of individual regression machine learning models, including Gradient Boosting, Tree, kNN, and AdaBoost. Additionally, an ensemble stack comprising these models is implemented using the Orange software, as illustrated in Fig. 1. Each of these models was fine-tuned with specific settings to maximize their predictive performance. A Gradient Boosting model is constructed with a total of 100 trees, employing a learning rate set at 0.1, and enforcing a maximum tree depth of 3. It utilizes all training instances and stops splitting nodes with a maximum of 2 instances. The AdaBoost model uses decision trees as its base estimator and consists of 100 individual estimators. For classification tasks, it employs the Samme.r algorithm, and for regression tasks, it uses the square loss function. The k-Nearest Neighbors (kNN) model employs 5 neighbors for predictions, employing the Euclidean distance metric and uniform weights. The Tree model is configured with specific pruning criteria: it requires a minimum of 2 instances in leaf nodes, at least 5 instances in internal nodes, and it imposes a maximum depth limit of 100. Additionally, in classification tasks, the model halts further splitting when the majority class within a node reaches a threshold of 95%. Lastly, these models are combined in a Stack ensemble, which includes the Gradient Boosting, AdaBoost, kNN, and Tree models as base learners. The stacking ensemble's aggregator uses the default Ridge regression method for combining predictions.

4 Results

The "Test and Score" widget in Orange serves as a pivotal tool for evaluating the performance of machine learning/ensemble learning models on a given dataset. This widget is essential for assessing the model's performance across various metric [16]. In the context of the proposed regression model, the performance criteria employed encompass Mean Squared Error (MSE), Root Mean Squared Error (RMSE), Mean Absolute Error (MAE), and the R^2 Score, which is also referred to as the coefficient of determination. The dataset was divided into 70:30 ratio for training and testing. The models are tested using 10-fold cross validation. In k-fold cross-validation, the dataset is partitioned into k separate and approximately uniform subsets. 'Fold' signifies the count of these distinct subsets. This division is done by randomly selecting data points without replacement from the learning set. The model is therefore trained

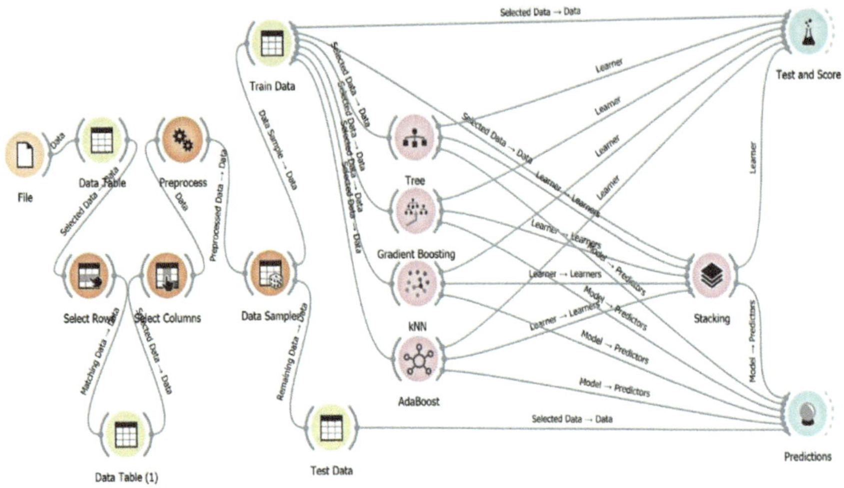

Fig. 1 Regression model for earthquake prediction

using k − 1 of these subsets, which collectively serve as the training set. Afterwards, the model is applied to the remaining subset which is called as the validation set, and is used to assess performance. Entire process is repeated until each of the k subsets has been used as the validation set. The cross-validated performance is determined by averaging the performance measurements obtained from these k validation sets [17]. Since this is a regression model, stratification during test and score is ignored.

The R-squared (R^2) value indicates the degree to which a model accurately predicts the dependent variable's outcome. As shown in Table 2, the R^2 score of 0.905 achieved for the stacking ensemble, indicates better performance compared to the other models. Given that lower values of MSE and RMSE indicate a higher level of accuracy in a regression model, it can be inferred that the proposed Stack ensemble model outperformed other models for predicting earthquakes using the provided dataset.

Table 2 Predictions performance score of different implemented algorithms

Model	MSE	RMSE	MAE	R^2
Stack	0.069	0.262	0.189	0.905
Gradient boosting	0.096	0.311	0.226	0.867
kNN	0.091	0.302	0.219	0.874
AdaBoost	0.072	0.268	0.187	0.901
Tree	0.139	0.372	0.262	0.809

5 Conclusion

This paper emphasizes the utilization of ensemble learning capabilities offered by machine learning algorithms, presenting a promising domain for researchers to explore and design models employing a variety of ensemble techniques. The application of ensemble learning contributes to enhanced model performance and a reduction in prediction errors. In this study, the Orange software platform was employed to implement and construct an ensemble model comprising four algorithms: Gradient Boosting, Tree, kNN, and AdaBoost. A dataset consisting of 28,872 events was employed for this research. Models were trained using cross-validation techniques and evaluated for prediction accuracy. Notably, the Stack ensemble demonstrated outstanding performance, achieving an R^2 score of 0.905. The proposed model showed good results on US Dataset in comparison to the other models that were considered. In the future, the same can be implemented to predict earthquakes in India.

References

1. Kumar, D.S.A., Kumar, A., Dhanraj, A., Thakur, A.: Earthquake prediction using machine learning **08** (2021)
2. Yousefzadeh, M., Hosseini, S.A., Farnaghi, M.: Spatiotemporally explicit earthquake prediction using deep neural network. Soil Dyn. Earthq. Eng. **144**, 106663 (2021). https://doi.org/10.1016/j.soildyn.2021.106663
3. Abdul Salam, M., Abdelminaam, Salama, D.-D.: Earthquake Prediction using Hybrid machine learning techniques. Int. J. Adv. Comput. Sci. Appl. **12**, 654–665 (2021). https://doi.org/10.14569/IJACSA.2021.0120578
4. Chin, T.-L., Huang, C.-Y., Shen, S.-H., Tsai, Y.-C., Hu, Y.H., Wu, Y.-M.: Learn to detect: improving the accuracy of earthquake detection. IEEE Trans. Geosci. Remote Sens. **57**, 8867–8878 (2019). https://doi.org/10.1109/TGRS.2019.2923453
5. Rahman, M.H.: Earthquakes don't kill, built environment does: evidence from cross-country data. Econ. Model. **70**, 458–468 (2018). https://doi.org/10.1016/j.econmod.2017.08.027
6. Asim, K.M., Martínez-Álvarez, F., Basit, A., Iqbal, T.: Earthquake magnitude prediction in Hindukush region using machine learning techniques. Nat. Hazards **85**, 471–486 (2017). https://doi.org/10.1007/s11069-016-2579-3
7. Asencio-Cortés, G., Morales-Esteban, A., Shang, X., Martínez-Álvarez, F.: Earthquake prediction in California using regression algorithms and cloud- based big data infrastructure. Comput. Geosci. **115**, 198–210 (2018). https://doi.org/10.1016/j.cageo.2017.10.011
8. Li, R., Lu, X., Li, S., Yang, H., Qiu, J., Zhang, L.: DLEP: a deep learning model for earthquake prediction. In: 2020 International Joint Conference on Neural Networks (IJCNN), pp. 1–8 (2020). https://doi.org/10.1109/IJCNN48605.2020.9207621
9. Xiong, P., Tong, L., Zhang, K., Shen, X., Battiston, R., Ouzounov, D., Iuppa, R., Crookes, D., Long, C., Zhou, H.: Towards advancing the earthquake forecasting by machine learning of satellite data. Sci. Total. Environ. **771**, 145256 (2021). https://doi.org/10.1016/j.scitotenv.2021.145256
10. Debnath, P., Chittora, P., Chakrabarti, T., Chakrabarti, P., Leonowicz, Z., Jasinski, M., Gono, R., Jasińska, E.: Analysis of earthquake forecasting in India using supervised machine learning classifiers. Sustainability **13**, 971 (2021). https://doi.org/10.3390/su13020971

11. Banna, Md.H.A., Ghosh, T., Nahian, Md.J.A., Taher, K.A., Kaiser, M.S., Mahmud, M., Hossain, M.S., Andersson, K.: Attention-based bi-directional long-short term memory network for earthquake prediction. IEEE Access **9**, 56589–56603 (2021). https://doi.org/10.1109/ACCESS.2021.3071400
12. Essam, Y., Kumar, P., Ahmed, A.N., Murti, M.A., El-Shafie, A.: Exploring the reliability of different artificial intelligence techniques in predicting earthquake for Malaysia. Soil Dyn. Earthq. Eng. **147**, 106826 (2021). https://doi.org/10.1016/j.soildyn.2021.106826
13. US EarthQuakes Data (2021–2022). https://www.kaggle.com/datasets/ayushv322/us-earthquakes-data20212022. Accessed 25 Sept 2023
14. Vaishnav, D., Rao, B.R.: Comparison of machine learning algorithms and fruit classification using orange data mining tool. In: 2018 3rd International Conference on Inventive Computation Technologies (ICICT). pp. 603–607. IEEE, Coimbatore, India (2018). https://doi.org/10.1109/ICICT43934.2018.9034442
15. Elshoush, H.T., Dinar, E.A.: Using Adaboost and stochastic gradient descent (SGD) algorithms with R and orange software for filtering e-mail spam. In: 2019 11th Computer Science and Electronic Engineering (CEEC). pp. 41–46. IEEE, Colchester, United Kingdom (2019). https://doi.org/10.1109/CEEC47804.2019.8974319.
16. Ljubljana, B.L.: University of test and score. https://orangedatamining.com/widget-catalog/evaluate/testandscore/
17. Refaeilzadeh, P., Tang, L., Liu, H.: Cross-validation. In: Liu, L., Özsu, M.T. (eds.) Encyclopedia of Database Systems. pp. 532–538. Springer US, Boston, MA (2009). https://doi.org/10.1007/978-0-387-39940-9_565

Event-Driven Stream Processing Using Window-Based Query Execution

Shruti Arora, Anushka Aggarwal, and Rinkle Rani

Abstract Event-driven stream processing has emerged as a fundamental paradigm for real-time data analysis in various domains, ranging from financial markets to IoT applications. Various stream processing frameworks have been developed in response to this demand. Among these, event-driven stream processing is critical because it continuously consumes incoming data stream while returning query results only when particular events occur. However, in the conventional query execution approach, even in the absence of relevant events, incoming data instances are continually filtered by queries, even though they do not yield any useful insights. Resultantly, it not only increases the system's computational burden but also wastes valuable system resources. To address this challenge, an intelligent event-driven stream processing approach that uses sliding window concept to buffer incoming data instances in key instances has been developed. The queued data instances are released and executed by ensuing operators when an event occurs. When certain queued instances within the window become obsolete owing to limited size of the window prior to the event occurrence, these immediately disappear from the window. As a result, system overhead decreases while throughput increases. Extensive CQL experiments have been conducted to validate the effectiveness of the proposed approach on Auction and Bidding Data streams.

Keywords Event detection · Stream query · Sliding window joins · CQL

S. Arora (✉) · A. Aggarwal · R. Rani
Department of Computer Science and Engineering, Thapar Institute of Engineering and Technology, Patiala, India
e-mail: sarora_phd18@thapar.edu

A. Aggarwal
e-mail: aaggarwal_be22@thapar.edu

R. Rani
e-mail: raggarwal@thapar.edu

B. Soni et al. (eds.), *Beyond Artificial Intelligence*, Lecture Notes in Networks and Systems 1326, https://doi.org/10.1007/978-981-96-4170-3_54

1 Introduction

A growing number of applications necessitate the near-real-time processing of endless streams of data in response to a large collection of continuous requests. In an era characterized by the explosive growth of real-time data streams from diverse sources such as social media, financial markets [1], and IoT devices [2], the efficient processing of streaming data has become imperative for organizations across industries. Streaming data analysis [3], offers the promise of timely insights, enabling rapid decision-making [4] and response to emerging trends and events.

Processing queries with low latency in data streams is crucial to make timely decisions. Stream processing systems must minimize query processing times to meet these requirements. Queries in data stream processing may involve complex event patterns, including sequences, time intervals, and aggregations. Therefore, designing query languages and engines that can express and efficiently process such patterns is challenging. Further, the queries may need to maintain state information over time to detect patterns or calculate aggregations. Henceforth, it is crucial to manage and update this state efficiently without introducing bottlenecks.

Data streams often contain noisy or incomplete data. Stream processing systems may need to incorporate data quality checks and cleaning mechanisms into query processing to ensure accurate results. To address, all these critical challenges a data processing paradigm that focuses on handling and analyzing data streams in real-time based on the occurrence of specific events is proposed which works on the principles of state-of-the-art event driven stream processing frameworks. In the proposed approach, event-driven system is designed to respond to specific triggers or conditions. These triggers can be defined using rules, thresholds, patterns, or other criteria. When a trigger is met, a corresponding action or computation is executed. In some cases, event-driven systems maintain state information across multiple events to perform tasks like session tracking or real-time aggregation. Stateful processing enables the system to remember and utilize historical context.

In the following sections, Sect. 2 provides an in-depth review of existing literature, offering insights into methodologies and techniques related to event based stream query optimisation. Section 3 offers a concise overview of the proposed novel approach. In Sect. 4, experimental results and a thorough analysis, including benchmark datasets, metrics, and comparative assessments is presented. The paper concludes in Sect. 5, results, comparative analysis with existing methodologies is discussed, along with directions for future research.

2 Literature Review

The landscape of query optimization in streaming data has evolved significantly in response to the growing importance of real-time data analytics and decision-making. This section provides an overview of key developments, methodologies, and trends in this field, highlighting both historical contributions and recent advancements. The history of query optimization in streaming data can be traced back to the early efforts in database management systems and continuous query processing. Early techniques focused on basic filtering and aggregation operations. However, as data streams became increasingly prevalent in various domains, the need for more sophisticated query optimization strategies emerged.

Early approaches to query optimization in streaming data primarily revolved around rule-based systems [5] and heuristic-based techniques [6]. These methods often struggled to handle the high-volume, high-velocity nature of data streams effectively. Furthermore, they had limited support for complex query structures and dynamic optimization.

Further, the advent of window-based processing introduced significant improvements. Time-based and count-based windows allowed for the aggregation of data within specified time intervals or over a fixed number of events. Techniques like Sliding Windows [7] and Tumbling Windows [8] became popular for managing continuous queries on data streams.

Incremental processing emerged as a key optimization technique, aiming to minimize redundant computations. By updating query results incrementally as new data arrives, these approaches significantly reduced computational overhead. Methods like Continuous Query Language (CQL) [9] and Data Stream SQL (DSQL) [10] played a pivotal role in advancing incremental processing.

Stream joins [11], a common operation in streaming data, have been a focal point of research. Techniques for optimizing join operations in streaming queries have evolved to handle various join types, including equi-joins, interval joins, and outer joins, while maintaining low-latency processing.

The CQL query language [12] provides the foundation for the proposed optimisation scheme for continuous queries in this paper. In this context, CQL serves as a powerful tool for expressing and executing complex queries that identify, capture, and analyze events of interest in real-time data streams by using the concepts of sliding windows. Sliding windows in CQL provide a temporal context for event detection queries. These windows can be adjusted to capture events occurring within specific time intervals, allowing for the identification of events that are time-sensitive or dependent on historical data. CQL provides mechanisms for managing sliding windows efficiently, including sliding window size, sliding step, and eviction policies. These settings allow users to balance resource consumption and event detection precision.

3 Proposed Methodology

This section is organised according to the processes involved in query optimisation and evaluation. The sequence of events in query processing are depicted in Fig. 1, and it is similar to the tried-and-true procedures used in conventional database

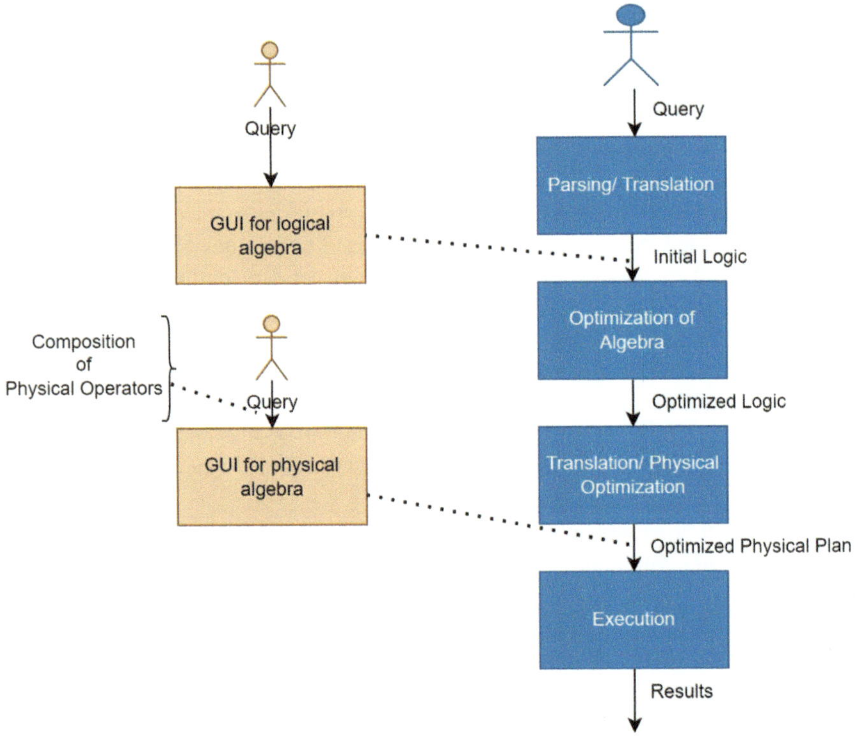

Fig. 1 Steps from the creation of the query to its execution

management systems. However, the proposed technique introduces few variations highlighted below:

1. Considering the continuous nature and sub-queries, specify the sliding windows to capture events over time and optimise the subplan of the sub-query.
2. Define event conditions and criteria within the queries. Determine the appropriate size and sliding step for the sliding windows.
3. Monitor and optimize the performance of the CQL queries and sliding windows. Fine-tune query conditions and window parameters for efficiency

A. Data source

Data sample utilised in this paper for experimental evaluation are taken from the Niagara project's NEXMark benchmark for data stream systems (Schema given in Fig. 2) [13]. An online auction's structure and query specifications are provided by the benchmark. The system is similar to that of eBay. There are countless auctions happening at once. The following actions proceed indefinitely: (i) Registration of new users, (ii) new articles listed (iii) Arrival of bids.

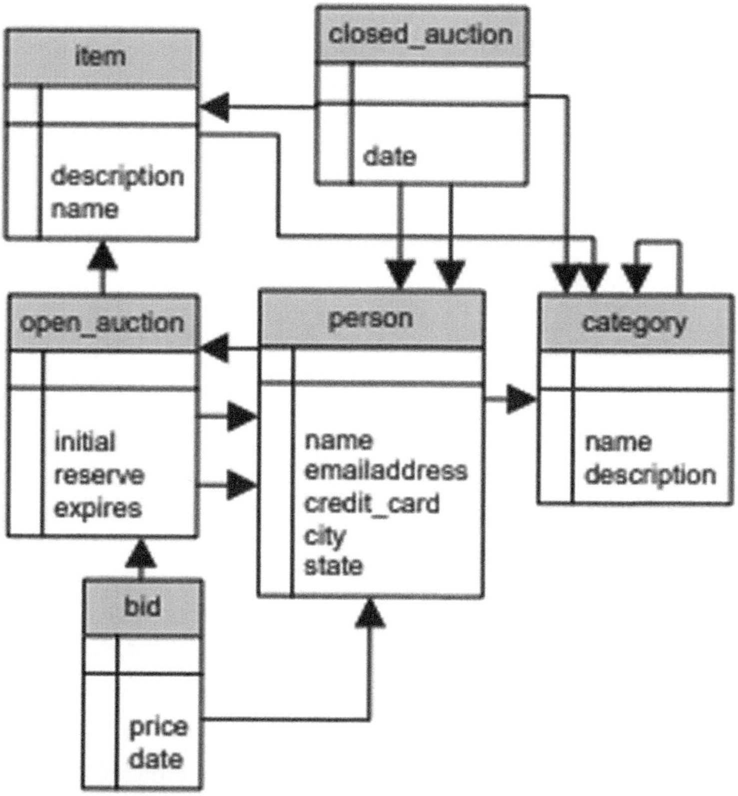

Fig. 2 Schema of NexMark Auction Data

B. Data stream parsing and translation

Users can manually create query plans using the operator algebra by using various operators. A minor expansion of regular SQL that includes a selection of the window constructs in continuous query language.

(a) Raw to logical stream

The $\varphi^{p \to q} : M_S^p \to M_S^q$ function converts an unprocessed stream M^p to its conceptual equivalent by following Eq. (1).

$$\varphi^{a \to b}(M^p) := \{(r, s, q) \in \lambda_s \times s \mid q = |\{(r, s) \in M^p\}| \wedge q > 0\} \quad (1)$$

Value-equivalent components of the unprocessed stream are incorporated into the logical one. Resultantly, the multiplicity q is equal to the degree of redundant instances (r) that are seen in in M^p at a particular instance of time (s).

(b) Raw to physical stream

Let N_p be a type R raw or unprocessed stream. A raw stream is passed to the function $\psi^{p \rightarrow q}: M_R^p \rightarrow M_R^q$, that outputs similar stream which is physical (processed). Each instance in the raw stream is transformed to a physical stream component using the following mapping given in Eq. (2).

$$\{r, s\} \rightarrow \{r, (s, s + 1)\} \tag{2}$$

C. Time-based sliding window operations

The proposed scheme focuses on time-based sliding window operator since continuous queries typically employ this kind of window. As per the sliding window requirements of SQL and CQL, count and partition windows have been integrated. S and W are the inputs required by the time-based sliding window where S is the logical stream and W is the size of window, time: $S_R^l \times W \rightarrow S_R^l$. A period of (application) time is represented by the window size W€R, $w > 0$, which is written as a subscript. To define the output stream over its input stream, the operator changes within a range of window size w.

D. Algebraic optimisation

A well-defined semantics is the foundation for any query optimisation. Logical algebra defines the logical relationships between conditions, and allows to express intricate patterns and constraints in the continuous queries. This section illustrates to what extent the proposed approach allows query optimisation.

The proposed technique introduces a timeslice operator, which takes snapshots of a logical stream. It can be defined mathematically as: υ: $(S_R^l \times R \rightarrow \wp(\beta R \times M)$. The function requires two arguments: a logical stream S^l and a time moment represented by the subscript s. The system calculates the current state of S^l at time s using function τ_t given in Eq. (3), which refers to the collection of all tuples that are valid at that specific moment.

$$\upsilon_t(S^l) = (\{r, q\} \in R \times M \mid \{r, s, q\} \in S^l) \tag{3}$$

In order to achieve optimisation in streaming queries, it is necessary for the framework to assess the efficiency of numerous common queries or fulfil a maximum number of query predicates. Common queries refer to search queries that encompass a wide range of search terms and yield many types of results.

Hence, our attention is directed towards classifying queries based on the categories of results and search keys in order to investigate the efficiency of retrieving a response. Additionally, it is imperative to consider the various procedures that impact the execution of queries.

Given the significance of time in stream processing, it has been decided to incorporate the snapshot-reducibility attribute into the proposed stream operator. This approach is used in the extensive research and advancements made in the field of temporal databases.

Further, some stream transformations are applied so that the query when fired gives output in real-time. It is possible to incorporate additional constraints for operations into the existing transformation rules. It can be assumed that the default window which is time-based with size of the window as 1 can be excluded. (Eq. 4).

$$\beta_{1}^{\text{time}}\left(S^{1}\right) = S^{1} \tag{4}$$

The reliability of this transformation criterion is immediately derived from the specification of the time-based sliding window by combining data from different streams or sources based on temporal conditions within time-based windows. Stateless operators help perform the join operation within each window. The transformation rules for logical streams are defined as follows:

$$\sigma_{p}\left(\beta_{w}^{\text{time}}\left(S^{1}\right)\right) = \beta_{w}^{\text{time}}\left(\sigma_{p}\left(S^{1}\right)\right) \tag{5}$$

$$\mu_{f}\left(\beta_{w}^{\text{time}}\left(S^{1}\right)\right) = \beta_{w}^{\text{time}}\left(\mu_{f}\left(S^{1}\right)\right) \tag{6}$$

$$\cup\left(\beta_{w}^{\text{time}}\left(S_{1}^{1}\right), \beta_{w}^{\text{time}}\left(S_{2}^{1}\right)\right) = \beta_{w}^{\text{time}}\left(\cup\left(S_{1}^{1}, S_{2}^{1}\right)\right) \tag{7}$$

The utilisation of prior window rules might yield benefits for a system due to several factors. Rule (5) describes that if the queue is positioned between window and filter operator and further the filter operator is executed after the window operator, it will result in a reduction of queue length. Hence, this conversion helps in cost optimization of window operator as the process significantly reduces the rate of the incoming data instances for the window.

In summary, modifications have the potential to be advantageous in the context of subquery sharing. In the scenario when a time-based window is implemented following a union operation, and a subsequent query is formulated that could utilise the capability of window on a single incoming stream. In such a case, rule (7) is applied and the window operation is relocated to a lower position within the union operation.

4 Experiments Performed

The results after firing the optimized continuous queries show that the data has been retrieved from data streams in real-time. Some of the queries that were fired on NEXMark data streams are discussed in this section.

(a) Conversion of currency

```
SELECT itemID, DToE(bid_price), bidderID
FROM Bid
```

This query converts all the incoming tuples for bids from US Dollars to Euros. DToE is our customised function. This query is fired without window option, hence default window will be applied.

(b) Find out all the bids that closed within 1 h

```
SELECT OpenAuction.* FROM OpenAuction oa
WINDOW (RANGE 60 MINUTES), ClosedAuction
ca WHERE oa.itemID = ca.itemID;
```

The given query above serves as a window based join in data stream. The input for this process consists of two streams, specifically the *OpenAuction* and the *ClosedAuction* stream. A sliding window with a duration of 1 h is established for the OpenAuction stream, while a default window is implemented for the *ClosedAuction* stream.

(c) *A sliding window based two-way join having a filter operation on* S^1, as mentioned in Eq. (7)

The data operators encompass the following: sequential window for S_1 and S_2, selection, two-way join, aggregation, and redundancy-preserving. The transformation rules are commonly exchanged by window operators and the join operation, whereas the selection operation is moved after join operation. In general, it is not possible to push the selection operation below the window operator due to the lack of commutativity between tuple-based windows and selection.

```
CREATE STREAM PresentRate AS
SELECT T.itemID, T.price, Q.sellerID AS Dealer
FROM ((SELECT itemID, bid_price AS rate
FROM Bid WINDOW(RANGE 60 MINUTES))
UNION ALL
(SELECT itemID, start_price AS rate
FROM OpenAuction WINDOW(RANGE 60 MINUTES)))
T,
ClosedAuction ca,
OpenAuction Q WINDOW(RANGE 60 MINUTES)
WHERE T.itemID = ca.itemID AND ca.itemID =
Q.itemID;
```

This CQL query will shows the use of sliding window joins with time. In addition, it returns the final settled price and it's dealer for each auction within 1 h.

5 Conclusion

Although there has been a rise in the number of techniques in the area of stream processing in recent years, there has been relatively limited research focus on establishing a robust query processing for a versatile stream algebra that can be applied broadly. To address this research gap, a comprehensive scheme is proposed for optimizing streaming queries for continuously arriving data instances, demonstrating the relation between the each query from query creation to it's output. Consequently, the proposed technique aims to integrate the sliding window concepts and query optimization in the data stream settings. The proposed algebra idenfies the critical events using CQL language in data streams.

The proposed scheme adapts existing stateless transformations from conventional big data frameworks to streaming frameworks, utilising the idea of snapshots of logical stream, from a theoretical standpoint. The query rules have been improvised by incorporating innovative suggestions that have been customized for the window operators.

In a nutshell, this study has successfully optimised the continuous streaming queries having stream joins and conceptually solid framework for fetching crucial events from data streams.

References

1. Edge, M.E., Falcone Sampaio, P.R.: A survey of signature based methods for financial fraud detection. Comput. Secur. **28**(6), 381–394 (2009). https://doi.org/10.1016/j.cose.2009.02.001
2. Bulut, A., Singh, A.K.: A unified framework for monitoring data streams in real time. In: Proceedings—International Conference on Data Engineering, pp. 44–55 (2005). https://doi.org/10.1109/ICDE.2005.13
3. Shukla, A., Simmhan, Y.: Model-driven scheduling for distributed stream processing systems. J. Parallel Distrib. Comput. **117**, 98–114 (2018). https://doi.org/10.1016/j.jpdc.2018.02.003
4. Arora, S., Rani, R., Saxena, N.: An efficient approach for detecting anomalous events in real-time weather datasets. Concurr. Comput.: Pract. Exp. **34**(5), e6707 (2022). https://doi.org/10.1002/cpe.6707
5. Peña, M., Biscarri, F., Guerrero, J.I., Monedero, I., León, C.: Rule-based system to detect energy efficiency anomalies in smart buildings, a data mining approach. Expert Syst. Appl. **56**, 242–255 (2016). https://doi.org/10.1016/j.eswa.2016.03.002
6. Nardelli, M., Cardellini, V., Grassi, V., Lo Presti, F.: Efficient operator placement for distributed data stream processing applications. IEEE Trans. Parallel Distrib. Syst. **30**(8), 1753–1767 (2019). https://doi.org/10.1109/TPDS.2019.2896115
7. Wang, W., Li, J., Dong, Z.D., Guo, L.: Processing sliding window join aggregate in continuous queries over data streams (2004)
8. Carney, D., Çetintemel, U.: Monitoring streams — a new class of data management applications. In: VLDB '02: Proceedings of the 28th International Conference on Very Large Databases, pp. 215–226 (2004)
9. Arasu, A., Babu, S., Widom, J.: The CQL continuous query language: Semantic foundations and query execution. VLDB J. **15**(2), 121–142 (2006). https://doi.org/10.1007/s00778-004-0147-z
10. Wang, H., Zaniolo, C., Luo, C.R.: ATLAS: a small but complete SQL extension for data mining and data streams. In: Proceedings 2003 VLDB conference, pp. 1113–1116 (2003)
11. Gulisano, V., Papadopoulos, A.V., Nikolakopoulos, Y., Papatriantafilou, M., Tsigas, P.: Performance modeling of stream joins. In: DEBS 2017—Proceedings of the 11th ACM International Conference on Distributed Event-Based Systems, 2017, pp. 191–202 (2017). https://doi.org/10.1145/3093742.3093923

12. Kharlamov, E., et al.: An ontology-mediated analytics-aware approach to support monitoring and diagnostics of static and streaming data. J. Web Semant. **56**, 30–55 (2019). https://doi.org/10.1016/j.websem.2019.01.001
13. Tucker, P., Tufte, K., Papadimos, V., Maier, D.: NEXMark—A benchmark for queries over data streams, October 2003. Technical draft (2003)

YouTube Video Performance: Leveraging Data Analysis and Forecasting with ARIMA Modeling for Optimal Engagement

K. S. Hemanth, R. Krishna Murthy, and B. Muralidhara

Abstract This study explore the potential of data analysis and modeling for precise YouTube video performance forecasting, recognizing YouTube as a pivotal platform for global content dissemination. Examining data from 6 YouTube channels, we employed ARIMA modeling to predict upcoming video performance based on key attributes such as title, publish date, likes, comments, views, and engagement metrics. Our model revealed a mean absolute error of 7.2% and a root mean squared error of 9.1%, underscoring the significance of various video characteristics in predictive accuracy. These findings offer actionable insights for content creators to enhance content and boost engagement. Moreover, the study underscores the necessity for advanced modeling techniques to achieve accurate YouTube video performance forecasting.

Keywords YouTube · EDA · Engagement metrics · YouTube data API · ARIMA modeling

1 Introduction

YouTube has become one of the most popular platforms for content creators to share their work with a global audience. However, with millions of videos uploaded every day, standing out from the crowd and ensuring the success of a video can be a daunting task. To optimize video performance and increase engagement, it is

K. S. Hemanth (✉) · R. Krishna Murthy
School of Computer Science and Applications, REVA University, Bangalore, India
e-mail: hemanth.ks@reva.edu.in

R. Krishna Murthy
e-mail: krishnamurthy.r@reva.edu.in

B. Muralidhara
Department of Computer Science, Bangalore University, Bangalore, India
e-mail: murali@bub.ernet.in

B. Soni et al. (eds.), *Beyond Artificial Intelligence*, Lecture Notes in Networks and Systems 1326, https://doi.org/10.1007/978-981-96-4170-3_55

essential for content creators to understand the factors that influence video success and predict the performance of their videos [1, 2].

The objective of this study is to explore the potential of data analysis and modeling for accurate YouTube video performance forecasting [3, 4]. Specifically, we aim to answer the research question: "How can data analysis and modeling be used to accurately forecast the performance of YouTube videos?" We propose a methodology that involves generating a YouTube data API, scraping of YouTube data using API, data cleaning and preprocessing, feature selection, Exploratory data analysis, and ARIMA modeling [5].

2 Literature Review

The scholarly research on the prediction of YouTube video performance has thoroughly examined a range of characteristics that impact the level of success a video achieves on the site. The criteria discussed in this context include both intrinsic content aspects, such as title, description, tags, and length, as well as engagement measures, such as views, likes, dislikes, comments, and shares [6]. Various modeling strategies have been used by researchers to identify the crucial aspects in forecasting video performance and to clarify the complex interconnections among these factors [7, 8]. Time series forecasting methods, such as ARIMA, have been widely used to accurately predict future video performance by leveraging past data [9].

Despite the wide range of modeling tools used in previous studies, there is still a need for more advanced methodologies that can effectively capture the intricate dynamics of video performance on the platform. A significant aspect to take into account is the possible influence of external variables, such as ongoing events, holidays, and popular subjects, on the success of videos [10]. Furthermore, taking into consideration user demographics and interests, together with comprehending the influence of recommendation algorithms in pushing films to viewers, has the potential to augment the precision of forecasting [11]. Hence, there exists an acknowledged need for more sophisticated modeling methodologies that can thoroughly tackle the intricacies inherent in video performance on the YouTube platform [12].

3 Methodology

3.1 Data Source

With the theme "Tech Review" in mind, we collected data on 6 YouTube channel, including Channel name, Subscribers, Total views, Total videos, and playlist ID. We also collected video data tracking multiple performance metrics such as Views, Likes, Comments, Publish date, video length etc. We leveraged Google Cloud platform to generate a YouTube data API and used it to scrape data for our analysis. We used data cleaning techniques such as removing duplicates and outliers and missing values. We then chose the most important features i.e., Views, Likes and Comments for video performance as they are considered as the engagement metrics in YouTube.

3.2 ARIMA Model

We trained an ARIMA model to forecast the performance of upcoming videos, with a focus on views as the primary performance metric. ARIMA (p, d, q) is called differential autoregressive moving average model, AR is autoregressive, MA is moving average, p and q are corresponding orders, and d is the number of times the time series becomes stationary. The model is a linear regression model that is used to track linear trends in stationary time series data where future values of time series are generated from linear functions observed in the past. The ARIMA (p, d, q) model essentially performs the d-order differential processing on the non-stationary historical data Yt to obtain a new stable stock history sequence Xt, fits the Xt to the ARMA (p, q) model, and then the original d-time differential restoration can obtain the predicted data of Yt. Among them, the general expression of ARMA (p, q) is shown in Eq. (1).

$$Y_t = \varphi_0 + \varphi_1 Y_{1-1} + \varphi_1 Y_{t-p} + \varepsilon_t - \theta_1 \varepsilon_{t-1} - \cdots - \theta_q \varepsilon_{t-q} \quad (1)$$

The first half of the equation is the autoregressive part, p is the autoregressive order, and φp is the autoregressive coefficient. The second half is the moving average part, q is the moving average order, θq is the moving average coefficient; Yt is the correlation sequence of the consumed stock data, and εt is the random error. Basically, this method has three phases: model identification, parameter estimation and diagnostic checking. For example, the ARIMA (1, 1, 1) model can be represented as follows.

$$Y_t = \varphi_0 + \varphi_1 Y_{t-1} + \varepsilon_t - \theta_1 \varepsilon_{t-1} \quad (2)$$

The ARIMA model is basically a data-oriented approach that is adapted from the structure of the data themselves.

To optimize the performance of the model, we used grid search to test different combinations of hyperparameters, including the autoregressive order (p), the degree of differencing (d), and the moving average order (q). The best parameters we found were of the order (1, 2, 1 which can be represented as follows:

$$Y_t = c + 2Y_{t-1} - Y_{t-2} + \phi_1 \varepsilon t_{-1} - \theta_1 \varepsilon_{t-2} + \varepsilon_t \quad (3)$$

where:

Yt is the time series variable at time t

ϕ1 is the autoregressive coefficient

$\theta 1$ is the moving average coefficient

εt is the error term at time t

c is a constant term.

Additionally, we used optimization techniques such as cross-validation to ensure that our model was robust and performed well on both the training and testing data. We also evaluated the model's performance using performance metrics such as mean squared error, mean absolute error and root mean squared error.

Finally, we developed a Streamlit application that allowed content creators to select the channel of their choice and receive a recommendation for the best date to publish the video based on historical trends and other factors.

3.3 Explratory Data Analysis

To gain a better understanding of the data and extract meaningful insights, we conducted exploratory data analysis (EDA) on the collected YouTube channel and video data. Considering the engagements metrics we analyzed the different aspects of 5 different channels (Beebom, Mrwhosetheboss, Unbox Therapy, Geeky Ranjit, Technical Guruji and Marques Brownlee) by comparing and contrasting their corresponding metrics. We chose the channel "Mrwhosetheboss" as part of EDA for video statistics since it's one of the most popular tech review channels. The data consisted of over 1500 videos, with performance metrics tracked over a period of 12 years. We analyzed the metrics of the videos to identify patterns and trends in the data.

We also examine the data to identify patterns, trends, and seasonality as part of time series EDA. We use visualizations such as time series plots, autocorrelation plots, and decompositions to identify these patterns. This EDA can help us understand the underlying dynamics of the time series, identify potential outliers or missing values, and inform the selection of appropriate modeling techniques.

Some of the questions to be answered with EDA:

i. What is the subscriber count for each channel and which channel has the highest number of subscribers?
ii. What is the total number of views for each channel and which channel has the highest overall views?
iii. What is the total number of videos posted for each and which channel has the greatest number of videos posted?
iv. How is the View versus Likes count data distributed for the Mrwhosetheboss channel?
v. What are the top 10 most viewed videos for the Mrwhosetheboss channel based on the Views?
vi. How many videos were posted by Mrwhosetheboss each month?
vii. In which month did Mrwhosetheboss post the highest and lowest number of videos?

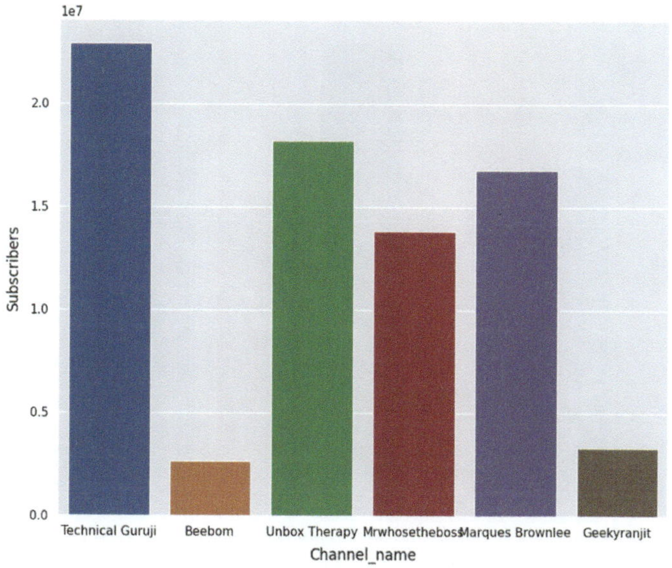

Fig. 1 Channels versus subscriber count of each channel

viii. What is the correlation between the various metrics (such as subscriber count, view count, and number of videos posted) for the Mrwhosetheboss channel?

We then performed some Time series EDA to gain insights into the patterns and relationships between the metrics over time. It involves visualizing and summarizing the data to understand its underlying characteristics, such as trend, seasonality, cyclicality, and irregularity. The analysis revealed that the data was non-stationary, with a clear increasing trend and seasonality. To make the data stationary, differencing was applied, which helped to remove the trend and seasonality from the dataset. After differencing, the time series data appeared to be stationary, with a constant mean and variance over time. This stationary data was then used as input to the time series forecasting model.

Channel Statistics (Figs. 1, 2 and 3)

Video Statistics (Figs. 4, 5, 6 and 7)

Time Series (Figs. 8 and 9)

4 Results

The results of this study were promising. While performing EDA we were able to extract meaningful insights with the help of simple visualizations. The ARIMA model was able to forecast the statistics of upcoming videos with a reasonable degree

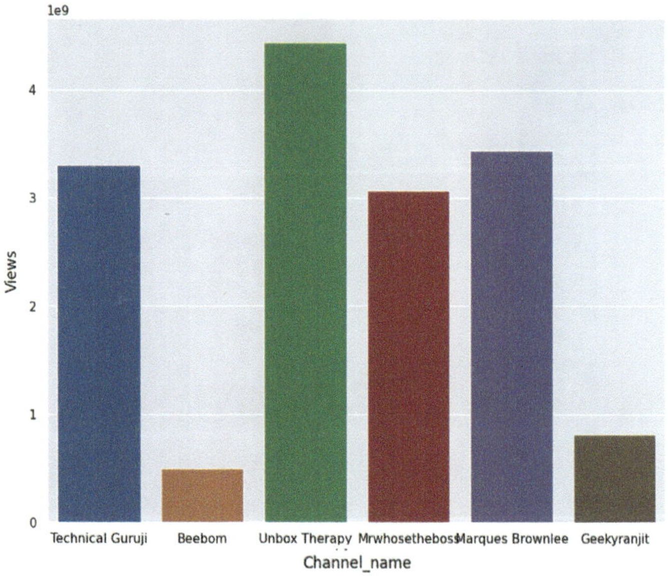

Fig. 2 Channels versus view count of each channel

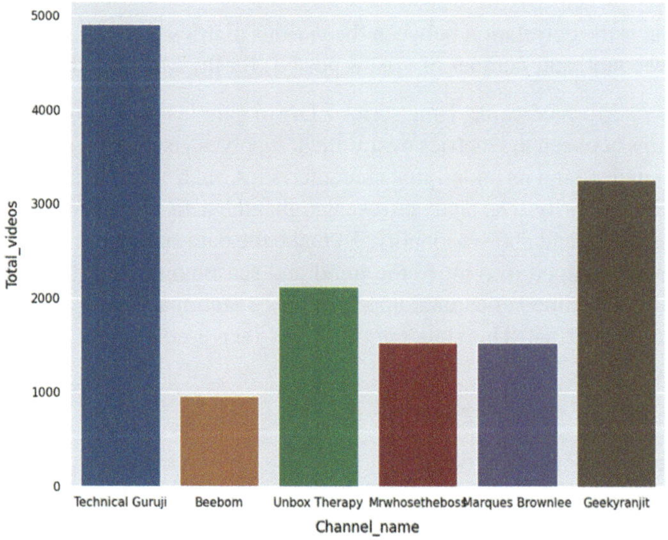

Fig. 3 Channels versus the total number of videos uploaded in each channel

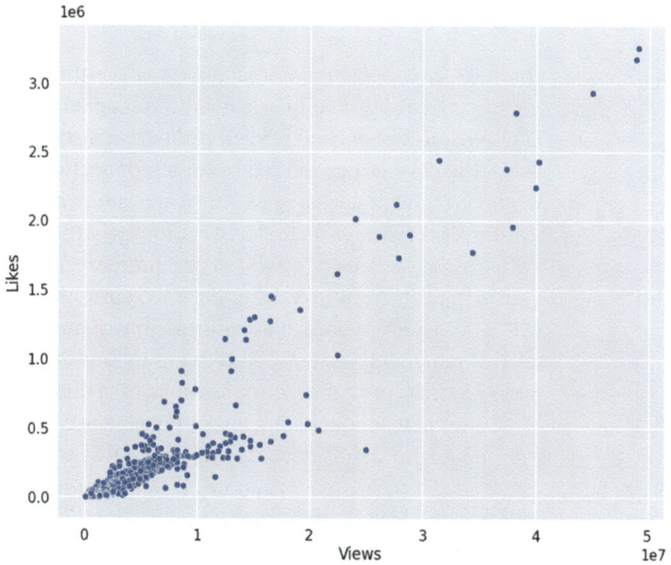

Fig. 4 Data distribution of views versus likes of Mrwhosetheboss

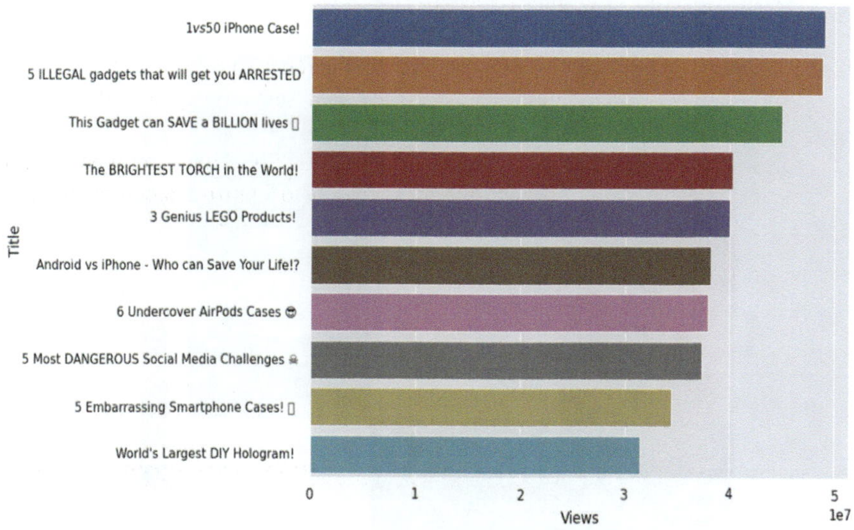

Fig. 5 Views versus videos uploaded by Mrwhosetheboss

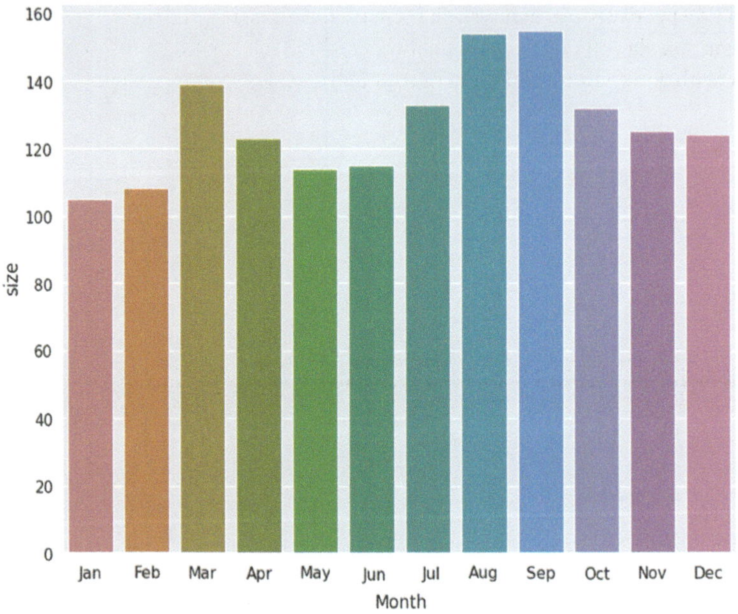

Fig. 6 Months versus the count of the videos uploaded by Mrwhosetheboss

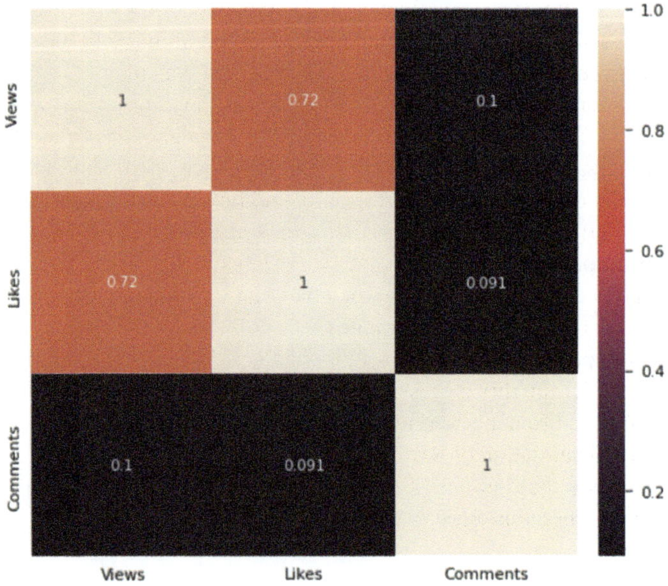

Fig. 7 Correlation between the engagement metrics of Mrwhosetheboss

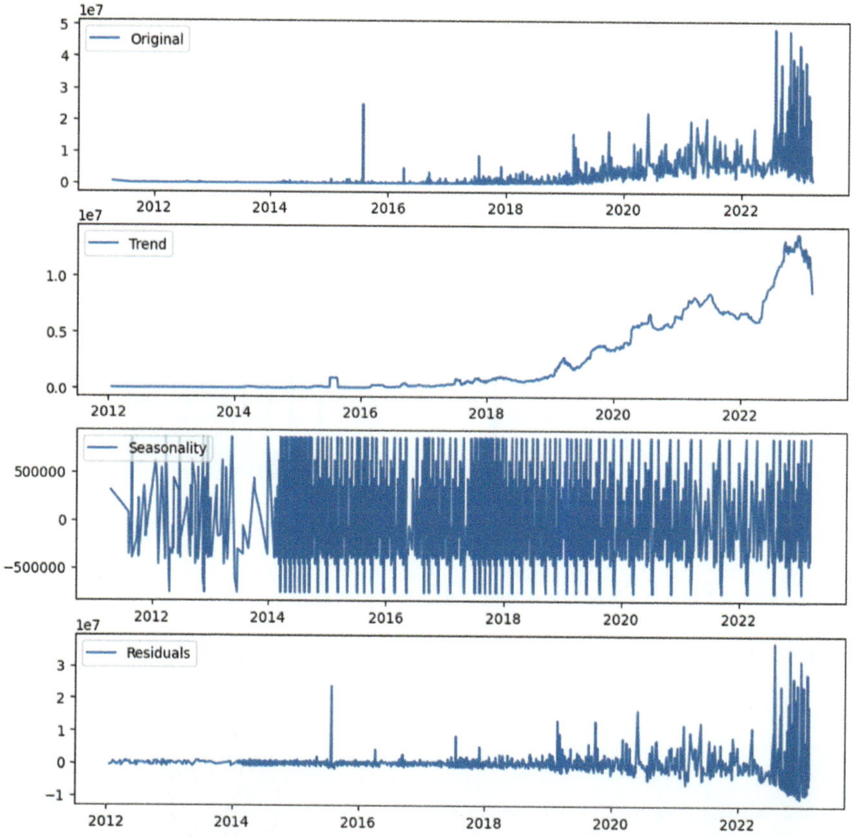

Fig. 8 Time series decomposition of views to understand trend, seasonality and the residuals of Mrwhosetheboss

of accuracy. Additionally, the model allowed users to interactively select a channel and number of days to predict, making it a useful tool for content creators who want to plan their video release schedule.

Some of the key insights gained from the EDA include (Tables 1 and 2).

Interpretations

- Even though some channels have a higher number of Subscriber count, they have fewer Views on their channel.
- There is a negative correlation between the Views and length of the videos, indicating that people are less likely to watch a video if it's too long.
- October typically sees a multitude of consumer electronics industry events, announcements, and new product releases; hence we see a spike in the videos being uploaded (Tables 3 and 4).

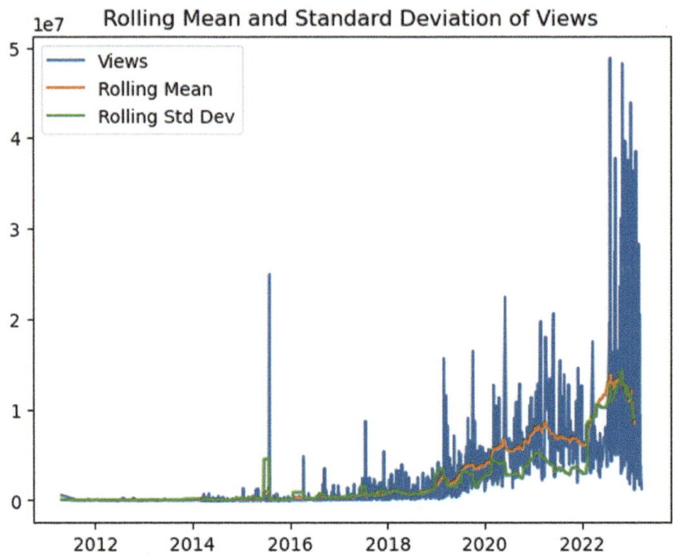

Fig. 9 Rolling mean and standard deviation of views over time

Table 1 Channel insights

a.	Highest subscriber count	Technical guruji
b.	Highest view count	Unbox therapy
c.	Greatest no. of videos posted	Technical guruji

Table 2 Mrwhosetheboss channel's videos insights

a.	Data distribution	As referred to Fig. 4
b.	Top 10 videos posted based on views	As referred to Fig. 5
c.	Count of videos posted in each month	As referred to Fig. 6
d.	Highest and lowest no. of video posted	Highest– September, October Lowest- January
e.	Correlation between views, likes and comments	Positive

Table 3 Augmented Dicky Fuller test for checking the stationarity of the data

ADF statistic value	−3.672286
p-value	0.004520
Indication	Time series is stationary and not a random walk

Table 4 Key performance metrics of the ARIMA model

Measures	Values
Mean Absolute Error (MAE)	350.42
Mean Squared Error (MSE)	100.0254
Root Mean Squared Error (RMSE)	317.23

5 Conclusion

The ARIMA model employed in this study has demonstrated notable predictive performance for forecasting YouTube video outcomes. The mean absolute error (MAE) of 350.42, mean squared error (MSE) of 100.0254, and root mean squared error (RMSE) of 317.23 collectively indicate the model's ability to accurately capture the dynamics of video performance. While the MAE reflects an average absolute difference between predicted and actual values, the MSE and RMSE underscore the overall accuracy and magnitude of prediction errors.

These metrics, when considered in conjunction, suggest that the ARIMA model is proficient in minimizing errors and providing reliable forecasts for key video performance indicators. However, it is imperative to interpret these results within the context of the specific dataset and the nature of the forecasting task. Furthermore, future research could explore the optimization of the model and its application to diverse datasets to enhance generalizability.

References

1. Cheng, X., Mehrdad, F., Ma, X., Zhang, C., Liu, J.: Understanding the YouTube partners and their data: measurement and analysis. China Commun. **11**(12), 26–34 (2014). https://doi.org/10.1109/CC.2014.7019837
2. Niture, A.A.: Predictive analysis of youtube trending videos using machine learning. Doctoral dissertation, Dublin Business School (2021)
3. Halim, E., Anindya, R., Hebrard, M.: The ımpact of motivation to watch YouTube, subjective norms, behavior control, ınformation success model to watching YouTube engagement. In: 2020 International Conference on Information Management and Technology (ICIMTech), Bandung, Indonesia, pp. 800–805 (2020). https://doi.org/10.1109/ICIMTech50083.2020.9211225
4. Gürsun, G., Crovella, M., Matta, I.: Describing and forecasting video access patterns. In: 2011 Proceedings IEEE INFOCOM, Shanghai, China, pp. 16–20 (2011). https://doi.org/10.1109/INFCOM.2011.5934965
5. Basch, C.H., Menafro, A., Mongiovi, J., Hillyer, G.C., Basch, C.E.: A content analysis of YouTube™ videos related to prostate cancer. Am. J. Mens Health **11**(1), 154–157 (2017)
6. Paek, H.J., Kim, K., Hove, T.: Content analysis of antismoking videos on YouTube: message sensation value, message appeals, and their relationships with viewer responses. Health Educ. Res. **25**(6), 1085–1099 (2010)
7. Mislove, A., Marcon, M., Gummadi, K.P., Druschel, P., Bhattacharjee, B.: Measurement and analysis of online social networks. In: Proceedings of the 7th ACM SIGCOMM Conference on Internet measurement, October 2007, pp. 29–42 (2007)
8. Baron, P.: YouTube's social media analytics as an evaluation of educational teaching videos. In: 2022 IEEE IFEES World Engineering Education Forum-Global Engineering Deans Council (WEEF-GEDC), November 2022, pp. 1–8. IEEE (2022)
9. Nochai, R., Nochai, T.: ARIMA model for forecasting oil palm price. In: Proceedings of the 2nd IMT-GT Regional Conference on Mathematics, Statistics and Applications, June 2006, pp. 13–15 (2006)
10. Barman, A.: Time series analysis and forecasting of covid-19 cases using LSTM and ARIMA models (2020). arXiv:2006.13852
11. Yenidoğan, I., Çayir, A., Kozan, O., Dağ, T., Arslan, C.: Bitcoin forecasting using ARIMA and PROPHET. In: 2018 3rd International Conference on Computer Science and Engineering

(UBMK), Sarajevo, Bosnia and Herzegovina, pp. 621–624 (2018). https://doi.org/10.1109/UBMK.2018.8566476
12. Wang, Y., Guo, Y.: Forecasting method of stock market volatility in time series data based on mixed model of ARIMA and XGBoost. China Commun. **17**(3), 205–221 (2020). https://doi.org/10.23919/JCC.2020.03.017

COVIDWavesSenti: Large Scale Twitter Dataset of Different COVID Waves

Jaspal Kaur Saini, Jyoti Maggu, and Pratishtha Verma

Abstract Covid-19 has led this entire world into a series of lockdowns and economic crises. Social media have been flooded with so many sentiments and emotions since the inception of Covid-19. People have taken social media such as Twitter, a platform to express their sentiments during Covid-19 times. As a part of this study, we collected data from twitter for majorly for three covid waves with timeline: first wave (March–April 2020), second wave (September–November 2020) and third wave (January–February 2021). We have collected around1.5 million tweets constituting three waves of Covid-19 worldwide. The contribution of this paper is two folded. One, we have presented a large scale twitter CovidWavesSenti dataset which can be utilized by policy makers to deal with such a situation in future or for any further research and development purposes. Secondly, we have also studied sentiments during different covid waves and found that 73% of tweets were negative during Covid first wave, which may be due to misinformation or fake news spread over twitter about covid. However, sentiment shifted to positive during the second wave as people started enjoying working at home. Our study supports the view that there is a need to develop a dynamic and agile public-oriented health information system to combat fake news from social media.

Keywords Data analytics · COVID · Sentiment analysis · Exploratory data analysis · Machine learning · Classification

---

J. K. Saini (✉)
Department of Information Technology, Dr B R Ambdekar NIT Jalandhar, Jalandhar, Punjab, India
e-mail: sainijk@nitj.ac.in

J. Maggu
Department of CSE, TIET, Patiala, Punjab, India
e-mail: jyotim@iiitd.ac.in

P. Verma
Department of CSE, NIT Kurukshetra, Kurukshetra, Haryana, India
e-mail: pratishthaverma@nitkkr.ac.in

B. Soni et al. (eds.), *Beyond Artificial Intelligence*, Lecture Notes in Networks and Systems 1326, https://doi.org/10.1007/978-981-96-4170-3_56

1 Introduction

Our lives took a huge turn in the beginning of 2020 when a virus, infamously known as COVID-19, started spreading across the globe at a much faster pace than any of us expected it to and affected the lives of every individual in this world in more ways than we could describe. The pandemic that began in the Wuhan district of China soon forced almost every nation in the world to enter a state of complete lockdown. Our normal lives took a huge break.

Due to these nationwide lockdowns, people in every country felt all kinds of emotions. Some were happy, some confused, some angry, some were ignorant towards COVID while others were frustrated, etc. These emotions themselves were not constant as they kept changing regularly as the time spent in lockdown increased. Nobody expected the lockdown to go on for such a long time and as the time progressed and the different waves of COVID kept affecting the world, people felt different emotions towards the pandemic. The Wordcloud in Fig. 1 represents the top 4000 words that were being used on the social media platform, Twitter in 2020 drawn using python wordcloud library.

This paper analyzed the different emotions of the world population towards the different waves of the COVID-19 pandemic and contrast the change in emotions of the world population towards different waves of COVID-19 pandemic with the help of analyzing the tweets shared by users across the globe on the Twitter.

Fig. 1 Wordcloud of top 4000 words used during Covid-19 on Twitter

The reason for choosing a social media platform is that as people were confined to their homes, they took to various social media platforms such as Facebook, Twitter, Instagram, Reddit, etc. to express the different kinds of emotions they were experiencing. Almost the entire world population used social media networks as an outlet for their emotions. People also used social media as an effective and a fast way to access information about the developments related to COVID-19 pandemic and its spread across the world.

As people came across any kind of news or updates regarding the pandemic, they expressed their views related to it there which could prove to be of great use to deeply analyze their emotions and feelings towards the pandemic. We used SNScrape to collect tweets of Twitter users worldwide for the first three waves of COVID-19 and then compared the changing emotions of the world population towards COVID during different waves.

2 Related Work

Various sources from numerous domains can be linked to the work done presented in this article. Text analysis is not a new concept. It has been used over the years to analyze reviews, filtering emails, speech sentiment detection, sarcasm detection and many more use cases [1, 2].

A study for analyzing customer feedback for a French Power Company utilized around 70,000 tweets made over a year [3]. Since the data is massive, critical insights are hidden and not observable in plain sight. LDA (Latent Dirichlet Allocation) was used to analyze the data and get the interesting insights [4]. Information systems used for designing human-trait identification systems, done using textual analysis also involve a similar study. Extracting the core emotion behind a social-media post also involves the use of linguistic and psychological analysis and this has been used in our research also. Various other applications of text analysis are e-mail filtering [5], irony detection [6], emotion detection [7–10].

Tracking twitter tweets has been involved in past epidemic analysis and crisis-situations. Widener and Li in their research paper concerning healthy and unhealthy food, showed that people in urban areas as well as suburbs tweet more than those residing in rural areas, using the spatial distribution of analyzed tweets [11]. Logistic Regression has been used to find that the tweets in the low-income areas were related to unhealthy food. Twitter data has been used in sentiment analysis of the Healthcare domain. Behavioral change and a change in linguistic style due to post-childbirth effects can also be detected by analyzing different tweets [12]. Authors in [13–16] presented various machine learning and data science approached to applied to different applications including big data in healthcare, information security, etc. Deep learning algorithms details are also explored by authors in [17] which can provide insights to utilize deep learning methods in healthcare applications also.

Concerning research regarding social media, Lopez et al. studied the various government policies put in use concerning the COVID-19 pandemic [18]. Different responses towards the popular policies were identified using tweets from different regions in various languages. Saire and Navarro showed that the number of infected people correlated positively with the number of tweets using the technique of data mining [19], they showed distribution (frequency, pattern) and determinants (causes, risk factors) of COVID-19. Kaila et al. took a random sample of 18,000 tweets based on COVID-19 and did COVID-19 topic modeling generating the top-10 topics from the sample taken for study [20]. Depoux et al. demonstrated that the sentiments, public behavior and the rumors that go around need to be detected and should be responded to as soon as possible by the relevant people concerned as they study showed that panic which emerged by people posting on social media is more likely to spread faster than that of COVID-19 [21]. Many other studies indicate to the same too [2, 22–25]. Sentiment embeddings, using different language (other than English), Natural Language Processing, Data Visualization are applied by researchers in [26–32]. In our articile data was collected using python SNscrape library [33]. Text mining and comparative studies of social media texts to mine different kinds of patters such as violent extremists, etc. are few trending and emerging approaches used in recent years [34].

3 Methodology

The flowchart in Fig. 2 depicts the proposed methodology followed to crawl data, Feature extraction and perform further exploratory data analysis.

3.1 COVIDSentiWave Data Collection and Labeling

Out of almost 1.5 million tweets scrapped from March to April 2020 (i.e. *First Wave*), and July to August (i.e. *Second Wave*) we included 1,000,000 unique tweets. COVIDSENTI was divided equally into two sub-datasets of COVIDSENTI-A, COVIDSENTI-B, respectively, of positive and negative sentiments. SNScrape, an official Python Twitter API library [33], was used to collect the dataset of tweets. SNScrape can be used to collect tweets of any time span which makes it very suitable for Data Collection.

(1) *Selection criteria:* COVIDWaveSenti contains two month's worth of tweets. We used only english tweets and scrapped only those tweets which were relevant/ related to Covid.
(2) *Labeling:* So that the data is explanatory in itself, labeling of all tweets was done. Each tweet was marked as positive or negative. For this task we utilized a library which helped us to label the emotion sentiment between positive and negative. The sentiment of the tweet was found by calculating it's polarity between 0

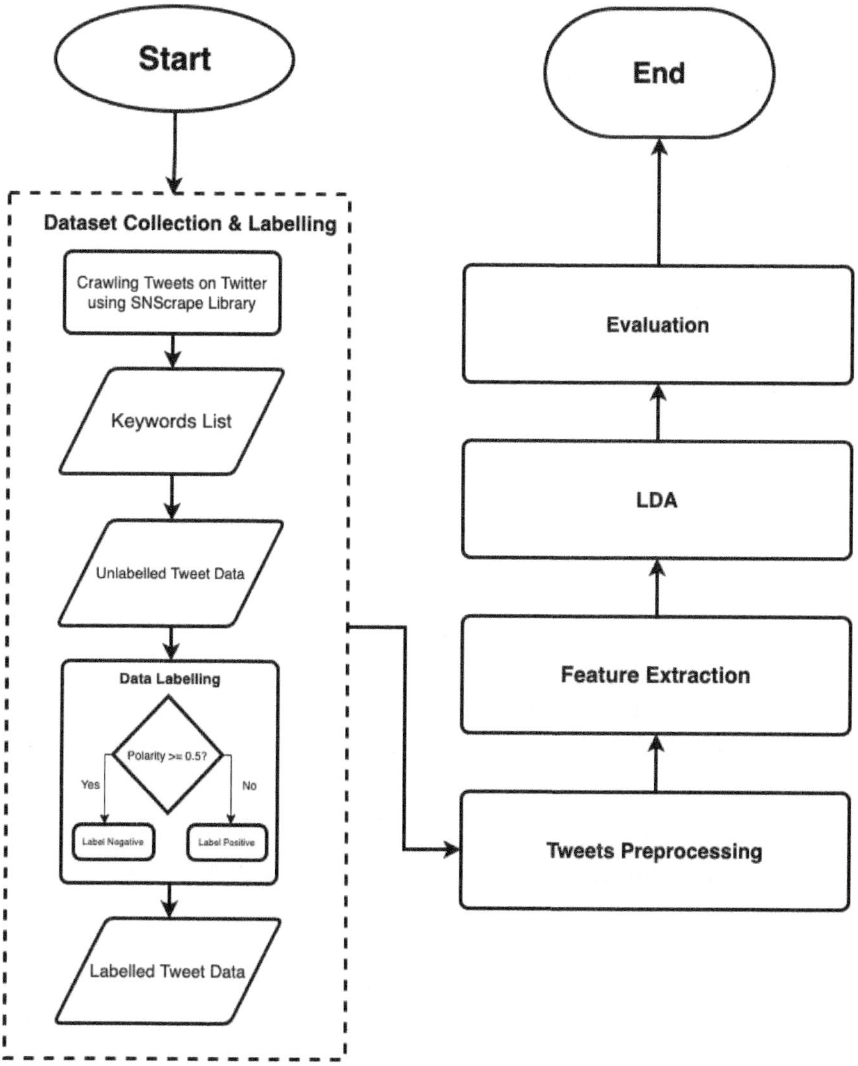

Fig. 2 Workflow

and 1. If the polarity came out as more than 0.5, the sentiment was marked as negative. Similarly, positive sentiment was when the polarity was less than 0.5.

3.2 Data Preprocessing

As raw tweets are usually short and unstructured, the first step of the analysis is to preprocess the data [16]. To perform that, the following series of techniques are applied for Data Cleaning:

(1) Twitter uses hashtags to represent topics, i.e., #COVID-19, #StayHome, #StaySafe, and #Coronavirus. In most cases, hashtags are unnecessary to sentiment and can affect the performance. So in our first step we removed all the hashtags.
(2) The second step is to convert each and every word into lower case as difference in casing may lead to too many words in the corpus.
(3) There are many words that are concatenated in other words, especially the words that are hashtagged such as "stayathomestaysafe" and "coronavirus," should be "stay home stay safe" and "coronavirus," respectively. Hence, we perform word segmentation to achieve this.
(4) We removed all the stopwords that were contained in the tweets, as stop words do not provide any important value to the sentiment analysis (Some words like for, is and etc. were found and removed).
(5) Same word may be found in different tenses or different verb forms so now we perform lemmatization on the words so that we have a unique word and can provide some value for the analysis we used nltk library for performing the same (The word wave can also be present as waves etc.).
(6) We now removed all the numeric and special characters from the tweets as they have no contribution to the sentiment.

3.3 EDA (Exploratory Data Analysis)

(1) *Keyword Trend Analysis*: More frequently used words were separated from the lesser frequently used one's from the corpus collected using Keyword trend analysis. This was done on preprocessed corpus of tweets. It was realized that people are using keywords like coronavirus cases, social distancing, the coronavirus outbreak, quarantine after the analysis was done on the dataset. The word cloud made up of most common words in the two classes namely positive and negative is also studied in subsequent sections.
(2) *Topic Modeling*: After performing Keyword trend analysis on the pre-processed data, we used the resultant data set and applied LDA on it to analyze the diverse topics present in it. LDA is a very popular topic modeling algorithm, which say that a different variety of topics comprise a piece of text. We determined the topics described by the distribution of words and the topics contained within the document using the LDA. While applying LDA, the number of topics was restricted to ten. We determined the topic distribution of the words contained inside the document and the distribution of topics shown by the distribution

of words. The word cloud of words within top six topics and distribution of top-seven dominant topics in the corpora using LDA is shown in Fig. 4.8.

(3) *Sentiment Comparison with COVID-19 Progression*: To understand the trends of positive and negative sentiments over time, we take help of the Sentiment comparison. As shown in Fig. 4.1, people expressed their negative sentiments more often as compared to positive sentiment in the duration of COVID-19 first wave. It is clear from the trend that the intensity and count of positive sentiment were low before mid of March 2020. As shown in Fig. 4.2, people expressed their positive sentiments more often as compared to negative sentiment in the duration of COVID-19s wave. The count and intensity of negative sentiments dropped and shifted towards positive emotion after mid of March 2020 as people began to backup lockdown and social distancing policies enforced by the authorities. This shift in intensity shows how the general masses accepted the COVID-19 pandemic because of getting accustomed to lockdown circumstances and various government policies being implemented.

3.4 Classification

To provide a comprehensive analysis, we used ML-based classifiers to gauge performance in the sentiment classification task. ML-based classifiers, such as naive Bayes (NB). For the classification of positive and negative tweets. We also used a kaggle dataset as training data and predicted multiple emotions present in the tweets namely joy, sad, fear and anger as a part of EDA.

4 Results

The pie chart below in Fig. 3 shows that the attitude of people towards the first wave of COVID-19 was mostly NEGATIVE. This pie chart in Fig. 4 shows that the attitude of people towards the second wave of COVID-19 was mostly POSITIVE (Figs. 5 and 6).

This pie-chart in Fig. 7 depicts the share between 10 major topics in which the tweets were distributed during first wave.

This pie-chart in Fig. 8 depicts the share between 10 major topics in which the tweets were distributed during second wave.

This bar chart in Fig. 9 depicts the classification among the 4 emotions namely sad, joy, fear and anger.

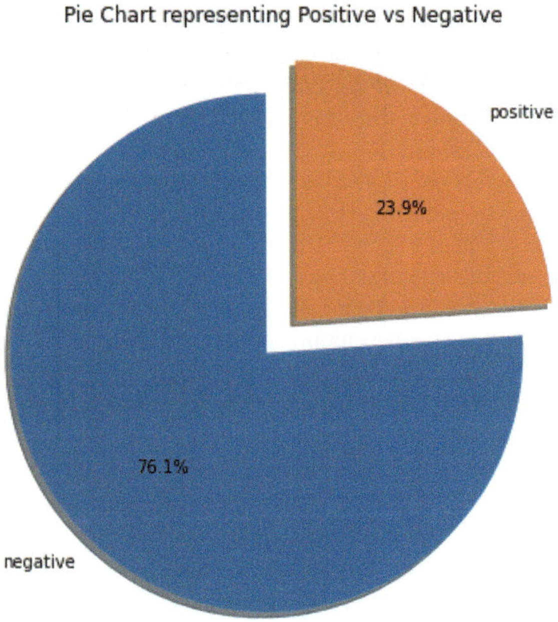

Fig. 3 Distribution of positive versus negative emotion in first wave

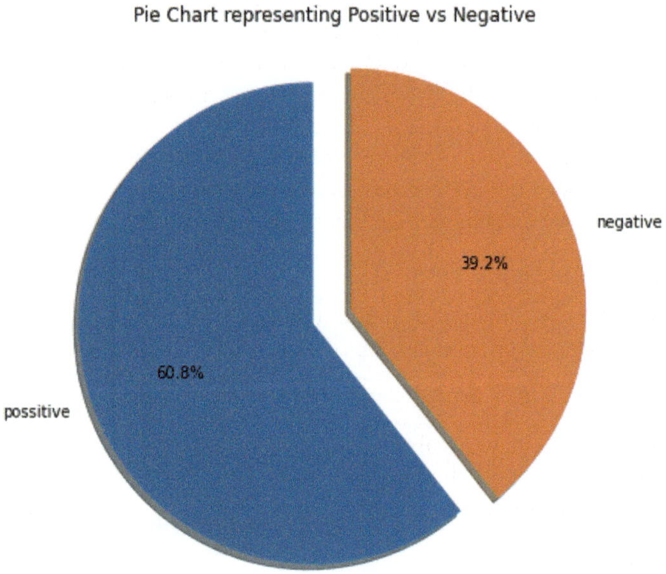

Fig. 4 Distribution of positive versus negative emotion in first wave

Fig. 5 Top 10 topics being discussed during first wave

	Topics
0	People getting vaccinated and covid tests bein...
1	Lockdown impact getting bigger day by day.
2	Covid workers during lockdown getting evicted.
3	Covid lockdown time coming back.
4	Another death in America during lockdown.
5	Covid lockdown helps school students's business.
6	New number of death cases in covid lockdown.
7	Covid lockdown protest case in Melbourne dring...
8	Mask really wanted during lockdown.
9	People get covid lockdown during coronavirus t...

Fig. 6 Top 10 topics being discussed during second wave

	Topics
0	American dying during covid lockdown.
1	Biden and Trump vaccination drives during Covi...
2	Hey, try this cool test app called ZOOM.
3	New death case numbers during covid lockdown.
4	New week first day in covid lockdown.
5	Like thinking in people developing during covi...
6	People need due help during covid lockdown.
7	Ex Coronavirus lockdown protest case in Melbor...
8	Poeple like to get back to covid lockdown times.
9	New world reveal by workers during covid lockd...

5 Conclusion

Covid-19 has shaken the health infrastructure and mental health status of every country around the world. We have taken the data from Twitter for three waves in our current work. We studied the emotion classification of difference waves of covid-19. We observed that people had more negative sentiments during first wave as people faces this deadly new virus for the first time. If we look at second wave

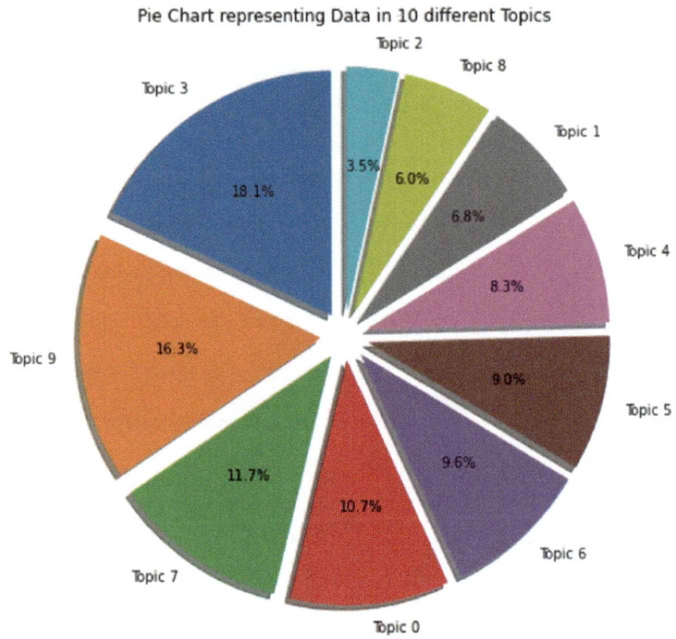

Fig. 7 Topic distribution in first wave

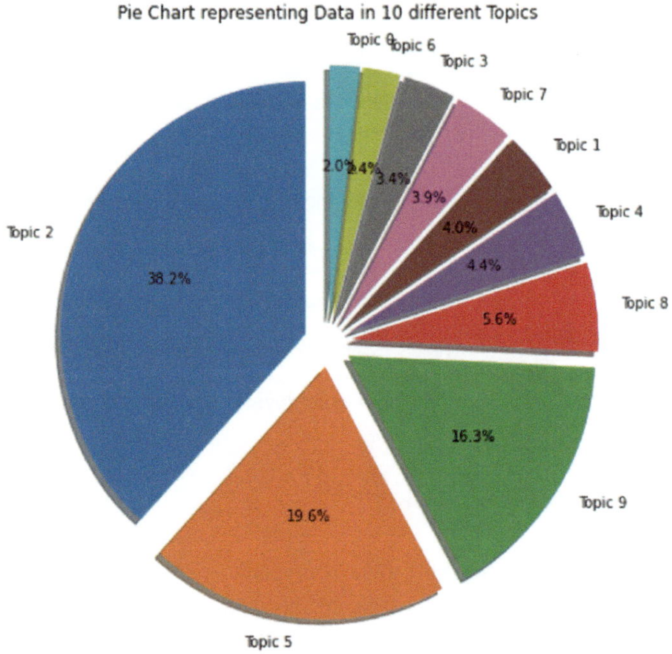

Fig. 8 Topic distribution in second wave

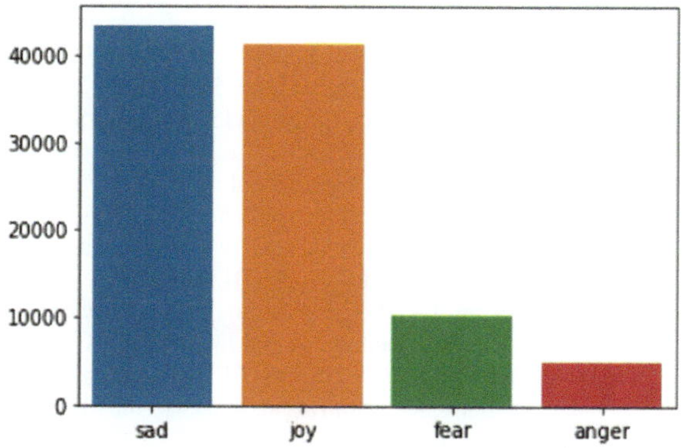

Fig. 9 Emotion classification b/w sad, joy, fear, and anger

sentiment, there is less negative sentiments as people were not scared and the health infrastructure and facilities were also improved, but there is need to develop more proactive and agile health information system in order to handle such type of virus in future. To facilitate research in this context, we have collected large scale of tweets from twitter which can be utilized by another research community for further research and development purposes.

References

1. Singh, R., Singh, R., Bhatia, A.: Sentiment analysis using machine learning technique to predict outbreaks and epidemics. Int. J. Adv. Sci. Res **3**(2), 19–24 (2018)
2. Samuel, J., Garvey, M., Kashyap, R.: That message went viral?! Exploratory analytics and sentiment analysis into the propagation of tweets (2020). arXiv:2004.09718.
3. Naseem, U., Razzak, I., Eklund, P., Musial, K.: Towards improved deep contextual embedding for the identification of irony and sarcasm. In: Proceedings of International Joint Conference on Neural Networks (IJCNN), July 2020, pp. 1–7 (2020)
4. Blei, D.M., Ng, A.Y., Jordan, M.I.: Latent Dirichlet allocation. J. Mach. Learn. Res. **3**, 993–1022 (2003)
5. Carreras, X., Màrquez, L.: Boosting trees for anti-spam email filtering (2001). arXiv:cs/0109015
6. Barbieri, F., Saggion, H.: Automatic detection of irony and humour in Twitter. In: Proceedings of ICCC, pp. 155–162 (2014)
7. Ahmad, T., Ramsay, A., Ahmed, H.: Detecting emotions in English and Arabic tweets. Information **10**(3), 98 (2019)
8. Carducci, G., Rizzo, G., Monti, D., Palumbo, E., Morisio, M.: TwitPersonality: computing personality traits from tweets using word embeddings and supervised learning. Information **9**(5), 127 (2018)
9. Jain, V.K., Kumar, S., Fernandes, S.L.: Extraction of emotions from multilingual text using intelligent text processing and computational linguistics. J. Comput. Sci. **21**, 316–326 (2017)
10. Kim, E.H.-J., Jeong, Y.K., Kim, Y., Kang, K.Y., Song, M.: Topicbased content and sentiment analysis of Ebola virus on Twitter and in the news. J. Inf. Sci. **42**(6), 763–781 (2016)
11. Widener, M.J., Li, W.: Using geolocated Twitter data to monitor the prevalence of healthy and unhealthy food references across the US. Appl. Geogr. **54**, 189–197 (2014)

12. De Choudhury, M., Counts, S., Horvitz, E.: Predicting postpartum changes in emotion and behavior via social media. In: Proceedings of SIGCHI Conference on Human Factors in Computing Systems, April 2013, pp. 3267–3276 (2013)
13. Goyal, Ishita, Amritanshu Singh, and Jaspal Kaur Saini. "Big Data in Healthcare: A Review." 2022 1st International Conference on Informatics (ICI). IEEE, 2022.
14. Saini, J.K., Bansal, D.: Detecting online recruitment of terrorists: towards smarter solutions to counter terrorism. Int. J. Inform. Technol. **13**, 697–702 (2021)
15. Saini, J.K., Saini, K.: New advance encryption standard to analyze encrypted image quality. Int. J. Comput. Appl. **74**(7), 1–5 (2013)
16. Goyal, T., Saini, J.K., Bansal, D.: Analyzing behavior of ISIS and Al-Qaeda using association rule mining. In: Proceedings of 2nd International Conference on Communication, Computing and Networking: ICCCN 2018, NITTTR Chandigarh, India. Springer, Singapore (2019)
17. Thakur, D., Saini, J.K., Srinivasan, S.: DeepThink IoT: the strength of deep learning in internet of things. Artif. Intell. Rev. 1–68 (2023)
18. Lopez, C.E., Vasu, M., Gallemore, C.: Understanding the perception of COVID-19 policies by mining a multilanguage Twitter dataset (2020). arXiv:2003.10359
19. Saire, J.E.C., Navarro, R.C.: What is the people posting about symptoms related to coronavirus in Bogota, Colombia? (2020). arXiv:2003.11159
20. Kaila, D.P., et al.: Informational flow on Twitter–corona virus outbreak– topic modelling approach. Int. J. Adv. Res. Eng. Technol. **11**(3), 128–134 (2020)
21. Depoux, A., Martin, S., Karafillakis, E., Preet, R., Wilder-Smith, A., Larson, H.: The pandemic of social media panic travels faster than the COVID-19 outbreak. J. Travel Med. **27**(3), Art. no. taaa031 (2020)
22. Huang, B., Carley, K.M.: Disinformation and misinformation on Twitter during the novel coronavirus outbreak (2020). arXiv:2006.04278
23. Aggarwal, C.C., Reddy, C.K.: Data Clustering: Algorithms and Applications. CRC Press, Boca Raton, FL, USA (2013)
24. El Zowalaty, M.E., Järhult, J.D.: From SARS to COVID-19: a previously unknown SARS-related coronavirus (SARS-CoV-2) of pandemic potential infecting humans—Call for a one health approach. One Health **9**, Art. no. 100124 (2020)
25. Gupta, V., Lehal, G.S.: A survey of text mining techniques and applications. J. Emerg. Technol. Web Intell. **1**(1), 60–76 (2009)
26. Rocha, G., Cardoso, H.L.: Recognizing textual entailment: challenges in the Portuguese language. Information **9**(4), 76 (2018)
27. Saeed, Z., et al.: What's happening around the world? A survey and framework on event detection techniques on Twitter. J. Grid Comput. **17**(2), 279–312 (2019)
28. Zhou, J., Yang, S., Xiao, C., Chen, F.: Examination of community sentiment dynamics due to COVID-19 pandemic: a case study from Australia (2020). arXiv:2006.12185
29. Ye, X., Li, S., Yang, X., Qin, C.: Use of social media for the detection and analysis of infectious diseases in China. ISPRS Int. J. Geo-Inf. **5**(9), 156 (2016)
30. Tang, D., Wei, F., Qin, B., Yang, N., Liu, T., Zhou, M.: Sentiment embeddings with applications to sentiment analysis. IEEE Trans. Knowl. Data Eng. **28**(2), 496–509 (2016)
31. Schild, L., Ling, C., Blackburn, J., Stringhini, G., Zhang, Y., Zannettou, S.: 'Go eat a bat, chang!': an early look on the emergence of sinophobic behavior on Web communities in the face of COVID-19 (2020). arXiv:2004.04046
32. Pépin, L., Kuntz, P., Blanchard, J., Guillet, F., Suignard, P.: Visual analytics for exploring topic long-term evolution and detecting weak signals in company targeted tweets. Comput. Ind. Eng. **112**, 450–458 (2017)

33. JustAnotherArchivist. (2025). snscrape (Version retrieved May 18 2025) [Computer software]. GitHub. https://github.com/JustAnotherArchivist/snscrapelink
34. Saini, J.K.: LSTM based deep learning approach to detect online violent activities over dark web. Multimed. Tools Appl. **83**(14), 42379–42390 (2023). https://doi.org/10.1007/s11042-023-17222-8

Classification of Codling Moth in Apple Orchards with Ensemble Learning

Kunzang Dolma and Mala Kalra

Abstract The Codling moth (Cydia pomonella) is a major insect pest that primarily infests fruit trees, especially apple and pear trees, causing substantial global economic and agricultural issues. Researchers have explored pest classification and detection methods. With the rising popularity of the machine learning across various research areas, it's potential for improving pest classification is increasingly recognized. This study introduces an ensemble learning-based method for identifying Codling moths through images, simplifying the categorization of orchards as either infected or uninfected. Besides reducing the workload and time of farmers, this approach can significantly mitigate Codling moth damage while preserving orchard productivity and health. In this research, the authors integrated a stacking ensemble model, amalgamating the functionalities of random forest, AdaBoost, and gradient boosting algorithms. The proposed model achieved 94% accuracy on field data. This research highlights the promise of advanced machine learning in addressing Codling moth-related challenges.

Keywords Codling moth · Invasive species · Image classification · Ensemble learning · Orchard management · Orange software

1 Introduction

The COVID-19 pandemic has significantly impacted lifestyles and dietary habits, fostering a renewed focus on health and nutrition. In this health-conscious era, fruits have become an essential component of a healthy diet. Consequently, the challenge of producing abundant high-quality fruits to meet the growing global population's needs has become paramount. However, fruit growers face a formidable obstacle in the form of pest invasions, leading to substantial losses in fruit production. The

K. Dolma (✉) · M. Kalra
Department of Computer Science and Engineering, National Institute of Technical Teachers Training and Research, Chandigarh, India
e-mail: dolma.kunzang@nic.in

B. Soni et al. (eds.), *Beyond Artificial Intelligence*, Lecture Notes in Networks and Systems 1326, https://doi.org/10.1007/978-981-96-4170-3_57

Fruit and Agriculture Organization (FAO) reported a 20–40% loss in fruit production worldwide in 2020 due to pests, resulting in staggering economic losses of around $220 billion [1]. Among these pests, the Codling Moth (CM) is a particularly menacing adversary, especially for stone fruits like apples, pears, apricots, and walnuts [2]. CM is challenging to control and classified as a quarantine pest, rendering infested fruits ineligible for export to prevent its spread. Infestations by CM degrade fruit quality, rendering them inedible and causing significant economic hardships for farmers.

The invasion of CM is a global concern. It is believed to have originated from Eurasia and has infiltrated numerous countries. CM poses a substantial threat in India, where fruit cultivation is a significant agricultural activity. Key fruit-cultivating states like Jammu and Kashmir, Himachal Pradesh, Arunachal Pradesh, and Ladakh have been affected. As the world's second-largest fruit producer, India boasts vast orchards covering upto 70 lakh hectares and yielding a total production of 86 lakh tons. Apple cultivation spans 3.11 lakh hectares, producing 19.15 lakh tons of apples [3]. However, CM has invaded Ladakh, potentially originating from the west–east border of Pakistan. This has led the government to impose restrictions on fruit transportation from Ladakh to other regions under SRO 397, dated 8th September 1981, issued under the Jammu and Kashmir Plant and Diseases Act 1973 [1, 3], directly impacting the rural economy. Efforts to manage and control CM must be proactive, as once it establishes itself, containment becomes exceedingly difficult and costly. While effective, the extensive use of pesticides and insecticides poses environmental and health hazards, including the risk of cancer, fetal death, genetic diseases, and severe respiratory ailments [1, 4]. Therefore, there is a need for innovative solutions to tackle CM infestations. In this research paper, a novel approach for Codling Moth (CM) classification in orchards is presented. The methodology employes ensemble learning, specifically a stacking approach. The ensemble model comprises Random Forest, Gradient Boosting, and AdaBoost methods. The stacking model combines the predictions of 'Gradient Boosting' and 'AdaBoost' as base learners, with 'Random Forest' acting as an aggregator to harness their collective strengths, thereby enhancing the overall model performance. The main findings of this research are as follows:

- To present an insight into existing research based on ML and DL techniques to address pest infestation.
- To propose an ensemble learning based approach for classification of CM from image dataset.
- To validate the effectiveness of proposed approach by comparing it with existing techniques considering precision, accuracy, F1-Score and recall as evaluation metrics.

The paper's structure can be outlined as follows. In Sect. 2 offers an insight into previous research within the field, setting the context. Section 3 provides a comprehensive explanation of the formulated methodology, and Sect. 4 presents the experimental setup and results achieved. In Sect. 5, we recap the main discoveries and highlight possible directions for future research and advancement.

2 Related Work

Several researchers have implemented different methodologies to address the concerns related to pest infestation in the orchards. They have worked on various ML and deep learning algorithms to predict critical scenarios. Rady et al. experimented on the GoldRush apples, which were brought from a farm in Georgetown and the study was conducted in two batches. They studied the effect of Hyperspectral Imaging (HIS) in the spatial region for visible and near-infrared. To get the spatial information, push-broom was used. The scanned images were classified for CM. Many classification methods were used, out of which linear discriminant analysis (LDA) and decision trees (DT) showed the best results with a classification rate of 82%. The study showed the potential of HIS for detecting CM without destroying the apples [5]. Sun et al. proposed an RTBnet model that could differentiate the red turpentine beetle from the other five bark beetles, China's most threatening pine species. A deep learning technique was employed to count the number of beetles in the conventional pheromone trap. For training, they have four RTBnets and two RetinaNets, out of which the RTBnet outperforms the other three. The authors utilized Keras and TensorFlow frameworks for designing RTBnet and achieved a precision of 74.6% for classification [6].

Remboski et al. proposed a solution to identify pests using digital image processing and ML. It consisted of three stages: segmentation of images, extracting features, and separating the region of interest from the background. A supervised learning approach was adopted, and four algorithms (Gaussian Naive Bayes, Support Vector Machine (SVM), Decision Tree and k-nearest neighbors (KNN), were evaluated. The findings revealed that SVM achieved the highest accuracy of 86.38%, surpassing the other algorithms [7]. Brunelli et al. implemented Deep Neural Network (DNN) model along with power efficient Internet of Things (IoT) sensor network. The model classifies the moth from other moth-like insects. They have embedded the system in a pheromone trap, which captures the images and classifies them. If a pest is found, it sends an alarm to the farmer, which will reduce the cost of human intervention for trap checking [8]. Kasinathan et al. suggested a model for identification and categorization of pests in agricultural fields. Their approach involved the examination of three distinct datasets, focusing on various attributes of insects, including shape, color, texture, Histogram of Oriented Gradients and Global Image Descriptor. To enhance the classification performance, the researchers employed both individual and combined machine learning models, with the ensemble model utilizing the majority voting scheme demonstrating the most robust performance. The results revealed an improvement in accuracy to 96.5%, 92.1%, and 92.3% for the Wang, Xie and Butterfly image dataset, respectively, through the implementation of the majority voting ensemble method [9]. Kasinathan et al. introduced methods for foreground extraction and figuring out the outlines (contour identification) to find insects in complicated backgrounds, utilizing three different datasets. They employed various machine learning methods and performance was evaluated using a rigorous 9-fold cross-validation process. Notably, the CNN model emerged as the top performer among all while achieving classification accuracy of 90% [10]. Tuda et al. utilized supervised machine learning techniques to categorize both insect species and gender. They based their classification on static images of specimens with mixed genders, concentrating their study on a stored product pest beetle and parasitoids. Notably, simple logistic regression, logistic model trees, and AdaBoost achieved high accuracy rates for sex and species classification, with rates ranging from 88.5 to 98.5%. Models combining object-based and pixel-based features performed best. Simple logistic regression exhibits robust and consistent performance regardless of the classification target. This research is among the earliest to apply machine learning for gender identification in insects using images, offering practical benefits for ecology and pest control [11].

Xiaoming Bi evaluated a cloud-based application for detecting pests in agricultural fields. The system used regression analysis and ML algorithms to classify pests with response time of just two seconds and an accuracy rate of 85%. The application provides end users with relevant information and a management strategy for dealing with pest infestations. The system's innovative approach and effectiveness make it a valuable tool for agricultural professionals [12]. Abbaspour-Gilandeh et al. a model was developed with sparse coding technique to build a dictionary with the intention of reducing duplication in the training set to speed up identification. A multilayer perception neural network took the constructed lexicon as input. For multiple apple pest species available in dataset, they received an average classification accuracy of 91% [13]. These studies underscore the remarkable progress in pest detection via ML, providing vital resources for early pest identification and crop safeguarding. Given the limited research on codling moth classification, these methods hold significant promise for addressing this specific challenge. Importantly, we intend to employ EL to classify codling moths in apples, further advancing the understanding and capabilities in pest management. These innovations herald more sustainable agriculture, bolstering food production and ecosystem well-being.

3 Proposed Methodology

This research paper introduces a method for identifying Codling Moths (CM) in orchards. The technique utilizes ensemble learning, specifically through a stacking technique. The ensemble model includes Random Forest, Gradient Boosting, and AdaBoost methodologies.

3.1 Data Acquisition and Augmentation

The authors in this study collected CM invasion related dataset from various areas of the Ladakh region which includes around 985 images of apples comprises of

Fig. 1 Images of healthy and infected apples

Table 1 Number of images before and after augmentation

Type of apples	Before augmentation	After augmentation
Healthy	600	1500
Infected	385	1500

385 infected and 600 healthy images. Figure 1 shows the sample images from the collected dataset of infected and healthy apples.

Data augmentation is a vital technique in computer vision and ML to enhance dataset diversity for improved model robustness and generalization. In this research, diverse transformations such as rotation (randomly rotated up to 20°), Horizontal Flip, colour Jitter, and Resize (Crop 500 × 500 pixels) are applied to original apple images to create a comprehensive dataset. These augmentations expanded the dataset, aiding robust training and model generalization, crucial for an improved performance with unseen apple samples. Table 1 displays the number of images depicting both healthy and unhealthy apples before and after augmentation.

3.2 Image Embedding

The process of image embedding in this study was executed through the utilization of the Inception V3 model, a renowned and robust deep learning architecture known for its exceptional performance in image recognition tasks. Inception V3, initially developed by Google, incorporates advanced convolutional neural network (CNN) structures, enabling it to effectively extract intricate and meaningful visual features from images [14]. With a dataset encompassing 3,000 images, the Inception V3 model was leveraged to extract intricate patterns and representations, laying a strong foundation for subsequent analysis and processing.

3.3 Feature Selection

In order to distil the most relevant and impactful features from the pool of extracted attributes, the researchers employed the widely acclaimed Chi-squared (χ^2) technique. This statistical method facilitated the prioritization of features based on their significance in relation to the specific classification task at hand. By evaluating the degree of association between each feature and the target variable, the Chi-squared method played a pivotal role in selecting the most influential features, thereby streamlining the dataset and enhancing the efficiency of the model [15].

At the onset, the dataset comprised a substantial 2048 features. However, post-application of the Inception V3 model and subsequent utilization of the Chi-squared technique, the researchers identified and retained a refined set of 944 features, ensuring an optimal balance between data comprehensiveness and computational efficiency.

3.4 Model Development

To enhance prediction accuracy, the authors have used Ensemble Learning, a potent approach in machine learning that combines multiple models. This strategy involves combining the predictions from AdaBoost, Random Forest, and Gradient Boosting. The stacking method is used to harness the collective strengths of these models. In the stacking model, the predictions generated by 'Gradient Boosting' and 'AdaBoost' serve as base learners. To aggregate these predictions effectively and enhance the overall model performance, 'Random Forest' is employed as the aggregator. The stacking technique allows the ensemble to capitalize on the complementary capabilities of these individual models, resulting in a more robust and accurate classification system.

3.5 Model Evaluation

To assess the effectiveness of the machine learning models, a comprehensive evaluation process was employed. This process involved utilizing a 10-fold cross-validation strategy to ensure robustness and reliability of the results. The model performance was further analysed using five important evaluation metrics: Accuracy, Area under the Curve (AUC), Recall, Precision and F1 Score. The mathematical equations of these metrics are given below in Eqs. 1–4 [16].

$$\textbf{Accuracy} = \frac{TP + TN}{TP + TN + FP + FN} \tag{1}$$

$$\textbf{Recall} = \frac{TP}{TP + FN} \tag{2}$$

$$\textbf{Precision} = \frac{TP}{TP + FP} \tag{3}$$

$$\textbf{F1 Score} = 2 \times \frac{(Recall \times Precision)}{(Recall + Precision)} \tag{4}$$

where: TN represents True Negative, TP denotes True Positive, FN stands for False Negative, and FP indicates False Positive.

4 Experiments and Results

This research study utilizes orange software [17–19] for implementing the suggested approach. Orange a versatile tool for machine learning and data analysis that can be operated through both Python scripting and visual programming. Different widgets serve as visual representations of functionalities. Widgets are placed on the canvas and their inputs and outputs are connected to create programs [20]. Figure 2 shows the execution of different steps of the proposed methodology using GARF stack model in orange software. It consist of individual models, namely Random Forest, Gradient Boosting, AdaBoost, and SVM. Ensemble learning techniques, specifically AdaBoost, Random Forest, and Gradient Boosting, were employed to combine their predictions. Stacking was utilized to enhance classification performance collectively. Configurations were optimized for each model: Random Forest (90 trees, max depth 3, 5 features/node), Gradient Boosting (100 trees, learning rate 0.102, max depth 4), AdaBoost (60 decision tree estimators, 'Samme.r' algorithm), and SVM (linear kernel, $C = 1.0$, $\varepsilon = 0.1$, tolerance 0.001). The stacking model utilized 'Gradient Boosting' and 'AdaBoost' as base learners, with 'Random Forest' aggregating their predictions for improved overall performance.

Table 2 shows the parameters set for different ML models used in the proposed approach.

The proposed approach is compared with different ML models to validate its performance. Table 3 presents the comparison of the results obtained. The adopted approach achieves AUC of 0.987, Accuracy, F1 score, and recall of 94% respectively.

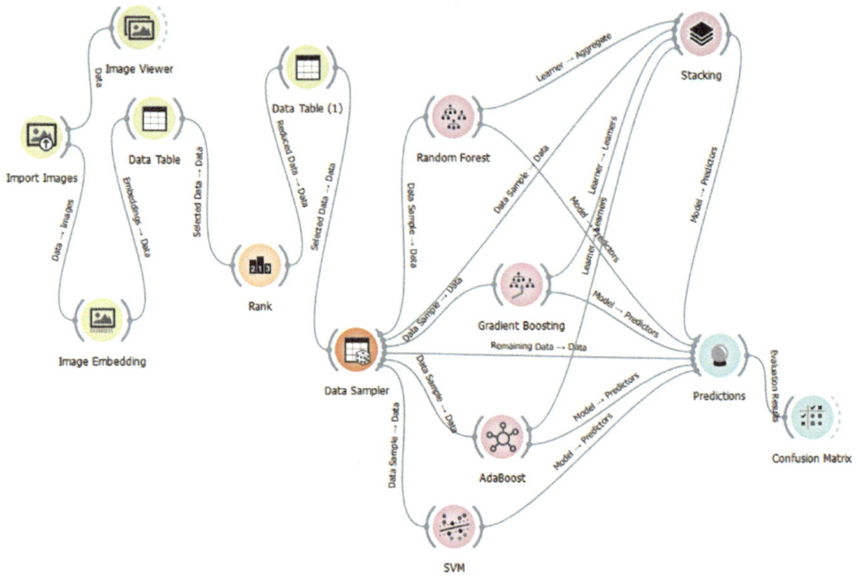

Fig. 2 GARF stack model build in orange

Table 2 Parameter settings for different ML models used in proposed model

Model	Parameters settings
Random forest	Number of trees: 90, Maximum tree depth: 3, Maximum features per node: 5
Gradient boosting	Number of trees: 100, Learning rate: 0.100, Maximum tree depth: 4
AdaBoost	Base estimator: Decision Trees, Number of estimators: 60, Algorithm: 'Samme.r', Loss function: Exponential
SVM	Kernel: Linear, Cost (C): 1.0, Regression loss epsilon (ε): 0.1, Numerical tolerance: 0.001

Table 3 Comparison of results of various models implemented on orange

ML model	AUC	Accuracy	Recall	F1 score	Precision
AdaBoost	0.813	0.813	0.813	0.813	0.817
SVM	0.791	0.693	0.693	0.693	0.693
Random forest	0.936	0.870	0.870	0.870	0.871
Gradient boosting	0.987	0.937	0.937	0.937	0.937
Proposed model	0.987	0.940	0.940	0.940	0.940

5 Conclusion

In conclusion, this research paper addresses the pressing issue of Codling Moth (CM) infestations in orchards, particularly apple orchards, and its detrimental impact on fruit production and the agricultural economy. The study introduces an innovative approach to CM classification using ensemble learning, specifically the GARF Stack model, which combines Random Forest, AdaBoost, and Gradient Boosting algorithms. This approach aims to simplify the categorization of orchards as infected or uninfected, ultimately reducing the workload and time required by farmers while mitigating CM damage. The experimental analysis shows a powerful method for precisely recognizing and categorizing CM-infested apples by leveraging the capability of ensemble learning and deep feature extraction. With an AUC of 0.98, accuracy, precision, recall, and F1-score of 94%, the stacking ensemble method has performed exceptionally well and has showed the potential of proposed methodology to transform pest management procedures. Orchard health, fruit output, and global agricultural future look brighter with these results, which will eventually be good for both farmers and ecosystems. As CM continues to threaten agricultural economies and food security, the integration of innovative technologies and machine learning into pest management becomes increasingly vital. This research serves as a significant step forward in the pursuit of environmentally friendly, economically viable, and highly effective solutions to combat CM infestations and similar agricultural challenges. By advancing the understanding and capabilities in pest management, the proposed study contributed to the broader mission of safeguarding crop yields, preserving orchard health, and sustaining global food production. In addition to the current focus on classifying CM in apples, the techniques employed in this study hold potential for extension to other stone fruits. Furthermore, the development of a real-time pest detection model emerges as a promising avenue for future investigation.

References

1. Karar, M.E., Alsunaydi, F., Albusaymi, S., Alotaibi, S.: A new mobile application of agricultural pests recognition using deep learning in cloud computing system. Alex. Eng. J. **60**, 4423–4432 (2021). https://doi.org/10.1016/j.aej.2021.03.009
2. Kuyulu, A., Genç, H.: Genetic diversity of codling moth Cydia pomonella L. (Lepidoptera: Tortricidae) populations in Turkey. Turk. J. Zool. **44**, 462–471 (2020). https://doi.org/10.3906/zoo-2006-4
3. Khan, A.A., Wani, A.R., Zaki, F.A., Nehru, R.K., Pathania, S.S.: Pests of apple. In: Omkar (ed.) Pests and Their Managemen, pp. 457–490. Springer Singapore, Singapore (2018). https://doi.org/10.1007/978-981-10-8687-8_14
4. Stobdan, T., Deen, M., Gupta, V., Raghuvanshi, M., Phunchok, T.: ApricotYellowTailMoth.pdf (2019). https://www.researchgate.net/publication/342550430_Integrated_management_of_yellow_tail_moth_Euproctis_similis_on_apricot_in_Leh_Ladakh_India_a_case_study_of_a_success_story

5. Rady, A., Ekramirad, N., Adedeji, A.A., Li, M., Alimardani, R.: Hyperspectral imaging for detection of codling moth infestation in GoldRush apples. Postharvest Biol. Technol. **129**, 37–44 (2017). https://doi.org/10.1016/j.postharvbio.2017.03.007
6. Sun, Y., Liu, X., Yuan, M., Ren, L., Wang, J., Chen, Z.: Automatic in-trap pest detection using deep learning for pheromone-based Dendroctonus valens monitoring. Biosyst. Eng. **176**, 140–150 (2018). https://doi.org/10.1016/j.biosystemseng.2018.10.012
7. Remboski, T.B., De Souza, W.D., De Aguiar, M.S., Ferreira, P.R.: Identification of fruit fly in intelligent traps using techniques of digital image processing and machine learning. In: Proceedings of the 33rd Annual ACM Symposium on Applied Computing. pp. 260–267. ACM, Pau France (2018). https://doi.org/10.1145/3167132.3167155.
8. Brunelli, D., Albanese, A., d'Acunto, D., Nardello, M.: Energy neutral machine learning based IoT device for pest detection in precision agriculture. IEEE Internet Things Mag. **2**, 10–13 (2019). https://doi.org/10.1109/IOTM.0001.1900037
9. Kasinathan, T., Uyyala, S.R.: Machine learning ensemble with image processing for pest identification and classification in field crops. Neural Comput. Appl. **33**, 7491–7504 (2021). https://doi.org/10.1007/s00521-020-05497-z
10. Kasinathan, T., Singaraju, D., Uyyala, S.R.: Insect classification and detection in field crops using modern machine learning techniques. Inf. Process. Agric. **8**, 446–457 (2021). https://doi.org/10.1016/j.inpa.2020.09.006
11. Tuda, M., Luna-Maldonado, A.I.: Image-based insect species and gender classification by trained supervised machine learning algorithms. Ecol. Inform. **60**, 101135 (2020). https://doi.org/10.1016/j.ecoinf.2020.101135
12. Bi, X.: Research and development of insect situation data intelligent cloud platform using geographic information system. J. Phys. Conf. Ser. **2033**, 012046 (2021). https://doi.org/10.1088/1742-6596/2033/1/012046
13. Abbaspour-Gilandeh, Y., Aghabara, A., Davari, M., Maja, J.M.: Feasibility of using computer vision and artificial intelligence techniques in detection of some apple pests and diseases. Appl. Sci. **12**, 906 (2022). https://doi.org/10.3390/app12020906
14. Szegedy, C., Vanhoucke, V., Ioffe, S., Shlens, J., Wojna, Z.: Rethinking the ınception architecture for computer vision. In: 2016 IEEE Conference on Computer Vision and Pattern Recognition (CVPR), pp. 2818–2826. IEEE, Las Vegas, NV, USA (2016). https://doi.org/10.1109/CVPR.2016.308
15. Singhal, R., Rana, R.: Chi-square test and its application in hypothesis testing. J. Pract. Cardiovasc. Sci. **1**, 69 (2015). https://doi.org/10.4103/2395-5414.157577
16. Rahmani, A.M., Yousefpoor, E., Yousefpoor, M.S., Mehmood, Z., Haider, A., Hosseinzadeh, M., Ali Naqvi, R.: Machine Learning (ML) in medicine: review, applications, and challenges. Mathematics **9**, 2970 (2021). https://doi.org/10.3390/math9222970
17. Elshoush, H.T., Dinar, E.A.: Using Adaboost and Stochastic gradient descent (SGD) algorithms with R and orange software for filtering e-mail spam. In: 2019 11th Computer Science and Electronic Engineering (CEEC), pp. 41–46. IEEE, Colchester, United Kingdom (2019). https://doi.org/10.1109/CEEC47804.2019.8974319
18. Ishak, A., Siregar, K., Aspriyati, Ginting, R., Afif, M.: Orange software usage in data mining classification method on the dataset lenses. IOP Conf. Ser. Mater. Sci. Eng. **1003**, 012113 (2020). https://doi.org/10.1088/1757-899X/1003/1/012113
19. Popchev, I., Orozova, D.: Algorithms for machine learning with orange system. Int. J. Online Biomed. Eng. IJOE **19**, 109–123 (2023). https://doi.org/10.3991/ijoe.v19i04.36897
20. Jovic, A., Brkic, K., Bogunovic, N.: An overview of free software tools for general data mining. In: 2014 37th International Convention on Information and Communication Technology, Electronics and Microelectronics (MIPRO), pp. 1112–1117. IEEE, Opatija, Croatia (2014). https://doi.org/10.1109/MIPRO.2014.6859735

Effective Advertisement Decision-Making Using Telecom Data Analytics

Aditya Bhardwaj, Divyansh Palia, Umesh Gupta, and Shreya Chatterjee

Abstract In today's data-driven and hyper-competitive business environment, making informed decisions is paramount for the success of advertising campaigns. Telecom data analytics has emerged as a valuable tool for businesses seeking to optimize their advertising strategies. Surveys have been conducted to understand whether the product or service has reached the target audience. Doing market research surveys is both time-consuming and expensive. Another possible solution to accomplish the task is to study the user pattern on existing customer data. In the telecommunication sector, such customer data is readily available to the company. This paper analyzed Singapore residents' data on the Singtel communications network. We have carried out exploratory results that predicts what cities of Singapore would be suitable for which network plans. Simulation environment using Python and Power BI has been setup to perform data cleaning, wrangling, and analysis operations. High competition among telecom companies compels them to introduce a new series of packages and discounts every month to seek customer satisfaction regularly to stay in the market. Such analytical insights are helpful for telecom companies to prepare customized and flexible network plans for their customers.

Keywords Singapore telecommunications analytic · Singtel dataset · Advertisement decision-making · Microsoft power BI

A. Bhardwaj (✉) · U. Gupta
School of CSET, Bennett University, Greater Noida, India
e-mail: aditya.cse@nitttrchd.ac.in

D. Palia
NCS @ Singtel, Singapore, Singapore
e-mail: divyanshpalia@outlook.com

S. Chatterjee
MiQ Digital, Andheri East, Mumbai, Maharashtra, India

B. Soni et al. (eds.), *Beyond Artificial Intelligence*, Lecture Notes in Networks and Systems 1326, https://doi.org/10.1007/978-981-96-4170-3_58

1 Introduction

In the telecommunication sector, the leading companies have covered most customers in developed countries. With the technological progress, it's challenging to rise amongst the competitors and to match the level of available plans. Companies are working hard to survive this competition and sustain this competitive market and plan on strategies to go ahead in this race. Companies used traditional marketing, surveying to work on their advertising and increase their customer base. Nowadays, companies are shifting to work on smart marketing to acquire customers, upsell existing customers, and increase the fall-off period. Smart marketing uses existing data by analyzing different parameters to sustain the present hold and reach new customers [1].

In this paper, we have used data analysis and visualization techniques to draw accurate area-wise details of Singapore, focusing on attaining the correct advertising strategy at any given location. Depending upon the area's needs, specific services can be advertised, for example, customer support services, network services, customized plans, etc. Data analytics can also predict consumer movement toward particular data plans. In this work, we have used Power BI for data analysis and Matplotlib for visualization [2].

- An effective advertisement decision making approach for telecom data analytic has been proposed.
- We have performed data analysis on SingTel exploratory data set using PowerBI tool formulating advertising strategies.
- By employing a simulation environment with Python and Power BI, this study provides valuable insights into suitable network plans for different cities in Singapore.
- Extensive simulations have been carried out to prepare customized and flexible network plans for a telecom company's customers.

The remaining sections are structured as: Sect. 2, we presents background and motivation of this study. In Sect. 3, proposed methodology with data set preparation, data grouping and simulation setup is presented. In Sect. 4 the obtained results are discussed with different visualization perspective. Finally, in Sect. 5, this article is concluded with open future research directions.

2 Background and Motivation of the Study

2.1 Background

Singapore telecommunications (SingTel) is one of the companies which has used smart marketing to keep coping up with rising competition. The mentioned technique has been used into effort for rolling out new plans. Though they were benefited from it but they were not performing data analysis for advertising. Smart marketing for better outreach support and reach could be an avenue to keep up with their existing customers [3]. SingTel provides services like phone, online security, online backup, tech assistance, streaming TV, internet and movies. Millions of subscribers use these services to obtain their day-to-day data needs. The database of such users can be utilised for data analysis and decision making. It has been shown that churn in the telecom business is frequently caused by a high level of customer effort. Contracts were cancelled by customers for the following reasons: Companies squandered their time (37% percent waited too long for their issue to be handled) by having to contact more than once 52% [4].

2.2 Motivation

Churn refers to percentage of customers who stopped getting your business products. Earlier researches has analysed the data available publicly for making customer churn predictions but since the data set was a global data set, specific city-wise information was not included [5]. Data analysis has been performed to predict the distribution of customers over different plans, similar data plans run by different companies. Companies advertising patterns, to name a few major telecom companies like SingTel, Hi! Telecom, etc. are using these research results to understand the customer needs and advertise appropriately. Churn being one of the biggest problems, earlier researches have shown that the average monthly churn rate among the top carrier in Singapore is 1 to 2%. Prices by far is one of the common reasons for this churn as it makes customers shift from existing provider to another provider. Telecommunication companies apply various data analysis models to figure out ways to offer some discounts, special offers or other gratifications to keep their customers engaged. Transactions on the internet, emails, movies, click streams, social networking services, are other sources companies use to advertise.

3 Proposed Methodology

In this work, we have performed data analysis on SingTel exploratory data set. The provided data was in raw and unstructured form. We used Power BI to clean and optimise it and further defined the parameters with the use of tools available in Power BI [6]. Next, we performed visualizations on structured data to gain knowledge helpful for formulating advertising strategies. Unfortunately, all of this complexity leads us down a rabbit hole of wasted time and missed opportunities since you spend more time seeking to understand and quantify churn rate than really exploiting the statistic to build your business and generate revenue development. Figure 1 shows the block diagram illustrating the methodology [7].

3.1 Data Set Preparation

SingTel exploratory dataset contains information of more than 3300 entities of its previous customers and their churned out data. Within the area of states and districts of Singapore we have divided the data into 20 parameters. With reference to day, evening and night calls, and other parameters as International calls. Scope for future can include Tourist SIM issued and calls made to specific country within the SingTel telecommunication system and services. It is also called the mobile procure data set which is often used with advertising agencies. A broad look to the data set and components are displayed [8].

SingTel used survey method for data collection where the questionnaire had 30 objective questions and 10 subjective questions. The survey was conducted on 3600 customers of SingTel industry in Singapore. A customer was defined as a person who is a consumer of some kind of network provider. Twenty minutes time was given to the participants to fill in the survey anonymously, and 3408 customers responded. Because no surveys were fully completed, 371 survey results were included in the analysis. These three research questions are addressed in this study. It proposes a data-derived small data strategy as the theoretical and methodological framework for answering the second and third research objectives. It then employs the provided method to examine the top. According to the research, the range of big data analytic primarily covers data mining, machine learning, data science, and systems in order of relevance [9]. Finally, this study investigates the theoretical, technological, and social ramifications of this discovery. The proposed technique and conclusions may be applicable to other data analytic and data science academics and practitioners.

3.2 Business Intelligence

Microsoft did not mention business intelligence until May 2017, when the tagline "Business Intelligence Like Never Before" first appeared on the top pages of leading business magazines. Business intelligence (BI) is a collection of processes, structures, and technology that convert unstructured data into meaningful and valuable information. It enables business users to make educated business decisions based on real-time data, allowing a company to stay ahead of its competition. Traditionally, the focus of BI technology options has been on essential functions such as reporting and analytic, but as those features become commoditize, a whole new set of possibilities has developed. According to Forrester's BI study, technology is advancing and organisations who are on the bleeding edge of these new trends may acquire a competitive advantage in their industry [13]. Combining machine learning and BI may have substantial impact on the insights gained from a company's available data, making BI a true game changer in helping organisations improve productivity, quality, customer service, and other metrics.

- **Data Sourcing**: Receiving the provided data from the MPI department of studies and research at SingTel.
- **Data Analysis**: Working on multiple factors and customizing various characteristics of numerical features and then explaining the concern for how to use them regarding the needed data.
- **Situation Awareness**: Peak awareness for including parameter statistics on non-numerical features which can cause analysis failure, mismanaged value count to have a look on distribution and proportional features.
- **Visualizations**: Visualizing given data into pictographs, pie- charts and bar graphs with various parametric comparisons.
- **Decision support**: Conclusions post analysis helps companies to make necessary decisions regarding their advertising strategies or data plan strategies.

3.3 Data Grouping

Data grouping allowed us to create summaries for display or analysis, while calculating total area charges or calls we implemented grouping techniques by applying group *dl* command. Filters are significantly used feature in Power BI. Filters can be used in query components, data models, and data visualisation components. At the Query level, two primary filters are utilised to find the needed data: *M functions* and *Query editor*. This eliminates the need to download and import data models.

Further, spark engine is used to represent each series of parameters, it helped analysts to understand the average cost each of the sector presented while purchasing the plan packs and how they can modify it further accordingly. In mathematical terms the grouping of the data helped us condensate the data into several groups and easier to record frequencies. Call detail records or CDR's contains all charging information and this data has a large amount of information and in addition, often the data is churned out in groups and data churning is implemented on specific relational database [10].

We form data Pivot Table and mark variables known as: Values—a list of variables for which statistics is calculated Index—a list of variables to group data by Aggfunc—statistics for groups e.g., mean, median, mode, etc.

3.4 Simulation Setup

We grouped the data by state and number of international calls per unit for analysis. Recorded results can be used to ensure efficient service in non-functional conditions. Dataset visualization is part of data analysis, a method of understanding data value through visual context performed after data modification. Today, visualizations are more useful in business intelligence and analysis across all industries. There are

Table 1 Grouped fields of the data set

Simulation parameters	Values
Account length	102.664596
Area code	437.817805
Number Mail Messages	5.115942
Total Day minutes	206.914079
Total Day calls	101.335404
Total international minutes	10.700000
Total international calls	4.1634561
Total international charge	2.889545
Customer service calls	2.229814
Churn	1.000000

several ways to visualize dynamic or interactive information, and datasets can be visualized through various forms of visual insight. This article focuses on the process model, working with Microsoft Power BI, the types of data sources the tool can use, and the various forms of visual insight and context associated with it. After adding the formula for customer churn rate (number of customers churn ÷ total number of customers at the beginning of the period) x 100, we finalized the forecast result. Table 1 shows the different parameter values for the collected dataset.

4 Results and Discussion

In this section, we have leveraged vast sets of telecom data to analyze and improve the effectiveness of advertising campaigns. We have used visualization tools such as Matplotlib to perform various representations. The analytical results are divided into different objectives, as explained in the following. The results of this research are presented in the following sections.

4.1 Extracting Data With International Calls

Extracting data from International calls involves gathering information from phone conversations between parties in different countries. Using data extraction to remain on top of data processing helps your team to access data more quickly [11]. This basic procedure of collecting and storing data makes it more visible to everyone in the organization who requires it. Reporting is a critical component that may be done with the help of business intelligence and data analytic technologies. On the other hand, the reports or visualizations generated change based on the type of business

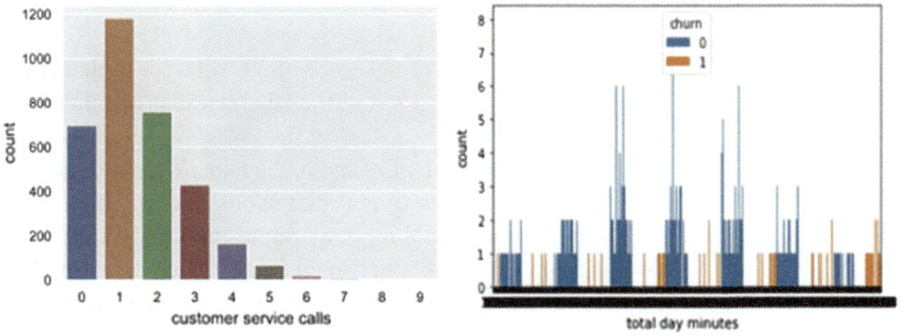

Customer services international calls data

Count of churn rate data

Fig. 1 Extracting data for international calls

data and parameters. Data analytics would be the finest choice when a corporation wishes to deal with current market trends and deliver ad-hoc reports. Data analytics is also important for firms forecasting future data patterns based on previous data. Otherwise, business intelligence is the solution if the client manages data housed in a data warehouse and generates reports by pulling data from the data generation and analytics campus of NCS and Singtel. Figure 1a, b show the count of data extracted for customer service calls.

4.2 Visualising Data For Different Parameters

The bars can be of two different types: horizontally or vertically. A vertical bar chart is also known as a column chart. The data for International calls could be used to boost the usage in lesser-used areas. We will use graphs to differentiate variables, such as age groups, classes, schools, etc. Bar graphs are the most efficient way to analyze data if there are too many categories to compare. With a good understanding of a proper bar graph, the advertising agency will clearly understand parametric stats and take notes about the market.

4.3 Visualizing International Calls Using Violin Plot

This violin plot is a hybrid of a boxplot and a kernel density plot that shows the peaks in the data. It is used to visualize the distribution of numerical data. Unlike boxplots, which can only display summary statistics, violin plots display summary statistics and the density of each variable [12].

The density plot's narrower sections represent values that occur less frequently. A violin plot shows the structure of a data collection and summarises it using five

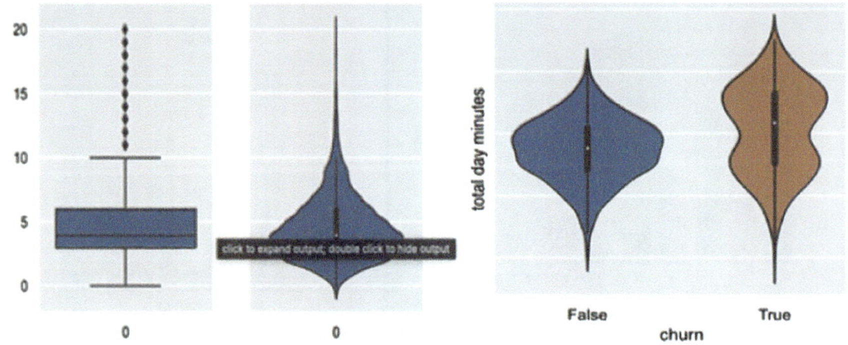

Violin plot on International bills

Churned data in violin bar format

Fig. 2 Visualizing international calls using violin plot

summary values: the lowest, first quartile, median, third quartile, and maximum. Figure 2a, b show the visualization of international calls and churned data using a violin plot.

4.4 Data Visualization with Correlation Matrix

The correlation matrix displays the correlation values used to calculate the degree of linear connection between each. If one variable rises while the other falls, the correlation value is negative [13]. The observable pattern in our example above is that all of the variables are significantly correlated with one another. The correlation matrix shows the correlation values used to quantify the degree of linear relationship between each pair of variables. The correlation matrix, also known as the correlation table, is an analytical tool that gathers correlation coefficients between an X-axis and a Y-axis where we locate various variables.

4.5 Data Visualization Using Scatter Plot

Each object in a dataset is represented by a point (or other mark) on two continuous, orthogonal dimensions in the standard scatterplot. The classic scatterplot design quickly becomes useless as data rises and complexity increases. As a result, several alternative scatterplot designs have been developed. While these designs may address scale, they are frequently tailored to certain data qualities and activities. Designers are given minimal information on choosing between design options [14]. We discuss how to examine analytical situations regarding task and data attributes to identify whether scatter plot designs apply to this study.

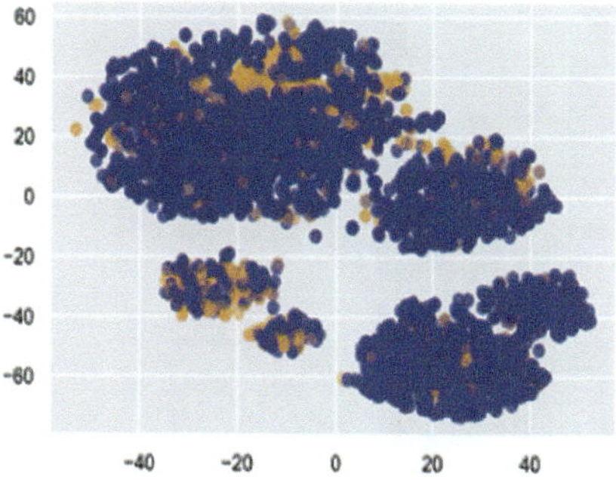

Concentrated data as partial scatter

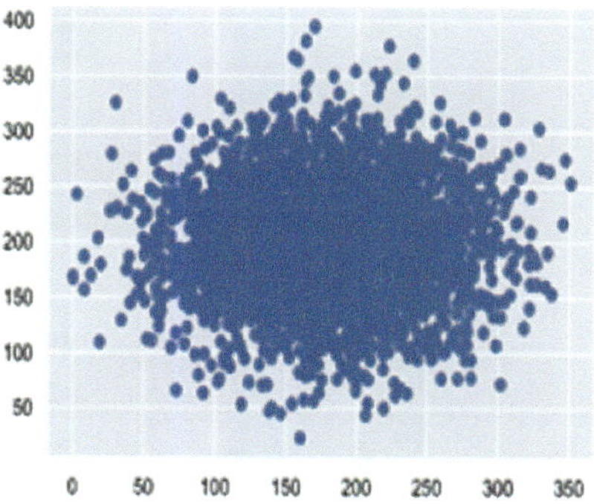

Scatter plot for day and night minutes

Fig. 3 Data visualization using scatter plot

We create a framework by gathering and abstracting scatter plot use cases from the literature. We gather model tasks completed with scatter plots to create an abstraction that helps us comprehend the scatter plot task space. We also identify various data attributes crucial to design, such as the number of objects. We survey scatter plot designs to group and cluster comparable design decisions to find the universe of feasible designs. We reason about the applicability of various designs using tasks and data attributes. A scatter plot is used to find out the relations between two variables. In the given case, we have used variable one as day minutes and variable two as night minutes. Using the color cyan, we can predict that there is a relative number of data that matches between the two; hence we can conclude with the given graph that the number of day and night calls made were relatively similar in numbers. Figure 3a and 3b show a visualization of the day and night data calls using a scatter plot.

4.6 Visualisation of Day, International, Evening and Night Calls Data

Visualized data with four parameters while using pictographs, which clarified the range of calls made during the day. Figures to be scaled and accuracy as per data. The benefit of using the given graphical representation is to have a rough vision but accurate ideation of the data. Customer outcomes are important for predicting a company's health since a high churn rate leads to a high revenue churn rate, which hurts the bottom line. The more clients you lose, the greater your turnover rate. Figure 4 shows the visualization of the day as per International, evening, and night calls.

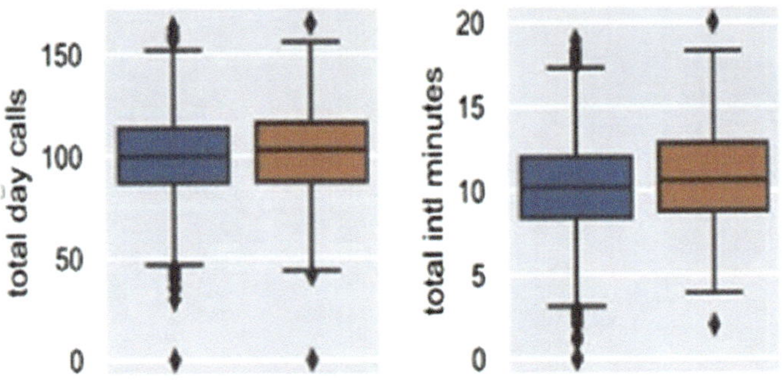

Fig. 4 Visualization of day, international, evening and night call data

5 Conclusion

The integration of telecom data analytics proves to be a pivotal asset for businesses navigating the complexities of today's data-driven and highly competitive business landscape. The study, focusing on Singapore residents' data from the Singtel communications network, exemplifies the potential of leveraging existing customer data to optimize advertising strategies. By employing a simulation environment with Python and Power BI, the research successfully performed data cleaning, wrangling, and analysis operations, providing valuable insights into suitable network plans for different cities in Singapore. The findings of this paper underscore the importance of data-driven decision-making in enabling telecom companies to adapt, innovate, and tailor network plans to meet the evolving needs of their customers in a rapidly changing market.

References

1. Arkin, P., Abushova, E., Bondarenko, V., Przdetskaya, N.: The concept of smart cities: prospects for the telecommunications business and the current trend in the development of modern society. In: Internet of Things, Smart Spaces, and Next Generation Networks and Systems: 20th International Conference, NEW2AN 2020, and 13th Conference, RuSMART 2020, St. Petersburg, Russia, pp. 308–317 (2020)
2. Bhardwaj, A., Krishna, C.: Performance evaluation of bandwidth for virtual machine migration in cloud computing. Int. J. Knowl. Eng. Data Min. **5**, 139–152 (2018)
3. Wu, S., Yau, W., Ong, T., Chong, S.: Integrated churn prediction and customer segmentation framework for telco business. IEEE Access. **9**, 62118–62136 (2021)
4. Ahmad, A., Jafar, A., Aljoumaa, K.: Customer churn prediction in telecom using machine learning in big data platform. J. Big Data. **6**, 1–24 (2019)
5. Hung, S., Yen, D., Wang, H.: Applying data mining to telecom churn management. Exp. Syst. Appl. **31**, 515–524 (2006)

6. Prashanth, R., Deepak, K., Meher, A.: High accuracy predictive modelling for customer churn prediction in telecom industry. In: Proceedings 13th Machine Learning and Data Mining In Pattern Recognition: 13th International Conference, MLDM 2017, New York, NY, USA, pp. 391–402 (2017)
7. Huang, B., Kechadi, M., Buckley, B.: Customer churn prediction in telecommunications. Exp. Syst. Appl. **39**, 1414–1425 (2012)
8. Keramati, A., Ghaneei, H., Mirmohammadi, S.: Developing a prediction model for customer churn from electronic banking services using data mining. Financ. Innov. **2**, 1–13 (2016)
9. Bhardwaj, A., Singh, V., Narayan, Y., Others Analyzing bigdata with hadoop cluster in hdinsight azure cloud. In: 2015 Annual IEEE India Conference (INDICON), pp. 1–5 (2015)
10. Effendy, V., Baizal, Z., Others Handling imbalanced data in customer churn prediction using combined sampling and weighted random forest. In: 2014 2nd International Conference on Information and Communication Technology (ICoICT), pp. 325–330 (2014)
11. Dahiya, K., Bhatia, S.: Customer churn analysis in telecom industry. In: 2015 4th International Conference On Reliability, Infocom Technologies and Optimization (ICRITO)(Trends and Future Directions), pp. 1–6 (2015)
12. Almana, A., Aksoy, M., Alzahrani, R.: A survey on data mining techniques in customer churn analysis for telecom industry. Int. J. Eng. Res. Appl. **4**, 165–171 (2014)
13. Peng, Z., Zhang, M., Li, K., Jin, R., Ren, B.: Speed-ANN: Low-Latency and High-Accuracy Nearest Neighbor Search via Intra-Query Parallelism (2022)
14. Hu, X., Yang, Y., Chen, L., Zhu, S.: Research on a customer churn combination prediction model based on decision tree and neural network. In: 2020 IEEE 5th International Conference On Cloud Computing And Big Data Analytic (ICCCBDA), pp. 129–132 (2020)

Employee Retention and Attrition Prediction Using Machine Learning. Case Study on ITES Company

Uttam Singh Chaudhary and Darpan Anand

Abstract Human Resources or Human Capital is the core component for the progress and success of any organization. The employees within an organization constitute its most valuable asset, as they are the driving force behind innovation, the delivery of products & services, and customer interactions. Organizations are investing in capacity building and upskilling their resources. Therefore, it is essential to retain the resources for the development and growth of any organization. The task of managing these kinds of situations is a major challenge for any human resource department. The proposed work aims to offer a solution to this significant issue. The paper presents a possible solution based on a case study of an ITES company. Machine learning, the popular technique is used to predict the possibility of attrition and is useful to get its possibility. It is very difficult to trust the feedback of the person willing to leave the company. Therefore the clustering is used through the k-means technique to handle the output of this problem. The results of the solution are very effective and efficient in giving the solution to the problem of attrition.

Keywords Machine learning · K-means clustering · Logistic regression · Support vector machine

1 Introduction

Employee happiness and engagement play a pivotal role in their commitment to the company and their contributions to its prosperity. The impact of employees on attrition and retention rates is indeed substantial. Employee retention, a critical process, focuses on maintaining high levels of employee satisfaction and engagement in their

U. S. Chaudhary
Sir Padampat Singhania Univesity, Udaipur, Rajasthan, India
e-mail: uttamsinghchaudhary3@gmail.com
URL: http://www.spsu.ac.in

D. Anand (✉)
Chitkara Institute of Engineering and Technology, Chitkara University, Rajpura, India
e-mail: darpan.anand.agra@gmail.com

B. Soni et al. (eds.), *Beyond Artificial Intelligence*, Lecture Notes in Networks and Systems 1326, https://doi.org/10.1007/978-981-96-4170-3_59

roles. This emphasis is justified by the considerable costs and disruptions associated with high employee turnover. When employees depart, companies must invest valuable time and resources in recruiting and training replacements. The consequences include diminished productivity, increased expenditures, and a detrimental effect on overall morale.

Crucial metrics that reflect a company's stability, productivity, and overall success are its attrition and retention rates, offering insights into the flow of employees within the organization. Attrition signifies the departure of employees, whether voluntary or involuntary, while retention signifies the company's effectiveness in maintaining employee engagement and satisfaction. For a company to thrive in the market, it is imperative to maintain low attrition rates. Understanding the concrete reasons that attract employees to an organization and, more importantly, keeping them as long-term assets is pivotal for the growth of any business or organization.

The primary objective of this paper is to assess the likelihood that new employees will stay with the company by analyzing past data. A significant portion of this research centers on verifying the authenticity of job satisfaction ratings submitted by employees. This is accomplished by initially selecting key attributes that influence employees' job satisfaction and subsequently forming clusters based on these attributes. These clusters are then compared with the original dataset's job satisfaction index, thereby enhancing the model performance to predict the false feedback of employee ratings in the original dataset. Machine learning techniques are subsequently employed to train a model for predicting of retention probabilities of new employees.

The motivation of this paper is to uncover the factors, affecting employee satisfaction and engagement, aiming to enhance working conditions and job satisfaction of the employees, Ultimately retaining invaluable talent in an organization. This research also provides the facility to HR department with insights that can enhance talent management programs, reduce costs, and facilitate informed decision-making concerning the workforce, thus contributing to the organization's overall success and stability. The process is technically driven therefore it has credibility, transparency and effectiveness.

A comprehensive analysis is also presented in this paper along with the analysis encompasses a range of aspects, including employee demographics, departmental distributions, attrition patterns, job roles, and experience levels, as well as an assessment of business travel trends. The dataset used for this research is the IBM HR Analytic Employee Attrition & Performance dataset from Kaggle. It is a fictional dataset created by IBM data scientists, containing various demographic, qualitative, and quantitative variables about individual employees. The dataset was downloaded as a csv file., containing various HR metrics commonly collected in contemporary organizations. These metrics cover aspects such as business travel, department, job role, job level, performance ratings, years at the company, years in the current role, and personal details, including age, education field, gender, and marital status, as well as monthly income, among others. In total, the dataset comprises 1,470 entries with 35 distinct attributes, shedding light on employee performance ratings and attrition behaviors within the company.

2 Literature Survey

ANALYZING EMPLOYEE ATTRITION USING DECISION TREE ALGORITHMS Alao D. & Adeyemo A. B have taken a dataset of 309 complete records of employees from Nigerian Institutions who left the institution between 1978 and 2006. The data revolve around demographic and job-related records of the employee to classify employees into attrition classes. The tools used were WEKA 3.6.7 and See5, popular ML software written in Java. Various classifiers of WEKA were used, and the results show that the Boosted SeeTree and the Boost SeeRule both had given a 74% accuracy, followed by the C4.5 (J48) and the CART (SimpleCart) classifier. The REPTree gave the lowest accuracy of 0.62. With predictive model, the best performance of accuracy has given that employee salary and length of services were major factors in determining the attrition of the institution [1].

Prediction of employee attrition using data-mining By R Shiva Shankar et al. contributed using the same dataset as IBM attrition by adding the approach of the Feature Selection method for preventing employee attrition. With the use of various classification models like Decision Tree, Logistic Regression, SVM, KNN, Random Forest, and Naïve Bayes they analyzed the dataset and presented their research [2].

Christopher Boom Hower et al. gave the answer to the big question, What Makes an Employee Quit? He did his work on publicly available data from the Office of Personnel Management, the Bureau of Labor Statistics, and IBM. They used Principal Component Analysis and classification methods. It was concluded with Logistic Regression that allowed them to simplify the model and predict employee quits with the highest accuracy of all their testing methods, achieving a greater than 74% success rate [3].

Analysing employee attrition using machine learning by Usha P. M. et al. as they conducted their research on the IBM attrition dataset using the Data Mining technique to understand the factors responsible for attrition of human resources using Weka, a data science tool for predictive analysis [4].

HR analytics: Employee attrition analysis using logistic regression by I. Setiawan et al. has shown that the data for the study were around four thousand employees, covering 261 days (one year working days) during 2015—the data period between January and December. In conclusion, to reduce the employee attrition rate, the company needs to improve the human resource department by evaluating the working environment, job satisfaction, employee workload, and interaction between manager and employee [5].

Machine Learning Approach for Employee Attrition Analysis by Dr. R. S. Kamath1 et al. analyzed the employee attrition status and the likelihood of retention-attrition of the employee by using the Kaggle dataset. They have used various prediction algorithms and evaluated the model's performance in terms of Error matrix and Pseudo R square, for estimation of error rate. Performance accuracy revealed that the Random Forest model can be effectively used for classification. Their analysis concludes that employee attrition depends more on employees' satisfaction level as compared to other attributes [6].

Predicting Employee Attrition Using Machine Learning Techniques by Francesca Fallucchi et al. focussed on the analysis of objective factors that make an impact on employee attrition, which could lead to the main cause of worker's decision to leave the company and further made a model that can predict the same for the future employee. The model was tested on a dataset of IBM analytics, which includes 35 features and 1500 samples. They concluded that the Gaussian Naïve Bayes classifier gives the best recall rate of 0.54. Results showed that monthly income, age, overtime, and distance from home were the main attrition variables [7].

IBM Employee attrition analysis by Shenghuan Yang et al. analyzed the dataset and reported some significant features affecting employee attrition rate. Then with K-means clustering they classified people into two categories and with the help of binary logistic regression, they concluded that the attrition rate of people who traveled frequently was twice then those who travelled rarely. They also said that people in HR have a higher tendency to resign from jobs. They also stated the model accuracy score. Random forest has given 84.56% accuracy and Logistic regression gives an accuracy of 88.43 respectively which shows Logistic regression fitted better and was considered more suitable for predicting [8].

Prediction of Employee Attrition Using Machine Learning and Ensemble Methods By Aseel Qutub et al. researched on IBM attrition dataset for improving their retention strategies and boosting employee satisfaction. They deployed several machine learning models namely: Decision tree, Random forest regressor, Logistic regressor, Adaboost Model, and Gradient Boosting classifier Models. They showed the results using first supervised model evaluation metrics and then with the Ensemble Model's Evaluation. With the ensemble technique, 86.73% accuracy was recorded with Adaboost and Random Forest Classifier. In a supervised manner, 88.43% accuracy was recorded By Logistic Regression [9].

The research done revolved around the factors that can lead to higher attrition rates in the company, and the implementation of ML algorithms added the service to predict the future attrition rates for the company and especially help HR with its rubrics for handling employee management. This research was focused not only on finding useful factors that lead to higher attrition rates but also on identifying the reviews or surveys covered by the leaving employee as either fake or not by considering major correlation factors making clusters with their similarity and then matching the rating with actual rating. Ultimately aligning the factors with Job satisfaction ratings for identifying fake reviews or not to improve the transition employed feedback system. Also, the implementation of the ML model was done for future prediction of employees leaving the job or not.

3 Experimental Setups

To analyze the dataset we need to do some experiments and apply various data science methods. so the process is explained in the upcoming section.

3.1 Dataset

Employee attrition is the internal data of the company, which is difficult to obtain, and some data has a certain degree of confidentiality, therefore this paper used the dataset disclosed by Kaggle. The sample size of the dataset is 1471 rows and 34 feature variables, namely: Age, Attrition, BusinessTravel, DailyRate, Department, DistanceFromHome, Education, EducationField, EmployeeCount, EmployeeNumber, EnvironmentSatisfaction, Gender, HourlyRate, JobInvolvement, JobLevel, JobRole, JobSatisfaction, MaritalStatus, MonthlyIncome, MonthlyRate, NumCompaniesWorked, Over18, OverTime, PercentSalaryHike, PerformanceRating, RelationshipSatisfaction, StandardHours, StockOptionLevel, TotalWorkingYears, TrainingTimesLastYear, WorkLifeBalance, YearsAtCompany, YearsInCurrentRole, YearsSinceLastPromotion, YearsWithCurrManager.

3.2 Machine Learning Methods

Machine Learning methods are computational methods that make the system learn and extract meaningful patterns from the data. Thus improving existing systems and deriving future predictions. There are various types of machine learning methods, each having different techniques for all possible kinds of data whether it is structured, unstructured, or semi-structured. Some of the major applications of machine learning techniques are used in real-life problems [10–13].

3.3 ML Workflow

The proposed model Fig. 1 highlights the workflow for machine learning. It begins by stating the research problem of predicting employee attrition and retention. This involves collecting relevant data and applying various data exploration techniques, such as plotting data visuals, to identify key factors and insights. By understanding the data, we can better comprehend what steps are necessary for data preprocessing, such as cleaning, eliminating null values, changing categorical variables, removing useless columns, and more. After preprocessing the dataset, we created a correlation matrix and graphical visualizations for feature selection. Then, we used Kmeans unsupervised techniques to identify fake employee ratings by clustering the data based on inherent similarities between data points. By comparing the job satisfaction ratings of employees with the clusters created by Kmeans, we were able to determine whether an employee's rating was genuine or not. Finally, we used various supervised ML techniques like decision tree, SVM, Random Forest, and Logistic regression to classify whether new employees will remain in the company or not.

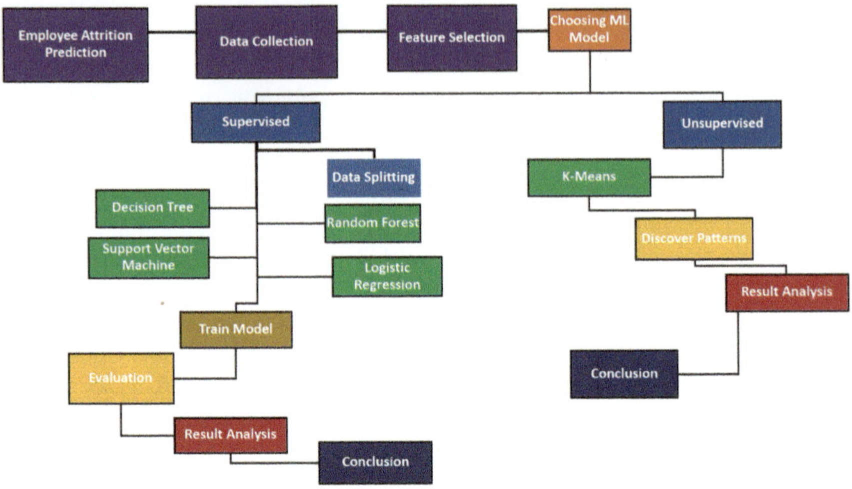

Fig. 1 ML workflow

K-Means clustering is a popular unsupervised machine learning technique that plays a crucial role in data analysis, pattern recognition, and segmentation. At its core, K-Means aims to organize unstructured data into meaningful clusters based on the inherent similarities between data points. Each data point is then assigned to the cluster whose centroid is nearest, typically measured using the Euclidean distance. The centroids are recalculated as the mean of all data points within each cluster. These steps are repeated iteratively until convergence, often defined by a maximum number of iterations or when the centroids no longer change significantly.

Decision trees are supervised Machine Learning models used for both classification and regression tasks. They provide a structured way of making decisions based on a series of hierarchical choices, similar to a flowchart. Decision Trees work in the following way:-

1. ROOT NODE: At the top of the tree, there's a root node that represents the initial decision or question based on a specific feature from the dataset.
2. INTERNAL NODES: These nodes follow the root and subsequent nodes. They represent decisions or conditions based on l node, branches extend to child nodes, each corresponding to possible out features that partition the data into subsets.
3. BRANCHES: From each intername of the decision or condition at the parent node.
4. LEAVES: The terminal nodes of the tree, called leaves, provide the final predictions or decisions. In a classification tree, each leaf represents a class label. In a regression tree, leaves contain numerical values.

Logistic regression is a widely used statistical model for binary classification tasks, where the goal is to predict one of two possible outcomes (e.g., yes/no, spam/

ham) based on input features. It's a fundamental algorithm in machine learning and has been extensively researched and applied in various domains.

Working: Hypothesis Function: Logistic regression models the probability of the binary outcome using a logistic (sigmoid) function. The logistic function maps the linear combination of input features to a value between 0 and 1.

Parameters Estimation: In logistic regression, the model learns the parameters (weights and bias) through an optimization process, typically using techniques like Maximum Likelihood Estimation (MLE) or gradient descent. The goal is to find the parameters that maximize the likelihood of the observed data.

Decision Boundary: Logistic regression finds a decision boundary in the feature space that separates the two classes. This boundary is typically a hyperplane. It assigns a probability to each input, and based on a threshold (usually 0.5), classifies the input into one of the two classes.

Support Vector Machine is a powerful machine learning algorithm used for both binary and multi-class classification tasks. It's known for its robustness and ability to handle complex datasets, and it has also been extensively researched and applied in various fields.

WORKING: Margin Maximization: SVM aims to find a hyperplane that maximizes the margin between the two classes in the feature space. The margin is the distance between the hyperplane and the nearest data points (support vectors) from each class.

Kernel Trick: SVM can handle non-linearly separable data by using a kernel function. This function implicitly maps the data into a higher-dimensional space where it becomes linearly separable. Common kernel functions include the linear, polynomial, radial basis function (RBF), and sigmoid kernels.

Regularization: SVM introduces a regularization parameter (C) to control the trade-off between maximizing the margin and minimizing the classification error. A smaller C value leads to a wider margin but may allow misclassifications, while a larger C value tries to minimize misclassifications at the expense of a narrower margin.

Support Vectors: SVM identifies a subset of data points called support vectors, which are the data points closest to the decision boundary. These points are critical for defining the margin and the decision boundary.

Random Forest The Random Forest algorithm is an ensemble learning method commonly used for classification and regression tasks. It belongs to the family of decision tree-based algorithms and is known for its robustness and high predictive accuracy. Random Forest builds multiple decision trees during the training phase and combines their predictions to achieve a more reliable and generalized model. The algorithm works as follows,

Bootstrapping: Random Forest employs a technique called bootstrapping, where multiple subsets of the original dataset are created by random sampling with replacement. Each subset is used to train an individual decision tree.

Random Feature Selection: At each node of a decision tree, only a random subset of features is considered for splitting. This introduces diversity among the trees

and prevents them from being overly correlated. The goal is to avoid overfitting and increase the robustness of the overall model.
Decision Tree Construction: A decision tree is constructed for each bootstrapped subset. The tree is grown by recursively splitting the data based on the selected features, aiming to maximize the information gain or decrease in impurity at each node.
Voting Mechanism: For classification tasks, each tree in the Random Forest independently predicts the class of a sample. The final prediction is determined by a majority vote among the trees. For regression tasks, the average prediction of all trees is taken as the final output.
Aggregation of Predictions: The predictions from individual trees are aggregated to form the overall prediction of the Random Forest. This process is robust and helps mitigate the risk of overfitting.
Parameter Tuning: Random Forest provides several hyperparameters that can be tuned to optimize its performance. Key parameters include the number of trees (n_estimators), the maximum depth of each tree (max_depth), and the number of features considered at each split (max_features).
Feature Importance: Random Forest naturally assesses the importance of each feature by evaluating how much each feature contributes to the improvement in prediction accuracy across all trees. This information can be used to identify the most influential features in the dataset [14–18].

3.4 Proposed Process

The workflow proposes the use of Machine learning techniques like EDA to have an exploratory view of data before making any assumptions as well as better understand patterns within the data, detect outliers or anomalous events, and find interesting relations among the variables. The use of various Machine Learning algorithms like Decision Tree, Logistic Regression, Random Forest, and Support vector Machine was introduced in order to build a model that could make the facts and assumptions clearer. We have made use of Unsupervised algorithms like Kmeans to make a comparison of existing employee ratings and machine learning formed clusters to verify the trueness of data and whether the ratings were actual or not.

4 Pre-data Analysis

- **Age Distribution and Gender Distribution**
- The accompanying bar graph Fig. 2a and b shows the Gender Distribution and Age Distribution in the company. It can be observed that most of the employees belong to the age group of 28 – 40 as the bars are at peak count at this range and there is a good number of female and male employees in the company. It's like

the ratio of 40:60, which shows a fair representation of opportunities inside the company as well.

- **Educational Background**

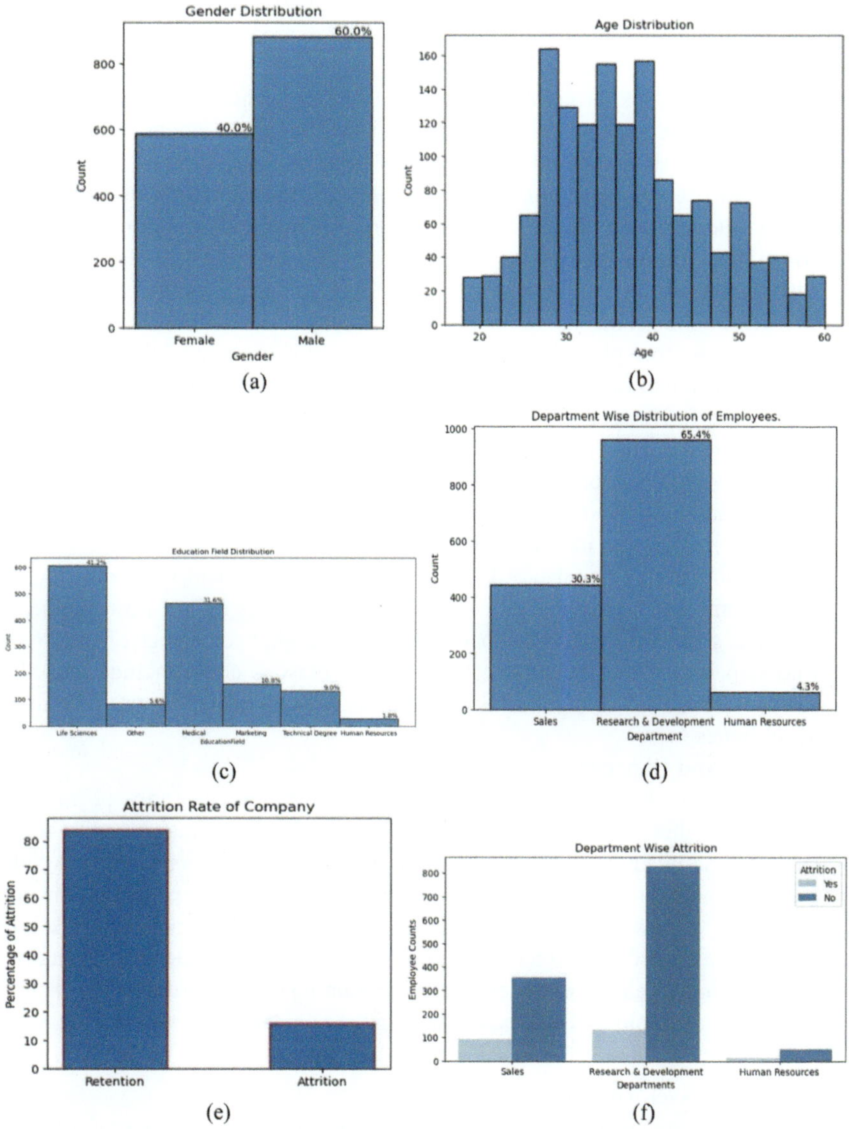

Fig. 2 Figures showing **a** gender distribution **b** age distribution **c** education field distribution **d** department wise employee distribution **e** attrition rate **f** department wise attrition rate

- The Following bar graph Fig. 2c shows the educational background of employees. It's very surprising that a Fortune 500 company, IBM, an IT company, has more than 50% of staff from non-technical backgrounds. It shows that 1000+ employees are from the Life Sciences and Medical fields, and the rest of the employees are from Marketing, HR, and technical degrees. A percentage of people are from another background also.
- **Department Wise Distributions**
- The accompanying bar graph Fig. 2d shows the department-wise distribution of employees. The Research & Development Department has a huge concentration of employees, as compared to the sum of both of Sales and HR departments, around 900+ employees. This shows the company has maintained its competitive level in the market by doing more and more research and development work to bring customer satisfaction.
- **Attrition Analysis**
- *Overall Attrition Analysis*
- The following bar plot Fig. 2e talks about the attrition rate of the company. It is impressive that the attrition rate of the company is around 15–16%, which is not very high. However, the company may aim to reduce it to less than 10% for better growth of the company and its employees. Although the retention rate is 84–85%, which is good enough, it shows that the company has provided proper job satisfaction to most of its employees.
- *Department-Wise Attrition Rates*
- The accompanying count plot Fig. 2f shows the department-wise attrition rate of the company. It reveals that even though R & D has the greatest number of employees, the Sales and HR departments have a higher percentage of employees who step back from their job roles. Possible reasons could include work-life balance, job satisfaction, monthly income, salary increments, late promotions, or travel issues.
- **Job Role and Experience**
- The accompanying bar plot Fig. 3a shows the different categories of job roles with their highest time in the current role. It shows that the Manager and Research Director had given 6+ years of service in their current role, followed by a Sales executive with 5+ years of experience. This shows all the implementations and executions of new programs are working with the best strategies undertaken by managers and research directors. Among all the job roles, Sales representatives have less than 2 years of experience. If more training is given to them then chances are there that employees will be retained in the company.

Business Travel Analysis

The accompanying bar plot Fig. 3b shows the different categories of job roles with their highest time in the current role. It shows that the Manager and Research Director had given 6+ years of service in their current role, followed by a Sales executive with 5+ years of experience. This shows all the implementations and executions of new programs are working with the best strategies undertaken by

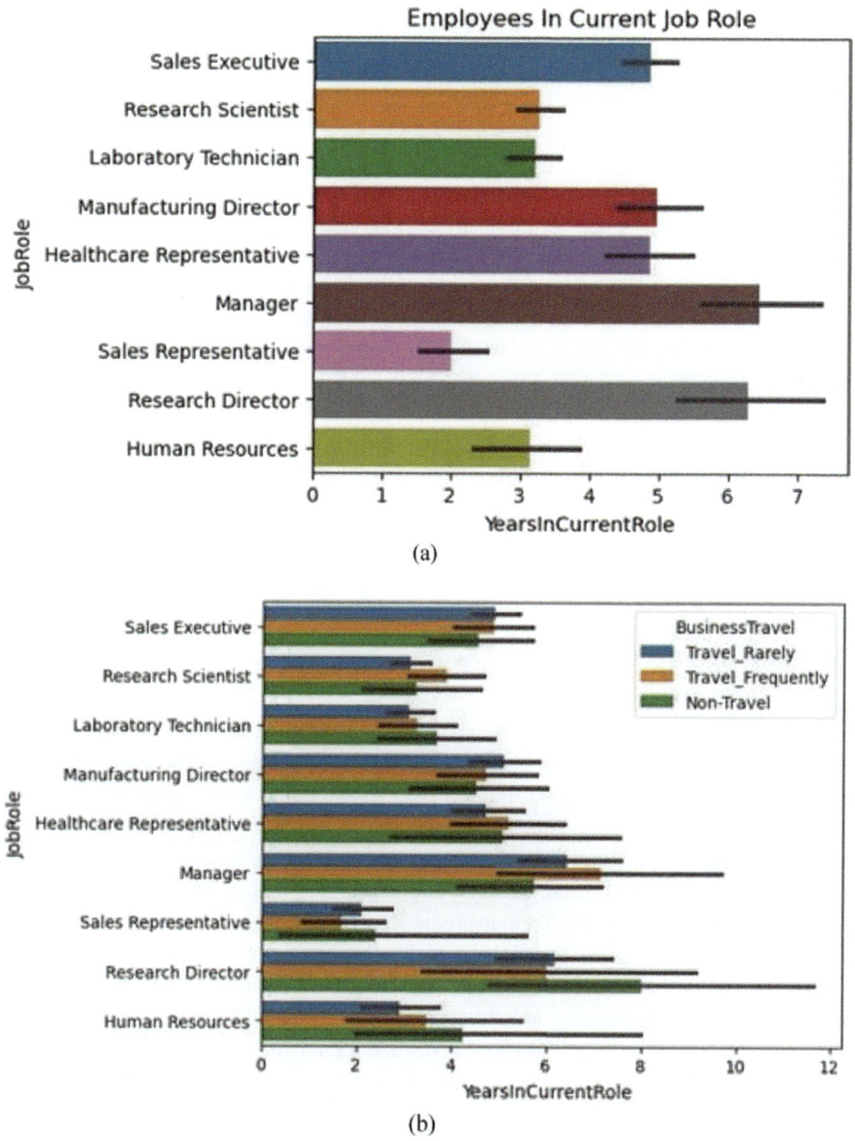

Fig. 3 **a** Job roles with years in current role **b** job roles with business travel

managers and research directors. Among all the job roles, Sales representatives have less than 2 years of experience. If more training is given to them then chances are there that employees will be retained in the company.

5 Results and Discussions

After conducting thorough research, utilizing a correlation matrix and various visualizations, five key factors are selected and these are: 'Education', 'Monthly Income', 'Percent Salary Hike', 'Performance Rating', and 'Years in Current Role'. Using k-means clustering Fig. 4a and b, we grouped the data points and compared them to the Job Satisfaction ratings given by employees. We found that 74% of the data points did not match the behavior of the clusters formed by k-means and the 'Job Satisfaction' ratings. We trained machine learning models using k-means Fig. 4c and d, with the decision-tree model achieving an accuracy of 75.85%, the logistic regression and support vector classifier models achieving an accuracy of 86.73%, and the random forest model achieving an accuracy of 82.65%. As a result, we conclude that the Logistic Regression and Support vector machine performed the best and can be used to predict an employee's chances of attrition in the company. Additionally, providing promotions at appropriate intervals, revising salary increment policies, hiring highly qualified employees, and prioritizing tech backgrounds can increase employee retention rates.

6 Conclusions and Future Scope

This paper focuses on the analysis of the employees' data related to job satisfaction and retention. To achieve our objective, the data collected for the experiment and research is analyzed, and various machine learning techniques to correlate with actual output. But, again there is a gap, and therefore, the clustering is applied for classification of the data. Accordingly, the data is labeled and further machine learning models are applied for prediction. our newly obtained data is feeded into machine learning algorithms such as Decision Tree, Logistic Regression, Random Forest, and Support Vector Machine. Logistic Regression and Support vector machine performed the best with an 86.73% accuracy score. Each model was evaluated based on its accuracy score as a performance metric. The dataset is very limited. Therefore, in the future, the model will be trained for versatile data not focused on a specific vertical. Some psychological parameters are also important to understand the actual problem. In the future, we are planning to incorporate these parameters in the prediction model.

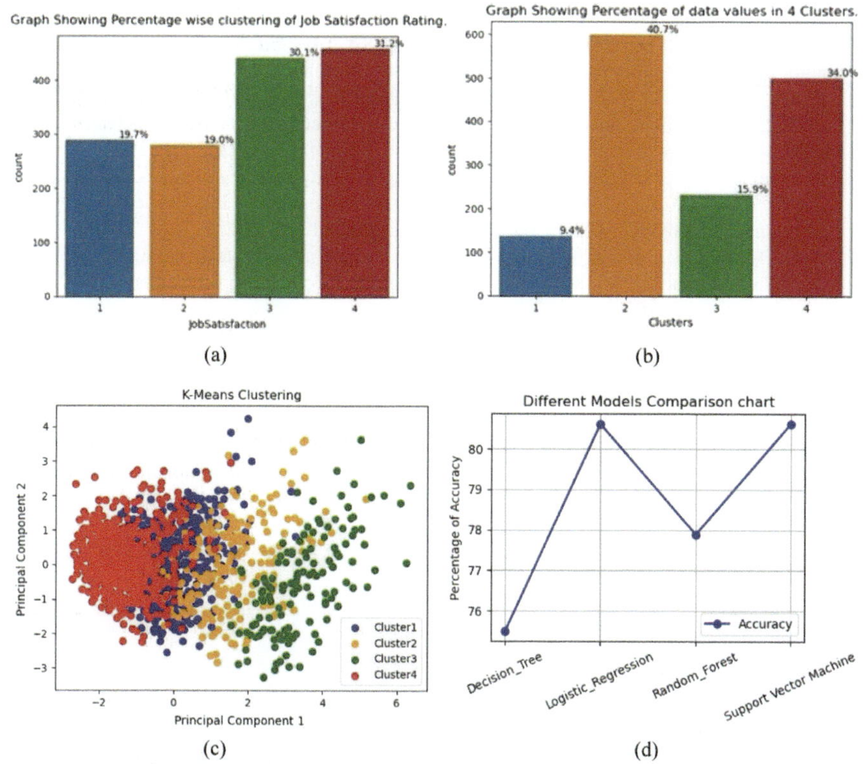

Fig. 4 **a** figure showing job satisfaction rating **b** figure showing k-means clusters formed groups **c** figure of scatter plot showing clusters formation by k-means **d** figure showing different models accuracy score

References

1. Alao, D., Adeyemo, A.B.: Analyzing employee attrition using decision tree algorithms. https://core.ac.uk/download/pdf/234697248.pdf
2. Shiva Shankar, R., Rajanikanth, J., Sivaramaraju, V.V., Murthy, K.V.S.S.R.: Prediction of employee attrition using datamining. https://ieeexplore.ieee.org/abstract/document/8541242/
3. Frye, A., Boomhower, C., Smith, M., Vitovsky, L.: Employee attrition: what makes an employee quit? Southern Methodist University, Stacey Fabricant, Penn Mutual Life Insurance Company. https://scholar.smu.edu/datasciencereview/vol1/iss1/9/
4. Usha, P.M., Balaji, N.V.: Analysing employee attrition using machine learning
5. Setiawan, I., Suprihanto, S., Nugraha, A.C., Hutahaean, J.: HR analytics: employee attrition analysis using logistic regression. https://iopscience.iop.org/article/10.1088/1757-899X/830/3/032001/meta
6. Kamath, R.S., Jamsandekar, S.S., Naik, P.G.: Machine learning approach for employee attrition analysis
7. Fallucchi, F., Coladangelo, M., Giuliano, R., William De Luca, E.: Predicting employee attrition using machine learning techniques. https://www.mdpi.com/2073-431X/9/4/86
8. Yang, S., Islam, Md.T.: IBM employee attrition analysis. Submitted 2 Dec 2020 (v1), last revised 1 Jan 2021

9. arXiv:2012.01286v6 [cs.CY]. Accessed 1 Jan 2021
10. Singh, G., Anand, D., Cho, W., Joshi, G.P., Son, K.C.: Hybrid deep learning approach for automatic detection in musculoskeletal radiographs. Biology **11**(5), 665 (2022)
11. Essah, R., Anand, D., Singh, S.: Assessment on crop testing based on IOT and machine learning. In: 2022 10th International Conference on Reliability, Infocom Technologies and Optimization (Trends and Future Directions) (ICRITO), pp. 1–6, October 2022. IEEE (2022)
12. Sharma, P., Anand, D., Kapoor, N.: Student's employability indexing using machine learning approach. In: 2022 Fifth International Conference on Computational Intelligence and Communication Technologies (CCICT), pp. 242–249, July 2022. IEEE (2022)
13. Kaur, A., Sahu, R., Anand, D.: Deep reinforcement base multi context routing in social IoT. In: 2023 International Conference on Circuit Power and Computing Technologies (ICCPCT), pp. 675–682, August 2023. IEEE (2023)
14. Mehta, V., Modi, S.: Employee attrition system using tree based ensemble method. https://ieeexplore.ieee.org/abstract/document/9689398/
15. Qutub, A., Al-Mehmadi, A., Al-Hssan, M., Aljohani, R., Alghamdi, H.S.: Prediction of employee attrition using machine learning and ensemble methods (2021)
16. Srivastava, D.K., Nair, P.: Employee attrition analysis using predictive techniques. https://link.springer.com/chapter/10.1007/978-3-319-63673-335
17. Comparative study of the machine learning techniques for predicting the employee attrition. https://ijrar.org/papers/IJRAR1903202.pdf
18. Chowdhury, A.H., Malakar, S., Seal, D.B., Goswami, S.: Understanding employee attrition using machine learning techniques

Cold Start Problem in a Recommender System: Application, Evaluation, Performance and Methods

Rajiv Kumar Nath and Tanvir Ahmad

Abstract The challenges that arise from the initial phase of recommender systems, known as cold start problems, pose a variety of obstacles during the setup and utilization of these systems. This is particularly evident when incorporating new products and welcoming new users to the platform. This hinders organizations from fully grasping the business worth of recommender systems, as they need to allocate resources for marketing and operations to attract fresh users and endorse novel products. Appreciation to recommender system researchers, who have conducted numerous studies to address cold start challenges. Nevertheless, there has been a scarcity of recent research focused on integrating these methodologies and algorithms. This study's extent is confined to a methodical examination of existing literature, excluding any experimental analysis for comparing and suggesting optimal strategies along with their applicable contexts in cold start situations.

Keywords Cold start · Approach driven · Data driven · New user · New item

1 Introduction

As the amount of information available to users continues to grow, it can be overwhelming and lead to decision fatigue. To address this problem, Recommender Systems have been developed to suggest items based on a user's preferences. These systems can be categorized based on the techniques used to filter items and the data sources used. As discussed, the three primary methods for filtering items in recommendation systems are content-based, collaborative filtering, and hybrid recommender systems [1, 2]. To lessen the cognitive load caused by an overwhelming amount of information, Recommender Systems propose suggestions to users based

R. K. Nath (✉) · T. Ahmad
Department of Computer Engineering, Faculty of Engineering & Technology, Jamia Millia Islamia, New Delhi 110025, India
e-mail: rajivknath@gmail.com

T. Ahmad
e-mail: tahmad2@jmi.ac.in

B. Soni et al. (eds.), *Beyond Artificial Intelligence*, Lecture Notes in Networks and Systems 1326, https://doi.org/10.1007/978-981-96-4170-3_60

on preferences provided by them. These systems can be categorized based on the methods used to filter items, or the type and origin of information employed. Recommender systems can be categorized into three main types on the basis of their filtering techniques: content-based, collaborative filtering, and hybrid recommender systems.

In addition to the three main types of recommender systems (content-based, collaborative filtering, and hybrid), there are several other types such as social network, knowledge-based, utility-based, context- aware, and demographic information-based recommender systems. Critique-based and constraint-based knowledge-based recommender systems are two new forms of knowledge-based recommender systems that use different data sources to develop the recommender system. Feedback from consumers is utilized as critiques in critiquing-based systems to produce item recommendations. What are the many types of knowledge-based systems?

Similarly, Recommender systems can be categorized depending on the technique used, kind or origin of data that they utilize, such as recommender systems based on social networks, expertise, utility, context, and demographics. Moreover, there are more recent forms of knowledge-based recommender systems, like critiquing based and constraint-based recommender systems. To generate suggestions, critiquing-based recommender systems use user comments as critiques, on the other hand, constraint-based recommender systems utilize restrictions produced from particular item characteristics and their association with user preferences.

Out of all the available types of recommender systems, the most commonly used and traditional methods are content-based and collaborative filtering techniques. Collaborative filtering algorithms provide top N item suggestions for users based on historical collaborative preferences between users and items. Depending on the entities involved in the interaction, recommender systems are classified into three types: user-based collaborative filtering, item-based collaborative filtering, and latent factor-based models that use matrix factorization. In user-based collaborative filtering, a user's item rating pattern is used to recommend items to other item-based collaborative filtering involves recommending items by on the other side, discovering neighboring users who have similar item rating patterns or preferences, identifies the closest neighbors of items and recommends items to users by considering their past item interactions and preferences.

Recommender systems assist in addressing the issue of excessive choices by suggesting items to users based on analysis of their preferences. These systems can be classified in numerous ways, including by the methods used to filter items (collaborative filtering, content-based, hybrid), or by the type or origin of data used (based on social networks, expertise, utility, context, and demographics). Critiquing-based recommender systems employ user feedback as critiques, whereas constraints derived from item attributes and user requirements are used in constraint-based recommender systems. The collaborative filtering and content-based approaches are the most commonly employed recommender systems. Collaborative filtering encompasses user-based, item-based, and matrix factorization-based latent factor models. Nevertheless, these techniques face challenges with new users and items, resulting in accuracy and quality issues. Content-based recommender systems be able to prevail

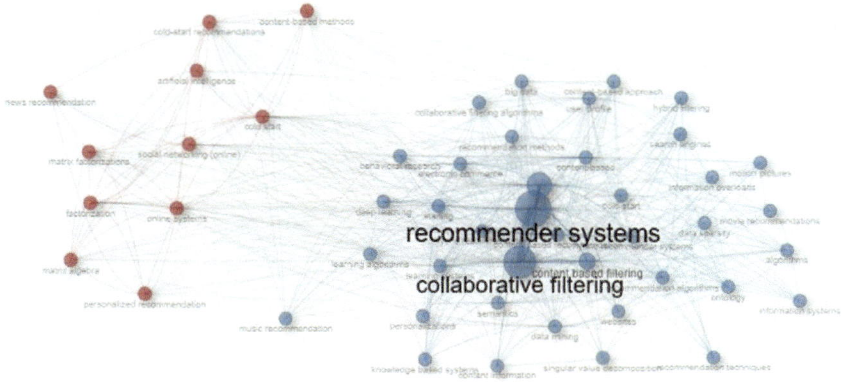

Fig. 1 Recommender system and their co-occurrence network

over these problems by examining item metadata, but they still encounter difficulties in providing dependable recommendations in the absence of user interaction data.

New users not finding recommendations useful can be a challenge to recommender systems' business value and may lead to reduced user engagement. To tackle this issue, researchers from academia and industry are actively developing new strategies. A method for enhancing recommendation accuracy and system efficiency in collaborative filtering is to focus on item-specific data rather than user data. This approach involves using item similarity rather than user similarity to generate recommendations. Advanced content-based methods that incorporate content, tagging, and cross-domain information have also been proposed. Additionally, context-aware recommender systems have been developed to enhance the performance of recommender systems by incorporating contextual information. These systems address challenges such as the cold start problem, new features or data types (for example, location, time, social information, similar groupings, and event details) (Fig. 1).

2 Objectives of the Research and Relevant Literature Review

The issue of cold start has an adverse effect on both the efficiency and the efficacy of recommendation systems, which has led researchers from both academia and industry to develop various algorithms and techniques. There has, however, been little research on integrating and consolidating these approaches into a comprehensive strategy for addressing the cold start problem. This study seeks to analyse existing works in this area (Fig. 2).

The author endeavored to address the issue of cold start by examining and contrasting four algorithms, but the study had some limitations [3]. It did not employ

Fig. 2 Summary of the article(s) used

a systematic literature review to cover all research methods in the field of cold start. Rather, the study focused only on four algorithms: MIPFGWS-CS [4], NHSM [5], FARAMS [6], and HU-FCF [7]. Using MovieLens and Jester datasets, the study examined the advantages and disadvantages of these algorithms and performed a comparative analysis of their effectiveness using MAE and RMSE values. The study's findings suggested that the NHSM algorithm was the most effective solution to overcome the challenges posed by the cold start problem.

The author conducted a comprehensive review of the literature to identify key components, strategies, and evaluation metrics associated with the cold start problem in recommender systems [8]. The article encompassed papers published from 2006 to 2016, but the authors did not employ a systematic literature review methodology. Therefore, some of the most recent research findings and potential approaches and algorithms may have been omitted, and not every application domain of recommender system may have been included. The authors admitted that the review may have overlooked certain domains, such as online shopping applications (Fig. 3).

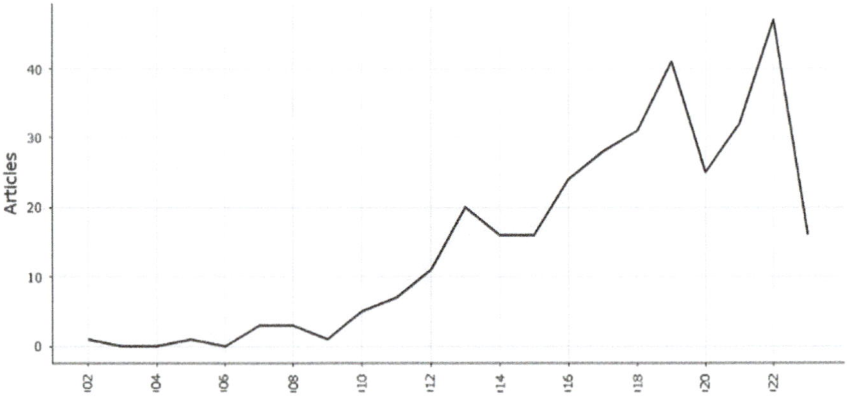

Fig. 3 Annual scientific production

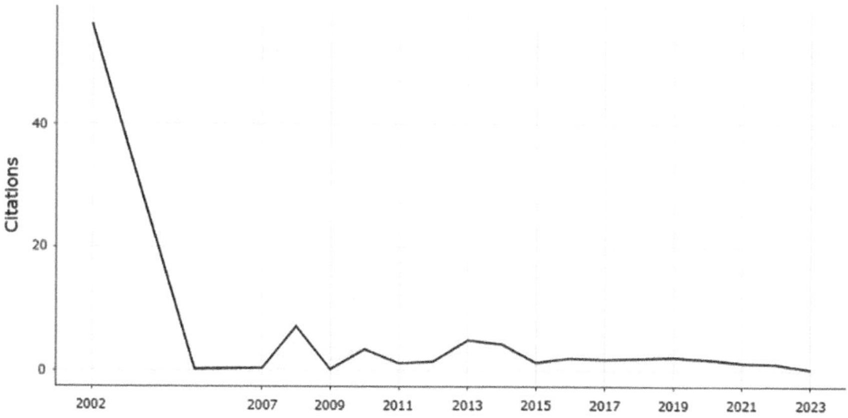

Fig. 4 Average citation per year

Social network data and auxiliary information are utilized to solve cold start issues in recommender systems [9]. However, this study recognizes that there is still a need for a more comprehensive review of the literature on this topic. Hence, the primary aims of this study are: (1) to perform a comprehensive analysis of all the techniques and algorithms that have been suggested for resolving the cold start issue, the authors conducted a systematic literature review and (2) to consolidate the different strategies used by researchers in terms of benchmark algorithms, types of information used, computing techniques, and their effectiveness in various domains. This study aims to overcome the limitations of previous research by providing a more comprehensive process of examining existing literature regarding the difficulty of the cold start in recommender systems (Fig. 4).

Paper is organized into different sections for clarity. Section 3 explains the methodology used for the systematic literature review. In Sects. 4 and 5 of the study, the review results are presented through statistical analysis and consolidation of information. In Sect. 5, the review qualitatively summarizes the various tactics and methodologies suggested by researchers to tackle the cold start problem. The synthesized approaches are discussed in greater detail in Sect. 6, and potential areas for future research are suggested. In conclusion, the Sect. 7 of the paper provides a summary of the key discoveries, contributions, and shortcomings of the study (Fig. 5).

3 Methods

This study has employed a methodical review of various research works to identify and analyze existing research on different techniques and methods proposed by scholars to address the issue of cold start in recommender systems. Scholars recommend using this approach as it is rigorous, systematic, and enables answering

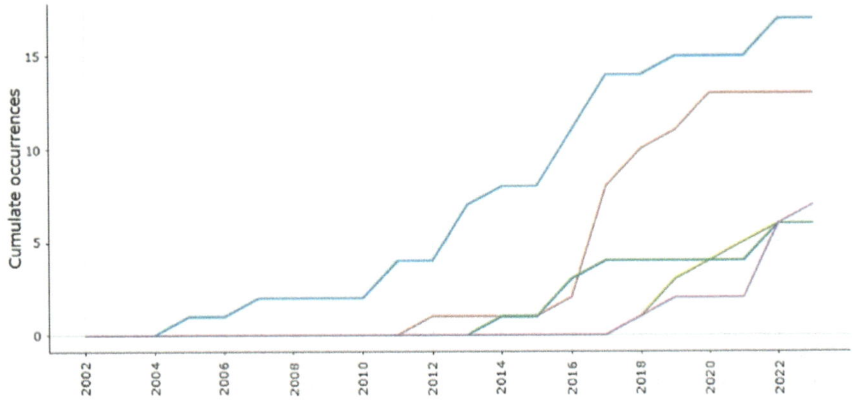

Fig. 5 Sources production over time

research questions and contributing to academic discussions [10]. In contrast, a standalone literature review only discusses the literature in a paper that does not have any empirical facets or aspects. The comprehensive review of relevant literature is characterized by a higher degree of comprehensiveness, organization, and methodology, which enhances its reproducibility by other scholars [11], the article highlights the significance of precisely defining the extent as the process of conducting a methodical and thorough examination of existing literature research question should not be too broad or too narrow. In this study, the focus was to examine the different methodologies and algorithms suggested in the research on recommender systems to tackle the challenges posed by cold start problems. To gather relevant literature, the researchers conducted a search using keywords such as "cold start," "recommender system," "recommendation system," and "collaborative filtering".

In order to narrow down the literature, the study employed specific criteria for inclusion and exclusion. The reviewers only considered articles written in English, as determined by their proficiency. To maintain the quality and reliability of the literature reviewed, the study focused only on peer-reviewed articles that were published in scholarly journals and conferences, while excluding other types of materials. The review was limited to articles published between January 2010 and December 2021; in order to cover the major research contributions this study reviewed the literature on the topic of addressing cold start problems in recommender systems within the last ten years. The articles that were gathered were reviewed through a two-step process, where the first step was to assess the title and abstract, and the second step was to evaluate the entire content of the article, to ensure that no duplicates were included in the final analysis.

4 Statistical Analyses of the Literature

The study included 66 articles that were published in 50 various scholarly journals that have undergone peer review. The data analysis demonstrates that the number of published articles on strategies to solve cold start problems has generally increased, with some exceptions in 2016, 2018, and 2021. In 2020, the number of published articles was the highest, indicating that more researchers are recognizing the importance of addressing cold start issues in recommender systems, as they have an impact on user engagement, which is driving more research in this area.

The upcoming section of the paper will provide a summary of the reviewers' discoveries from the literature. Summarizing literature is a useful way to arrange existing research to establish a classification, establish concepts or meta-theories, and propose study agendas or questions for prospect study. Aim of this paper is to create research groups based on the summarized information, which can be utilized in subsequent sections for discussion, thereby offering a more comprehensive understanding of the topic.

5 Qualitative Syntheses of the Literature

The paper provides a detailed overview of different methods that have been proposed to discourse cold start issues in recommender systems, after reviewing 66 selected articles. The approaches are categorized into two main groups: data-oriented approaches and technique-oriented approaches. Data oriented approaches focus on utilizing various kinds of information and their connections considering both users and objects features to mitigate a fresh user and a fresh item issues. Researchers have proposed novel methods for addressing cold start problems by using information regarding the whereabouts of individuals, data from social networks, data indicating trustworthiness, and data from various fields.

Approach-driven strategies suggest solutions based on creating new algorithms, employing a blend of techniques, and utilizing various recommender system approaches. The group of strategy-based methods can be categorized into five groups: approaches based on deep learning, matrix factorization, hybrid recommender systems, innovative methods in collaborative filtering, and innovative methods in content-based recommender systems. These strategies present solutions by introducing novel algorithms, leveraging a blend of techniques, and implementing diverse approaches within the realm of recommender systems. These approaches provide innovative solutions, such as clustering, KNN, genetic algorithms, using semantic relationships, semantic tags, and deriving personality characteristics to alleviate cold start problems.

5.1 Data Based Methods

Researchers have proposed data-oriented strategies to Researchers have proposed data-oriented strategies to address the challenges posed by cold start problems in recommender systems. These approaches leverage different forms of data, including social network data, trust data, location data, and cross-domain data. Social network data has been used to improve the rating profiles of users who lack sufficient ratings and trust information. The author presented an approach that suggests collecting user profile data extracted from social media platforms, specifically focusing on social tagging activities, to enhance the system semantically [12]. Other researchers had utilized social network data, using social interests and social influence as a means of addressing cold start issues in recommender systems. For example, [13] proposed a random walk algorithm that considers various social network features to effectively tackle the problem of cold start users and enhance the precision of recommendations [14].

Proposed the "Reputation-based Trust-Aware Recommender System (RTARS)," This method leverages social information to assess user reputation and predict simulated ratings, by doing so, it addresses the challenges associated with cold start and data sparsity, while also enhancing the quality, reliability, and diversity of the recommender system.

To solve the problem of new users encountering cold starts, researchers have suggested the amalgamation of social network data with other data attributes as a recommended approach. In their work, the authors introduced a user similarity detection engine (USDE) as a proposed solution that extracts social information from personal smart devices to cluster users based on contextual and similarity information, subsequently, this user similarity detection engine (USDE) is utilized within the recommendation algorithm to deliver personalized recommendations [15–17]. Additionally, Ref. [18] utilized social media data to construct a behavioral profile, which can then be employed for user classification purposes. This classification was then used to make predictions for these users through the use of classification trees and random forests, aimed at addressing cold start problems.

To address cold start issues and get better performance, correctness of recommender systems, researchers have utilized cross-domain approaches. Introduced Cross-Domain Recommender Systems (CDRS), that create a user profile based on ratings and find a neighborhood by using cross-domain computation in cases where user ratings are unavailable [19–22]. The approach utilizes trust, influence, and reputation assessment to tackle the challenges of limited data and commencing from the beginning in the desired domain, leading to enhanced accuracy metrics with more precise and lower absolute error values. The authors of Ref. [23] introduced a technique called consistent information transfer (CIT), which aims to ensure that the knowledge consistency between user and item groups is maintained. The approach also transfers information and learning methods across domains to produce recommendations. The CIT-based collaborative filtering recommender system (CDRS) has proven to be useful in addressing cold start problems, as demonstrated in both research studies and real-world applications such as Smart Bizseeker. Reference [24] presented CrossRec, a recommendation system that operates across domains and relies on social big data from multiple sources. The system employs association rule mining based on social tags to produce recommendations. This method not only addresses cold start issues faced by users but also offers a wider range of recommendations compared to algorithms for collaborative filtering that leverage information fusion across different domains. Furthermore, CrossRec outperforms such algorithms in terms of effectiveness.

Introduced two new cross-domain recommendation models called PLSR-CrossRec and PLSR-Latent, they rely on regression analysis using partial least squares [25]. In cold start situations of the target domain, these models utilize ratings from the source domain to forecast ratings for new users. A solution to the desired domain's cold start problem was introduced in Refs. [26, 27] through the proposal of a CDLFM model. The model leverages the user rating behaviour from the source domain and conducts matrix factorization. This model takes into account the process of creating a mapping of features for users and their associated users in the neighborhood using boosted decision trees and neural networks with multiple layers. In addition, [28] suggested a clustering-based matrix factorization approach that utilizes data obtained from supplementary domains to improve recommendations when dealing with cold start situation. Finally, the model acquires useful priors and resolves cold start issues [29].

5.2 Strategies Based on Approach

Strategies derived from the approach entail employing novel techniques, algorithms, and their combinations to address the cold start problem.

5.2.1 Deep Learning Based Methods

The application of deep learning techniques in recommender systems has been leveraged to enhance precision and quality while tackling the challenges posed by the cold start problem. Reference [30] proposed SemRe-DCF, a recommender system that uses a denoising autoencoder based on deep learning to produce recommendations that take into account item similarities in detail, utilizing semantic descriptions of items and auxiliary user data. This approach is highly accurate and effective, successfully addressing the cold start problem. Furthermore, Ref. [31] introduced Deep-MINE, a comprehensive deep learning model that integrates data from various sources, including product images, descriptions, and reviews, to generate personalized recommendations. The model also features a cognition layer that considers the various cognitive styles of consumers to enhance the accuracy of recommendations, particularly when faced with the cold start problem.

Reference [32] proposed MSIR, presents a solution for tackling the cold start challenges in service recommendation. Reference [33] have proposed a recommender system for items based on a two-stage neural network architecture that uses a method for predicting ratings in sparse user-item matrices that involves a cold start item rating model based on denoising autoencoder and a collaborative filtering predictor based on neural networks. Reference [34] have proposed two recommendation models, SADE and time SVD++, using collaborative filtering and deep learning neural networks, in that order, to anticipate ratings in situations where there is little to no data available. Reference [35] have introduced the Meta-User2Vec model, which uses metadata embeddings and item representations to solve cold start problems, have proposed ML2E to generate desirable initial embeddings for new items and forecast new users' preferences. Reference [36] have introduced the ConTS model, the approach employs the Thompson sampling framework to create informal suggestions when dealing with new or unexplored scenarios. This approach is used to produce recommendations in situations where there is little to no data available [37].

5.2.2 Matrix Factorization

In order to improve the caliber of suggestions and address the issue of cold start users, matrix factorization techniques have been adopted by many researchers in recommender systems. These methods break down the user-item interaction matrix into smaller dimensions and utilize the resulting latent features to produce recommendations, which have been demonstrated to be more successful than rudimentary collaborative filtering methods [38]. Similarly, in Ref. [39], matrix factorization techniques of collaborative filtering were utilized, along with item-attribute and leveraging item similarity data to enhance recommendation quality and address cold start challenges, particularly in relation to new or unexplored items. A matrix factorization technique for cross-domain recommender systems, which factors in items while accounting for their varying content across different domains. The method employs Linked Open Data to extract items from three separate domains and form a knowledge graph, enabling the identification of item similarities across domains. This method improves both the precision and variety of recommendations, while also addressing cold start user problems [40].

A matrix factorization model was suggested that integrates item-side information to deal with the issue of cold start items. Their approach involves incorporating a two-stage soft-cluster embedding algorithm into the collaborative matrix factorization framework [41]. The authors also introduced a new evaluation metric called Cold Items Precision (CIP) to assess the performance of recommender systems in cold start situations. They compared their proposed algorithm with three deep learning-based algorithms and found that it outperformed them in terms of the CIP score.

Reference [42] Recommend a content-based approach that uses user tags and item keywords to generate a tag- keyword relation matrix. They use 3-factor matrix factorization to extract user interests and item correlations to make recommendations. The proposed solution involves using a weighted bipartite modularity index to divide users and items into groups, and then implementing localized low-rank matrix factorization to predict ratings for new users, effectively addressing the cold start problem in user-based collaborative filtering [43–46]. Improve matrix factorization by introducing a regularization weight on the norms of latent factors for both users and items [47, 48]. Propose a technique named local representative-based matrix factorization, which utilizes decision trees to generate clusters of users for producing recommendations to new users [49]. Reference [50] proposes mrf-MF, which integrates user attributes and latent features into a Markov random field prior to matrix factorization in order to create recommendations and resolve cold start difficulties.

5.2.3 Hybrid Recommender Systems

To address the issue of cold start problems, scholars have discovered that hybrid recommender systems that merge collaborative filtering and content-based techniques can be successful. Various techniques have been proposed to enhance the performance of these systems. One such technique is the Hybrid Features Selection Method (HFSM) by [51]. HFSM employs a clustering algorithm for content-based recommendations that merges statistical and semantic features of items. Following that, the clusters are processed by a regular collaborative filtering module as well as a hybrid module that combines a linear blend of content-based and collaborative filtering approaches. This method has been observed to enhance the efficiency of recommender systems and resolve the problems of cold start.

Reference [52] proposes a hybrid e-learning recommender system that generates recommendations using ontology and sequential pattern mining. Their technique comprises establishing ontology of learning resources, computing similarity scores between ontology and domain knowledge-based resources, producing top N learning items using collaborative filtering, and applying to obtain the ultimate suggestions, apply sequential pattern mining to the top N learnt items. Implementing sequential pattern mining, ontology, and domain knowledge approaches aids in overcoming cold start issues and boosting recommendation precision.

A novel movie recommender system was developed in Ref. [53] to effectively address the cold start difficulty associated with new items. The authors of [54] put forth a hybrid recommender system called "DISCOVER" that leverages social network investigation and contextual similarity analysis to suggest venues to scholars.

According to [55], the cold start problem for TED talks was addressed by combining collaborative filtering and content-based methods that employed item-based similarity. The algorithm's recommendations were assessed using techniques based on keywords and semantic vector spaces. The research findings indicated that the utilization of semantic vector space-based methods significantly enhanced the accuracy and quality of recommendations when dealing with cold start scenarios. They employed demographic information and cosine similarities between users to generate precise recommendations, effectively resolving the cold start user problem [56]. The authors of reference [57] developed an e-tourism travel package recommendation system that incorporates a hybrid approach.

5.2.4 Innovative Methods in Collaborative Filtering

Researchers have been working on improving collaborative techniques and introducing new techniques for tackling issues related to cold start. One such technique is Div-clustering, proposed by [58], which uses active users in social networks to enhance data clustering, improve recommendations, to overcome the issue of cold start, various methods have been proposed. Other methods include bipartite network representation; a method was developed to address the cold start problem for new items, along with a collaborative filtering approach that takes into account soft ratings and community preferences [59] developed a group recommender system that uses singular value decomposition to identify similar attributes among users and items. Their approach performs well in cold start situations by filling in missing attributes through clustering and similarity patterns. The authors also proposed a computationally scalable approach by introducing a back-fitting algorithm to alternating least squares. This method avoids the need for huge matrix operations or rigorous memory storage.

The authors of [60] have created a recommendation algorithm named BLI (GA), which employs both past rating data and semantic information to generate personalized recommendations. Reference [61] developed two item-based similarity measures, based on asymmetric correlation and relative interconnection, to take care of the cold start problem in recommendations. These measures are combined with an enhanced prediction algorithm, proposed a method that combines item attribute information and a historical rating matrix to predict cold start user preferences and generate recommendations using matrix decomposition. Reference [62] proposed a weighted cosine-based approach that considers contextual information of items' attributes to modify session-based recommender systems, a covering reduction collaborative filtering method, this method aims to minimize the number of redundant elements in covering-based rough sets and create decision classes for new users. The authors also suggested techniques for maximizing coverage, including Max-coverage and Category-exploration. Reference [63] suggested a strategy for recommending Points-of-Interest (POI) in cold start situations, which involves gathering POIs using crowd sourcing and suggesting the top-K POIs that include new and pre-existing POIs.

5.2.5 New Methods in Content-Based

In order to handle the issue of cold start, many researchers have integrated various item feature spaces, personality characteristics, and semantic relationships with traditional content-based filtering models.

Reference [64] have proposed an algorithm called ECSN, which combines social network information with a content-based approach to improve recommendation accuracy. When there is no interaction data available for a new user, the recommendation system generates suggestions by computing the preference score based on the interactive information of the user's social circle, which is a solution to the cold start problem.

Other researchers have used semantic information and tagging patterns to enhance user and item profiles and solve cold start problems. Reference [65] created an algorithm that sorts tagging patterns into relevant patterns based on user preferences to generate recommendations. They also suggested a semantic-based recommendation system that uses external knowledge bases like WordNet and Wikipedia to improve user-generated tag patterns to semantic-based profiles. These profiles are then augmented with a semantic spread mechanism to address cold start problems with greater precision and accuracy.

A model named Transitive Semantic Relationships (TSR) was created by [66] to tackle the issue of the cold start problem in big data, where there is a scarcity of labeled data. The TSR model utilizes text content from both users and items, and also incorporates a few labeled examples to establish relationships and combat data sparsity. This approach is especially useful in fields where comparable content implies comparable relationships among items.

6 Discussion and Suggesting Possible Areas for Future Investigation

Recommendation systems may experience a decline in accuracy due to problems with Data Sparsity and Cold Start. Data Sparsity arises when there are insufficient item ratings, while Cold Start arises when the system is unable to create profiles for new users or items. Cold Start is more difficult to handle because it can result in reduced user involvement. Cold Start issues tend to affect Collaborative Filtering systems more than Content-Based systems, but the latter typically have lower levels of accuracy. As a result, many researchers have proposed various methods and algorithms to tackle these issues. The progression of recommender systems from Web 1.0 to Web 3.0 has helped overcome the limitations of earlier systems. Web 2.0 systems rely more on social information and trust data to tailor recommendations, whereas Web 3.0 systems use hybrid ensemble algorithms that incorporate diverse data types such as location and real-time health information. Recent progress in computing, as well as machine and deep learning algorithms, has also contributed to tackling these difficulties.

This review article aims to provide an overview of the various techniques and algorithms proposed in previous studies to take care of the cold start issue in recommender systems. The review identifies two main types of strategies: data and approach-driven. Strategies employing Machine Learning and Deep Learning have been employed to handle the challenges posed by new users and new items. These strategies include

KNN algorithms, clustering, matrix factorization, profile expansion, trust relation derivation, and soft clustering.

Hybrid recommender systems that work across multiple domains have also been suggested, utilizing methods like information transfer and context awareness. These methods demonstrate superior performance in handling the challenges of cold start problems.

This study has identified and documented the techniques and algorithms employed by authors, standard algorithms meant for benchmarking, and evaluation metrics used for assessing performance. Standard collaborative filtering algorithms are frequently used as benchmark algorithms, while performance evaluation metrics like MAE, Precision, Recall, and RMSE are commonly employed to gauge algorithm effectiveness. Further investigation is necessary to establish which benchmark algorithms and metrics are suitable for evaluating algorithm performance in cold start scenarios.

The article has identified the areas of interest that researchers have focused on in relation to the cold start problem. This information could be useful for researchers working on the cold start problem in specific domains, as they can refer to previous research to inform their own work. Moreover, the study proposes that there is a shortage of research in domains related to travel, tourism, food, and hospitality, which creates an opportunity for researchers to propose new algorithms to enhance recommender system performance in these areas. Finally, the study suggests that researchers could explore the use of recommender.

7 Conclusions

In conclusion, this review paper provides a comprehensive compilation of different strategies and algorithms developed by researchers in the past 21 years (2002–2023) to overcome the challenges of new user and new item cold start problems in recommender systems. This study's primary contribution is its consolidation of 328 articles but only 66 journal articles focused on recommender systems and their proposed solutions to reduce cold start problems. This review is a valuable resource for future researchers and practitioners to explore existing research on cold start issues and identify algorithms and methods that can be employed to alleviate these challenges. However, the study acknowledges that it has limitations and did not conduct comparative or experimental analysis to determine the most effective strategies and their corresponding application contexts. This limitation indicates that there is a possibility for future research by scholars in the recommender system field to investigate the cold start domain more deeply.

References

1. Abdullah, N.A., Rasheed, R.A., Nasir, M.H.N.M., Rahman, M.M.: Eliciting auxiliary information for cold start user recommendation: a survey. Appl. Sci. **11**(20), 9608 (2021)
2. Abel, F., Herder, E., Houben, G.J., Henze, N., Krause, D.: Cross-system user modeling and personalization on the social web. User Model. User-Adap. Inter. **23**, 169–209 (2013)
3. Ahmadian, S., Afsharchi, M., Meghdadi, M.: An effective social recommendation method based on user reputation model and rating profile enhancement. J. Inf. Sci. **45**, 607–642 (2019)
4. Aksnes, D.W., Langfeldt, L., Wouters, P.: Citations, citation indicators, and research quality: an overview of basic concepts and theories. SAGE Open **9**(1), 2158244019829575 (2019)
5. Alhijawi, B., Kilani, Y.: Using genetic algorithms for measuring the similarity values between users in collaborative filtering recommender systems. In: IEEE/ACIS 15th International Conference on Computer and Information Science (ICIS), Okayama, Japan, pp. 1–6 (2016)
6. Alhijawi, B., Kilani, Y.: A collaborative filtering recommender system using genetic algorithm. Inf. Process. Manag. **57**, 102310 (2020)
7. Anwaar, F., Iltaf, N., Afzal, H., Nawaz, R.: A hybrid framework to integrate content embeddings in recommender systems for cold start items. J. Comput. Sci. **29**, 9–18 (2018)
8. Bahrani, P., Minaei-Bidgoli, B., Parvin, H., Mirzarezaee, M., Keshavarz, A., Alinejad-Rokny, H.: User and item profile expansion for dealing with cold start problem. J. Intell. Fuzzy Syst. **38**, 4471–4483 (2020)
9. Bi, X., Qu, A., Wang, J., Shen, X.: A group-specific recommender system. J. Am. Stat. Assoc. **112**, 1344–1353 (2017)
10. Bobadilla, J., Ortega, F., Hernando, A., Alcal: Improving collaborative filtering recommender system results and performance using genetic algorithms. Knowl.-Based Syst. **24**(8), 1310–1316 (2011)
11. Bobadilla, J., Ortega, F., Hernando, A., Gutiérrez, A.: Recommender systems survey. Knowl.-Based Syst. **46**, 109–132 (2013)
12. Cai, M., Gong, Z., Li, Y.: Fuzzy prototype classifier based on items and its application in recommender system. Int. J. Comput. Intell. Syst. **10**, 1016 (2017)
13. Chalyi, S., Leshchynskyi, V., Leshchynska, I.: Method of forming recommendations using temporal constraints in a situation of cyclic cold start of the recommender system. EUREKA Phys. Eng. **4**, 34–40 (2019)
14. Chen, C.C., Wan, Y.H., Chung, M.C., Sun, Y.C.: An effective recommendation method for cold start new users using trust and distrust networks. Inf. Sci. **224**, 19–36 (2013)
15. Chen, H.H., Chen, P.: Differentiating regularization weights - a simple mechanism to alleviate cold start in recommender systems. ACM Trans. Knowl. Discov. Data **13**, 1–22 (2019)
16. Chen, L., Pu, P.: Critiquing-based recommenders: survey and emerging trends. User Model. User-Adap. Inter. **22**, 125–150 (2012)
17. Choi, S.M., Jang, K., Lee, T.D., Khreishah, A., Noh, W.: Alleviating item-side cold start problems in recommender systems using weak supervision. IEEE Access **8**, 167747–167756 (2020)
18. Deldjoo, Y., Dacrema, M.F., Constantin, M.G., Eghbal-zadeh, H., Cereda, S., Schedl, M., Ionescu, B., Cremonesi, P.: Movie genome: alleviating new item cold start in movie recommendation. User Model. User-Adap. Inter. **29**, 291–343 (2019)
19. Singh, S., Malik, S., Nath, R.K.: Lane and curve detection using deep learning. In: 2021 3rd International Conference on Advances in Computing, Communication Control and Networking (ICAC3N), Greater Noida, India, pp. 331–333 (2021)
20. Ebesu, T., Fang, Y.: Neural semantic personalized ranking for item cold-start recommendation. Inform. Retr. J. **20**, 109–131 (2017)
21. Felfernig, A., Burke, R.: Constraint-based recommender systems: technologies and research issues. In: 10th International Conference on Electronic Commerce (ICEC) '08 Innsbruck. ACM Press, Austria (2008)
22. Feng, J., Xia, Z., Feng, X., Peng, J.: RBPR: a hybrid model for the new user cold start problem in recommender systems. Knowl.-Based Syst. **214**, 106732 (2021)

23. Ferdaous, H., Bouchra, F., Brahim, O., Imad-eddine, M., Asmaa, B.: Recommendation using a clustering algorithm based on a hybrid features selection method. J. Intell. Inform. Syst. **51**, 183–205 (2018)
24. Fernández, D., Formoso, V., Cacheda, F., Carneiro, V.: High order profile expansion to tackle the new user problem on recommender systems. PLoS ONE **14**, 1–16 (2019)
25. Fernández-Tobías, I., Cantador, I., Tomeo, P., Anelli, V.W., Noia, T.D.: Addressing the user cold start with cross-domain collaborative filtering: exploiting item metadata in matrix factorization. User Model. User-Adapt. Interact. **29**, 443–486 (2019)
26. Ghavipour, M., Meybodi, M.R.: Stochastic trust network enriched by similarity relations to enhance trust-aware recommendations. Appl. Intell. **49**, 435–448 (2019)
27. Guan, Y., Wei, Q., Chen, G.: Deep learning based personalized recommendation with multi-view information integration. Decis. Support. Syst. **118**, 58–69 (2019)
28. Guo, G., Zhang, J., Thalmann, D.: Merging trust in collaborative filtering to alleviate data sparsity and cold start. Knowl.-Based Syst. **57**, 57–68 (2014)
29. Guo, X., Yin, S.C., Zhang, Y.W., Li, W., He, Q.: Cold start recommendation based on attribute-fused singular value decomposition. IEEE Access **7**, 11349–11359 (2019)
30. Han, D., Li, J., Yang, L., Zeng, Z.: A recommender system to address the cold start problem for app usage prediction. Int. J. Mach. Learn. Cybern. **10**, 2257–2268 (2019)
31. Hasan, M., Roy, F.: An item–item collaborative filtering recommender system using trust and genre to address the cold-start problem. Big Data Cogn. Comput. **3**, 1–15 (2019)
32. Nath, R.K., Rasool, A., Gangwar, N., Jha, H.: Human activity recognition with video classification. In: 2022 4th International Conference on Advances in Computing, Communication Control and Networking (ICAC3N), Greater Noida, India, pp. 1010–1014 (2022)
33. Hernando, A., Bobadilla, J., Ortega, F., Gutiérrez, A.: A probabilistic model for recommending to new cold-start non-registered users. Inform. Sci. **376**, 216–232 (2017)
34. Hong, D.G., Lee, Y.C., Lee, J., Kim, S.W.: Crowdstart: warming up cold-start items using crowdsourcing. Expert Syst. Appl. **138**, 112813 (2019)
35. Hu, L., Cao, L., Cao, J., Gu, Z., Xu, G., Yang, D.: Learning informative priors from heterogeneous domains to improve recommendation in cold-start user domains. ACM Trans. Inform. Syst. **35**, 1–37 (2016)
36. Jeevamol, J., Renumol, V.G.: An ontology-based hybrid e-learning content recommender system for alleviating the cold-start problem. Educ. Inf. Technol. **26**, 4993–5022 (2021)
37. Ji, K., Shen, H.: Addressing cold-start: scalable recommendation with tags and keywords. Knowl.-Based Syst. **83**, 42–50 (2015)
38. Nath, R.K., Ahmad, T.: Content based recommender system: methodologies, performance, evaluation and application. In: 2022 4th International Conference on Advances in Computing, Communication Control and Networking (ICAC3N), Greater Noida, India, pp. 423–428 (2022)
39. Kim, H.N., El-Saddik, A., Jo, G.S.: Collaborative error-reflected models for cold-start recommender systems. Decis. Support. Syst. **51**, 519–531 (2011)
40. Kolahkaj, M., Harounabadi, A., Nikravanshalmani, A., Chinipardaz, R.: A hybrid context aware approach for e-tourism package recommendation based on asymmetric similarity measurement and sequential pattern mining. Electron. Commer. Res. Appl. **42** (2020)
41. Li, C.T., Hsu, C.T., Shan, M.K.: A cross-domain recommendation mechanism for cold-start users based on partial least squares regression. ACM Trans. Intell. Syst. Technol. **9**, 1–26 (2018)
42. Li, S., Lei, W., Wu, Q., He, X., Jiang, P., Chua, T.S.: Seamlessly unifying attributes and items: conversational recommendation for cold-start users. ACM Trans. Inform. Syst. **39**, 1–29 (2021)
43. Liu, H., Hu, Z., Mian, A., Tian, H., Zhu, X.: A new user similarity model to improve the accuracy of collaborative filtering. Knowl.-Based Syst. **56**, 156–166 (2014)
44. Liu, S., Meng, X.: A location-based business information recommendation algorithm. Math. Probl. Eng. **1**, 345480 (2015)
45. Ma, Y., Geng, X., Wang, J.: A deep neural network with multiplex interactions for cold-start service recommendation. IEEE Trans. Eng. Manag. **68**, 105–119 (2021)
46. Mansoury, M., Shajari, M.: Improving recommender systems' performance on cold-start users and controversial items by a new similarity model. Int. J. Web Inform. Syst. **12**, 126–149 (2016)

47. Masood, M.A., Abbasi, R.A., Maqbool, O., Mushtaq, M., Aljohani, N.R., Daud, A., Aslam, M.A., Alowibdi, J.S.: Mfs-lda: a multi-feature space tag recommendation model for cold start problem. Program **51**, 218–234 (2017)
48. Mazumdar, P., Patra, B.K., Babu, K.S.: Cold-start point-of-interest recommendation through crowd sourcing. ACM Trans. Web **14**, 1–36 (2020)
49. Mirbakhsh, N., Ling, C.X.: Improving top-n recommendation for cold-start users via cross-domain information. ACM Trans. Knowl. Discov. Data **9**, 1–19 (2015)
50. Misztal-Radecka, J., Indurkhya, B., Smywiński-Pohl, A.: Meta-ser2vec model for addressing the user and item cold-start problem in recommender systems. User Model. User-Adapt. Interact. **31**, 261–286 (2021)
51. Mobasher, B., Burke, R., Bhaumik, R., Williams, C.: Toward trustworthy recommender systems: an analysis of attack models and algorithm robustness. ACM Trans. Internet Technol. **7**, 23 (2007)
52. Movahedian, H., Khayyambashi, M.R.: Folksonomy-based user interest and disinterest profiling for improved recommendations: an ontological approach. J. Inf. Sci. **40**, 594–610 (2014)
53. Movahedian, H., Khayyambashi, M.R.: A semantic recommender system based on frequent tag pattern. Intell. Data Anal. **19**, 109–126 (2014)
54. Natarajan, S., Vairavasundaram, S., Natarajan, S., Gandomi, A.H.: Resolving data sparsity and cold start problem in collaborative filtering recommender system using linked open data. Expert Syst. Appl. **149** (2020)
55. Nguyen, P., Wang, J., Kalousis, A.: Factorizing lambdamart for cold start recommendations. Mach. Learn. **104**, 223–242 (2016)
56. Nguyen, V.D., Sriboonchitta, S., Huynh, V.N.: Using community preference for overcoming sparsity and cold-start problems in collaborative filtering system offering soft ratings. Electron. Commer. Res. Appl. **26**, 101–108 (2017)
57. Nie, D. C., Zhang, Z. K., Dong, Q., Sun, C., Fu, Y.: Information filtering via biased random walk on coupled social network. Sci. World J. (2014); Nouh, R.M., Lee, H.H., Lee, W.J., Lee, J.D.: A smart recommender based on hybrid learning methods for personal well-being services. Sensors, (Switzerland) **19**
58. Ojagh, S., Malek, M.R., Saeedi, S.: A social-aware recommender system based on user's personal smart devices. ISPRS Int. J. Geo-Inform. **9**, 519 (2020)
59. Okoli, C.: A guide to conducting a standalone systematic literature review. Commun. Assoc. Inf. Syst. **37**, 879–910 (2015)
60. Paleti, L., Krishna, P.R., Murthy, J.V.: Approaching the cold-start problem using community detection based alternating least square factorization in recommendation systems. Evol. Intel. **14**, 835–849 (2021)
61. Pan, R., Ge, C., Zhang, L., Zhao, W., Shao, X.: A new similarity model based on collaborative filtering for new user cold start recommendation. IEICE Trans. Inf. Syst. **E103D**, 1388–1394 (2020)
62. Pappas, N., Popescu-Belis, A.: Combining content with user preferences for non-fiction multimedia recommendation: a study on ted lectures. Multimed. Tools Appl. **74**, 1175–1197 (2015)
63. Peng, F., Lu, J., Wang, Y., Xu, R.Y.D., Ma, C., Yang, J.: N dimensional Markov random field prior for cold-start recommendation. Neurocomputing **191**, 187–199 (2016)
64. Pereira, A.L.V., Hruschka, E.R.: Simultaneous co-clustering and learning to address the cold start problem in recommender systems. Knowl.-Based Syst. **82**, 11–19 (2015)
65. Polohakul, J., Chuangsuwanich, E., Suchato, A., Punyabukkana, P.: Real estate recommendation approach for solving the item cold-start problem. IEEE Access **9**, 68139–68150 (2021); Pradhan, T., Pal, S.: A hybrid personalized scholarly venue recommender system integrating social network analysis and contextual similarity. Future Gener. Comput. Syst. **110**, 1139–1166 (2020)
66. Puthiya Parambath, S.A., Chawla, S.: Simple and effective neural-free soft-cluster embeddings for item cold-start recommendations. Data Min. Knowl. Discov. **34**, 1560–1588 (2020)

Malwares for Numerous Platforms: A Deep Learning

Vinisha Sumra, Naveen Malik, and Sandip Kumar Goyal

Abstract Nearly everyone relies on smart devices such as Android, iOS and Windows for daily activities. This dependence on such devices paves the way for the malware developers to achieve their ambitions. Next generation malwares are even more complicated than those of customary malwares. These days, researchers are coming forward with deep learning rooted malware detectors. This paper throws light on numerous platforms such as iOS, Android and Windows and their corresponding malware detector systems. Analysis of these detector systems proves that every platform needs a unique way to deal with the malwares targeted on that platform. The comparison will navigate the future researchers in an appropriate direction.

Keywords Malware · Detection · Android · Windows

1 Introduction

1.1 A Subsection Sample

In neoteric years, nearly everyone has been utilizing the Internet for routine tasks. Reason behind such large scale usage is that the Internet has spread its feet in almost all the activities such as marketing, health allied transactions, online banking and social interactions. Owing to such speedily growth of the Internet, criminals have kicked off perpetrating crimes on this platform instead of the real world. Delinquents are normally embedding malicious programs to set afloat cyber attacks to the sufferer machines. Any program or software which deliberately brings about malicious payloads on target machines including computer networks, smartphones and computers is contemplated as malware [1]. With the shoot up in usage of applications and network systems, numerous malwares are also raising their heads to attack the devices and networks. Malware is a program that coerces your scaffolding to

V. Sumra (✉) · N. Malik · S. K. Goyal
MM Deemed To Be University Mullana, Ambala 133203, India
e-mail: Vinishamalik317@gmail.com

B. Soni et al. (eds.), *Beyond Artificial Intelligence*, Lecture Notes in Networks and Systems 1326, https://doi.org/10.1007/978-981-96-4170-3_61

attain something that a mugger intends it to do [2]. Malware is an inconceivable most contemplative coercion for the Internet and Distributed Systems [3]. There are numerous types of malwares such as worm, virus, rootkit, trojan horse and ransomware. Different malware families are mapped out to take hold of the victim machine in numerous ways such as authorizing remote code execution, vandalizing the target system and pilfering the sensitive data etc. Customary malwares were delineated for unexacted purposes which makes their detection easier. This kind of malware is called Conventional or Customary malware. However, now a days, the malware that can be executed in kernel mode and is more cataclysmic and denser to detect than customary malware can be referred as Next-generation malware. This type of malware can easily detour safeguarding software that is executing in kernel form such as antivirus and firewalls etc. Normally, customary malware relies on one process and does not utilize knotty techniques to camouflage itself. On the contrary, next generation malware utilizes multitudinous different processes at once and embeds few muddled techniques to bury itself and become tenacious in the device. Next generation malware can set afloat more noxious attacks such as earmarked and indefatigable and more than one kind of malware is embedded in the attack [1]. The collation between Next Generation and Customary malware is presented in Table 1.

As per the business and scientific reports, nearly one million malicious programs are originated each single day. Owing to such huge numbers, cybercrime may harm the world economy by nearly six trillion dollars yearly [4].

Pivoted on the attributes, mobile malware can be cataloged under the following heads:

- Botnet
- Virus
- Spyware
- Rootkits
- Ghost push
- Worm
- Trojan.

A botnet may be contemplated as a station of stratagems that are adulterated with a virus that furnishes the assailant the facility to be remotely guided by them. Botnets

Table 1 Next generation and customary malware

Collation attribute	Next generation malware	Customary malware
Composition	Multitudinous processes	Generally single process
Seclusion techniques	Yes	No
Type of attack	Earmarked	General
Detection process	Difficult	Easier
Aimed devices	Numerous devices	General computers
Augmentation	Each replica is different	Each replica is akin

are pronounced for a diligent security pitfall on the Internet and these are expressly designed for codified malefaction performing attacks to grasp mazuma.

Virus is expressed as a program code that can engender dupes of it. Contrastive copies of a virus gaze at infecting other programs by tethering itself with the target file.

Spyware is the type of malware that carries off its intent by stockpiling, utilizing and mushrooming a utilizer's peculiar information without the utilizer's acquiescence. It is chiefly consigned into four heads: trojans, system monitors, tracking cookies and adware.

Rootkits confound malignant utilizer-space programs and files, immobilize anti-viruses and firewalls. They may also plonk Trojans [5].

Ghost push is a program that contaminates the Android OS (Operating System) by ineluctably seizing root ingress, downloading malicious program, transfiguring it to system application and then overlooking root access.

Trojan is a malware that appears to bestow a few valuable assignments but instead it billets a malicious file or program. Moreover a Trojan is always in need of user reciprocation to be triggered off [6].

This paper throws light on the investigation of numerous malware detection techniques intended for mobile, android and iOS etc. The rest of the paper is structured as follows: Sect. 2 revolves around analysis of mobile malware detection techniques whereas Sect. 3 comprises deep insight into windows malware detection. Section 4 concludes the paper.

2 Mobile Malware Detection

Utilization of mobile devices has escalated in everybody's daily tasks. Ticket booking, bank transactions, social networking, online gaming, online learning and many more on the list are entangled with mobile devices such as tablets, smartphones and laptops. As per the statistical data, the economy hinged on smartphones is more advanced than PCs [7]. Assailants keenly pick out mobile devices in comparison to PCs because mobile devices accommodate more privileged information of utilizers [8–11]. Malware detection for Android and iOS is further elaborated as follows.

2.1 Android Garrisoned Malware Detection

The Android platform has been the most menacing since 2019 [12–14]. Trojans inhabit a prime part in the dispersal of android hinged malwares in 2019, squarely 93.93%. "Hiddad" named android trojan is also occupying a seat in the top 10 android trojans. It baffles utilizers and debilitates security software while remaining camouflaged in Google play store applications [15]. Static analysis and detection techniques focus on static features such as intents and permissions whereas dynamic detection techniques aim at utilization of network traffic features and system calls. Hybrid detection techniques amalgamate the static and dynamic features to attain better accuracy [16, 17].

AndroDialysis is a grading model hinged on the intents of android device. It focuses on the execution time binding message system which comprises the processes of application quarried from the malicious files or programs. Performance is evaluated utilizing K-fold cross validation by Bayesian network [18]. This model utilized permissions and attained an enhanced detection ratio [19]. Static analysis techniques do not aspire to contaminate a device and pose no standard manifesto for assemblage of information during execution in a confined environment. Nevertheless, code obfuscation methods cannot be circumvented by static techniques [20–24].

FlowDroid [25] involves execution of Inter Component Communication (ICC) analysis over and above the configuration programs and code of the testing app. This model dwindles false positives in the result. By amalgamation of features from static and dynamic analysis, advanced methods for classification of malwares are persuaded [26]. Machine Learning techniques are embedded for stratification goals in which a feature set is retrieved by utilizing both static and dynamic analysis for amicable malicious and normal applications [27].

BRIDEMAID framework exploits dynamic analysis to detect malign behavior during execution, in which numerous phases of surveillance are endured on applications, devices and behaviors of utilizers [28]. Andromaly framework is considerate about dynamic features for stratification of malwares [29]. Numerous attributes are surveiled such as packet count through the network, utilization of the Central Processor Unit, battery drainage and the number of live processes. Multiple deep learning techniques also contribute to the malware detection [30].

Sandbox system alloys static and dynamic techniques [31]. Execution is performed in sandpile environment and perilous activities such as remote server association and file open processes are probed. An augmentation of the Android OS is put forward as TaintDroid [32]. Third party malicious applications are efficient in digging up the confidential information. Aside from, surveillance is put forward for a real time application environment. Few of the android garrisoned malware assessment systems are presented in Table 2.

2.2 *iOS Based Malware Detection*

Mac OS is generally safer than windows and android [13]. Above 100 malware samples were programmed to splinter MacOS in 2019. Owing to no malware systems, it is climacteric to focus on Mac OS. Sensitive information larceny is carried out by more perceptible malwares in iOS devices. As per the report presented by Shannon, a spyware named watering hole was delineated in 2019. It possesses an aptness to clasp sensitive information such as GPS location and iMessage pictures [39]. Table 3 throws light on comparison of few iOS malware detection systems as follows.

Table 2 Android hinged malware assessment systems

Research paper	System/Model name	Approach
Zhou et al. [33]	DroidRanger	Behavior and heuristic grounded approach
Lindofer et al. [34]	Andrubis	Hybrid Analysis approach over Dalvik VM
Zheng et al. [35]	SmartDroid	Hybrid analysis approach to unmask UI triggers of android applications
Hei et al. [36]	HAWK	Amalgamation of android relationships and entities among behavior
Zhu et al. [37]	SEDMDroid	Bootstrapping and principal component analysis
Zhou et al. [33]	DroidRanger	Behavior and heuristic grounded approach
Gao et al. [38]	GDroid	Graph convolutional network to segregate the android malwares

Table 3 Collation among few iOS malware detection systems

Research paper	Approach for categorization	Attributes	Malware type
Lindofer et al. [40]	iHoneypot	Mac OSX	Mac OS grounded malware
Cimitile et al. [41]	oneR approach	Opcode Histogram	Mac OS
Bojjagani et al. [42]	VAPTAi	Mobile banking	Mac OS and Android
Husainiamer et al. [43]	Phylogenetic concept	Mobile susceptibility	Mac OS

3 Windows Malware Detection Systems

Above 600 official cases were jotted down in one year regarding the menace in the soundness of the system [13]. Targets for assailable attacks reached to 357 in number. Windows Server 2019 and Windows Server 2016 witnessed prominent security dereliction. During the first quarter of 2020, Windows 7 and Windows 10 occupied the utilization segment of 30 and 51.68% respectively [13, 15].

Cybercriminals are putting Windows Operating System susceptibilities on the peak of available targets. Numerous detection techniques are administered for Windows malware. Table 4 puts forward a few of Windows Malware Detection mechanisms.

Table 4 Windows hinged malware detectors

Research paper	Segregation approach	Malware type	Attributes
Shamsul et al. [44]	Hybrid wrapper and SVM wrapper approach	Windows	Way of behaving during execution environment
Irshad et al. [45]	Random forest, naive Bayes and support vector machine classifiers	Windows	Cuckoo named malicious file analysis tool
Gandeva et al. [46]	Hybrid analysis approach	Windows	Malware actions
Yang et al. [47]	Event tracing for windows	Windows	RAT Logging
Bander et al. [48]	Fine behavioral detection for malwares	Windows	Directory of contingent files
Nir et al. [49]	Support vector machine	Windows	Vigorous learning

4 Conclusion

Being dynamic instead of being susceptible takes us one step ahead than the malware threats. These days, evolving malware detection models are stretching, specifically when merchandising with the next generation malware. Newfangled circumvention techniques have sanctioned the progression of next generation malware. However, malware detection systems hinged on deep learning diminish the flaws of customary methods. This paper dispenses systematized analysis of malware detection utilizing deep learning methods. Present-day techniques to detect malware on iOS, Android, and Windows are reconnoitered and contrasted. This study will maneuver the researchers in the pertinent path to design the diminution techniques for both customary and complicated malwares.

References

1. Aslan, O., Samet, R.: A comprehensive review on malware detection approaches. IEEE Access **8** (2020)
2. Bhattacharya, A., Goswami, R.T.: DMDAM: Data mining based detection of android malware. In: Manadal, J.K., Satapathy, S.C., Sanyal, M.K., Bhateja, V. (eds.) Proceedings of the First International Conference on Intelligent Computing and Communication, pp. 187–194. Springer, Singapore (2017)
3. Souri, A., Asghari, P.P., Razaei, R.: Software as a service based CRM providers in the cloud computing: challenges and technical issues. J. Serv. Sci. Res. 219–237. JSSR (2017)
4. Morgan, S.: 2019 cybersecurity almanac: 100 facts, figures, predictions and statistics. Cisco and Cybersecurity Ventures. https://cybersecurityventures.com/cybersecurity-almanac-2019
5. La Polla, M., Martinelli, F., Sgandurra, D.: A survey on security for mobile devices. IEEE Commun. Surv. & Tutor. **15**(1), 446–471 (2013)
6. Malik, V., Malik, N., Goyal, S.K.: Analysis of android malwares and their detection techniques. In: Fourth International Conference on Parallel, Distributed and Grid Computing (PDGC). IEEE (2016)

7. Digitaltrends. https://www.digitaltrends.com/android/smartphone-sales-exceed-those-of-pcs-for-first-time-applesmashes-record/
8. Fasano, F., Martinelli, F., Mercaldo, F., Santone, A.: Energy consumption metrics for mobile device dynamic malware detection. Procedia Comput. Sci. **159**, 1045–1052 (2019)
9. Ciobanu, M.G., Fasano, F., Martinelli, F., Mercaldo, F., Santone, A.: A data life cycle modeling proposal by means of formal methods. In: Proceedings of the 2019 ACM Asia Conference on Computer and Communications Security, pp. 670–672. ACM (2019)
10. Martinelli, F., Mercaldo, F., Santone, A.: Social network polluting contents detection through deep learning techniques. In: 2019 International Joint Conference on Neural Networks, IJCNN, pp. 1–10. IEEE (2019)
11. Gandotra, E., Bansal, D., Sofat, S.: Malware analysis and classification: a survey. J. Inform. Secur. **5**, 56–64 (2014)
12. Garg, S., Baliyan, N.: Comparative analysis of android and iOS from security viewpoint. Comput. Sci. Rev. **40**(2021), 1574–0137 (2021)
13. Av-test. https://www.av-test.org/fileadmin/pdf/security_report/AV-TEST_Security_Report_2019-2020.pdf
14. Razgallah, A., Raphael, K., Sylvain, H., Kobra, K.: A survey of malware detection in Android apps: recommendations and perspectives for future research. Comput. Sci. Rev. **39**, 1574–0137 (2021)
15. Gopinath, M., Sethuraman, S.C.: A comprehensive survey on deep learning based malware detection techniques. Comput. Sci. Rev. **47** (2023)
16. Malik, V., Goyal, S.K., Malik, N.: Android malware detector based on hybrid features. In: GCAIA Conference, Jaipur (2021)
17. Malik, V., Goyal, S.K., Malik, N.: A hybrid model for android malware detection. Int. J. Innov. Technol. Explor. Eng. (IJITEE) **8**(12), IJITEE (2019). ISSN: 2278–3075
18. Feizollah, A., Anuar, N.B., Salleh, R., Suarez-Tangil, G., Furnell, S.: Andro-dialysis: analysis of android intent effectiveness in malware detection. In: Comput. Secur. **65**, 121–134 (2017)
19. Canfora, G., Mercaldo, F., Visaggio, C.A.: A classifier of malicious android applications. In: Proceedings of the 2nd International Workshop on Security of Mobile Applications, in Conjunction with the International Conference on Availability, Reliability and Security (2013)
20. Mercaldo, F., Nardone, V., Santone, A., Visaggio, C.A.: Ransomware steals your phone. Formal methods rescue it. In: International Conference on Formal Techniques for Distributed Objects, Components, and Systems, pp. 212–221, Springer (2016)
21. Cimitile, A., Mercaldo, F., Nardone, Santone, A., Visaggio, C.A.: Talos: no more ransomware victims with formal methods. Int. J. Inform. Secur. **17**, 1–20 (2017)
22. Mercaldo, F., Nardone, V., Santone, A., Visaggio, C.A.: Hey malware, I can find you! In: 2016 IEEE 25th International Conference on Enabling Technologies: Infrastructure for Collaborative Enterprises, WETICE, pp. 261–262. IEEE (2016)
23. Canfora, G., Sorbo, A.D., Mercaldo, F., Visaggio, C.A.: Obfuscation techniques against signature-based detection: a case study. In: 2015 Mobile Systems Technologies Workshop, MST, pp. 21–26. IEEE (2015)
24. Martinelli, F., Mercaldo, F., Nardone, V., Santonre, A., Sangaiah, A.K., Cimitile, A.: Evaluating model checking for cyber threats code obfuscation identification. J. Parallel Distrib. Comput. **119**, 203–218 (2018)
25. Arzt, S. Rasthofer, S., Fritz, C., Bodden, E., Bartel, A., Klein, J., Traon, Y.L., Octeau, D., McDaniel, P.: Flowdroid: precise context, flow, field, object-sensitive, and lifecycle-aware taint analysis for android apps. ACM SIGPLAN NOT **49**(6), 259–269 (2014)
26. Lindofer, M., Neugschwandtner, M., Platzer, C.: Marvin: efficient and comprehensive mobile app classification through static and dynamic analysis. In: 2015 IEEE 39th Annual Computer Software and Applications Conference (COMPSAC), vol. 2, pp. 422–433, IEEE (2015)
27. Faiella, M., LaMarra, A., Martinelli, F., Mercaldo, F., Saracino, F., Sheikhalishahi, M.: A distributed framework for collaborative and dynamic analysis of android malware. In: 2027 25th Euromicro International Conference on Parallel, Distributed and Network-Based Processing, PDP, pp. 321–328. IEEE (2017)

28. Martinelli, F., Mercaldo, F., Saracino, A.: Bridemaid: an hybrid tool for accurate detection of android malware. In: Proceedings of the 2017 ACM on Asia Conference on Computer and Communications Security, pp. 899–901. ACM (2017)
29. Shabtai, A., Kanonov, U., Elovici, Y., Glezer, C., Weiss, Y.: Andromaly: a behavioral malware detection framework for android devices. J. Intell. Inform. Syst. **38**(1), 161–190 (2012).
30. Malik, V., Malik, N., Kumar, S.: An inclusive analysis on deep learning hinged malware detection techniques. In: 5th International Conference Artificial Intelligence and Speech Technology, AIST 2023, Communications in Computer and Information Science, vol. 2268, pp. 417–425. Springer (2024)
31. Blasing, T., Schmidt, A.D., Batyuk, L., Camtepe, S.A., Albayrak, S.: An android application sandbox system for suspicious software detection. In: Proceedings of the 5th International Conference on Malicious and Unwanted Software (2010)
32. Enck, W., Gilbert, P., Han, S., Tendulkar, V., Chun, B.G., Cox, L.P., Jung, J., McDaniel, P., Sheth, A.N.: Taintdroid: an information-flow tracking system for realtime privacy monitoring on smartphones. ACM Trans. Comput. Syst. (TOCS) **32**(2) (2014)
33. Zhou, Y. Wang, Z., Zhou, W., Jiang, X.: Hey, you, get off of my market: detecting malicious apps in official and alternative android markets. In: Proceedings of the Network and Distributed System Security Symposium, NDSS (2012)
34. Lindofer, M., Neugschwandtner, M., Weichselbaum, L., Fratantonio, Y., Veen, V.V., Platzer, C.: Andrubis - 1,000,000 apps later: a view on current android malware behaviors. In: Proceedings of the 3rd International Workshop on Building Analysis Datasets and Gathering Experience Returns for Security, BADGERS (2014)
35. Zheng, C., Zhu, S., Dai, S., Gu, G., Gong, X., Han, X., Zou, W.: Smartdroid: an automatic system for revealing UI-based trigger conditions in android applications. In: Proceedings of the 2nd ACMWorkshop on Security and Privacy in Smartphones and Mobile Devices, SPSM, pp. 93–104. New York, NY, USA (2012)
36. Hei, Y., Yang, R., Peng, H., Wang, L., Xu, X., Liu, J., Liu, H., Xu, J., Sun, L.: HAWK: rapid android malware detection through heterogeneous graph attention networks. IEEE Trans. Neural Netw. Learn. Syst. (2021)
37. Zhu, H., Li, Y., Li, R., Li, J., You, Z., Song, H.: SEDMDroid: an enhanced stacking ensemble framework for android malware detection. IEEE Trans. Netw. Sci. Eng. **8**(2), 984–994 (2021).
38. Han, G., Shaoyin, C., Weiming, Z.: GDroid: android malware detection and classification with graph convolutional network. Comput. Secur. **106**, 0167–4048 (2021)
39. Shannon, W.: Mobile malware and exploitation amongst biggest cyber threats for 2020. In: Security Brief Asia (2020). https://securitybrief.asia/story/mobile-malware-and-exploitation-amongst-biggest-cyber-threats-for-2020
40. Lindofer, M., Miller, B., Neugschwandtner, M., Platzer, C.: Take a bite-finding the worm in the apple. In: 2013, 9th International Conference on Information, Communications, and Signal Processing, pp. 1–5, ICICS. IEEE (2013)
41. Cimitile, A,. Martinelli, F., Mercaldo, F.: Machine learning meets iOS malware: Identifying malicious applications on apple environment. In: ICISSP (2017), pp. 487–492 (2017)
42. Bojjagani, S., Sastry, V.N.: VAPTAi: A threat model for vulnerability assessment and penetration testing of android and iOS mobile banking apps. In: 2017 IEEE 3rd International Conference on Collaboration and Internet Computing, CIC (2017), pp. 77–86 CIC (2017)
43. Husainiamer, M., Saudi, M.M., Ahmad, A.: Classification for iOS mobile malware inspired by phylogenetic: Proof of concept. In: 2020 IEEE Conference on Open Systems, pp. 59–63. ICOS (2020)
44. Huda, S., Islam, R., Abawajy, J., Yearwood, J., Hassan, M.M., Fortino, G.: A hybrid-multi filter-wrapper framework to identify run-time behaviour for fast malware detection. Future Gener. Comput. Syst. **83**, 193–207 (2018)
45. Irshad, A., Maurya, R., Dutta, M.K., Burget, R., Uher, V.: Optimization for run time analysis of malware in windows operating system using machine learning approach. In: 2019 42nd International Conference on Telecommunications and Signal processing, pp. 255–260. TSP, Budapest, Hungary (2019)

46. Satrya, G.B., Cahyani, N.D.W., Andreta, R.F.: The detection 8 type malware botnet using hybrid malware analysis in executable file windows operating systems. In: Proceedings of the 17th International Conference on Electronic Commerce, 2015, ACM (2015)
47. Yang, R., Chen, X., Xu, H., Cheng, Y., Xiong, C., Ruan, L., Kavousi, M., Li, Z., Xu, L., Chen, Y.: RATScope: recording and reconstructing missing rat semantic behaviors for forensic analysis on windows. IEEE Trans. Depend. Secure Comput. (2020)
48. Alsulami, B., Srinivasan, A., Dong, H., Mancoridis, S.: Lightweight behavioral malware detection for windows platforms. In: 2017 12th International Conference on Malicious and Unwanted Software, MALWARE, pp. 75–81. IEEE (2017)
49. Nissim, N., Moskovitch, R., Rokach, L., Elovici, Y.: Novel active learning methods for enhanced PC malware detection in windows OS. Expert Syst. Appl. **41**(13), 5843–5857 (2014)

Pushing the Boundaries of Mortality Prediction: Advancing High-Risk Sepsis-III Patient Care Through Cutting-Edge Deep Learning Techniques

Deepak Mahto, Sunil K. Singh, Sudhakar Kumar, Manya, Rashmi, Varsha Arya, Kwok Tai Chui, and Brij B. Gupta

Abstract Sepsis-III is a widely accepted medical framework introduced in 2016 to precisely define sepsis. It employs a clinical approach that emphasizes organ dysfunction resulting from an exaggerated response to infection. When organ dysfunction is severe and co-occurs with an elevated risk of mortality, it may be classified as septic shock. High-risk sepsis-III patients pose a challenging problem in accurately predicting mortality, necessitating the development of precise tools for timely intervention. This paper performs a comprehensive comparative analysis of machine learning techniques, including Logistic Regression, XGBoost Classifier, LGBM Classifier, and Deep Learning techniques like Artificial Neural Networks, primarily aimed at predicting mortality in sepsis-III patients. Leveraging the "Refined MIMIC-III 30-day mortality prediction of sepsis-3 patients" dataset from IEEE Dataport, the study systematically evaluates these models, with an emphasis on the deep learning model, Artificial Neural Network. Key metrics, such as accuracy and AUC Score, are utilized for a comprehensive assessment for the identification of the most effective model for mortality prediction. The LGBM classifier model gives the best predictive results providing accuracy 0.9 and AUC score 0.76. The significant contributions involve enhanced mortality prediction in sepsis-III patients, offering a comprehensive comparison of machine learning techniques. The work systematically evaluates models using key metrics, including accuracy and AUC score, to identify optimal predictive performance. The future refinement of study focuses on the ANN-based methods by combining PCA with hyperparameter tuning. This study provides a feasible approach for the betterment of sepsis healthcare systems.

D. Mahto · S. K. Singh · S. Kumar (✉) · Manya · Rashmi
Chandigarh College of Engineering and Technology, Chandigarh, India
e-mail: sudhakar@ccet.ac.in

V. Arya · K. T. Chui
Hong Kong Metropolitan University, Hong Kong, SAR, China

B. B. Gupta
Department of Computer Science and Information Engineering, Asia University, Taichung, Taiwan

University of Economics and Human Science, Warsaw, Poland

B. Soni et al. (eds.), *Beyond Artificial Intelligence*, Lecture Notes in Networks and Systems 1326, https://doi.org/10.1007/978-981-96-4170-3_62

Keywords Sepsis-III · Machine learning · Logistic regression · XGBoost · LGBM classifier · Artificial neural network

1 Introduction

Sepsis is a condition characterized by a combination of physiological, pathological, and biochemical abnormalities triggered by an infection. The documented prevalence of sepsis is on the rise, likely attributable to aging populations with a higher burden of comorbid conditions. According to a recent scientific publication, it was approximated that in 2017, there were 49 million reported cases of sepsis, resulting in 11 million sepsis-related fatalities on a global scale. This alarming statistic amounted to nearly 20% of all recorded deaths worldwide. Sepsis exhibits a disproportionate impact on vulnerable demographic groups, including newborns, expectant mothers, and people residing in resource-limited environments. An estimated 85.0% of all sepsis cases and associated fatalities are concentrated within these specific segments of the population. Half of the annual 49 million sepsis cases are observed in pediatric patients, leading to 2.9 million fatalities, a significant proportion of which can be averted through timely diagnosis and proper clinical intervention [1]. On a global scale, it is estimated that among every 1000 women undergoing childbirth, 11 women encounter infection-related severe organ dysfunction or fatal outcomes.

Sepsis-III is defined as a life-threatening condition that results in severe dysfunction of organs due to an imbalanced host reaction to infection. According to WHO, Sepsis is termed a critical medical condition where the body's immune system exhibits an excessive response to an infection. This highly immune reaction leads to significant damage to the body's tissues and vital organs. Bacterial infections are the primary cause but other infectious agents such as viruses, parasites, and fungi, can also be responsible for this condition [1]. Sepsis is differentiated from a typical infection by the presence of dysregulated host response, accompanied by organ dysfunction [2]. Managing sepsis necessitates a comprehensive medical approach, encompassing the administration of antimicrobial agents, intravenous fluids, and vigilant monitoring to ensure timely and effective treatment [3, 4].

The work focuses on a clinical prediction model to find the predictive outcomes for refinement in healthcare systems and better preventive measures against sepsis. The various machine learning algorithms like XGBoost classifier, LGBM classifier Logistic regression, Linear SVC, and artificial neural network are used and the accuracies of these models are compared to find out the best resulting model. The significant contributions of work are:

(i) Enhanced Mortality Prediction: This study contributes to the field by addressing the challenging problem of accurately predicting mortality in high-risk sepsis-III patients.

(ii) Comprehensive Comparative Analysis: This research provides a thorough comparative analysis of a wide range of machine learning and deep learning techniques [5].
(iii) Utilization of Key Metrics: This work employs essential evaluation metrics, including accuracy and AUC Score, to systematically identify the most effective model for mortality prediction.

Section 2 of this paper is about Literature review which includes previous studies on Sepsis-III patients' mortality prediction. The experimental setup and methods are described in Sect. 3, and it expands on the dataset, and methodology. Section 4 highlights the results found and comparative analysis, whereas Sect. 5 presents the limitations of the results obtained and future scope of the models. Section 6 presents the conclusion of the study.

2 Literature Review

It is imperative to conduct a comprehensive survey of existing research endeavors within this domain to develop effective machine learning [6–9] and deep learning models [10, 11]. A survey on deep learning techniques can found in [12–15]. Deep learning has indeed gained significant traction in the healthcare industry [16–20] in recent years. Dealing with imbalanced data in classification [21, 22] tasks is a common challenge in machine learning. Deep learning methods can be effective for addressing this issue. This section of the paper summarizes numerous earlier studies on Sepsis-III and mortality prediction of patients diagnosed with Sepsis-III.

In their 2020 study, Nianzong Hou and colleagues created a machine learning approach using XGBoost to predict 30-day mortality in sepsis-3 patients with the MIMIC-III dataset [23, 24]. They assessed its performance against traditional models like logistic regression and SAPS-II to determine its effectiveness. The limitation of this study is that no additional investigation into the database was conducted, potentially resulting in the oversight of critical variables.

In 2023, Jin Wang and colleagues investigated the link between myocardial enzymes and one-year mortality. They created a predictive nomogram that combined myocardial enzymes and clinical features, surpassing LDH, SOFA, and SAPS II scores in forecasting one-year mortality [25]. LDH levels were found to be related to one-year mortality. The paper is limited to providing generalizability as it excludes people with missing data on myocardial enzymes. By including a larger and more diverse group of people, the potential impact of bias can be reduced. Jiyoung Shin et al. (2021) in their research included a diverse range of predictive variables, encompassing demographic attributes and a comprehensive set of physiological indicators derived from electronic health records (EHR). These variables are employed to forecast mortality among patients diagnosed with sepsis-III [26, 27]. To address the absence of specific physiological variables, the study employs the Multivariate Imputation by Chained Equations (MICE) method. It's noteworthy that the MICE method assumes a missing-at-random pattern, which may not precisely mirror the actual physiological values observed in clinical practice. This assumption represents one of the study's limitations.

Yong Li et al. (2023) in their studies introduced a new deep learning model, DGFSD, based on refined core indicators from clinical electronic medical records, offering more accurate mortality risk assessment for sepsis patients, outperforming traditional methods and improving clinical diagnosis efficiency [28, 29]. The limitation is that data only contains (EMRs) of patients in the US but does not have EMRs from other countries.

In 2021, Inger Persson and her team created an effective machine learning algorithm for sepsis prediction. They harnessed convolutional neural networks to forecast sepsis onset in adults (age 18+) based on 20 variables [30]. The study lacks prospective validation, faces challenges related to bias and external validation, and does not assess the clinical or economic impact of algorithm integration.

Jau-Woei Perng et al. (2019) focused on predicting mortality among suspected infected patients in emergency departments. Deep learning algorithms were compared with other methods and sepsis screening tools, and the Convolutional Neural Network plus SoftMax demonstrated the highest accuracy [31, 32]. The limitation is slower processing time, which can lead to delays for doctors and patients in obtaining accurate results.

In 2020, Yao and colleagues aimed to create a mathematical model for forecasting in-hospital mortality in postoperative sepsis patients. They achieved improved performance using the XGBoost model over stepwise logistic regression [33]. A limitation of this study is its single-center, retrospective design employing the XGBoost model, which may raise concerns about overfitting and constrain the assessment of vital variables.

Hanyin Wang et al. revealed significant performance disparities when applying the predictive model to different demographic subgroups, particularly among Asian, Hispanic, and Spanish-speaking patients. These findings underscore the impact of social determinants on sepsis diagnosis and prediction, highlighting the necessity for adaptable diagnostic systems to accommodate diverse patient populations and ensure accurate assessments [34, 35].

A summary of all literature work is presented in Table 1.

3 Experimental Setup and Methods

3.1 Dataset

For research, the "Refined MIMIC-III 30-day mortality prediction of Sepsis-III patients" dataset [36], sourced from IEEE Dataport, served as the primary data source. This dataset contains data from 4240 patients who were suffering from Sepsis and admitted to ICUs [37]. The dataset contains a wide range of data,

Table 1 Comparative analysis of various works done in the field of mortality prediction in sepsis patients

S. No.	Author	Year	Technique used
1	Hou et al. [23]	2020	Logistic regression model (AUC 0.819), SAPS-II score prediction model (AUC 0.797), and XGBoost model (AUC 0.857)
2	Wang et al. [25]	2023	Univariate logistic regression and multivariate logistic regression named LDH model (AUC for training set 0.773 and testing set 0.750)
3	Shin et al. [26]	2021	L1- and L2-regularized logistic regression (LR), random forest (RF), L1- and L2-regularized support vector machine, XGboost and Multi-Layer Prediction and CNN
4	Yao et al. [28]	2023	DGFSD model with an accuracy of 80.73%
5	Persson et al. [30]	2021	Convolutional Neural Network with AUROC 0.90 and AURPC 0.62
6	Perng et al. [31]	2020	CNN with softmax with an accuracy of 89.02% within 72 h and 81.79% within 28 days
7	Yao et al. [28]	2012	XGBoost (AUPRC 0.418) and stepwise logistic regression (AUPRC 0.280)

including the patient's laboratory results, vital signs, fluid balance, etc.[38]. The dataset comprises various columns, the first three providing miscellaneous information. The column thirtyday_expire_flag, represents the outcome of interest, while all remaining columns are feature columns [23]. Furthermore, this dataset also encompasses continuous physiological data obtained from bedside monitors. This offers valuable perspectives on the treatment of critically ill patients and can be harnessed to discern trends and patterns in patient care [39]. The feature description of the dataset used is presented in Table 2 and plotted in Fig. 1 as per these features and their mortality prediction percentage.

3.2 Different ML Models Used

In this study, the models used are Logistic regression, XGBoost model, LGBM model, Linear SVC model, and artificial neural network. All these models have their own specific accuracies and AUC values [40]. The primary aim is mortality prediction of sepsis-3 patients using features like sofa, lactate, respiration rate, any chronic disease like diabetes, and many other health factors.

Logistic regression, a fundamental machine learning technique within the realm of supervised learning, serves as a pivotal tool in predicting the mortality of sepsis patients [25]. It leverages a formula that models the probability of a categorical

Table 2 Feature description of the dataset used

S. No.	Feature	Datatype	Description
1	Sofa	Int64	The Sequential Organ Failure Assessment (SOFA) is a scoring system for assessing and monitoring organ dysfunction's severity
2	lods	Int64	Logistic Organ Dysfunction System, is another scoring system particularly useful for evaluating the risk of mortality in these patients
3	Lactate_ mean	Float64	Monitoring changes in lactate levels over time, including calculating the mean or average lactate level, can provide valuable information about the progression of sepsis and the patient's response to treatment
4	Lactate_ min	Float64	Minimum or lowest recorded level of lactate in a patient's blood
5	Lactate_ max	Float64	Elevated lactate levels can be indicative of inadequate oxygen delivery to cells
6	Aniongap_ min	Float64	The Lowest recorded anion gap helps assess the severity of metabolic disturbances and changes in acid–base balance
7	Aniongap_ max	Float64	An elevated anion gap may signal more significant metabolic acidosis and can be an indicator of a more severe condition

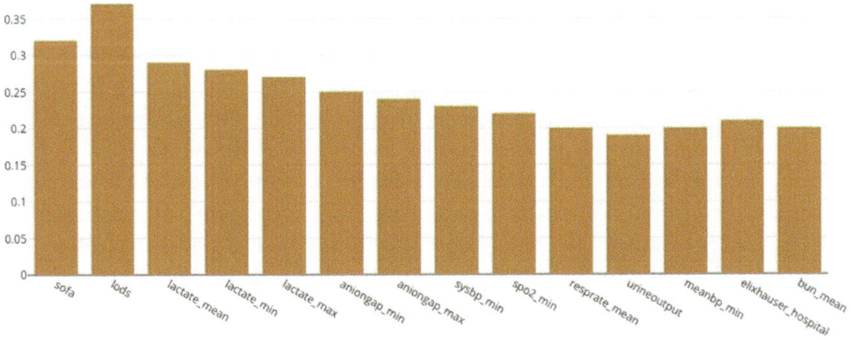

Fig. 1 Mortality prediction is highly dependent on the top fifteen features

outcome based on independent variables, with a specific application to the prognosis of sepsis patients.

In the research, XGBoost, a meticulously optimized gradient boosting framework is used. This approach, deeply rooted in the principles of ensemble learning, empowers us to significantly enhance our predictive capabilities when it comes to determining the mortality of sepsis patients. LightGBM, introduced by Microsoft, serves as an efficient gradient-boosting framework which is known for its remarkable fast speed. For proficiency management of extensive datasets, LightGBM is excellent as its root lies in histogram-based learning. It is an ideal candidate for our research into sepsis patient mortality prediction. It can easily handle large health datasets using complex formulas with limited memory requirements.

Along with the above two, the research also involves the Linear Support Vector Classifier. It serves best for the substantial volumes of data. By using linear kernel functions, this classifier applies a practical approach to addressing the challenge of mortality prediction in sepsis cases. Artificial Neural Networks (ANNs) are a key component of our research efforts. ANNs are known for their ability to deal with complex data correlations. These models, driven by weighted inputs and activation functions, delve into the intricacies of health data [41]. They enhance our comprehension of sepsis management and enhance patient outcomes by forecasting sepsis patient mortality using various health factors.

4 Experiments and Comparative Analysis

4.1 Implementation of Different ML and DL Techniques

A systematic approach is employed to evaluate the performance of a variety of machine learning models for mortality prediction in Sepsis-III patients. The primary source of data for this study is the "Refined MIMIC-III 30-day mortality prediction of sepsis-3 patients" dataset, obtained from IEEE Dataport [36]. Data preprocessing is conducted with meticulous attention to detail, guided by the necessity to guarantee data quality and the preparedness of the models in the specific domain of sepsis patient mortality prediction. Addressing the prevalent issue of missing data, advanced imputation techniques are thoughtfully applied to estimate missing values, thereby preserving dataset integrity and mitigating potential bias introduction. Additionally, the transformation of categorical data into numerical formats is undertaken. Given the common use of categorical variables in the healthcare domain for representing diagnoses, treatments, and patient attributes, a methodical encoding of these variables is executed (Fig. 2).

Model validation, a critical step in the context of sepsis patient mortality prediction, is meticulously approached by partitioning the dataset into distinct training and testing subsets. This division facilitated a rigorous assessment of predictive model performance on previously unseen data. Furthermore, feature scaling is implemented to standardize the data, ensuring fair and unbiased comparisons among various model algorithms. Subsequently, a suite of machine learning models is applied, including Logistic Regression, XGBoost Classifier, LGBM Classifier, Linear SVC, and Deep Learning models, such as Artificial Neural Networks, to predict sepsis-III patient mortality. Each model underwent rigorous training and testing, and their performance is evaluated based on various metrics, prominently including accuracy, AUC Score, and F1 score.

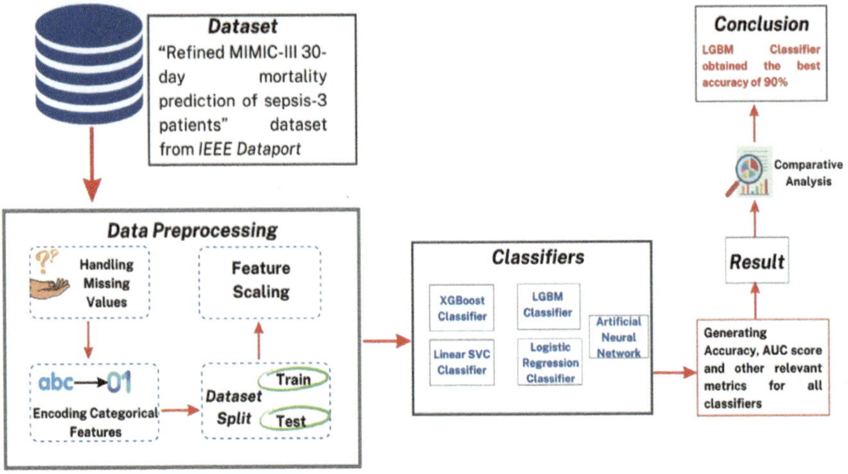

Fig. 2 Framework for the development of mortality prediction model

4.2 Comparative Analysis of Various Techniques

In addition to the machine learning models employed in our research, we conducted a comparative analysis of model performances, assessing their accuracies and area under the curve (AUC) scores using the same dataset and feature set. Comparative analysis of various machine learning and deep learning models is depicted in Fig. 3.

From Table 3, it is inferred that the LGBM Classifier Demonstrates exemplary performance with an accuracy of 0.90, an AUC of 0.76, and an impressive F1 score of 0.89. This model excels in predicting sepsis patient mortality, leading the pack.

XGBoost Classifier delivers robust performance, boasting an accuracy of 0.89, an AUC of 0.75, and a formidable F1 score of 0.88. Its consistent and high-quality results establish its significance in this predictive task. Logistic Regression offers reliable results, featuring an accuracy of 0.88, an AUC of 0.72, and a solid F1 score of 0.87. This model showcases commendable capabilities in the realm of sepsis mortality prediction. Linear SVC exhibits noteworthy performance with an accuracy of 0.88, an AUC of 0.71, and an impressive F1 score of 0.86. Its ability to accurately forecast sepsis patient mortality is evident. The Artificial Neural Network model provides an accuracy of 0.78, an AUC score of 0.61, and the lowest F1 score which is 0.37. The fine accuracy and AUC score encourage further betterment of the model.

Notably, the LightGBM (LGBM) classifier emerged as the top performer, boasting an impressive accuracy of 0.90 and a substantial AUC score of 0.76. This underscores its efficacy in accurately predicting the outcomes under consideration. On the opposite end of the spectrum, the Quadratic Discriminant Analysis model exhibited the lowest accuracy, registering a value of 0.20. Additionally, we observed that the Dummy Classifier, Label Spreading, and Label Propagation models collectively presented the lowest AUC scores, each scoring 0.50. These contrasting results offer

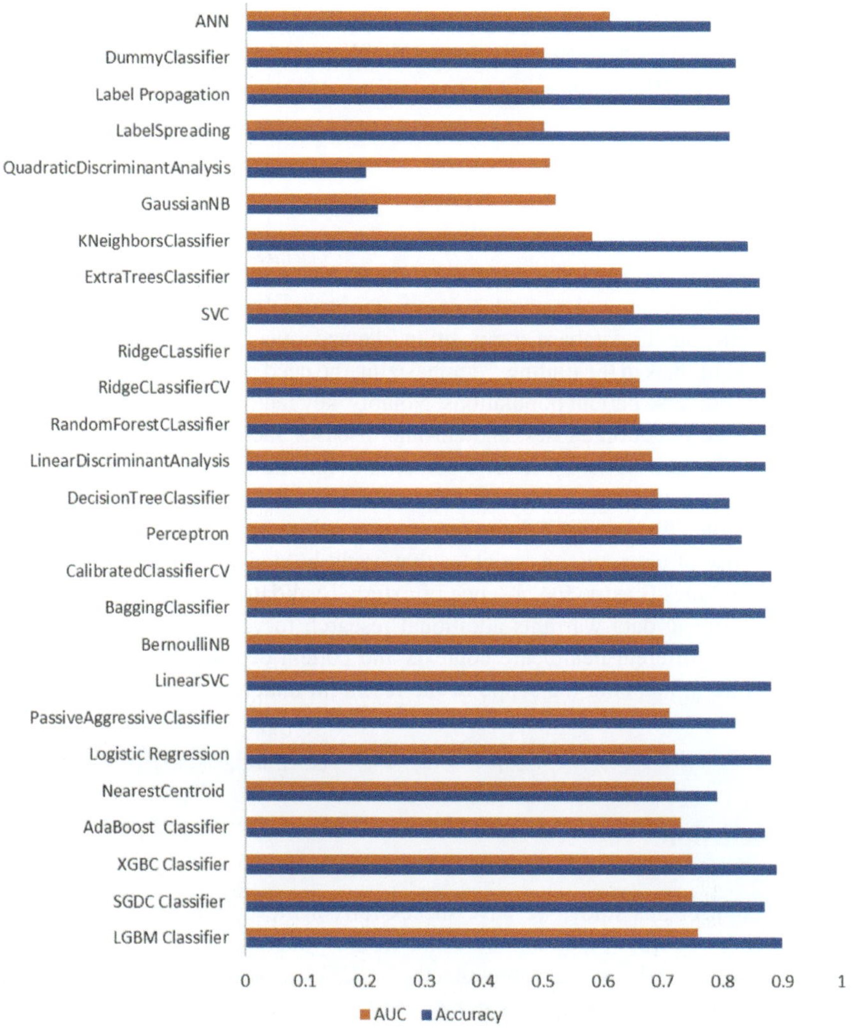

Fig. 3 Comparative analysis of various machine learning and deep learning models

Table 3 Accuracy, Area Under Curve (AUC) and F1 score of models employed

S. No.	Model	Accuracy	AUC	F1 score
1	LGBM classifier	0.90	0.76	0.89
2	XGBoost classifier	0.89	0.75	0.88
3	Logistic regression	0.88	0.72	0.87
4	Linear SVC	0.88	0.71	0.86
5	Artificial neural network	0.78	0.61	0.37

valuable insights into the varying predictive capacities of the models tested in our study, aiding in the selection of the most suitable model for the specific healthcare context under investigation.

5 Limitations and Future Scope

The proposed model, aimed at deepening our understanding of sepsis-3 outcomes, offers valuable insights tempered by several noteworthy limitations. Firstly, the predominantly single-database composition, skewed toward one ethnic group, raises the potential for bias in the findings, emphasizing the need for caution in generalizing the results to diverse patient populations. Notably, key factors such as genetics, pregnancy, specific infection causes, and chronic diseases such as Chronic Obstructive Pulmonary Disease (COPD), heart disease, and chronic kidney disease, were not incorporated into the analysis, which may limit the model's comprehensiveness.

Furthermore, the study's relatively narrow exploration of the database leaves the possibility of unexamined variables that could hold significance for predicting sepsis-3 outcomes. Lastly, the absence of external validation using an independent dataset or clinical data warrants additional research to verify the model's performance across different healthcare contexts and patient demographics. These limitations, while influencing the scope of the research, underscore areas for potential future investigations and the importance of careful interpretation of the findings.

6 Conclusion

In conclusion of this study, we have successfully achieved our goal of predicting the mortality of sepsis-3 patients by using various machine learning models. With the comparative analysis, it is found that the LGBM Classifier model gives the best-desired outcome on the basis of accuracy, AUC score, and F1 score in comparison to other models. Along with this, the fine accuracy result of the ANN model inspires exploring further optimization through hyperparameter tuning. By considering the work mentioned in our study, it can be concluded that our research introduces a pioneering machine learning model poised to make a profound impact on healthcare, with a particular emphasis on sepsis-3 patients. The model focuses on predicting mortality among sepsis patients, utilizing an array of critical health factors.

The future scope of this ongoing research, emphasis on ANN model refinement along with PCA. The artificial neural network (ANN) model has demonstrated a current accuracy of 0.8. To further refine its predictive capabilities, the next step involves exploring the potential of hyperparameter tuning. This meticulous fine-tuning process offers the promise of optimizing the model's performance and achieving even more favorable results. Concurrently, considering the integration of Principal Component Analysis (PCA). PCA serves as a valuable dimensionality reduction technique, which, when combined with hyperparameter tuning, can significantly enhance model efficiency and reduce computational complexity. By merging these approaches, we aim to establish a comprehensive strategy for improving model performance, setting the stage for more robust and accurate predictions in our research endeavors.

The potential application of this model offers the promise of more effective and proactive healthcare strategies focusing on early prediction, precise diagnosis, and a comprehensive assessment of sepsis severity. This research endeavor holds the potential to redefine the landscape of sepsis management and contribute to improved patient outcomes, thus serving as a significant advancement in the field of healthcare.

References

1. World Health Organization. Sepsis. World Health Organization. https://www.who.int/health-topics/sepsis
2. Huang, B., Liang, D., Zou, R., Yu, X., Dan, G., Huang, H., Liu, H., Liu, Y.: Mortality prediction for patients with acute respiratory distress syndrome based on machine learning: a population-based study. Ann. Transl. Med. **9**(9), 794 (2021). https://doi.org/10.21037/atm-20-6624
3. Stevenson, E.K., Rubenstein, A.R., Radin, G.T., Wiener, R.S., Walkey, A.J.: Two decades of mortality trends among patients with severe sepsis: a comparative meta-analysis*. Crit. Care Med. **42**(3), 625–631 (2014). https://doi.org/10.1097/CCM.0000000000000026
4. Tembhurne, J.V., Almin, M.M., Diwan, T.: Mc-DNN: fake news detection using multi-channel deep neural networks. Int. J. Semant. Web Inf. Syst. (IJSWIS) **18**(1), 1–20 (2022)
5. Aggarwal, K., Singh, S.K., Chopra, M., Kumar, S., Colace, F.: Deep learning in robotics for strengthening industry 4.0.: opportunities, challenges, and future directions. In: Robotics and AI for Cybersecurity and Critical Infrastructure in Smart Cities, pp. 1–19 (2022)
6. Singh, I., Singh, S.K., Singh, R., Kumar, S.: Efficient loop unrolling factor prediction algorithm using machine learning models. In: 2022 3rd International Conference for Emerging Technology (INCET), pp. 1–8, May 2022. IEEE (2022)
7. Peñalvo, F.J.G., Maan, T., Singh, S.K., Kumar, S., Arya, V., Chui, K.T., Singh, G.P.: Sustainable Stock Market Prediction Framework Using Machine Learning Models. Int. J. Softw. Sci. Comput. Intell. (IJSSCI) **14**(1), 1–15 (2022)
8. Gupta, S., Agrawal, S., Singh, S.K., Kumar, S.: A novel transfer learning-based model for ultrasound breast cancer image classification. In: Computational Vision and Bio-Inspired Computing: Proceedings of ICCVBIC 2022, pp. 511–523. Springer Nature Singapore, Singapore (2023)
9. Khade, G., Kumar, S., Bhattacharya, S.: Classification of web pages on attractiveness: a supervised learning approach. In: 2012 4th International Conference on Intelligent Human Computer Interaction (IHCI), pp. 1–5, December 2012. IEEE (2012)
10. Singh, I., Singh, S.K., Kumar, S., Aggarwal, K.: Dropout-VGG based convolutional neural network for traffic sign categorization. In: Congress on Intelligent Systems: Proceedings of CIS 2021, vol. 1, pp. 247–261, July 2022. Springer Nature Singapore, Singapore (2022)
11. Gupta, A., Singh, S.K., Gupta, B.B., Chopra, M., Gill, S.S.: Evaluating the sustainable COVID-19 vaccination framework of india using recurrent neural networks. Wirel. Pers. Commun. 1–19 (2023)
12. Ren, P., Xiao, Y., Chang, X., Huang, P.Y., Li, Z., et al.: A survey of deep active learning. ACM Comput. Surv. (CSUR) **54**(9), 1–40 (2021)
13. Khoudja, M.A., Fareh, M., Bouarfa, H.: Deep embedding learning with auto-encoder for large-scale ontology matching. Int. J. Semant. Web Inf. Syst. (IJSWIS) **18**(1), 1–18 (2022)

14. Vijayakumar, P., Rajkumar, S.C.: Deep reinforcement learning-based pedestrian and independent vehicle safety fortification using intelligent perception. Int. J. Softw. Sci. Comput. Intell. (IJSSCI) **14**(1), 1–33 (2022)
15. Li, S., Qin, D., Wu, X., Li, J., Li, B., Han, W.: False alert detection based on deep learning and machine learning. Int. J. Semant. Web Inf. Syst. (IJSWIS) **18**(1), 1–21 (2022)
16. Shankar, K., Perumal, E., Elhoseny, M., Taher, F., Gupta, B.B., El-Latif, A.A.A.: Synergic deep learning for smart health diagnosis of COVID-19 for connected living and smart cities. ACM Trans. Internet Technol. (TOIT) **22**(3), 1–14 (2021)
17. Hammad, M., Alkinani, M.H., et al.: Myocardial infarction detection based on deep neural network on imbalanced data. Multim. Syst. 1–13 (2021)
18. Nguyen, G.N., Le Viet, N.H., Elhoseny, M., Shankar, K., et al.: Secure blockchain enabled Cyber–physical systems in healthcare using deep belief network with ResNet model. J. Parallel Distrib. Comput. **153**, 150–160 (2021)
19. Afify, M., Loey, M., Elsawy, A.: A robust intelligent system for detecting tomato crop diseases using deep learning. Int. J. Softw. Sci. Comput. Intell. (IJSSCI) **14**(1), 1–21 (2022)
20. Anil, B.C., Dayananda, P., Nethravathi, B., Raisinghani, M.S.: Efficient local cloud-based solution for liver cancer detection using deep learning. Int. J. Cloud Appl. Comput. (IJCAC) **12**(1), 1–13 (2022)
21. Srivastava, A.M., Rotte, P.A., Jain, A., Prakash, S.: Handling data scarcity through data augmentation in training of deep neural networks for 3D data processing. Int. J. Semant. Web Inf. Syst. (IJSWIS) **18**(1), 1–16 (2022)
22. Hasib, K.M., Towhid, N.A., Islam, M.R.: Hsdlm: a hybrid sampling with deep learning method for imbalanced data classification. Int. J. Cloud Appl. Comput. (IJCAC) **11**(4), 1–13 (2021)
23. Hou, N., Li, M., He, L., Xie, B., Wang, L., Zhang, R., Yu, Y., Sun, X., Pan, Z., Wang, K.: Predicting 30-days mortality for MIMIC-III patients with sepsis-3: a machine learning approach using XGboost. J. Transl. Med. **18**(1), 462 (2020). https://doi.org/10.1186/s12967-020-02620-5
24. Suganya, G., Premalatha, M., Geetha, S., Chowdary, G.J., Kadry, S.: Detection of COVID-19 cases from chest X-rays using deep learning feature extractor and multilevel voting classifier. Int. J. Uncertain. Fuzziness Knowl.-Based Syst. **30**(05), 773–793 (2022)
25. Wang, J., Fei, W., Song, Q.: One-year mortality prediction for patients with sepsis: a nomogram integrating lactic dehydrogenase and clinical characteristics. BMC Infect. Dis. **23**, 668 (2023). https://doi.org/10.1186/s12879-023-08636-8
26. Shin, J., Li, Y., Luo, Y.: Early prediction of mortality in critical care setting in sepsis patients using structured features and unstructured clinical notes. In: 2021 IEEE International Conference on Bioinformatics and Biomedicine (BIBM), Houston, TX, USA, pp. 2885–2890 (2021). https://doi.org/10.1109/BIBM52615.2021.9669822
27. Burtenshaw, P., Finlayson, B., El-Abed, O., Palmer, C.: The DEEPSAL project: using the past for local community futures in Jordan. Conserv. Manag. Archaeol. Sites **21**(2), 69–91 (2019)
28. Yao, R.Q., Jin, X., Wang, G.W., Yu, Y., Wu, G.S., Zhu, Y.B., Li, L., Li, Y.X., Zhao, P.Y., Zhu, S.Y., Xia, Z.F., Ren, C., Yao, Y.M.: A machine learning-based prediction of hospital mortality in patients with postoperative sepsis. Front. Med. **7**, 445 (2020). https://doi.org/10.3389/fmed.2020.00445
29. Haddad, B.M., Dodge, S.F., Karam, L.J., Patel, N.S., Braun, M.W.: Locally adaptive statistical background modeling with deep learning-based false positive rejection for defect detection in semiconductor units. IEEE Trans. Semicond. Manuf. **33**(3), 357–372 (2020)
30. Persson, I., Östling, A., Arlbrandt, M., Söderberg, J., Becedas, D.: A machine learning sepsis prediction algorithm for intended intensive care unit use (NAVOY Sepsis): proof-of-concept study. JMIR Form. Res **5**(9), e28000 (2021). https://doi.org/10.2196/28000
31. Perng, J.-W., Kao, I.-H., Kung, C.-T., Hung, S.-C., Lai, Y.-H., Su, C.-M.: Mortality prediction of septic patients in the emergency department based on machine learning. J. Clin. Med. **8**(11), 1906 (2019). https://doi.org/10.3390/jcm8111906
32. Srivastava, G., Kavitha, V., Vimaladevi, M., Parthiban, N.: The hydrological impact of tropical cyclones on soil moisture using a sensor based hybrid deep learning model. Acta Geophys. **70**(6), 2933–2951 (2022)

33. Wang, H., Li, Y., Naidech, A., et al.: Comparison between machine learning methods for mortality prediction for sepsis patients with different social determinants. BMC Med. Inf. Decis. Mak. **22**(Suppl 2), 156 (2022). https://doi.org/10.1186/s12911-022-01871-0
34. Sahoo, S.R., Gupta, B.B.: Multiple features based approach for automatic fake news detection on social networks using deep learning. Appl. Soft Comput. **100**, 106983 (2021)
35. Kline, A., Wang, H., Li, Y., et al.: Multimodal machine learning in precision health: a scoping review. NPJ Digit. Med. **5**, 171 (2022). https://doi.org/10.1038/s41746-022-00712-8
36. Yoo, J.: Refined MIMIC-III 30-day mortality prediction of sepsis-3 patients. IEEE Dataport (2022). https://doi.org/10.21227/qrhd-a469
37. Peng, S., Li, W., Ke, W.: Association between red blood cell distribution width and all-cause mortality in unselected critically ill patients: analysis of the MIMIC-III database. Front. Med. **10**, 1152058 (2023). https://doi.org/10.3389/fmed.2023.1152058
38. Luo, C., Duan, Z., Xia, Z., et al.: Minimum heart rate and mortality after cardiac surgery: retrospective analysis of the Multi-parameter Intelligent Monitoring in Intensive Care (MIMIC-III) database. Sci. Rep. **13**, 2597 (2023). https://doi.org/10.1038/s41598-023-29703-9
39. Zhang, H., Tian, W., Sun, Y.: The value of anion gap for predicting the short-term all-cause mortality of critically ill patients with cardiac diseases, based on MIMIC-III database. Heart & Lung **55**, 59–67 (2022). https://doi.org/10.1016/j.hrtlng.2022.04.004, ISSN 0147-9563
40. Chopra, M., Singh, S. K., Sharma, S., Mahto, D.: Impact and usability of artificial intelligence in manufacturing workflow to empower Industry 4.0. In: International Conference on Smart Systems and Advanced Computing (SysCom 2021) (2022)
41. Su, Y., Guo, C., Zhou, S., et al.: Early predicting 30-day mortality in sepsis in MIMIC-III by an artificial neural networks model. Eur. J. Med. Res. **27**, 294 (2022). https://doi.org/10.1186/s40001-022-00925-3

Enhancing Review Authenticity: A Linguistic and Behavioral Approach for Detecting Deceptive Online Reviews Using Machine Learning Models

Ishtveer Singh Billing, Sudhakar Kumar, Sunil K. Singh, Manraj Singh Gill, Brij B. Gupta, and Varsha Arya

Abstract Online customer reviews are invaluable tools for making informed decisions, yet this utility is affected by generation of Fake Reviews. This problem equally concerns both consumer and business fields. Existing Methods are based on linguistic attributes. With the incoming of various open-source LLMs, the distinction between machine and human generated text is fading. In this research, we observed both linguistic and behaviour-based detection systems provide better accuracy. We adopted a hybrid approach using Semantic Analysis and supervised ML approach on Yelp Review dataset. Various useful attributes were used/formulated to form new parameters. Semantic Analysis has been used to form certain attributes in the dataset to include the linguistic behaviour of review. Our study is grounded on comparative analysis of 10 different supervised ML models. Comprehensive evaluation of models is performed using various metrics. The best achieved accuracy was 94.362% with ROC AUC of 0.982, precision of 0.915, recall of 0.978 and F1 Score of 0.945. The detailed experimental analysis shows the feasibility and effectiveness of proposed system.

Keywords Fake review detection · Semantic analysis · Supervised machine learning · Model comparison · Model evaluation · Feature synthesis · Natural language processing

I. S. Billing (✉) · S. Kumar · S. K. Singh · M. S. Gill
Department of Computer Science and Engineering, Chandigarh College of Engineering and Technology, Chandigarh, India
e-mail: ishtveer@gmail.com

B. B. Gupta
Department of Computer Science and Information Engineering, Asia University, Taichung, Taiwan

V. Arya
Department of Business Administration, Asia University, Taichung, Taiwan

B. Soni et al. (eds.), *Beyond Artificial Intelligence*, Lecture Notes in Networks and Systems 1326, https://doi.org/10.1007/978-981-96-4170-3_63

1 Introduction

Online reviews are customer or user opinions shared on digital platforms like social media, e-commerce sites, and travel booking websites. These reviews influence consumer choices, trust, and a company's reputation, offering valuable insights. They cover various topics, from product quality to customer service, with a range of sentiments. Businesses use them for marketing and customer engagement. In the digital age, online shopping highlights the importance of reviews, which impact the reputation of products, services, and businesses.

False online reviews [1] are a major issue, misleading consumers and damaging trust in legitimate businesses and review sites. A growing field of research focuses on detecting fake reviews, and we contribute by analysing review, reviewer, and product/service behaviour.

Consumers must be aware of fake review warning signs, emphasizing transparency in review collection and verification. However, the validity of online reviews is a constant worry in the digital age that it can be jeopardized by phony or fraudulent content. It's a constant struggle for review platforms [2] and regulatory bodies to deal with fraudulent reviews and ensure the validity of online input. To combat this, review systems employ user reporting, human moderation, and detection methods. Some countries consider legal restrictions. When exploring the effects of digital transformation on firms' innovation performance [3] and review authenticity in the context of heterogeneous resources, a workflow scheduling algorithm [4] plays a crucial role in optimizing resource allocation and enhancing overall efficiency.

We propose a hybrid model which encompass behaviour and linguistic based deceptive online review detection systems to enhance review authenticity utilizing a supervised machine learning approach [5–9] and deep learning [10, 11]. Our research not only offers a robust hybrid model for fake review detection but also evaluates the strengths and weaknesses of these models, contributing to the broader research in this field. The significant contributions of the research work are:

(a) The method of detecting fake reviews using hybrid model which encompass behaviour and linguistic based deceptive online review using various ML and deep learning techniques.
(b) Incorporation of semantic analysis to form certain attributes in the dataset to include the linguistic feature of review.
(c) In order to assess different ML and deep learning models' performance in relation to the classification of online reviews, we rigorously assessed and compared them in our research.

The rest of the paper is arranged as follows: In Sect. 2, we delve into the research pertinent to fake review detection. Section 3 of this document elucidates dataset description. Section 4 presents the suggested approach. Section 5 gives models evaluation and comparison and provides a discourse on them. The paper's closing remarks can be found in Sect. 6.

2 Literature Review

Elmogy et al. [12] proposed a supervised machine learning approach to identify fake reviews and compares the performance. In Kalaivani et al. [13], a Naive Bayes algorithm is used as an effective method for detecting fake reviews using machine learning. Zaki et al. [14] proposed a node embedding approach using machine learning classifiers for accurate detection of fake reviews which uses a graph-based machine learning approach with explainable AI. Li et al. [15] proposed a deceptive review detection method based on multidimensional feature construction and ensemble feature selection. Sentiment Analysis and its linking with SVM and Naïve Bayes was proposed by Singh et al. [16]. Although they followed text-based classification of movie reviews. Spam Filtering used by Gupta [17] proposed a classification algorithm. Most of the previous works around the Yelp Restaurant Review Dataset [18, 19] has been on the text-based review classification for fake and authentic reviews. Detecting phishing websites [20], malware [21], and compromised social network accounts [22] using semantic features involves leveraging natural language processing (NLP) and machine learning classification techniques [23]. Semantic features focus on the meaning and context of text data. Texture classification [24] for semantic analysis in Natural Language Processing (NLP) typically involves techniques like instance-based [25] or contextual matching [26, 27]. These approaches are used to understand and analyze the texture, or the structure and context, of text data (Table 1).

3 Dataset Description

3.1 Dataset Collection

The dataset used in our research was Yelp Restaurant Review Dataset. The official source of dataset is the yelp website which has open-source review dataset available to download in different formats on the website: https://www.yelp.com/dataset [19].

3.2 Dataset Characteristics

The dataset available in SQL format has 3 different relations—reviewer, restaurant and review. All of the parameters have more than 40 attributes combined. The dataset has 7,88,471 different reviews in the review relation and 16,941 different reviewers has reviews present in this dataset. A total of 2,42,652 restaurants were reviewed in the given dataset.

Preprocessing SQL Dataset. Using SQL queries, the data was extracted using python scripts and converted into .csv format for easy handling of the dataset. We

Table 1 Comparative analysis of related work

Ref. No.	Proposed approach	Dataset(s) used	Algorithm used	Result(s) obtained
10	Direct application of ML models on the dataset including the user behaviour analyis	Yelp restaurant review dataset	KNN, Naive Bayes (NB), SVM, logistic regression and random forest	Best F-score 82.40%(using KNN)
11	Usage of Naive Bayes classifier to classify reviews	Unknown	Naive Bayes classifier	Unknown
12	Node embedding approach to detect online fake reviews by creating distance matrix and graphs	Deceptive opinion spam corpus and YelpChi datasets	Support vector machines	98.44% accuracy, 98.44% precision, 98.44% recall, and 98.44% F1-score
14	SentiWordNet approach with different variations of linguistic features, scoring schemes and aggregation thresholds for sentiment classification	Cornell sentiment polarity dataset, polarity dataset v1.0	Naive Bayes and SVM	81.07% accuracy with Naive Bayes classifier
18	Applying 16 ML Models upon 10 semantic features that represent the most effective features for detection of phishing websites	Huddersfield university dataset (R M. Mohammad, Thabtah and McCluskey 2012)	Gradient boosting classifier and random forest classifier	Best accuracy around 97% and lowest around 81%
27	An improved version of min-min algorithm for workflow scheduling	Not used	Min-min and ELBMM algorithms	Not applicable

have extracted 26,959 unique review values in the .csv file with various parameters such as (the name and definition of these parameters is expected to be implied): reviewID, reviewerID, restaurantID, reviewDate, rating, reviewContent, restaurentLocation, yelpJoinDate, reviewCount, firstCount, usefulCount, reviewLength, restaurantRating, flagged (label to tell authentic or fake review).

3.3 Feature Engineering

Feature engineering plays a pivotal role in the project. Some new introduced parameters enhanced the accuracy of ML models being trained. We trained Random Forest Model on the just extracted dataset from the SQL format and we achieved an accuracy of 86.671% and after feature engineering we now have a final accuracy of 89.855% after doing feature engg. It shows that feature engineering and new introduced features has a significant impact on accuracy of the models.

From the existing parameters, we deduced new ones as follows:

Review Deviation. Review deviation is a novel feature created to capture the extent to which individual review deviates from the typical or expected behaviour of the reviews. It is derived from restaurant rating, review length and sentiment in accordance with the features of restaurant.

Consider $S_{mean}(r)$ as the mean rating of a restaurant and normalizing according to 5-star rating scale, the review deviation score on a scale of [0, 1]. For a given ri

$$\mathrm{RD}(\mathrm{r_i}) = \frac{|r_i - S_{mean}(r)|}{4}$$

Content Similarity. This parameter is deduced with the help of semantic analysis that ranges the similarity of the text of the review with the comparison of the restaurant features. It measures the similarity of the content posted in the review with the context and is a good parameter to detect fake reviews. Also, a lot of spammers reproduce the same content to many different restaurants. It would also help in detecting spammers. To take the average similarities of all the reviews for particular user, we would use the cosine similarity. This can be used to measure the reviewer behaviour.

The content similarity score is also from the scale of [0,1] of a reviewer a(ri) attributed to review r_i, is the average of the similarities of each review $r_j \in R_{a,j}$.

$$CS(a(r_i)) = Avg\left(\frac{\sum_{z=1}^{|R_{a,j}|} \cos ine\left(r_j, r_z\right)}{\left|R_{a,j}\right|}\right), j \neq z$$

Max Review Count. Max Review Count or Peak Review Count or Max. Number of Reviews (MNR) is the parameter that measures the number of reviews on a given day/ max number of reviews posted on any day by a user.

We also scale this parameter data between 0 and 1, where 1 represents the reviewer's maximum number of reviews on a single day.

Mathematical Formula:

$$MRC = \frac{R_i * D}{Max\{R_i\}}$$

where R_i = No. of reviews by the reviewer.

Average Review Count. We have two different parameters yelpJoinDate and reviewCount that could be used to find the avgReviewCount feature. A lot of spammers have a lot of reviewCount per day and this is a very useful feature to catch spammers and by finding outliers that have a very high review postage per day.

Mathematical Formula:

$$Avg.\ Review\ Count = \frac{N(R_i)}{N(R_i * D)}$$

i.e., No. of total reviews/No. of days on platform.

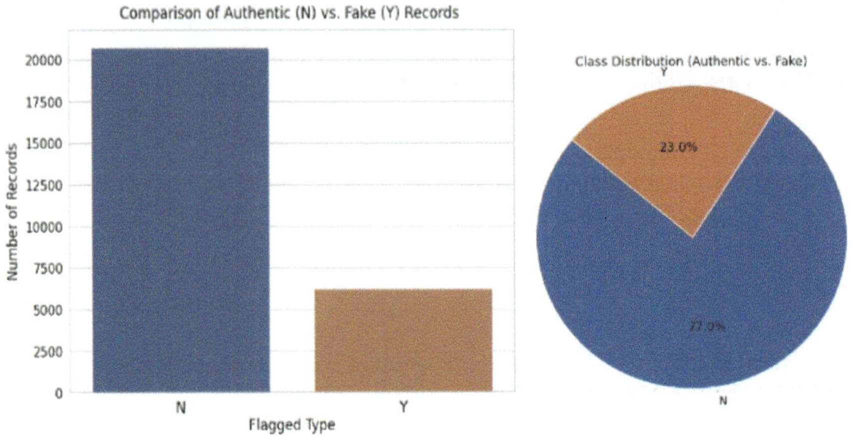

Fig. 1 Distribution of classes in our dataset with a class imbalance having 77% authentic reviews and 23% fake reviews

3.4 Dataset Size and Distribution

The dataset we have extracted in the .csv format has 26,959 different tuples and it has the records divided into two different classes in the label parameter—authentic and fake reviews.

We have 20,752 authentic reviews in our dataset and 6,206 fake reviews as labelled in the 'labelled' parameter. This leads to class imbalance in our dataset which could possibly lead to decrease in accuracy in ML models. Some models like Random Forest are not affected by such imbalance and thus on this imbalanced data, Random Forest Model has performed the best with the accuracy of 87.314% without the use of sampling and scaling (Fig. 1).

Now to counter this imbalance, we would apply scaling and sampling which is further discussed as in next Sect. 3.5 (Scaling and Sampling).

3.5 Scaling and Sampling

Feature Scaling. Scaling is one of the most fundamental preprocessing steps, especially as we are dealing with numerical features as in our case. It normalizes and standardize all the features over a chosen range which we applied to transform our data to have a mean of 0 and a standard deviation of 1. Our goal is to ensure that no single parameter dominates the learning process. Although, the changes we found in accuracy of most of the ML models was rather quite nominal but this technique was applied to help models in converging more efficiently.

Sampling. Addressing class imbalance was a vital part in our research and we experimented with a lot of sampling techniques of oversampling, under sampling,

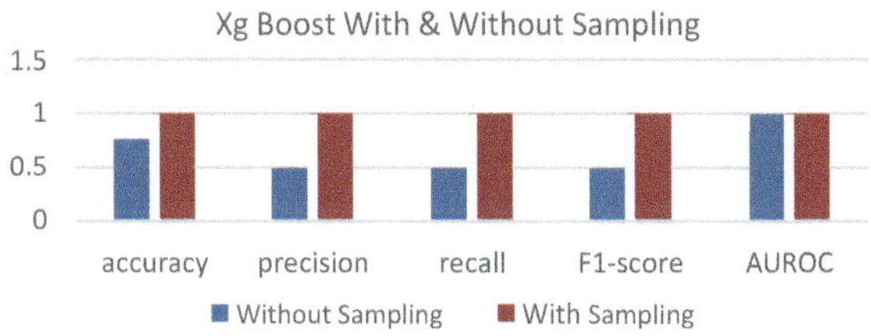

Fig. 2 The difference between XGBoost model before and after random oversampling

probability sampling, mean sampling etc. As a part of our research, we found Random Oversampling to be providing best results in our case. It had most significant results in the best model we trained—XGBoost model whose accuracy increased from 70.580 to 94.362% after applying Random Oversampling.

After oversampling, the minor class 'fake' was produced randomly matching 'authentic' values at 20,752 now dataset having total 41,504 values (Fig. 2).

These both were our main parts of dataset preprocessing along with label encoding the label class parameter which has 'N' as authentic values which we encoded to '0' and 'Y' as fake values encoded to '1' using in-built LabelEncoder().

4 Hybrid Model

In our study, we introduced a unique hybrid model designed to review authenticity detection. We used Semantic analysis in order to find the values of features such as Content Similarity and Review Deviation. These parameters would be further used in machine learning model to detect fake reviews. Both of these parameters are scaled from 0 to 1 and gauge the similarity of content posted by the reviewer with the context of reviewed place.

4.1 Semantic Analysis

In the semantic analysis component of our hybrid model, we use pre-build libraries for our research as our main focus was on the ML models and their optimization. For NLP, we used spaCy library and NTLK NLP library for text processing and similarity analysis.

We matched the textual reviews from Yelp datasets in the review table with the attributes of restaurant such as location, type, food attributes that would tell us about restaurant and useful to see whether the review is similar or not.

As stated above the mathematical model employed in this process is cosine similarity measure. It accessed the semantic similarity and provides the values in range of 0–1, where 0 indicates no similarity and 1 indicates a perfect match (Fig. 3).

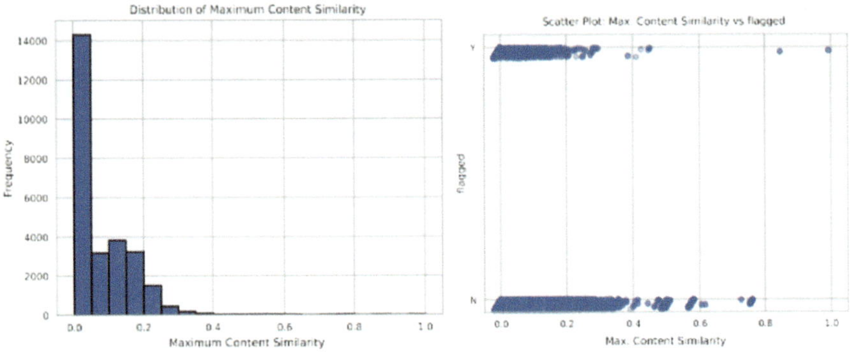

Fig. 3 Distribution of content similarity feature over the dataset and relation of MCS with flagged parameter

4.2 ML Models

In our hybrid model, we have seamlessly integrated traditional ML models that are useful for numerical data classification along with semantic analysis. ML models are trained from the contextual insights gained from semantic analysis model. The training of each of these models are done with the help of K-fold Cross Validation as it gives a better evaluation of the models rather than split testing. We trained and tested 10 different ML models to find the best one for fake review detection. The evaluation of all these models is presented in detail in the next Sect. 5 (Fig. 4).

4.3 Use of Hybrid Model

From our reserach and previous work on this topic, we reached the conclusion that both linguistic of the review and behavioural features of the reviwer play conclusive role in fake review classification. Thus, the major benefit of using a hybrid model was to incorporate both of these featues as we used combination of semantic analysis and feature engineering to derive certain features from the textual data of the review. Use of hybrid model increases our input features but provides more better and accurate classification. Along with that, mostly popular open-community platforms has user behaviour paramteres on their backend which would server as additional parameters and could be input directly into the Hybrid Model to create a more robust detection system for the particular platform. Thus, this hybrid model makes use of two of the most important aspects of the review—What is written in the review? And Who has written the review? These two parameters as stated in title of this research are equally contributing factors making the use of a hybrid model an evident and efficient approach to solve the problem.

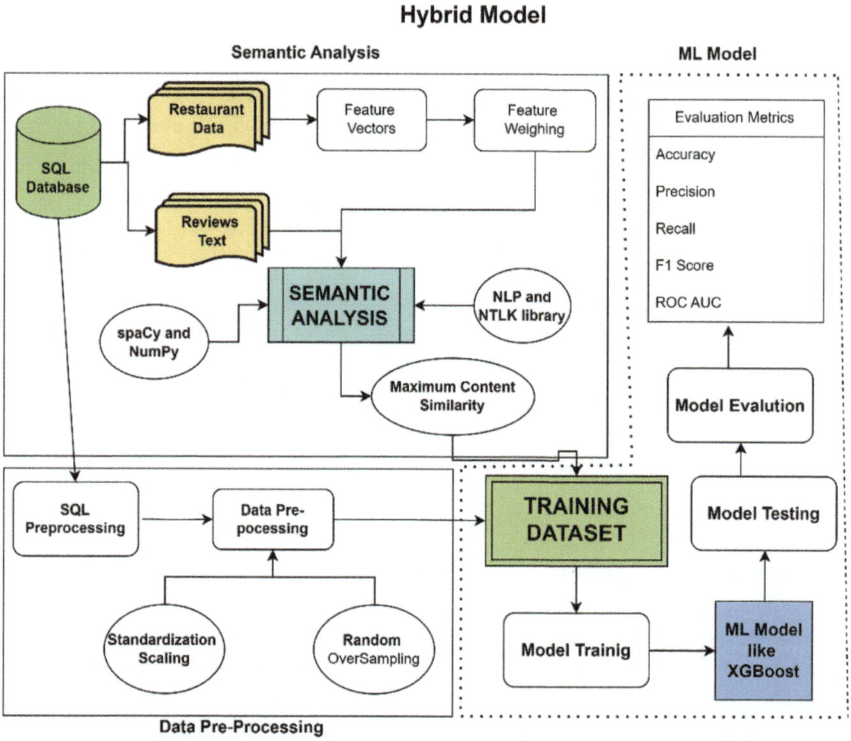

Fig. 4 Diagrammatic representation of hybrid model and data preprocessing

Proposed Algorithm
1. **Input:** *In* = reviewText, restaurantData, reviewData, reviewerData
2. **Ouptut**: *Out* = 0 ← authentic reviews, 1 ← fake reviews
3. for review in reviewData //Semantic Analysis
4. reviewData ← dropna() // Drop null value tuples
5. New_Data ← reviewData U reviewerData U restaurantData //All required parameters
6. Mcs ←similarity(reviewText, restaurantData)
7. New_Data ← to_csv(Mcs) // Adding MCS
8. // ML Classification // Random Oversampling New_Data ← RandomOverSampler(New_Data)
9. New_Data ← StandardScaler(New_Data) // Standarized Scaling
10. Label ← LabelEncoder(New_Data, 'flagged') // Label Encoding
11. model ← classifier(New_Data, Label) // Applying Classification
12. model.fit(train) model.Kfold() cross_val_score(model)
13. model.fit.summary()

5 Models Evaluation and Comparison

In our research, we rigorously evaluated and compared various ML models to access their performance in context on online review classification. As evaluation was our main motive, we also employed a diverse set of evaluation metrics as discussed in next Sect. 5.1.

5.1 Evaluation Parameters

The parameters we used are:

Accuracy. Accuracy means the proportion of correctly classified instances among all instances.

Precision. Precision here accesses the accuracy of positive fake review predictions made by model. It is ratio of true positive (fake) over the sum of true positive and false positive.

Recall. Recall also known as Sensitivity evaluates model's ability to identify all relevant instances within a dataset. It is ratio of true positive over sum of true positives and false negatives.

F1 Score. F1 Score is the harmonic mean of precision and recall, providing a balance between two metrics. Good to check performance on imbalanced datasets.
ROC Curve. ROC Area Under Curve (AUC) quantifies a model's ability to distinguish between positive and negative classes across various probability thresholds. Higher ROC AUC indicates superior discrimination power.

5.2 Models Evaluation

For each evaluation the evaluation metrics are calculated using K-fold Cross Validation. To generate a Confusion Matrix along with ROC Curve, the model is then tested on a standard split of 20% for each model.

Linear Support Vector Machines. Linear SVM is a classification algorithm that aims to find hyperplane that best separates data points into different classes. In our project, we applied scaling as SVM gives best results when data is scaled. All the data was numerical so the performance or SVM is as follows (Table 2 and Fig. 5).

Table 2 Performance of all ML models on our dataset

Model	Accuracy	Precision	Recall	F1 score	ROC AUC
Linear support vector machines	85.314	0.643	0.827	0.722	0.921
Decision tree classifier	85.738	0.694	0.691	0.693	0.797
Random forest classifier	89.855	0.787	0.773	0.780	0.956
Naïve Bayes classifier	71.494	0.449	0.984	0.616	0.905
Logistic regression	85.793	0.692	0.701	0.696	0.912
Artificial neural network	88.557	0.757	0.749	0.753	0.941
XG boost	**94.362**	**0.915**	**0.978**	**0.945**	**0.982**
K-nearest neighbour	77.373	0.516	0.454	0.483	0.809
Light BGM	91.524	0.820	0.813	0.817	0.968
SVM with RGB kernel	82.105	0.613	0.652	0.627	0.883

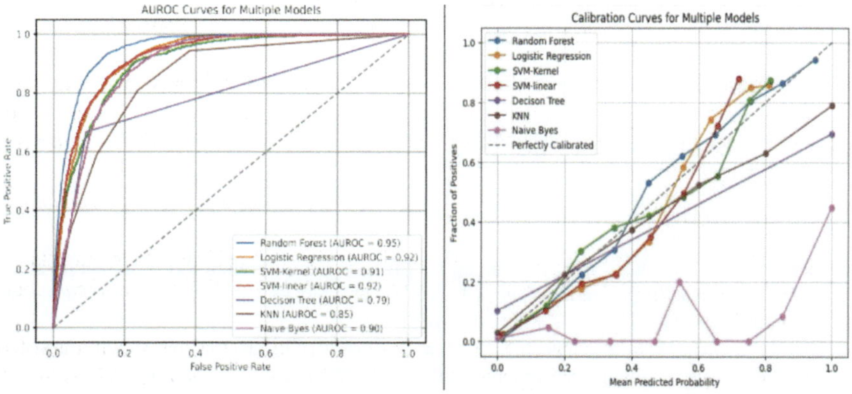

Fig. 5 Left: comparison of ROC curve of models evaluated. Right: calibration curve of the models providing comparative analysis

6 Conclusion

In this study, we propsed a hybrid model to tackle the increasing İssue of fake reviews that captures multiple aspects of the review. Along with that, we conducted a comparative study of multiple machine learning models to classify the reviews based on parameters generated using the hybrid model and address the issue of fake review detection. Our analysis revealed that while other models exhibit average results, the ensemble models—combining XGBoost, LightBGM along with Random Forest Classifier and deep learning model—ANN outperforms other individual models. From individual models Random Forest Classifier performed really well. Additionally, we incorporated 13 different features including the content similarity feature that helps to capture the review context accurately with the use of semantic analysis. These features are independent of any certain category of products and are based on universal features of a review. The proposed model could be used along with multiple platforms such as product services, video platforms like YouTube or community disccusion platforms like Reddit etc. The hybrid model achieved remarkable results with the best model having an accuracy of 94.362% with a F1 score of 0.945 and ROC AUC of 0.982. As online reviews play an increasingly pivotal role in decision-making of the consumers and authentic reviews are a challenge to maintain for product platforms, our findings provide valuable insights for fake review detection whether for businesses or for thse online platforms. With this study, we aim to maintain review authenticity and trustworthiness.

References

1. Habib, A., Asghar, M.Z., Khan, A., Habib, A., Khan, A.: False information detection in online content and its role in decision making: a systematic literature review. Soc. Netw. Anal. Min. **9**, 1–20 (2019)
2. Xiang, Z., Du, Q., Ma, Y., Fan, W.: A comparative analysis of major online review platforms: implications for social media analytics in hospitality and tourism. Tour. Manag. **58**, 51–65 (2017)
3. Li, S., Gao, L., Han, C., et al.: Exploring the effect of digital transformation on firms' innovation performance. J. Innov. Knowl. **8**(1), 100317 (2023)
4. Bisht, J., Vampugani, V.S.: Load and cost-aware min-min workflow scheduling algorithm for heterogeneous resources in fog, cloud, and edge scenarios. Int. J. Cloud Appl. Comput. (IJCAC) **12**(1), 1–20 (2022)
5. Singh, I., Singh, S.K., Singh, R., Kumar, S.: Efficient loop unrolling factor prediction algorithm using machine learning models. In: 2022 3rd International Conference for Emerging Technology (INCET), pp. 1–8, May 2022. IEEE (2022)
6. Peñalvo, F.J.G., Maan, T., Singh, S.K., Kumar, S., Arya, V., Chui, K.T., Singh, G.P.: Sustainable stock market prediction framework using machine learning models. Int. J. Softw. Sci. Comput. Intell. (IJSSCI) **14**(1), 1–15 (2022)
7. Mengi, G., Singh, S.K., Kumar, S., Mahto, D., Sharma, A.: Automated Machine Learning (AutoML): the future of computational intelligence. In: International Conference on Cyber Security, Privacy and Networking, pp. 309–317. Springer International Publishing, Cham (2021)
8. Raza, A., Ullah, N., Khan, J.A., Assam, M., Guzzo, A., Aljuaid, H.: DeepBreastCancerNet: a novel deep learning model for breast cancer detection using ultrasound images. Appl. Sci. **13**(4), 2082 (2023)
9. Khade, G., Kumar, S., Bhattacharya, S.: Classification of web pages on attractiveness: a supervised learning approach. In: 2012 4th International Conference on Intelligent Human Computer Interaction (IHCI), pp. 1–5, December 2012. IEEE (2012)
10. Aggarwal, K., Singh, S.K., Chopra, M., Kumar, S., Colace, F.: Deep learning in robotics for strengthening industry 4.0.: opportunities, challenges and future directions. In: Robotics and AI for Cybersecurity and Critical Infrastructure in Smart Cities, pp. 1–19 (2022)
11. Singh, I., Singh, S.K., Kumar, S., Aggarwal, K.: Dropout-VGG based convolutional neural network for traffic sign categorization. In: Congress on Intelligent Systems: Proceedings of CIS 2021, vol. 1, pp. 247–261. Springer Nature Singapore, Singapore (2022)
12. Elmogy, A.M., Tariq, U., Ammar, M., Ibrahim, A.: Fake reviews detection using supervised machine learning. Int. J. Adv. Comput. Sci. Appl. **12**(1) (2021)
13. Kalaivani, P., Raj, V.D., Madhavan, R., Naveen Kumar, A.P.: Fake review detection using naive bayesian classifier. In: 2023 International Conference on Sustainable Computing and Smart

Systems (ICSCSS), Coimbatore, India, pp. 705–709 (2023). https://doi.org/10.1109/ICSCSS57650.2023.10169838
14. Zaki, N., Krishnan, A., Turaev, S., Rustamov, Z., Rustamov, J., Almusalami, A., Ayyad, F., Regasa, T., Iriho, B.B.: Node embedding approach for accurate detection of fake reviews: a graph-based machine learning approach with explainable AI (2023)
15. Li, S., Zhong, G., Jin, Y., Wu, X., Zhu, P., Wang, Z.: A deceptive reviews detection method based on multidimensional feature construction and ensemble feature selection. IEEE Trans. Comput. Soc. Syst. **10**(1), 153–165 (2022)
16. Singh, V.K., Piryani, R., Uddin, A., Waila, P.: Sentiment analysis of movie reviews and blog posts. In: 2013 3rd IEEE International Advance Computing Conference (IACC), Ghaziabad, India, pp. 893–898 (2013). https://doi.org/10.1109/IAdCC.2013.6514345
17. Gupta, S., Chhabra, A., Agrawal, S., Singh, S.K.: A comprehensive comparative study of machine learning classifiers for spam filtering. In: International Conference on Cyber Security, Privacy and Networking, pp. 257–268, September 2021. Springer International Publishing, Cham (2021)
18. Luca, M.: Reviews, reputation, and revenue: the case of Yelp.com., 15 March 2016. Harvard Business School NOM Unit Working Paper (12-016) (2016)
19. Asghar, N.: Yelp dataset challenge: review rating prediction (2016). arXiv:1605.05362
20. Almomani, A., Alauthman, M., Shatnawi, M.T., Alweshah, M., Alrosan, A., et al.: Phishing website detection with semantic features based on machine learning classifiers: a comparative study. Int. J. Semant. Web Inf. Syst. (IJSWIS) **18**(1), 1–24 (2022)
21. Gaurav, A., Gupta, B.B., Panigrahi, P.K.: A comprehensive survey on machine learning approaches for malware detection in IoT-based enterprise information system. Enterp. Inf. Syst. 1–25 (2022)
22. Yen, S., Moh, M., Moh, T.S.: Detecting compromised social network accounts using deep learning for behavior and text analyses. Int. J. Cloud Appl. Comput. (IJCAC) **11**(2), 97–109 (2021)
23. Cvitić, I., Peraković, D., Periša, M., et al.: Ensemble machine learning approach for classification of IoT devices in smart home. Int. J. Mach. Learn. Cybern. **12**(11), 3179–3202 (2021)
24. Chen, X., Li, J., Zhang, Y.F.: Multidirectional gradient feature with shape index for effective texture classification. Int. J. Semant. Web Inf. Syst. (IJSWIS) **18**(1), 1–19 (2022)
25. Barbosa, A., Bittencourt, I.I., Siqueira, S.W., Dermeval, D., Cruz, N.J.: A context-independent ontological linked data alignment approach to instance matching. Int. J. Semant. Web Inf. Syst. (IJSWIS) **18**(1), 1–29 (2022)
26. Gu, J., Vo, N.D., Jung, J.J.: Contextual Word2Vec model for understanding Chinese out of vocabularies on online social media. Int. J. Semant. Web Inf. Syst. (IJSWIS) **18**(1), 1–14 (2022)
27. Sarivougioukas, J., Vagelatos, A.: Fused contextual data with threading technology to accelerate processing in home UbiHealth. Int. J. Softw. Sci. Comput. Intell. (IJSSCI) **14**(1), 1–14 (2022)

GPSR Compliance

The European Union's (EU) General Product Safety Regulation (GPSR) is a set of rules that requires consumer products to be safe and our obligations to ensure this.

If you have any concerns about our products, you can contact us on ProductSafety@springernature.com

In case Publisher is established outside the EU, the EU authorized representative is:

Springer Nature Customer Service Center GmbH
Europaplatz 3
69115 Heidelberg, Germany

Batch number: 10091386

Printed by Printforce, the Netherlands